High Speed Pulse Technology

VOLUME I

HIGH SPEED PULSE TECHNOLOGY

By Frank B. A. Früngel
IMPULSPHYSICS ASSOCIATION
HAMBURG, GERMANY AND ZÜRICH, SWITZERLAND

VOLUME I

CAPACITOR DISCHARGES - MAGNETOHYDRODYNAMICS
X-RAYS - ULTRASONICS

1965

ACADEMIC PRESS • New York and London

ACADEMIC PRESS INC.
111 Fifth Avenue, New York, New York 10003

United Kingdom Edition published by
ACADEMIC PRESS INC. (LONDON) LTD.
Berkeley Square House, London W.1

LIBRARY OF CONGRESS CATALOG CARD NUMBER: 65-16665

PRINTED IN THE UNITED STATES OF AMERICA.

Foreword

After some ten years or more of contact with Dr. Frank Früngel, at numerous technical meetings and through his writings, I am now glad to see all his ideas, devices, and thoughts published in an English edition.

His excellent German book "Impulstechnik" (Akademische Verlagsgesellschaft Geest & Portig K.-G., Leipzig, 1960) has not had widespread use in the English-speaking world, partly owing to the language barrier problem. Now this translation of the latest edition will open a new door for many into the "impulse world" of Dr. Früngel.

Anyone with a technical education in physics or electrical engineering will see Dr. Früngel's impulse technique at work in this monograph. He approaches his attack on the problems like a capacitor discharge. For years he has been storing up the energy (ideas) in his brain, now the discharge moment has arrived. Let us hope that the efficiency of idea conversion from his brain to those of the readers will be high.

As I thumbed through the massive pile of galley proofs of this monograph, I kept thinking, "What impulse phenomena has he missed?" The subjects covered by his impulse methods range from deep-sea sound sources to cloud height measurement, as well as the conventional (and the unconventional) electronic flash method of producing and controlling light flashes. Has he considered impulsing the solar system, or our galaxy, or the universe? Does he consider the sun as an impulse device? Everyone knows that the sun must have started at some time and will eventually go out like an ordinary flash lamp. Let us hope that it will not explode as some parts of the universe seem to do at, fortunately for us, infrequent intervals of time or in a few spaces.

This monograph should be an inspiration to the new generation of eager electrical engineering and physics students who, at this moment, are preparing for their life efforts. My advice to students is, "Read but do not completely believe everything of the present-day presentation of facts." All should be put to test in the laboratory and in the field. The older generation, such as Dr. Früngel and myself, only cover some of the facts, others remain to be discovered, and these may be the most important ones. "What for example" ask the students. I reply, "If I knew I would immediately try to discover them myself, and so would Dr. Früngel."

One cannot read and study the book without getting many new ideas. This monograph, published in two volumes, will be a classic for years to come. Congratulations and thanks to Dr. Frank Früngel for completing it!

Cambridge, Massachusetts

HAROLD E. EDGERTON
Massachusetts Institute of Technology

Preface

Capacitive discharges are more and more frequently used in various technical fields in order to produce energy pulses of various forms. The physicist or engineer designing equipment and apparatus will, therefore, feel the need to become more thoroughly acquainted with this special field.

This monograph deals with the theoretical and engineering problems which arise in the capacitor discharge technique.

After twenty years of practical experience in the technique of capacitor discharges, the author attempts to present a comprehensive monograph based upon the scattered literature existing in this field. This is a comparatively ambitious attempt since capacitor discharge techniques range from high-speed photography to nuclear physics and from pulse welding to acoustics and meteorology, not to mention the numerous fields of high-voltage applications.

This work also contains new and still unpublished work of the author. These passages are represented in more detailed form than those for which relevant literature already exists. Available Russian literature was also incorporated.

With regard to measuring system standards the author deviated from the general norm and at times employed the old cgs system used for mechanical engineering terms, e.g., *t* instead of megapond, in order to facilitate readability for the practical engineer.

Most of the book was translated from the second German edition. This also refers to drawings and illustrations. Consequently, it became necessary to retain certain German symbols, e.g., U for the voltage instead of E as is usually employed in English-speaking countries. In this first English edition, new material was incorporated using the customary symbols.

I wish to thank my colleagues M. and H. Voege, H. G. Patzke, and O. H. Bock, who translated the monograph, and W. Thorwart, H. Alberti, and H. Wegmann, who assumed responsibility for the high-speed photography and the technical accuracy and compilation of the tables and diagrams.

FRANK FRÜNGEL

Contents

D. Conversion of capacitor energy into current impulses

E. Conversion of capacitor energy into voltage impulses

F. Conversion of capacitor energy into x-ray flashes and neutrons

G. Conversion of capacitively stored energy into heat

H. Conversion of capacitively stored energy into magnetic fields

I. Conversion of capacitively stored energy into acoustic impulses

J. Material working by high-frequency capacitor discharges (spark erosion)

Contents of Volume II

A. The capacitor, its characteristics and its applications in a discharge circuit

1. Capacitors in Discharge Circuits in General and the Equivalent Circuit Diagram*

The idea of storing weak electric currents for subsequent discharges in the form of intensive current pulses dates back to the time of the early experimental discovery of electricity. Charging of a sphere by static or frictional electricity, experiments with conductors, for example with the electrophor, and storing of electricity in the Leyden jar are steps toward the development of electric condensers or, more precisely, capacitors.** Constructions somewhat similar to the Leyden jar can be found in modern designs such as the ceramic cup capacitor.

Capacitance is present between any two adjacent conductors. A capacitor consists basically of two parallel metal plates separated by a dielectric material, such as air, gas, paper, ceramic, or oxide film.

When a voltage is applied across the plates the capacitor will become charged and the amount of charge will depend mainly on the voltage. The capacitance of the capacitor is the ratio of the charge acquired to the voltage applied or

$$C = Q/V$$

where Q = charge in coulombs (or ampere-seconds), V = voltage in volts, C = capacitance in farads.

The farad is the unit of capacitance and a capacitor has a capacitance of 1 farad if its potential rises by 1 volt when it receives a charge of 1 coulomb. This unit is inconveniently large for practical usage and the submultiples microfarad (μf, one-millionth of a farad) and micromicrofarad ($\mu\mu$f) or picofarad (pf, one-million-millionth of a farad) are normally used.

Besides the electrical definition there is recognized the physical unit of capacitance in accordance with the cgs system. This is defined as the capacitance of a sphere having a radius of 1 cm and is known as 1 cm capacitance.

---

* Chapter A, Sections 1–4, is reprinted by permission from G. W. A. Dummer, "Fixed Capacitors," Vol. III of "Radio and Electronic Components," pp. 9–27, 35–36, Sir Isaac Pitman & Sons, London (1956).

** For the complete bibliography of fixed capacitors, see reference [182].

The units of 1 pf and 1 cm are interrelated as follows:

$$1 \text{ pf} = 0.9 \text{ cm}, \qquad 1 \text{ cm} = 10/9 \text{ pf}$$

The energy of the charge is stored up as electrostatic energy in the dielectric and is equal to $CV^2/2$. If this energy is absorbed at a uniform rate over a time t the power required is

$$P_{\text{av}} = \tfrac{1}{2}\,(CV^2/t)$$

where P_{av} = average power in watts, C = capacitance in farads, V = voltage in volts, t = time in seconds.

Under ac conditions the rate at which energy is supplied to and returned by the capacitor is

$$P_{\text{av}} = 2\pi CV^2 f$$

where f = frequency in cycles per second and V = rms voltage.

When a direct voltage is applied to a capacitor, the electric field within the dielectric is displaced and the bound electric charges are polarized or displaced from their normal position of equilibrium. Work is, therefore, done in charging the capacitor. The work done, in joules or watt-seconds, which is available as stored potential energy, is

$$J = \tfrac{1}{2}\,QV, \quad \text{or } \tfrac{1}{2}\,CV^2, \quad \text{or } Q^2/2C$$

where J = joules or watt-seconds, C = capacitance in farads, V = voltage in volts, Q = coulombs or ampere-seconds.

The fundamental formula for the capacitance of two flat parallel plates separated by a dielectric material is

$$C = (\epsilon A \times 10^{12})/(4\pi d \times 9 \times 10^{11})$$

which becomes

$$C = 0.0885\,\epsilon A/d$$

where C = capacitance in micromicrofarads (pf), ϵ = permittivity, or dielectric constant,* A = area of one plate in square centimeters, d = distance between plates in centimeters.

When more than one plate is used, $(N - 1)$ is added to the top line where N is the number of plates. Due to edge and fringe effects, this formula is not strictly true: the capacitance should be slightly higher and the following corrections should therefore be added to the plate dimensions:

For straight edges, add $0.44d$ to the sides.

For circular edges, add $0.11d$ to the radius.

---

* The recommended symbol (B.S. 1991:1954) for permittivity, ϵ (Greek epsilon), will be used throughout this book. The symbol κ (Greek kappa) was formerly used for this quantity and is still currently used in capacitor work, together with the contraction "high-k" for "high permittivity."

The equivalent circuit of a capacitor may be represented as in Fig. A1-1, where C = capacitance of capacitor, R_s = resistance due to leads, plates, and contacts, R_p = resistance due to the dielectric and case material, L = inductance of leads and plates of the capacitor.

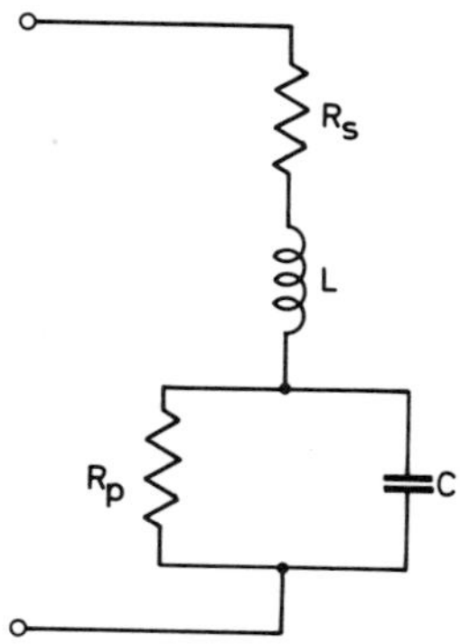

FIG. A1-1. Equivalent circuit of a capacitor. C = capacitance of capacitor; R_s = resistance due to leads, plates, and contacts; R_p = resistance due to the dielectric and case material; L = inductance of leads and plates of the capacitor. By permission from "Fixed Capacitors," G. W. A. Dummer, Vol. III of "Radio and Electronic Components," Pitman, 1956.

The behavior of the capacitor at various frequencies is described later in this chapter (Section 2f). It is important to remember that capacitance is never constant, except under certain fixed conditions. It changes with temperature, frequency, and age, and the capacitance value marked on any capacitor strictly applies only at room temperature and at low frequencies.

2. Characteristics of Dielectric Material

a. *Permittivity (Dielectric Constant)*

Dielectric materials used for radio and electronic capacitors can be grouped into the following five main classes:

(1) Mica, glass, low-loss ceramic, etc: Used for capacitors from a few pf to a few hundred pf.

(2) High-permittivity ceramic: Used for capacitors from a few hundred pf to a few tens of thousands of pf.

(3) Paper and metallized paper: Used for capacitors from a few thousand pf up to some μf.

(4) Electrolytic (oxide film): Used for capacitors from a few μf to many μf.

(5) Dielectrics such as polystyrene, polythene, polytetrafluorethylene, etc.: Range of use from a few hundred pf to many pf.

Many factors affect the dielectric properties of a material when it is used in a capacitor, among them being the permittivity, power factor, leakage current, dielectric absorption, dielectric strength, and operating temperature, and these are discussed briefly below.

The permittivity, dielectric constant, or specific inductive capacity of any material used as a dielectric is equal to the ratio of the capacitance of a capacitor using the material as a dielectric to the capacitance of the same capacitor using vacuum as a dielectric. The permittivity of dry air is approximately equal to one. A capacitor with solid or liquid dielectric of higher permittivity than air or vacuum can therefore store ϵ times as much energy for equal voltage applied across the capacitor plates. A few typical figures for capacitor dielectrics are shown in the tabulation.

Material	Permittivity ϵ
Vacuum	1.0
Dry air	1.00059
Polythene, polystyrene, etc.	2.0 to 3.0
Impregnated paper	4.0 to 6.0
Glass and mica	4.0 to 7.0
Ceramic (magnesium titanate, etc.)	Up to 20
Ceramic (titania)	80 to 100
Ceramic (high-ϵ)[a]	1000 upwards

[a] The high permittivity in these high-ϵ ceramic capacitors comes from the fact that the electric charges in the molecular structure of the material are very loosely bound and can move almost freely under the polarizing voltage, resulting in high total capacitance.

Dielectrics can be classified in two main groups: polar and nonpolar materials. Polar materials have a permanent imbalance in the electric charges within the molecular structure. The dipoles within the structure consist of molecules whose ends are oppositely charged. These dipoles therefore tend to align themselves in the presence of an alternating electric field (if the frequency is not too high). The resultant oscillation causes a large loss at certain frequencies and at certain temperatures.

In nonpolar materials the electric charges within the molecular structure are balanced and the dipoles do not rotate under applied fields although they may distort. No sharp loss peaks with frequency and temperature therefore exist.

Polyvinylchloride (PVC) is an example of a polar material, having a permittivity of 10 at low frequencies, falling to 3.0 or 4.0 at a few megacycles per second, whilst polystyrene is an example of a nonpolar material, having a permittivity of approximately 2.5 both for direct current and for alternating currents up to many thousands of megacycles per second.

An exceptional material having special properties is high-permittivity ceramic, which is a nonpolar material in that it maintains its permittivity for frequencies up to many thousands of megacycles per second, but has very high induced polarization effects. Under voltage stress, the molecular structure distorts to such an extent that it becomes extremely sensitive to temperature, mechanical pressure, and applied voltage. Under these circumstances, the permittivity becomes very high.

b. *Losses in Dielectric Materials*

Losses occur due to current leakage, dielectric absorption, etc., depending on the frequency of operation. For a good nonpolar dielectric the curve relating loss with frequency takes the approximate shape given in Fig. A2b-1.

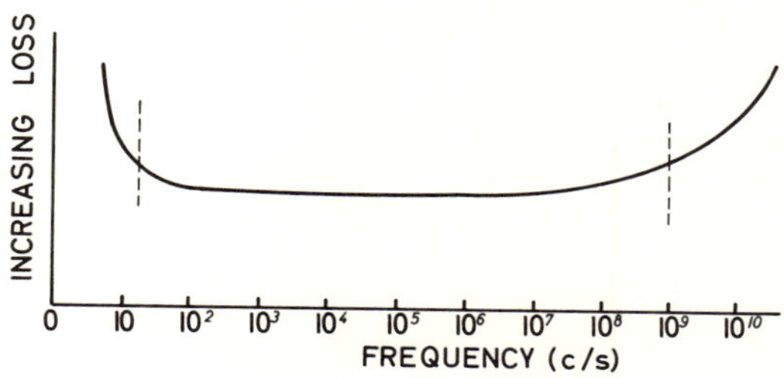

FIG. A2b-1. Loss/frequency curve for a nonpolar dielectric. By permission from "Fixed Capacitors," G. W. A. Dummer, Vol. III of "Radio and Electronic Components," Pitman, 1956.

For a polar material the loss-frequency curve may be shown approximately as in Fig. A2b-2.

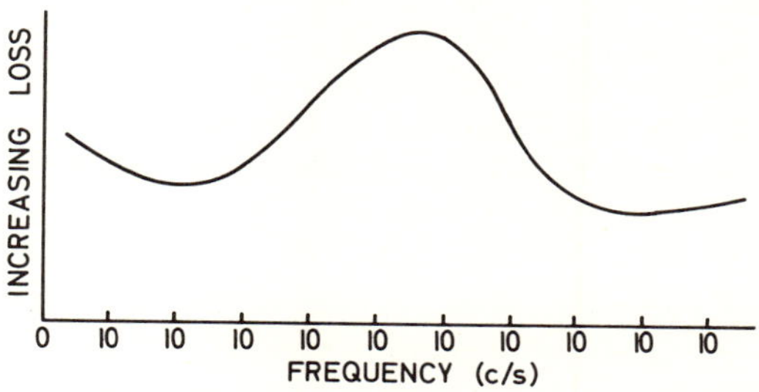

FIG. A2b-2. Loss/frequency curve for a polar dielectric. By permission from "Fixed Capacitors," G. W. A. Dummer, Vol. III of "Radio and Electronic Components," Pitman 1956.

The variation of permittivity with frequency is negligible so long as the loss is low. Increased losses occur when the process of alignment cannot be completed, owing to molecular collisions, and in these regions there is a fall in permittivity. Viscous drag in the molecular structure limits the frequency at which full alignment can be carried out. If the applied frequency is comparable with the limiting frequency, losses will become high.

Equivalent circuits showing series and parallel loss resistance can be given, but are greatly dependent on the system of measurement at any particular frequency. The important criterion is the ratio:

$$\frac{\text{Power wasted per cycle}}{\text{Power stored per cycle}}$$

This is the power factor of the material and for good dielectrics it is independent of frequency.

Power factor, phase angle, and phase difference are defined as shown in Fig. A2b-3. When alternating current flows in a capacitor, the current leads the voltage by a little less than the theoretical 90° (angle ϕ in Fig. A2b-3).

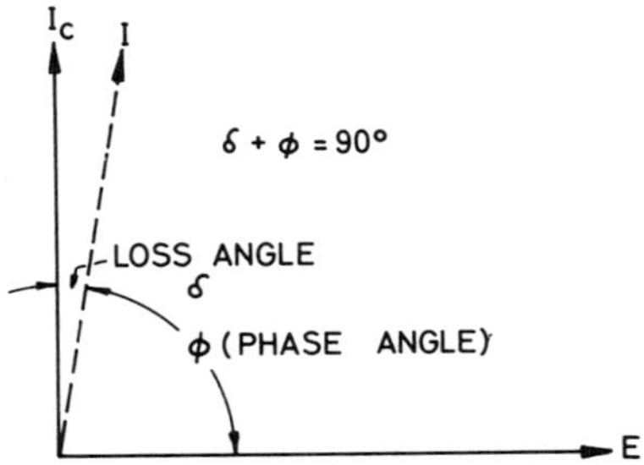

FIG. A2b-3. Vector diagram for capacitor. By permission from "Fixed Capacitors," G. W. A. Dummer, Vol. III of "Radio and Electronic Components," Pitman, 1956.

This is the phase angle and the complement of this angle (δ) is termed the loss angle. The cosine of the phase angle or sine of the loss angle is the power factor. The dielectric loss is therefore $EI \cos \phi$, or $EI \sin \delta$. The loss angle is normally so small that $\tan \delta$ can be taken as $\sin \delta$: It is more convenient to use $\tan \delta$, rather than $\sin \delta$ or $\cos \phi$, as it is easier to measure. In a perfect capacitor with no dielectric loss $\delta = 0$. The power factor is

$$\cos \phi = 2\pi f \times RC \times 10^{-9}$$

where f = frequency in kilocycles per second, R = equivalent series resistance in ohms, C = capacitance in micromicrofarads. The power factor can also be represented as the ratio of the power loss in the dielectric to the product of the applied voltage and current, i.e.

$$\frac{\text{Total power loss in watts}}{\text{Volts (rms) applied} \times \text{amperes (rms) flowing}}$$

The power factor therefore can be calculated from

$$\cos \phi = W \times 10^6/2\pi fCE^2$$

where W = total power loss in watts, C = capacitance in microfarads, E = voltage in volts, f = frequency in cycles/second. The Q of a capacitor is the reciprocal of the power factor. It may alternatively be defined as the ratio of the pure reactance to the effective resistance.

If a capacitor were completely free from dielectric absorption the initial charging or polarization current when it was connected to a dc supply would be

$$i = (E/R)e^{-t/CR}$$

where i = current flowing after a time t, E = applied voltage, and R = capacitor series resistance, and the polarization current would die off asymptotically to zero. If R is small, this takes place in a very short time and the capacitor is completely charged. In all solid-dielectric capacitors it is found that, after a fully charged capacitor is momentarily discharged and left open-circuited for some time, a new charge accumulates within the capacitor, because some of the original charge has been "absorbed" by the dielectric. This phenomenon is known as dielectric absorption. A time lag is thus introduced in the rate of charging and of discharging the capacitor.

Dielectric absorption is due to the finite time taken to displace the bound charges in the dielectric from their normal positions because of the viscous forces which resist their motion. Polarization time varies with different dielectrics—the polarization may be almost instantaneous or it may take many hours. Even within the same dielectric a few electrons or ions only become free to move after a period which may be seconds or even days. The effect is exceedingly complex, and, for example, in the case of an impregnated-paper capacitor, the polarization times of paper and of impregnant are different. There are, in addition, many types of polarization which must be taken into account.

The effect of dielectric absorption is to reduce the capacitance as the frequency of operation is increased and also to cause unwanted time delays in certain pulse circuits or in circuits requiring rapid charge and discharge characteristics.

c. *Leakage Currents and Time Constants of Capacitors*

Losses due to leakage currents when a capacitor is being used on direct current prevent indefinite storage capacity being realized, and the charge acquired will leak away once the source is removed. The time in which the charge leaks away to $1/e$, or 36.8% of its initial value, is given by RC, where R is the leakage resistance and C is the capacitance. If R is measured in megohms and C in microfarads the time constant is in seconds. This can also be expressed as megohm-microfarads or as ohm-farads. Some typical time constants for various dielectrics used in capacitors are

Polystyrene, several days.

Impregnated paper, several hours.

Tantalum-pellet electrolytic capacitors, one or two hours.

High-permittivity capacitors (ceramic), several minutes.

Plain-foil electrolytic capacitors, several seconds.

It should be borne in mind that below capacitance values of about 0.1 μf the time constant is generally determined by the structure, leakage paths, etc., of the capacitor assembly itself rather than the dielectric material. Leakage current increases with increase of temperature (roughly exponentially) In good dielectrics at room temperature it is too small to measure, but at higher temperatures the current may become appreciable, even in good dielectrics.

d. *Insulation Resistance*

The insulation resistance of a dielectric material may be measured in terms of surface resistivity in ohms or megohms, or as volume resistivity in ohm-centimeters. Surface and volume resistivities are measured using apparatus similar to that shown in Fig. A2d-1. The specimen is mounted in mercury

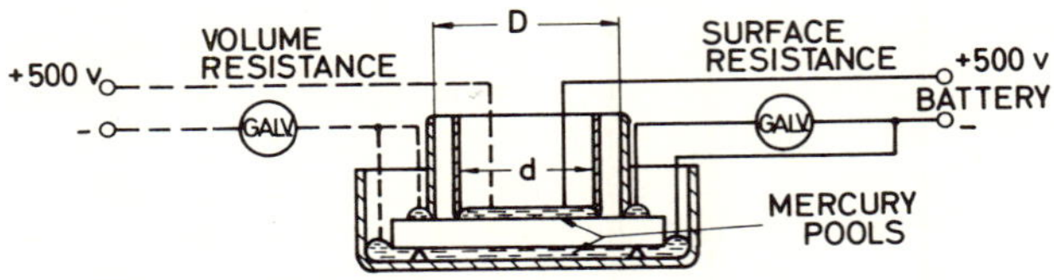

FIG. A2d-1. Typical arrangement of apparatus for surface and volume resistivity test. By permission from "Fixed Capacitors," G. W. A. Dummer, Vol. III of "Radio and Electronic Components," Pitman, 1956.

and a potential difference of 500 volts dc is connected across the inner and outer metal rings with mercury electrodes to the top surface. Typical surface insulation resistance figures for some good dielectrics measured in this way (to B.S. 771:1954) are given in Table A2d-1. Often a figure of ohm-farads is quoted (see paper-dielectric capacitors), as the insulation resistance varies with capacitance, particularly in the higher capacitance values.

TABLE A2d-1

TYPICAL SURFACE INSULATION RESISTANCE VALUES

Dielectric	Surface resistivity (ohms)	Volume resistivity (ohm–cm)
Polythene	4×10^{14}	10^{19}–10^{20}
Polystyrene	10^{16}	10^{17}–10^{19}
Polytetrafluorethylene (PTFE)	3.5×10^{13}	10^{16}
Polyethylene terephthalate (Terylene or Melinex)	10^{13}	10^{13}

Volume resistivity is measured similarly between the top and bottom surfaces, as shown on the left side of Fig. A2d-1. Typical figures for volume resistivities are shown in Table A2d-1. It is important to remember that the insulation resistance of good dielectrics in a capacitor (glass, mica, etc.) may be reduced by any encasing material having a lower insulation resistance, such as the phenol resins, particularly under conditions of high humidity or temperature.

e. *Dielectric Strength*

The ultimate dielectric strength of a material is determined by the voltage at which it breaks down. The stress in kilovolts per inch (or volts per mil) at which this occurs depends on the thickness of the material, the temperature, the frequency and the waveform of the testing voltage, and the method of application, etc., and therefore comparisons between different materials should ideally be made on specimens equal in thickness and under identical conditions of measurement. The ultimate dielectric strength is measured by applying increasing voltage, through electrodes fitted with guard rings, to a specimen with recessed surfaces (to ensure that the region of maximum stress shall be as uniform as possible). Preparation of the specimens is important and their previous histories should be known. Dielectric-strength figures for most capacitor dielectrics are given in Tables A3-1 and A3-2.

A practical limit to the dielectric strength of a material which is convenient to use is the discharge inception voltage, above which progressive breakdown occurs. This is usually much lower than the ultimate dielectric strength for short applications of voltage. Above the discharge inception voltage, a form of corona is created by the discharges and progressive deterioration takes place. It is an advantage that tests for discharge inception voltage are usually nondestructive tests, as corona produces "noise" which can be detected and measured.

The dielectric strength of a material is always reduced when it is operated at high temperatures or if moisture is present. Few materials are completely homogeneous and breakdown may take the form of current leak along certain small paths through the material; these become heated and cause rapid deterioration, or flashover along the surface and permanent carbonization of the surface of organic materials. Inorganic materials such as glass, ceramic, and mica are usually resistant to this form of breakdown. The time for which the voltage is applied is important; most dielectrics will withstand a much higher voltage for brief periods. With increasing frequency the dielectric strength is reduced, particularly at radio frequencies, depending on the power factor, etc., of the material.

f. *The Effect of Frequency on Dielectric Materials and Capacitor Assemblies*

At very low frequencies and also at very high frequencies there is an increase of loss which sets a limit to the practical use of a capacitor with any given dielectric. At very low frequencies various forms of leakage in the dielectric material have time to become apparent, such as dc leakage currents and long time-constant effects, which have no effect at high frequencies. At very high frequencies some of the processes contributing to dielectric polarization do not have time to become effective and therefore cause loss. These losses may be simply and approximately represented as in Fig. A2f-1.

At very low frequencies the circuit is entirely resistive, all the current passing through the shunt resistance (dc leakage resistance, etc.). At very high frequencies the current passes through the capacitance C but all the volts are dropped across the series resistance and again the circuit is lossy. This series resistance may be due to the resistance of the capacitor leads, the silvering (in the case of silvered mica or ceramic capacitors), contact resistances, etc., in the capacitor assembly itself. These limit the upper frequency independently of the dielectric material used. Similarly, leakage across the case containing the dielectric may limit the lower frequency so that not all the useful range of the dielectric itself may be realized.

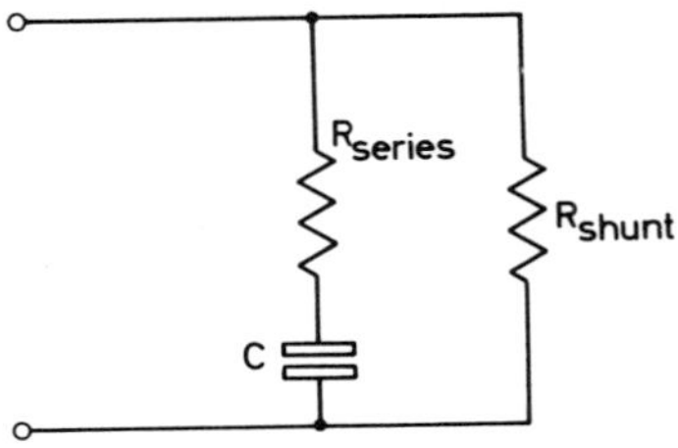

FIG. A2f-1. Losses in a capacitor. By permission from "Fixed Capacitors," G. W. A. Dummer, Vol. III of "Radio and Electronic Components," Pitman, 1956.

The chart in Fig. A2f-2 shows the approximate usable frequency ranges for capacitors with various dielectric materials. The construction of the capacitor assembly will affect the frequency coverage to some extent, so that the chart should be regarded as a guide only.

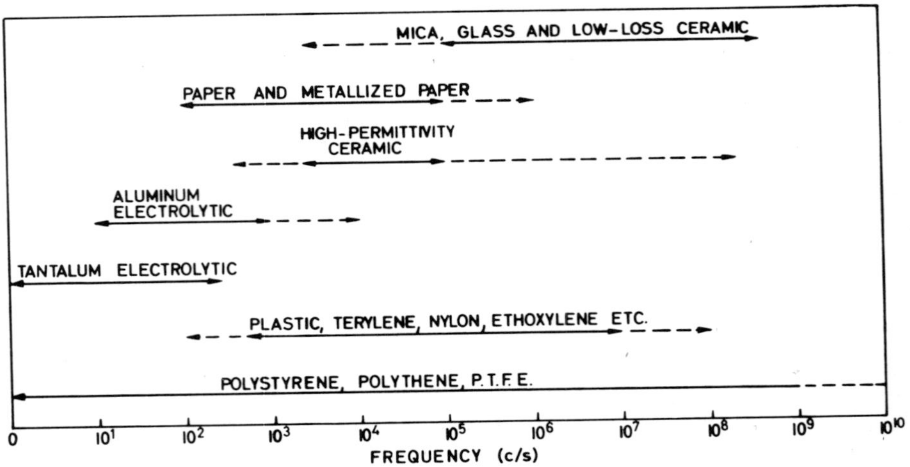

FIG. A2f-2. Frequency coverage of different classes of capacitors. By permission from "Fixed Capacitors," G. W. A. Dummer, Vol. III of "Radio and Electronic Components," Pitman, 1956.

The current in a capacitor when alternating voltage is applied is given by

$$I = 2\pi fCV \text{ amp}$$

where C = capacitance in farads, V = voltage in volts, f = frequency in cycles per second, and the reactance is given by

$$X_c = -\tfrac{1}{2}\pi f C \Omega$$

An ideal capacitor would have entirely negative reactance but the losses, described previously, due to dielectric case and leads, preclude this. In addition, inductance is also present in varying amounts and therefore as the frequency is increased the inductive or positive reactance increases, and above a critical frequency the capacitor will behave as an inductor. At the resonant frequency the impedance of the capacitor is controlled by its effective resistance, which in turn is made up of the losses described. Every capacitor will resonate at some given frequency (depending on its construction) and, having inductance and resistance, will exhibit a complex impedance, capacitative in one range of frequencies, resistive in another, and inductive in still another.

3. Tables of Properties of Dielectric Materials Used for Capacitors

Tables A3-1, A3-2 and A3-3 give an indication of the main properties of most of the materials used as capacitor dielectrics. Most of the recently developed dielectric materials are included.

As the loss-frequency curves for most good dielectrics are flat over most of the frequency band, rising at very low and very high frequencies (see Table A3-1) it is considered necessary to give only the upper and lower limits of frequency for these good materials. In the tables, therefore, only these figures are given unless there are marked variations. The temperature limits over which the dielectrics can operate are also given, but it is emphasized that these figures are a guide only, as variations in composition, etc. may affect the working temperature ranges of the materials. As so many plastic materials have been developed, a separate table of properties for insulating plastic materials is given in Table A3-2, including some materials not employed as dielectrics but often encountered in electrical work.

Even today pulse capacitors of higher energy level generally use paper as dielectric material. It is only recently that special methods of fabrication engaging electron microscopes and employing modern means of impregnation under high vacuum conditions led to qualities of dielectric paper which comply with nearly any demand. The aim of the designer was to obtain paper material with great dielectric strength and power factor. It is remarkable that paper material [924] has been developed with a guaranteed breakdown voltage of 130 volts per μ (μ, one-thousandth of a millimeter), corresponding to 3300 volts per mil., with a power factor of approximately 2.3×10^{-3} which is practically constant between 30 and 100°C. It increases slightly to lower and more rapidly to higher temperatures. Another type of paper called TUKO W [924] holds a breakdown voltage of 120 volts per μ, independent of its layer thickness. It has a power factor of 2.2×10^{-3} which is nearly constant between 20 and 100°C. Figure A3-1 shows approved values for thickness of paper and the required number of layers plotted against the operating voltage, which differs for ac and dc.

TABLE A3-1

PROPERTIES OF SOME CAPACITOR DIELECTRIC MATERIALS

Material	Loss (at room temperature): Limiting frequency of operation		Loss: Power factor (loss angle tan δ)	Permittivity (over operating frequency range)	Dielectric strength (volts per mil) (breakdown)	Temperature limits (°C)		Remarks (and some (registered) trade names)
	Approx. min.	Approx. max.				Approx. min.	Approx. max.	
1	2	3	4	5	6	7	8	9
Paper Capacitor tissue (Kraft)	Poor below 100 c/s (Can be dc)	1 Mc/s	0.01 to 0.03	4.5	500 to 1,000[b] (Depends on impregnant)	No limit but impregnant freezes	85 to 100	Capacitor properties depend greatly on impregnant
Mica (Ruby)	Precision work 100 c/s (Can be dc)	10,000 Mc/s	0.0005	5.4	1,000	No limit	+ 200	Properties vary according to source of origin
Ceramic (Low-permittivity types) Magnesium silicate	100 c/s[a]	10,000 Mc/s	0.001	5.4 to 7.0	200 to 300	No limit	+ 150 to 200	Steatite, Tempradex, Frequentex, etc.
(Medium permittivity types) Titania Rutile (TiO_2)	500 c/s[a]	5,000 Mc/s	0.001	70 to 90	100 to 150	No limit	+ 120	Faradex, Condensa, (Has large temperature coefficient, approx. −750 ppm/°C)[c]
(High-permittivity types) Titanate	1,000 c/s[a]	1,000 Mc/s	0.01	Approx. 1,000 to over 7,000	100	− 100	+ 120	Characteristics dependent on temperature. Sharp max. in permittivity at 120°C. Permittivity depends on volts. Below 120°C hysteresis occurs and loss dependent on volts. Other high permittivity materials exist with different peaks.

Glass								
(Soft lead-soda)	200 c/s	10,000 Mc/s	0.001	6.5 to 6.8	< 500	No limit	+ 200	
(Hard, borosilicate)	100 c/s	10,000 Mc/s	0.001	4.0	< 500	No limit	+ 200	
Quartz (Fused)	100 c/s	>10,000 Mc/s	0.0002	3.8	1,000	No limit	+ 300	
Plastics								
Polyethylene	dc	10,000 Mc/s	0.0002	2.3	1,000	Becomes brittle − 40	+ 70	Polythene, Alkathene, Telcothene
Polystyrene film	dc	10,000 Mc/s	0.0003	2.3	1,000	− 40	+ 65	Distrene, Lustron, Styron, Trolitul
Polytetrafluorethylene (PTFE) shaved film	dc	10,000 Mc/s	0.0001	2.0	1,000 to 2,000	− 60	+ 285	Fluon, Teflon
Polymonochlorotri-fluorethylene (PCTFE) film	dc	10,000 Mc/s	0.01 to 0.05	2.3 to 2.8	3,000 to 5,000	− 195	+ 180	Kel-F, Hostaflon
Polyethylene terephthalate film	Poor between say 0.1 to 10 Mc/s		0.01 at l.f. 0.001 at h.f.	3.1	4,000 to 5,000	− 40	+ 130	Terylene or Melinex. Could replace paper at l.f.
Paraffin wax	dc	10,000 Mc/s	0.0001	2.2	1,000	No limit but hardens	60	Capacitor impregnant, coil potting, etc.
Liquid insulants								
Paraffin	dc	10,000 Mc/s	0.001	2.2	1,000	No limit but solidifies	50	Ozokerite (slightly plastic). Used for coil potting
Transformer oil	dc	1 Mc/s	0.001	2.4	1,000	− 40	120	
Silicone oil	dc	10,000 Mc/s	0.0001	2.8	1,000	− 50	250	
Nitrobenzene	dc	10,000 Mc/s	0.0005	40	50 to 1,000	Freezes at + 3 then low permittivity	150	Note high permittivity. Cannot be used on dc because of leakage. Has large negative temperature coefficient. Has specialized use as capacitor dielectric

[a] Can be dc but slow capitance change possible.
[b] Special papers for high-energy pulse capacitors can stand 3000 volts per mil. For further information see [924].
[c] ppm/°C ≡ parts per million per degree Centigrade.

TABLE A3-2

CHARACTERISTICS OF VARIOUS

Material	Usual form	Method of molding	Permittivity (at 1 kc/s)	Power factor (at 1 kc/s)	Breakdown strength (volts/mil)
Phenol formaldehyde (PF)	Molding powder; paper, cloth and glass laminated sheeting, rod and tubing. Usually brown or black in color	Compression	4 to 7	0.01 to 0.2	250 to 400
Urea formaldehyde (UF)	Molding powder, paper and cloth laminated sheeting. Full range of colors available	Compression	7 to 8	0.02 to 0.04	300 to 400
Melamine formaldehyde	Molding powder. Usually gray-brown in colour. Laminated sheeting	Compression	6 to 9	0.015 to 0.036	300 to 400
Polymethyl-methacrylate	Molding granules, sheeting. Transparent or coloured	Compression, injection, blowing, and extrusion	3 to 3.2	0.04	380 to 390
Cellulose acetate	Molding powder, film and sheeting. Transparent and colored	Compression, injection, and extrusion	3.5 to 7	0.01 to 0.06	250 to 365
Polystyrene	Molding granules, rod and film. Transparent or colored	Compression, injection, and extrusion	2.45 to 2.65	0.0001 to 0.0005	500 to 700
Polyvinyl-chloride (PVC)	Sheeting, tubing, and pastes. All colours. Rigid and flexible grades available	Injection, extrusion, casting, dipping, and coating	6 to 7	0.012	400

PLASTIC INSULATING MATERIALS

Common uses	(Registered) trade names	Remarks
Cabinets, knobs, valve holders and bases, terminal blocks, tag-boards, fuse holders, instrument cases, aerial mountings, etc.	Bakelite, Mouldrite, Nestorite, Elo, Paxolin, Tufnol, Pirtoid, Rockite, Beckacite, Delaron, Ashlam, etc.	The inclusion of fillers such as paper, cloth, mica, asbestos, nylon, etc., considerably modifies the properties. Special grades are available for electrical purposes; these possess low water absorption and high resistance to tracking
Cabinets, knobs, instrument cases, name plates, etc.	Bakelite, Mouldrite, Beetle, Rockite, etc.	Seldom used for electrical but extensively for decorative purposes
Plugs and sockets, switch parts, etc.	Melamine	Highly resistant to arcing and tracking
Transparent canopies, scales, and radomes, etc.	Perspex, Diakon, Lucite, Plexiglass, etc.	Easily fabricated and bent into complex shapes
Cabinets, knobs, scales, etc.	Cellomold, Celastoid, Tenite, etc.	Non-inflammable film material
Coil formers, aerial insulators, rf plugs and sockets, capacitor film, etc.	Lustrex, Styron, Styroplas, Distrene, Trolitul, etc.	A high-quality electrical insulator and dielectric material; suffers from low heat resistance (about 65°C)
Wire coverings, sleeving, tubing, etc.	Welvic, Vinatex, Vinatube, Vinylite, Geon, Vibak, etc.	Properties considerably modified by included plasticizers. Some PVC compounds are co-polymers with vinylidene, acetate, butyrate and other resins

TABLE A3-2

Material	Usual form	Method of molding	Permittivity (at 1 kc/s)	Power factor (at 1 kc/s	Breakdown strength (volts/mil)
Polyethylene	Molding granules, rod and film. White greasy feeling	Injection and extrusion	2.3	0.0001 to 0.0002	1,000
Polytetrafluor-ethylene (PTFE)	Molding granules, film. White, greasy feeling	Dry pressing followed by heat curing	2.0	<0.0001	480
Polychlortri-fluorethylene (PCTFE)	Molding granules, film. White, translucent	Injection, extrusion, and compression	2.8	0.024	2,500
Polyamide	Molding powder, film or monofila-ment. All colors. Horn-like appearance	Compression, injection, and extrusion	3.6	0.02	300 to 400
Polyester	Liquid syrups. Usually clear or amber colored. May be dyed all colors	Casting, pouring, and impreg-nating (usually glass fabric)	3 to 5	0.01 to 0.05	250 to 450
Ethoxyline (Epoxy)	Powder or viscous liquids. Amber or brown in color	Casting, pouring	4	0.03	320

continued

Common uses	(Registered) trade names	Remarks
Rf cable dielectrics, plugs, sockets, aerial mounts and terminations, capacitor film, etc.	Polythene, Alkathene, Telcothene	A high-quality electrical insulator and dielectric material available in a variety of grades. Low-grade numbers should be used for maximum heat resistance. Many contain polyisobutylene as a plasticizer. Anti-oxidants are often included
Rf cable dielectric, aerial mounts, etc.	Fluon, Teflon	The best plastic dielectric material discovered to date, withstands temperatures up to 250°C. Very difficult to mold into intricate shapes. No known adhesive will stick to it. Very expensive to produce
Rf cables, aerial mounts	Kel-F, Hostaflon	Almost as good as PTFE but may be molded by conventional methods. Very expensive to produce
Coil formers, etc.	Nylon	May be used at temperatures up to 135°C. Very tough and resistant to abrasion. High water absorption may produce unstable electrical and mechanical properties
Radomes, component protection, potted circuits	Marco, Paraplex, Laminac, Nuron, etc.	Polymerized by means of a catalyst and sometimes an accelerator. May be cured either at room or elevated temperatures. Properties are modified by fillers—mica, slate, glass, chalk, etc. being commonly used
Component protection, potted circuits, adhesives	Araldite, Epikote, etc.	Polymerized in conjunction with a hardener—usually an amine substance. Excellent adhesive properties to metal, glass, ceramic, etc. May be filled with mica, slate, sand, etc., to reduce thermal expansion

TABLE A3-3

SUMMARY OF THE ELECTRICAL CHARACTERISTICS OF FIXED CAPACITORS

Capacitor type	Capacitor stability (after climatic tests) (%)	Normal manufacturing tolerance on capacitance (%)	Best manufacturing accuracy (%)	Permittivity (ϵ)	Power factor (at 1 kc/s)	Temperature coefficient (ppm/°C)	Maximum capacitor temperature for long life (°C)
Impregnated-paper (rect. metal-cased and tubular capacitors)	5	± 20	5	approx. 5	0.005 to 0.01	+ 100 to 200	100
Metallized-paper tubular capacitors	5	±25	5	5	0.005 to 0.01	+150 to 200	Normal 85; special 125
Moulded stacked-mica capacitors	2	± 20 or ± 10	5	4 to 7	0.001 to 0.005	± 200	Up to 120; depends on casing
Moulded metalized-mica capacitors	1	± 10	± 2	4 to 7	0.001 to 0.005	± 60	Up to 120; depends on casing
Glass-dielectric capacitors	1	± 10	(± 1 down to 10 pf)	approx. 8	0.001	+ 150	200
Glaze-dielectric capacitors	1	± 10	1	5 to 10	0.001	+ 120	150
Ceramic-tubular, normal-ϵ capacitors	1	± 10	± 1	6 to 15 80 to 90	0.001	+ 100 −30 −750 (according to mix)	150

Ceramic tubular, high-ε capacitors	20	± 20	5 (low values) 1 (high values)	1,500 and 3,000 (may be higher)	0.01 to 0.02 (varies with temperature, etc.)	−1,500, varies Nonlinear	100
Polystyrene-film, tubular and rectangular, capacitors	> 1	± 20	5	2.3	0.0005	−150	60–70
Polyethylene terephthalate (Melinex) capacitors	5	± 20	5	2 to 5, depending on frequency	0.01 to 1 kc/s, varies with temperature and frequency	Varies with temperature	130
Electrolytic, (normal) capacitors	10	− 20 to + 100	10	—	0.02 to 0.05	+ 1,000 to 2,000 approx.	70–85
Electrolytic, (tantalum-pellet) capacitors	5	± 20	10	—	0.05	+ 100 to 200	125
Electrolytic, tantalum-foil capacitors	10	± 10	5	—	0.05	+ 500	85
Precision-type air-dielectric capacitors	± 0.01	—	0.01	1	0.00001	+ 10	20

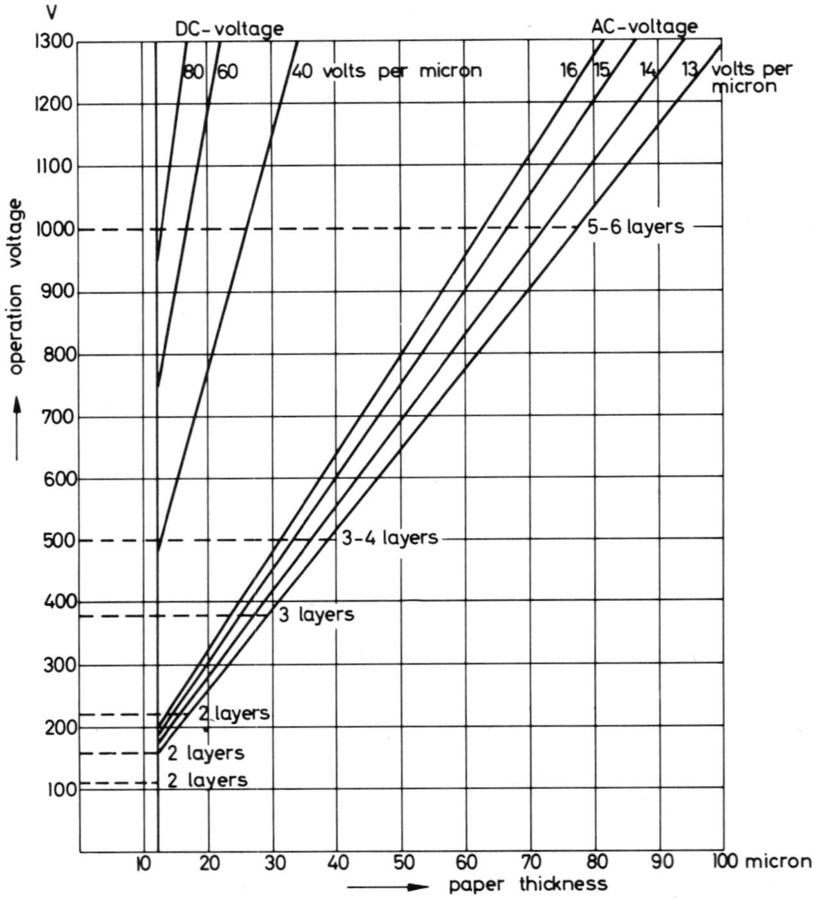

FIG. A3-1. Number of paper layers and paper thickness as functions of ac and dc operation voltage.

4. Specifications and Standards of Capacitors

a. *American and Canadian Specifications*

American and Canadian Specifications on fixed capacitors may be of interest and are given below:

American R.M.A. Standards

- REC-118: Fixed paper-dielectric capacitors in tubular non metallic cases.
- TR-113A: Metal-encased fixed paper-dielectric capacitors for dc application.
- REC-115A: Standard fixed mica-dielectric capacitors.

REC-107A: Standard ceramic-dielectric capacitors.
S-417 (M4/571 to 574): Wet electrolytic capacitors.
S-418 (M4/591 to 598): Dry electrolytic capacitors.

b. *I.E.C. and R.C.S.C. Specifications*

The I.E.C.* is an international body coordinating the various industrial standards, represented through the B.S.I. International. Specifications for components are now being drawn up and approved. Some of these specifications are:

Color code for ceramic-dielectric capacitors.
First draft specification for electrolytic capacitors.
Specifications for fixed paper capacitors for dc.
Specifications for type 1A and 1B ceramic-dielectric capacitors.
Draft specification for aluminum electrolytic capacitors for general purpose application.

There are two main sets of specifications in general use in the United Kingdom, those prepared by the Radio Components Standardization Committee (R.C.S.C.) for service components, and those prepared by the Radio Industry Council (R.I.C.) for commercial components, Specifications are also available from the British Standards Institution (B.S.I.) and the International Electrotechnical Commission (I.E.C.). Some specifications affecting fixed capacitors published by the R.C.S.C. are given below.**

RC 130: Guide on capacitors, fixed.
RCS 131: Specification for capacitors, fixed, paper-dielectric (dc rated).
RCS 132: Specification for capacitors, fixed, mica-dielectric.
RCS 133A: Specification for capacitors, fixed, ceramic-dielectric, (normal permittivity).
RCS 133B: Specification for capacitors, fixed, ceramic-dielectric, (high permittivity).

c. *Color Codes*

1. American Color Codes.

American color codes for fixed capacitors may be of interest and are given below.

Mica-dielectric Capacitors

The standard American R.T.M.A. color marking (R.C.M.), R.T.M.A. Standard REC-115A, uses a six-dot system positioned as shown in Fig. A4c-1.

---

* I.E.C. Specifications are published by The Central Office of the I.E.C., 39 Rout de Malagnou, Geneva, Switzerland.

** These are obtainable from The Radio Components Standardization Committee, 77/91 New Oxford Street, London, W.C.1.

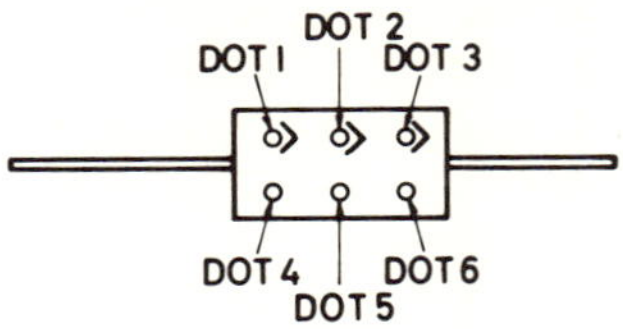

FIG. A4c-1. American six-dot mica capacitor color coding. By permission from "Fixed Capacitors," G. W. A. Dummer, Vol. III of "Radio and Electronic Components," Pitman, 1956.

Dot 1. White indicates R.T.M.A. molded mica-capacitor color coding. Black indicates JAN color coding (see following paragraph). Any other color indicates old R.M.A. six-dot system.

Dot 2. First digit of capacitance value (see Table A4c-1).

Dot 3. Second digit of capacitance value (see Table A4c-1).

Dot 4. Class of capacitor (see REC-115A).

Dot 5. Tolerance (see Table A4c-1).

Dot 6. Multiplier for capacitance value (see Table A4c-1).

TABLE A4c-1

AMERICAN COLOR CODE FOR MICA CAPACITORS

Color	Numeral	Multiplier	Tolerance (%)	Class
Black	0	1	20	A
Brown	1	10	—	B
Red	2	100	2	C
Orange	3	1,000	3	D
Yellow	4	10,000	—	E
Green	5	—	5	—
Blue	6	—	—	—
Violet	7	—	—	—
Gray	8	—	—	I
White	9	—	—	J
Silver	—	0.01	—	—
Gold	—	0.1	10	—

The color code for ceramic capacitor is shown in Table A4c-2. The temperature coefficient is indicated by the end color. The first and second color markings indicate the first and second digits of the value of the capacitance in micromicrofarads. The third color marking indicates the decimal multiplier of the value of the capacitance and the fourth color marking indicates the capacitance tolerance as shown in Table A4c-2.

TABLE A4c-2

AMERICAN COLOR CODE FOR CERAMIC CAPACITORS

Position	End	Nos. 1, 2	No. 3	No. 4 Capacitance tolerance	
Color	Temperature coefficient (ppm/°C)	Significant figures	Decimal multiplier	$C > 10$ pf (%)	$C < 10$ pf (pf)
Black	0	0	1	± 20	
Brown	− 30	1	10	± 1	± 0.1
Red	− 80	2	10^2	± 2	—
Orange	− 150	3	10^3	± 2.5	—
Yellow	− 220	4	10^4	—	—
Green	− 330	5	—	± 5	± 0.5
Blue	− 470	6	—	—	—
Violet	− 750	7	—	—	—
Gray	+ 30	8	0.01	—	± 0.25
White	General-purpose capacitor	9	0.1	± 10	±1.0

Capacitance Change
Silver, + 25% (Class 4 capacitors).
Silver, + 25% to 50% (Class 5 capacitors).

2. German Color Codes

In Germany it is the practice to color-code ceramic capacitors according to type and not to capacitance values. High-permittivity capacitors, for instance, are colored throughout a dark olive green. A proposed German color code based on temperature coefficient is given in the tabulation.

Temperature coefficient (ppm/°C)	Color
+ 100	Red
0	Orange
− 100	Light green
− 220	Dark green
− 330	Light blue
− 470	Dark blue
− 750	Violet

5. Faults Which May Occur in Fixed Capacitors*

One of the most common faults in fixed capacitors is the deterioration of the insulation resistance caused by the penetration of moisture through incomplete sealing. Although many types are now well sealed it is a difficult problem to seal a capacitor completely and thoroughly at low cost, and without creating production difficulties. The deterioration occurs mainly under arduous service conditions, but can occur under normal conditions in such places as India. In ordinary domestic radio, room and valve heat reduce the effect. Varying humidity and temperature cause moisture vapor to be sucked in, either because of molecular penetration of the materials used or along the connecting leads to the electrodes. This vapor is trapped, remains as moisture, and, in turn, lowers the insulation resistance, causes loss, and, in some cases, breakdown of the component.

In low voltage impregnated-paper capacitors the main faults which have occurred have been due to high leakage currents and short circuits. Of several probable causes of high leakage currents in this class of capacitor the commonest is deterioration of the impregnant. This use of impure impregnants gives rise to many failures, one cause of deterioration of the impregnant being that resin leaks in during soldering, in the course of manufacture. In some equipments ambient temperatures higher than normal have been used and, with a paper dielectric, deterioration of the paper itself, caused by dehydration, becomes rapid above 70°C. Short circuits and burn-outs have probably been due to excessive voltage surges, in which the dielectric breaks down at its weakest spot. The progressive deterioration of the impregnant would be a contributory cause. The date of manufacture of this type of capacitor is important, as old capacitors may suffer from aging effects during storage.

In molded silvered-mica capacitors failures have occurred due to the slow building-up of conducting material with time across the dielectric and cases have occurred where the silver has migrated over the surface of the mica, causing low insulation resistance followed by voltage breakdown. This may be caused by corona effects from excessive alternating voltage or it may be due to the presence of electrolytic impurities in combination with moisture inside the molded case.

Silvered-ceramic capacitors may suffer from similar growths or migration of silver in time away from the silvered electrode areas, and in some cases actual breakdown has followed the path of one of these growths through a flaw in the ceramic.

In electrolytic capacitors, difficulties due to corrosion caused by leakage of electrolyte through the case are now rare; they were due mainly to faulty assembly, and this problem has been overcome. High leakage currents are

---

* Chapter A, Section 5, is reprinted by permission of the publisher from G. W. A. Dummer, "Fixed Capacitors," Vol. III of "Radio and Electronic Components," pp. 149–152, Sir Isaac Pitman & Sons, London (1956).

often caused by depolarization during storage and old capacitors of this type should be reformed before use.

The electrolyte is usually the main cause of failure in aluminum oxide electrolytic capacitors. At high temperatures drying out of the electrolyte eventually produces a high power factor and loss of capacitance. At low temperatures the electrolyte freezes and causes a high power factor and low capacitance. Any corrosion effects are generally attributable to the electrolyte. Surges over the working voltage in this type of capacitor should be avoided in the design of equipment, by using, where necessary, indirectly heated rectifier valves, delay switches, etc., otherwise scintillation will occur with consequent breakdown of the capacitor.

Leaks and the unwanted production of pulses have caused difficulties with high-voltage capacitors. Mechanical leaks can be avoided by proper sealing of the can, but the spontaneous pulsing and noise generation by capacitors is a fundamental problem which is discussed in the following paragraphs.

Pulsing and random noise generated in capacitor units may have very varied origins, and only a little is known about these, especially infrequent noise pulses which would not be noticed in the ordinary tests which capacitors have been subjected to in the past. Certain sources of noise have been well established during such tests, and it is possible that they may persist under other conditions in causing random pulses, and thus be important sources of error in counting circuits, although for normal purposes the component is completely satisfactory.

Five possible sources of noise or random pulses are considered here.

1. *Internal discharge*. The best known of these normal sources is the incipient internal discharge which preceeds normal breakdown. This phenomenon is due to local ionization under high stress when there are any air occlusions or voids. Under ac excitation, this leads to a new discharge twice per voltage cycle: its onset is an indication that the component has been stressed beyond the safety limit, and the detection of this noise forms the basis of the E.R.A. nondestructive test. Under dc, a pulse of current may be followed by a quiescent period during which stress is built up again by leakage or by general relaxation of the dielectric until there is another flash.

It is quite feasible that such local pulses might be built up under dc stresses considerably lower than those at which the noise becomes appreciable in normal tests, owing to the indefinite time available for such stresses to accumulate. Noise caused in this way, however, may be expected to fall off if the stress is reduced by connecting elements in series for a given voltage.

In badly designed or faultily assembled components, the existence of incomplete leakage paths, or of exposed points and edges at which corona and sparking can form, will obviously be bad sources of noise; these can be regarded as extreme and removable causes of incipient noise.

2. *Scintillation or flutter*. This is a possible source of random noise in radio frequency components of silvered-mica or silvered-ceramic type. It consists

of random jumps in capacitance, and therefore of charge stored. It is attributed to incomplete adhesion of the silver to the material of the dielectric so that, when there is thermal expansion or mechanical shock, complete areas are jolted in or out of circuit. This source of noise is not be expected in large paper-dielectric capacitors, where foil electrodes are usual, except in capacitors having inserted tab contacts. These often make poor contact when carrying small currents, and should not be used if freedom from noise is important. It can, however, occur in mica and ceramic elements, although rarely in modern types. Correct metallizing and, in some cases, electroplating, overcomes the fault. From its nature, however, it cannot be assumed that a component will not become liable to the fault as a result of rough usage—probably any such tendency would be brought out by subjecting a sufficient number of specimens to stringent high- and low-temperature cycling.

3. *Dielectric after-working.* At very low rates of occurrence, it is possible that the effects of dielectric after-working may produce noise pulses, and unfortunately this is most marked with good dielectrics such as polystyrene and quartz. The stressing of the whole material is not released at once, but regions are left in a stressed condition, which is relaxed by processes of a viscous nature but is taken up in the structure, thus resorting some of the original polarization. A similar process may be expected to occur during charging and the temporary release of such stresses may be represented by impulses of voltage. Liquid dielectrics might be used to reduce this effect.

4. *Polymerization.* It is known that in such plastic materials as polythene, polymerization is incomplete and will continue with time so that eventually stresses may be built up which may possibly be released either as a slow variation or as an incipient pulse.

5. *Radiation.* The action of light, or of radioactive particles and x-rays, on all the good dielectrics (polythene, quartz, mica) produces ionization, and is therefore a potential source of noise, especially in components of small capacitance. It is of interest that a single crystal of anthracine, a pure hydrocarbon, is a standard form of β-ray counter—the scintillations being photoelectrically magnified. Hence, if components of these materials are exposed to radiations, some noise may be expected as a result, and there can be no protection from such noise except by adequate shielding and disposition of the apparatus. In ferroelectrics there is the analog of the Barkhausen effect. This is well established and produces large voltage pulses.

The above list of sources of noise may not be exhaustive and mechanical movements of electrodes, etc., due to thermal expansion overcoming friction, may cause pulses to be generated, either by change of capacitance, or contact resistance or by tribo-electric effects, but it is a summary of the principal causes of trouble which, from present limited knowledge, may be expected to cause noise. The summary shows especially the causes of noise at low frequencies which would not have been detected in ordinary capacitor tests.

6. Basic Combinations of *RCL* Elements in the Circuitry

This section is concerned with some of the more basic configurations which may be constructed from resistors, capacitors, and inductors. Many useful circuits each contain but one energy storage element, that is, either one inductor or one capacitor. This type of circuit will be studied in detail. The transients in these circuits will be initiated by a switching process. Also to be considered in this chapter are simple forms of the series and parallel *RLC* circuits.

a. *Single Capacitor Linear RC Circuit**

A restricted *RC* circuit will be considered first. This circuit will consist of but one capacitor and a two-terminal network which contains the switching system and any number of constant voltage sources, constant current sources, and linear resistors.

Regardless of the manner in which the capacitor is connected into the system, it can be represented as shown in Fig. A6a-1. For any particular condition, the two-terminal network can be replaced by a constant voltage source in series with a resistor. As given by Thevenin's theorem, this voltage source will have a magnitude and polarity equal to that of the open-circuit voltage of the two-terminal network. The series resistor will have a resistance equal to that looking back from the capacitor terminals when the voltage sources are short circuited and the current sources are open circuited.

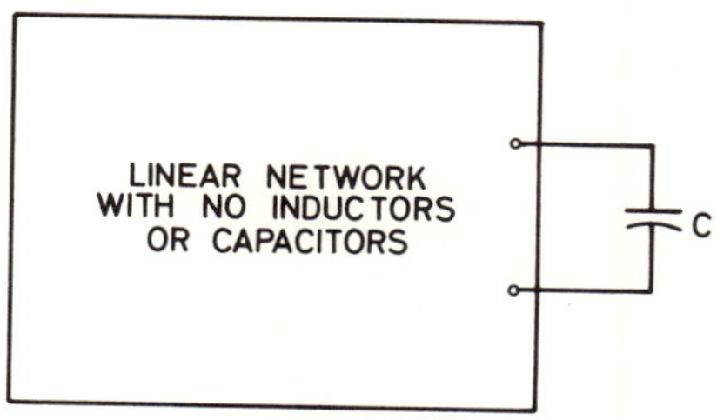

FIG. A6a-1. Restricted *RC* circuit. By permission from "Recurrent Electrical Transients," L. W. von Tersch and A. W. Swago, Prentice-Hall, 1953.

Let the two-terminal network, before the switching operation takes place, be represented by E' and R' as shown in Fig. A6a-2a. The network after switching is shown in Fig. A6a-2b. When the circuits are switched at $t = 0$, the capacitor voltage is equal to E_0. This may or may not be equal to E' depending upon the duration of time the circuit of Fig. A6a-2a has been in

---

* Chapter A, Sections 6a and 6b, is reprinted by permission from L. W. von Tersch and A. W. Swago, "Recurrent Electrical Transients," pp. 18–27. Prentice-Hall, New York (1953).

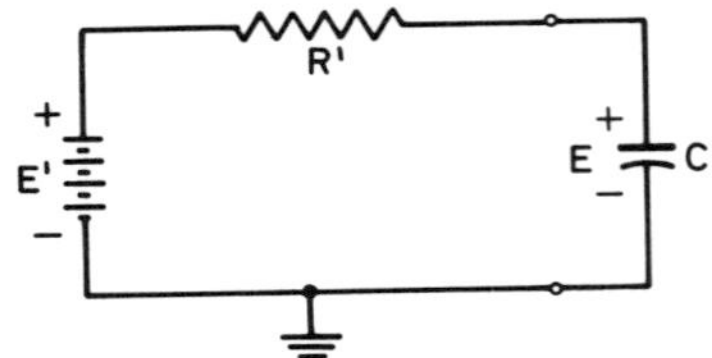

FIG. A6a-2a. *RC* circuit before switching. By permission from "Recurrent Electrical Transients," L. W. von Tersch and A. W. Swago, Prentice-Hall, 1953.

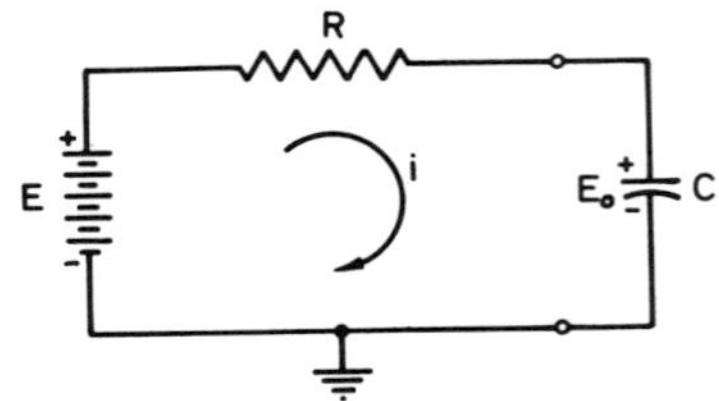

FIG. A6a-2b. *RC* circuit after switching. By permission from "Recurrent Electrical Transients," L. W. von Tersch and A. W. Swago, Prentice-Hall, 1953.

existence. The following equation can be written from Fig. A6a-2b, assuming the capacitor voltage at $t = 0$ to be E_0.

$$E = iR + \frac{1}{C}\int_0^t i\,dt + E_0 \tag{A-1}$$

Differentiating all terms and dividing by R,

$$\frac{di}{dt} + \frac{i}{RC} = 0 \tag{A-2}$$

or in D operator form,

$$\left(D + \frac{1}{RC}\right) i = 0 \tag{A-3}$$

This has the solution

$$i = Ae^{-t/RC} \tag{A-4}$$

Evaluating Eq. (A-1) at $t = 0$ will give the required boundary condition

$$E = Ri(0) + E_0 \text{ or } i(0) = (E - E_0)/R \tag{A-5}$$

where $i(0)$ is the current at $t = 0$.

Therefore $A = (E - E_0)/R$ and

$$i = \frac{E - E_0}{R} e^{-t/RC} \tag{A-6}$$

The voltage across the capacitor

$$E_c = E_0 + \frac{1}{C}\int_0^t \frac{E - E_0}{R} e^{-t/RC} dt \tag{A-7}$$

$$= E_0 + (E - E_0)(1 - e^{-t/RC}) \tag{A-8}$$

Equation (A-8) is effective only so long as the circuit of Fig. A6a-2b is unchanged. Further switching in the circuit can be accommodated by reapplying the technique above to the new equivalent circuits. The product RC which appears in Eqs. (A-4) and (A-8) is known as the time constant of the circuit; it must have the dimension of time and will be expressed in seconds if R has the units of ohms and C has the units of farads. Other combinations are shown in Fig. A6a-3.

R	C	TIME CONSTANT
MEGOHMS	MICROFARADS	SECONDS
OHMS	MICROFARADS	MICROSECONDS
MEGOHMS	MICROMICROFARADS	MICROSECONDS

FIG. A6a-3. Time constant combinations for *RC* circuit. By permission from "Recurrent Electrical Transients," L. W. von Tersch and A. W. Swago, Prentice-Hall, 1953.

The changing part of the capacitor voltage is seen to be

$$(E - E_0)(1 - e^{-t/RC})$$

while the changing part of the capacitor current is

$$[(E - E_0)/R]e^{-t/RC}$$

The terms

$$e^{-t/RC} \text{ and } 1 - e^{-t/RC}$$

appear in all single-capacitor RC transient problems. The value of each of these expressions can be plotted as a function of t/RC to provide two very useful curves. These curves are shown in Fig. A6a-4.

From Fig. A6a-4a it is apparent that after a period of time equal to one time constant, the voltage across the capacitor will have traversed 63% of the total change which would be experienced if the circuit were indefinitely undisturbed. Similarly, a check of the current equation and Fig. A6a-4b shows that the current will also have changed 63% of the total possible change during the same period of time. Increments of 86 and 95% are shown for elapsed time periods equal to two and three time constants, respectively. For most practical applications, the transient can usually be assumed to have

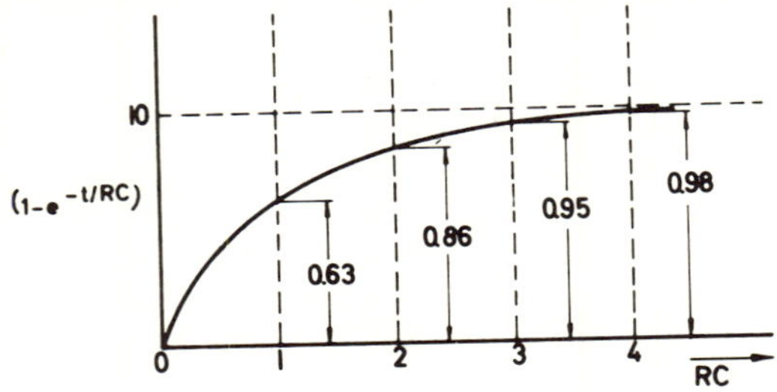

FIG. A6a-4a. Plot of $(1 - e^{-t/RC})$. By permission from "Recurrent Electrical Transients," L. W. von Tersch and A. W. Swago, Prentice-Hall, 1953.

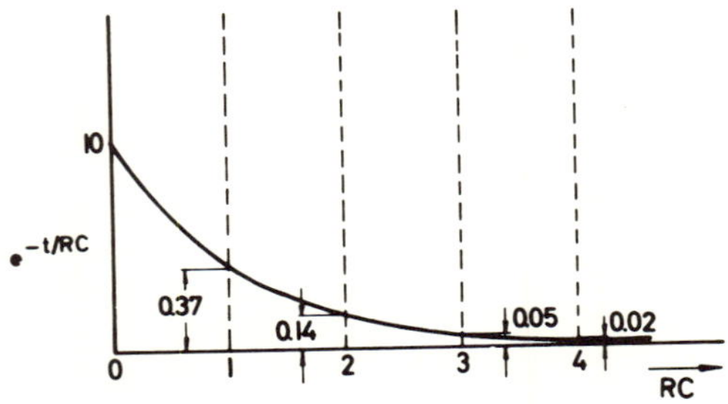

FIG. A6a-4b. Plot of $e^{-t/RC}$. By permission from "Recurrent Electrical Transients," L. W. von Tersch and A. W. Swago, Prentice Hall, 1953.

disappeared after a period of time equal to four time constants, since 98% of the eventual change has taken place.

If the two-terminal network of Fig. A6a-1 contains only constant voltage sources, constant current sources, and linear resistors, several general statements may be made. All voltage and current changes in the system will be of a general form similar to that of Eq. (A-8), that is,

$$e \text{ (or } i) = k_1 + k_2 (1 - e^{-t/RC}) \tag{A-9}$$

A form similar to that Eq. (A-6) can be considered to be a specific case of the general form of Eq. (A-9). This equation is not restricted to representation of the voltage across, or the current through, a single element. For example, the equation might give the voltage across a particular series-parallel combination, or the total current through some other combination.

When, owing to some switching operation, a sudden change takes place in the circuit, any desired voltage or current can be obtained by determining the constants k_1 and k_2 along with the time constant RC. Assume that in Fig. A6a-1 some internal combination of resistors is to be switched at $t = 0$, and that some particular voltage is to be determined for the time interval following the switching.

The first step is to determine the voltage in question just after the switching has taken place ($t = 0^+$). The voltage across the capacitor cannot change instantaneously; therefore if the voltage on the capacitor is known just prior to the time of switching, it is known for the instant subsequent to the switching. Thus to find any voltage or current immediately after the switching, the capacitor can be replaced by a constant voltage source, symbolically a battery, equal to the capacitor voltage before the switching. The circuit is then solved as a simple dc circuit. The magnitude of the voltage under consideration will be the k_1 of Eq. (A-9).

The second step is to find the voltage in question under the conditions which would exist if the transient were ever completed. This does not imply that the transient will be completed, but only that the terminal conditions are of interest in calculating the transient. The capacitor can be considered as an open circuit at this time, and the resulting circuit again solved for voltage relationships. The value of the voltage obtained in this way is the sum of k_1 and k_2. The two points which have just been found will then be connected by an exponential curve having the time constant RC.

Example: The switch in Fig. A6a-5 has been closed for a long time. It is then opened for 1000 μsec, after which it is reclosed. The voltage from point A to ground shall be determined.

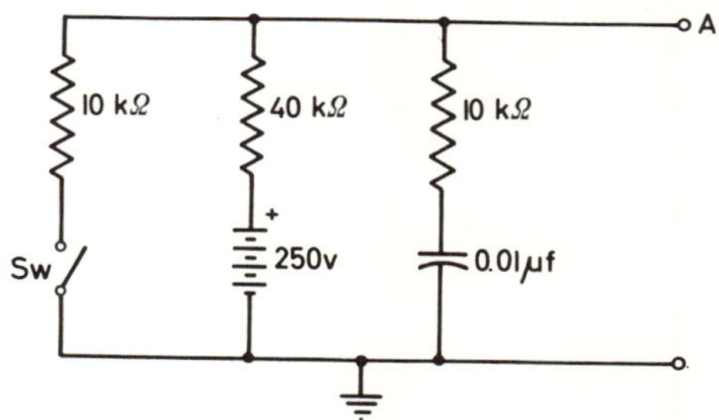

FIG. A6a-5. Example of *RC* circuit. By permission from "Recurrent Electrical Transients," L. W. von Tersch and A. W. Swago, Prentice-Hall, 1953.

Since the switch has been closed for a long time, the original voltage across the capacitor may be found.

$$E_c = \frac{10{,}000 \times 250}{10{,}000 + 40{,}000} = 50 \text{ volts}$$

The switch will now be opened. To find the voltage from A to ground immediately following the switching operation, the capacitor will be replaced by a 50 volt battery as shown in Fig. A6a-6.

If the switch were to stay open for a long time, the capacitor voltage would become 250 volts. The time constant of the exponential curve which connects the 90 and 250 volt points is

$$RC = (50)\,(10)^3\,(0.01)\,(10)^{-6}$$

$$= 500 \times 10^{-6} \text{ sec} = 500\ \mu\text{sec}$$

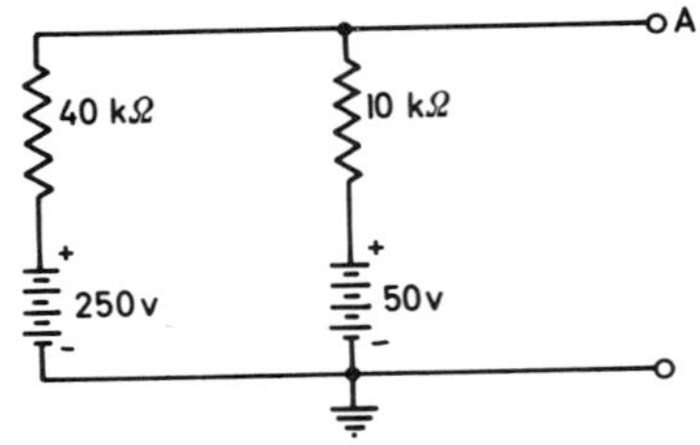

FIG. A6a-6. Equivalent circuit with switch open. By permission from "Recurrent Electrical Transients," L. W. von Tersch and A. W. Swago, Prentice-Hall, 1953.

The voltage from A to ground is then of the form

$$e_A = k_1 + k_2\,(1 - e^{-t/RC})$$

where $k_1 = 90$ volts, and $k_2 = 250 - 90 = 160$ volts. Then

$$e_A = 90 + 160\,(1 - e^{-t/RC})$$

This voltage is sketched in Fig. A6a-7. The complete exponential between 90 and 250 volts would exist only if the switch remained open. Since the switch is reclosed at $t = 1000$ μsec, the potential at point A at this time will be

$$e_A = 90 + 160\,(1 - e^{-1000/5000}) = 228 \text{ volts}$$

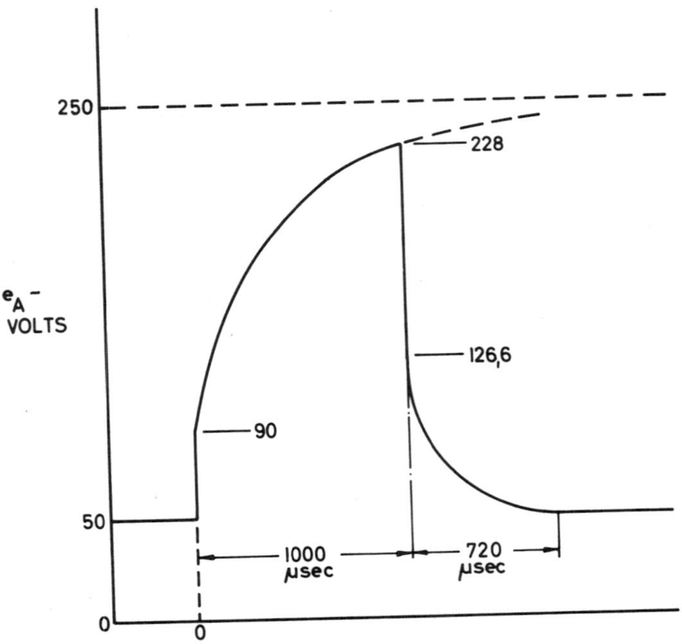

FIG. A6a-7. Wave forms for circuit of Fig. A6a-5. By permission from "Recurrent Electrical Transients," L. W. von Tersch and A. W. Swago, Prentice-Hall, 1953.

In order to solve the circuit again after the switch is reclosed, it is necessary to know the capacitor voltage at the time of closing. Summing the currents at anode A gives

$$\frac{228 - E_c}{10{,}000} + \frac{228 - 250}{40{,}000} = 0$$

$$E_c = 222.5 \text{ volts}$$

Immediately after the switch is reclosed the circuit is that of Fig. A6a-8 and the value of e_A will again be calculated:

$$\frac{e_A - 225.5}{10{,}000} + \frac{e_A}{10{,}000} + \frac{e_A - 250}{40{,}000} = 0$$

$$e_A = 126.6 \text{ volts}$$

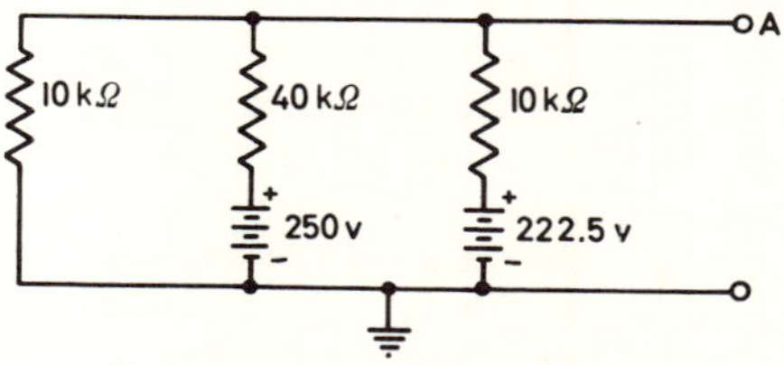

FIG. A6a-8. Equivalent circuit immediately after switch is opened. By permission from "Recurrent Electrical Transients,' L. W. von Tersch and A. W. Swago, Prentice-Hall, 1953.

A new exponential change will now occur. The starting point is 126.6 volts. At $t = \chi$, the voltage will have returned to the 50 volt value which existed before any switching occurred. Over this period of time, the exact equation is

$$e_A = 50 + (126.6 - 50) \exp [-(t - 1000 \ \mu\text{sec})/R'C]$$

It is usually easier to shift the time axis, that is, to pick a new time zero, than to carry along the more complicated exponent of the previous equation. Rewriting the equation on this basis.

$$e_A = 50 + 76.6^{-t/R'C}$$

The resistance in the time constant is marked with a prime to indicate that it is not the same resistance that was used in the original time constant. However, R′ is still the resistance looking into the circuit from the capacitor terminals, or

$$R' = \frac{(10{,}000)\,(40{,}000)}{50{,}000} + 10{,}000 = 18{,}000 \text{ ohms}$$

The voltage from A to ground will go through 98% of the change to its final value (50 volts) in approximately four time constants, or 720 μsec.

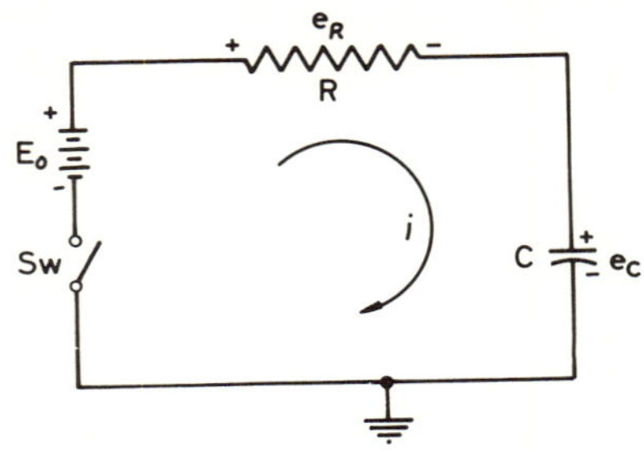

FIG. A6b-1. Nonlinear *RC* circuit. By permission from "Recurrent Electrical Transients," L. W. von Tersch and A. W. Swago, Prentice-Hall, 1953.

b. *Single Capacitor, Nonlinear RC Circuit* [184]

By conventional graphical methods any circuit containing one capacitor, any number of current sources, voltage sources, or linear and nonlinear resistors can be resolved into the circuit of Fig. A6b-1.

A completely general system would require the resistance R to be a function of i (or of the voltage e_R), and the capacitance C to be dependent upon the terminal voltage e_c (or the charge q on the capacitor). However, most low dielectric capacitors can be considered to be essentially linear devices, that is, C is independent of the terminal voltage or charge. As already indicated in Section 1, a great variety of nonlinear resistors may be encountered, particularly in the study of vacuum tubes and various forms of semiconductors.

The nonlinear resistance R could be treated in a number of ways. One method might consider the voltage e_R as a power series in i or

$$e_R = a_0 + a_1 i + a_2 i^2 + \ldots . \tag{A-10}$$

For the particular case of the linear resistor,

$$a_0 = a_2 = a_3 = \ldots . = a_n = 0, \text{ and } e_R = a_1 i$$

It is also possible to use a Fourier series to represent the volt-ampere characteristic. However, both of these methods result in cumbersome and difficult differential equations, particularly when discontinuities appear in the derivative of the volt-ampere characteristic. For these reasons, an approximate graphical method is favored.

The resistor R in Fig. A6b-1 is described by the volt-ampere characteristic shown in Fig. A6b-2. The switch is closed at $t = 0$. Let Δt_1 be an increment of time following the switching. If Δt_1 is taken to be very small, it can be assumed that the current during the Δt_1 period is essentially constant at Δi_1. If there is no initial charge on the capacitor, the capacitor voltage at the end of Δt_1 will be

$$\Delta e_{c1} = \Delta t_1 \, \Delta i_1 / C \tag{A-11}$$

Also at this time,

$$E_0 = \Delta e_{c1} + \Delta e_{R1} \tag{A-12}$$

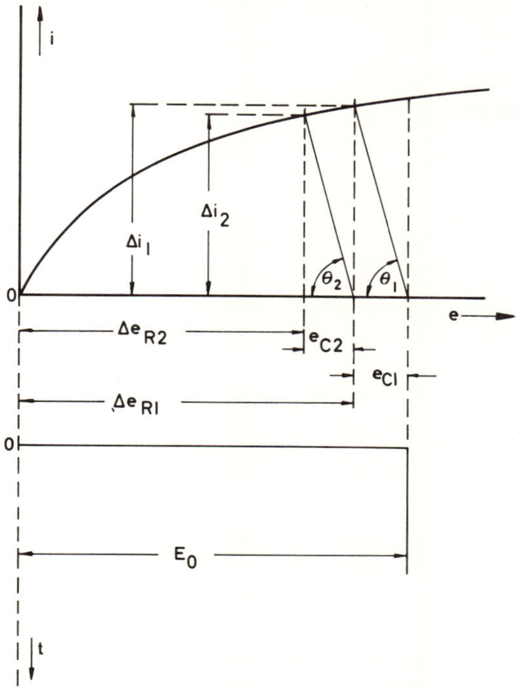

FIG. A6b-2. Graphical construction for nonlinear *RC* circuit. By permission from "Recurrent Electrical Transients," L. W. von Tersch, and A. W. Swago, Prentice-Hall, 1953.

With reference to Fig. A6b-2

$$\tan \Theta_1 = \frac{\Delta i_1}{\Delta e_{c1}} = \frac{\Delta i_1 C}{\Delta t_1 \Delta i_1} = \frac{C}{\Delta t_1} \qquad \text{(A-13)}$$

Thus for any chosen increment of time, Δt_1, the current Δi_1 can be found by calculating tan Θ_1, and then laying off a line from E_0 making an angle Θ_1 with the base line.

To find the second increment of current, the effective supply voltage can be considered to be $E_0 - e_{c1}$. If equal increments are used, $\Theta_1 = \Theta_2$, and Θ_2 can be laid off from the $E_0 - e_{c1}$ point.

The process is now continued until the current and voltage increments become arbitrarily small. Values of i, e_R, and e_c can be plotted as functions of time. The accuracy of the system will depend upon the size of the time increments which are used. It is not necessary to use equal increments, although this usually simplifies the solution.

If the magnitude of the supply voltage is an arbitrary function of time, instead of remaining constant as was assumed above, the same technique can still be used.

Example: The current and voltage in the circuit of Fig. A6b-3 will be found

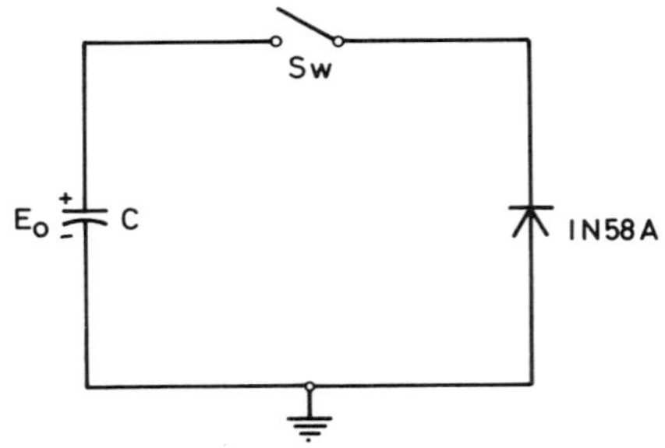

FIG. A6b-3. Example of nonlinear *RC* circuit. By permission from "Recurrent Electrical Transients," L. W. von Tersch, and A. W. Swago, Prentice-Hall, 1953.

by the graphical process which has been discussed. The capacitor has an initial voltage of 100 volts. The switch is closed at $t = 0$. Although the capacitor is discharging through the back resistance of the diode, this resistance is sufficiently low to allow the capacitor to be discharged in less than 1000 μsec.

7. Calculation and Layout of Capacitors for Discharge Operation

In nonoscillatory discharges through low ohmic loads, the resistances of the capacitor layers on the one hand, and of the load on the other hand, form a voltage divider. When, for example, a capacitor having its own resistance of 1 milliohm is discharged through a load of 9 milliohms, the ohmic resistance loss amounts to 10%. Even at a large loss angle of $\tan \delta = 10^{-2}$ in the dielectric, the loss during an aperiodic discharge the duration of which extends through one half-wave or only $\frac{1}{2} \times 10^{-2}$ in the present example is negligibly small. Besides the essential aim of obtaining low self-resistance in an impulse discharge capacitor, another quality factor is low inductance. Low inductance is desirable because the impulse current branches off energy equal to $\frac{1}{2} LI^2$, which harms the primary impulse even when this energy is not lost but is

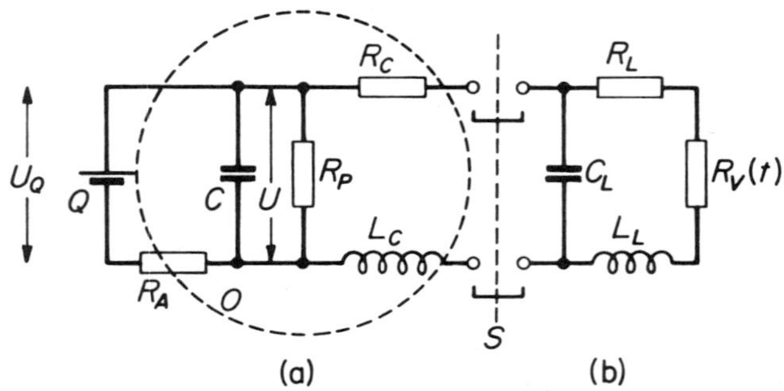

FIG. A7-1. Equivalent circuit diagram of capacitive energy storing means with impulse discharge loads. (a) Electrical values within the capacitor up to its terminals. (b) Electrical values of the external discharge circuit.

stored and becomes effective during subsequent discharges. Nevertheless, this inductance reduces the first peak value of the impulse current.

The mechanical stability of the layers of an impulse capacitor is also of primary importance. During each charge and discharge, the layers are affected by electrostatic attraction and release in proportion to the compressibility of the dielectric. Moreover, all other degrees of mechanical freedom of the layers are strained. Plastic changes, at first elastic, later persistent, occur in the capacitor structure. Finally, fatigue failure manifests itself in a spontaneous rupture at a mechanically weak point in the dielectric. In practical capacitors, mechanical layer movement announces itself by a shock sound during each discharge.

Figure A7-1 shows the equivalent circuit diagram representing a capacitance connected for discharge operation. The symbol designations are as follows:

C = storage capacitance
U = charging voltage
R_p = leakage resistance
R_c = effective total resistance between terminals
L_c = effective total inductance between terminals
O = effective surface of the capacitance, i.e. effective cooling surface
$\tan\delta = R\omega C$ = loss angle, ω = angular frequency of all C and L in the whole circuit
R_A = charging resistance
Q = charging power source of voltage U_Q
f = discharge frequency, i.e. operating frequency of discharge switching means S
C_L = harmful total capacitance of the internal conductors of the load
L_L = harmful inductance of the internal conductors of the load
$Z = \sqrt{L_L/C_L}$ = impedance of the internal conductors of the load circuit
R_L = resistance of the internal conductors
$R_V(t)$ = effective beneficial resistance of the load, mostly time dependent
I = average effective current during the discharge cycle

General definition of the efficiency:

$$\eta = \frac{\int_0^t I^2 R_V(t)\,dt}{\sum_0^t \frac{1}{2}CU^2} \qquad \left[\frac{\text{watt-seconds}}{\text{watt-seconds}}\right]$$

Limit cases:

1. R_V, for example, in ionization chambers, only under the condition of discharge carries within the measuring space.
   Desirable: time constant of self-discharge

$$ts = R_p \times C \to \infty\text{ ; therefore } R_p \to \infty$$

   Obtainable: 3-foil capacitor, $ts \approx 2000$ years [1]
   Without influence: All other values.

2. $R_V \rightarrow 0$, for example, in heaviest current spark discharges. Nearly without influence: R_p, C_L, as long as $C_L \ll C$. Of great influence: R_C, L_C, R_L, L_L.
   These values should be made as nearly as possible to zero by proper design.
3. Load having predetermined R_V between 1 and 100 ohms: Here, owing to the unavoidable conductors, the impedance must be made

$$Z = \sqrt{\frac{L_L}{C_L}} \approx (R_L + R_V)$$

   in order to assure freedom from reflex phenomena.
4. Load adapted to assure discharge progression within the aperiodic limits:

$$(R_V + R_L + R_C) = \sqrt{\frac{4(L_C\, L_L)}{C + C_L}}$$

   For best efficiency there should be:

$$R_L + R_C \ll R_V$$

   For highest energy peak: $L_C \ll L_L$, obtainable about 10^{-8}h = 10 cm.

For short-time or high-speed discharges, particularly single discharges, the temperature rise of the capacitors is practically without consequence. It is, however, of importance to withdraw the charged or stored energy instantaneously and completely in a power impulse of extremely short duration. The discharge time t and the energy $\frac{1}{2}CU^2$ give, when a rectangular curve is assumed, a power impulse of

$$N = \frac{\frac{1}{2}CU^2}{t} \qquad \text{[watts]}$$

wherein t (if we disregard an eventual time-extending influence of the impulse object which, in an extremely intense spark, can be assumed to be zero) depends only on C and the inductance of the whole discharge circuit. With reference to Fig. A7-1, a and b:

$$L = L_c + L_L$$

The half-cycle is then

$$t = \pi\sqrt{LC} \qquad N = \frac{\frac{1}{2}CU^2}{\pi\sqrt{CL}}$$

Consequently, with C predetermined, the achievable power is proportional to $1/\sqrt{L}$

In order to bring L as close as possible to zero, the capacitors are constructed as shown in Fig. A7-2. Around a tube, 1, there is tightly wound a metal foil and dielectric, 2. Both terminals 3 and 4 of the winding foils lie free. Contact discs, 6, are arranged at each end and pressed tightly against the terminals by a center bolt, 5. It is also possible to solder the discs to the foils or to vaporized end pieces, or to join the same by metal spray (metalization method).

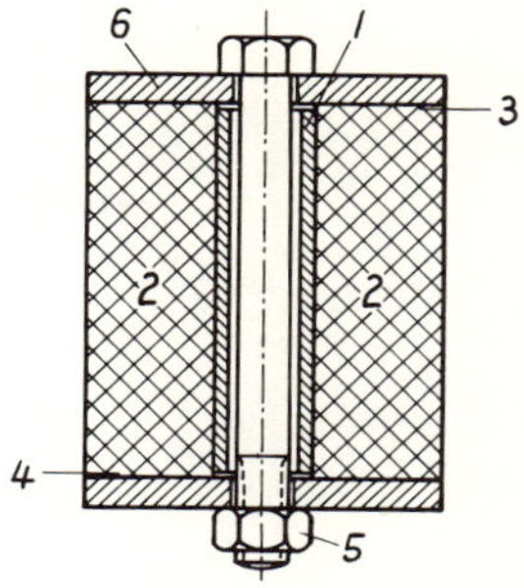

FIG. A7-2. Design of an inductance poor wind capacitor (for explanation of reference numerals, see text).

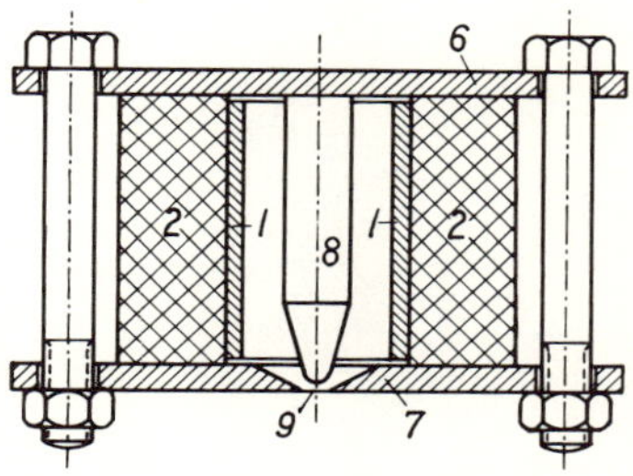

FIG. A7-3. Design of a capacitor with coaxial projection of the rear electrode, as used for point-spark generation. Highest instantaneous power output (for explanation of reference numerals, see text).

Figure A7-3 shows a discharge system composed of such capacitors. With such construction inductances as low as 0.01 μh are obtainable. In an example, assuming $C = 0.1\ \mu$f, $U = 10$ kv:

$$N_{\text{peak}} = \frac{\frac{1}{2}CU^2}{\pi\sqrt{LC}} = \frac{\frac{1}{2} \times 10^{-7} \times 10^8}{\pi\sqrt{10^{-8} \times 10^{-7}}} = \frac{\frac{1}{2} \times 10^8}{\pi\sqrt{10^{-1}}} \approx 50 \text{ megawatts}$$

at an energy (output time) of only 5 watt-sec the discharge time would be

$$t = \pi\sqrt{LC} \approx 10^{-7} \text{ sec}$$

This time is hypothetical as there exists no load having a resistance that could consume the energy during this half-wave period. Spark resistances grow with the life of the spark. At this point, however, the capacitor energy output possibilities will be considered first. A practical commercial capacitor having very little inductance and capable of high impulse energy output is shown in Fig. A7-4. It is a capacitor battery feeding a spark gap. Three cylindrical polystyrol-wound capacitors are arranged around an axis. Each capacitor consists of compressed superimposed single windings, and the inductance, as inherent in the cylindrical form of every capacitor, is here effectively compensated through a bifilar jacketing of the whole capacitor by a pair of copper cylinders.

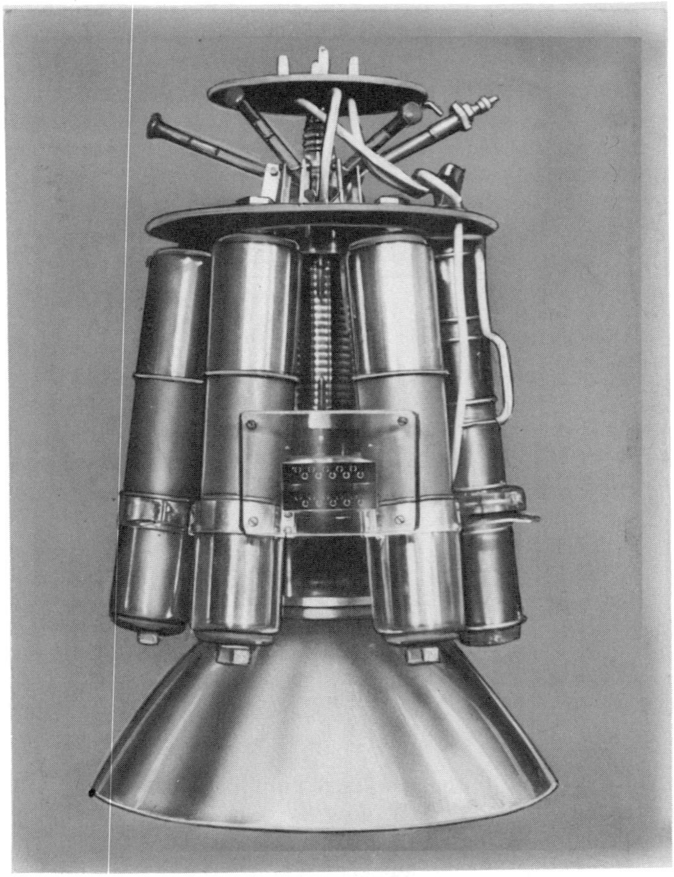

FIG. A7-4. Polystyrol capacitors in a spark-gap device of low inductance, 0.03 μh, **peak** load 30–40 megawatts (FF Impulsphysics Inc.).

Several capacitors of this type can be flanged to a common conductor plate. The capacitors operate here at 10 kv charging voltage, 6 kamp peak impulse current achieved by direct discharge through a spark chamber at about 0.6–1 ohm spark resistance. The maximum apparent output is therefore 60 megavolt-amp. The effective output can be determined only from oscillographic traces of current and voltage and multiplication of both curves. Such procedure results in about 30–40 megawatts. For large peak currents at high voltage and for the construction of discharge devices having the lowest possible inductance, aluminum-foil–wound paper, or polystyrol-foil, capacitors have gained widespread acceptance, The latter are mostly paraffin impregnated under vacuum. The wide aluminum contact surfaces, that are formed by compression of the layers, are connected with the terminal of an adjacent capacitor by exactly regulated pressure. Experience has shown that such units can withstand current peaks up to 2000 amp even at high pulsing rates such as 50,000 per sec.

It is not so much the small loss angle of the material as the extremely low resistance of the capacitor elements that is responsible for the low total losses. Mechanical strengthening of the capacitor against electrostatic stresses can be increased by proper impregnation. The measured losses during cyclical discharge operation are lower in impregnated capacitors than in those which have not been properly vacuum impregnated.

8. The Capacitor with Large Energy Content for Powerful Single Discharges

In the field of nuclear research, the need often arises for a large quantity of energy to be utilized in an extremely short time. Peak powers in the order of 2000 megawatts are not uncommon. The usual solution to this problem lies in the use of energy storage capacitors charged to a relatively high voltage and discharged in a very short time through the required load. To accomplish the rapid discharge, the internal inductance and resistance of the capacitor must be held to a minimum.

A second requirement for low-loss, low-inductance capacitors occurs in research programs requiring tank capacitors in high power rf generators. Both the energy storage and the rf tank requirements are met by essentially the same capacitors.

Nearly all manufacturers of high-voltage capacitors have developed a line of low-inductance capacitors to meet such requirements. The design and construction of such capacitors incorporates new techniques and materials not found in normal capacitor design. Coaxial line principles are utilized to provide the unusually low inductance.

The finished capacitor can be compared to a coaxial transmission line of low impedance and short length in which care is taken to maintain a low ratio of outer to inner conductor diameters in order to minimize the inductance.

Generally, capacitors of this type are custom designed to meet the requirements of specific applications. Requests for information should include the following data:

a. Microfarad rating.
b. Peak charging voltage on capacitor and maximum reversal voltage, if any.
c. Minimum ringing frequency or maximum internal inductance.
d. Ambient temperature range.
e. Type of termination (terminal arrangement).
f. Dimensions (space available) and mounting arrangement.
g. Ripple voltage and frequency (if in excess of 10% of charging voltage).
h. Duty cycle (discharge cycle in terms of discharges per second; time on charge, time off). A sketch of approximate voltage-time curve is desirable.
i. Life expectancy, number of discharges or equivalent hours on charge.
j. Magnitude of peak discharge current.

The smallest capacitor discharge circuit with pinching sparks was devised by H. Fischer [761] with an energy of only 0.05 watt-sec. This was made possible because of an extremely small circuit inductance, in the millimicrohenry range.

Normally, for generation of pinching spark discharges with self-contraction of the plasma at highest temperatures, the energy storage of the feeding capacitor battery must be between 5 and 1500 kj. Its inductance must be extremely low so that the development of the highest power peak in the impulse (within $\frac{1}{4}$ cycle after initiation of the discharge) is assured within the shortest time. Capacitors are available for such purposes. Their self-resistance should not be higher than a few microhms for the whole battery. Since in some applications there are very often several hundred impulse capacitors connected in parallel, the portion for each element shrinks to less than 1 milliohm. A similar statement can be made with regard to inductance. In the parallel connection of, for instance, 500 elements of 0.2 μh each, the total inductance within the capacitor itself is 1/2500 μh, a value which shows that the problem of obtaining low inductances in parallel-connected capacitors lies not in the capacitors themselves but in the arrangement of the connections. Bifilar construction of the bus bars is necessary. At the capacitor end the main problem is the realization of extremely low resistance and dynamic rigidity.

Since unidirectional currents attract one another, it is important that by proper design and construction of the capacitor winding connections, an arrangement is assured that prevents contraction of the current path within the capacitor. With increasing current magnitude there would be an increase in inductance due to cross-sectional contraction. The forces within a capacitor battery under discharges similar to short circuits can be realized by the following consideration.

An impulse voltage capacitor of modern design contains about 50 to 100 joules per liter volume. By assumption of nonresistance, there is no primary possibility of heat generation in an ideal discharge case. Therefore the total energy is set free mechanically through dynamic, i.e., magnetic contraction forces, in each capacitor and between adjacent capacitors, as well as through the spontaneous relaxation of the dielectric and through electrostriction of material. One hundred joules are equivalent to 10 kg-meters, so that 10 kg-meters per liter have to be absorbed. If this energy conversion takes place within only 1 μsec, there results an output peak of 10^7 kg-meters per second per liter of capacitor volume, or roughly 10^5 hp, and this during 1 μsec. These forces are comparable to the explosion of a small projectile within the capacitor. The problem can best be overcome by constructing the capacitor to be so rugged and solid—for example, with layers produced by the

---

FIG. A8-1. Side elevation and cross section of an impulse capacitor (Hydrawerk, Berlin). Nominal capacitance 10 μf, charging voltage 102 kv, 52 kw-sec, self-inductance about 0.15 μh, amplitude of impulse discharge current when discharging into an undamped and inductance free discharge circuit:

$$U \times \sqrt{\frac{C}{L}} = \text{about } 830{,}000 \text{ amp.}$$

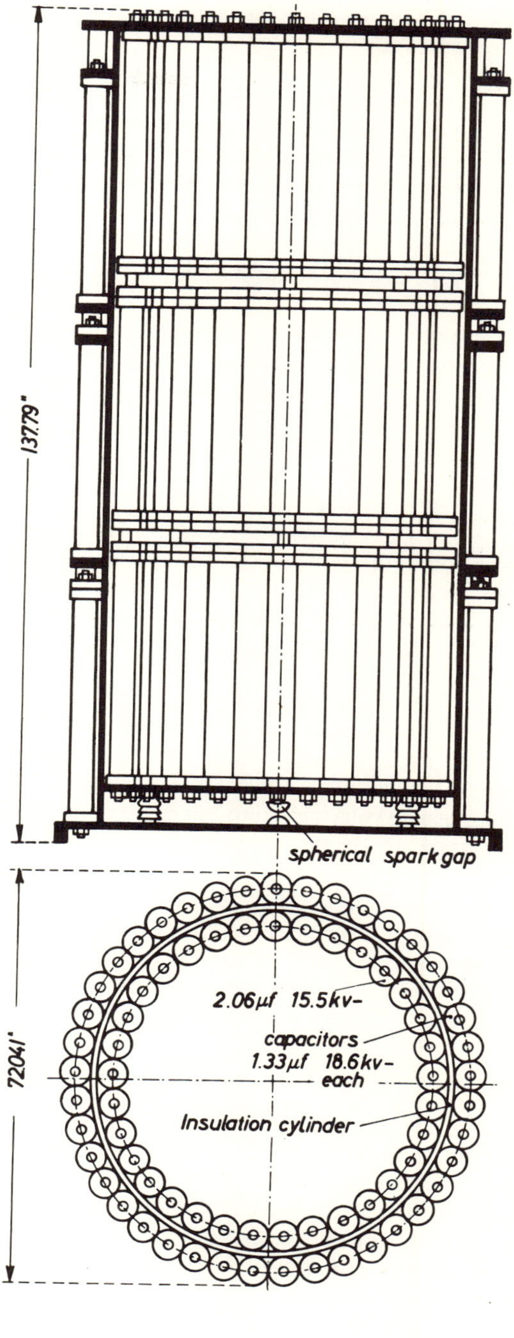
137.79"
spherical spark gap
72041'
2.06μf 15.5kv-
capacitors
1.33μf 18.6kv-
each
Insulation cylinder

metallization method and by subsequent turning—that it can withstand the impact of a small projectile striking its layers with the outer armor of the sheet metal housing removed.

Figure A8-1 shows the arrangement of a capacitor battery for 10 μf and 102 kv charging voltage at 52 kw-sec having a self-inductance of only 0.15 μh. The low self-inductance is achieved by proper geometrical arrangement of the component elements. The manufacturer (Hydrawerk-Berlin) claims an amplitude of impulse discharge current, with discharge into an undamped and inductance-free discharge, circuit of

$$I = U\sqrt{\frac{C}{L}} = \text{about } 830{,}000 \text{ amp}$$

The self-inductance given refers to the tube system only, exclusive of the inductance of the face plates, which leads to the discharge half-spheres.

The loads of such impulse capacitor batteries are mostly special spark chambers. They often carry the additional burden of auxiliary magnetic fields. Application for pure magnetic field production to observe nuclear physics processes creates similar problems. (For instance, the generation of 500,000 oe during a few microseconds requires 30–50 kj, as will be explained later.

In 1956, in London, Kurchatov [2] observed nuclear fusion in intense current spark discharges. For the first time neutron emission from spark discharges was demonstrated experimentally. The result has been intense work in many widely separated centers on the development of apparatus with large capacitor batteries.

The assembling of especially dimensioned capacitor batteries for high-energy single pulses from commercial components is the economical way of producing this rather expensive apparatus. This is because there is hardly a demand below 50 kj.

The design of the usual American high-energy storage capacitors is similar. One type, for instance, made by Tobe-Deutschmann has 0.85 μf, 120 kv, and 120 nh self-inductance. This capacitor will deliver $\frac{2}{3}$ megamp with $\frac{1}{4}$ μsec rise time. The unique low-inductance terminal insulator geometry permits connection using a coaxial return sheet, with adequate creepage for capacitor operation under oil, high vacuum, or pressure. A coaxial return sheet consists of two coaxial cylinders, a small distance apart. It has the same characteristics as sandwich lines.

9. The Impulse Capacitor in Cyclic Charging Operation

For longer lasting series of high-energy discharge impulses in rapid succession the thermal requirements, which present no problem in single discharge operation, must be considered. Generally, during charging in the charging resistance R_a of the capacitor, there is converted into heat the same quantity of energy as is charged into the capacitor in the form of electrical energy, namely $\frac{1}{2}CU^2$ (Fig. A9-1).

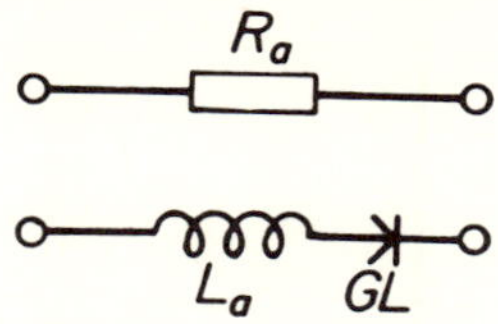

FIG. A9-1. Charging resistance and charging choke.

This thermal loss can be reduced by exchanging R_a for an inductance or choke coil with a blocking rectifier. Dimensioning is dependent on the highest frequency f to be encountered in service.

For resistance charging, it is a requirement for this frequency that the charging potential lie only a little below the voltage of the feeding high voltage source. For instance, 13.5% corresponding to $1/e^2$ where $e = 2.718\ldots.$

$$U \geqq 0.84\ U_q$$

When this condition is met, the time constant of charging $t_a = R_a \times C$ will be about equal to one-half of the shortest time spacing between discharges, since after one time constant.

$$U = \left(1 - \frac{1}{e}\right)U_q, \qquad t_a = \frac{1}{2f}, \qquad R_a = \frac{1}{2fC}$$

and after two time constants

$$U = \left(1 - \frac{1}{e^2}\right)U_q \approx 0.84\ U_q$$

If R_a is replaced by a charging choke L_a and a rectifier GL, the ohmic resistance losses of the charging rectifier are changed to an inductive storage of $\frac{1}{2}LI^2$ with the effect that the capacitor C is charged to a higher potential. The total charge is

$\frac{1}{2}CU^2$	+	$\frac{1}{2}LI^2$	=	CU_e^2
Capacitor charge through resistor		Storage in charging choke		Discharge

The voltage U_e that can be expected by resistance-free choke operation can be $\sqrt{2}$ times higher than without it, but such operation makes necessary a reverse-flow-preventing blocking rectifier GL. The resistance of the rectifier minimizes the value of U_e in the forward direction by a corresponding resistance drop. The choke coil must be dimensioned in accordance with Fig. A9-2. Thus for one-half of the highest operative discharge frequency,

$$L = \frac{1}{\left(2\pi.\frac{f}{2}\right)^2 \times C} \sim \frac{1}{10f^2C}$$

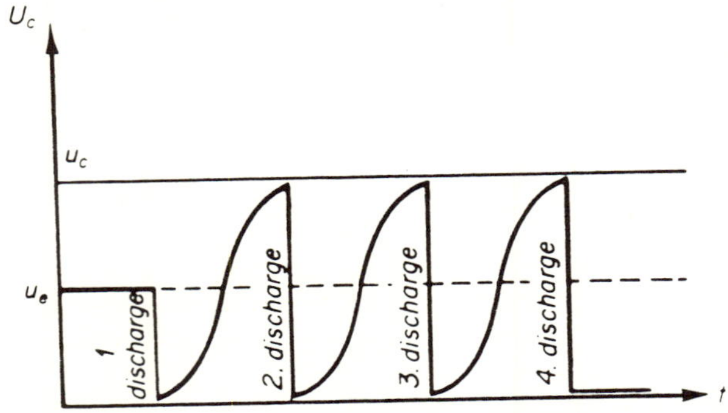

FIG. A9-2. Voltage drop U_c at the capacitor during choke-charging of a discharge circuit.

The rectifier must be proportioned for short time peak currents to be expected as follows:

$$\frac{1}{2} \times LI^2 = \frac{1}{2}CU^2 \qquad I_{\max} = \sqrt{\frac{CU^2}{L}}$$

The mean value of the rectifier current is given by the capacitor charge $Q_{As} = CU$ and the opertive discharge frequency f; therefore $I = fCU$. The blocking potential of the rectifier must be U_e.

The permissible total resistance of the choke-rectifier combination will depend on the consideration whether the additional cost is justifiable as against a resistance R_a, when, for instance, the thermal losses could be reduced to one-third. It is thus an economic or a weight problem. As long as a resistance-poor rectifier with a choke coil will be cheaper than a high-potential power source that has been derated for the output increase necessary for resistor operation, a rectifier will be of advantage.

Example:

A capacitance

$$C = 0.5\ \mu\text{f} = 5 \times 10^{-7}\ \text{farads}$$

as discharge capacitor shall be charged and discharged at a sequence frequency of $f = 10{,}000$ per second and a potential of $U = 10$ kv. (Here only charging is of interest.) For this purpose there is necessary a charging energy of:

$$A_L = \tfrac{1}{2}CU^2 = \tfrac{1}{2} \times 5 \times 10^{-7} \times 10^8 = 25\ \text{watt-sec}$$

and the charge is

$$Q = C \times U = 5 \times 10^{-7} \times 10^4 = 5 \times 10^{-3}\ \text{amp-sec}$$

The average or mean current becomes

$$I = f \times Q = 10^4 \times 5 \times 10^{-3} = 50\ \text{amp}$$

In charging through an ohmic resistance the required charging resistor is

$$R_a = \frac{1}{2fC} = \frac{1}{2 \times 10^4 \times 5 \times 10^{-7}} = 100 \text{ ohms}$$

The voltage of the charging current source—in order to meet above conditions of $U_c = (1 - 1/e^2)\, U_q$—must be

$$U_q = \frac{10^4}{1 - (1/e^2)} \approx 12 \text{ kv}$$

The effective output of the capacitor during the periodic discharge sequence is

$$N = f \times \tfrac{1}{2}CU^2 = f \times A_L = 10^4 \times 25 \text{ watt-sec/sec} = 250 \text{ kw}$$

The current source has to be capable of delivering

$$U_q \times I = 12 \text{ kv} \times 50 \text{ amp} = 600 \text{ kw}$$

Consequently, the efficiency of capacitor charging through an ohmic resistance becomes

$$\eta = \frac{25}{60} = 42\%$$

For choke-coil-rectifier charging there must be

$$L = \frac{1}{10f^2C} = \frac{1}{10 \times 10^8 \times 5 \times 10^{-7}} = \frac{1}{500} = 2 \text{ mh}$$

$$I_{\max} = \sqrt{\frac{CU^2}{L}} = \sqrt{\frac{5 \times 10^{-7} \times 10^8}{2 \times 10^{-3}}} \sim 160 \text{ amp}$$

In order to meet the requirement that $R_a = 30\%$ of the value in purely ohmic resistance charging, the self-resistance of the choke-rectifier combination must be

$$R_{GL} + R_L \leqq 30 \text{ ohms}$$

Example of choke:

Single layer air cylinder coil

$$L = \frac{\pi^2 \times D^2 \times n^2}{l} \times K \times 10^{-9} \text{ [henrys]} \qquad K = \text{form factor here about } 0.8$$

Wire size, 2 mm = 0.2 cm diameter, $l = \frac{1}{5} \times n$

$$L = \frac{10D^2 \times n^2}{\frac{1}{5} \times n} \times K \times 10^{-9} \text{ [henrys]} \qquad (\pi^2 \sim 10)$$

$$L = 50D^2 \times n \times 0.8 \times 10^{-9} \text{ [henrys]}$$

$$L = 40D^2 \times n \times 10^{-9} \text{ [henrys]}$$

If it is assumed that $L = 2 \times 10^{-3}$ henrys, it will be

$$D^2n = \frac{2 \times 10^{-3}}{40 \times 10^{-9}} = \frac{1}{2} \times 10^5$$

When a diameter of 20 cm is assumed, $D^2 = 400$ and

$$n = \frac{1 \times 10^5}{2 \times 400} = 125 \text{ turns}$$

The resistance

$$R_L = \frac{\rho \times l^*}{q}$$

at

$$l^* = D\pi \times n = 0.2 \times \pi \times 125 = 25\pi \text{ [meters]}$$

$$\rho = 0.017 \left[\frac{\text{ohm millimeters}^2}{\text{meter}}\right]$$

$$q = 1 \times \pi \text{ [millimeters}^2\text{]}$$

becomes:

$$R_L = \frac{0.017 \times 25\pi}{1 \times \pi} \left[\frac{\text{ohm mm}^2 \text{ meter}}{\text{mm}^2 \text{ meter}}\right] = 0.42 \text{ ohm}$$

that is, negligibly small relative to the total resistance of $\leqq 30$ ohms, as assumed in the present example.

The selenium rectifier to be used in the charging circuit should have, at 10 kv blocking voltage, a rating for 170 amp impulse peak current and for an average direct current of

$$I = f \times C \times U = 50 \text{ amp}$$

and should have a forward resistance below 30 ohms. For only short time intermittent duty of the arrangement described above, the thermal loading in watts is less important than the limitation of this resistance. That this condition is met must be assured by experimental measurement on the element itself.

Example:

Selenium rectifier, blocking voltage 30 volts per element. Number of elements for 10 kv, 333. Allowable forward resistance at 170 amp (to be measured oscillographically) per disc element = 30/333 = 1/10 ohm. Such a selenium rectifier would be rather expensive.

In the case of a glow (tube) rectifier, which has very little resistance owing to its constant voltage drop (ionization potential of 10–15 volts), the cathode must be designed for 170 amp load and 10 kv blocking potential in the anode-cathode path.

For intermittent short time duty, the commercial hydrogen thyratron Type 522 meets this condition.

The voltage of the charging source U_q can now be lower than in the case of pure resistance charging.

$$U_q = \frac{1}{\sqrt{2}} U_c + I \times R_{GL}$$

For 10 kv charging voltage and 30 ohms rectifier resistance there are required $U_q = 0.71 \times 10^4$ (volts) + 50 (amp) × 30 (ohms) = 7100 + 1500 = 8600 volts.

The efficiency is now higher.

The output of the capacitor at $f = 10^4$ per sec is again 250 kw. From U_B there have to be supplied:

$$8.6 \times 10^3 \text{ volts} \times 50 \text{ amp} = 430 \text{ kw}$$

Therefore

$$\eta = \frac{250}{430} = 58\%$$

as against 42% in a pure R_a circuit.

Since the cost of a rectifier can be considerable, it is clear that in short-time devices this improvement in efficiency from 42 to 58%, that is, by 16%, may not be worth while. Normally, therefore, the simpler resistance charging method described first, with a somewhat larger current supply unit, will be preferable. Investigation would be necessary to show whether the difference between 600 kw and 430 kw—that is, 170 kw more in the high-voltage charging device—would be a costlier investment for a planned duration of operation than a 170 amp, 10 kv rectifier.

The circuit with a charging choke and rectifier, results, in contrast to resistance charging as evidenced by Fig. A9-2, in a slow rate of current rise after every discharge. Thus it has the effect of a quenching action on the discharge (quenching choke). Since the apparatus technique generally requires relatively wide operating frequency ranges, the combination of glow valve (rectifier tube) and charging choke is used—for instance, in stroboscopes—almost exclusively. Because of the heavy peak currents the only devices that can be used are gas-filled rectifier tubes or thyratrons which function as rectifiers.

10. Periodic Discharge Operation

In periodic, or cyclic impulse, discharge operation at high outputs considerable heating occurs within the capacitor, even at low dielectric losses.

Even when the combined true ohmic and dielectric losses in an impulse capacitor total only 1%, this, in a unit of 250 kw rating, amounts to 2.5 kw in heat to be dissipated. If cooling is to be effected by air, it is evidently easier to bring the cooling air in contact with the dielectric in a system of several capacitors connected in series or parallel than in a large single unit.

Guided by this experience, one may obtain an arrangement of a capacitor system that possesses particularly good cooling conditions and needs only a very short time interval for establishment of temperature equilibrium. The required capacitance consists of flat capacitors of known make. At the plane of their largest surface expansion they are joined to form slot discs. Several such discs are combined into a rib system. It is well known that so-called rib systems are most suitable for good heat dissipation into flowing air. This heat dissipation becomes even more effective when the single ribs are provided with slots. Because usually every capacitor has its own cup, it is readily possible to join flat capacitors physically to form such a rib system with slotted ribs. Utilization of the aforementioned flat capacitors gives, at a right angle to the cooled surfaces, a relatively short path for the outward-streaming heat flow. Thereby, a temperature equilibrium corresponding to the prevailing load condition in the unit, is reached quickly.

The arrangement just described has proven its value, particularly in impulse voltage generators having compressed air spark gaps as periodic switching means [3]. For such purposes it is recommended that air which has been cooled by relaxation in the outlet nozzles of the spark gap be allowed to pass through the capacitor system.

Two examples of the described construction are illustrated in Figs. A10-1a and A10-1b. Figure A10-1a shows a front elevation and a longitudinal section of such a capacitor system with a radial rib arrangement which is especially suitable for application in high-voltage phase shifters. Figure A10-1b shows a front elevation and a longitudinal section of a capacitor system with longitudinal ribs particularly suitable for impulse voltage generators. In Fig. A10-1a the capacitors, 1, 2, 3, 4, 5, and 6, are secured radially around an insulation tube, 7. They form in combination a disc-like cooling rib with wedge-shaped

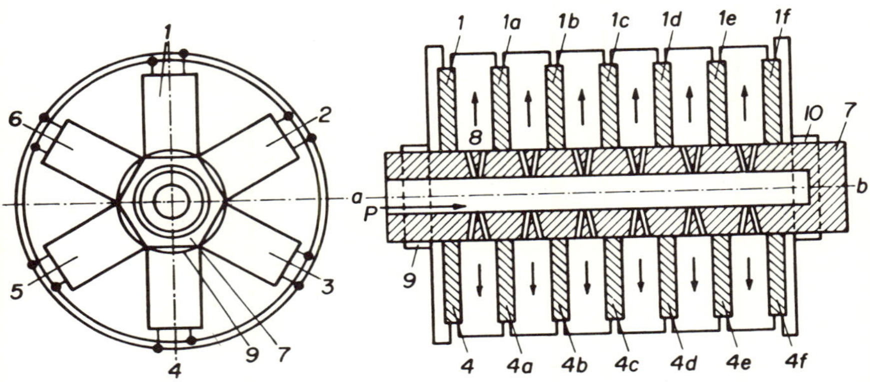

FIG. A10-1a. Air-cooled capacitor batteries of small inductance. The connection leads are fed into a pair of coaxial cylinders.

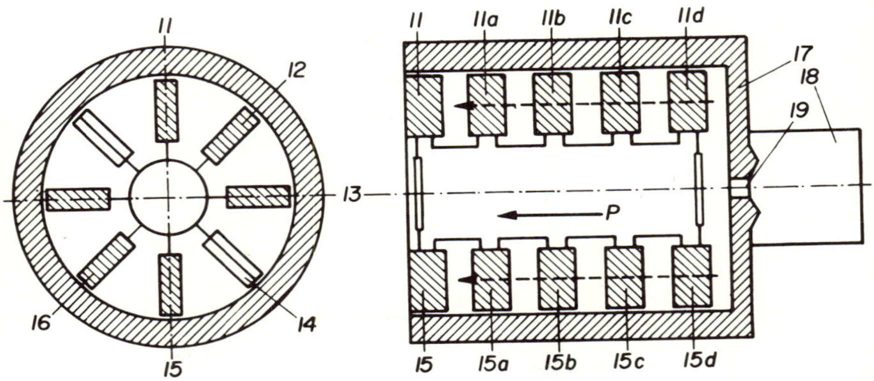

FIG. A10-1b. Another design of an air-cooled capacitor with series connection of the capacitor elements.

slots. As shown in the longitudinal section, a number of such cooling ribs are arranged in equispaced relation on the insulating tube, 7, and are thereby combined into a cooling rib system. The capacitors, 1, 1a, 1b . . ., form thereby a chain in series. The several chains, or longitudinal rows, can be connected with one another in parallel. In this manner the most favorable insulation conditions result because the capacitors of each rib row are practically of similar potential. The cooling air enters, in the direction shown by the arrows P, the interior of the insulating tube, 7, which at its periphery is provided with outlet nozzles, 8. If desirable, the whole system can be arranged to revolve continuously about its axis, a-b. In such an arrangement, the insulating tube, 7, serves as a shaft and as carrier of the slip rings, 9 and 10. When the system is revolving, cooling is more effective and the capacitors do not heat above allowable limits even under extremely heavy loads.

Another constructive modification is illustrated in Fig. A10-1b. Here the capacitors, 11, 12, 13, 14, 15, and 16, are arranged radially within an insulating tube 17. The plan of their largest areal expansion lies parallel to the axial section of this tube. As can be discerned from the longitudinal section, the successively arranged chains of capacitors, 11, 11a, 11b, . . ., form a multi-slotted cooling rib in the longitudinal direction of the tube, 17. To the end of the tube, 17, there is directly attached a compressed air spark gap (see Figs. B6c-1 and B6c-2) from whose housing, 18, there escapes the expanding (and thereby cooled) compressed air through the nozzle, 19, into the tube, 17. The capacitors 11, 11a, 11b, . . ., are again connected in series within the several rows running parallel to the axis, a-b, of the carrier, 17. The rows in turn are connected in parallel with one another. In this way one obtains a very compactly built device of high voltage and current capability. The thermally removable energy rate can be derived with help of conventional formulas used for calculation of heating radiators in freely circulating air [4].

For every short series of discharges, as a rule, the thermalic heat capacity of the capacitor battery proper suffices for ballistic consumption of the Joule's and ohmic heat that has been set free during the discharge.

An example may serve to give a better insight in this phenomenon: At an average specific heat for a capacitor in the order of 0.3 calories per gram and degree C, it is proper to assume that there can be dispersed 1.2 joule per gram of capacitor weight per degree temperature rise. At a normally allowable temperature rise of 100°C for the first operating periods, this will amount to a total of 120 joules in heat dispersion.

A capacity battery for a high-speed discharge sequence, as shown in Fig. A7-4, has a weight of about 3 kg and a capacitance of 0.2 μf at 10 kv charging voltage (that is, for 10 joules energy). When the battery is employed in a high-frequency flashing device at, for instance, 5000 discharges per second, and if the losses by resistance, loss angle, and dynamics of the discharge current path in the capacitor are assumed to be roughly 2%, then it is possible, owing to the weight of 3 kg and in accordance with above assumption, to invest thermally during a discharge series

$$3000 \text{ grams} \times 1.2 \text{ joules/(grams} \times {}^\circ\text{C)} \times 100^\circ\text{C} = 360 \text{ kj}$$

At 5000 discharges per second and 10 joules, there are converted in the discharge circuit 50 kw. Thus, at the assumption of 2% losses, the losses in the capacitor amount to 1 kw. Therefore, for 100 degrees of temperature rise, the capacitor could be operated at the specified load for 360 sec without having to resort to external cooling.

In impulse generators for light flash production it is generally possible to operate without external cooling when film exposures are taken at a maximum rate of, for instance, only 10^4 flashes. In contrast thereto, for experiments with impulse modulation of optical rays it is practically always necessary to cool continuously.

Extreme heating conditions occur when a capacitor battery is employed in a quenching or blow-spark discharge circuit for generation of high frequency for longer durations—for instance, in installations for hardening processes. An installation, calculated, designed, and executed by the author in 1948, has 20 kw deliverable hf output at 10,000 sparks per second and 3 joules in the

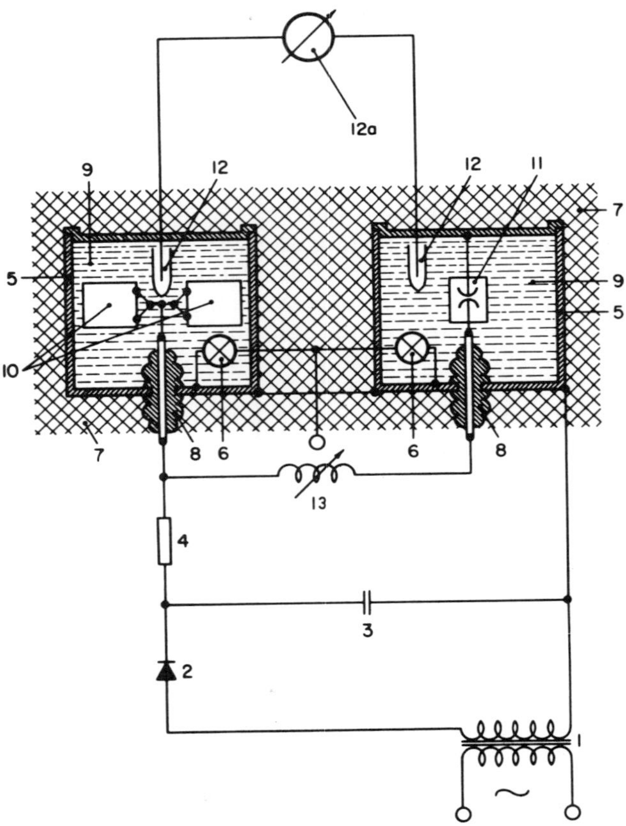

KEY
1. line transformer
2. rectifier
3. capacitor
4. charge resistor
5. calorimeter vessel
6. external heater for calorimeter fluid
7. thermal insulation
8. insulator
9. calorimeter fluid (oil)
10. capacitor elements under test
11. spark-gap for periodical discharge of capacitors 10 through a risetime varying inductance 13
12. thermopiles
12a. quotient measuring system indicating loss factor or efficiency
13. variable inductance

FIG. A10-2. Simple calorimetric measuring setup for total loss measurement of a capacitor in discharge operation.

capacitor battery at about 65% efficiency. For a calorimetrically measured capacitor loss in ceramic cup capacitors (resistance heating, not loss angle) it was found that there had to be carried off, continuously, 2% of 30 kw, or 600 watts. It was, of course, possible to work a few minutes without cooling owing to the thermal ballistic storage capability of the unit. Thereafter it was necessary to resort to oil cooling. With ceramic elements it presents no problem to install the battery with connections in a circulating oil system that has been designed in accordance with the thermosyphon principle.

Unfortunately, it is practically impossible to calculate the losses in impulse operation of a capacitor, because the following values, each of which in itself is often indeterminable, enter into the calculation: impulse wave form, Fourier's frequency content, loss angle for the different frequencies, skin effect in the capacitor, resistance coefficient for every current density during the impulse, static and electrostrictive coefficients of the capacitor construction used, and acoustic damping effects of the cooling medium.

In contrast, the experimental evaluation is rather simple. For that purpose the capacitor battery or the single element is set in a calorimeter as shown in Fig. A10-2. The calorimeter may be improvised from a room that has been air sealed and calibrated by incandescent lamp heating. The application of a simple oil calorimeter is also possible. The test subject is suspended in oil. In the latter case it is often possible to put the load, e.g., a spark chamber, into the same calorimeter and connect both thermocouples to a cross coil instrument which measures the quotient of both couple readings. In such manner, a direct calibration of the over-all efficiency in percent is possible. This reduces the time required for deriving extensive measuring data, e.g., with changing inductances or various gas fillings in the spark chamber as parameters. The saving in time may readily compensate for the more elaborate primary calibration.

11. Capacitor Battery for Continuously High Output in Discharge Operation (Impulse Welding Operation)

In impulse welding machines operating on production conveyors, as is the case in the manufacture of wire mesh for concrete construction, the impulse welding transformers have to receive the energy impulses for welding at the rate of one per second (corresponding to the feed of the wire mats). Operation can be continuous in three shifts per day (theoretically at the rate of 86,400 welds per day). In practice, however, owing to interruptions through loading, and so forth, only 30,000–40,000 welds are performed per day. Impulse rates of this order must be continuous for years of service and must be reliable in two respects, namely: there must be a low loss rate, and an unexpected break down of one capacitor must not cause a shutdown or an interruption of production, except for very short intervals. A capacitor battery of that kind that has served satisfactorily for years is described here briefly. Technical data:

100 single elements, 40 μf each, 5 kv

Total energy, 50,000 watt-sec = 50 kj

Discharge sequence, 1 per sec

Charging output of the rectifier, 100 kv-amp (selenium rectifier with leakage reactance transformer, without charging resistance, no effective watt losses)

Discharge impulse form resembling a 50 cycle half-wave

In order to enhance the reliability of operation, the battery has been subdivided into the mentioned 100 elements. Premature failure of one unit results in a scarcely detectable drop of only 1% in the output of the machine. Each single capacitor is fused by an upward arched link of fuse wire. In case of a breakdown of one element, the current of the remaining 99 elements follows a path through the single fuse link of the defective element and evaporizes it. Undestroyed energy in the form of mechanical burst energy within the defective capacitor would result in the soiling of the installation, as well as in an overly-rapid discharge of the sound elements. In order to destroy as much energy as possible the fuses have been formed of iron wire having a great mass melting heat. (The spark flow from the blowing fuse link is guided by magnetic action in a desired direction.) A discharge energy of 1 kw-sec is capable of melting 1 gram of iron. For the desirable energy consumption there are needed 10 grams or more wire material. In cases where the arc formed by blowing of a fuse is quenched magnetically by proper shaping of the fuse wire and leads, it is quite natural that less wire mass will suffice. Final decision concerning fusing can be made only by experimenting.

Natural cooling of the battery (at 1% estimated loss, each capacitor sets free about 5 watts in heat) is sufficient in open installations where air can circulate freely around the capacitor battery.

FIG. A12-1. Drawing of a pocket dosimeter according to Früngel, Kristensen, and Ambrosen [1]. (I) Cross section through the measuring capacitor in closed condition; (II) cross section through the measuring capacitor after opening and inserting of the contact plug for reading the charging decrease; (III) measuring capacitor with outer screening which at the same time serves to increase the wall thickness to eliminate softer radiations; (IV) fitting of measuring capacitor between support plates which are of a material suitable to write on. Key: 1–Aluminium cup, 2–base plate, 3–upper connecting plate, 4–packing, 5–cover, 6–holes for mounting key, 7–bore holes for mounting, 8–perforations in the soft plastic cover 5, 9–spring washer and grove, 10–shim, 11–screws for the contact membrane, 12, 13–contact rivet of precious metal, 15–hollow contact body of the insulated cover of the measuring capacitor, 16–drying agent pill, 17–contact surface, 19–thread for screwing the contact body into the measuring capacitor, 20–measuring capacitor, 21–radius of the discharge space in which the ionizing radiation forms charge carriers, 22–shim, 23–wall thickness, 24–effective discharge space, 25–lower limitation disc with graphite coating, 26–support sleeve of the measuring capacitor, 27–shim to prevent squeezing of the dielectricum, 28–detrimental discharge space which cannot be calibrated but must be kept as small as possible, 29–contact ring for neutral connections of the grounded cover, 30–grounded (neutral) capacitor foil, 31–bore hole for mounting key, 32–sleeve for the measuring device, 33–highly insulated mounting plate, 34–measuring contacting pin for charge checking, 35–protective coat, 36–sealing material respectively special paper for recording the indicated charge differences between the various capacitor recharges, 37–control and support plate with notes.

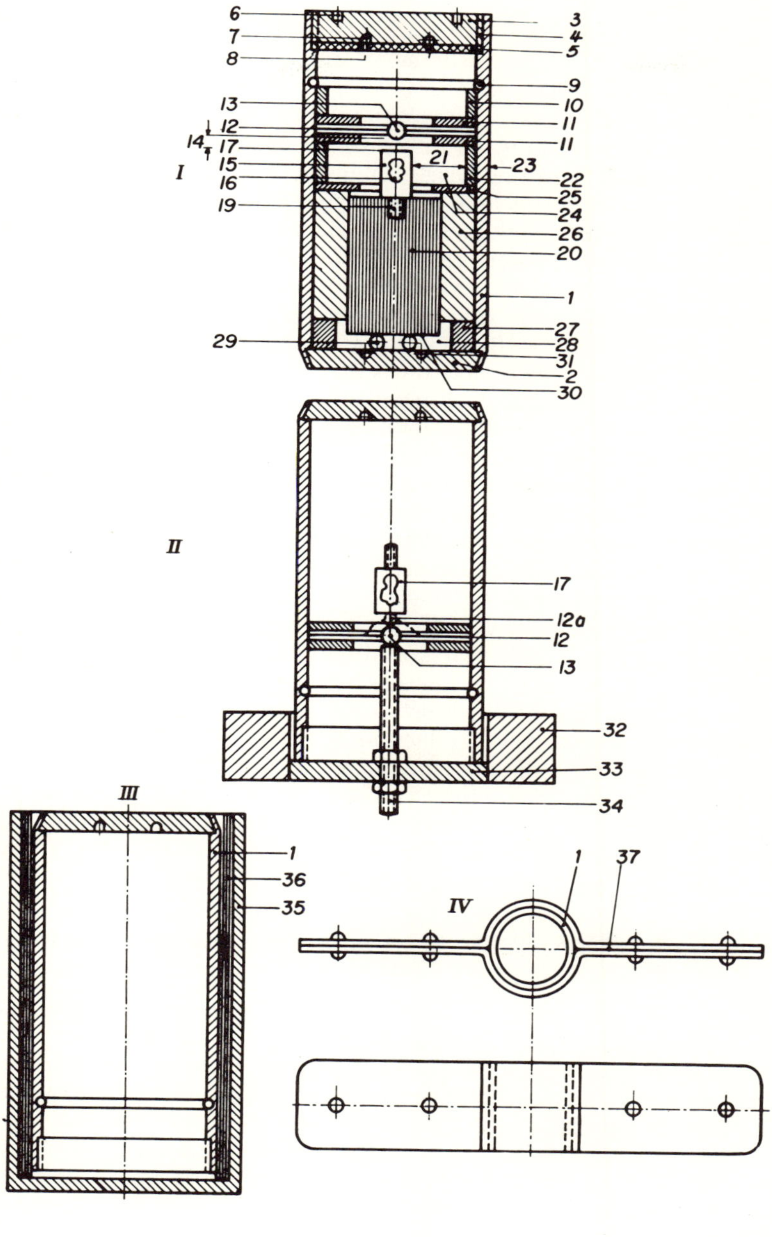
I
6
7
8
13
12
14
17
15
16
19
29
3
4
5
9
10
11
11
21
23
22
25
24
26
20
1
27
28
31
2
30
II
17
12a
12
13
32
33
34
III
1
36
35
IV
1
37

12. Capacitors Having Extremely High Insulation Values

In the field of radiation measurement, one of the most exact measuring instruments is the so-called chamber dosage meter (*Kammer dosimeter*). Here a very heavily insulated measuring capacitor is charged to a standard potential. After a measuring interval it is checked for its charge loss. If the measuring intervals are extended to one year—for instance, by yearly reading of the x-ray unit dose at buildings and in open areas—the insulation resistance of the capacitor must be so high that during the measuring year no falsification is introduced through inherent capacitor losses. Therefore, half-value times of 50–100 years are desirable. At a capacitor value of 1000 pf, a practical size for measuring radiation values between 10 and 300 x-ray units, the self-resistance must be 3.2×10^{18} ohms for a half-value time of 100 years. Ambrosen [1] obtained half-value times of up to 2000 years by utilizing two polystyrol foils with intermediate mylar foil as a dielectric. The unavoidable holes in each foil, which make multilayer insulation necessary in high-voltage capacitors, here become effective loss resistances in this low-voltage operation. In the foil combination referred to here, there is built up empirically around each imagined hole a potential protective wall by several day-long polarization treatments at a few hundred volts. The wall prevents further current flow. The insulation resistances thus obtained are so high that, under the influence of continuously acting cosmic rays (the discharging action of which is higher than the rest resistance by 1–2 powers of 10), they can only be extrapolated. About 2000 years, or in the present example more than 6×10^{19} ohms or 6×10^{22} ohms/pf, can thus be achieved.

Figure A12-1 shows the construction of a dosage meter (Kristensen and Früngel) for inexpensive mass production. When a measurement is to be made, a contact pin pierces a diaphragm; the hole thus caused closes itself after removal of the pin. The protective hull of the capacitor contains the paper on which the recorded values are printed. Such dosage meters have been built with capacitances of 10, 30, 100, 300, and 1000 pf and proved sufficiently well sealed to be buried for ground research work.

13. Capacitor Lines with Constant Impedance per Unit of Length

Frequently it is necessary to generate electrical pulses with very short rise times for the simulation of scintillation pulses for photomultiplier arrangements, the analysis of photosignals, impedance measurements, testing of photomultiplier tubes and light-sensitive semiconductors, and measurements with magnetic materials.

In answer to this challenge, in developmental work which took almost 10 years, the USAF Research Laboratories in Bedford [767], under the direction of Dr. Heinz Fischer, developed and refined a special spark-gap diode for the nanosecond range. Provisions are made for the connection of a voltage divider, permitting the selection of any desired pulse voltage.

These Fischer capacitors have been described in the *Journal of the Optical Society of America* [471]. In contrast to the earlier spark-gap techniques, the so-called Fischer Nanolite is capable of producing extremely short rise times down to 1 nsec. This has been accomplished by the capacitor design which resembles electrically a length of transmission line with an inductance of about 1 nh. Because it is similar to a cable with very thin insulation, each amount of capacitance corresponds to an equivalent amount of inductance. An inductance of 2 nh with 4 kv operation voltage has about 0.005 μf, another type has 0.00065 μf with 1 nh. The following diagram, Fig. A13-1, shows the discharge curve of such a capacitor measured by a photodiode.

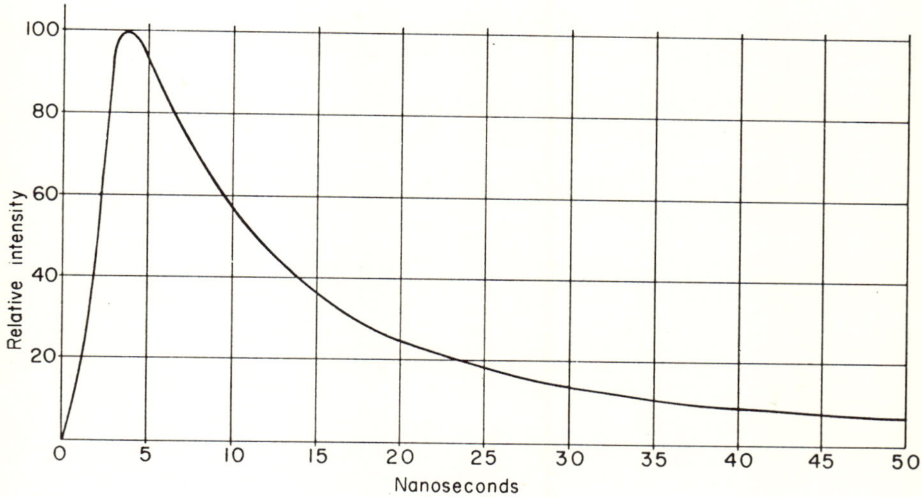

FIG. A.13-1. Discharge of the Fischer capacitor. Photodiode: FW-114a, 2kv, S-4 response. Oscilloscope, EGG Model 707. C = 650 pf; U = 4.1 kv; gap = 0.97 mm; 10–90% rise time, 1.9 nsec, half-width, 9.2 nsec, 10% width, 36 nsec.

Basically it should be possible to realize the Fischer principle with any discharge capacity. However, the fabrication of the coaxial capacitors—which in addition to the required electrical capacity have to be adjusted to the time constant impedance—and the outside shell of extremely low inductance require high precision workmanship. Therefore this manufacturing process is economical only for a limited number of types and in correspondingly larger quantities.

The above mentioned capacitor with 0.005 joule energy is suitable for the half-time measurements on semiconductors, photomultiplier tests, and other photoelectric research work as well as for the generation of electrical pulses with a rise time of 2 nsec and a voltage of about 4 kv.

The spark discharge of the Fischer-Nanolite is a true pinch discharge. The peak current occurs at the pinch condition. The "tail" of the light emission corresponds to the subsequent plasma expansion during the low current or zero-current phase. When the Fischer Nanolite is operated at high repetition rates, special care should be taken so that the electrodes do not begin to glow and the Teflon gap insulation does not get too hot, otherwise the current duration increases and the life of the capacitor is shortened. The life expectancy of this capacitor at low repetition rates of for instance 50/sec is exceptionally long, up to 10^7 discharges.

B. Switching means

1. Thyratrons

To select a switching means, one must know whether the discharge of the energy-spending capacitor is aperiodic or oscillatory. In both cases it is essential to know the peak current, the total number of coulombs carried during the discharge, and the charging voltage of the capacitor. The switching means described in this chapter are adapted to meet the requirements mentioned.

The thyratron generally is suitable only where an oscillating discharge is impossible or can be dispensed with. However, it can also be used to cut off undesirable oscillations in a discharge circuit at the first passing of the current through zero, thus making oscillation impossible. Two essential values of the thyratron are its cut-off potential and its peak emission current. In cases of repeated discharge the average allowable cathode current is of interest, and in devices for continuous operation the guaranteed life expectancy must be considered.

a. *The Cold-Cathode Thyratron*

The cold-cathode thyratron includes a nonheated cathode. Adjacent to it, or concentric within its aperture, there is mounted an ignition electrode. Facing the cathode in spaced relation and of different configuration is the anode. Cold-cathode thyratrons have the great advantage that they need no heater. This advantage becomes acute when the cathode terminal of the thyratron is at a high potential and a heater transformer would increase the cost of the switching project. The cold-cathode thyratrons should also be considered when the switching operation must occur at an instant which cannot be predetermined—for example, in twilight switches or in devices without basic current consumption, or where a heater would be objectionable under locally given conditions. The field of application for the cold-cathode thyratron is expanding rapidly today. The classic cold-cathode thyratron withstood an inverse voltage of about 300 volts at an allowable peak cathode emission current of about 0.2 amp. In recent years, however, new developments have continually entered the market. Today the upper limits of the cold inverse voltage lie in the range of a few kilovolts, and the maximum peak current at the cathode in the kiloampere range. Basically it can be said in regard to the cold-cathode thyratron that, owing to the delay between initiation of the ignition spark and the discharge proper, it is not possible to obtain such precision in starting as can be expected in hot-cathode thyratrons. The reason for this is that before ignition free electrons are not present in the tube in sufficient quantity, as is the case in the hot cathode. These electrons must first be produced at the cathode after initiation of the ignition discharge. Special constructions of cold-cathode thyratrons are available in which the discharge itself is the sole purpose. This discharge is lengthened in capillary fashion to become visible as an intense light flash with an energy output corresponding to the energy of a charged capacitor fed into the tube. Most of the discharge lamps of the so-called glow-lamp stroboscopes and the xenon-filled stroboscope lamps are actually cold-cathode thyratrons. Moreover, the photo flash lamps designed with 20–300 joules flash energy belong to the same species. Forming relatively high ohmic loads, such stroboscopic flash lamps can also be used with relatively high blocking voltage, a few kilovolts, as switching tubes for heavy currents. However, the wall-stabilized glow discharge, which is chiefly responsible for the light output, is burdened with a high voltage drop. As a result, the switching circuits employing such stroboscopic flash tubes generally have low efficiencies. However, when only a current impulse is required, and a predetermined number of coulombs, depending on the capacitance of the capacitor, has to be supplied to the load, the efficiency is of little consequence, since a high voltage loss in the tube can be tolerated. An application of this sort presents itself in small magnetizing devices in which only one current pulse is passed, without oscillation, through a small coil.

The thermally limited energy of lamps lies in the range of 50 to 10,000 joules. The allowable average load is between 1 and 100 watts. The inverse voltage, which hereby becomes effective at the tube before ignition, is 300–500 volts in the smallest types, in larger ones, about 2.5 kv. Ignition in flash lamps can be capacitively effected from the outside through the glass. Within the envelope in photo flash lamps, there are always only the two main electrodes. Ignition is caused through a helix wound around the tube or through a cathode-adjacent metallization energized by a steep voltage pulse.

The cathode in most cases is prepared with a barium oxide paste in order to reduce the electronic work function (emission potential) and thereby facilitate ignition. Connecting with wrong polarity can destroy the lamp at the cathode side.

The discharge duration of photo flash lamps is 1/10,000 to 1/100 sec. A tube for 250 joules at 500 volts and 10 millisec flash duration delivers consequently 50–100 amp as a current pulse. The cold-cathode thyratrons have been perfected during recent years to a high degree. They can perform several thousand switching operations per second as in rapid counters, or they can stand ready for action over the years without change in their data, as in twilight switching devices, because dischargeless times have no effect on the life of the tubes.

b. *The Heated Thyratron*

In the field of heated thyratrons, development over the years has brought forward essentially three distinct types characterized by different fill-gases. The mercury vapor filled tubes with a slight admixture of argon have the highest life expectancy—over one year in continuous service.

The temperature dependency of the argon admixture is low, whereas with a pure mercury vapor filling it is high. Such tubes are commercially fabricated for guaranteed peak currents up to 100 amp and for inverse voltages up to 5 kv, and in special cases up to 10 amp and 20 kv respectively. A rough dimensioning rule is that about 100 watts heater energy will render 10 amp current emission. For frequently repeating discharges at a sequence frequency below 1000 per second and with average discharge currents below the aforementioned limits a proper selection for a nonoscillating case would be the mercury vapor filled thyratron. It also has the advantage of needing very little ignition energy. If delays on the order of 10^{-5} sec between ignition pulse and main discharge are permissible, the discharge energies for ignition can be as low as a few microwatts. However, when short delay intervals are requested and deviation cannot be more than some 10^{-7} sec an energetic ignition pulse must be imparted to the grid of the tube. This pulse must be strong enough to supply an essential ionization contribution for plasma ionization between cathode and grid. For a rough estimate the grid ignition energy needed for shortest ignition delay can be derived from the product of gas pressure, volume between grid and cathode, and ionization potential. The latter can be assumed to be 100; the number of particles under customary fill-gas pressure, about 10^{16}–10^{17} per cubic centimeter [6]; and the volume, depending on tube type, 1–10 cm^3. From the product of these three values and the elementary ionization loading the required ignition pulse energy is obtained. After evaluation, one obtains energies in the range from 0.1 to 10 milliwatt-sec. This energy, of course, has to be imparted to the grid in the form of an instantaneous voltage pulse, for instance, within 10^{-7} sec, in order to assure a rapid initiation of firing. By means of the measurement techniques described by Kröbel [7a] it is possible to reduce deviation to a value far below 10^{-7} sec.

The inert gas filled thyratron has properties similar to the mercury vapor filled one. The present trend is to replace the heavy and temperature-dependent mercury vapor with xenon. In its discharge mechanism xenon resembles mercury vapor with respect to quantity of electrons and variety of excitation states. But it has a much lower temperature dependency. A slight drawback is that tubes with xenon filling have lower inverse voltage values. An important construction advantage is their posture independency, i.e., they do not have to be mounted base down. Such mounting is necessary with mercury vapor tubes, unless their filling is under-critical, because at cooling of the tube particles of condensed liquid mercury can be deposited upon the cathode. This will markedly shorten the life of the tube. At present, almost all tubes which were formerly fabricated with mercury vapor filling are commercially obtainable with xenon filling. The life expectancy of xenon-filled tubes is about 5000 hours, and that of mercury vapor filled tubes is at the maximum about 30,000 hours when mercury is injected in excess, i.e., more than required for the vapor filling.

Tubes having 2–3 kv inverse voltage and about 80 amp permissible peak current at a few amperes permissible average current are common. A numerical example may serve for illustration: A capacitor of 3000 pf capacitance and 3 kv charging voltage is discharged at a frequency of about 1000 per second into the inductance of the primary winding of an impulse transformer, e.g., to produce ignition sparks. Then the average current will be

$$I = Q \times f = (3 \times 10^{-9})\,(3 \times 10^{3})\,(10^{3}) = 9 \text{ milliamp}$$

It appears that a small tube would be sufficient. When the primary inductance of the impulse transformer is assumed to be 100 μh, one obtains as peak current

$$L \times I^2 = C \times U^2,\ I = \sqrt{\frac{(C)\,(U^2)}{L}} = \sqrt{\frac{(3 \times 10^{-9})\,(9 \times 10^{6})}{10^{-4}}}$$

$$= \sqrt{270} = 16.4 \text{ amp}$$

It is therefore necessary to select a tube good for 17 amp continuously repeatable cathode peak current at 9 milliamp average cathode current and a positive inverse voltage of 3 kv. The deionization time of this tube must be $t = 10^{-3}$ sec because $f = 1000$ discharges per second.

c. *Operating Principle of Thyratron Tubes* [9]

By the term "thyratron" there is meant a discharge chamber in which are arranged a cathode, one or several grids, and an anode, and which is filled with an inert gas or a metal vapor. The gas filling distinguishes this tube in its function from the high vacuum tubes used in the fields of communication and entertainment. The varying characteristics due to different gases will be discussed later.

Disregarding the control grid, we may say that the discharge in the tube progresses so that the electrons emitted from the cathode assume high velocities under the influence of the supplied anode voltage. These velocities are sufficient to ionize the gas atoms in their path, i.e., to split them into one electron and one positively charged ion. The ions are responsible for a considerable part of the charge conveyance within the tube (on account of their positive charge, in direction anode→cathode) and effect a neutralization of the space charge. The result is that the anode current becomes larger by several numerical orders than that in a corresponding high vacuum tube.

If a grid is interposed between cathode and anode, and a sufficiently high negative control potential is imposed on this grid, no current flows through the tube when an anode voltage is applied. However, when the negative potential at the grid is reduced to a certain value, or when it is made somewhat positive, the full anode current begins to flow instantaneously at a certain grid potential. The tube has fired, and a light arc forms between cathode and anode. The voltage drop is practically independent of the current value. It depends on the kind of fill-gas and its pressure. Moreover, the pressure in mercury vapor filled tubes is, to a large degree, temperature dependent so that the arc voltage drop lies in the range of 16 to 32 volts. The losses occurring in the tube (arc voltage, average anode current) are therefore relatively low.

d. *Control Characteristic and Anode Current Value*

The relationship between anode and grid voltage of the tube required for firing, is generally represented graphically by a curve of critical control characteristics. There are tubes having negative, positive, and transgressing characteristics, depending on whether the grid control potential required for firing lies in the negative region, in the positive, or in the boundary between both. Figures Bld–1, Bld–2, and Bld–3 show the three possibilities for tubes having mercury vapor filling.

In the range at the left of each curve the tube remains nonconducting. At the right it becomes conducting. It should, however, be clearly understood that a tube which has become conducting cannot be extinguished by making

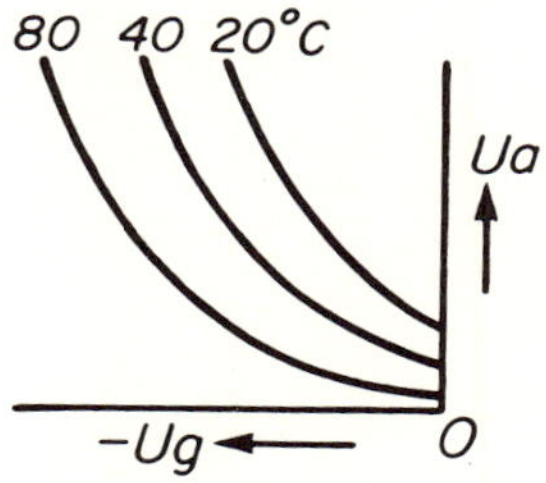

FIG. B1d-1. Negative control characteristic of mercury vapor filled thyratron tubes.

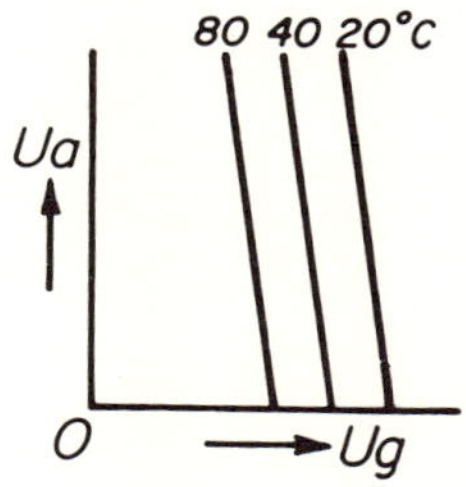

FIG. B1d-2. Positive control characteristic of mercury vapor filled thyratron tubes.

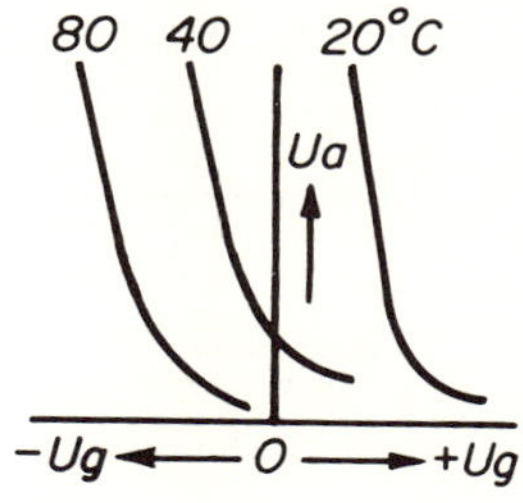

FIG. B1d-3. Transgressing control characteristics of mercury vapor filled thyratron tubes. Rising operating temperature shifts the curve into the range of more negative control voltage. *Ua* = anode voltage; *Ug* = control grid voltage.

the control voltage more negative. This is not the case with high-vacuum tubes, in which the flow of anode current can be stopped any time by the negatively controlled grid. A cloud of positive gas ions which envelops the grid causes it to lose its control ability. The discharge can no longer be influenced by it. The anode current flows until the voltage between anode and cathode drops below a certain value, namely the arc voltage drop value. The deionization time, which counts from extinguishing of the light arc or cessation of current flow, can be shortened only by such actions as are described in Section 3 of this chapter.

e. *Structural Assembly of the Thyratron Tube*

Figure B1e-1 shows the construction of a thyratron tube. The envelope consists of heavy wall glass to resist rough operating conditions in industrial installations. The anode is of metal or graphite. The form must be such that good heat dissipation, to avoid secondary emission, is assured. The cathode is provided with an oxide layer and is capable of delivering an appreciable emission current. It can be heated directly or indirectly. Its design deviates

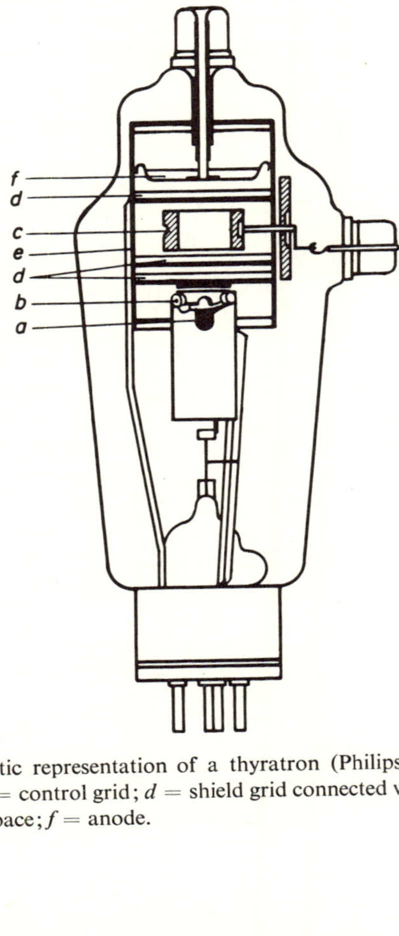

FIG. B1e-1. Schematic representation of a thyratron (Philips Tetrode). a = cathode; b = cathode shield; c = control grid; d = shield grid connected with the cylindrical shielding e of the discharge space; f = anode.

from that of a cathode in a high-vacuum tube. Since the electron paths in gas filled tubes need not necessarily be straight lines, it is possible to give the cathode, e.g., the form of a helix. Such shape considerably reduces losses due to heat conduction and radiation.

The time needed for preheating varies, with style, form, and size of the cathode, between a few seconds and several minutes. Tubes with mercury filling cannot be put into operation until a sufficient quantity of mercury has been evaporated by heating to produce the necessary pressure in the tube. During this interval there must be no load in the anode circuit of the tube.

The control grid generally has an annular or screenlike shape. It is made of graphite or metal. Thyratrons having a negative control characteristic are provided with an annular control grid having a relatively large inner opening. Tubes with a positive control characteristic have a system of perforated, screenlike metal plates as grids. It is evident that the increased shielding obtained in the latter case results in firing at more highly positive grid voltages.

As previously mentioned, there are tubes which have a shield grid arranged on both sides of the control grid. Its purpose is to shield the control grid and limit the current flowing through this grid during operation as much as possible. This construction is an advantage in some applications. Thyratrons with shield grids make possible the operation of the control grid in connection with a very high ohmic grid leak resistor. With capacitive or photoelectric control of the ignition impulse this measure is often necessary.

f. *Current and Voltage Characteristics of Gas Filled Rectifier Tubes*

Since the mercury thyratron or the inert gas filled tube forms the most widely used electronic switching means in impulse engineering, a more complete description of essential basic dimensioning rules will now be given.

Every thyratron is in principle a gas filled rectifier tube in which, by means of a control grid, current passage can be prevented at will. Therefore, it appears to be proper to discuss briefly the fundamentals of gas filled rectifier tubes.*

According to the type of filling, there exist three groups of rectifier tubes: high vacuum tubes, metal vapor tubes (mercury, cesium), and inert gas tubes (argon, xenon, krypton, neon, helium, etc.).

The gas filling lends the characteristic properties to the tube. These characteristics and the resultant operating behaviors will be discussed here. Concepts that are essential for the understanding of some phenomena will be explained with graphs.

There are glow cathode tubes with gas filling as well as high vacuum tubes in practical use. The difference in behavior between the two groups can best be ascertained by contrasting the characteristic curves U_a versus I_a of two diodes of similar size (Fig. B1f-1). In the high vacuum tube (RG 62) the

* The remainder of Section 1f is taken from a commercial paper of the Brown Boveri Company by R. Hübner and from Weicker [117].

internal voltage increases considerably with rising current value. It remains low and is practically independent of the current in the gas tube (DQ2). A high voltage drop, however, results in the electrons bombarding the anode with great force and heating it. Therefore, thermal considerations limit the current handling capabilities of high vacuum tubes. A means of increasing the current handling capability of a high vacuum tube is to improve the heat radiation of the anode by enlarging its surface. This, however, is possible only to a certain degree because of the unreasonable increase in tube size. In contrast, in the gas filled tube the anode surface can be small with respect to the current load. This fact makes it possible to build gas filled tubes for considerably larger current outputs (at present up to 25 amp). Since a lower voltage drop (10–20 volts) causes less energy loss, gas filled tubes operate at considerably higher efficiencies than high vacuum tubes, the voltage drop of which can be as high as a few hundred volts owing to high internal resistance.

The internal resistance (R_i) is defined as the change in voltage drop for a certain change of anode current; thus $R_i = \Delta U_a/\Delta I_a$. In high vacuum tubes R_i lies in the range of a few 100 ohms (in RG 62, for instance, 200 ohms). In gas filled tubes this quotient hardly changes, so that $R_i \approx 0$. This peculiarity has to be considered in the operation of gas filled tubes. During a short circuit in the load wiring, the current limitation otherwise present in the inner tube resistance is missing. Destruction of the tube will occur if devices such as quick acting circuit breakers, soft network transformers, or protective resistors are not provided in the anode circuit of each rectifier tube.

In the mercury vapor filled rectifier tubes a negative internal resistance is possible. Under heavy currents a summation of the pulse energies of the electrons striking a gas atom may take place resulting in the arc voltage dropping below the ionization potential. With heavy currents, such tubes can still burn and conduct at an arc voltage of only 5 volts. Therefore, in the parallel connection of tubes, there must be a series resistor in the lead of each tube. This causes, with respect to the arc voltage, a relatively high ohmic voltage and counteracts the possibility of a negative R_i. Example: Peak current 10 amp, ionization potential 13 volts, voltage drop at series resistance $2\ U_i = 26$ volts, i.e., 2.6 ohms.

The short circuit surge current, as given in the technical data sheet of a tube, must not be surpassed in the event of unexpected short circuits. Its duration must not be longer than 0.1 sec. If possible the tube should never be burdened up to this surge current (in the TQ 2/6, for instance, 600 amp) because it can carry it for only a few seconds without damage. The purpose of this data figure is to guide the designer in providing proper means in the application circuit to protect the tube in case of a short circuit.

As shown in the diagram, Fig. B1f-1, in gas filled tubes there is also an abrupt increase in voltage drop (arc voltage drop) at the saturation point (I_{ap}) of the cathode emission current. Therefore, at the instant when the cathode is loaded to the limit of its emission capacity, a heavier current is possible only through higher ionization. However, in order to produce higher ionization work, the electrons must be accelerated faster. The voltage drop between cathode and

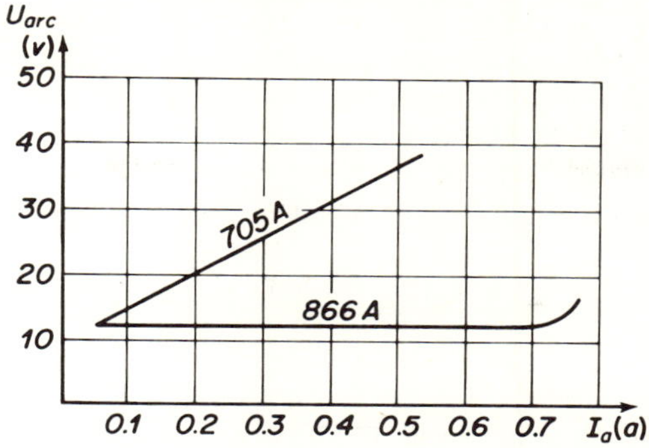

FIG. B1f-1. Internal voltage drop (arc voltage) U_{arc} in dependency of anode current I_a in a high-vacuum tube (RG 62=705A), and in a gas-filled rectifier tube (DQ 2=866A).

anode is raised. The result is increased heating and dangerous bombarding of the cathode with highly accelerated, returning, positive gas ions. These can rapidly destroy the cathode. There is for each gas filling, when an oxide cathode is employed, a critical voltage drop (disintegrating voltage, at the surpassing of which the danger of cathode destruction is present). With mercury vapor this voltage is about 22 volts; with neon it is 28 volts. During the life of the tube the internal voltage drop increases. In inert gas tubes there is also gas consumption, since during operation inert gas ions are shot into the microcrystalline crevices of the cathode and become lodged in the structure. This means that they are lost for further performance.

This effect may occur not only when the upper anode peak current value (I_{ap}) is surpassed, but also in the case of under-heating. Operating the cathode at a low temperature (Fig. B1f-2) prevents the oxide cathode from emitting

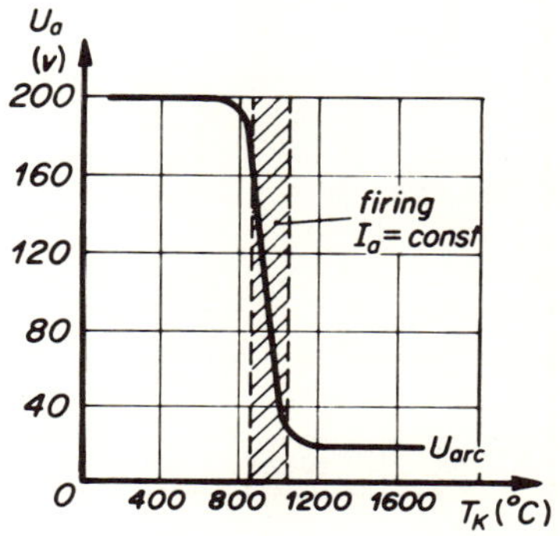

FIG. B1f-2. Voltage trace of a gas discharge path in relation to cathode temperature T_k. Only after bringing the cathode up to the prescribed operating temperature, so that it can deliver the necessary energy for the work function (emission work) of its material, does the internal voltage drop to the low value of the arc voltage, that is, about 10–12 volts in Hg vapor tubes, 14–16 volts in argon and xenon tubes, and 20 volts in neon tubes.

a sufficient quantity of electrons. Therefore, the following precautions should be observed in experimenting and operating with gas filled tubes:

a. Adhere strictly to the prescribed preheating time and the heater voltage tolerances. Never under-heat!

b. Adhere to the prescribed momentarily permissible anode peak current (I_{ap}), limiting it to the short interval given by the integration time (never more than a few seconds).

By interposing a grid one can convert the rectifier tube just described into a thyratron.

g. *The Firing Characteristic of a Thyratron* [117]

For every anode voltage value there exists a certain value of grid voltage at which a thyratron fires or becomes conductive. This value is U_{gc} = critical grid voltage. If these firing points at various anode voltages are connected in a graph, there results the plot of a certain thyratron type U_{ap} (or U_a) = f (U_g) (see Fig. B1g-1).

For higher U_a this curve follows the linear relationship $U_g \approx 1/D \times U_a$. For low U_a it departs from the straight line and can even enter the positive area, so that here positive grid voltage become necessary for firing.

In this range, at first a predischarge between cathode and grid occurs, the grid acting as anode. The charge carriers produced by this action travel to the anode to start firing (breakdown) between cathode and anode (see Section 1b of this chapter).

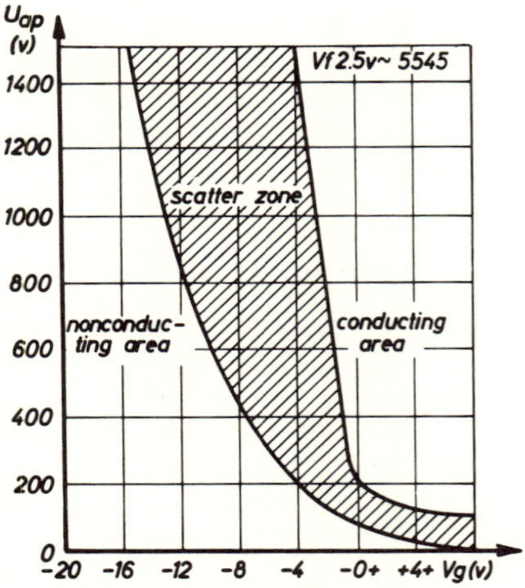

FIG. B1g-1. Firing characteristic of a thyratron (TX 2/6). $U_{ap} = f$ (U_g), U_{ap} = peak value of the applied anode ac voltage (in the rare case of operation with anode dc voltage the latter would be designated by U_a). At an anode ac voltage of 440 volts, corresponding to a peak value of 600 volts, the tube would positively fire at a negative grid potential of 2 volts (depending on tube type and external conditions, possibly at −9 volts). The cross-hatched area represents the variable range from tube to tube.TX216 equivalent to 5545.

Since for practical use it is not necessary to build thyratron tubes with close tolerances, manufacturers allow certain variations from tube to tube. These are represented by the (cross-hatched) variable range of the characteristic. At the left of this range the tube does not fire (no breakdown area); at the right, the tube will positively fire (certain-breakdown area).

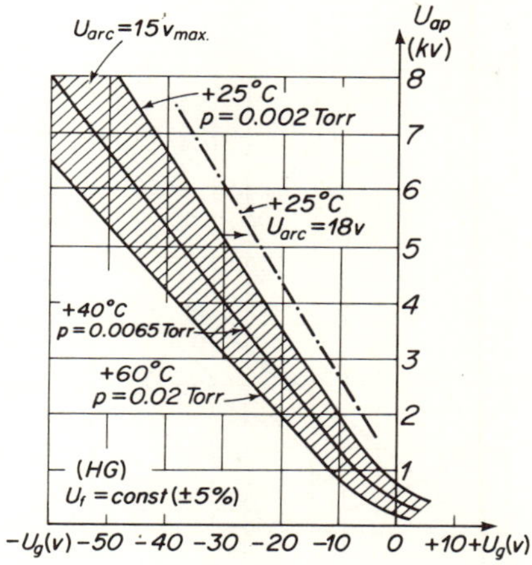

FIG. B1g-2. Firing characteristic of a mercury vapor thyratron at various Hg temperatures and Hg vapor pressures (p). Highest temperatures require highest gas pressures and shift the curve toward the left, i.e., at given U_a the tube will fire at a higher grid potential or earlier than at lower temperatures. The temperatures entered are the ones of the condensed Hg (measureable by thermoelements in a region a few millimeters above the tube socket on the glass envelope). Higher heater temperatures shift the curves toward the left; higher arc voltages (arc drops), toward the right.

Within the cross-hatched areas lie not only variations from tube to tube but also allowable heater voltage fluctuations permissible to $\pm$ 5% as well as out-of-phase conditions between heater and anode ac voltage. Besides, shifts of the curves occur owing to increased internal resistance near the end of the life of the tube. With progressing time of use, the curves shift toward the right as shown in the graph. Tubes having mercury vapor filling are furthermore very dependent on the vapor pressure and the ambient temperature. At higher temperatures the curves shift toward the left (Fig. B1g-2). This temperature dependency of mercury tubes reduces their usefulness (use at lower ambient temperatures than about + 10°C is generally not feasible) as against inert gas filled tubes, which can be operated in ranges from about − 50°C to + 70°C.

h. *Temperature Characteristic of a Mercury Vapor Tube* [117]

Figure B1h-1 shows the relationship among mercury vapor pressure p (Torr), inverse voltage U_{inv} (kilovolts), and the internal voltage drop U_{arc} (volts) (also known as arc voltage drop, or arc drop) versus the temperature of the condensed mercury T_{Hg} (°C). From this representation it is apparent that rising temperatures increase vapor pressure, and reduce inverse voltage. In contrast, lower temperatures increase the arc drop and may lead to destruction of the cathode. The theoretically possible increase in inverse voltage at lower temperatures up to the maximum value recorded in the tube data sheet is not obtainable owing to voltage problems.

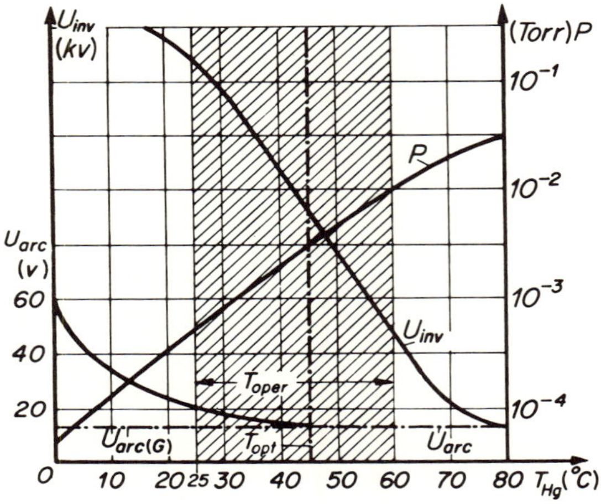

FIG. B1h-1. Mercury vapor pressure p (Torr), inverse voltage U_{inv} (kv), and arc voltage U_{arc} (volts) as functions of the Hg temperature T_{Hg} (°C). T_{oper} = permissible operating range of the mercury vapor tube; T_{opt} = recommended preferable Hg temperature; U_{arc} (G) = arc voltage (drop) of a new tube.

i. *Ionization and Deionization Characteristics* [117]

Ionization time is defined as the time between the instant the critical grid voltage is first reached and the time of complete firing. During this time the voltage between anode and cathode decreases to the normal operating value of the arc voltage (Fig. B1i-1). This time is not constant but depends on several factors. It can vary from fractions of one μsec up to several hundred of μsec. The figures given in the tube data are only rough approximations [7a].

The ionization time depends not only on the curve shape and the amplitude of the grid voltage initiating firing, but also on the tube design and construction and the kind of gas used. Whereas high negative grid voltages lengthen the ionization time, t_i, short positive pulses (up to +40 volts) may shorten it down to 0.1 μsec (see Section 1b of this chapter). The application of grid overvoltages, that is, voltages above U_{gc}, makes possible the ulilization of inert gas filled triodes in blocking circuits and oscillographs, where very short t_i are necessary. The graphs represented in Fig. B1i-2 are furthermore dependent on the gas pressure and kind of gas.

The heavier the molecules of a gas, the slower they move and the longer it takes until sufficient ions are produced to initiate firing. Since mercury atoms are heavier than xenon atoms, it is clear that tubes with mercury vapor filling have longer ionization times than inert gas filled ones, although in the latter the higher gas pressure acts unfavorably in that respect. The shortest t_i have thyratrons with hydrogen filling, as mentioned in Section 1o of this chapter; they are therefore well suited for higher frequencies (radar).

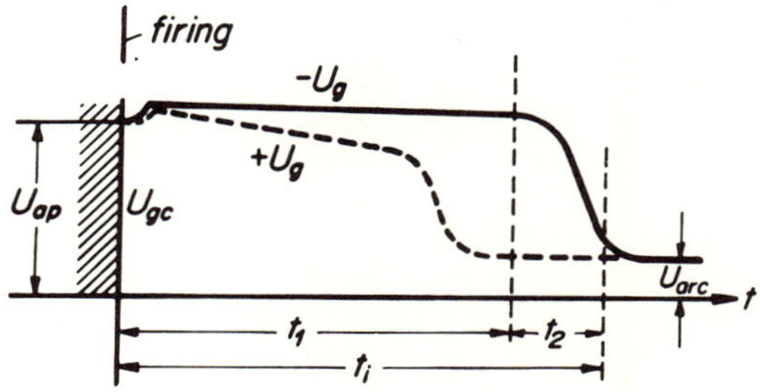

FIG. B1i-1. Representation of the progression of ionization t_i in a mercury vapor thyratron for positive and negative grid voltage. $t_i = t_1 + t_2$; t_1 = time lag, t_2 = breakdown period; U_{ap} = peak anode voltage (e.g. line voltage applied to a nonfired (cross-hatched) tube); U_{gc} = highest grid voltage at which firing is initiated; U_{arc} = normal operating arc voltage; t_1 results from the travel time of the ions to the cathode; t_2 is the time during which the electron space charge cloud in the vicinity of the cathode is depleted and an ion space charge is again built up.

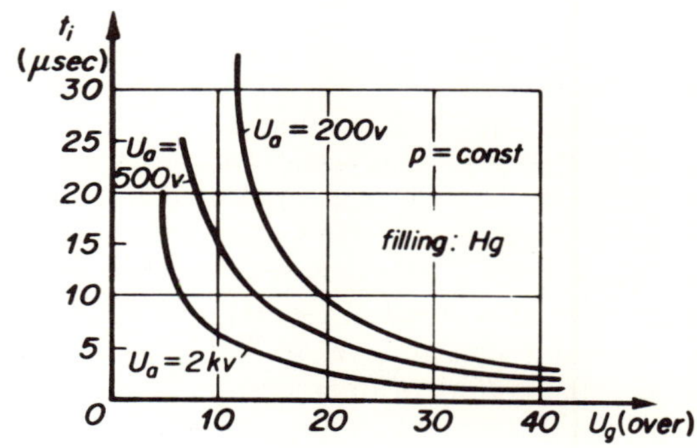

FIG. B1i-2. Influence of various anode and grid voltages on the ionization time t_i of a mercury vapor thyratron under the particular condition p = constant. This curve is greatly influenced by gas pressure and the kind of gas used.

The deionization time (t_d) can be defined in two sets of terms:

a. In diodes it is defined as the time which elapses from the end of firing ($I_a = 0$) until the instant when all ions are reconverted into gas atoms, i.e., when the prefiring conditions have been reestablished.

b. For thyratrons it is usual to establish a shorter waiting time, namely, from the end of firing until the instant at which the grid has regained control and is ready to block a new firing cycle. This time is also known as recovery time.

For tubes with mercury filling t_d is about 1000 μsec, t_i about 10 μsec. For tubes with xenon filling t_d is about 50 μsec, t_i about 10 μsec. Deionization time (t_d) depends on various factors which affect its duration considerably. Therefore, it is also impossible to furnish accurate figures for t_d. Tube design as well as grid voltage and anode current shortly before extinction of the arc are of influence.

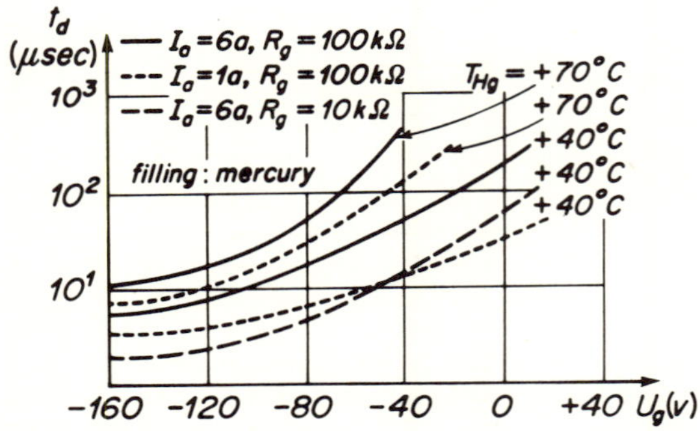

FIG. B1i-3. Dependence of the deionization time t_d of a mercury vapor triode on the negative grid voltage (U_g) during the tube conducting time, on the grid resistance (R_g), the average anode current (I_a), and the mercury vapor temperature (T_{Hg}). For tubes having xenon filling, t_d is of about 33% shorter duration.

A reduction in the duration of t_d can be achieved by enlarging the grid surface, by means of several small holes (instead of a large one), or by arranging the grid relatively close to the anode, etc. High anode voltage and high negative grid potential as well as low gas pressure act toward a shorter t_d. A large anode current prolongs this time (Fig. B1i-3). Magnetic fields and the electrode capacitance are also of influence. With an inductive load the anode current flows during a portion of the negative half-wave and can, in controls with phase-shifted alternating current, prolong t_d considerably. The most favorable condition exists when U_a and U_g pass simultaneously through zero at the time of interruption, so that the whole half-wave time can be utilized. In tubes with mercury filling, a high mercury temperature with its higher vapor pressure prolongs the deionization time.

k. *Grid Current Characteristics*

The magnitude of grid current I_g flowing in the grid circuit of a thyratron depends on the polarity of the grid relative to the cathode, on the impedance of the grid circuit, and on the anode current.

The prevailing functions of the grid current are illustrated in Fig. B1k-1. Before firing ($I_a = 0$) there will flow only a very small grid current when the grid voltage is sufficiently low. This small current is due to insulation defects and grid emission.

When the negative grid potential is reduced or made positive ($U_g = 1$ volt, for example), a grid current I_g of a few microamperes will begin to flow. This current can rise with increasing positive grid potentials up to several hundred microamperes. This behavior of the thyratron is somewhat similar to that of a high vacuum tube. As soon as the tube fires, I_g changes its direction, owing to the positive ion flow away from the grid ($-I_g$). It does

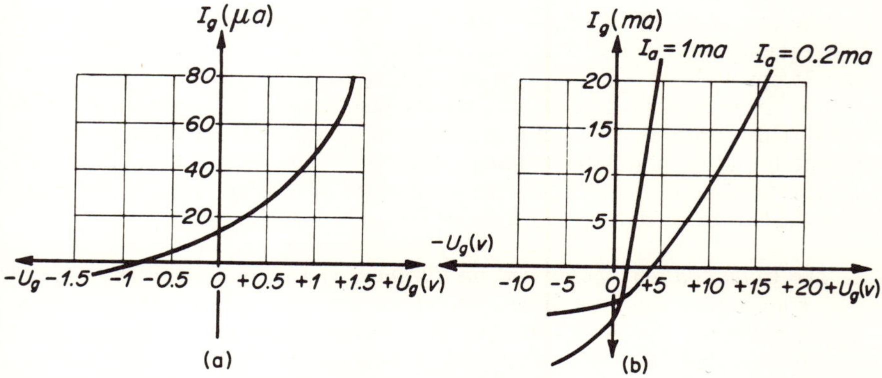

FIG. B1k-1. Relationship between grid current, grid voltage, and anode current: (a) in a nonconducting tube ($I_a = 0$); the corresponding grid current is caused by passage of negative ions to the grid which is surrounded by positive ions. (b) in a conducting tube, here a small gas-filled triode with positive anode voltage U_a.

not return to the original direction until $U_g \approx 0$. I_g can reach a magnitude of several milliamperes, since the grid now assumes the function of an anode. This current must be limited by use of a grid resistor, so that the rated grid current, as stated in the tube data sheet, cannot be surpassed. Otherwise sputtering and grid overheating may result. The value of R_g must be at least 1 kohm, but preferably not higher than 0.1 megohm. Otherwise there is danger that, owing to negative grid current, an undesirable forward shift of the firing point will occur in phase control applications. In tetrodes, in which the grid is replaced by two electrodes, this consideration applies to the grid which is closest to the anode.

In operation at ac grid potentials the grid becomes alternatively negative and positive. In phase control operation the phase relationship between grid and anode changes and as a result the representations of the diagrams change. For ac voltages the arithmetical mean values apply.

1. *Dependence of Anode Current and Gas Filling on the Frequency* [117]

Tubes with mercury filling, owing to their larger and more sluggish atoms and the resulting longer ionization and deionization times, can be operated only at frequencies in the range from 25 to 150 cps. It is possible to extend this range above the upper limit with inert gas or hydrogen filled tubes.

Since anode load and temperature fluctuate rhythmically with the impulse sequence frequency, this frequency affects the permissible loading of the anode. At a frequency $f < 25$ cps, the temperature fluctuations of the anode are rather small, owing to heat inertia. The anode temperature depends chiefly on the arithmetical average (root mean square, rms, value) of the anode current I_a. Under given conditions this current may have to be reduced at $f > 150$ cps. At a frequency $f < 25$ cps, it is the magnitude of the peak currents I_{ap} that influences temperature rise of the anode. At this lower impulse frequency the anode temperature will fluctuate in synchronism with the frequency. If one operates a gas filled tube at a lower sequence frequency than 25 cps without reducing I_{ap} considerably, the anode will overheat and, with an inert gas filled tube, excessive consumption of the inert gas will occur. This consideration holds true, naturally only in the case of intense current impulses of relatively long duration, such as in the operation of impulse welding devices and in the generation of strong, long-lasting magnetic fields, not in devices working in the μsec range.

m. *The Commutation Factor and the Problem of Gas Consumption*

In contrast to mercury tubes, in which the vapor can be continually replenished from the large reservoir of condensed mercury, inert gas tubes do not have a reserve to draw upon. Their gas content is gradually absorbed in the course of their operating life. If the tube manufacturer did not provide for some protection against gas loss, the life expectancy of gas filled tubes would be relatively short. Thanks to modern production methods, this gas consumption can be effectively reduced. Three methods have been found for lengthening tube life:

a. Introduction of a sufficient quantity of gas into the envelope. However, this necessitates a larger tube.

b. Increase of gas pressure. However, the inverse voltage is decreased.

c. Gas replenishment from a built-in reservoir (glow helix with H_2-saturated texture).

For inert gas thyratrons the second method is preferable. In conjunction with proper shielding (Fig. B1m-1) and an effective grid action, favorable life expectancies can be achieved.

Because of the high gas pressure, which is usually higher than in mercury tubes, the maximum inverse voltage of an inert gas thyratron is generally only about 15,000 volts (in contrast to maximum 22,000 volts with mercury filling).

That it is possible to select the method leading to volume enlargement is shown by the example of an inert gas filled diode, Type DX 2 in which an inverse voltage of 10,000 volts is allowable. The relationship between gas pressure and electrode spacing is governed by Paschen's law (Fig. B1m-2).

Paschen's law states that the breakdown voltage in gas filled discharge chambers is constant when the product of electrode spacing d and gas pressure p is constant ($d \times p = K$). When this product becomes too small it is detrimental

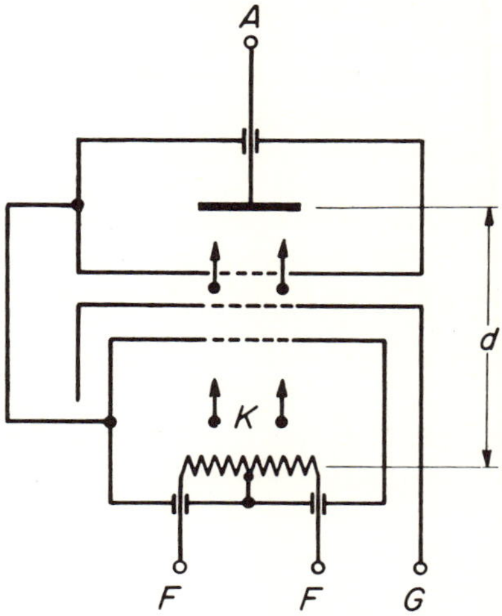

FIG. B1m-1. Schematic representation of the arrangement in an inert gas thyratron. A = Anode of material having a high work function, e.g., graphite or sheet nickel that has been coated with a layer of carbon; K = cathode mostly oxide; anode and cathode surrounded by shield sheets between which the grid sheet is inserted. All sheets slotted or perforated, to allow electrons to pass to the anode; passage holes, however, as small as possible in order to prevent passage of ions into the anode as much as possible. To obtain a small product pd it is essential to make d as small as possible, whereby d designates the average electron path from cathode to anode. There are tubes available with anode–cathode distances of about 1 cm. G = control grid. F = heater filament.

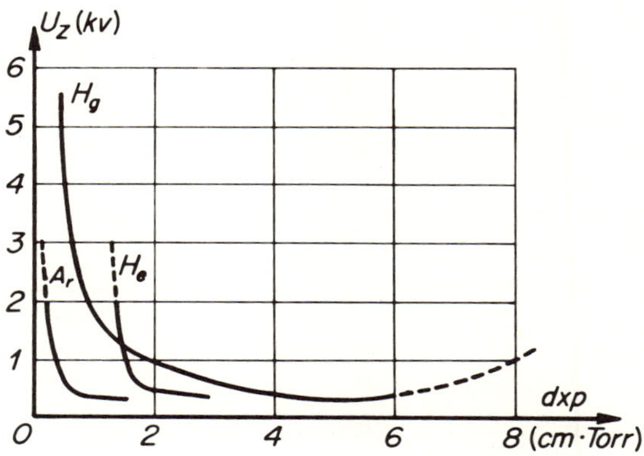

FIG. B1m-2. Dependency of firing potential of a gaseous discharge on relative electrode spacing for various gases.

to the discharge and must be compensated for by an increase in anode voltage, in order to accelerate the gas ions enough in the space between cathode and anode to produce a sufficient number of collisions of electrons and ions. Although it is possible to reduce gas consumption to a minimum through proper design, it appears proper to give the tube user, through additional electrical data, some advice as to precautions against gas consumption. Adherence to the maximum values of the following is essential: frequency limits, average and peak anode current, inverse voltage, and commutation factor. Whenever a current-conducting tube is suddenly shut off, U_a increases rapidly to its negative value. There is not sufficient time for the positive ions in the discharge chamber to deionize themselves. They are

TABLE B1m-1

COMPARISON OF TWO RECTIFIER TUBES WITH Hg AND Xe FILLING

	DQ 2	DX 2
Height/width	155/51 mm	158/51 mm
Ambient temperature limits	+ 10 . . . + 60°C	− 75 . . . + 90°C
Max. inverse voltage	10 kv	10 kv
Max. average anode current	0.25 amp	0.25 amp

accelerated by the high negative voltage now effective at the anode and enter the latter. Here they are neutralized and remain in the anode as gas atoms (xenon atoms, for example). They are then lost for further action in the tube.

The commutation factor (K.F.) in inert gas filled tubes is defined as the product of anode current drop (amp/μsec) and the immediately following negative anode voltage rise (volts/μsec) (see also Section 1o of this chapter).

This factor is not affected by the tube but by the circuit in which it operates. A higher K.F., as given in the tube data, reduces the life expectancy of the tubes. Measuring of K.F. is preferably accomplished by a two-beam oscilloscope. From the two traced curves $I = f(t)$ and $U = f(t)$, there are read the current I (in amperes) 10 μsec before commutation and the time t_2 which elapses after commutation until the anode voltage in the following nonconducting phase has reached 200 volts. Then K.F. = 20 I/t_2.

For the inert gas thyratron TX 2/6 the K.F. is maximum 130. For mercury vapor thyratrons this value is of no interest, since a sufficient reservoir of mercury is always available.

Occasionally the danger of surpassing K.F. exists, for example, in circuits in which the anode potential changes very rapidly from its positive to its negative value as can be the case with inductive loads (inductively stored energy) in multi-phase rectifiers. It is possible to reduce K.F. by inserting

R–C links as a series connected arrangement of resistance and capacitance in parallel to the thyratron (Fig. B1m-3). As in a low pass filter, these snubbing circuits prevent rapid voltage and current changes from becoming effective in the tube (Fig. B1m-4). However, these *R–C* links may impose supplementary costs and energy losses or may cause additional heating.

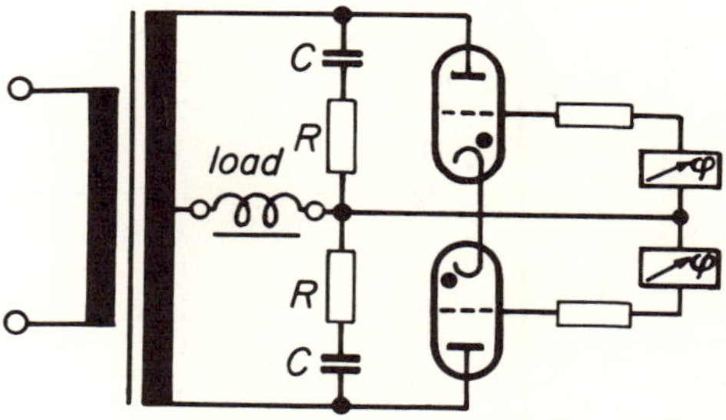

FIG. B1m-3. Schematic diagram of a single-phase center-tap circuit with inductive load and parallel connected phase correction links (*RC* links).

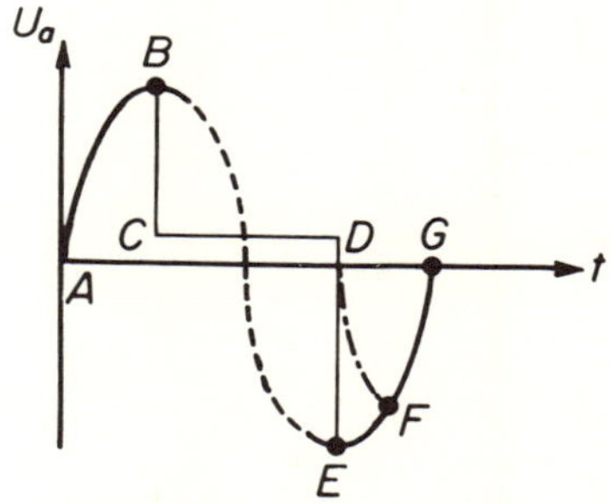

FIG. B1m-4. Trace of voltage in a tube connected as in Fig. B1m-3. The counter emf of the inductive load keeps up a positive anode voltage and forces the tube to remain conducting after the line voltage has become negative (*CD*). At the instant of becoming extinct (D), the voltage drops rapidly to a negative value (*E*) with the result of a high commutation factor which is only minimized by stray inductance of the line transformer. By insertion of the *RC* link, the drop is softened and the voltage trace assumes the shape indicated in the figure as a dash-dot line.

In all larger rectifying devices with controllable loads, controlled thyratrons are extensively used. They are indispensable in high voltage rectifiers for capacitor charging. Therefore, their designs, construction and behavior have been described at great length, although they are not directly involved in the technique of switching means in capacitance discharge circuits.

A very careful consideration of the commutation factor is indispensable in the design of long-lived impulse switching means, for example, in production setups in which continual tube change is impossible. The inductances in discharge circuits should always be damped as much as possible.

n. *Characteristic Data of a Few Industrial Mercury and Xenon Thyratron Tubes*

TABLE

THE MOST IMPORTANT LISTED TECHNICAL DATA OF

Tube type		TQ 2	TQ 1/2	TQ 2/3
Heater voltage[a]	H(±5)%	2.5	2.5	2.5
Heater current, about	amp	7	7	12
Min. preheating time	sec	30	30	30
Filling		Hg	Hg + A	Hg + A
Ambient temperature range	°C	+ 15 ... + 50	− 40 ... + 70	− 40 ... + 70
Ionization time, about	μsec	10	10	10
Deionization time, about	μsec	1000	1000	1000
Highest operating frequency	cps	150	150	150
Grid-anode capacitance	pf	1.8	1.8	2.5
Grid-cathode capacitance	pf	5	5	16
Internal voltage drop (arc drop), about	volts	15	15	15
Peak rated anode voltage in inverse and forward directions	volts	2500	2000	2000
Maximum rated anode current:				
(a) average value Ia, peak value	amp	1	1.6	3.2
(f ≧ 25 cps) I_{ap}	amp	4	20	40
Single short-circuit impulse (max. 0.1 sec)[b]I	amp	25	120	500
(b) Max. rated grid current, average Ig	amp	0.01	0.1	0.25
Max. negative grid potential:				
tube non conducting	volts	−500	−500	−500
tube conducting	volts	−10	−10	−10
(a) Integration time (Ia)	sec	15	15	15
(b) Integration time (Ig)	sec	0.02	0.02	0.02
Weight (unpacked)	grams	70	70	230
Weight (packed)	grams	260	260	1000
Anode connection		HF 506 709 P1 NB 863 820 P3	HF 506 709 P1 NB 863 820 P3	HF 506 709 P1 NB 863 820 P3
Socket		HF 402 587 P3 HF 402 587 P4	HF 402 587 P3 HF 402 587 P4	HK 400 927 P1 NB 861 730 P2
Base		A4—10	A4—10	A4—18

[a] Cathode oxide, directly heated; with exception of TQ 2/25, which is indirectly heated.

[b] The circuit has to be designed so that in case of short circuit no current of greater magnitude can flow. These data are therefore not to be assumed as operating values.

[c] For operating values (Ia) at higher frequencies contact manufacturer.

B1-n1

SOME INDUSTRIAL THYRATRONS (BROWN BOVERI)

TX 2/3	TQ 2/6	TX 2/6	TQ 2/12	TQ 2/25
2.5	2.5	2.5	2.5	5
12	22	22	33	36
30	30	30	120	300
Xe	Hg + A	Xe	Hg + A	Hg + A
− 55 . . . + 70	− 40 . . . + 70	− 55 . . . + 70	− 55 . . . + 70	− 40 . . . + 70
10	10	10	10	10
40	1000	40	1000	1000
400[c]	150	400[c]	150	150
1	3.5	1	4.5	5.4
50	22	50	25	11
15	15	15	15	15
1500	2000	1500	2000	2000
3.2	6.4	6.4	12.5	25
40	80	80	150	300
560	1000	1120	1500	3000
0.2	0.25	0.2	0.5	1.0
−250	−1000	−250	−1000	−1000
−10	−10	−10	−10	−10
15	15	15	30	30
0.02	0.02	0.2	0.02	0.02
250	380	280	790	1665
1400	1500	1400	3600	8500
HF 506 709 P1		HF 506 709 P1	HK 400 913 P1	HK 400 913 P1
NB 863 820 P3	NB 400 202 P1	NB 863 820 P3	NB 400 202 P1	NB 400 202 P1
HK 400 927 P1	HK 400 927 P1	HK 400 927 P1	—	—
NB 861 730 P2	NB 861 730 P2	NB 861 730 P2	—	—
A4—18	A4—18	A4—18	—	—

TABLE B1n-2

CHARACTERISTIC DATA OF A FEW INDUSTRIAL MERCURY AND XENON THYRATRON TUBES

Type	Heater voltage (volts)	Heater current (amp)	Permissible max. anode current (amp)	Average anode current (amp)	Inverse voltage (volts)	Anode voltage (volts)	Arc voltage arc-drop (volts)	Preheating time (minutes)
AEG								
S 0.5/12 im	5	12	12	3	—	500	10–20	3
S 5/1 i	4	3	1	0.3	—	5000	10–20	1
S 7.5/2.5 d	5	10	2.5	0.8	—	7500	10–20	5
S 15/5 d	5	19	5	1.6	—	15000	10–20	5
S 15/40 i	5	20	40	10	—	15000	10–20	10
Philips								
PL 17	2.5	5	2	0.5	5000	2500	16	0.1
PL 57	5	4.5	15	2.5	1000	1000	16	5
PL 105	5	10	40	6.4	2500	2500	16	5

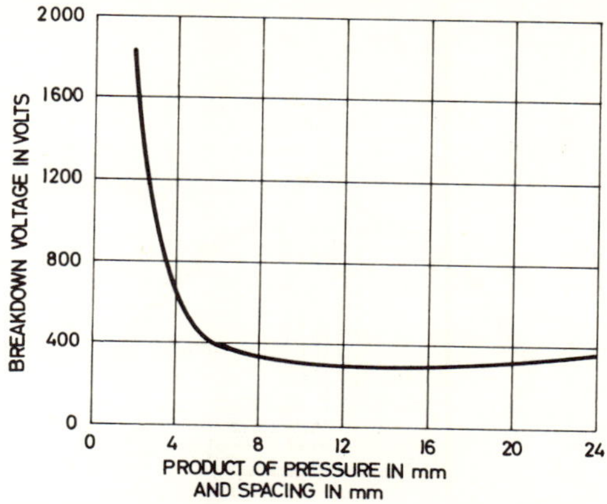

FIG. B1o-1. Plot of breakdown voltage in hydrogen as a function of the product of spacing and pressure. By permission from The hydrogen thyratron, K. J. Germeshausen, *in* "Pulse Generators," McGraw-Hill, 1948.

o. *General Characteristics of Hydrogen Filled Thyratrons* [186]*

The use of hydrogen in place of mercury vapor or a rare gas as the filling for thyratrons [668] appears to have some real advantages.

The first discussion of hydrogen thyratrons is given by Drewell [668]. The primary advantage of hydrogen in a thyratron, is the high fall of cathode potential that can occur without injury to the oxide cathode. The destruction voltage, or voltage corresponding to the ion velocity at which destruction of an oxide cathode sets in, is about 30 volts for mercury ions and has about the same order of magnitude for the rare gases, including helium. For hydrogen the destruction voltage is about 600 volts, or 20 times as great. This fact is of considerable importance in the design of a thyratron for operation at high voltage and high pulse currents.

A second advantage of hydrogen filled tubes is the short deionization time that is obtained. In any given circuit this deionization time is about one-tenth that of mercury, argon, or xenon filled thyratrons.

A serious problem connected with the use of hydrogen in thyratrons is gas cleanup, or the disappearance of the gas during operation. In addition to the use of materials of carefully controlled purity, it is important to prevent contamination resulting from improper cleaning or handling of the parts.

Because of gas cleanup, the gas pressure is raised as high as possible by making the spacing between the anode and grid much smaller than in the more conventional thyratron designs. The purpose of the small spacing can best be explained by referring to Fig. B1o-1, which is a plot of the breakdown voltage in hydrogen versus the product of spacing and pressure, the familiar Paschen curve of breakdown in gases.

* Chapter B, Section 1o, is reprinted by permission from K. J. Germeshausen, The hydrogen thyratron, *in* "Pulse Generators," McGraw-Hill, New York, pp. 336–339 (1948).

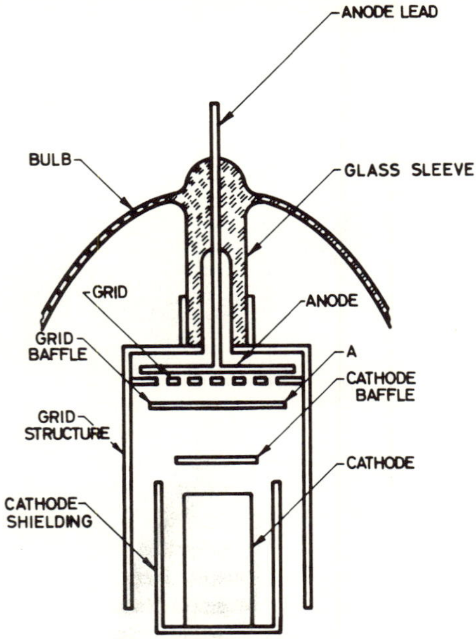

FIG. B1o-2. Structure of a hydrogen thyratron. By permission from The hydrogen thyratron, K. J. Germeshausen, *in* "Pulse Generators," McGraw-Hill, 1948.

A characteristic design is shown in Fig. B1o-2 and B1o-3. Long path discharges between the anode lead wire and the outside of the grid structure are prevented by surrounding the lead by a glass sleeve that fits tightly into a collar attached to the grid structure. From Fig. B1o-2 it is apparent that the distance between the upper part of the anode lead inside the glass sleeve and the grid structure is considerably greater than the shortest spacing from grid to anode. Breakdown over this relatively long path is prevented partly by the small spacing between the glass sleeve and the anode lead, and partly by the fact that the field between the anode lead and the grid structure in this region is in such a direction that the acceleration of ions or electrons over the long path is prevented.

Hydrogen thyratrons differ from the more common thyratron types in that they are designed with a positive control grid characteristic. In order to trigger the tube it is necessary to drive the grid to a positive voltage sufficient to draw grid current between the grid and the cathode. Two considerations dictated the positive control characteristic. First, positive control simplifies the trigger circuit, since, for most applications, no negative line supply is required. Second, the extensive baffling between the anode and cathode that is associated with positive control minimizes grid emission difficulties.

In order to obtain a positive control characteristic, the cathode is completely shielded from the anode field in the manner shown in Fig. B1o-2. The grid itself is a perforated, or mesh, section situated just below the anode,

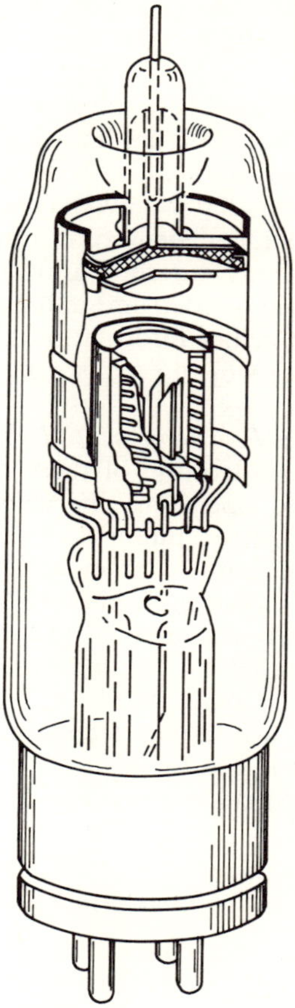

FIG. B1o-3. Cutaway perspective view of the 5C22 hydrogen thyratron. By permission from The hydrogen thyratron, K. J. Germeshausen, *in* "Pulse Generators," McGraw-Hill, 1948.

and beneath it is a grid baffle, which is attached to the grid. In such a structure the anode field does not extend beyond the point marked A, and there is no opportunity for the anode field to act upon electrons emitted from the cathode, most of which are confined to the region inside the cathode shield and baffle structure. In order to trigger the tube, or to start conduction, it is necessary to draw current between the grid and the cathode. This current produces electrons and ions in the region outside the cathode shield structure some of which reach the point A. As soon as the electron density at A is high enough, the anode field is able to produce ionization in the region above the grid baffles, and breakdown takes place. During the initial stages of the breakdown, the anode field is unable to penetrate into the grid cathode space. This glow discharge raises the grid to a high positive potential, which quickly ionizes the grid cathode region to an extent sufficient to bring the grid back almost to the cathode potential. The entire breakdown process described above occurs in 0.02 to 0.07 μsec, which is thus the ionization time.

For extremely high-power application it is useful to fill the thyratrons with deuterium instead of hydrogen. This gas has the same chemical characteristics as hydrogen. L. E. Algar *et al.* [945] have designed such a deuterium filled thyratron for a peak power of 200 megawatts and an average load of 150 kw. Because of the lower mobility of deuterium ions the breakdown voltage is higher than that of hydrogen. Deuterium allows operation up to 40 kv. The cathode has a heated surface of 1000 cm^2 and emits a current of 10,000 amp at 5-μsec pulse duration. Further information about deuterium filled thyratrons is given by Cook *et al.* [946] and Armstrong *et al.* [947].

p. *Characteristic Data of a Few Industrial Hydrogen Thyratrons*

Tube characteristics and operation of three often used smaller hydrogen thyratrons and their basic operating characteristics are shown in Table B1p-1. The circuit of a thyratron pulser is given in Fig. B1q-11. For higher current ranges up to 2400 amp a larger selection of industrial hydrogen thyratrons is available as shown in Table B1p-2.

The latter types especially are used in the technique of pulsing big klystrons. The impulse delivered by the hydrogen thyratron becomes transformed up to 250 kv with a pulse width of 7 μsec. The klystron tubes are often used in the particle accelerator technique for feeding the accelerator plates which are mounted in the torus.

Owing to the relatively small average thermal load the radiation cooling normally is sufficient.

TABLE B1p-1

OPERATING CHARACTERISTICS OF THE 3C45, 4C35, AND 5C22 HYDROGEN THYRATRONS

Characteristics	3C45	4C35	5C22
Maximum forward anode voltage (kv)	3	8	16
Maximum inverse anode voltage (kv)	3.0	8.0	16.0
Maximum value of inverse anode voltage for 25 μsec after pulse (kv)	1.5	2.5	5.0
Maximum anode pulse current (amp)	35	90	325
Pulse power into resistance load (kw)	50	350	2500
Maximum average anode current (ma)	45	100	200
Cathode: indirectly heated, unipotential			
Heater power (watts)	15	40	65
Heater voltage (volts) ($\pm$ 7.5%)	6.3	6.3	6.3
Maximum pulse duration (μsec)	6	6	6
Grid bias (volts)	0	0	0
Minimum trigger voltage (volts)	150	150	150
Tube weight (grams)	64	194	273
Tube life at maximum ratings (hours)	500	500	500

TABLE B1p-2

OPERATING CHARACTERISTICS OF SOME LARGE HYDROGEN THYRATRONS

Characteristics	Type GL 7390	Type GL 7890	Type Z 5212
Peak current (amp)	2000	2400	2000
Anode voltage (kv)	33	40	50
Average current (amp)	4	4	8
Peak power output (megawatts)	33	48	50
Average power output (kilowatts)	60	70	200
Anode dissipation factor	30×10^9	55×10^9	400×10^9
Rms current (amp)	72	72	126

TABLE B1p-3

OPERATING CHARACTERISTICS OF SOME LARGE CERAMIC HYDROGEN THYRATRONS

Characteristics	Type 7322/1802[a]	Type GL7390[b]	Type GL7890[b]
Maximum peak anode voltage (kilovolts)	25	33	40
Maximum peak cathode current (amp)	1000	2000	2400
Average cathode current (amp)	1.25	4.0	4.0
Peak output power (megawatts)	12.5	33	48
Average output power (kilowatts)	15	60	44
Anode dissipation factor (Product of the peak forward anode voltage, pulse repetition rate, and peak anode current)	12.5×10^9	30×10^9	55×10^9

[a] Manufactured by EGG.
[b] Manufactured by GE.

A group of ceramic sealed hydrogen thyratrons was developed by the firms EGG* and GE** for switching service in radar modulators in airborne devices to carry high peak currents and withstand very high voltages. Table B1p-3 shows some characteristic data.

The above-mentioned series of big thyratrons can now handle a number of switching problems for which formerly only spark gaps were suitable.

q. *Circuitry and Applications of Hydrogen Thyratons* [190]

The hydrogen thyratron was developed primarily as a switch tube for line type, high-powered pulse generators. The typical circuit for this type

* Edgerton, Germeshausen & Grier, Boston.
** General Electric Co.

of service is shown in Fig. B1q-1. It consists basically of the following: (1) A pulse-forming artificial transmission line, characterized by a two-way transit time, τ, and an impedance, Z_0; (2) an energy absorbing load, R_L, considered as a pure resistance in this work; (3) the thyratron trigger system; (4) a charging system for restoring energy to the line during the interpulse interval; and (5) the thyratron.

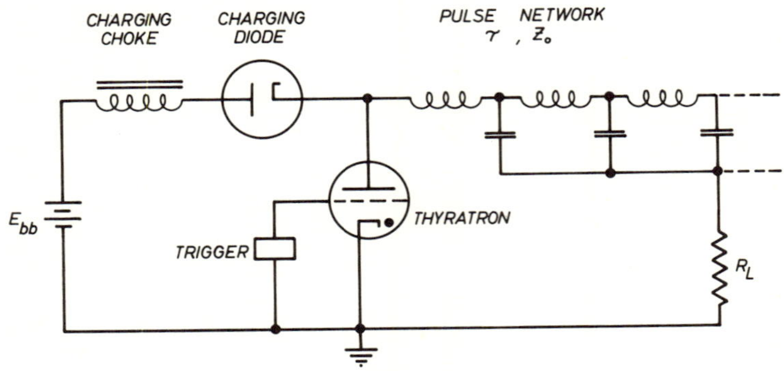

FIG. B1q-1. Hydrogen thyratron pulse-generating circuit.

The circuit functions basically as synchronized relaxation oscillator. When the circuit is triggered, the energy stored in the pulse line capacity during the interpulse period is discharged through the load and the thyratron. The peak pulse current is given by $e_p/(R_L + Z_0)$ where e_p is the peak network voltage and the pulse duration is determined by the two-way transit time of the pulse network.

The thyratron is extinguished following the current pulse by maintaining the anode at less than a critical reignition voltage for a period greater than the recovery time. This is accomplished through the details of the charging system and facilitated by the presence of inverse voltage on the network. This latter is given ideally by

$$e_x = \frac{e_p\,(R_L - Z_0)}{R_L + Z_0}$$

As far as the thyratron switch is concerned, the circuit must perform three basic functions: (1) Close the switch; (2) allow sufficient recovery time for the switch to open; and (3) maintain the switch at its proper operating ratings at all times.

The data were obtained with the circuit shown in Fig. B1q-2 using a 4C35 thyratron. Its nominal ratings are as shown below.

Peak forward voltage	8 kv
Peak anode current	90 amp
Repetition rate	2800 cps
Pulse length	0.5 μsec

The anode cathode discharge of a hydrogen thyratron is initiated by striking an arc in the grid cathode region of the tube by application of a positive voltage (trigger pulse) to the grid. Figure B1q-2 shows the basic circuit used to study this process. A substantially flat voltage pulse, which rises in a time which is quite short compared to the rate of growth of the grid current, is applied to the grid at zero time.

The grid current is measured by the voltage drop across a 100 ohm terminated viewing cable which constitutes practically all of the grid drive source impedance. The driving thyratron has an effective resistance of only about 15 ohms. Figure B1q-3 shows the resulting current and voltage pulse forms.

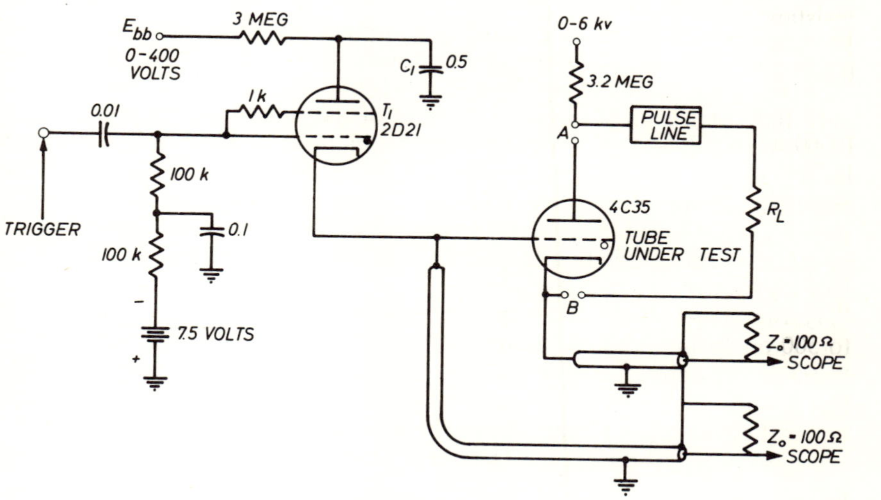

FIG. B1q-2. Basic circuit used for studying the grid cathode breakdown process of the 4C35.

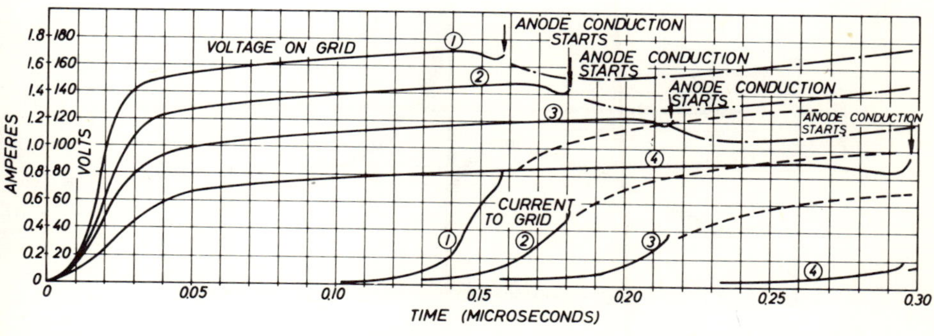

FIG. B1q-3. Grid current during breakdown of the cathode grid space and cummutation to the anode as functions of time at various voltages. Anode voltage 3 kv.

The time at which the grid current becomes apparent and its rate of growth thereafter are functions of the amplitude of the applied trigger voltage. After the grid current has built up to a certain value, which appears to depend on the applied trigger voltage, the arc is commuted to the anode and the tube as a whole breaks down.

The dotted portions of the curve indicate the course of grid current in the absence of anode potential. The current here rises until limited by the source resistance, the applied voltage, and the steady state arc drop. More detailed analysis of the data of Fig. B1q-3 indicates that anode conduction occurs at a fixed time delay after the grid current reaches a certain critical value. For the case shown, where the anode voltage was 3 kv, this current was about 0.060 amp and time delay about 0.03 μsec.

The magnitude of this critical current provides some index of the order of trigger source impedance required to obtain minimum delay times. That is, the source impedance need only be low enough to pass this critical current with negligible voltage drop. Thus it was found that the delay time increased only 10% (0.016 μsec) upon an increase of the source impedance from 100 to 500 ohms. It is quite probable that this increase in delay time was caused mostly by the grid-cathode capacitance of the thyratron. This amounts to about 20 $\mu\mu$f and thus causes a 0.01-μsec time constant delay in the application of grid voltage for 500 ohms source impedance. It was not found possible to measure this capacity effect directly because of the loading effect of the viewing system.

Figure B1q-4 shows the triggering situation for an applied trigger of the minimum form specified in the JAN ratings of the tube. In this case the open-circuit applied trigger was 140 volts, the source impedance was 1500 ohms, and the rise rate was approximately 120 volts/μsec. The slow rate of application of grid voltage causes the grid current to build up at a slower average rate. In addition, the 1500 ohm source impedance is large enough to

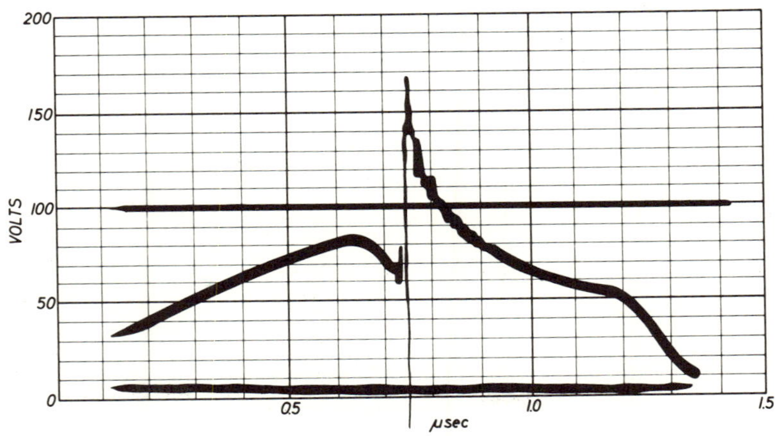

FIG. B1q-4. Grid voltage with 6 kv applied anode voltage during application of a JAN specified trigger pulse having a peak open circuit value of 140 volts.

further reduce the grid voltage at current levels near the critical value. Both of these factors account for the 0.7 μsec delay time for these conditions as contrasted to the delay times of 0.2 μsec or less, obtained with rapidly rising low impedance triggers.

A further trigger system design factor becomes apparent when one is considering trigger requirements as a function of tube life. It has been found that the trigger requirements increase after about 1000 hours life. Accompanying this is a marked deterioration of the cathode coating. In spite of this, however, it has been found possible to obtain satisfactory operation for as much as an additional 1000 hours providing a sufficiently large voltage was applied to start the discharge. After several minutes operation under these conditions, the cathode became sufficiently heated (due to ion bombardment and i^2R losses in the coating) to allow the trigger to be returned to its normal voltage. This effect was simulated by studying the trigger current as a function of cathode temperature on a new tube. The results are shown in Fig. B1q-5. At lower cathode temperatures, which correspond to lower cathode activity, the rate of increase of grid current is decreased, requiring increased trigger voltages to achieve the same anode breakdown delay time.

When the grid cathode arc is commuted to the anode, the elements of the discharge circuit come into play. These include, besides the pulse line, load resistance, and the thyratron, the associated stray capacities and inductances inherently present. The initial effect of the circuit is on the rate of build up of cathode current. The rate at which this current increases in relation to the

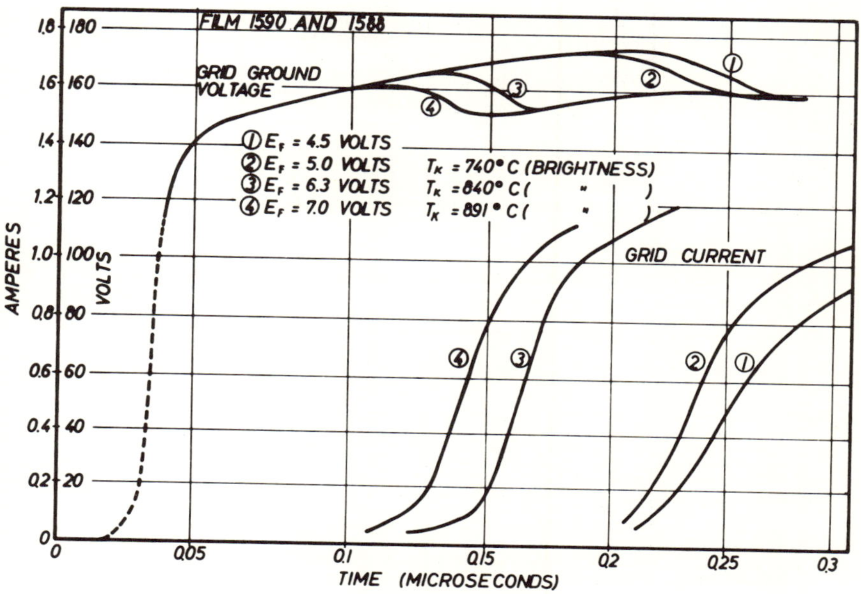

FIG. B1q-5. Grid current as a function of time during breakdown of the grid cathode space; cathode temperature as parameter.

rate of fall of anode potential bears directly on the tube dissipation and thus is an important parameter to control.

Ideally, the rate of rise of current is determined by the pulse line characteristics, that is, the number of network sections and the pulse length. In a practical circuit, however, stray capacities are quite apt to change the rise of current markedly from that expected from the network properties. Those of principal importance in this respect are the capacities from thyratron anode to cathode. These generally originate in the elements of the charging circuit.

Figure B1q-6 indicates the initial rise of cathode current in a circuit where no corrective means were taken to isolate or eliminate these normally present capacities. Figure B1q-7 shows a single sweep oscillograph of the current pulse under the same conditions except that the charging circuit was physically disconnected from the discharge circuit after the charging interval. It will be noticed that the major part of the oscillations have been eliminated by this means. The slight remaining oscillation was found to be caused by a resonance of the first section network inductance.

The stray capacitance from anode to cathode in the original circuit amounted to 60 $\mu\mu$f, most of which was due to the charging diode filament transformer. The lead inductance (approximately 0.5 μh) connecting this capacity to the anode accounts for the 30 mc oscillation. This stray capacity

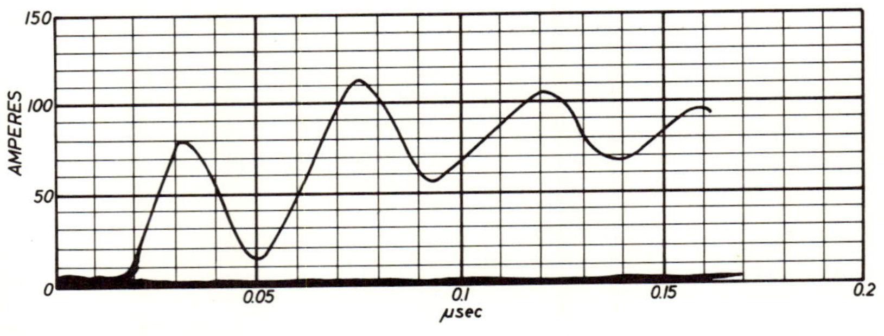

FIG. B1q-6. Current pulse obtained on equipment as first assembled.

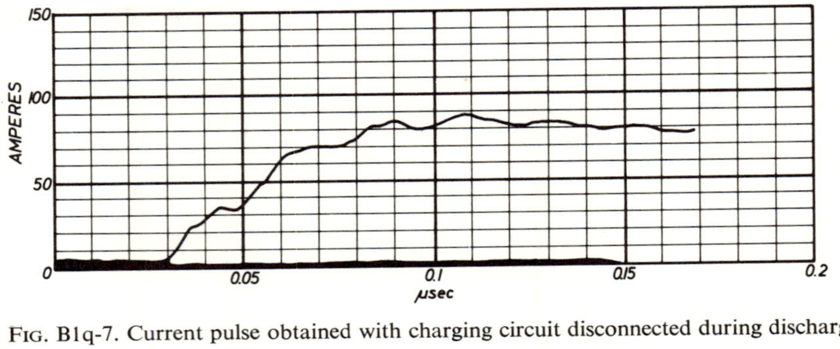

FIG. B1q-7. Current pulse obtained with charging circuit disconnected during discharge.

effect was eliminated in a more practical manner by the use of isolating circuit elements. Wire-wound inductive resistors were found to work particularly well for this.

Figure B1q-8 shows the current pulse with a 125 ohm *line-cord* (Mallory YO-25 resistor cord) inductive type resistor in series with a 160 ohm composition resistor, isolating the charging circuit. This combination fully damped the oscillation. For most applications, the power dissipated in these isolating resistors is less than a few percent of the average power switched. Also, they are sufficiently low in resistance as to have negligible effect on the charging voltage wave forms.

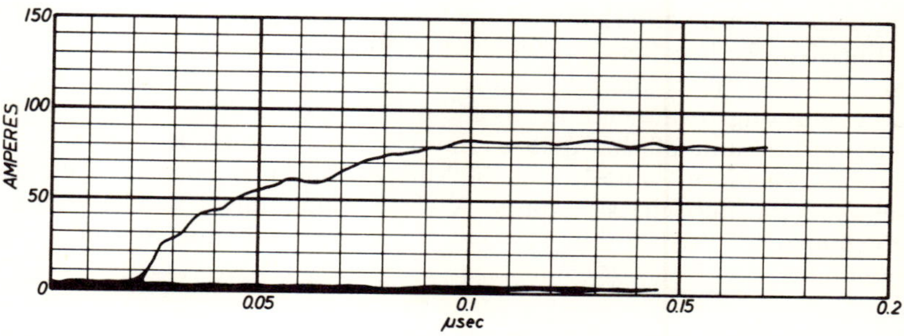

FIG. B1q-8. Current pulse obtained with 160 ohms added to line cord resistor. Rate of rise is 1150 amp/μsec.

Following the rise portion of the pulse, the current attains a peak value given by

$$i_p = \frac{e_p}{R_L + Z_0}$$

When the load resistance is less than the line impedance, the arc will extinguish at a time equal to the two-way transit time of the line as the tube voltage drops below the value necessary to support ionization (approximately 50 volts). If the load impedance is greater than the network impedance, one will observe an output which will decrease in steps of τ_p duration until the network voltage has fallen below the minimum potential necessary for sustained ionization. It will be shown later that this type of operation with its lack of inverse voltage on the thyratron anode imposes a severe recovery problem.

Some rather severe problems are encountered in properly measuring the discharge conditions. This is due to the very short time scales involved in the changes of voltage and current in the thyratron and the associated circuit. For example, the oscillographs of rise of cathode current shown in the preceding section were obtained by recording the voltage drop across a 1 ohm current viewing resistor. Thus a peak deflection of 90 volts (about $1\frac{1}{4}$ inch deflection on the oscilloscope was used) was obtained for the 90 amp current pulses shown. However, the rates of rise of current in the uncorrected circuit had maxima on the order of 4000 amp/μsec. At this rate of rise an inductance on the order of 10^{-9} henrys in the viewing resistor would cause a perceptible error in the observed voltage. Thus, resistors having less than this order inductance are required for error-free current viewing.

A further requirement is that the effective skin depth of the resistor must be greater than the resistor cross section for the highest frequency components of the pulse. Components up to 100 mc have been observed under certain circuit conditions.

A number of satisfactory resistor configurations have been designed by Evans Signal Laboratories and other agencies. Two that have been found simple to construct and that have the frequency response required are shown in Fig. B1q-9. The first is a spoked arrangement of 1 watt composition resistors. The rim connecting to the chassis serves as one terminal and the hub connecting to the cathode as the other. The oscillographs shown were taken with a unit of this type containing ten resistors. In connecting the resistors, no more than $\frac{1}{16}$ inch of lead was allowed between the resistor itself and the hub chassis points. A close fitting shield would reduce the inductance still further, but was not found necessary for the rates of rise encountered.

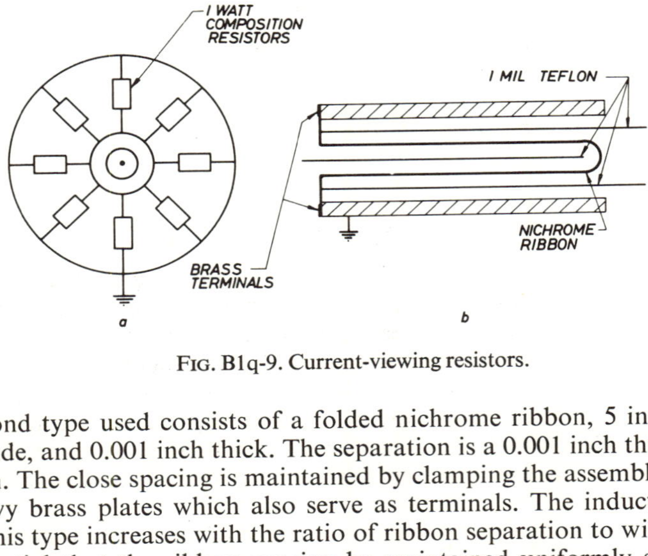

FIG. B1q-9. Current-viewing resistors.

A second type used consists of a folded nichrome ribbon, 5 inches long, 1 inch wide, and 0.001 inch thick. The separation is a 0.001 inch thick ribbon of Teflon. The close spacing is maintained by clamping the assembly between two heavy brass plates which also serve as terminals. The inductance of a unit of this type increases with the ratio of ribbon separation to width. Thus, it is essential that the ribbon spacing be maintained uniformly small. The requirements for large width and the limited resistivities and foil thickness available restrict this type of construction to resistors having less than about $\frac{1}{2}$ ohm resistance. The dimensions given were for a $\frac{1}{4}$ ohm unit. The spoked type is limited in dissipative abilities because of the use of commercial composition units. To secure more than 25 watts capacity would require an unwieldy number of resistors.

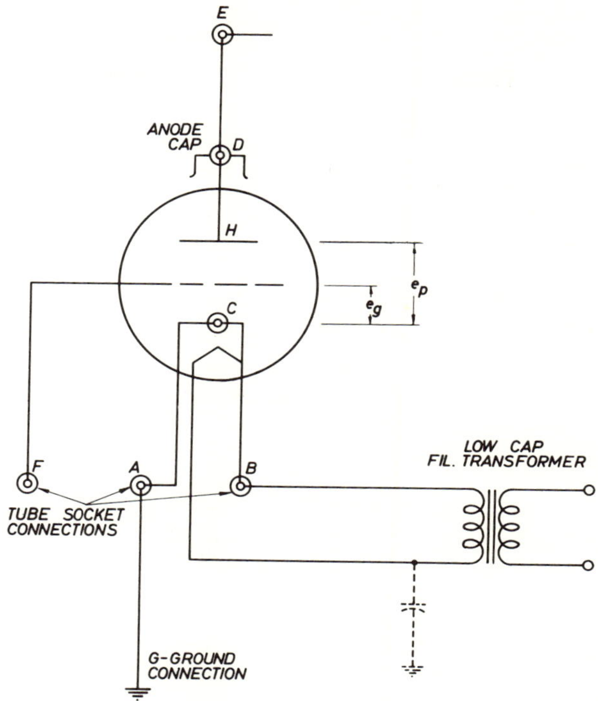

FIG. B1q-10. Sketch of typical thyratron wiring arrangement showing voltages primarily affected by stray inductive effects.

A further instrumentation problem caused by the high rate of current growth is encountered in measuring the voltages across the tube elements. The measurement of these voltages at the start of anode conduction is made difficult by the effects of the very rapidly rising currents flowing in the grid and cathode leads. The JAN specified rate of current rise for the 4C35, 1000 amp/μsec will produce a voltage drop of 100 volts in the inductance of a straight wire of 0.04 inch diameter and $3\frac{1}{2}$ inches long (approximately 0.1 μh). Since these dimensions correspond to the length of lead connecting and supporting the various electrodes inside the 4C35, it is apparent that, unless potential leads not carrying current are available for measurements, very serious errors may result. Even if these potential leads are available, the mutual inductance between the loop formed by them and the loop formed by the load, network, and anode and cathode must be carefully minimized. Figure B1q-10 is a sketch of a typical thyratron wiring arrangement showing voltages which would be primarily affected by stray inductive drops. E-H and C-G (G being ground) represent lead lengths, and thus inductances, from the actual tube anode and cathode respectively to the points to which voltages would normally be referred.

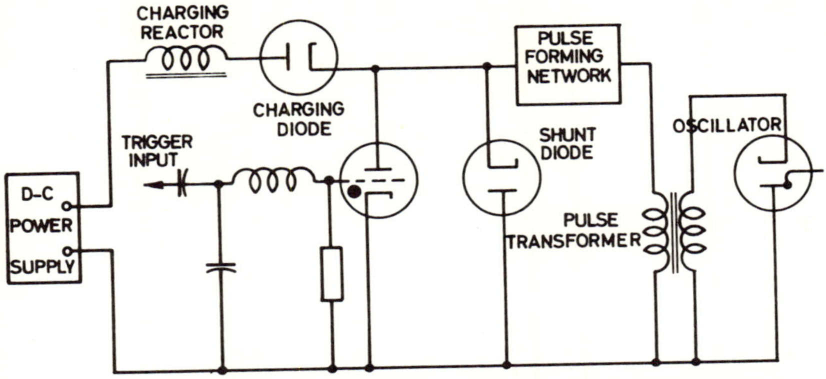

Fig. B1q-11. A typical hydrogen-thyratron pulser circuit. Reprinted from The hydrogen thyratron, K. J. Germeshausen, *in* "Pulse Generators," McGraw-Hill, 1948.

In the following paragraphs some examples of applications are shown:* Hydrogen thyratrons are designed to work primarily into a load impedance of 50 ohms and, in the case of the 3C45 and 4C35, maximum pulse power is obtained at this level. The 5C22 supplies its maximum pulse power at 25 ohms; however, the average power is the same as at 50 ohms. Any desired load impedance may be used, provided that none of the maximum tube ratings are exceeded. In general, both pulse and average power are less if a load impedance of other than the optimum value is used. (See Fig. B1q-11.)

The life of the thyratron depends greatly on the operating voltage, current, and recurrence frequency. A typical life of about 500 hours can be obtained at the maximum voltage and current ratings, with recurrence frequencies of 2000 pps for the 3C45 and 4C35 and 1000 pps for the 5C22. The length of life is considerably increased by reducing any of these parameters. In general, the operating level of thyratron pulsers should be kept somewhat below the maximum tube ratings. A much longer tube life can be obtained by operating the tubes well below these ratings, for instance, 4C35's operated at 4 kv and 45 to 50 amp pulse current usually give 2500 to 3000 hours of satisfactory service. Improvements in manufacturing methods are constantly increasing the life expectancy, which may soon be at least 1000 hours at full rating.

Hydrogen thyratrons may be mounted and operated in any position, but care must be taken lest the life be reduced by gas cleanup or puncture of the bulb. Gas cleanup is accelerated if the bulb temperature is lowered by subjecting the tube to forced cooling, or if intense rf fields are present to cause the ionization of the gas. Corona extending to the glass surface causes erratic operation and, eventually, puncture of the bulb.

It is often possible to satisfy switch requirements for which no single thyra-

---

* The remainder of Chapter B, Section 1q, is reprinted from K. J. Germeshausen, The hydrogen thyratron in "Pulse Generators," McGraw-Hill, New York, pp. 341–344 (1948).

tron is suited by the operation of two tubes in series or in parallel. Parallel operation provides twice the load current, while series operation provides twice the load voltage. There are a number of possible circuits, two of which are shown in Figs. B1q-12 and B1q-13.

In aparallel operation, the principal problem that arised is to secure proper division of the current between the two tubes. When the anodes of the two tubes are connected directly, the tube that breaks down first reduces the anode voltage of the second tube so rapidly that it may not break down at all. The best way to avoid this difficulty is to use a balanced reactor in the thyratron-anode circuit, a device that has been used with gas-tube rectifier circuits. The reactor is wound so that the fields of the two halves are in opposition, with very close coupling between the two. When the triggered tube is broken down, the full anode voltage appears across its half of the reactor, producing an equal and opposite voltage in the other half of the reactor which abruptly raises the anode voltage of the second tube to twice its previous value.

The grid-to-anode capacitance of the second tube is such that the grid is pulled positive by the transient voltage, thus triggering the second tube. Until the currents in the two halves of the reactor are equal, there is considerable reactance in each anode circuit; however, when they become equal the

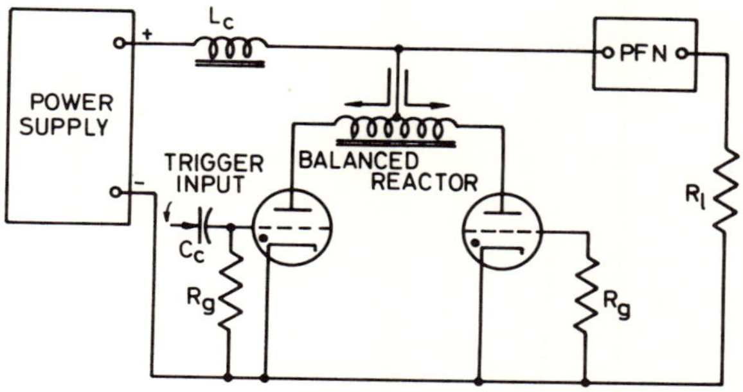

FIG. B1q-12. Parallel operation of thyratrons. Reprinted from The hydrogen thyratron, K. J. Germeshausen, *in* "Pulse Generators," McGraw-Hill, 1948.

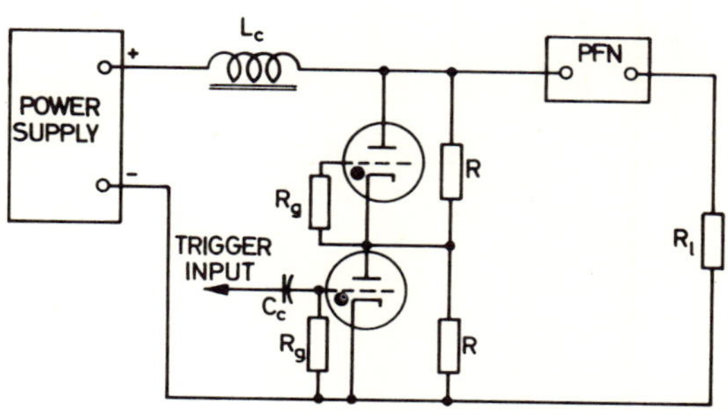

FIG. B1q-13. Series operation of thyratrons. Reprinted from The hydrogen thyratron, K. J. Germeshausen, *in* "Pulse Generators," McGraw-Hill, 1948.

the net reactance is negligible. Any tendency of one tube to draw more current than the other unbalances the reactor and raises the anode voltage of the lagging tube until the anode currents are again equal.

A practical circuit for the operation of two thyratrons in series is shown in Fig. B1q-13. Equal division of the voltage across the two thyratrons is obtained by means of the voltage divider comprising the resistor R. The minimum value of the resistors R is limited by the excessive drain on the power supply and by the effect on the peak network voltage. The maximum value is limited by the effect of the thyratron capacitance on the proper division of voltage. When the lower tube is triggered, the cathode potential of the upper tube is abruptly lowered to ground, but the potential of the grid of the upper tube tends to stay fixed because of the grid-anode capacitance. As a result, a positive voltage is applied to the grid of the upper tube, which triggers it. By this means both tubes are rendered conducting.

The major tube characteristic affecting the performance of the anode circuit is the variation of tube drop with time. It may be divided into two regions, an initial, or ionization period, when the drop is relatively high, and a conducting period during the pulse when the tube drop is relatively

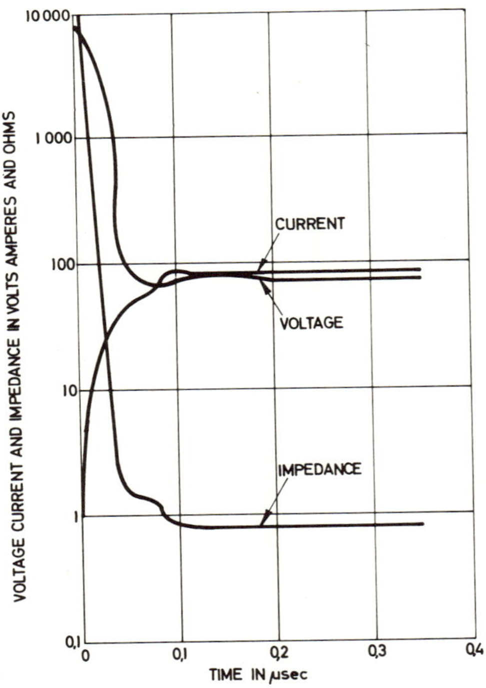

FIG. B1q-14. Tube drop, current, and impedance as functions of time for a 4C35 hydrogen thyratron. $V_N = 8$ kv; $I_p = 90$ amp; $\tau = 1.1$ μsec. Reprinted from The hydrogen thyratron, K. J. Germeshausen, *in* "Pulse Generators," McGraw-Hill, 1948.

low. In Fig. B1q-14 is shown the relation between the tube drop, current, and impedance, as a function of time. The data for the curves were obtained from oscillographic records of the operation of a 4C35 in a 50 ohm pulser circuit. The portion of the curves to the left of 0.05 μsec represent conditions in the tube during the time that the gas is being ionized. This section of the curves is called the ionization period. The portion of the curves to the right of 0.05 μsec, occurring after the gas is fully ionized, is the conducting period during which the tube drop remains relatively constant.

Ionization time and the shape of the voltage-time curve during ionization are almost entirely independent of the external circuit; that is, they are a function of tube characteristics only, mainly because of the high anode voltage employed in pulser applications, which insures that any electrons present are accelerated sufficiently to ionize the gas.

R. J. Armstrong *et al.* [947] apply deuterium filled thyratrons with 40 kv operation voltage up to 10,000 amp and 5-μsec pulse duration as switching means in heavy duty plasma-discharge equipment. This tube allows 150 kw average power. (See also Section 1o of this chapter.)

2. The Ignitron

In the hot cathode thyratron the cathode, when heated, is constantly emitting electrons, and control is exercised by allowing the electrons to initiate ionization only when conduction is wanted.

In the ignitron, on the other hand, the cathode is a cold pool of mercury from which, initially, there is practically no emission. When an arc current is desired, steps are taken to form an electron-emitting spot on the mercury surface. An arc then develops between the anode and the cathode with a low voltage drop, usually a few volts higher than for a hot cathode tube of comparable size. The ignitron is thus a trigger control device [193]. If a suitable current is passed through the ignitor and the mercury—for example, a pulse of current produced by discharging a capacitor by means of a thyratron, as shown diagrammatically in Fig. B2-1—a cathode spot forms at the point of immersion. If the anode is sufficiently positive an arc is established to it. The timing is very precise, since the formation of the cathode spot can take

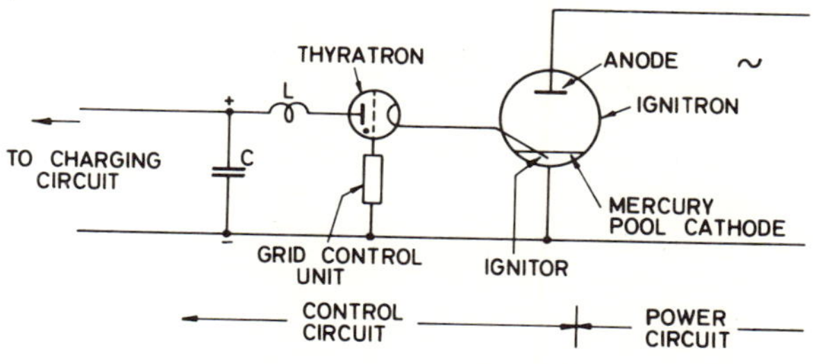

Fig. B2-1. Ignitron control circuit, elementary diagram. Current pulse passes through ignitor when capacitor C is discharged by thyratron; timing governed by thyratron control circuit.

place within a very short time, depending on the current pulse employed, and less than a microsecond if desired.

The function of the ignitor is to initiate the arc and so to start current flow at the desired time. In an ac circuit the conditions are similar to those for a thyratron. The arc goes out when the anode voltage falls to zero and must be reignited in each cycle. By means of the ignitor it is possible to select the starting time in the cycle. By varying this time, phase control of the current is obtained.

The operation of the ignitor is illustrated diagrammatically in Fig. B2-2. The ignitor itself is commonly in the form shown in Fig. B2-3. A slender rod of semiconductor material is bonded to a connector, generally with an intermediate shank of graphite.

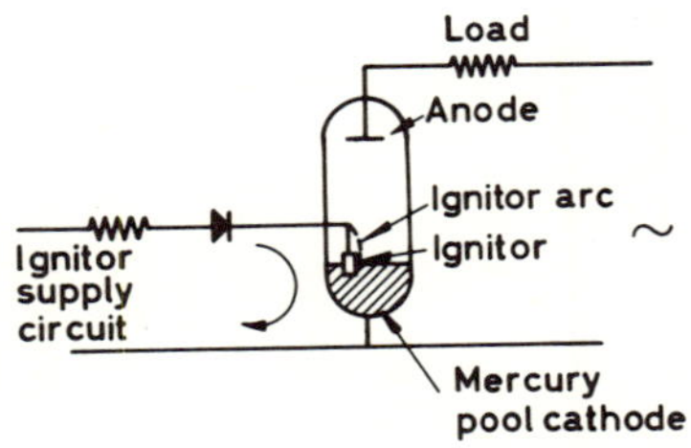

FIG. B2-2. Diagram illustrating the function of the ignitor.

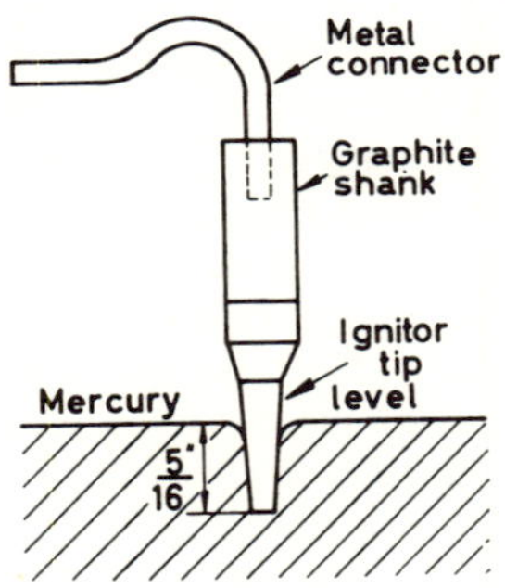

FIG. B2-3. Typical semiconductor ignitor.

The term semiconductor in this case signifies that the material has a moderate resistance. This varies in different designs, between about 20 ohms and a few hundred ohms. The resistance quoted is measured by an ohmmeter between the connector and a pool of mercury into which the ignitor projects to a specified depth. The resistance, which includes the contact resistance between the ignitor and the mercury, varies with the depth of immersion.

When current flows in the ignitor cathode circuit in the direction shown by the arrow, a discharge takes place at the point of immersion. This triggers an arc between the anode and the cathode (if the anode is positive). The initiating discharge takes place in the narrow tapering space between the ignitor surface and the convex meniscus of the mercury. Because of the resistance, there is a potential drop along the ignitor. As a consequence the discharge spreads up the rod and the latter is bypassed by a discharge from the ignitor connector direct to the mercury, as shown in Fig. B2-2. The main arc may be initiated, even if this spreading is prevented by screening all of the ignitor and its connectors above the mercury level so that the ignitor discharge is confined to the meniscus space. In practice, however, the bypassing discharge is desirable; it helps the rapid formation of an arc from the main anode to the cathode.

If the ignitor current were passed in the opposite direction, with the ignitor negative, the resulting discharge at the point of immersion would have its cathode spot on the ignitor; this would lead to erosion of the ignitor material and early failure. It is important with this type of ignitor that ignitor current should flow only in one direction. This is best ensured by the inclusion of a rectifier in the ignitor circuit, as shown in Fig. B2-2. If a controller rectifier such as a thyratron is included, the initiation of ignitor current, and hence of the main arc, can be timed very precisely by the grid control circuit.

Originally cerite was the material for the ignitor electrode. Now many modifications of semiconductors are used by manufacturers. The life expectancy of the device is given by the rate of consumption of this half-conductor electrode. Even under this limitation, life will last for a few 5×10^7 ignition impulses. For capacitor discharge circuits, which are always operated with direct current, the simple circuit of Fig. B2-4 can be employed.

During ignition a luminous spot is formed at the pin in the mercury that is visible to the eye. From this initial emission the discharge progresses. The advantage of the ignitron is its high current capacity, which reaches 10,000 amp in larger units. A drawback, for technological reasons since the chambers are mostly of iron or steel, is that it is very difficult to allow an inverse voltage of more than 3 kv. Owing to the steel shell, the cathode can be cooled very effectively. A generally accepted construction is to arrange at the outside

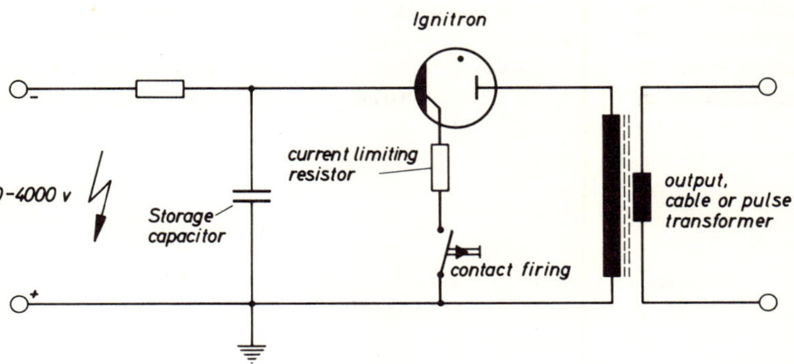

FIG. B2-4. Schematic circuit diagram of an ignitron for a capacitor discharge ignited by contactor.

of the ignitron a cooling jacket that has been brazed to the iron chamber. This cooling jacket serves to conduct the loss heat from the cathode to the outer atmosphere. Air cooled rib coolers are used for smaller units (e.g., Type 5550).

Calculation of the heat losses is relatively simple. The average current is multiplied with the ionization voltage of the mercury. When, for instance, there are to be converted 100 amp at about 10 volts ionization voltage, the heat loss is 1000 watts, which must be dissipated. This loss is independent of the total voltage converted in the device.

There are, of course, some ignitrons of special design that have a glass instead of a steel chamber which can withstand high inverse voltages above 10 kv. By changing from steel to glass one sacrifices the good cooling condition otherwise obtained by brazing of the cooling jacket to the steel chamber. On account of this sacrifice, the cathode in a glass envelope has little opportunity for cooling. Ignitron tubes have been built with a glass envelope which at about 20 kv inverse voltage can carry a peak current of 5000 amp. However, they can consume only a rated average cathode current of 2 amp, corresponding to about 20 watts average cathode output. Such tubes serve very well in laboratories and test stands as switching means. Ignitrons in steel chambers have gained general acceptance in the field of capacitor discharge practice, in impulse welding, in large magnetizing devices, and for switching of large capacitor batteries at 2–3 kv charging voltage (e.g., Raytheon impulse welding machines; impulse welding machines "Impulsa" and "Vitessa" of FF Impulsphysics Corp).

3. The Multivibrator [184]*

Multivibrators are the most essential circuit means for triggering other switching elements, such as thyratrons, spark gaps, trigatrons. For low current condenser discharging they may be useful directly.

Perhaps the easiest way to begin a study of multivibrators is by comparison with the Eccles–Jordan circuit. This trigger circuit was found to have two stable operating conditions. If, in this fundamental trigger circuit, one coupling resistor is replaced by a capacitor, and the bias is also removed from that side, a completely different form of operation will result. Such a circuit is shown in Fig. B3-1b with the original trigger circuit given in Fig. B3-1a. In the original trigger circuit, a drop in potential at one plate caused the opposite grid potential to drop to a low value. This value was usually below cutoff or at least sufficiently low to prevent any further regeneration. Thus a stable state resulted, with the grid maintaining its high negative voltage. This type of operation is represented by the sketch of Fig. B3-2a. In the modified circuit of Fig. B3-2b, the drop in potential of plate P_1 causes a corresponding drop in the grid potential of tube T_2. However, the grid circuit of T_2 is now not a simple resistance voltage divider, but is a resistance-

* The following 11 paragraphs are reprinted from L. W. von Tersch and A. W. Swago, "Recurrent Electrical Transients," Prentice-Hall, New York, pp. 320–332 (1953).

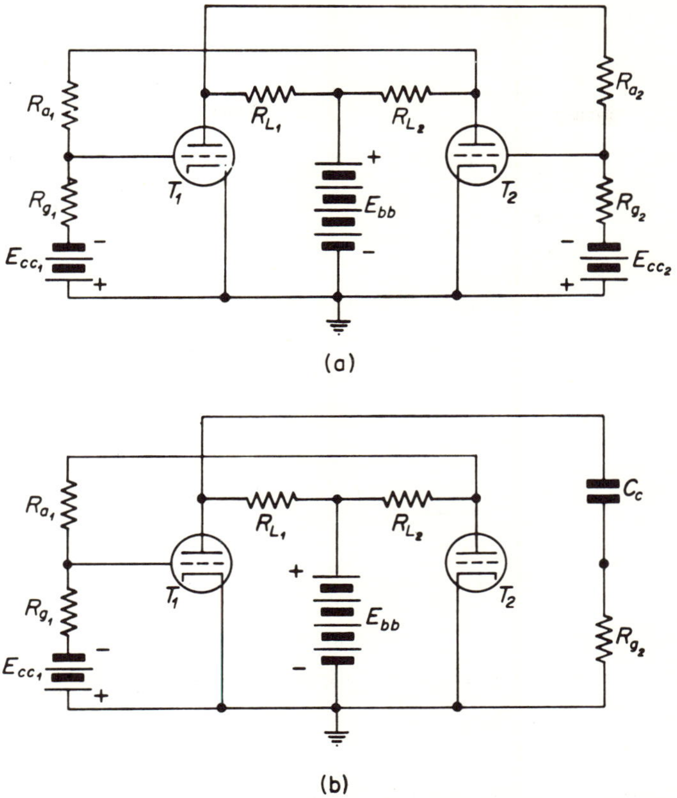

FIG. B3-1. (a) Eccles–Jordon trigger circuit. (b) Single-cycle, plate-coupled multivibrator. By permission from "Recurrent Electrical Transients," L. W. von Tersch and A. W. Swago, Prentice-Hall, 1953.

capacitance circuit, so that the grid voltage of T_2 returns toward zero as the coupling capacitor charges. When the grid voltage comes up to the cutoff value, T_2 will conduct slightly.

This will lower the plate potential of T_2 and the grid potential of T_1, whereupon the familiar cumulative action will occur, with conduction being transformed back to T_2. The new circuit thus has but one stable condition. If some external signal should cause conduction to be transferred to T_1, the circuit will eventually return, without any further external disturbance, to the condition with T_2 conducting. The time required for this return will depend upon the effective time constant of the circuit, the voltage applied to the circuit, and the tube characteristics.

The circuit of Fig. B3-1b is called a plate-coupled, single-cycle multivibrator. If both coupling resistors in the basic Eccles–Jordan circuit are replaced by capacitors, a free-running, plate-coupled multivibrator will result. This is the circuit originally described by Abraham and Bloch. The

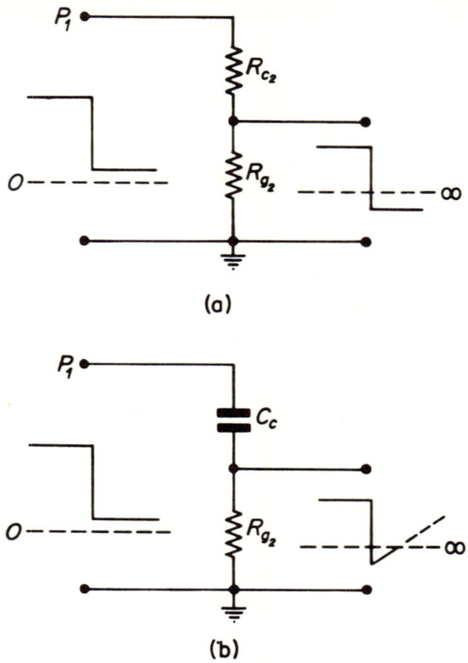

FIG. B3-2. (a) Resistance coupling in trigger circuit. (b) Capacitance coupling in a multivibrator. By permission from "Recurrent Electrical Transients," L. W. von Tersch and A. W. Swago, Prentice-Hall, 1953.

circuit can be considered a combination of two single-cycle multivibrators with operation alternating back and forth between the two semistable states. No external signal is needed to cause this alternation, although one may be used to influence the rate of change.

Each basic form of multivibrator also has numerous modifications. Some of the more important of these modifications and their uses will be discussed in detail in the following sections.

The clearest model is the single-cycle, plate-coupled multivibrator. The Eccles–Jordan trigger circuit as modified into a single-cycle, plate-coupled multivibrator is shown in Fig. B3-3. This circuit will have but one stable state in which T_2 is conducting and T_1 is nonconducting. Under these conditions the circuit of Fig. B3-4 may be drawn. If the circuit of Fig. B3-3 and Fig. B3-4 has been at rest for a long time, the capacitor C_{c2} will have a potential of E_{bb} across its terminals. The term E_{cc1} will be treated as a positive quantity, thus if R_{g1} is returned to a potential which is negative with respect to ground, a negative number must be used in place of E_{cc1}.

For the grid voltage of T_1 the following equation is obtained:

$$(e_{c1})_0 = \frac{R_{g1}(e_{b2})_0 + R_{c1}E_{cc1}}{R_{c1} + R_{g1}}$$

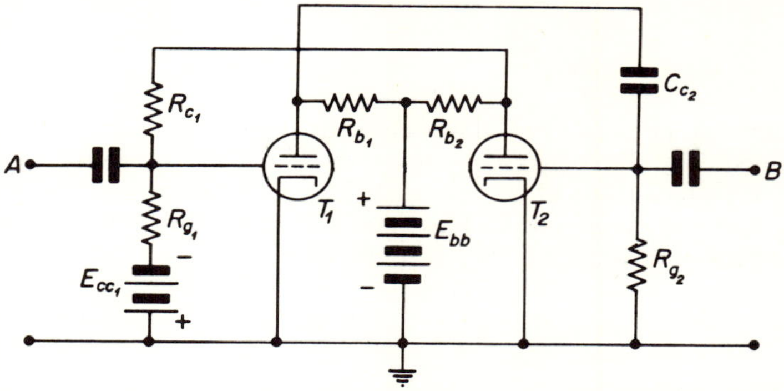

FIG. B3-3. Single-cycle, plate-coupled multivibrator. By permission from "Recurrent Electrical Transients," L. W. von Tersch and A. W. Swago, Prentice-Hall, 1953.

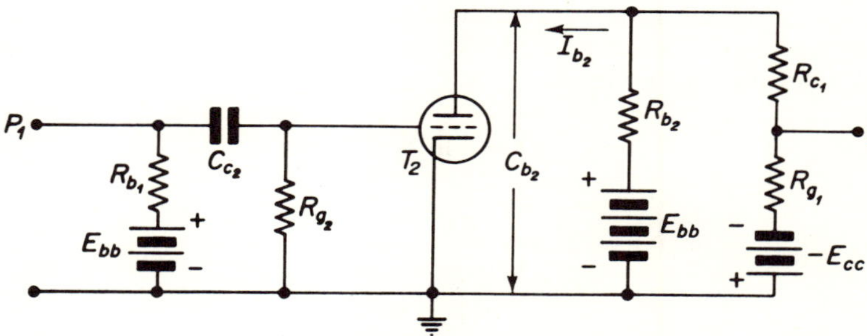

FIG. B3-4. Circuit equivalent to that of Fig. B3-3 when T_2 is conducting and T_1 is cut off. By permission from "Recurrent Electrical Transients," L. W. von Tersch and A. W. Swago, Prentice-Hall, 1953.

and the plate voltage is

$$(e_{b1})_0 = E_{bb}$$

In order that the original assumptions be justified $(e_{c1})_0$ must be negative and sufficiently great in magnitude to keep T_1 cut off for a plate voltage of E_{bb} volts.

The circuit can now be triggered in either of two general ways. A positive pulse can be applied to point A, or a negative pulse can be applied to point B. The positive pulse at A must be sufficiently large to bring the grid of T_1 to a point just above the cutoff potential in order that regeneration can occur.

The amplitude of the negative pulse needed to trigger the system can only be approximated, but will be nearly that value which will cause a change of voltage at the plate of T_2 such that

$$\frac{R_{g1}}{R_{g1} + R_{c1}} \Delta e_{b2} = e_{co1} - (e_{c1})_0$$

or

$$\Delta e_{b2} = \frac{R_{g1} + R_{c1}}{R_{g1}} [e_{co1} - (e_{c1})_0]$$

The equivalent circuit of Fig. B3-4 can be used in conjunction with the plate characteristics of T_2 to find the change in grid voltage needed to cause a change in plate voltage of Δe_{b2} volts. It is possible to trigger the circuit if $(e_{c1})_1$ is negative, but a more stable action will result if $(e_{c1})_1$ is made positive.

The effective resistance in the grid circuit of T_2 is

$$R = \frac{\bar{r}_{p1} R_{b1}}{\bar{r}_{p1} + R_{b1}} + R_{g2}$$

The time constant is

$$\tau = \left(\frac{\bar{r}_{p1} R_{b1}}{\bar{r}_{p1} + R_{b1}} + R_{g2} \right) C_{c2}$$

The duration of the time interval during which T_2 is cut off and T_1 is conducting can now be calculated. A sketch of the grid voltage of T_2 is shown in Fig. B3-5. From the figure,

$$\Delta t = \tau \ln \frac{(e_{c2})_1}{e_{co2}}$$

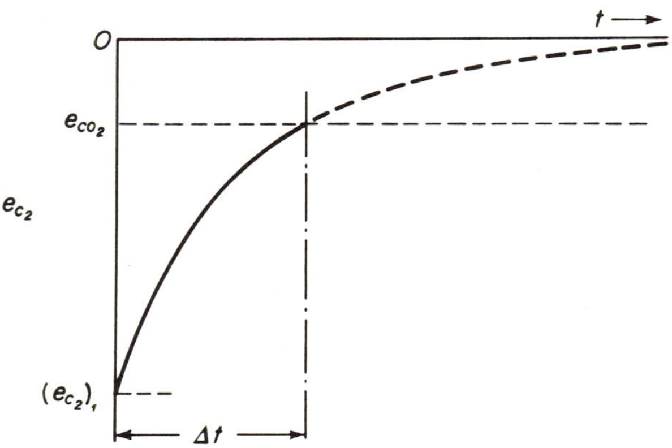

FIG. B3-5. Grid voltage of T_2 when T_2 is cut off. By permission from "Recurrent Electrical Transients," L. W. von Tersch and A. W. Swago, Prentice-Hall, 1953.

During the time Δt, the grid voltage goes through $[(e_{c2})_1 - e_{co2}]/(e_{c2})_1$ of its total charge. Thus, the plate potential of T_1 will go through the same fraction of its total change. The final value of e_{b1} is that found from the application of E_{bb} and R_{b1} to the plate characteristics. If the final plate voltage of T_1 and the cutoff potential of T_2 are known, the final capacitor voltage may also be determined.

Several modifications of the circuit in Fig. B3-3 are possible, with some

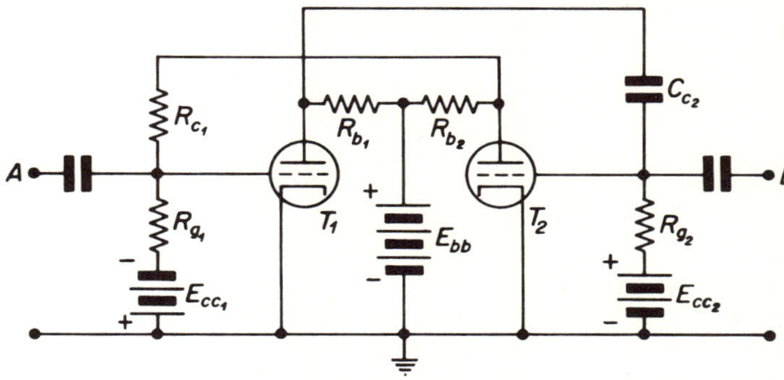

FIG. B3-6. Single cycle, plate-coupled multivibrator with positive grid return. By permission from "Recurrent Electrical Transients," L. W. von Tersch and A. W. Swago, Prentice-Hall, 1953.

of these modifications having definite practical value. One of these is shown in Fig. B3-6. The grid of T_2 is now returned through the grid resistor R_{g2} to a positive source E_{cc2}, instead of to ground as was done in Fig. B3-3.

Such circuits generally consist of two triode tubes with interconnected cathodes. In smaller tubes, these are accommodated in a common glass envelope. In a widely applied multivibrator circuit there is taken from the anode of the first tube a positive or negative voltage impulse which is fed to the grid of the second tube. The consequence of this connection is that, at the common cathode resistance, there occurs a current impulse that is unidirectional with the original current flow. This creates a feedback that results in a current increase, depending on circuit connection, either in the first or in the second multivibrator tube up to the point of emission. A measure of the efficiency of such multivibrator circuits is the emission current of one of the two tubes. The other tube can be a weaker type since it serves only to control the first one and the inverse voltage, i.e., the permissible cold anode voltage without current load, since before relaxation occurs the control tube is currentless. In the form of combination tubes there are available double three-pole systems which can produce within allowable temperature rise limits about 1 amp emission current at 500 watts impulse output at any desirable frequency. When two separate tubes are employed it is advantageous to consider strong-emission special tubes having fast rise times, e.g., such tubes as are used chiefly in the ultra high-frequency field.

In radar equipment hard tube pulsers are used for feeding the transmission tube. These have high heater capacity and emission, e.g. 10 kv and 10 amp, thus 100 kw. Circuitry with such components is necessary in impulse control devices for triggering hydrogen thyratrons with extraordinarily steep-fronted pulses at the grid. Moreover, it is possible to give the trigger pulse the favorable shape referred to in the previous paragraph, e.g., a fast positive rise time with a subsequent negative longer-lasting but also energy-strong "tail" (see Sections 1b and 1n in this chapter). Triggering problems in question here can be solved in practice only with correspondingly designed multivibrator circuits. A multivibrator itself requires for triggering only very weak pulses. The small multivibrator types of the ECC series can, in properly designed circuits, be made sensitive enough to respond to a pulse of only 10 mv, whereas even more substantial circuits generally need approximately 1 volt control pulses for release. Moreover, since the control voltage can be applied through a high-resistance circuit, a number of otherwise necessary amplifier stages may be discarded when employing a multivibrator circuit. Another advantage of this circuit is that it is possible to trigger a multivibrator synchronized with a certain event through only the low preamplifaction provided by a standard amplifier tube. It must be observed, however, that by triggering of a multivibrator, owing to the steep voltage decay, a considerable number of disturbing voltages and oscillations can be released in the switching network which, fortunately, dampen out very quickly.

However, since, in cyclic impulse operation the interval between subsequent pulses is generally of such duration that the following impulse arrives at the grid late enough to find a disturbance-free atmosphere, it is safe to combine standard amplifier units even with high-powered multivibrator circuits without resorting to complicated shielding. In every instance, however, one should make certain by proper measurements that the igntion pulse of the multivibrator with all its disturbances has decayed before another pulse occurs.

Figure B3-7 shows an example of an ignitron-controlled capacitor circuit in which the ignitron is triggered by a thyratron which in turn is triggered

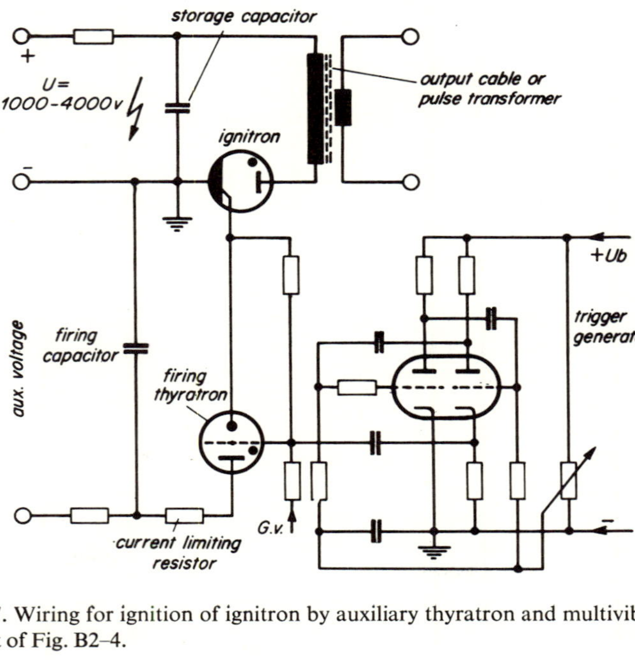

FIG. B3-7. Wiring for ignition of ignitron by auxiliary thyratron and multivibrator in the basic circuit of Fig. B2–4.

by a multivibrator. Particular attention is called to the current limiting resistors which must be adapted to the rated operating voltage and the rated cathode current.

It appears proper to mention here the Schmitt trigger as a special design multivibrator. This circuit requires, independent of the pulse sequence frequency, uniform magnitude of the triggering pulses, whereas a capacitance-coupled multivibrator requires for its start a constant steepness of the voltage rise at the grid.

The art of perfecting multivibrators has advanced rapidly. The aim of the circuitry of the new transistors is to reach the subnanosecond rise-time region in order to increase the calculation speed of computers. But also in use are special nuvistor tubes with which subanosecond speed is available. (Nuvistors are tubes which are manufactured under high-vacuum conditions using ceramic insulation material only. They have parallel flat cathodes, grids, and anodes.) A short survey of the most modern circuitry of tubes, tunnel diodes, and transistors is given in [862].

4. Symmetrical Switch Tubes with Mercury Filling

In impulse welding machines, just as in all other capacitor discharge circuits, it is important to switch a charged capacitor upon a load, e.g. an impulse welding transformer, at any arbitrary instant. For such switching operation heavy magnetic starters are normally used. Ignitron switching, employed in various installations, cannot be used for large capacitor impulse welding machines because the inverse voltage of such ignitrons would have to be above 3000 volts. Under the full-load condition of the impulse transformer current values of some 1000 amps can readily develop. Ignitron switching tubes of such capacity would be very costly, particularly so considering that under oscillating discharge conditions one ignitron would be required for the positive and one for the negative current impulse. With oscillations of longer duration, longer than one negative and one positive half-cycle, other control and firing circuits of considerable complexity would be required. In contrast to devices operating at network frequency, the frequency of oscillation in capacitor discharge circuits adjusts itself, in accordance with the equations governing frequencies, to the capacitance and leakage inductance of the connected transformer as well as its load. Therefore, control and ignition elements would have to function at such nonconstant frequency. Relieving the difficulties pointed out it is possible to use as a switch a simple metal vapor switching tube that is filled, for example, with mercury. When a sufficiently high ion concentration is produced at the cathode of such a tube, there can be passed through the tube with very little switching loss the charge of the capacitor battery in an aperiodic or oscillating energy impulse at any arbitrary instant. A short description of a few versions of switch tubes follows.

In Fig. B4-1 a capacitor, 1, has been charged to a high potential through lines, 2, before the discharge process now to ensue. The discharge will take place through a chamber, 3, into the primary winding of a transformer, 4,

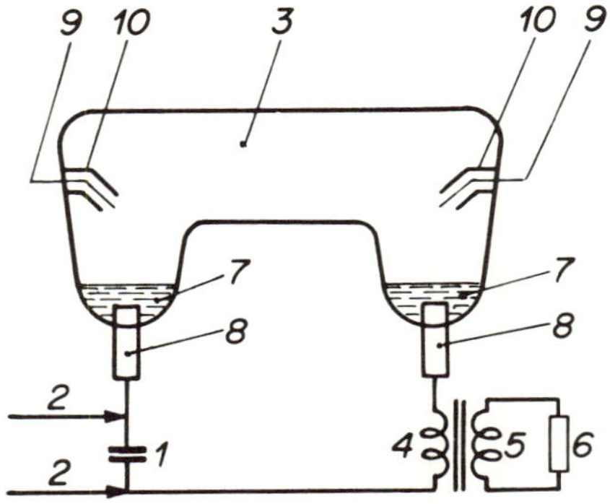

FIG. B4-1. Switch tube with 2 Hg electrodes and spark gap ignition.

which normally has a secondary, 5, connected to a load, 6. The load may be a work piece to be welded. When mercury is employed as filling, the simplest form of a switch tube, as shown in Fig. B4-1, has two meniscuses, 7, which are in electrical contact with the terminals, 8, of large cross section that have been sealed into the wall of the chamber. The average output N_k that has to be conducted to the cathode can be calculated as follows: When U_i, in accordance with the cathode drop, is the ionization voltage, then $N_k = U_i \times I$. Here, just as in the ignitron, it is not the current peak but only the average current which is of interest.

For effecting a discharge, in contrast to the ignitron method, an ignition electrode system, 9-9, is used. The ignition electrode lead-ins are surrounded by glass sleeves, 10, in order to prevent a creepage discharge to the outer glass shell surface of the chamber, 3. A high-voltage ignition impulse is applied for firing. This impulse can be a transformed capacitor discharge or the direct discharge of an auxiliary capacitor. With such a strong voltage pulse it is possible, without ignitron-like dipping of an ignition electrode in the mercury, to obtain, by a true gaseous discharge, sufficient ionization in the chamber to fire the main discharge between the meniscuses. A single ignition spark at one of the electrodes often suffices.

Usually, however, the discharge arcs throughout several half-waves until oscillation decays, since the terminals are symmetrically arranged. In cases where the second half-wave is of lower energy so that it can be dispensed with during the main discharge, it is possible to use a tube as shown in Fig. B4-2.

In such a tube it is again essential that the ignition spark be established in the gap between 9 and 7. Here, however, it is possible to select for the first half-wave a graphite anode, 11, passing through a heavy lead-in bushing, 12.

It has been found beneficial to abandon the semiconductor pin of classical origin and to devise an open spark gap as indicated in Fig. B4-2. This gap then has to be ignited through an impulse transformer as mentioned before. Such spark gap arrangements have, if properly dimensioned, a longer life

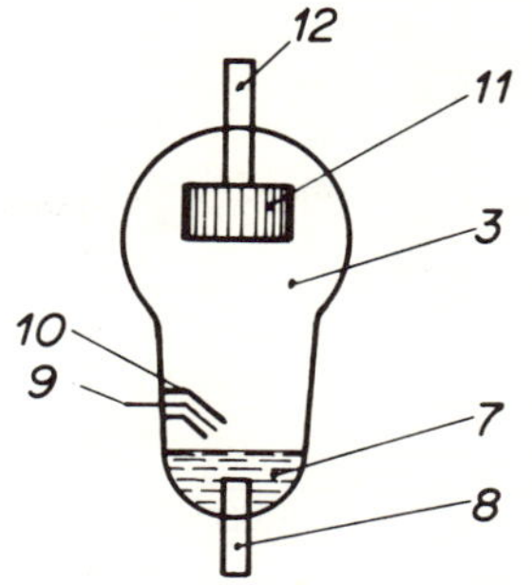

FIG. B4-2. Switch tube with Hg cathode and graphite anode.

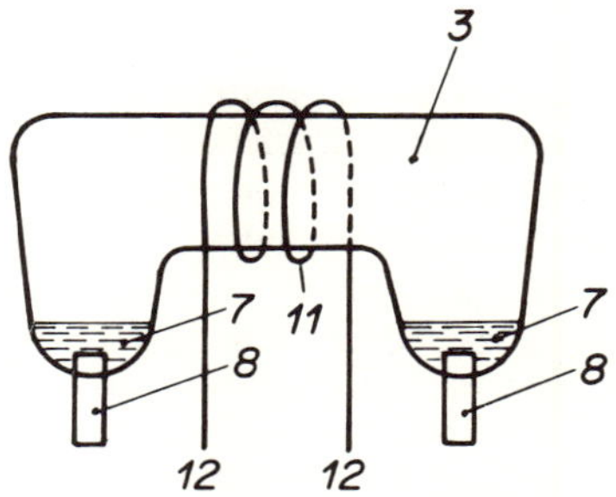

FIG. B4-3. Inductron (Duratron) with 2 Hg electrodes and ignition coil.

than the ignitron semi-conductor pins. Thus the useful life of switch tubes can be considerably increased by resorting to inductive external ignition. Capacitive ignition is also possible. For that purpose, a metal foil is put around the glass close to the meniscuses. Ignition is effected by a steeply rising or high-frequency oscillating voltage impulse. Adjacent to the capacitive ignition layer there should be interposed an insulation capacitor, because it has been found beneficial to tube life to keep even the smallest dc components away from the glass in order to prevent glass electrolysis.

A particularly simple example of inductive ignition is depicted in Fig. B4-3 (by Kluge and Früngel). In order to simplify the design of otherwise costly switch tubes, it is possible to ignite by a high-frequency oscillation from a coil, 11–12, instead of by an electrode-initiated ignition spark main discharge. In this connection the soundness of the following thought has proven itself: In the well-known glow emitter for the evacuation of vessels, a luminescence can be observed during preheating and treatment action of the high frequency operated glow emitter. Such luminescence is actually an electrodeless discharge that can now be utilized for ignition of the main discharge in a metal-vapor switch tube. To achieve this, an oscillating auxiliary discharge must be sent through coil 11. This discharge has a much greater susceptance during the ignition than is usual in the oscillatory coil

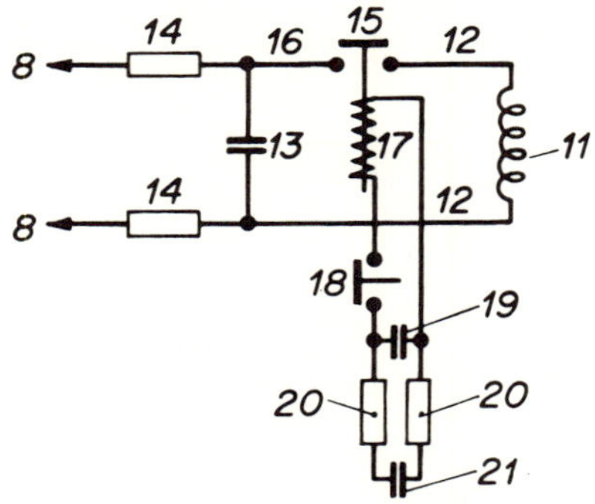

FIG. B4-4. Circuit for producing the ignition pulse for the coil in Fig. B4-3.

of a glow emitter. Glow emitters operate generally at an output of 2 kw and at a 50-fold increased susceptance factor, i.e., at about 100 kva. By dimensioning a discharge capacitor and an induction coil that has been wound around the tube so that the product $N = U^2\omega C$ gives a susceptance of about 100 kva or more, one obtains a forceful gas discharge in the tube and an arc breakthrough between 7 and 7. When a capacitor discharge is actuated in the manner set forth here it is advisable to provide means against premature release. The circuit shown in Fig. B4-4 is useful for that purpose.

The energy storage capacitor, 13, provided for oscillating discharge through the coil, is charged through charging resistors 14 from a high-voltage line, 8–8. It has been found advisable to use the same high-voltage source that is connected to the terminals of the switch tube. In this manner it is possible to complete the weld or to release the charge of the capacitor only when full potential exists at the switch tube, i.e., only after the capacitor is fully recharged. The capacitor discharge of 13 passes by way of 16 through coil 11. The coil, 17, of contact 15 can thereby be energized by closing another switch contact, 18, which may be a manual push button, to release a discharge from capacitor 19 into this coil. This ensures a forceful closing of contact 15. It is advisable to recharge capacitor 19 by means of charging resistors, 20, from a current source, 21, which can be a charged larger capacitor, in such a manner that the time constant between resistor 20 and capacitor 19 is identical with the time constant of capacitor 13 and resistor 14. This prevents premature discharge of capacitor 13 into oscillatory coil 11.

Difficulties at low operating voltages at switch tube 3 can be overcome by the circuit illustrated in Fig. B4-5. If voltages in the range of only 2 to 3 kv are applied at the tube, 3, the oscillations in coil 11, because of the low field intensity in the tube, will be considerably stronger than with 4 to 6 kv.

The ignition of a gaseous discharge is easier the closer the prevailing voltage is to the rupture voltage. But to obtain positive ignition and a high kva value, in the oscillating discharge coil 11, even under the adverse condition of low voltage, the Marx circuit can be employed. In this circuit the capacitors 131, 132, and 133 are being charged in parallel through the resistors

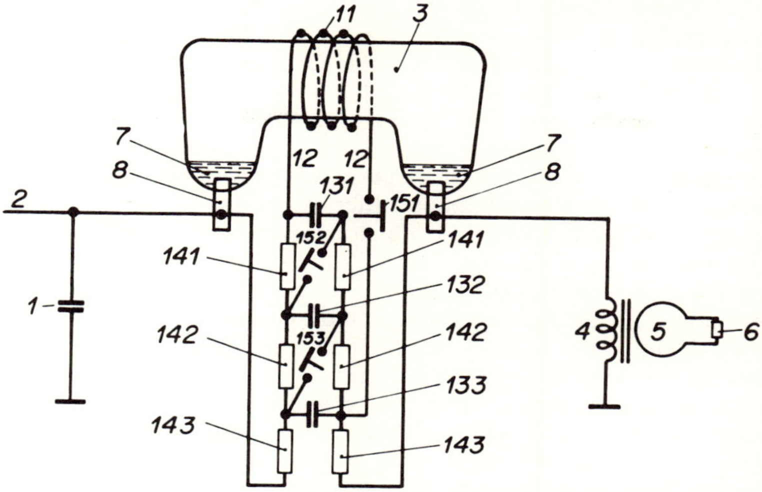

FIG. B4-5. Inductron (Duratron with cascade connection in the ignition circuit of an impulse welding machine). The ignition circuit draws its voltage from the operating voltage of the switch tube (Früngel-Kluge).

141, 142, and 143, respectively. They are in this case also charged by the same voltage that is applied to the terminals, 8–8, of the switch tube. To start ignition the switch contacts, 151, 152, and 153, are simultaneously closed. With a voltage of only 2 kv at the tube, 6 kv is obtained at the coil, 11–12, for the oscillation by this three-fold switching. Within a voltage range between 1.5 and 6 kv and a capacitor value for item 1 of between 50 μf and 5000 μf it is possible to achieve reliable ignition with such a voltage tripling arrangement. The oscillating wattless energy in the coil during ignition is considerable.

Example:

Capacitance of the ignition capacitors: 3 units, 10^{-9} farads each; charging voltage 3 kv.

During discharge in series connection the capacitance becomes

$$\tfrac{1}{3} \times 10^{-9} \text{ farads at } 9 \text{ kv}$$

The peak current is then

$$\frac{1}{2} CU^2 = \frac{1}{2} LI^2; \qquad I = \sqrt{\frac{CU^2}{L}}$$

With an inductance of 100 μh for the ignition coil, the susceptance becomes:

$$N_B = U \times \mathrm{I} = U^2 \sqrt{\frac{C}{L}} = 81 \times 10^6 \times \sqrt{\frac{10^{-9}}{3 \times 10^{-4}}} \approx 15 \times 10^4$$

$$\approx 150 \text{ kva}$$

This simple initial ignition results in a very considerable oscillation excitation within the plasma.

The arrangement shown is also suitable for all capacitor discharge circuits with relatively low resistive loads. It works with potassium or sodium instead of mercury as fill-gas. Mercury, however, is the least temperature dependent. For an operating inverse voltage of 5000 volts at 5000 amp nonoscillating or oscillating impulse current load with oil cooling, the device needs only chamber sizes of about 2 liters capacity.

At high average loads, e.g. at impulse sequences of 100 kw-sec/sec (100 kw), under adverse conditions, several kilowatts in loss heat are produced within the switch tube. Therefore, it is advisable to provide for operation under oil as illustrated in Fig. B4-6. The oil in turn is cooled by rib coolers with water. In order to keep the vapor pressure, and thereby the ignition conditions, in the tube constant, the temperature must be maintained at a constant level (e.g., 40°C). Temperature and pressure, however, must not surpass a certain limit. Otherwise the breakthrough voltage may become dangerously low. During the starting period, and a short time thereafter, temporary heating of the tube may be necessary for the above reasons. After the operating temperature has been reached, the heating may be changed to cooling in order to carry off the continuously generated operating heat. If the switching device is cooled by circulating water, which in industrial plants generally has a temperature of 25–40°C, too low a starting temperature is practically excluded beforehand.

FIG. B4-6. Internal arrangement of an inductron switching. The cascade impulse capacitors can be seen subdivided at the left, middle, and right of the illustration, above the inductron tube with the ignition coil.

Because of their long life, inductively ignited diodes are referred to as "duratrons."

5. Fixed Spark Gaps

a. *Measuring Spark Gaps with Two Electrodes*

Spark gaps appear to be the earliest switching means for capacitor discharges. The techniques for measuring high capacitor discharge potentials and high ac voltages are dependent upon them. They form the basis of standard high-potential measuring methods. It is well known that in a sphere spark gap the breakdown or sparking distance depends on the diameter of the spheres and the air pressure. There are tabulations available from several sources in which measured potentials and corresponding sphere diameters are listed (see Tables B5a-1 and B5a-2). The measuring potentials are referred to normal barometric air pressures and have to be converted to values of acutally prevailing pressures. Throughout wide limits, within the validity range of Paschen's law, the breakdown voltage varies in linear direct proportion to the air pressure. The simplest form of switching mechanism is the connection of a capacitor with a load through an interposed fixed spark gap. When the capacitor is charged to a potential that surpasses the breakdown potential of the spark gap (which can be adjusted to within 3% or even more accurately), the sparkover occurs. The spark in the gap now has a spark resistance that was first calculated by Toepler. The resistance of the spark R_F according to Toepler, which is sufficiently accurate for practical purposes, is

$$R_F = \frac{k \times l}{Q}$$

wherein $k_{\text{air}} = 0.8 \times 10^{-3}$ (see Chapter M), l is the length of the spark in centimeters and Q is the number of ampere-seconds transmitted through the spark. A spark which discharges a 1 μf capacitor charged to 30 kv through a spark gap of about 1 cm length would have a resistance of

$$R_F = \frac{0.8 \times 10^{-3} \times 1}{10^{-6} \times 3 \times 10^4} = 8/300 \text{ ohm} \approx 27 \text{ milliohms}$$

From this example it becomes apparent that the spark resistance is negligibly low in comparison with the usual resistances of loads into which capacitors are discharged. The widespread belief that a spark is an undesirable source of high losses is thus not justified. A modification of the solid state or self-igniting spark gap is an arrangement in which one sphere of the gap is made movable with respect to the other sphere. The movable sphere is advanced toward the other by a pneumatic or electromagnetic actuator until breakdown occurs or the electrodes strike each other. In rough use under heavy loads with very high current peaks the spheres may weld together. In order to avoid this it is advisable to provide a small circular recess in the sphere and to fill it, level to the sphere surface, with tungsten so that tungsten will strike against tungsten. Even under current impulses of many thousand

Table B5a-1

Initial Voltages of Symmetrical Sphere Spark Gaps[a, b]

Spark width (cm)	Sphere diameter in cm: 1.11	2	2.54	5	6.25	6.66	7.5	10	12.5	15	25	Plane Plates
	Peek	Weicker	Peek	Weicker, Estorff, Töpler	Peek	Peek	Estorff	Estorff, Weicker, Töpler	Peek	Weicker, Estorff, Töpler	Peek	
0.05	2.9	—	—	—	—	2.65	—	—	2.65	—	2.65	2.65
0.1	4.95	—	—	—	—	4.60	—	—	4.60	—	4.60	4.47
0.2	8.70	—	8.32	—	—	8.09	—	—	8.04	—	8.04	7.96
0.3	12.0	—	11.6	—	—	11.3	—	—	11.3	—	11.3	11.3
0.4	15.0	—	14.8	—	—	14.4	—	—	14.4	—	14.4	14.4
0.5	17.9	—	17.9	—	17.5	17.5	—	—	17.4	—	17.4	17.4
0.6	20.6	—	21.0	—	—	20.5	—	—	20.4	—	20.4	20.3
0.7	23.3	—	24.0	—	—	23.5	—	—	23.3	—	23.3	23.2
0.8	25.6	—	27.0	—	—	26.4	—	—	26.2	—	26.2	26.1
0.9	27.7	—	29.8	—	—	29.4	—	—	29.1	—	29.0	28.9
1	29.5	32.0	32.5	32.4	32.4	32.2	32.1	32.1	32.0	31.8	31.8	31.7
2	—	52.1	—	59.3	59.7	—	59.7	60.0	60.0	60.0	60.5	59.6
3	—	64.2	—	80.7	82.4	—	83.8	85.4	85.7	86.2	87.3	87.0
4	—	71.9	—	97.7	102	—	104	108	110	110	113.5	114
5	—	76.8	—	111	117	—	122	128	131	133	138	140
6	—	—	—	122	129.5	—	136	146	151	154	162	166
7	—	—	—	131	140	—	149	162	169	174	185	192
8	—	—	—	139	150	—	160	176	185	191	205	217
9	—	—	—	145	158	—	170	187	200	207	227	242
10	—	—	—	150	165	—	178	199.5	213	222	247	266
11	—	—	—	—	—	—	185	209	225	236	266	290
12	—	—	—	—	—	—	192	218.5	236	249	283	—
13	—	—	—	—	—	—	198	227	246	260	300	—
14	—	—	—	—	—	—	203	234	255	271	316	—
15	—	—	—	—	—	—	208	241	265	282	331	—
16	—	—	—	—	—	—	—	247	—	291	—	—
17	—	—	—	—	—	—	—	252	—	300	—	—
17.5	—	—	—	—	—	—	—	—	284	304	367	—
20	—	—	—	—	—	—	—	—	299	—	398	—
22.5	—	—	—	—	—	—	—	—	—	—	424	—
25	—	—	—	—	—	—	—	—	—	—	451	—
30	—	—	—	—	—	—	—	—	—	—	492	—
40	—	—	—	—	—	—	—	—	—	—	555	—

[a] Initial voltage with two equal spheres at symmetrical potential distribution in normal atmospheric air, barometric pressure 750 mm Hg, temperature 20°C, voltages in kilovolts.

[b] Dc voltage values. For effective ac voltage, divide by $\sqrt{2}$. These values are valid for initial voltage. If under limited current supply (influence machine) a complete breakdown does not occur the voltages given refer to the beginning of the discharge.

TABLE B5a-2

INITIAL VOLTAGE OF SYMMETRICAL SPHERE SPARK GAPS, ONE SPHERE GROUNDED[a, b]

Spark width (cm)	Sphere diameter in cm									
	0.5	1	2	2.5	5	6.25	12.5	25	37.5	
	Heyd-weiller	Heyd-weiller	Heyd-weiller	Orgler	Müller, Weicker	Peek	Peek	Peek	Chubb and For-tesque	Plane plates
0.01	1.10	1.02	0.993	—	—	—	—	—	—	—
0.02	1.60	1.53	1.52	—	—	—	—	—	—	—
0.03	2.07	2.02	1.97	—	—	—	—	—	—	—
0.04	2.50	2.46	2.38	2.26	—	—	—	—	—	—
0.05	2.92	2.87	2.79	2.67	—	—	—	—	—	2.65
0.06	3.32	3.26	3.18	3.07	—	—	—	—	—	—
0.08	4.08	4.03	3.96	3.86	—	—	—	—	—	—
0.1	4.87	4.80	4.74	4.61	—	—	—	—	—	4.47
0.2	8.51	8.37	8.28	8.20	—	—	—	—	—	7.96
0.3	11.5	11.7	11.6	11.5	—	—	—	—	—	11.3
0.4	14.0	14.8	14.7	14.6	—	—	—	—	—	14.4
0.5	15.9	17.5	17.7	17.6	17.5	17.4	17.4	—	—	17.4
0.6	17.3	20.0	20.7	—	20.4	—	—	—	—	20.3
0.7	18.4	22.0	23.5	—	23.2	—	—	—	—	23.2
0.8	19.3	24.0	26.2	—	26.1	—	—	—	—	26.1
0.9	19.9	25.5	28.6	—	28.9	—	—	—	—	28.9
1	20.5	27.1	31.0	—	31.8	31.9	31.9	31.7	—	31.7
1.1	—	—	—	—	34.5	—	—	—	—	—
1.2	—	—	35.0	—	37.1	—	—	—	—	—
1.3	—	—	—	—	39.9	—	—	—	—	—
1.4	—	—	—	—	42.5	—	—	—	—	—
1.5	—	—	—	—	45.0	45.3	45.8	46.0	—	—
2	—	—	—	—	56.7	58.9	59.5	59.7	—	59.6
3	—	—	—	—	74.6	78.4	85.0	86.0	87.2	87.0
4	—	—	—	—	87.4	92.4	108	111	113	114
5	—	—	—	—	96.6	105	129	136	139	140
6	—	—	—	—	104	114	146	160	164	166
7	—	—	—	—	109	121	162	182	188	192
8	—	—	—	—	114	126	175	203	212	217
9	—	—	—	—	—	131	186	223	235	242
10	—	—	—	—	—	135	197	241	256	266
12	—	—	—	—	—	—	214	274	296	—
15	—	—	—	—	—	—	231	314	351	—
17.5	—	—	—	—	—	—	244	342	392	—
20	—	—	—	—	—	—	253	365	426	—
22.5	—	—	—	—	—	—	—	385	—	—
25	—	—	—	—	—	—	—	402	—	—
30	—	—	—	—	—	—	—	430	—	—
40	—	—	—	—	—	—	—	468	—	—

[a] Initial voltages with two equal spheres, of which one is grounded, in normal atmospheric air, barometric pressure 760 mm Hg, temperature 20°C, voltages in kilovolts.

[b] Dc voltage values. For effective ac voltage, divide by $\sqrt{2}$. With one sphere grounded, the danger of (electrostatic) influence is much greater than with symmetrical voltage distribution. For this reason the values of this table are less reliable than the values of Table B5a-1.

amperes and some 10–100 kv, such switching spark gaps with tungsten facings operate reliably [10].

This form of capacitor discharge switch will be preferred in cases where large energy values have to be switched in slow sequence frequencies. The thermal load would otherwise become too high and would result in some portions of the device (for instance, the tungsten facings) glowing. Thus thermoemission would defeat the very purpose of the spark gap mechanism. Besides, a very high sequence frequency can hardly be reached by purely mechanical means. Motor-driven interrupters can handle up to 500/sec [11]. The accuracy of ignition timing is dependent on the precision of the mechanism. With precise mechanical control it is possible to switch accurately within about 5 μsec.

Example:

Peripheral speed of a switch tube, 100 meters/sec = 10^5 mm/sec; operating voltage, 8 kv corresponding to about 2 mm sparking distance.

With $\pm$ 0.5 mm deviation of spark initiation, there results

$$\text{a timing accuracy} = \frac{\pm 0.5 \text{ mm} \times \text{sec}}{10^5 \text{ mm}} = \pm 5 \ \mu\text{sec}$$

A variant of this mechanical switch, good for single arbitrary impulse release or contact making, is an arrangement in which the spark gap is expanded in a manner similar to the lock of a rifle and is arrested in this position for release by an electromagnetic trigger at the desired instant. A strong spring then speeds both switch spheres or switch conductors toward each other. This manner of switching has an additional advantage for use in the release of single discharge impulses. In the untensioned condition a positive short circuit exists in the high-voltage portion of the discharge cycle and recharging of the capacitor battery is only possible after deliberate opening or separating of the impact switch contacts. The reliability of such spring-tensioned electromagnetic release switches for high voltage work is practically unlimited. They can switch impulse loads on the order of 1000 megawatts without trouble. Tensioning the switch and setting it for the next action is usually done by hand (e.g., through a high-voltage insulating nylon cord).

A high-speed diode-switching technique is made possibe by the use of pinching sparks. A spark with a rapid increase of its magnetic field can be compressed by its own magnetic field. This results in higher plasma conductivity and extremely high temperature. An example of the use of the radiation of the gap itself is given by Fischer [797, 799]. The Teflon-guided spark prevents nearly any damage or burning of the electrodes.

b. *General Considerations for Switching Spark Gap Design* [187]*

The three fundamental considerations in the design of enclosed gaps are

* Chapter B, Section 5b, is reprinted by permission from J. R. Dillinger, General considerations for gap design, in "Pulse Generators," McGraw-Hill Book Company, New York (1948), pp. 316–332.

(1) the choice of a suitable gas or mixture of gases, (2) the choice of suitable electrode material, and (3) the determination of a satisfactory electrode geometry.

Contrary to demountable spark chambers, these gaps may be called tubes. Deionization characteristics and chemical activity are of major importance in the choice of a gas. After conducting the required pulse, the gas must be ionized at a sufficiently high rate in order that the gap will be able to hold off the network voltage that is applied following the pulse. In other words, the reignition voltage of the gap at any time after the pulse must be greater than the applied network or switch voltage at that time. The gas molecules should also have no metastable states in which electrons may be trapped during the discharge, and from which they would subsequently be released, causing a premature breakdown of the gap. During the required life of the tube, no constituent of the gas within the gap should be absorbed by or combine with any elements inside the tube in quantities large enough to affect the operating characteristics seriously.

Because of erosion of electrodes in air, a gas mixture of approximately 95% argon and 5% oxygen was used to fill the gap envelope. The oxygen maintains an oxide coating on the electrodes in the sparking area that is sufficient to limit the electrode erosion encountered during the required life of the tube. However, as the oxygen is thus used up by chemical combination, the operating characteristics of the tube change. The need of a satisfactory switch for pulsers of higher power than those in which the tube operated satisfactorily resulted in investigations [192] of enclosed gaps filled with various gas mixtures. The principal characteristic sought for in these investigations was stability of switch performance over an extended period of operation. This work included the study of gas mixtures such as argon and helium, argon and oxygen, helium and oxygen, and helium, argon, and oxygen. As a result of other work [189] there was evolved a hydrogen-argon gas mixture that was found to have good deionization characteristics and to be stable with life. At first hydrogen alone was tried, but it was found that, after a few hours operation at low pulse currents, the tube drop during the discharge was abnormally high and fluctuated from pulse to pulse, causing undesirable amplitude jitter in the output pulse. It was believed that this behavior did not exist during the first few hours of operation because of the presence of impurities in the hydrogen that aided the transfer from a glow to a low voltage arc discharge. When argon was added to the hydrogen, the tube drop was found to remain low and constant throughout the life of the tube. Therefore, mixtures of approximately 80% hydrogen and 20% argon are commonly used. The argon acts as a stable impurity in that it performs the same function throughout life, as did the impurities in the tube during the first few hours before they were cleaned up. The current at which this high tube drop appeared could be reduced by increasing the percentage of argon, within limits. This high voltage drop did not appear with tubes filled with 20% argon when conducting 1-μsec 100-amp pulses, but it did appear after operating these tubes for 25 to 50 hours with 1-μsec 50-amp pulses. Apparently, the above percentage of argon was not sufficient to effect the transfer from a glow to arc discharge at a pulse current of 50 amp without the help of the impurities present during the first few hours. Later work showed that no argon was needed at pulse currents of about 200 amp and above.

The choice of suitable electrode materials proved to be somewhat more difficult than the choice of a gas. It was found that all the tested materials eroded rapidly in the hydrogen-argon gas mixture. Studies of the erosion of several materials and combinations of materials when operated in a hydrogen-argon gas mixture show that the rate of erosion of the cathode depends somewhat on its geometry. Cathode erosion appears to be directly proportional to pulse current at a given pulse duration but is not directly proportional to pulse duration at a given pulse current. For example, at a constant pulse current, the erosion at pulse durations of 2 μsec and 5 μsec is about 2.7, and 8.3 times, respectively, that obtained at 1 μsec.

Since all cathode erosion studies have not been made with electrodes of the same geometry, or at the same pulse duration, it is not possible to present the results in concise tabular form. The erosion rates of tungsten, molybdenum, and aluminum by weight were approximately the same, and much less than that of either gold or carbon.

Aluminum was chosen as the cathode material in one major type of series spark gap because the material sputtered off the cathode was found to be deposited on the anode in such a way that the gap spacing stayed approximately constant under limited operating conditions. Experiments have also been made with several sintered materials containing various oxides that have low work functions. It was found that cathodes made by sintering 5% barium oxide, strontium carbonate, or beryllium oxide with a mixture consisting of 95% tungsten and 5% nickel or copper for a binder had an erosion rate of about 0.025 times that of tungsten or molybdenum alone [674]. These materials were not used as the cathode in series gaps because of the development of the sintered iron-sponge mercury cathode, which was superior to them.

Some older designs have concentric cylindrical electrodes, the outer one being an aluminum cathode. By providing a large cathode area, the change in the electrode spacing produced by cathode erosion can be minimized. Another design uses a mercury cathode. In order to immobilize the mercury, it is held in the interstices of a sintered honeycomb structure of iron powder, called an "iron sponge." The design is such that within the operable power range only mercury is eroded from this cathode.

The ratio of the diameter of the outer aluminum electrode, or cathode, to that of the inner electrode, or anode, is made approximately equal to e, the base of natural logarithms. With this ratio, the highest possible breakdown voltage that does not cause the formation of corona before breakdown for a given outside diameter is obtained, and a minimum change occurs in this breakdown voltage for a slight change in the size spacing of the electrodes.

Some conclusions concerning the effects on cathode erosion and anode buildup of variations in recurrence frequency, pulse current, pulse duration, gas pressure, and gap spacing are stated here. These conclusions are based on data taken for gaps having 0.200 inch and 0.330 inch spacings, similar to the Bell Telephone Laboratories. Types WX 3226 and WX 3240 gaps respectively, and filled with 80% hydrogen and 20% argon to total pressures in the range of 68 to 175 cm of mercury.

For a given gap spacing, pulse duration, and gas pressure, the cathode erosion and anode buildup per pulse were found to be directly proportional to pulse current in the range 100 to 400 amp.

For a given gap spacing, pulse current, and gas pressure, the cathode erosion and anode buildup per pulse increase much more rapidly than the pulse duration in the range 0.9 to 5 μsec. As a result the maximum allowable pulse current for a given gap design decreases rapidly with increasing pulse duration.

The ratio of anode buildup to cathode erosion per pulse was found to remain constant for a given gas pressure and gap spacing, even though the pulse current varied from 100 to 400 amp, the pulse duration varied from 0.9 to 5 μsec, and the recurrence frequency varied from 200 to 1200 pps corresponding to a variation from 25×10^{-8} to 560×10^{-8} mg in the mass of aluminum eroded from the cathode per pulse.

For a given gas pressure, pulse current, and pulse duration, an increase in gap spacing was found to decrease the amount of material eroded from the cathode per pulse, and the ratio of anode buildup to cathode erosion per pulse. A tube with large spacing may therefore be operated satisfactorily at higher values of pulse current and longer pulse duration than a tube with smaller spacing.

For a given gap spacing, pulse current, and pulse duration, an increase in gas pressure was found to increase the cathode erosion per pulse slightly, and to increase the ratio of anode buildup to cathode erosion per pulse appreciably. The increase in cathode erosion is perhaps due to a decrease in the size of the cathode spot with increasing gas pressure, which increases its temperature, and thus increases the amount of material evaporated from the cathode per pulse. Because the temperature along the arc column is high and the gas density is therefore low, and because a higher gap-filling pressure causes a region of higher gas density to surround the arc column, it is believed that the higher pressure aids in funneling aluminum vapor from the cathode across to the anode. The higher pressure thus concentrates the distribution of aluminum vapor from the cathode spot in the direction of the anode.

After the spacing of gaps has been chosen with regard to anode buildup and dissipation, the breakdown voltage can be adjusted to the desired value by changing the gas pressure. The operation of gaps filled with 80% hydrogen and 20% argon has been satisfactory over a range of pressures from 50 to 250 cm of mercury.

Nominal ratings for each of the seven industrially developed gaps are given in Table B5b-1. The 1B34 is a WX 3226 filled to a pressure of 75 cm of mercury, and the 1B41, 1B45, and 1B49 and WX3240 gaps filled to pressures of 90, 150, and 110 cm of mercury, respectively.

The tube drop is negligible in case of high voltage switching and lies in the region of 40–400 volts. It was found to be independent of the gas mixture,

TABLE B5b-1

NOMINAL RATINGS FOR VARIOUS FIXED SPARK GAPS

	I_p (amp)		τ (μ sec)						$I_p \times \tau$ (coulombs
Type	Max.	Min.	Max.	Min.	f_r (pps)	$I_{p\ av}$ (ma)	V_{SG} (kv)	$f_r \times \tau$	per pulse)
1B29	30	20	0.75	0.4	2100	—	1.4	—	—
1B22	75	50	0.75	0.4	1100	75	2.1	—	—
1B34	200	75	2	0.25	2000	160	3.5	0.001	2×10^{-4}
1B31	300	30	5	0.25	1600	150	4.0	0.001	3.8
1B41	450	100	5	0.5	2000	210	5.0	0.001	7.5
1B49	450	100	5	0.5	2000	210	5.5	0.001	7.5
1B45	450	100	5	0.5	2000	210	6.5	0.001	7.5

the gas pressure, the pulse current, and the recurrance frequency within the limits of error of these measurements, and within the range over which these quantities were varied. The gas mixture was varied from 100% hydrogen to 100% argon, the pressure from 68 to 150 cm of mercury, the pulse current from 80 to 240 amp, and the recurrence frequency from 200 to 2000 pps. The tube drop is primarily dependent on the electrode spacing and pulse duration.

The voltage across a gap during the first part of the discharge has been studied with the use of a capacitance voltage divider. By multiplying the values of tube drop and pulse current for each increment of time during the pulse, a curve can be obtained showing the rate at which energy is dissipated in a single gap during the pulse. Figure B5b-1 shows such a curve for a WX 3226 gap operated in series with two other gaps at a pulse duration of

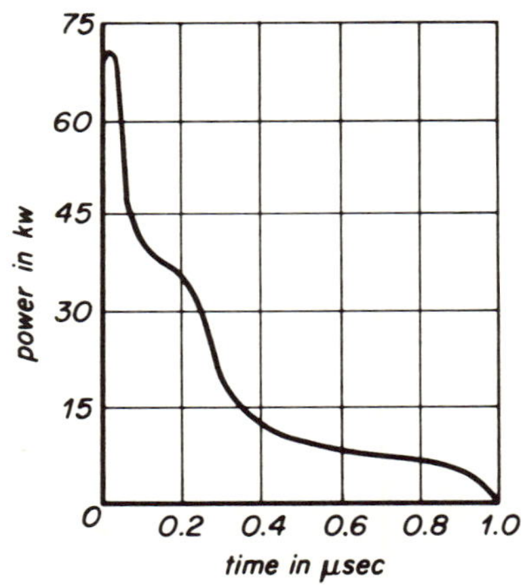

FIG. B5b-1. Power dissipated in one WX 3226 gap as a function of time for a single pulse. By permission from General considerations for gap design, J. R. Dillinger, *in* "Pulse Generators," McGraw-Hill, 1948.

0.9 μsec and a pulse current of 80 amp. Integrating this curve, a value of 15.4×10^{-3} joules is obtained as the energy dissipated per pulse in one of these gaps. A measurement of the average dissipation by means of a calorimeter gave a value of 15.5×10^{-3} joules per pulse. A comparison of other values obtained by integrating power versus time curves with those obtained calorimetrically also shows good agreement.

The dissipation per pulse has been found to be directly proportional to pulse current and to gap spacing. By arbitrarily assuming that an equation for dissipation per pulse involves terms including τ and τ^2, the following empirical relation has been developed:

$$D = (0.703 \times 10^{-6} + 0.420\,\tau - 0.0233 \times 10^{6}\tau^{2})\, SI_p$$

where D is the dissipation per gap in joules per pulse, τ is the pulse duration in seconds, S is the gap spacing in mils, and I_p is the pulse current in amperes. This equation agrees reasonably well with experimental data over the range of 0.9- to 5.0-μsec pulse duration, 0.200 to 0.440-inch spacing, and 80- to 500-amp pulse current [678].

The iron-sponge mercury-cathode gap [674, 679] is an improvement over the cylindrical-electrode aluminum-cathode gap in at least three respects: (1) The range remains constant during life; (2) the time jitter can be maintained at a value of the order of magnitude of 1% of the pulse duration; and (3) as wide an operating range can be obtained with two of these gaps as with three of the cylindrical-electrode aluminum-cathode gaps. For satisfactory operation, however, the reverse current must be kept very low.

The anode of this gap is a molybdenum rod with a diameter of approximately 0.060 inch. The cathode is mercury that is immobilized by an iron sponge. This iron sponge contains about 60% void space and is made by compressing iron powder into a kovar cup and sintering under appropriate conditions. After subsequent heat treatment in a hydrogen atmosphere the sponge of a typical gap holds about 9 cu cm of mercury when fully saturated. The mercury that is evaporated from the surface of the cathode during operation condenses and runs down the walls back into the sponge. Because of the surface tension of the mercury, a film of mercury is maintained over the surface of the iron sponge which prevents the erosion of the iron. In order to minimize the time jitter, the gaps are usually filled with 100% hydrogen. Radium salts are generally omitted in order to increase the operating range and to further minimize jitter.

Some specifications for these tubes are given in Table B5b-2. Each quantity has the same significance as in Table B5b-1. In columns containing more than one figure, the first represents the maximum and the second the minimum rating. The value of V_{SG}, the nominal operating voltage per gap, has been obtained from data for the operation of two gaps in series. For two Fe-I gaps operated at 15.5 kv, 290 amp, 300 pps, and 2 μsec the voltage range of 10.5 to 22.5 kv is constant for more than 500 hours of operation, with V_{start} equal to 5.5 kv. The required trigger voltage is 20 kv, and the time jitter is less than 0.02 μsec at voltages above 14 kv, provided that the rate of rise of trigger voltage is 55 kv/μsec or greater. For gaps filled to

TABLE B5b-2

RATINGS FOR TWO IRON SPONGE MERCURY CATHODE GAPS

No.	I_p (amp)	τ (μsec)	f_r (pps)	$I_{p\,\text{av}}$ (ma)	V_{SG} (kv)	τf_r	τI_p (coulombs per pulse)	$f_r I_p$ (amp × sec^{-1})
1B42	300–80	6.1–0.25	2000	0.250	5.0	0.0012	1.28×10^{-3}	40×10^{4}
Fe-1	750–100	5.5–0.5	1000	0.450	7.5	0.0011	1.5×10^{-3}	25×10^{4}

higher gas pressures, ranges of 13 to 29 kv for two gaps have been maintained during more than 500 hours of operation at 24.5 kv, 925 amp, 250 pps, and 2 μsec.

Time jitter can be reduced by improvements in the tube design. The elimination of argon from the gas filling, a reduction in the diameter of the anode, and a decrease in the gap spacing have all been found effective in reducing time jitter. It is possible to eliminate argon from these tubes because they are intended for operation at pulse currents of 100 amp and higher. A reduction in the anode diameter decreases the jitter, but increases the detrimental effects of reverse current in the tubes. The most effective way of reducing time jitter is to decrease the spacing, and to increase the gas pressure accordingly in order to obtain the desired gap breakdown voltage. However, the maximum gain that can be obtained by this method is limited because, for a gap with small spacing, the breakdown voltage increases linearly with gas pressure in the region of low values, but for high gas pressures, this voltage is not affected appreciably by a change in pressure.

As a result of caliorimetric measurements made at the Bell Telephone Laboratories [670] on iron-sponge mercury cathode gaps operated over a wide range of pulsing conditions, the following empirical formula has been established expressing the dissipation D in joules per pulse per gap in terms of the gap parameters and the pulsing conditions:

$$D = 5.7 \times 10^{-7} I_p S + (40 + 3.9 \times 10^{-2} p^{0.4} S) I_p \tau$$

where I_p is the pulse current in amperes, S is the gap spacing in mils, p is the gas pressure in inches of mercury, and τ is the pulse duration in seconds. This equation has been established from data covering the following ranges of parameters: spacing 40 to 350 mils, gas pressure, 28 to 50 inches of mercury, pulse duration, 1 to 6 μsec, pulse current, 45 to 1070 amp. The dissipation in watts per gap for any application may be obtained by multiplying the value of D by the recurrence frequency. This equation does not include the energy contributed by the trigger, which can usually be neglected, but which can be measured independently and added if necessary.

Dissipation is an important consideration not only from the standpoint of circuit efficiency, but also in determining whether or not forced cooling is necessary when operating mercury/cathode gaps. Since mercury vapor has poor deionization characteristics, there is some upper limit to the partial

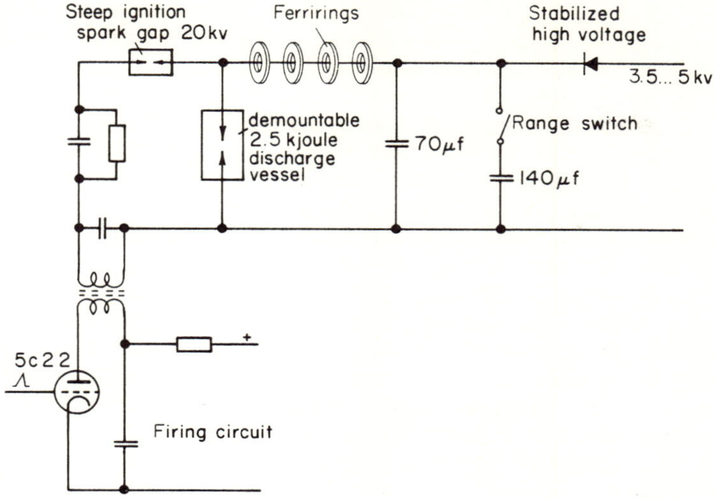

FIG. B5b-2. Diode as heavy current, high-voltage switch in a circuitry for producing high-intensity light flashes.

pressure of mercury, and thus to the operating temperature of the gap, above which the poor deionization characteristics of the mercury override the good characteristics of the hydrogen.

Früngel and G. Roeder [807, 856] have found a new method of firing two electrode spark gaps over a voltage range from several hundred volts to several kilovolts in a specific time interval. The method operates on the principle of automatic variation in series inductance.

The trigger circuit, Fig. B5b-2, is placed in parallel to the main discharge circuit, consisting of the capacitor bank and the flash tube. A fast trigger pulse is fed in by the pulse transformer. Its rise time is further shortened by the help of an auxiliary spark gap. The main discharge line is covered with ferrite rings giving a total inductance of 100 μh, thus blocking it against the trigger pulse which has to pass through the spark lamp. The main discharge follows immediately. Its high current saturates the ferrite rings and reduces their inductance to about 1 μh, which keeps the length of the discharge below 200 μsec. A trigger pulse of 40 kv with 0.1 μsec rise time triggers safely and permits use of a rather low voltage for the main discharge. Low voltage at a given electrode distance and gas pressure means high discharge current with correspondingly low content of double ionization processes. Instead of using the spark itself for light emission, the same principle can be used for serial resistances and other consumers of electrical power.

c. *The Ignited Three Electrode Spark Gap*

Whenever it is desirable to time the ignition of the sphere spark gap accurately, the so-called three-electrode spark gap can be used as a switching means. In its simplest form it consists of two spheres spaced at double the distance of the natural sparking distance of the desired capacitor discharge ignition voltage. Midway between these two spheres there is arranged a small insulated metal pin which is adjusted by very high-ohmic resistances to be at the average potential between the two spheres. At such a potential, the pin is unable to initiate a discharge by its presence. If a pulse is applied to the pin from a small capacitor and this pulse disturbs the potential symmetry there is immediately a spark breakdown from the pin to the other sphere. Since the pin, owing to its small ignition capacitor, does not represent an energetic discharge path, the discharge immediately is carried to the sphere electrode and the main breakdown occurs. It is thus possible to obtain twice the spark resistance which corresponds to the doubled sparking distance. Ignition mechanisms of this type are basically limited in their timing accuracy because of the delay caused by the ionization time required in the space between pin and sphere before spark breakdown can take place. Every ionization process depends on spacing, since an ionized channel requires a buildup time which is roughly proportional to the sparking distance (see Table B5c-1). Concerning ignition delay as a function of electrode material, see Strigel [12].

TABLE B5c-1

IGNITION DELAY IN SIMPLE POINT IGNITION

Capacitor voltage (kv)	Gap to be ignited (cm)	Delay time (μsec)
4–8	0.3	0.01–0.1
15–20	1.0	0.1–1
150	10.0	1–10

d. *The Three Electrode Spark Gap Design* [391-397, 189]*

In order to build a satisfactory three-electrode gap vessel for triggering or switching purposes, it is necessary to have an electrode material such that its erosion rate at the desired operating conditions is negligibly small in the gas to be used. Since a gap of this type has a somewhat more complex electrode configuration than does a two-electrode gap, much less electrode erosion can be tolerated. Such three-electrode spark gaps often are called trigatrons. Normally it is filled with 95% argon and 5% oxygen to pressures of about 1 to 6 atm, depending on the desired operating voltage and on the particular tube design. Typical operating conditions are 8 kv, 1200 pps, and 1 μsec in a 70-ohm line-type pulser. The anode and cathode are of molybdenum and the trigger pin is tungsten. The anode sleeve was found effective in making the time jitter less than 0.1 μsec.

Oxygen is used in a trigatron principally to maintain an oxide coating

* The following 6 paragraphs are reprinted by permission from J. R. Dillinger, The three-electrode fixed spark gap, *in* "Pulse Generators," McGraw-Hill Book Company, New York, 1948, pp. 332–335.

on the surfaces of the electrodes and thereby to limit the rate of electrode erosion, since there are effectively two cathodes in the tube when it is operating. In this tube, the principal discharge does not start between the anode and cathode after the tube is triggered, but takes place from the cathode to the trigger pin to the anode. Thus, the trigger pin must serve as both an anode and cathode, and is therefore doubly subject to erosion. Because of the small size of the trigger pin and the other electrodes, this erosion must be minimized. The oxygen is also needed in these gaps to quench metastable atoms of argon after the discharge.

The lifetime of a trigatron allows up to 400 hours of operation. After the lifetime, this tube was found to fail to deionize at frequent intervals, agreeing with British experience. Failure of this tube is not due to electrode erosion, for it operates satisfactorily after being pumped and refilled. The failure may be due to the formation of nitrogen peroxide, which is known from experience with rotary gaps to have poor deionization characteristics, or it may be due to a reduction in the amount of oxygen to a value that is insufficient to quench metastable atoms of argon after the discharge. This reduction is caused by the combination of oxygen with the molybdenum and tungsten electrodes during operation.

The time jitter in a trigatron type CV 125 (Fig. B5d-1) at 12 kv, 800 pps, and 0.9 μsec in a 50-ohm circuit remains less than 0.09 μsec throughout life, a value that is much less than that for cylindrical-electrode aluminum-cathode gaps. The life of series gaps under the above conditions is considerably longer than that of the CV 125, and can exceed 1500 hours. Also, the deionization characteristics of the series gaps are better than those of the trigatrons, as is evidenced by the greater decrease in V_M with increasing recurrence frequency for the trigatron than for the series gaps.

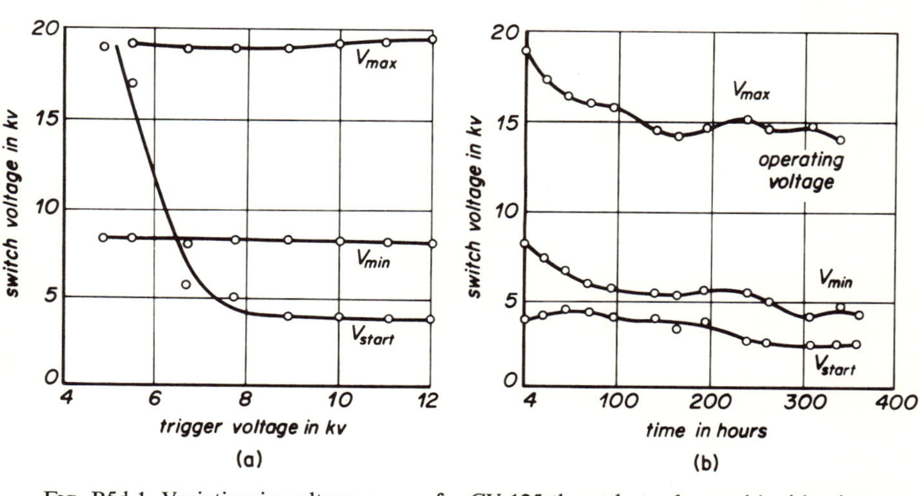

FIG. B5d-1. Variation in voltage range of a CV 125 three-electrode gap (a) with trigger voltage (new tube), and (b) with time. By permission from The three-electrode fixed spark gap, J. R. Dillinger, *in* "Pulse Generators," McGraw-Hill Book Company, 1948.

It would be desirable to combine the advantages of the hydrogen or hydrogen-argon gas filling, which is stable with life, with those of a three-electrode geometry of the trigatron type. Unfortunately, the erosion rate of possible electrode materials in the absence of an oxide coating on the surface is usually high.

An attempt has been made to overcome this difficulty by using electrodes made by sintering tungsten and barium oxide together, with nickel as a binder. Some three-electrode tubes have been constructed with these materials, but sufficient tests have not yet been made to indicate the stability of their characteristics during life.

In case the high pulse voltage that must be applied to the ignition point in order to assure positive ignition is in any way objectionable, there exists another arrangement in which the cathode in its spherical or semispherical shape is provided, at the closest point to the anodic sphere, with a hole in which the ignition pin is concentrically placed (see Fig. B5d-2). It is expedient to let the ignition pin project somewhat (about 2 mm). The pin is then connected by way of a high ohmic resistance with the cathode. A negative pulse of such potential is applied thereto that a sparkover occurs between the cathode pin and the surrounding hole rim. This ignition spark, which with a hole diameter of 10 mm and a pin diameter of 3 mm has to bridge a distance of only 3.5 mm, can be produced with customary ignition coil voltages of 15–20 kv. The spark leads to electron emission at the rim of the hole and simultaneously to considerable field disturbance between the

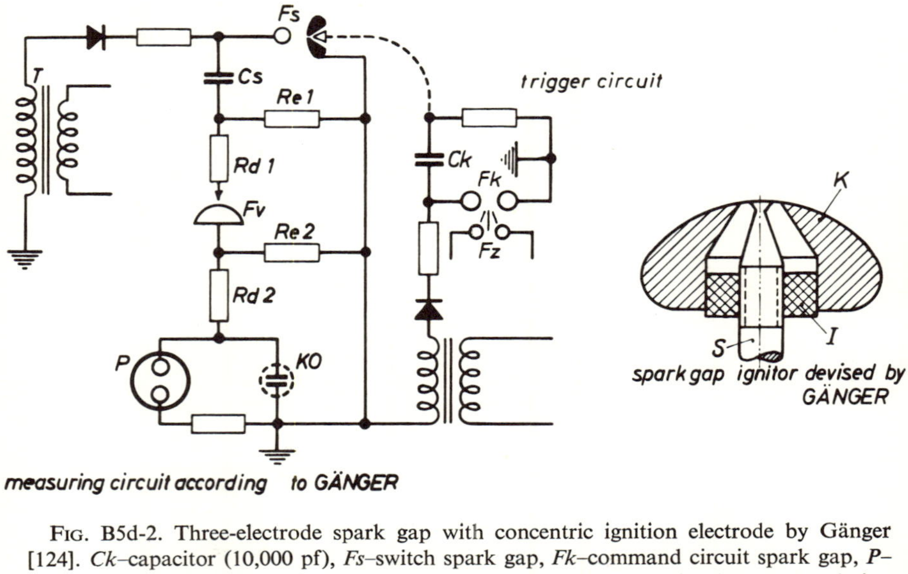

FIG. B5d-2. Three-electrode spark gap with concentric ignition electrode by Gänger [124]. *Ck*–capacitor (10,000 pf), *Fs*–switch spark gap, *Fk*–command circuit spark gap, *P*–work piece under test, *Rd*1 + *Rd*2–damping resistors, *K*–spherical calotte; *I*–Insulation insert, *S*–copper pin, *Cs*–capacitor (2 μf), *Fv*–distributor spark gap, *Fz*–time circuit spark gap, *KO*–oscilloscope, *Re*1 + *Re*2–discharge resistor.

two spheres. The spheres in this example must have such a spacing that they are apart by a maximum of 40% and a minimum of 10% above the natural sparking distance.

With such relationships it is possible to initiate positive breakdown even at great sparking distances with weak-current auxiliary voltage pulses. This ignition method, which can also be practiced by the use of a storage battery and an automotive ignition coil at the ignition pin, is therefore very popular in high-voltage laboratories. A wiring diagram is shown in Fig. B5d-2. (The ignition pin can eventually be arranged in the rear of the electrode hole when an ignition energy of some watt-seconds can be discharged through the pin. The ensuing detonation causes an acoustic injection of ions through the hole into the main discharge space.)

e. *Heavy-Duty Three-Electrode Spark Gap*

Whenever electrodes are subjected to considerable burn-off, e.g. at high impulse frequency, high coulomb value, or long operating times, or when it is required to obtain an extremely accurate timing within a jitter of not more than in the order of 10^{-7} sec by means of controlled spark gaps, one should consider the possibility of anodic ignition as devised by the author [152]. Figure B5e-1 shows a dimensioned design drawing of such a spark chamber.

The construction makes use of pin electrodes instead of sphere electrodes and arranges around the anode an ignition annulus which is dished toward the point of the anode. The hole of the annulus resembles the hole of the previous example but serves here to let the anode pin project. The electrodes face each other at a distance which corresponds, just as in Gänger's arrangement, to about 1.1 to 1.4 times the normal sparking distance of the prevailing discharge voltage. For proper adjustment it is advisable to arrange the cathode pin, 3, that is not surrounded by an ignition electrode, so that it can slide.

Ignition is now effected at the ignition annulus surrounding the anode, by a steep ignition pulse. As a result, during the instant before breakthrough of the ignition spark an extraordinarily high field intensity occurs. This emits very high-frequency ultraviolet radiation into the discharge space [13]. The travel distance of this very high-frequency ultraviolet radiation determines the limit of usefulness of this system. In air, this distance can be assumed to be about 5 mm. The sparking distance that can be positively ignited is about twice the radiation distance of this radiation. It is therefore impossible to increase the electrode spacing above 10 mm under normal atmospheric pressure. With needle-point electrodes, the sparking distance is greater than that of sphere electrodes, or, conversely, their breakdown voltage is lower. Ten mm spacing is good for about 10 to 15 kv voltage to assure safety from self-breakdown.

Through the electromagnetically caused very high-frequency ultraviolet emission (wavelength $\lambda \approx$ 500–1000Å) the ignition is practically inertia-free. One-half of the space between cathode and anode is thereby so strongly ionized that a vertical discharge plug extends from the cathode into this plasma. This results in a clearly centered spark. The light emission of such

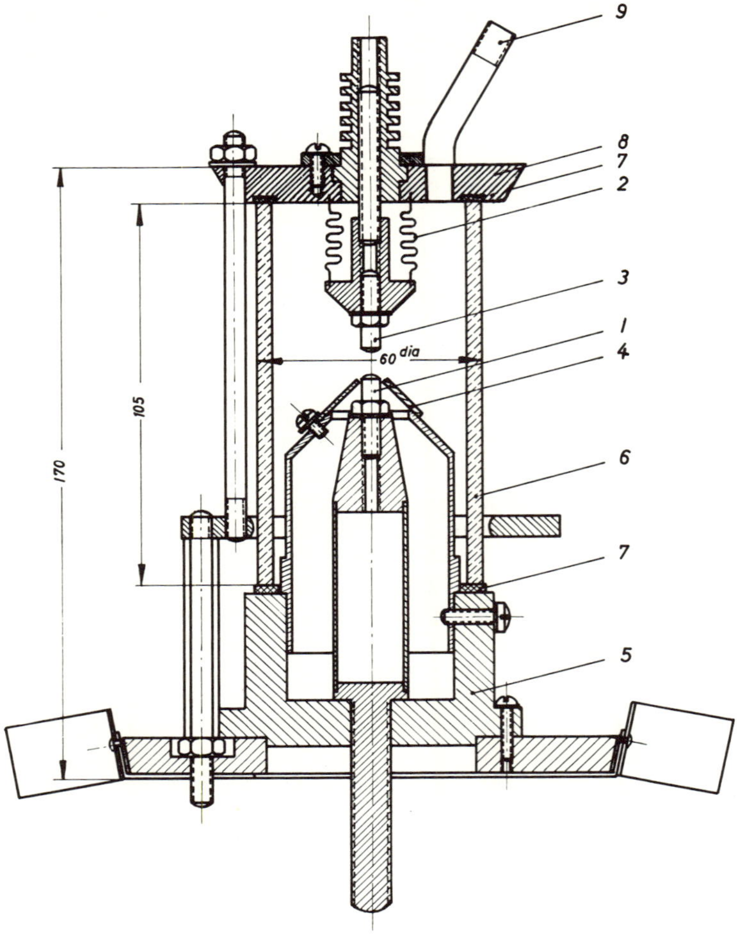

FIG. B5e-1. Cross-sectional view of a demountable ignition spark gap device for inert gas filling and discharge currents of maximum 6000 amp at 4–9 kv operating voltage (Früngel). 1–Anode pin, 2–metallic bellows, 3–adjustable cathode pin, 4–ignition voltage annulus, 5–ceramic bottom portion, 6–glass cylinder, 7–sealing gaskets, 8–cover plate, 9–inlet valve.

sparks will be discussed in more detail later. It is possible to operate such discharge systems in air as well as in inert gases. The latter lend the advantage of nonoxidation and thereby extend the life of the components considerably. The current peaks are very high. It is possible to supply external loads with peak currents up to 10^4 amp at an operating voltage between 4000 and 9000 volts, corresponding to the peculiarity of this ignition system as above discussed. The impulse output at a resistive load is consequently $\leqq 10^8$ watts. The advantage of almost any desirable accuracy in ignition timing is somewhat clouded by the necessity for close voltage limitation. Of benefit is the accessibility for ready replacement of the tungsten pins which serve as the main electrodes. Some outlay in equipment for producing and controlling the ignition pulses in such three-electrode switching spark gaps is necessary in order to gain the full benefit of their usefulness. It is possible to employ a hydrogen thyratron of high rating that is triggered at its grid by a multivibrator of high rating. When a minimum switching sequence is observed, say 10–100/sec, and certain residual ions remain in the spark gap, it will be noticed that in this electrode arrangement the average time lag between ignition sparking and main discharge varies much less than 3×10^{-8} sec. This type of switching is insensitive to oscillations and does not self-extinguish. If one tries to increase the physical size of such electrode systems proportionally and to operate them above 10 kv, a reliable ignition can still be obtained but the accuracy is decreased. Inaccuracy occurs with voltages between 20 and 100 kv. This is not much better than in ignition devices with half-sphere cathode and centered pin ignition at the cathode. Since the inert gas argon has about two to three times the sparking distance of air, it is possible to operate such electrode systems advantageously under 2–3 atm filling pressure. The higher pressure results in a steeper breakthrough (more particles) and better convection cooling of the electrodes.

The thermal loading limit of this kind of spark gaps lies at about 250 watts for a $\frac{1}{4}$ liter chamber volume and with a hard glass enclosure, if no forced cooling of any kind is applied. The spark resistance lies between 1 and 0.1 ohm, so that with an external resistance between 1 and 10 ohms the useful load can be 1–10 kw in continuous operation. For certain applications a filling of hydrogen may be advisable because this gas gives rise to a lower spark resistance on account of a smaller Toepler's spark constant. Since the discharge chambers for use as switches are demountable—such demountable construction is widely used in stroboscope lamps—refilling of the chamber through a common automobile tire valve presents no problem. Their use circumvents the trouble arising from the relatively short life of hydrogen thyratrons in sealed envelopes employed for similar switching operations.

A glass-metal triggered spark gap is manufactured by EGG. This glass-metal triggered spark gap, Model GP-11, is designed to replace hydrogen

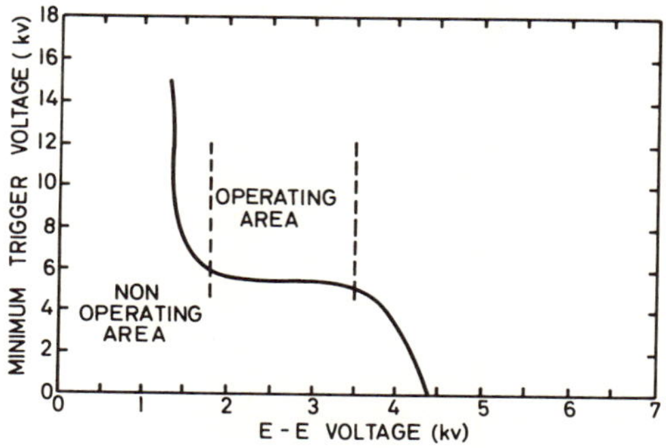

FIG. B5e-2. Typical GP–11 operating characteristics.

thyratrons in electronic crowbar high energy switching applications. The common advantages of this spark gap are: high holdoff voltage, high peak current capability, no filament power requirement, low energy triggering, and shock and vibration resistance.

Use of the GP-11 eliminates many of the disadvantages encountered in using hot cathode hydrogen thyratrons and mercury vapor ignitrons in crowbar and high energy switching applications. These disadvantages are: limited high voltage holdoff, limited peak current, filament power requirement, close temperature regulation, restricted mounting position, complex triggering, large size, high cost, and relatively short life.

In addition to this gap, EGG manufactures a variety of spark gaps which operate at voltages ranging from 2 kv at 50,000 amp to 100 kv at 2,000,000 amp (when immersed in oil).

Figure B5e-2 shows the operating characteristics and Fig. B5e-3 a typical crowbar circuit with the GP-11 spark gap.

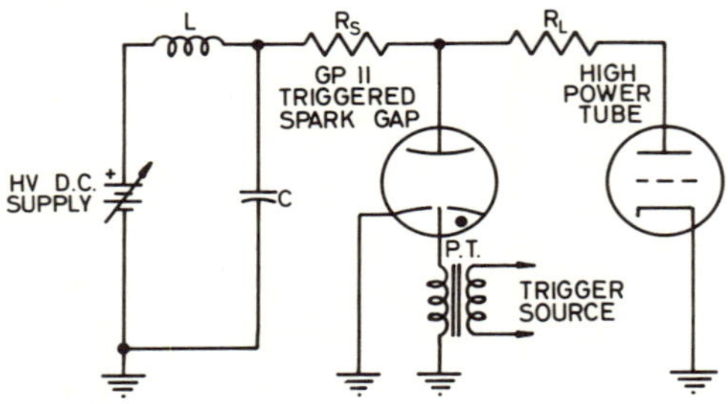

FIG. B5e-3. Typical crowbar circuit with GP–11.

A bigger type of the same general design is intended for 25 kv operation. Life tests performed on this GP-12 showed no sign of failure after 1000 shots at a peak current of 10,000 amp and 20 kv electrode voltage, and at 10 kv low energy trigger pulse. Another life test was performed at 14 kv from electrode to electrode, a 1 μf capacitance discharge, and a peak current in excess of 10,000 amp; more than 50,000 shots were obtained at this condition. The life test figures stated above for peak current are extremely conservative and operation at peak currents greater than 10,000 amp have already been performed.

The specifications of this spark gap should be mentioned because of the characteristic properties of such switching circuits. The operating range of electrode-to-electrode voltage shows a minimum cutoff of 9 kv and a maximum hold-off of 25 kv. The trigger pulse voltage must have 10 kv minimum. The allowable peak current is greater than 10,000 amp. The unavoidable delay time is 0.5 μsec when operation is at 80% of maximum hold-off voltage.

An application of such a spark gap in low-inductance feeding of plasma-discharge devices is described by Foitzig and Reizuch [810]. It is designed for mounting in coaxial return sheets and has a self-inductance of 20 nh. It allows switching up to 300 amp at 35 kv operating voltage, and has a practical lifetime of up to 10^5 discharges.

f. *Spark Gaps for Rapid High-Voltage Switching*

Research work sometimes requires the explosion of a wire for such purposes as generating plasma in a shock tube or generating hypersonic waves in a wind tunnel. The spark-gap switch to be discussed explodes a wire by discharging a bank of capacitors charged to 30 kv.

A basic requirement of this switch is that it must close in less than 10 millimicroseconds. The greatest problem encountered in using spark gaps for the switch was that of reducing the jitter time (the maximum statistical variation in firing time). When spark gaps that have a large jitter time are used in parallel, the first to fire may prevent some or all of the rest from firing. This is because the first gap to fire applies 30 kv to the load, and, as this load is also connected to all the load terminals of the unfired gaps, the net voltage across them drops to zero and they remain unfired.

The spark-gap switching circuit of Fig. B5f-1 uses 111 spark gaps. Thyratron V fires section 1, which contains a single spark gap. This gap triggers ten spark gaps in section 2. Each gap in section 2 triggers ten gaps in the final stage, section 3. The spark gaps of section 3 switch one hundred 0.03-μf

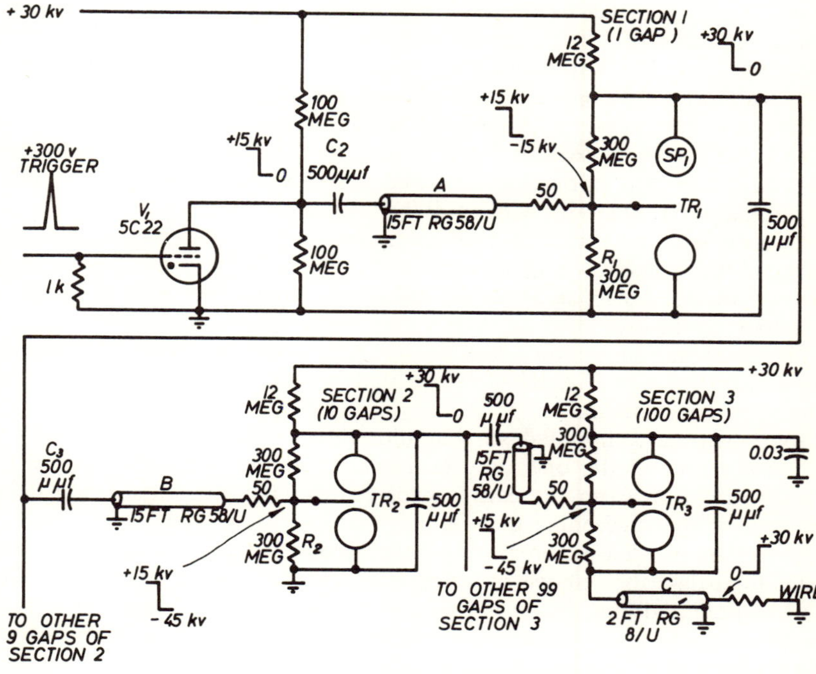

FIG. B5f-1. Spark gap switch. Only one of the 10 gaps of section 2 and one of the 100 gaps of section 3 are shown.

capacitors (one of these capacitors is shown), which are charged to 30 kv, through the wire load.

Coaxial cable sections A, B, and C couple each spark gap to the preceding stage that furnishes the trigger pulse for the gap. The center conductor of each coaxial cable is initially at + 15 kv.

When a trigger pulse to thyratron V switches capacitor C_2 to ground, capacitor C_2 discharges through resistor R_1. Simultaneously, cable section A forms a pulse of 15 kv. Discharge of capacitor C_2 and the pulse from cable section A produce a 30 kv pulse across resistor R_1. This pulse drops the potential of the spark-gap trigger electrode, TR_1, to −15 kv. Since the gap between the upper sphere SP_1 and trigger electrode TR_1 has a potential of 45 kv, it breaks down.

As soon as the upper half of the gap breaks down, trigger electrode TR_1 goes to + 30 kv. As the second half of the gap now has double its normal voltage, it breaks down. Breakdown time for 50 kv/cm, which is obtained with 30 kv across the electrodes, is about 8 milliμsec.

The ten capacitors that couple the first section to the second section (one such capacitor is represented in Fig. B5f-1 by C_3) now discharge from +30 kv to ground potential. The pulse formed by cable B and the discharge of capacitor C_3 impress 60 kv across resistor R_2. Spark gap trigger electrode TR_2 drops to −45 kv and the first half of the spark gap of section 2 breaks down. With 75 kv across the half gap, which has a spacing of 0.57 cm, a voltage gradient of 130 kv/cm appears and breaks down the gap in about 0.45 milliμsec. Breakdown in the other nine first-half gaps of section 2 occurs almost simultaneously and in the same way.

Breakdown processes in the second half of the spark gaps of section 2 and in the spark gaps of section 3 are similar to the processes that have been described. The bank of one hundred 0.03-μf capacitors in section 3 completes its discharge through the wire in 100 milliμsec.

As each section takes about 10 milliμsec to break down, the total time between the breakdown of thyratron V_1 and the start of the discharge through the load wire is about 30 milliμsec. Jitter time of each section is less than 1 milliμsec.

The spark gap in section 1 and the ten gaps in section 2 are similar to the spark gap shown in Fig. B5f-2 except that no effort is made to make them coaxial in construction.

Spark gaps are mounted so that ultraviolet radiation from spark gaps that fire earlier in the operating sequence irradiates succeeding gaps. The intense radiation from the earlier gaps reduces the statistical firing delay, or jitter, of the succeeding gaps.

It may be possible to break down the second half of the spark gap even faster than described above by using the circuit shown in Fig. B5f-3. Here the low-potential end of the gap is connected to trigger source by a short length of cable and a blocking capacitor, which in turn is connected to ground by a similar piece of cable.

Similar nanosecond spark gaps are used in the field of high-speed photography.

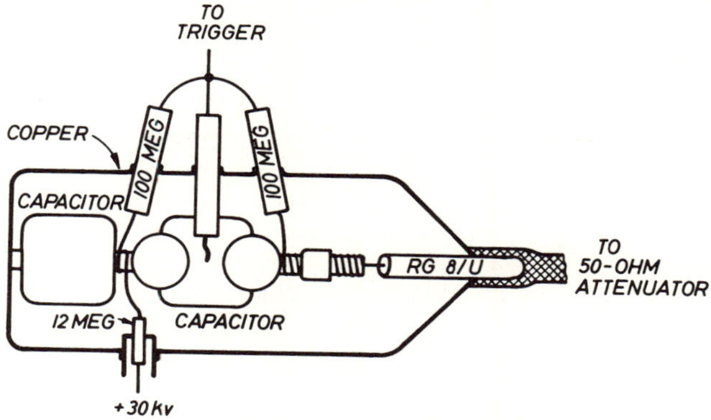

FIG. B5f-2. This experimental spark gap is similar to those used in sections 1 and 2 of Fig. B5f–1.

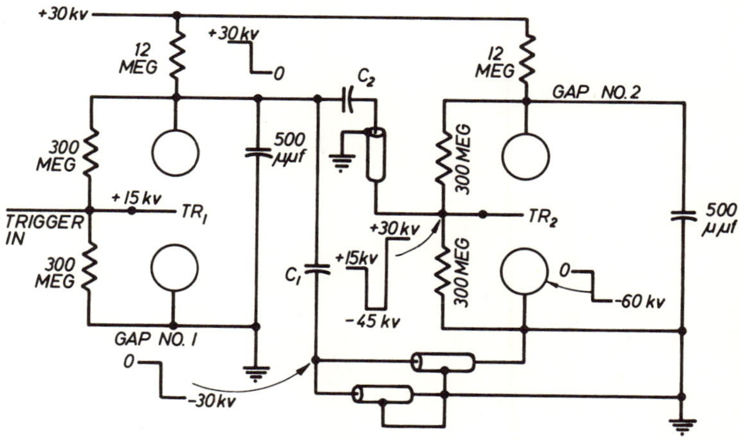

FIG. B5f-3. Improved spark gap switch which breaks down in less than 1 milli μsecond.

H. Fischer has developed nanosecond spark gaps which have a pinching spark character [471, 767, 799]. Smallest types of this gap, which are combined with coaxial return-sheet capacitors, show a subnanosecond rise time.

g. *On the Triggering of Cable Spark Gaps* [399]

Spark gaps which are built into a cable can be exactly triggered by generating charge carriers and by increasing at the same time the breakdown voltage. An experimental setup of this mode of triggering showed a jitter of 1×10^{-8} sec and the spark gap adjustment proved to be rather uncritical.

The term cable spark gap refers to a spark gap situated inside a coaxial cable and interrupting the inside conductor of said cable while inoperative, but effecting a connection when operative. Such special spark gaps are employed when heavy voltages in the region from 10 to 100 kv have to be

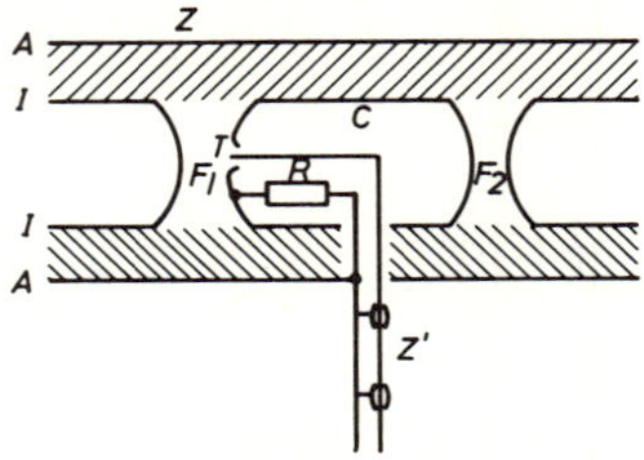

FIG. B5g-1. Cable spark gap with trigger circuit.

switched at extremely short times and with great accuracy. Such switching problems occur when voltage pulses are generated and fed through a pulse cable or when pinch discharges are produced by means of a low induction capacitor battery.

Figure B5g-1 shows such a cable spark gap, F_1, which is supposed to switch a voltage U.

It is arranged in a coaxial cable (A = exterior, I = interior conductor) having a characteristic impedance Z. Left from F_1 the cable is charged and a voltage U is supposed to be at the electrodes of F_1. When F_1 is triggered a voltage pulse of $U/2$ enters the cable at the right-hand side of F_1.

In order to switch the spark gap as accurately as possible (low jitter) the trigger pulse as such must perform two tasks, viz. produce free charge carriers and increase the voltage at the spark gap. The above illustration shows how this can be achieved. The cable at the right-hand side of F_1 is interrupted by a further spark gap, F_2. Thus the end between F_1 and F_2 in fact constitutes an additional small capacitance C. A negative trigger pulse having roughly a voltage of $-U/10$ is fed over the trigger cable with an impedance Z' and an ohmic resistance R to the trigger spark gap T, effects an immediate sparkover, and produces at R a voltage which charges the capacitor C. Since this charging is effected by a negative voltage, the voltage between the electrodes of F_1 is increased.

The values for R and Z' of this trigger circuit must be adjusted in such a way that C can be charged very quickly yet they must also be such that there is no substantial energy loss from the main cable Z, i.e. $R \approx Z' \approx 10Z$.

The spark gap F_2 must be adjusted so that the breakthrough is just effected by the voltage pulse $U/2$ and the distance from F_1 to F_2 must be such that there is no coupling between these spark gaps—for steep wave fronts approximately 10 cm.

A spark gap which was triggered in this way gave perfect performance with a jitter of 1×10^{-8} sec. The electrical data were as follows:

Breakdown voltage of F_1	65 kv
Charging voltage U	38 to 65 kv
Characteristic impedance Z	15 ohms
Impedance Z'	150 ohms
Resistance R	150 ohms

Experiments with this cable spark gap further showed that a decrease of the charging voltage U to almost half the value of breakthrough voltage of F_1 did *not* result in increased jitter. These results indicate a still unknown phenomenon which also makes the setting of F_1 uncritical. The experiments described above were carried out at the laboratories of the FF Impulsphysics Corporation/Impulsphysik A.G. The mentioned spark gap was built into a delay cable in order to obtain rectangular voltage pulses.

On closely observing the energy conditions of the mode of triggering one finds that the fed-in trigger energy is first of all used up for the production of plasma and UV radiation. Thereby the resistance of the trigger spark gap drops down so much that the adjustment of trigger spark gap and trigger cable becomes so poor that hardly any energy can enter the plasma. At the same time, however, the charging of C causes an increase of the spark gap voltage which again constitutes an effective feed-in of triggering energy.

h. *Abstracts of Additional Studies on Spark Discharges*

J. M. Meek [391], Department of Electrical Engineering, Liverpool University, shows the growth mechanism of spark discharges. He gives an account of the present theoretical picture of the spark mechanism together with a summary of the main features of the growth of corona and spark discharges as observed experimentally using photographic, photoelectric, oscillographic, and cloud chamber methods. The theoretical explanation of sparking phenomena is still uncertain. One group explains all observed phenomena in terms of the Townsend theory. Another group uses streamer theory and Townsend theory to account for the observed phenomena. The streamer theory is based primarily on considerations of space charge fields in electron avalanches and streamers which attract, in the case of a positive streamer, secondary avalanches into the tip of the streamer. Secondary avalanches are most likely initiated by photoelectric ionization. It is possible to regard the streamer theory as a special case of the Townsend theory in which secondary ionization occurs in the gas and not at the cathode. In the case where the field at the cathode is sufficiently high (e.g. uniform field) photoemission from the cathode may facilitate the growth of the spark. In nonuniform fields (positive point to plate corona) experiments appear to show that the cathode processes do not affect the growth of the spark. The discharge is initiated by a positive streamer (leaderstroke) which grows from anode to cathode. The head of the leader stroke is a luminous zone, its luminosity increasing to a bright central core, forming on its path a filamentary ionized track which bridges the gap and leads to complete breakdown. Similar characteristics have been observed in the growth of lightning discharges and it is again thought that these must be explained in terms of ionization processes which are gas dependent and not electrode dependent.

L. H. Fischer [392], College of Engineering, New York University, New York, describes formative time lags of spark breakdown. Formative time lags of spark breakdown in uniform fields have been investigated. The measurements have been made in the neighborhood of breakdown threshold with electrode spacing from 0.3 to 3.0 cm and pressures from 100 to 800 mm

Hg and yielded buildup times of the order of 10 to 100 μsec. These results show that Townsend buildup preceded breakdown in air, N_2, and H_2 within 2% above the breakdown voltage and in argon by overvoltages up to 100%. From the variation of time with pressure, electrode spacing and overvoltage, and the knowledge of Townsend's first ionization coefficient, it is possible to study the nature of the preceding Townsend discharge. In air, N_2, and H_2 it has been identified as a photoelectric emission from the cathode caused by photons, formed in the gas in conjunction with avalanches. In argon there is also a delay in the transmission of radiation through the gas. In oxygen a formation of negative ions plays a role in delaying the discharge. Positive point to plane corona in air does not show such long buildup times and so another mechanism is responsible. Some measurements of time lags in glow discharges will be discussed next.

Some properties of spark channels were measured by J. D. Craggs [393]. For some years Craggs has been studying in Liverpool the properties of spark channel discharges by the use of rectangular current pulses. The properties of afterglows can be studied without complications to residual currents. It was hoped that information of basic interest on various processes, for example radiative recombination, could be obtained with discharges of this type. It is believed that this is still possible, but detailed studies stress the complicated nature of the discharge.

The discharge can be divided into three periods: (1) the initial shock wave expansion period which will be briefly described with reference to Russian and American work; (2) the period of current flow, when a quasiequilibrium may be established depending on the pulse length; experiments on excitation temperatures, ion densities, drift currents and voltage gradients will be described; (3) the afterglow; measurements of ion densities and excitation temperatures are given with some discussion of their implications. Finally, the spark channels carrying currents of several hundreds of kiloamperes are described.

The internal radiation and nature of discharges were the subject of a paper published by A. von Engel [394] of the Clarendon Laboratory, Oxford.

In the study of the nature of discharges in gases the emission of light and the shape and size of various parts of the discharges have been observed; then the electric parameters and chemical changes have been investigated; and, finally, the elementary processes have been traced to explain working and starting conditions. The original idea that simple collisions between electrons and molecules had to be considered as the only production and multiplication process appeared to be inadequate to explain the phenomena fully without including cumulative processes as a correction. It is necessary to include the effects of radiation from the beginning. The idea that internal radiation (i.e. vacuum ultraviolet) plays an important role goes back to the beginning of this century, but has been little appreciated.

By means of examples, the role of internal radiation in the ionization in the gas itself and through its effect on conducting and insulating walls are described. In recent years the photoelectric yields of surfaces have become known. However there are no data available on the quantum yield from a swarm of electrons through a gas. A new method of obtaining information about excitation coefficients is described.

R. W. Warren and L. B. Loeb [394], University of California, Berkeley, California, present a description of field measurements in glow discharges with a refined electron beam probe and automatic recording. An electron beam of 0.1 mm diameter and an energy of 20 kv, or more, traversing a glow discharge tube of 11 cm diameter, produced a spot on a fluorescent screen, which was viewed by two photomultiplier tubes which controlled the current through Helmholz coils compensating the deflection of the beam. By means of this null method fields of from 0.2 to 1000 volts/cm could be measured, after calibration. Measurements have been made in He, A, N_2, H_2, and air in glow discharges with currents of from 0.1 to 10 ma. The curves of field strength versus distance to the cathode differ from the curves obtained by earlier workers. They show a sharp rise at the edge of the negative glow continuing parabolically at high currents later to rise more sharply near the cathode. The curves are similar for different gases and electrodes, except for the pressures and currents at which particular curves appear. The curves from the edge of negative glow toward the cathode reduce to one single curve characteristic of the pressure of each gas. Dividing them by the current density to the $\frac{2}{3}$ power shows parallel characteristic changes with reducing pressure even for different gases. The data obtained appear to indicate fields in the dark space, which are determined by ion drift velocities, varying as (E/p) $\frac{1}{2}$. They yield drift velocities of plausible magnitude. At very low pressures, the curves differ from those anticipated using reasonable drift velocities, indicating the presence of ionization in the gas. Indications are available that ionization by positive ions was active near the cathode when $E/p >$ 10,000 volts/cm/mm Hg. Fields in striations have been determined.

Time studies of ionization processes are the subject of a paper published by C. Grey Morgan and J. M. Sommerville [396], University of Wales, University College of Swansea, Swansea.

The temporal growth of ionization currents in gases under impulse electric fields has been measured. The practical problem involved is to measure the time from the liberation of initiatory electrons to the attainment of a given current, due to the primary ionization process (α) and secondary processes, including positive ion ionization (β), photo ionization (η), and electrode effects such as positive ion impact ($\alpha\ \gamma$), photons (δ), and excited atoms (ϵ). The relative importance of the various processes has been investigated, using the analysis given by Davidson (1953), based on the solution of continuity equations for particles and photons. For hydrogen at low pressures there appear to be two secondary processes the γ and δ processes. At high values of X/p, γ appears to be predominant. At lower values δ is predominant. Neither process is wholly exclusive. From this work it follows that under impulse conditions the mechanism working is the same as under steady conditions. No new process is involved. The growth of current depends also on surface effects of the electrodes.

In Germany, H. Raether [397], Institut für angewandte Physik, Hamburg, observed the electron avalanche and its development.

A single electron avalanche can be detected by the following methods:

(1) By means of a Wilson cloud chamber. The drift velocity of the electrons is determined from the length of the tracks and the time of the applied voltage. Their thermal velocity is ascertained from the diameter of the head of the avalanche.

(2) By measuring the potential drop in a series resistance. The drop is composed of a rapid electron component and slow ion component. From the latter the ion drift velocity in the avalanche can be derived. The statistical fluctuations of the pulse height are in agreement with theoretical considerations.

(3) By detecting the light quanta emitted from the head of the avalanche with a multiplier. If one enlarges the multiplication exp αd the following events occur. The number of ions moving to the cathode and the number of light quanta increase. The space charge field increases. The first event starts the spark by means of alternating avalanches (Townsend mechanism). With the described pulse method the time necessary for the formation of secondary avalanches can be studied and from this the formative time of the spark. The second process becomes important when exp αd becomes e^{20}, where the streamer mechanism starts.

H. S. W. Massey [398], University College, London, described fundamental primary processes in gas discharges. Many different types of reaction play an important part in gas discharges. The probability of ionization and excitation of atoms and molecules by electron impact, the mobility of electrons and ions in different gases, and the rate of recombination of electrons with positive ions are of vital importance. In certain gases metastable atoms are very important and the processes producing and destroying them must be regarded as basic for present purposes. In electronegative gases where, in certain cases, the negative charge is carried by ions rather than electrons, the understanding of these discharges must be based on a knowledge of the formative and destructive processes and of the recombination of positive and negative ions. Transfer of excitation or ionization on impact between atoms or molecules must sometimes be considered. In intensive discharges in diatomic and polyatomic gases, dissociation in the gas will occur (in H_2 dissociation can be complete). It is then necessary to have available the rates of ionization, excitation, and other processes involving unusual radicals. An outstanding challenge is offered to devise measuring methods of effective cross section for such collisions involving unsaturated chemical radicals. The processes to be discussed are electron impact, recombination of electrons and negative ions with positive ions, diffusion of positive ions in gases, and formation and destruction of metastable atoms and negative ions.

6. Moving Spark Gaps

a. *The Impact Switch High-Voltage Switch for Capacitor Discharge Circuits of Impulse Welding Machines*

Impulse welding devices requiring discharge durations of 1/1000 to 1/30 of a second require a special type of switch. Such devices expend energy at the level of 1 to 100 kw-sec with a current density of a few kiloamperes. The switch described in this section has proved reliable in such situations. It operates in air and avoids sparking during cut-in almost completely. In capacitor welding machines it is a well-known practice to release the charge of a capacitor through a suitable switch into the primary of a welding transformer. In its secondary the transformed impulse is fed to the load in the form of the weld to be produced.

Such switching operations have the disadvantage of heavy sparking during the switching-on operation. This is because a sparkover occurs before the contacts engage owing to the high voltage. Discharge of the capacitor is often over before the contacts close. It is apparent that such action is detrimental to the welding process.

If a voltage is applied to two electrodes at a fixed distance, there is a time lag before sparkover occurs. This time lag is between 10^{-7} and 10^{-4} sec and is dependent on the structure of the electrode material. Theoretically it would thus be possible to increase the velocity of the approaching contacts to such a degree that contact is made within a time of 10^{-7} or 10^{-8} sec, before a switching spark can develop. Such a velocity, however, cannot be achieved practically since the cost would be prohibitive. Besides, there is a limit to velocity from the viewpoint of suitability of material.

However, if the electrodes are of a metal that is inclined to form an electron-hampering protoxide layer, e.g., copper, the sparkover will be delayed for a short time interval, say 10^{-4} sec. The explanation is that field electrons and photoelectrons are captured in the protoxide layer owing to the semiconductor effect.

It is possible to design a high-voltage switch for capacitor discharge circuits which practically prevents the development of switch sparks in cyclically recurring switching actions. The solution of the problem is that (Fig. B6a-1) a movable bridging contact, 6, of highly oxidizing metal is actuated at a sequence frequency in synchronization with the welding cycle

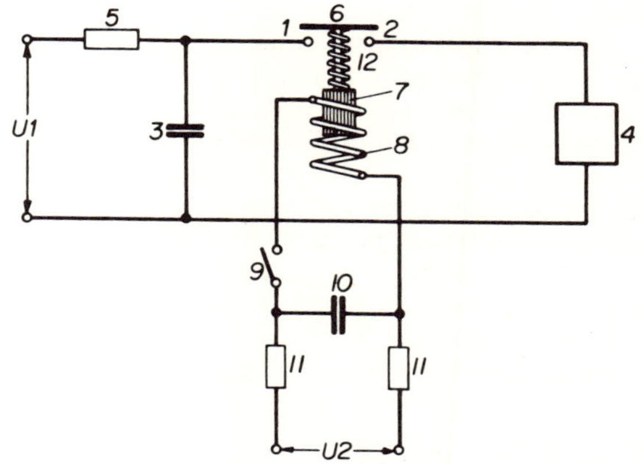

FIG. B6a-1. Capacitor discharge circuit with impact switch.

by means of current pulses derived from an auxiliary circuit, 10-8. It is moved so rapidly against the fixed counter contacts, 1–2, that no contact sparking can occur before contact is made. It is then rapidly opened at a predetermined speed by a spring, 12. The current pulse required for moving the contact bridge, 6, is derived from the discharge of an auxiliary capacitor, 10. The pulse energizes the coil, 8, pulling an iron core, 7, to which contact 6 is secured, into this coil. The charging time constant, 10–11, of auxiliary capacitor 10 by way of resistors 11 is longer than the charging time constant of main capacitor 3 by way of its resistor, 5. Moreover, the force of recoiling spring 12 as well as the electrical values of resistors 11 for charging of the auxiliary capacitor are so dimensioned that the current flowing through these resistors is not sufficient to hold the iron core in the coil against the force of the spring.

After every switching operation the oxide coating of the cathode, which is only a few molecules thick, renews itself on the air so that the switch remains always in operational condition.

The spontaneous striking movement of the contacts can also be effected purely mechanically or pneumatically. In experimenting with torus discharges impact switches are employed for operating loads up to 10^5 amp at 20 kv for the discharge of huge capacitor batteries.

b. *The Rotary Spark Gap*

J. V. Lebacqz and H. J. White [185] present a brief discussion of the sparking mechanism at atmospheric pressure and describe different kinds of rotary spark gaps. A spark is initiated between two electrodes by raising the voltage to, or above, the static breakdown voltage. In general, if the voltage applied to a gap is 2 to 3 times the static breakdown voltage, and if some initial ionization is provided, the time lag of the gap can be reduced to from 10^{-9} to 10^{-7} sec.

The rotary spark gap principle often is used in impulse setups for cyclic operation (see Section 4, Chapter E), and also for feeding high-voltage pulses into X-ray tubes (see Section 5, Chapter F); see also ref. [250].*

A rotary spark gap consists of a set of moving electrodes rotating in front of one or more fixed electrodes. The minimum spacing is adjusted so that the corresponding static breakdown voltage is smaller than the voltage required on the network, and the motion of the electrodes is synchronized in such a way that the maximum network voltage is obtained a short time before the minimum spacing is reached. Although the final design of successful rotary gaps has been achieved primarily by trial and error, there are some fundamental factors to be kept in mind, and some experimental data that can be used as a guide. These factors are the gap geometry, the nature of the gas and electrode materials, and the number of electrodes.

The gaps were built either with parallel electrodes having sufficient overlap

---

* The remainder of Chapter B, Section 6b, is reprinted by permission from "Pulse Generators," pp. 276–282 and 286–292, Glasoe and Lebacqz, McGraw-Hill Book Co., New York, 1948.

to increase the effective area which the spark may strike, or the electrodes are in two planes normal to the axis of rotation. The moving electrodes are radial and the fixed electrode makes an angle of 20 to 40 degrees with the radius, so that the gap geometry is essentially unaltered as the electrode wears away slowly from its tip.

Simple considerations of the voltage gradient between two moving cylindrical electrodes with parallel axes may be used to bring out some of the limitations of a gap of this type. Consider two electrodes (Fig. B6b-1) of equal diameter, d, and having a minimum spacing a. For any distance, s, between the axis of the moving electrode and its position corresponding to minimum spacing, the breakdown distance between the two electrodes is given by

$$D = \sqrt{s^2 + (d + a)^2} - d$$

For a rotary gap, the angular velocity of the moving electrode is constant. Then, if t_0 denotes the time at which $D = a$, the distance can be expressed as a function of angular velocity, ω, and radius, r, of the circumference of the center of the electrode. Then $s = \omega r\,(t_0 - t)$. This equation is correct only to the same approximation that $\omega\,(t_0 - t) \approx \sin \omega\,(t_0 - t)$, for usual rotary gap practice, the angle over which the spark may take place is always so small that the error introduced is negligible. Hence, the average voltage gradient between electrodes can be expressed as

$$E = \frac{V_N}{D} = \frac{V_N}{\sqrt{s^2 + (d + a)^2} - d}$$

$$= \frac{V_N}{\sqrt{\omega^2 r^2\,(t_0 - t)^2 + (d + a)^2} - d'}$$

where V_N is the network voltage.

For reasons explained later, resonant charging is nearly always used for rotary gap pulsers. In this case, the network voltage is so nearly constant near the discharge point that it may be assumed constant without introducing appreciable error. An exact discussion of the sparking phenomena requires a knowledge of the maximum electric field between electrodes for every position of the moving electrode. In the following discussion it is assumed that the voltage gradient E is uniform between the electrodes. It is also assumed that a spark may occur at any time after the electric field exceeds 30 kv/cm, and occurs instantaneously if the electric field reaches a value of 70 kv/cm. These assumptions are obviously far from correct, and are not intended to give a quantitative solution to the problem of gap design. If made consistently for various conditions, however, they do lead to results that are in accord with experience, and are therefore sufficiently accurate to indicate the expected trends of the phenomena. For instance, it is shown later that the experimental average gradient before static breakdown for a given gap varies as a function of spacing, and is higher than 30 kv/cm at the smaller spacings. The introduction of this variation in the minimum

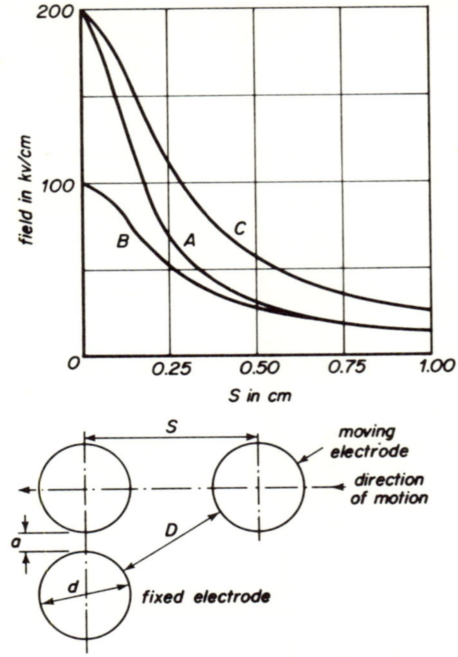

FIG. B6b-1. Average field between rotary gap electrodes. Curve A is for $V_N = 10$ kv, $a = 0.05$ cm; curve B is for $V_N = 10$ kv, $a = 0.10$ cm; and curve C is for $V_N = 20$ kv, $a = 0.10$ cm. By permission from The rotary spark gap, J. V. Lebacqz, *in* "Pulse Generators," McGraw-Hill, 1948.

average gradient in the discussion at this point would result in many additional complications without altering the conclusions.

For any given values of applied voltage, electrode dimensions and minimum spacing, and velocity of the moving electrode, a time $t = t_1$ can be found at which the voltage gradient reaches 30 kv/cm. Similarly, a time $t = t_2$ may be defined as the time at which $E = 70$ kv/cm. The spark then occurs between t_1 and t_2. Thus, $(t_2 - t_1)$ corresponds to the maximum time jitter that may be expected because of uncertainty in the initiation of the gap discharge. As an example, consider the hypothetical gap of Fig. B6b-1, and assume that the electrodes are 0.25 cm in diameter. The average gradient has been plotted as a function of s for several values of V_N and a. Then, with the assumptions made above, a spark may take place at any point between $s = 0.25$ and $s = 0.5$ cm for curve A, between $s = 0.18$ and $s = 0.43$ cm for curve B, and between $s = 0.41$ and $s = 0.85$ cm for curve C. To express these results as a function of time, some electrode velocity must be assumed. Taking $r = 10$ cm and a rotor speed of 60 revolutions per second, the linear velocity of the moving electrode is

$$\omega r = 2\,\pi \times 60 \times 10 = 3750 \text{ cm/sec}$$

Then, the maximum time jitter can be estimated as 0.25/3750 sec or 65 μsec for curves A and B, and about 115 μsec for curve C.

The time jitter to be expected is inversely proportional to the relative speed of the electrodes. Thus, in order to decrease the inherent time jitter caused by the statistical nature of the sparking process, it is theoretically sufficient to increase the steepness of the curve showing the variation of field with spacings or time within the range of voltage gradients that is most apt to produce a spark. One method of achieving the desired result is to increase the speed of the moving electrode. For instance, referring to Fig. B6b-1, it is seen that the time jitter of the gap with a minimum spacing of 0.10 cm for an applied voltage of 20 kv is almost twice its value for an applied voltage of 10 kv.

If the moving electrode velocity is held constant, the inherent time jitter of a gap that has an electrode diameter and minimum spacing that are twice those of the gap considered in the previous example and that operates at 20 kv is still much greater than that of the gap operating at 10 kv.

Another very important consideration in gap design is that of deionization. As pointed out previously, the gap is not required to interrupt the main pulse current. The network, however, starts to recharge immediately after the pulse, and care must be taken that its voltage does not exceed the breakdown voltage of the gap at any time before the end of the desired interpulse interval. Since the air in the vicinity of the electrodes is highly ionized by the passage of the spark current, it is safe to assume that there is very little time lag before breakdown, even for the minimum voltage gradient. If losses are neglected and the network is assumed to discharge completely, the voltage across the gap may be expressed as

$$v_N = \frac{V_N}{2}(1 - \cos \omega_0 t)$$

for dc resonant charging, and the average field is

$$E = \frac{v_N}{D}$$

where D is defined as before. As an example, consider the gap of Fig. B6b-1, curve A, $V_N = 10$ kv, and assume that the spark has taken place when $E = 40$ kv/cm. If the pulse recurrence frequency is 600 pps.

$$v_N = 5(1 - \cos 600\pi t),$$

and the electrode velocity is 3750 cm/sec. Thus, the voltage gradient across the gap can readily be plotted as in Fig. B6b-2 for various positions of the moving electrode. For the particular values of parameters chosen, there is no danger of the gap restriking and establishing a short circuiting arc across the power supply. If the same gap were used at $V_N = 20$ kv, however, the network discharge might take place at $s = 0.7$ cm. Then, the voltage built up on the network at the time of minimum spacing between the electrodes would be about 610 volts, corresponding to an average gradient greater than 12 kv/cm. Because of the nonuniformity of the field, which can be caused

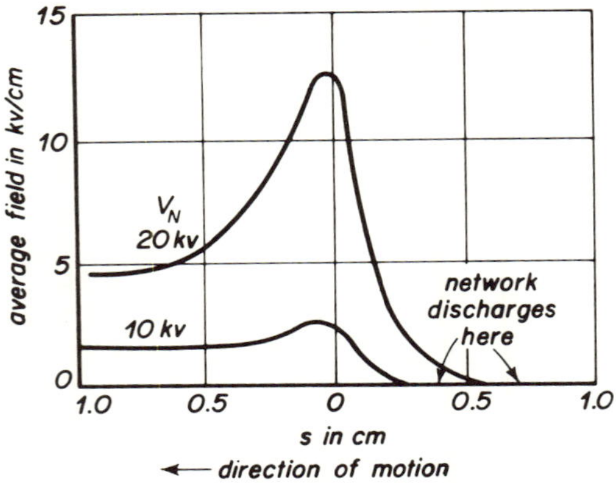

FIG. B6b-2. Average field between electrodes during the beginning of a charging cycle for $d = 0.25$ cm, $a = 0.05$ cm, $f_r = 600$ pps, $\omega r = 3750$ cm/sec, where d, a, and s are shown in Fig. B6b-1. By permission from The rotary gap, J. V. Lebacqz, *in* "Pulse Generators," McGraw-Hill, 1948.

by rough electrodes (resulting from long operation), the gap may be expected to restrike under these conditions. The average voltage gradient as a function of pin spacing has been plotted for this case in Fig. B6b-2.

Experience has shown that high-power gaps work satisfactorily with an average gradient of between 60 and 70 kv/cm corresponding to minimum spacing (Fig. B6b-1), and that low-power gaps may be built and operated with minimum spacing that would correspond to average gradients of between 200 and 300 kv/cm if the gap did not break down before that point were reached. These figures are only indicative of the range that has been used, and each case has to be considered separately, depending on the required recurrence frequency, the time jitter, the type of charging circuit, and the type of load.

Ventilated rotary gaps using tungsten electrodes have been used in the great majority of radar applications, although they cannot be used in airborne radar sets because of the change in breakdown voltage with pressure. Sealed rotary gaps have been used in one airborne radar set, for which the life requirements are much less severe than for ground-based or shipborne sets. In practice, a convenient unit for rate of wear is milligrams per ampere-hour. Experimental data obtained both at the Bell Telephone Laboratories and the Radiation Laboratory show that, for pulse currents from 40 to 170 amp the rate of cathode wear in air varies between about 2 and 6 mg/amp-hour, 3 or 4 mg/amp-hour being the median value. In nitrogen or gas mixtures that do not contain oxygen, the rate of wear is 10 to 20 times greater. The order of magnitude of the rate of anode wear in air is only 1/5 that of the cathode.

Figures B6b-3 and B6b-4 show the relationship between pulse power and average current and the diameters of the moving and fixed electrodes, obtained

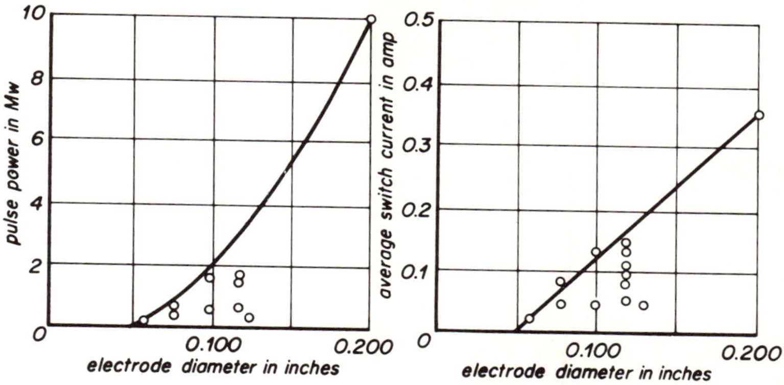

FIG. B6b-3. Diameter of the moving electrode for satisfactory gap operation and life. By permission from The rotary spark gap, J. V. Lebacqz, *in* "Pulse Generators," McGraw-Hill, 1948.

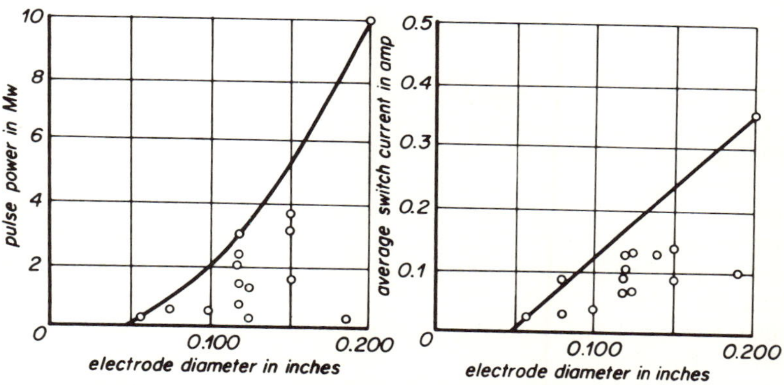

FIG. B6b-4. Diameter of the fixed electrode for satisfactory gap operation and life. By permission from The rotary spark gap, J. V. Lebacqz, *in* "Pulse Generators," McGraw-Hill, 1948.

from gap designs that have given satisfactory operating life. In these gaps, the fixed electrode is the anode and, as there are always several moving electrodes, the wear is divided between them.

The values indicated by the curves can be regarded as the maximum safe operating power and current for any given electrode diameter, as determined from present experience, that allows a satisfactory gap life.

The number of fixed and rotary electrodes, N_f and N_r, is related to the angular velocity, ω, of the rotor (in revolutions per second) and the required pulse recurrence frequency, f_r, by the relation,

$$f_r = N_f N_r \omega$$

This relation applies only if the respective positions of fixed and rotating electrodes are such that the minimum spacing between electrodes is never reached simultaneously by more than one fixed and one rotating electrode. If, for instance, there are six rotor and six stator electrodes all spaced 60 mechanical degrees apart, the total number of sparks per revolution is six, the same number that would be obtained for one, two, or three stator electrodes equally spaced. If, on the other hand, there are five rotor and three stator electrodes equally spaced, the number of sparks per revolution is 15, and $f_r = 15\omega$.

In general, because of the mechanical difficulties involved, only one fixed electrode is used, except for special applications such as the Marx circuit. In such cases, it is necessary to obtain two or more sparks simultaneously, and the number of fixed electrodes must therefore be at least equal to the total number of discharge paths desired.

Of all the possible gap configurations, several have been used extensively and have proved the most satisfactory. In all of these gaps, the electrodes were tungsten rods, so the types of gaps are classified by their geometry.

Figure B6b-5a shows a perspective view of the parallel pin rotary gap with sparking sector. A gap of this type was used extensively for a pulse-power range of 500 to 3000 kw. The clearance between the rim of the rotor disk and the sparking sector is usually from 0.010 to 0.050 inch, corresponding to a static breakdown voltage between 1 and 5 kv, approximately. Hence, little time delay occurs once the gap between the pins has broken down, and most of the network voltage appears between the pins until this gap is broken down, since capacitance between them is very much smaller than that between the rim and ground. An important advantage of the parallel pin gap is that its operation is relatively independent of the exact amount of electrode overlap, which in turn means that close end play tolerances are not required of the motor driving shaft.

Figure B6b-5b shows a variation of the preceding design called the parallel-pin double gap. In this case, the rotor must be insulated for the full network voltage, and not just for the breakdown voltage of the sparking sector. The rotary pins extend on either side of the insulating rotor, and spark to two fixed pins. A gap of this type is especially useful when using a Marx circuit pulser because, by using as many pairs of fixed pins as there are stages in the Marx circuit, the need for more than one rotor is avoided. It is well suited for use in high-voltage circuits.

The radial pin gap is commonly used for handling high powers. The fixed pin normally wears away rapidly, but it can be set at an angle, as indicated in Fig. B6b-5c, in order to maintain very nearly constant gap geometry despite the erosion. For powers in excess of 3 megawatts, the minimum gap spacing can be made greater than 0.05 inch. As a result, the motor end play need not be kept to unusually small values; however, it is clear that, at small minimum spacings, the shaft end play must be very accurately controlled. A gap of this type has been successfully used in a 10 megawatt, ac, charging pulser in which the gap current exceeds 1000 amp. It was conservatively estimated that the fixed electrode would not need replacement in less than 1000 hours.

Figure B6b-5d shows the opposing pin gap, which was used in early

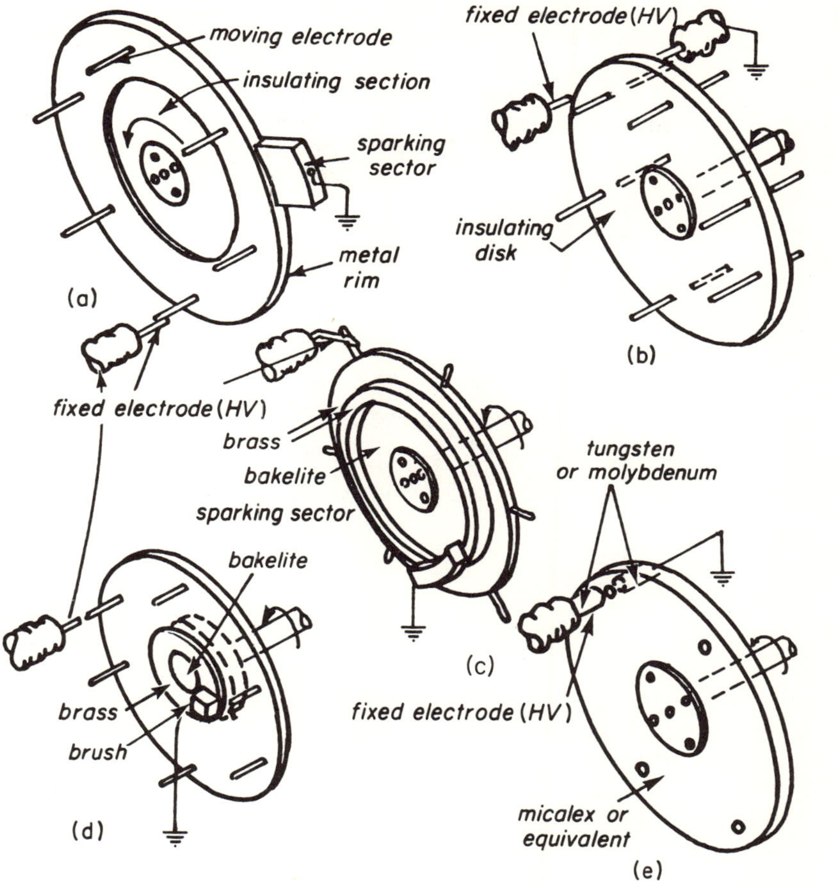

FIG. B6b-5. Types of rotary gaps: (a) Parallel pin gap with a sparking sector; (b) parallel pin gap, double gap with an insulating rotor; (c) radial pin gap with a sparking sector; (d) opposing pin gap with a grounding brush; (e) gap with holes in insulating disk. By permission from The rotary spark gap, J. V. Lebacqz, *in* "Pulse Generators," McGraw-Hill, 1948.

experimental sets. This gap was unsuitable for long service because of the rapid change in geometry with erosion. Except for very low powers, the grounding brush has also been abandoned because of the rapid deterioration of the brush and slip ring surface.

A gap in which both electrodes are stationary, and where breakdown is obtained by varying the dielectric constant and dielectric strength of the gap space, is shown in Fig. B6b-5e. It has a Micalex disk rotating between two fixed massive electrodes. A series of holes drilled in the Micalex causes the gap characteristics to change as the holes pass between the electrodes, in such a way that a spark takes place at that time. The life of a gap of this type is limited by the eventual deterioration of the insulating disk, which is eaten away by the sparks.

The minimum spacing between electrodes is normally adjusted so that the gap is overvolted by a factor of 2 to 10 at the point of closest approach. As a result, the spark always takes place before the closest approach is

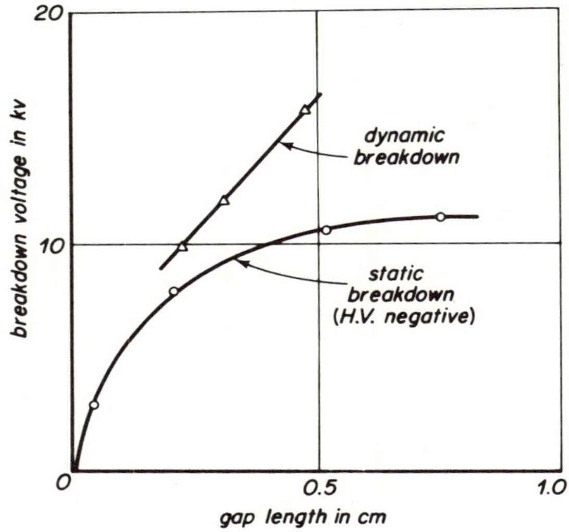

FIG. B6b-6. Breakdown characteristics of the rotary gap used in a 1-mw pulser. By permission from The rotary spark gap, J. V. Lebacqz, *in* "Pulse Generators," McGraw-Hill, 1948.

TABLE B6b-1

DESIGN CHARACTERISTICS OF ROTARY SPARK GAPS

Type of charging	Dc resonant charging					Dc charging
Electrical ratings voltage range (kv)	7–16	5–>9	12–22	12–22	5–>9	8–13
Peak current (amp)	160	>90	220	220	>90	130
Normal operating voltage (kv)	12	8	≈18	≈18	8	13
Normal operating current (amp)	120	80	180	180	80	130
Normal average current (ma)	45	100	130	130	70	70
Pulse recurrence frequency (pps)	410	825	800	800	750,375	580
Pulse duration (μsec)	0.9	1.5	0.9	0.9	$1\frac{1}{8}$, $2\frac{1}{4}$	0.9
Pulse power (kw)	600	360	1500	1500	300	700
Pins; no. of fixed pins	1	1	2	2	2, 1	1
No. of rotary pins	7	14	10	14	4	10
Diam. of pins (inches)	0.100	0.188	0.125	0.150	0.125	0.120
Motor speed (rpm)	3500	3535	4800	3400	5600	3450

realized, and the uncertainty in the time of breakdown produces time jitter. In order to reduce time jitter to a minimum, it is desirable to have the steepest part of the average field spacing characteristic (Fig. B6b-1) occur at those fields at which breakdown is most likely to occur, or at about 40 kv/cm (see Fig. B6b-6). If sufficient ionization is available, usually as a result of preceding sparks, the time jitter of the gap can be held to less than 10 μsec. This figure depends on the relative velocity of the electrodes, and decreases as the velocity increases.

Finally, Table B6b-1 is a short tabulation of characteristics for designing of rotary spark gaps.

c. *The Blast Spark Gap* [677a, 191]

In order to utilize very high peak currents and relatively high voltages at a high rate of discharge sequence, above 1000/sec, Marx has introduced the so-called blast spark gap. The feature of this spark gap is that a stream or blast of air or gas is continuously introduced into the gap between a pair of sphere-shaped electrodes facing each other in spaced relation. In practical constructions, air at high velocity is usually introduced at both electrodes into a cylinder surrounding the spark gap system. The electrodes have axial center bores through which the air or gas escapes axially to both ends of the spark gap. By such means every discharge between the electrodes is driven by the gas flow toward the axial center of the gap and from there to both sides in a flow direction through the center bores into the space behind the spherical faces of the electrodes. Any ionization is thus blown away pneumatically. Owing to this blow action every successive spark finds a virgin gap without ionization. With such switching devices it is possible to reach spark sequences up to 50,000/sec. These blast spark gaps have found acceptance in the field of high-frequency generation by capacitor discharges. In this application each capacitor discharge excites an oscillating circuit which gradually damps out. The discharge energy can be utilized for heating a work piece. For production expediency, one desires the next starting impulse to follow as soon as the energy of a previous spark has died away. When one is operating with such apparatus—for example, at an oscillator frequency of 300 kilocycles/sec—for heating a work piece, and after having designed the oscillation circuit for a decrement of 0.1, it is safe to figure a sufficiently diminished oscillation within 1/30,000 sec. A spark sequence of 30,000/sec is then necessary. If each discharge converts only 1 watt-sec into thermal energy, it will still be an output of 30 kw. With a special modification of a construction having a concentric ignition electrode, it is possible to control the discharge. Such a construction is rugged and easily maintainable for use as a switching means at any desirable rate of discharge.

With respect to peak currents, the rules of the previous chapter apply. These point out that some thousand amperes can be switched intermittently for a few microseconds at intervals. In regard to peak currents it must be remembered that the ion path preceding the discharge must have enough time to develop. It can be said that the dimensional limit of blast spark gaps lies at about 2-cm sphere spacing and at about 60-kv capacitor voltage.

(a)

(b)

(c)

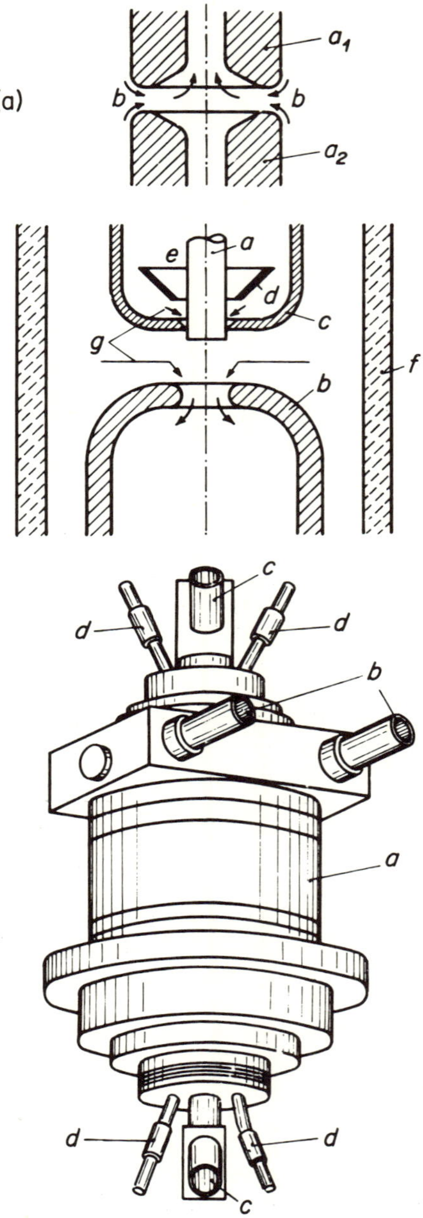

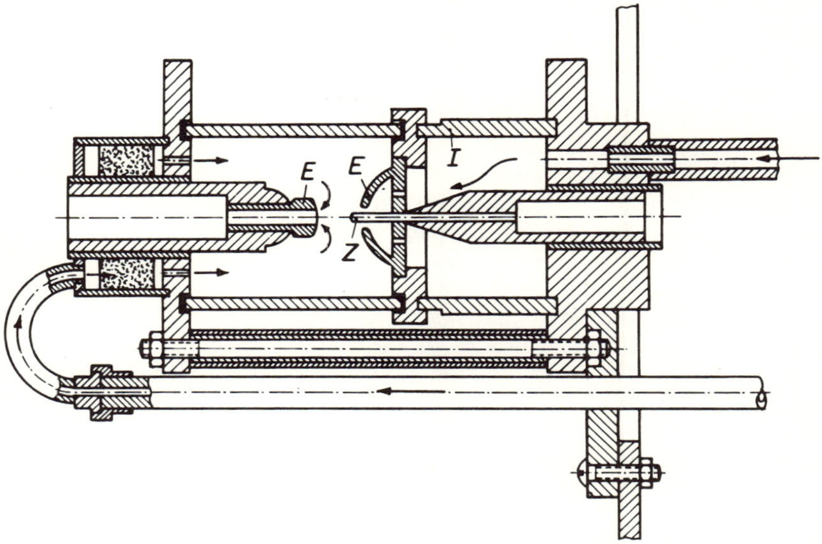

FIG. B6c-2. Impulse spark gap with air blast provision. E–main electrode; Z–ignition electrode.

Most of the devices in operation today are for capacitor voltages in the range from 10 to 15 kv and for switching sequences of about 10,000/sec. They are used frequently for blast spark operated high-frequency generators in the field of induction hardening [15]. Hereby the capacitor is simply charged by an ac voltage. The polarity of the discharge is unimportant for the heating process. The structural details of a blast spark gap device with indication of air flow direction are depicted in Fig. B6c-1a, b, and c and a controllable blast spark gap with ignition electrode is represented in Fig. B6c-2. The dimensional scale is about 1:2, i.e., practical units are about twice the size shown. In every case where compressed air is available, the blast spark gap is the proper selection for periodic switching of average energy impulses at high sequence frequencies in continuous operation.

More economical than the usual industrial compressed air supplies (6 atm) are medium-pressure blowers of 0.5–0.8 atm and air volume rates of 10 cubic meters per minute (Prokorny, Machine works Ärzen).

FIG. B6c-1. The blast spark gap of Marx [15].

(a) Scheme of the electrodes a_1 and a_2–electrodes; b = direction of compressed air flow.

(b) Controlled impulse spark gap. *a*–pin electrode, *b*–umbrella-like counter electrode, *c*–ignition electrode, *d*–preionization electrode, *e*–pressure chamber for ignition air flow, *f*–pressure cylinder, *g*–direction of compressed air flow.

(c) Side view of a blast spark gap device. *a*–pressure cylinder, *b*–compressed air inlet, *c*–compressed air outlet, *d*–cooling water connection.

d. *Quick-Response Spark Gaps with Air Stream for High Sequences* [191]

Generally, it is possible to use simple spark gaps for switching on discharge circuits. For heavy outputs, however, designs differing from the well-known conventional sphere spark gaps become necessary. This is because heavy loads cause rapid burnoff and fast consumption of the electrodes. They become practically useless within a short time.

The methods of spark gap construction now to be described assure uniform continuous operation with spark gap arrangements designed to produce voltage pulses with high-frequency spark sequences. When there is impressed on a spark gap device a steep-fronted voltage pulse 20% higher than the static breakdown voltage of the gap, there exists a time lag between the impression and initiation of the sparkover. This time lag depends on the surface structure of the electrodes or the state of their surface oxidation. Metals having a low work function (low resistance to electron emission) cause short time lags. Aluminum and magnesium are metals of low work function causing time lags on the order of 10^{-7} sec. Other metals, for instance copper, have the tendency to cover themselves with a thin coating of protoxide or oxide which has the peculiarity of hampering or even completely preventing electron emission from the metal surface. With electrodes of such metals several μsec may elapse before sparkover occurs at the given overvoltage. On the other hand, metals of this kind have, according to experience, the longest life expectancy in spark gap arrangements for continuous operation with heavy loads. They are therefore used in practice as electrode material for blast-spark high-frequency generators. In such generators, there is such a small interval between subsequent sparks that the copper finds no time to cover itself with a coating of protoxide. In consequence, the mentioned time lag in spark firing is noticeable only at the first ignition, and is absent thereafter. However, when such a generator must be switched on and off very often, considerable overvoltage surges develop each time during switching-on operation of the high-voltage transformer. Owing to the protoxide layer, these lead to delayed sparkover. Such overpotentials burden the high-voltage capacitor to twice the rated voltage, i.e., up to the breakdown limit, before the spark gap responds. With such voltages, the development of a breakdown in the solid dielectric of the capacitor needs only a few seconds and is therefore more speedy than the breakthrough in the slow-acting spark gap. In order to eliminate such capacitor breakdowns, it was necessary in the early days to overdimension the capacitor battery in such spark gap systems, so that it could safely withstand voltages in the order of switching-on voltage surges. The cost of capacitors of similar capacitance increases more or less with the square of the voltage rating. If a capacitor is dimensioned for safe operation with switching-on voltage surges of twice the rated operating voltage—and such surges have been measured in blast-spark high-frequency generators—such an increase in cost of the installation results that, in spite of many advantages, it is hardly competitive with other devices such as high-frequency transmitter tube generators. The capacitor is only one of the circuit elements that is endangered by these undesirable overvoltage surges.

When small capacitors are employed, the way is open to provide the copper electrode, blast spark gap with an auxiliary safety spark gap. This consists of a metal that causes no time lag, say, aluminum. The result of such an arrangement is that, when at switching-on an overvoltage surge occurs, the auxiliary spark gap responds with a subsequent forming of an arc. When this happens the setup must be switched off and on again repeatedly until by accident switching-on takes place at the instant when the line voltage passes through zero and consequently no voltage surge occurs. In difficult applications, such as the inductive surface hardening of steel, the way described above is impossible, since it would result in a work piece reject. In such situations, the use of overdimensioned capacitor batteries is the only solution today.

The overvoltage surge phenomenon does not occur in a spark gap arrangement when a metal with a low burning-off quality and sluggish spark initiation is combined with a metal of good ignition qualities. In Fig. B6d-1 numerals 1–1 represent the electrodes of a spark gap that face each other in spaced relation and are arranged in a stream or blast of gas or air entering at 2 and leaving in direction 3 coaxial with the spark gap.

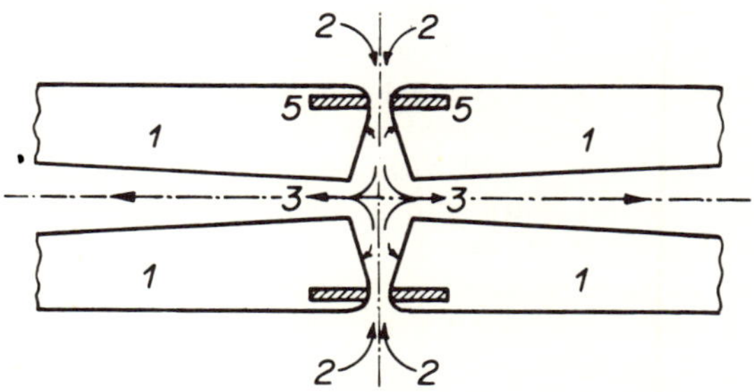

FIG. B6d-1. Direction of air flow in a blast spark gap that assures quick sparkover response (by Früngel). 1–copper alloy of great heat capacity, heat conductivity, and burn-off resistance; 2–direction of entering stream of air; 3–directon of outflow of air; 4–opposite, traveling surfaces for sparks with increasing spacing towards the axis; 5–inserted metal rings of low-work-function material, e.g. aluminum alloy, for spark start with no time lag.

The closest spacing of the electrodes is at 4. Here, as a rule, the sparkover begins. There are inserted radially out from this closest spacing a larger or smaller number of pins, 5, of another material, say, aluminum. Whenever an overpotential occurs, the first spark forms between the two active surfaces of the pins, 5–5. It is driven immediately by the blast of compressed air to the working point, 4. Subsequent sparks form only between 4 and 4 so that burning off at 5 does not take place.

A modified arrangement giving similar results is shown in Fig. B6d-2. Here a ring, 2, is put around each one of the electrodes, 1–1. It consists of ignition-willing metal, i.e. metal of low work function. The ring is so arranged and shaped that the smallest spacing at (2–2) 3–3 is somewhat larger than the closest spacing at (1–1) 4–4: for instance, 3.5 mm ring spacing

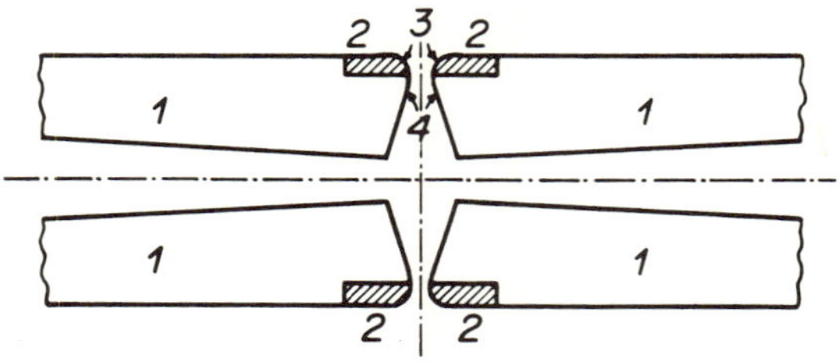

FIG. B6d-2. Modification of the spark gap with aluminum ring around the active portion of the electrodes for continuous operation in blast-spark high-frequency generators.

at 3, as against 3.0 mm electrode spacing at 4. If one selects equal spacings, each spark starts at 2, because of the more favorable work function here, and will cause a high burning-off rate at 2.

This will result automatically in a higher electrode consumption at 3 which would continue until the closest spacing would be at 4. Such burning off would result in a progressive change of spark gap characteristic. It is advisable to produce the illustrated relationship of Fig. B6d-2 beforehand to have a stable condition from the beginning. Since every electrical discharge grows, preferably, from the cathode, in dc operated spark gap arrangements it is generally only necessary to provide the cathode with a low time lag auxiliary metal. In dc operated electrodes it is therefore not necessary that both electrodes be provided with low work function metal because time lag–free response of the spark gap at overpotentials is also assured when only the cathode is made of, say, aluminum, and the respective anode is of any other conducting material. From the practical production standpoint, however, concentric arrangements are preferable.

Another field of application for the spark gaps just described is in rotary high-voltage rectifiers and inverters. In order to prevent inconsistencies in the operation of such equipment it is important to obtain a spark discharge as soon as possible after the voltage has passed through zero or after an electrode approach. Thus a metal of low work function would by analogy be employed for the first initiation of sparkover. The main arc discharge continues to take place between the burnoff-resistant electrodes as is usual in prior devices for that purpose.

For lightning-protective devices, where a quick acting spark gap is desirable, an arrangement as shown in Fig. B6d-3 can be employed. Here the electrodes, 1–1, of high temperature–resistant metal are provided with rings, 2, of low work function metal inflicting only a negligible time lag on the lightning discharges. After the discharge has started between the rings, 2, the thermal load is primarily carried by the high temperature–resistant main portion of the electrodes. This is owing to the self-contracting effect of the arc, whenever the lightning intensity of the stroke is high enough to develop an arc. For lightning protection it is customary to insert such a gap between lightning rod and ground. Similar arrangements are of benefit in the technique of the so-called cathode drop protectors for the protection of overhead

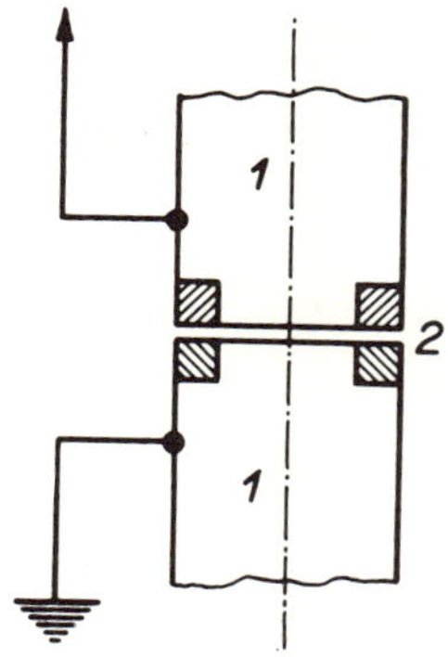

FIG. B6d-3. Modification of the spark gap shown in Fig. B6d-2 having parallel faces at close spacing for lightning protection of low voltage transmission lines.

transmission lines against lightning strokes, which, in fact, are atmospheric capacitor discharges of heavy currents.

In conclusion it can be said that the quick-response spark gap arrangement always has electrodes of two different metals, one set of which serves for inertialess initiation of the first sparkover while the other set has burnoff-resistant qualities for continuous operation. Under certain conditions, it may be feasible to consider replacing one of the metals by carbon or another nonmetallic conductor.

In continuous operation as a spark gap in a Marx blast spark high-frequency generator such as a spark gap with 3 mm electrode spacing, 40 mm diameter, and 6 cubic meters of blast air can convert 35 kw power at an output of 25 kw high-frequency energy. Rate of sparks $\approx 10^4$/sec, discharge energy per peak $\approx$ 3.5 joules. After about 50–100 hours of service it has been found advisable to recondition the electrodes on a lathe in order to equalize small inhomogeneities. (See schematic circuit diagram, Fig. C3-1.)

Sparks which have developed into heavy current arcs can be quenched additionally by the application of magnetic fields. A. C. MacKellar [878] and E. J. G. Beeson [792] describe an example of magnetic quenching of arcs. Especially in the light signal technique, e.g., lighthouse arc lamps, this technique allows an advantageous motionless operation. With magnetic switching, a substantially square-wave pulse of light of shorter duration can be obtained, the radiation being of daylight quality. Measurements have shown that flash energies up to 15,000 watt-sec are possible, with light outputs approaching 1.5×10^6 lumens.

e. *Pneumatic Switching Elements*

For switching on and off capacitor discharge circuits having compressed air–operated circuit elements, e.g. blast spark gap devices, it is quite logical to utilize compressed air for actuation of the switches. As an example of such pneumatically actuated switches there is the blast spark gap described

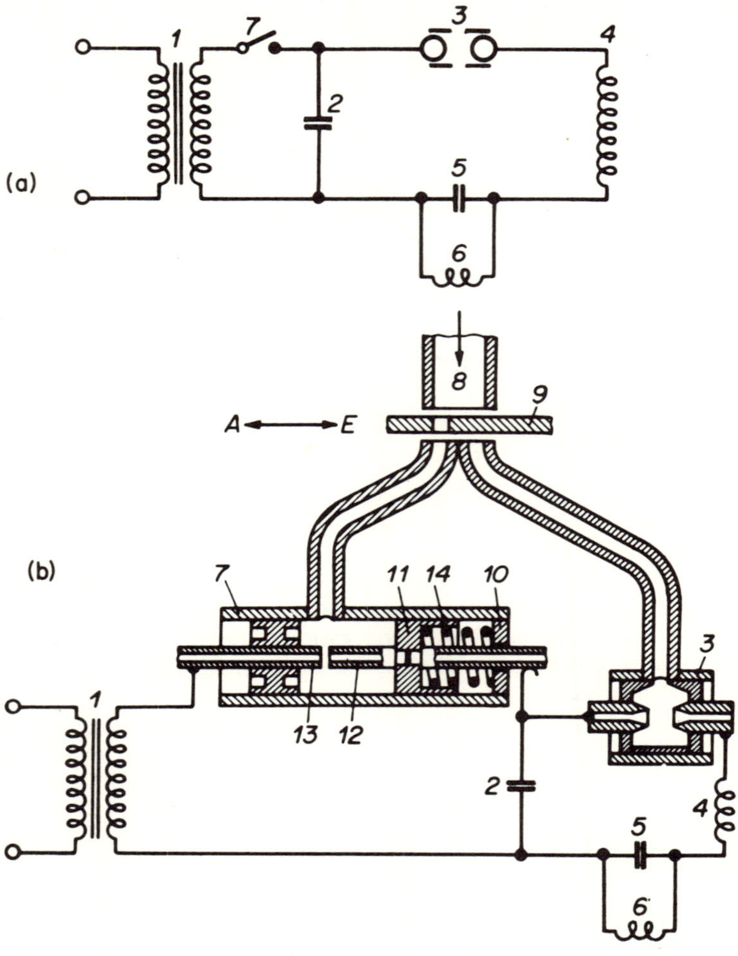

FIG. B6e-1. Pneumatic switching of blast spark gaps. a–schematic circuit diagram; b–example of structure.

in the previous chapter and illustrated in the schematic diagram Fig. B6e-1a.

For this purpose an impulse capacitor, 2, is charged periodically from a high-voltage transformer, 1. The capacitor, upon reaching the breakdown potential of the spark gap, 3, discharges through inductance 4 into the load circuit, 5–6, and creates a damped high-frequency oscillation there.

The blast spark gap, 3, is blown out after every sparkover in order to achieve as high a spark sequence frequency as possible.

With high-frequency loads of 20 kw and higher, there enter difficulties in periodic triggering of discharges for inductive hardening applications. Such switching requires large contactors at the primary side of the transformer and incurs a high burnoff rate at their contacts. The application of high-voltage switches at point 7 in Fig. B6e-1a in the form of expansion switches is too costly for the purpose at hand. It is therefore advisable to replace, in accordance with Fig. B6e-1b, the switch, 7, by a switching device that can feed the compressed air selectively either to the blast spark gap or to a switching chamber similar to a blast spark gap. Under the influence of compressed air this effects switching off of the high voltage passing through its movable electrodes.

This switching chamber operates as follows. The switch proper is accommodated in tube 10 in which there slides a piston, 11, that carries the electrode, 12, which by means of spring, 14, is resiliently pressed against electrode 13. Current passage from 13 to 12 is thereby assured. When compressed air is admitted to the cylinder, 10, it pushes piston 11 to the right and separates the electrodes, 12 and 13. A high-voltage arc develops between them. The compressed air can also escape through the axial bore of the electrodes, 12 and 13, and blows, as previously explained, the arc into these bores. The arc is cooled and extinguishes during the passage of the current through zero. Switching off the setup is thus accomplished. General control is by means of slide valve 9. In position A of this valve, compressed air can enter only cylinder 10. This keeps the installation switched off. In position E compressed air can enter only the spark gap chamber, 3. The switch, 7, receives no pressure. The electrodes close and the setup is switched on. Actuation of slide valve 9 is most conveniently accomplished by electromagnetic means. With this construction, one can take advantage of the fact that burnoff of the electrodes in the blast spark gap device is at a minimum when the electrode temperature does not drop below 200°C. If in the switched-off condition of the setup air still flows through the spark gap, 3, the electrodes will cool further and at switching on it will be necessary first to reestablish the operating temperature which is a prerequisite for stable operation. In an arrangement as illustrated in Fig. B6e-1b the spark gap, 3, receives compressed air only when it is switched on.

In consideration of the fact that the time during which an arc burns between electrodes 12 and 13 is very short with respect to the total on-time, the switching gap, 7, can be rather small, since a short circuit can never occur. All that can happen is passage of the secondary current given by the design of the setup, say, in the range from 1 to 10 amp. By proper design and economical construction the cost of pneumatic switches has been reduced, so that today it is only a fraction of that of large low voltage contactors.

7. The Lightning Protector

A switching means that has gained such considerable recognition that it should be mentioned here is the so-called lightning protector.

This spark switch serves the purpose, when it is installed parallel to high-voltage overhead transmission lines, of conducting to the ground extraordinary overvoltages caused by lightning strokes and at the same time extinguishing the arc carried over to the operating line voltage.

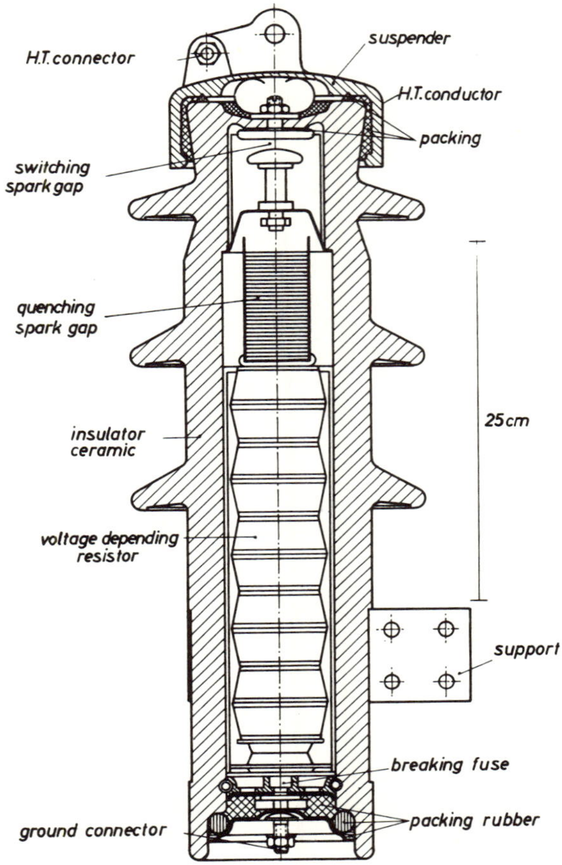

FIG. B7-1. Internal construction of a lightning protector.

Figure B7-1 shows a lightning protector in section. An auxiliary spark gap, which is set to respond when a predetermined overvoltage occurs, is interrupted by the traveling wave caused by the lightning stroke. This steep-fronted breakdown subsequently breaks down the quenching spark gap. In order to prevent a massive short circuit, there is interposed at the ground side of the device a potential-dependent sintered resistor, the resistance of which at operating line voltage may be a few megohms but decreases with rising voltage rather steeply. The overvoltage surge is thus carried immediately to ground. At an ac voltage phase angle close to zero the quenching spark gap cuts off the current at a point below the minimum current value required for maintaining the arc. The spark in the main spark gap also extinguishes.

These so-called lightning protectors are of very rugged construction. In laboratory work they can be employed as voltage fuses and can be useful in single impulse networks for impulses of a certain minimum value. As a result of mass production, they are rather reasonable in cost even for higher voltage ratings.

8. The Actuation of Magnetic Contactors at Predetermined Phase Angles

Sometimes it becomes necessary to close a contact at a correct phase angle, for instance, when a capacitor as phase shifter is charged by ac voltage and has continuously alternating potentials. The following switching technique is possible in such cases. It can be used in general for switching of capacitor discharges by electromagnetically operated switching devices.

When a coil of an electromagnetic switch is energized with the rated energizing voltage, the magnetic field gradually grows owing to voltage and coil inductance, and leads finally to closing of the switch contacts. The time that elapses between the instant of applying the coil voltage and the instant of contact making varies considerably. Particularly with ac energizing of the magnet there can be encountered variations of between 50 and 250 millisec. These variations depend largely on the phase position of the voltage at the instant of switching on. However, when a capacitor is discharged through the energizing coil of an electromagnetic switch, it is possible to convert a proper uniform quantity of capacitor energy into kinetic energy actuating the movable system of the switch, and to achieve thereby a phase-correct switching action.

A circuit diagram for such a switching method is illustrated in Fig. B8-1. The capacitor, 1, is charged by way of a resistor, 2, from a dc voltage source. When contact 3 is closed the capacitor discharge effects a response in electromagnet 4 which attracts armature 6 and thereby closes contacts 5. On a time basis the progression is as depicted in Fig. B8-2, where time is plotted as the abscissa.

The switch, 3, is closed at instant *a* and remains closed until *c*. The contact, 5, closes at instant *b*. The distance *a–b* is (at a constant charge of capacitor, 1, and with the same type of switch) dependent only on the mechanism of the switch and can thus be considered constant. When at instant *c* the switch, 3, is opened, contacts 5 open within a tolerance time *d*. There is active, besides the constant mechanical pull for opening the contacts, the nonconstant residual magnetism of core 4. Owing to small variations in residual magnetism, depending on the incidental air gap between core and armature, which is subject to

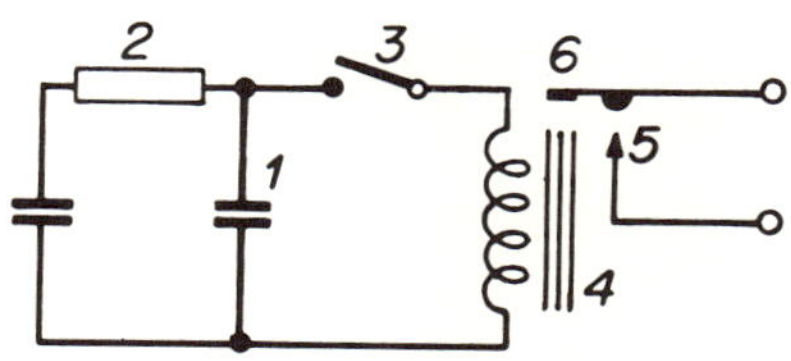

Fig. B8-1. Schematic diagram for contact closing at a predetermined phase angle.

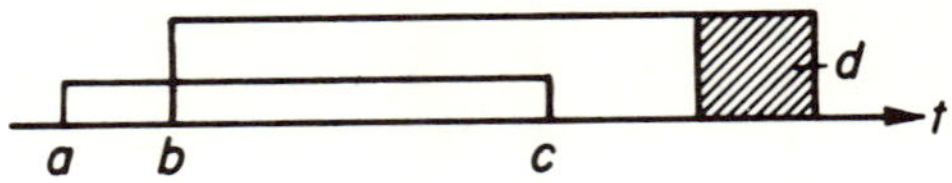

FIG. B8-2. Time program for contact closing of a relay-controlled phase-correct switching device.

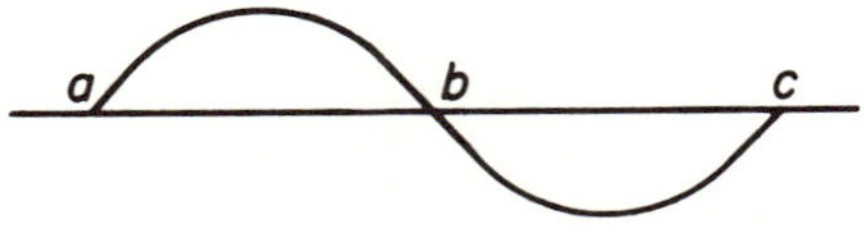

FIG. B8-3. Definition of zero value passing of a sine wave oscillation.

minute changes, the force of the contact opening spring will at an earlier or later instant be greater than the force of residual magnetism and will open the contacts. The tolerance range of contact opening cannot be reduced in this switching arrangement.

In Fig. B8-3 the problem may be to obtain contact making of contacts 5 (Fig. B8-1) at a predetermined point of the sine wave curve shown in Fig. B8-3. By means of an electromagnetic contact maker as shown at, 1, in Fig. B8-4, the contact tongue, 2, is moved periodically between the contacts, 3 and 4, when the coil of this relay is energized through a phase shifter, consisting of the capacitor, 5, and a resistor, 6. This is controlled for predetermined phase position by a constant line voltage drop at resistor 7. This drop is stabilized by the iron-hydrogen resistor, 7a. Figure B8-6 shows the sine wave progression *a* of the ac voltage drop across resistor 7 (Fig. B8-4). The contact tongue, 2, is, during the interval *b*, in contact with 3. During interval *c* it is in contact with 4. The time diagram shown in Fig. B8-6 corresponds to an oscillographic trace of a device designed for 50 cps control. Duration and time position of the intervals *b* and *c* can be adjusted by the magnitude of the excitation voltage as well as the phase position within

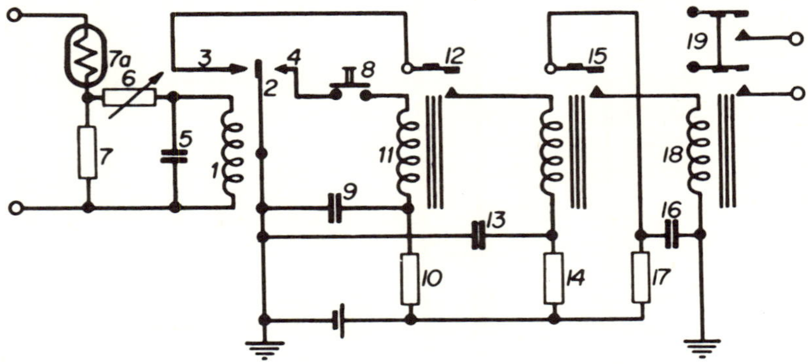

FIG. B8-4. Complete relay and contactor circuit diagram for phase-correct switching on of contacts 19 at predetermined phase angle, e.g., at zero passing of the line voltage wave.

certain limits, for practical purposes $\pm$ 30° phase angle. The light contact tongue, 2, is thereby the phase correction, controlling, switching organ proper. For actuation of electromagnetic switch 18–19 at a desired phase angle, switching contacts 8 must be closed (Fig. B8-4).

Closing these contacts, 8, effects that at the next engagement between 2 and 4 the charged capacitor, 9, which has been recharged by way of resistor 10 from a dc charging source, energizes relay coil 11. This coil closes its contacts, 12, at a constant delay time given by the distance *a–b* in Fig. B8-2. The contacts, 12, are now connected with contact 3 in such a manner that during the short engagement time *b* (Fig. B8-6) capacitor 13, which has also been charged from the dc source by way of resistor 14, closes the contacts 15 at an instant given by phase position of contact tongue 2. It is thus possible to close contacts 8 arbitrarily at any instant. This first effects, during the next contact making 2–4, a response at relay 11–12; then, during the short contact closing of 2–3, a phase-true contact which times the actuation of relay 14–15 in accordance with interval *b* (Fig. B8-6). Contact closing at 15 can now be utilized to discharge the energy of a capacitor, 16, which has been charged by way of resistor 17, into the energizing coil, 18, of the electromagnetic contactor for closing its contacts, 19.

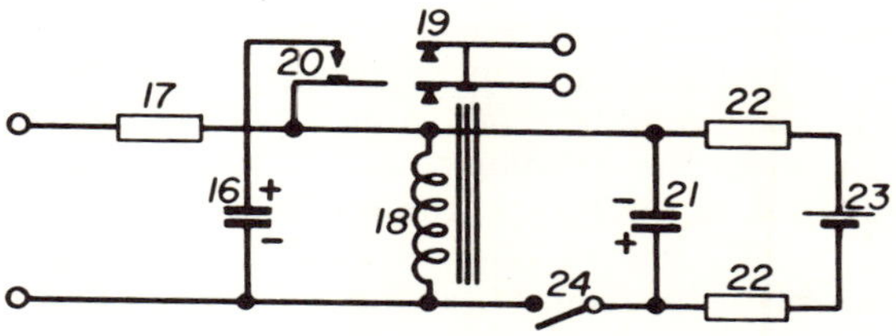

FIG. B8-5. Auxiliary circuit for phase-true switching-off through compensation of residual magnetism in the contactor by means of a dosed capacitor discharge.

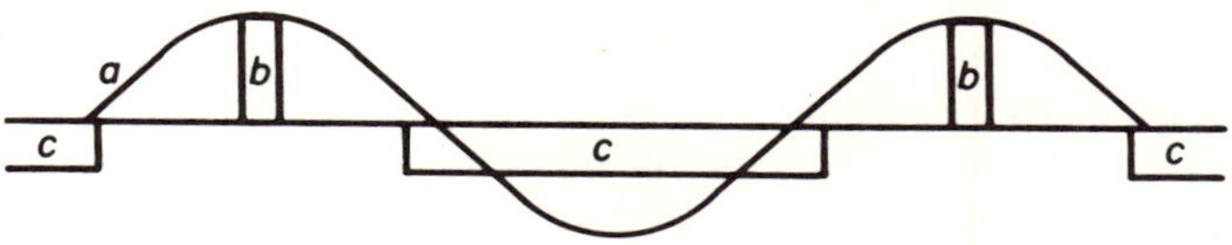

FIG. B8-6. Adjusting scheme in contact making for the intervals *b* and *c* relative to the ac voltage *a*.

The time lag between contact closing at 15 and contact closing at 19 is again, as illustrated in Figs. B8-1 and B8-2, constant. When, for instance, this time lag is 10 millisec and the time lag between engagement of 2–3 and response of relay 14–15 is 2 millisec, that is, a total of 12 millisec, it is necessary to advance initial contact making *b* in Fig. B8-6 by this amount through proper adjustment of the phase shifter 5–6. At 50 cps line frequency the duration of a full wave is 20 millisec and corresponds to 360 degrees. After every 10 millisec, corresponding to 180 degrees, there exist similar absolute voltages. Therefore, it is necessary to effect an advance of 2 millisec corresponding to a 36 degree phase angle for the actuation of contact closing 2–3, in order that phase angle–correct switching on will take place at the proper time. For switching at the instantaneous maximum voltage value, another advance of 90 degrees would be necessary. It is, of course, possible to obtain any other time advance by other settings of the phase shift member 6–5, Fig. B8-4, so that contact making of the work contacts, 19, occurs positively at the predetermined and preadjusted instant, after the release contact, 8, is actuated. In order to prevent the relays 11–12 and 14–15 from dropping out it is advisable to provide each one of these relays with a holding contact which, after closing, bridges contacts 2–3 and 8 in relay 11–12, and contact 12 in relay 14–15.

In order to separate contact 19 at a predetermined phase position and thereby prevent a scattering zone, *d* (Fig. B8-2), it becomes necessary to take steps toward quenching the magnetic field of the coil. Corresponding steps are depicted in Fig. B8-5 where the switching magnet, 18, is kept closed by direct current supplied through resistor 17, once the capacitor 16 has been switched off by an auxiliary contact 20, after closing of switch 18–19. It is obvious that in certain cases a dc-fed coil, 18, with series resistors, 17, can only be applied when phase-true switching is not required. For phase-true contact opening of contact 19, a capacitor, 21, which is fed by way of resistors, 22, from a dc voltage source, 23, is connected by way of contact 24 to the magnet coil, 18. The polarity symbols in Fig. B8-5 indicate the condition with switch 24 open. The holding-circuit direct current, flowing through resistor 17 and coil 18, induces in the holding magnet a certain magnetic flux. By selection of a proper value of capacitance in capacitor 21, which can be experimentally determined, at the instant of contact making in 24, there is a momentary overpowering of the discharge current of capacitor 21, the current in circuit 17–18. This is sufficient to temporarily demagnetize the core and reverse polarity. During this demagnetized interval, the relay armature return spring will open the contacts 19. The returning current in circuit 17–18 is, after discharge of capacitor 21, able to remagnetize the core. This magnetization is insufficient to attract the armature and reclose contact 19. The instant of demagnetization is determined by the characteristics of capacitor 21 and coil 18 and is constant. Thus the time lag between action of contacts 24 and the instant of demagnetization of coil 18 is also constant. Return of the armature and the opening of contact 19 also takes a constant time interval. This depends on the spring force and inertia of the armature with respect to the contacts. By means of this instantaneous quenching of the magnetic field, the scattering interval *d* (Fig. B8-2), which cannot otherwise be reduced, can be avoided. This is because residual magnetism will die away in a negligibly short time and only the constant armature return spring acts at this instant. The slowly returning field of the circuit 17–18 (Fig. B8-5) finds the relay armature at considerable distance from the core so that it cannot cause disturbance. When the circuit (Fig. B8-4) is used in such a way that contact 15 takes the function of contact 24 (Fig. B8-5), it is possible to obtain a phase angle–correct switching-off with the simplicity of Fig.

B8-4. For phase angle–correct switching-on and switching-off two circuits are necessary (Fig. B8-4). One, by means of contact making at 15, switches on phase-true. The other (Fig. B8-5), by contact making at 24, switches off phase-true. Both sets of contacts make and break, in accordance with the relay arrangement (Fig. B8-4, items 1–14, of which two are used), at the phase angles prescribed by the phase shifter, 5–6. Several construction components, e.g., the rectifier or the voltage stabilization for the periodic contact maker, can be used for both circuit arrangements in common.

Practical installations of this kind proved that switching on of 50 kva contactors with scattering times of about $\pm$ 5 degree phase angle are possible. Switching off in accordance with Fig. B8-5 and with an optimum value capacitor, 21, can be effected with an average scattering of $\pm$ 7.5 degree phase angle. These results have been achieved with commercial electrical circuit components. The accuracy of the control portion (items 1 to 15, Fig. B8-4), without the relatively heavy contactor system, 18–19, is considerably higher. The average is a $\pm$ 1 degree phase angle.

As periodic contact makers, polarized telegraph relays have proved reliable. It is often possible to dispense with a separate compensation current winding by setting the contact tongue asymmetrically with respect to the stationary contacts. Adjustment of the contact maker requires some skill and it is advisable to employ an oscilloscope in connection with an adjustment pattern. Since all contacts of the periodic contact maker carry current only during the switching on and switching off pulses, their rate of contact burnoff is practically unmeasurable. After more than 100,000 switching operations, there was not an observable timing change in contacting action.

The condensors feeding the relay and contactor coils vary, at constant charging voltage, in their capacitance values. They must, however, contain somewhat more energy than required for relay armature actuation. The capacitors for quenching the residual magnetism in the relay core have to be more critically adapted. To that end it is necessary to determine the inductance L at the holding current I of the relay coil by ac measurement $L = U/\omega I$.

The ohmic resistance can be neglected. In this case the residual magnetism-quenching capacitor must hold a capacitive energy equal to the inductive energy of the relay coil:

$$\frac{1}{2} CU^2 = \frac{1}{2} LI^2, \qquad C = \frac{LI^2}{U^2}$$

9. Quenching Spark Gaps

a. *General Applications of Quenching Spark Gaps*

The quenching spark gap appears to be the earliest form of a switching means for cyclic discharge of capacitors [197–201]. Such gaps were first created to replace the two-electrode explosive sparks in wireless communication and for supplying to the antenna in rapid succession energy impulses that dampen out in an oscillatory circuit. Through introduction of the so-called quenching spark gap it was possible in these early days to increase

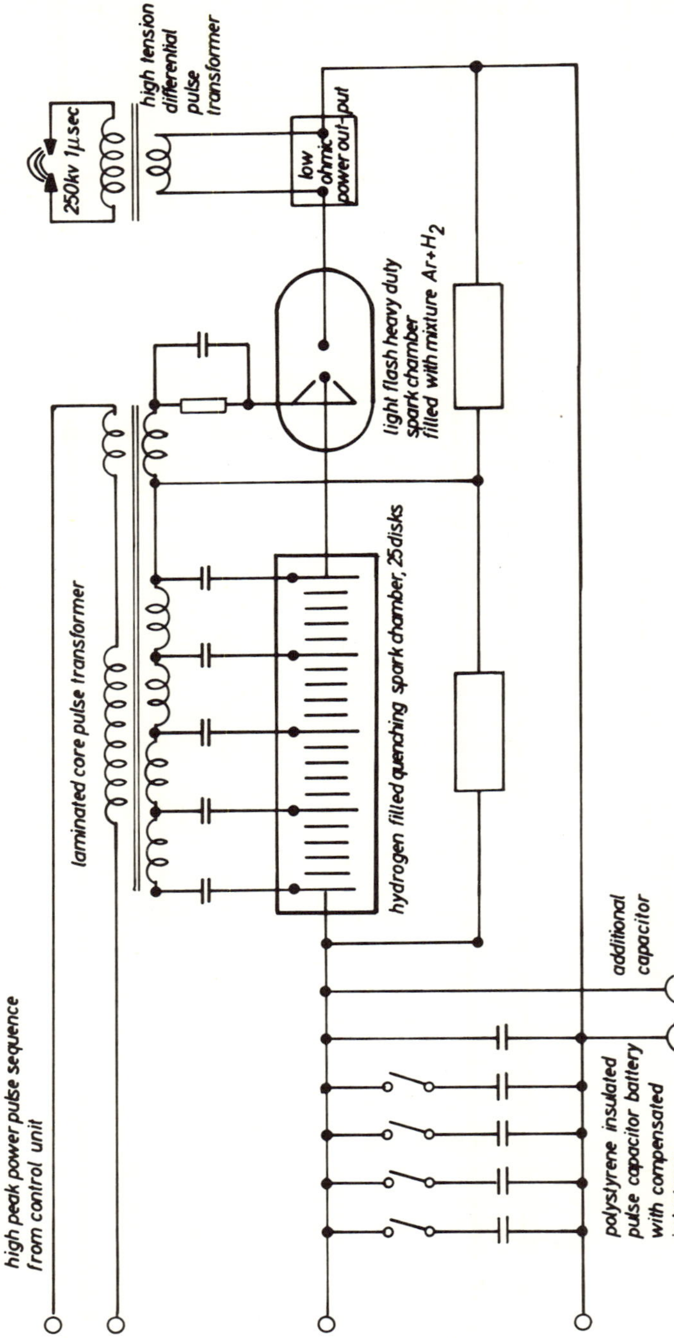

FIG. B9a-1. Schematic diagram of a circuit for impulse frequencies from 0 to 50,000/sec with quenching spark gap controlled by impulse ignition.

Example: impulse transformer giving 250 kv secondary voltage for producing spark series which travel with a stream of air and serve for photography of flow patterns.

the transmitter energy up to a few kilowatts, and thus to increase the radiating range correspondingly. The quenching spark gap is based on the phenomenon that in a spark gap of extraordinarily small sparking distance the small volumes of plasma lying between the electrodes reach a temperature favorable to the maintenance of good conductivity very soon after the beginning of discharge. They then offer a good thermal equilibrium in heating and cooling. As soon as the ionization energy drops, through discharge of the capacitor, the cooling effect of ion-striking electrons and dissociated atoms outweighs the energy supplied by the circuit and the discharge ceases. For the positive control of high switching voltages there are arranged, insulated from one another, a great number of flat quenching spark discs at spacings of about 50 to 200 micrometers. Figure B9a-1 shows the schematic diagram of a quenching spark gap used in a light flash producer for photographic purposes.

The ignition and quenching voltage of each disc can be assumed to lie between 600 and 300 volts. With 25 discs, the self-breakdown voltage would be 12 kv and quenching would take place at 6 kv. The discs are supported by mica interlays.

With particularly refined constructions of this kind having plane, parallel tungsten discs and very small capacitors on the order of 200 pf, and at operating voltages of about 10 kv, it was possible to trigger powerful output loads between 200 and 500 watts at a frequency of about 30 megacycles. Analogous, high-frequency generators were used in ultrahigh-frequency therapy until the advent of high-powered and long-lived transmitter tubes.

Figure B9a-2 is a schematic diagram of a commercial ultrahigh-frequency therapy apparatus (Sanitas A.G., Berlin) which operates with the type of quenching spark gaps just described. Systems of this sort, somewhat modified

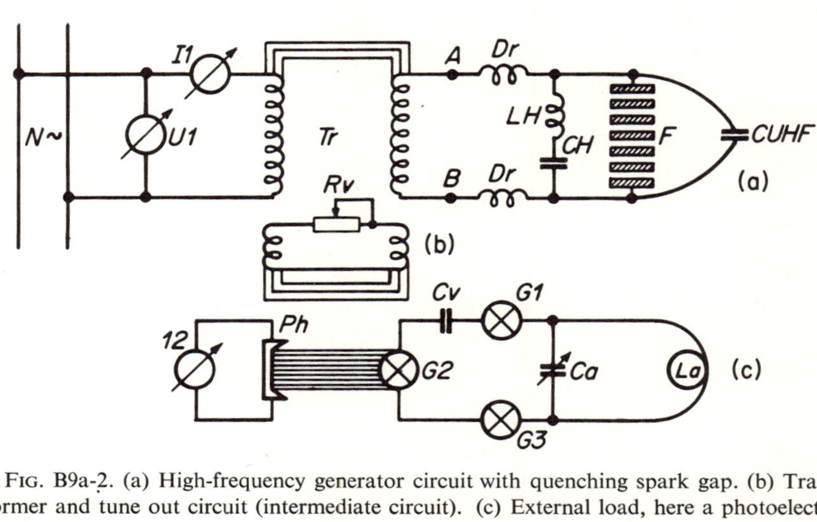

FIG. B9a-2. (a) High-frequency generator circuit with quenching spark gap. (b) Transformer and tune out circuit (intermediate circuit). (c) External load, here a photoelectric wattmeter.

for lower frequencies of about 1 megacycle are still in use today for electro-surgical instruments. They make use of tungsten quenching spark gaps for excitation of the oscillatory circuit, owing to recognition of their insensitivity to line voltage fluctuations and the possibility of disregarding heater and pre-heating time.

With the systems referred to above, there have been obtained low energy discharge switching operations up to one million per second. By making the quenching spark gap of copper and providing for continuously clean metallic surfaces by operation in an oxide-reducing hydrogen atmosphere, it is possible to increase the output rate. Copper, as an extraordinarily good heat conductor, provides for quick quenching. Hydrogen as fill-gas lends high ion mobility which is favourable to higher spark sequence frequencies. With such a copper quenching spark gap having 25 discs, it is possible to reach 2000 watts continuously at the load, be it a resistance or an oscillatory circuit, with such capacitor discharges following one another in rapid succession. A simple version of this operation is obtained by immersing the whole spark gap in spirit alcohol gas instead of hydrogen. A gas-tight metal container as a chamber for the quenching spark gap has been successful. Before the container is closed, a thin layer of boiling spirit alcohol is poured on the container bottom. Evaporating alcohol displaces the air in the container. The carbon atoms have practically no disturbing influence in the ionized spirit alcohol gas. Only the easily movable hydrogen ions effect the conduction mechanism. With perfect container sealing this mode of operation requires absolutely no maintenance. In the spirit alcohol reservoir there is sufficient operating gas for extended operating times. During operating intermissions spirit alcohol condenses again without overpressure from the outside.

The author discovered, while developing high-speed filming devices, that such quenching spark gaps can be made controllable. It may appear contradictory to try to control the mechanism of a quenching spark gap whose main objective it is to serve for rapid extinction of sparkovers at the electrodes. However, since these spark gaps are in themselves a rapidly working switching means for spark sequences, the thought occurred that the quenching spark gap might lend itself to precise ignition. It was found that this was possible by feeding powerful ignition pulses to a larger number of electrodes —for 20 copper electrodes, about 3–4 ignition electrodes. The pulse energy should be about 200 kw at a pulse duration of about 0.2 μsec. A separate supply is furnished to each one of the 3 or 4 ignition electrodes through individual impulse transformer windings. Such energy injection is sufficient to release with certainty a subsequent main discharge without fear of accidental breakthrough in the intervals between ignitions in the discharge system.

Within the permissible temperature limits of the whole system, which corresponds to a continuous load of 100 watts maximum, 10^4 to 10^5 and with good efficiency even 10^6 sparks in succession may be utilized with a scattering accuracy of within about 0.1 to 0.2 μsec. Control of the discharges is effected by a powerful hydrogen thyratron which in turn is controlled by a multivibrator. It is also possible to obtain time modulation, e.g., impulse spacing for program transmission of speech, etc. Such a transmission system modulated by a controlled hydrogen quenching spark gap results, with frequency modulation at the receiving end, from the standpoint of distortion factor in no more distortion than can be found in wire transmission. This benefit results from the fact that all momentary disturbances are limited to single impulses. These at the high basis frequency of, for instance, 30,000/sec lie below other noise disturbance values and carry little weight.

The controlled hydrogen quenching spark gap is an extremely reliable switching means for capacitor discharges in the range of 6000 to 12,000 volts charging voltage and peak currents between 500 and 5000 amp at impulse sequence frequencies in the range of from 1 up to about 300,000/sec [487]. Depending on the average output, continuous operation is often possible. Without output limitation, within the thermal storage capacity of the structure, impulse discharge series of from a few megawatts, to about 50 megawatts impulse energy for a single impulse, with impulse durations of less than 1 μsec and an average output at the load up to 50 kw, are possible. Load resistance values down to 0.5 ohm can be provided with energy pulses at about 50% efficiency. Efficiencies in the order of 70% at about 1–5 ohms load resistance can be assured. Even by best adaptation one can hardly get less than 10% losses within the quenching spark gap. Because of the difficulties encountered in cooling, the output limits lie at a few hundred watts. In continuous operation it is possible to feed only a few thousand watts to the load. The great metal mass, which, e.g., can be 1 kg for the total quantity of the electrode system, serves as a thermal reservoir. Thus it is possible to meet, by proper construction, the thermal requirements for a given total energy of an impulse discharge series. One kilogram of copper has heat capacity of 90 calories per degree centigrade and, at a permissible temperature rise of 150°C, the system can absorb $90 \times 150 = 13{,}500$ calories, or, in electrical units, about 50,000 watt-sec. Assuming an efficiency of 70%, it is possible to convert in one discharge series a total energy of about 120,000 watt-sec independent of the spark sequence frequency. Considering spark frequency, there would be possible (for instance, with 10,000 discharges of 1 watt-sec each), an operating duration of 12 sec. At longer operation times and 200°C temperature rise for the system, some radiation cooling through the glass or quartz cylinder becomes effective; this can be figured as being 250 watt-sec/sec.

b. *Quenching Spark Gaps as Trigger Elements in High-Speed Cinematography*

The generation of energy pulses of high voltage and high currents is necessary where point-shaped spark light sources flash at high repetition rates, e.g., in illumination of reflector systems in interference and schlieren photography. It is also required for energizing the primary of impulse transformers at high pulsing rates and predetermining energy per pulse, in order to generate at the secondary either high-frequency, high-voltage sparks or x-ray flashes. Also, for triggering Kerr cells at high repetition rates there is the possibility of energy feeding by means of such impulse transformers.

Research into the parameters relating to pulse energy, pulsing rate, number of spark gaps, design of spark gaps, and suitable filling gases was the subject of paper [196]. These paramaters defy calculation since the influences present in quenching spark gaps are so complex and varied that even analogous computation methods will not suffice for these purposes. Some of these parameters are: different sparkover distances in relation to the plasma temperature, cooling coefficient by the parallel electrodes, deionization time in a low-pressure discharge zone (the thickness of the discharge zone layer mostly varies between 0.5 and 0.1 mm), combined effect of a multiple of such spark gaps in series and effect of the consumer with its impedance on the backfeed of the system.

For these reasons an empirical method was adopted. To begin with, air spark gaps with quenching spark gaps were examined with electrode distances of 0.15 and 0.3 mm and between 1 and 4 quenching spark gaps.

Figure B9b-1 shows the experimental circuitry and the results obtained. Two capacitors of 0.03 and 0.066 μf were used as discharge capacitors. They were discharged over the quenching spark gap into a resistor of 0.2 ohms. The size of this resistor roughly corresponds to the smallest resistance of high-pressure rare gas sparks. The size of the capacitor corresponds to that of ultrarapid flasher units.

An attempt was made to determine the smallest charging resistor at which the spark changed into a continuous arc, since this would clearly set the limit of the system. The resistance together with the capacitance determines the flashing rate.

Figure B9b-1 shows that the most effective range of the quenching spark gaps lies between 1 and 4 elements since above that number the curves flatten out. One can further observe that the smaller distance of 0.15 mm with four spark gaps and 0.066 μf permits a charging resistor of 570 ohms compared to 1800 ohms at 0.3 mm. As was to be assumed the quenching effect increases with decreasing distance. If, however, one compares the permissible resistance with the capacitances, then one finds that with identical quenching spark gaps the resistances show a behavior directly inverse to that of the capacitances. Since the *RC* product gives the internal oscillation of the spark gap circuit, the quenching spark gap as such and not the size of the capacitance determines the frequency limit. This is explained by the theory that it is immaterial to deionization what number of spark channels may have been in existence for obtaining sufficient conductivity since their deionization proceeds at the same rate. It thus follows that the quenching spark gap as such determines the *RC* product.

Finally the curves of Fig. B9b-1 show the relating time constants. With four spark gaps, 40 μsec has been achieved. This would correspond to a possible sparking rate of 25,000.

Figure B9b-2 shows an experimental circuit for investigation into the operation of quenching spark gaps containing a larger number of discs (in this case 25). The results show that the largest number of spark gaps, which theoretically would permit a very high flashing rate, becomes impractical when the internal resistance is too high.

electrode width: $\delta = 150\,\mu m$

$U_{st} \approx 7.0$ kv				
	$C_L = 0.030\,\mu f$		$C_L = 0.066\,\mu f$	
n	R_L min	T	R_L min	T
1	7,5 kΩ	230 μsec	3,4 kΩ	220 μsec
2	2,2 kΩ	66 μsec	0,92 kΩ	61 μsec
3	1,4 kΩ	42 μsec	0,65 kΩ	43 μsec
4	1,2 kΩ	36 μsec	0,57 kΩ	38 μsec

electrode width: $\delta = 300\,\mu m$

$U_{st} \approx 7.0$ kv				
	$C_L = 0.030\,\mu f$		$C_L = 0.066\,\mu f$	
n	R_L min	T	R_L min	T
1	9,5 kΩ	290 μsec	5,3 kΩ	350 μsec
2	4,7 kΩ	140 μsec	3,0 kΩ	200 μsec
3	3,7 kΩ	110 μsec	2,4 kΩ	160 μsec
4	3,3 kΩ	99 μsec	1,8 kΩ	120 μsec

measuring diagram

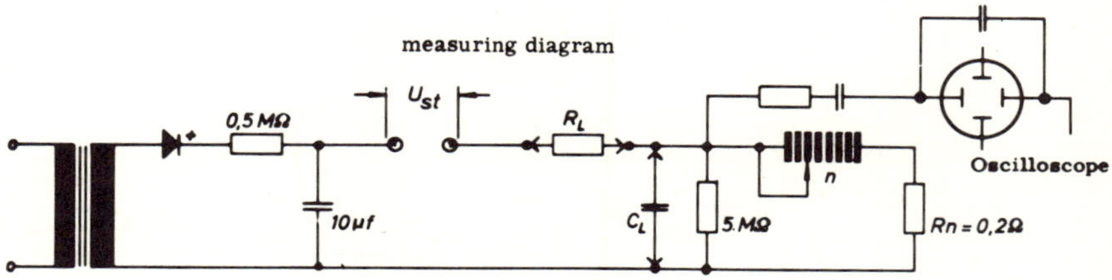

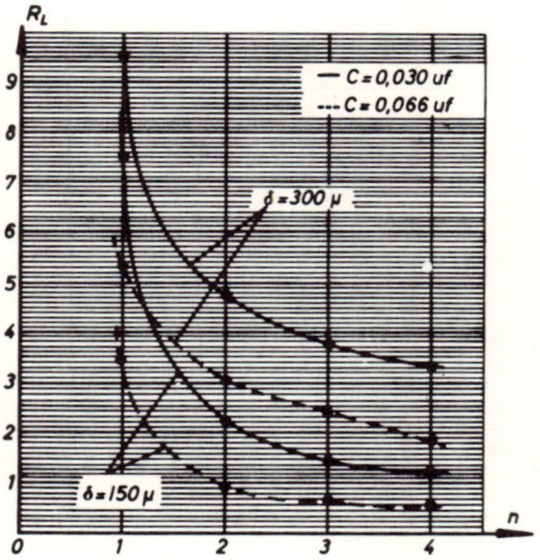

smallest charging-resistor R_L (kΩ) vs. number of quenching spark gaps n.

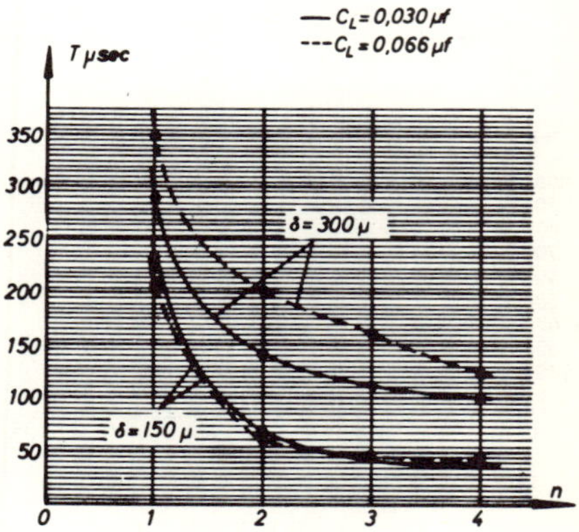

smallest intervals (μs) between two pulses vs. number of quenching spark gaps n.

in series with the quenching spark gaps is a resistor R_n to simulate the light emitting spark.

FIG. B9b-1. Influence of capacitance and charging resistor on the quenching rate of a number of spark gaps.

For measuring, resistors of various sizes corresponding to the spark resistances were wired to a calorimeter. The quenching spark gap was wired into another one. Here, the sparking rate as such was of no interest. The measuring results and the curves obtained are as follows:

With increasing consumer resistance, effectiveness in the beginning shows a linear increase up to approximately 1.5 ohm. It then remains relatively

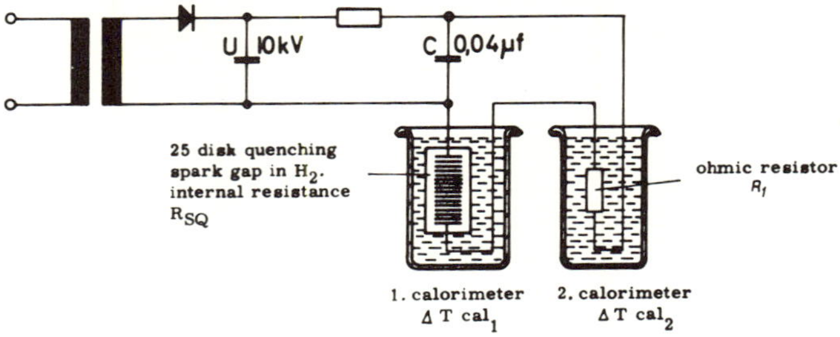

MEASURING CIRCUIT

R_1 (Ω)	$\Delta Tcal_1$ (°C)	$\Delta Tcal_2$ (°C)	Q = ENERGY TOTAL (kws)	$\eta\,\% = \frac{\Delta Tcal_1}{\Delta Tcal_1 + \Delta cal_2} \cdot 100$	$R_{SQ} = \frac{Tcal_2}{Tcal_1} R_1(\Omega)$	t = R·C μsec
15	23	5	31	82.0 %	3	0,72
5	50	9	66	85.0 %	0,9	0.24
1.5	27	10	41	73.0 %	0.55	0,08
0.5	25	31	62	44.7 %	0.6	0,04

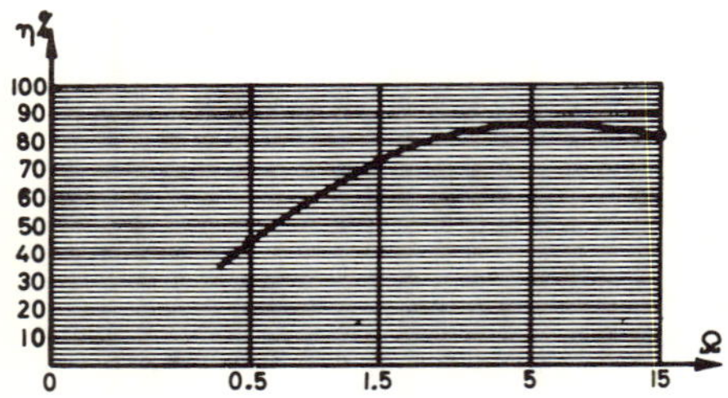

Efficiehcy under load, represented by an ohmic resistor R_1 (Ω) in a discharge circuit C = 0.04 uF, U = 10 kV in series with a 25 dish H_2 filled quenching spark gap , disk spacing δ = 0.15 mm. For comparison: a spark in rare gas atmosphere would in this discharge circuit have in average resistance between 0.6 and 2 Ω

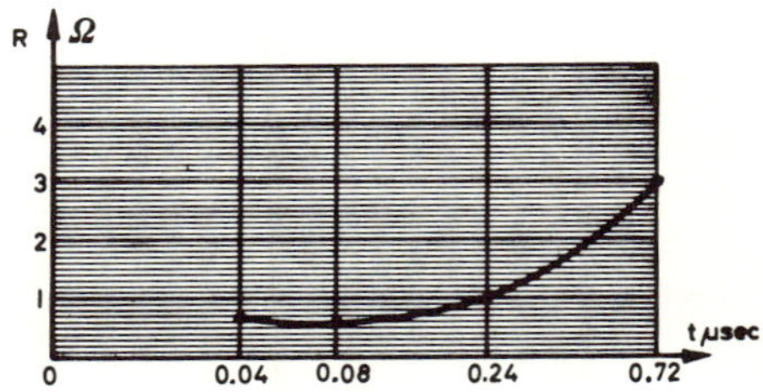

Internal resistance of a quenching spark gap (25 disks, γ = 0.15 mm, H_2) in relation to the pulse duration t. (t = R · C corresponding to the discharge circuit's RC-constant)

FIG. B9b-2. Efficiency and internal resistance of a 25-disc quenching spark gap as a function of spark resistance and the time constant in the discharge circuit.

constant at 80%. There is a tendency to fall with resistors larger than approximately 10 ohms. At such large resistances the discharges require too much time. The thermal ion losses during the discharge time as such make themselves felt and the internal consumption of the quenching spark gap as such increases. The logical conclusion is that quenching spark gaps should only be used in those discharge circuits where the resistance of the spark is somewhere between 1 and 10 ohms.

The curve in the right-hand table of Fig. B9b-2 shows the internal resistance of the quenching spark gap as a function of the time constant. The resistance shows a steep increase as soon as the discharge burns for more than approximately 1/4 μsec. The reason for this is, as already mentioned, that the deionization starts during the discharge. The free path of H_2 molecules at 1 atm of pressure is much shorter. A hydrogen molecule would travel approximately 1500 meters/sec, i.e. 1.5 mm in 1 μsec. It is a striking analogy that the quenching effect sets in after exactly this time, as can be seen from the curve.

In the case of the Strobokin flash lamp for which these preliminary measurements were performed, the current has risen to its maximum at 0.1 μsec, i.e., at roughly 1 ohm spark resistance. At 0.1 μsec operation is exactly at the limit of maximum efficiency.

A commercial equipment for producing high-power pulses with a high repetition rate is the so-called Strobokin [798]. The basic Strobokin equipment consists of a high-voltage power supply, one control unit for frequency setting, and one pulse generator with the "Quenchotron," a new tube manufactured especially for this equipment. The "Quenchotron" is a multielectrode hydrogen gas discharge tube for 10 kamp peak current, 12 kv peak inverse voltage, and a repetition rate of more than 300,000/sec. Its inductance is as low as 10 nh, therefore, it allows feeding of short 40-megawatt impulses into low ohmic electrical consumers, e.c. spark gaps, pulse transformers, and other transducers. Its metal mass is sufficiently large so that a 100-kjoule burst of discharge raises the temperature only about 100°C.

C. Line conductors

1. The Influence of Impedance

During every instant that a capacitor is discharging, in single or cyclic discharges, into a load by switching means, as discussed in Chapter B, the electrical discharge energy must pass the connecting conductors. In general, too little attention is paid to these connecting lines in discharge circuit design. An essential property of a line is its characteristic impedance Z.

$$Z = \sqrt{\text{Longitudinal resistance} \times \text{transverse (cross) resistance}}$$

or, as another equation:

$$Z = \sqrt{\frac{R + i\omega L}{G + i\omega C}}$$

where R = resistance of unit length of line, ω = angular frequency, $2\pi f$, L = inductance of that length, C = capacitance of that length, G = transverse (cross) resistance of the unit length between conductors and $i = \sqrt{-1}$

Impulses, according to Fourier's analysis, also contain frequencies below 20 kilocycles, but, since these have negligibly small amplitudes, it is permissible for all practical purposes to figure in all impulse discharge circuits with pure impedance. Since for high frequencies

$$\omega \times L \gg R \text{ and } \omega \times C \gg G$$

the impedance for high frequencies is

$$Z = \sqrt{\frac{L}{C}}$$

A line can be designed for either two different fundamental objectives: It can give either the highest efficiency or the highest output, i.e., a maximum of power at the load. Highest efficiency of the line will be necessary when matching transformers or the like convert the output energy pulse. Under such conditions and with very short connections the impedance of the line must be considerably lower than the load resistance; otherwise line impedance presents a bottleneck. Then even a powerful impulse cannot reach the load because it is unable to pass the line impedance with sufficient current magnitude. If the impedance of the line is considerably lower than the load resistance the bottleneck problem is absent. However, then the problem of partial reflection arises: The energy at the end of the line is reflected in a superimposed oscillation that may result in stationary waves on the line. In considering the energy alone, however, the line impedance is not very important. This is different in the impulse technique, where the impedance matching problem is regularly encountered when a maximum of power is desired at the output end of the line at the load. Here, in contrast to electric power engineering, the objective is to strive not for highest efficiency but for optimum power at the load, which in the ideal case of lossless conductors means a maximum efficiency of 50%.

$$R_i = Z = R_a$$

Thus, matching is obtained when the internal transmitter resistance R_i is equal to the impedance Z of the line (but the sign of the connected resistance phase angle is reversed). Moreover, the load resistance R_a at the end of the line must be equal to Z. The most favorable condition can thus be obtained when the internal resistance of the source is equal to the impedance of the connecting line, and this in turn is equal to the load resistance. Whenever the load resistance differs from Z, stationary waves occur on the line. The result is unequal and higher average current density as well as increased damping due to radiation losses in the line. Moreover, the arriving energy is reduced by mismatched conditions at the load. However, all such discrepancies play a considerable part only at pronounced mismatches. In the following sections a few characteristic line systems for discharge circuits will be discussed.

2. The Overhead Line for High and Highest Voltages with Large Conductor Radius

Overhead lines* generally have impedances in the range from 300 to 600 ohms against earth. Whenever the load has a high impedance—for instance, a transformer with sufficiently large matched primary—the transmission of energy impulses through such line systems presents no problem. It is only necessary that the radius of the conductors be large enough to avoid corona losses and eventual wild flashovers. Overhead lines of this type will normally be installed in laboratories for discharge stands of more than 50 kv capacitor voltage. With voltages of this order it is difficult to conduct the energy by other leads, e.g. concentric cables. If, however, at the end of such an overhead line, i.e. at the load, a sparkover is forced to occur—for instance, by connecting a standard wave of 1 μsec rise time and 50 μsec decay time—this sparkover is by no means an impedance like load. On the contrary, such a sparkover has only a few ohms spark resistance. At its occurrence, stationary waves develop on the conductor portion between capacitor battery and load. After actuation of the switching mechanism a direct connection of the conductor with the feeding capacitor battery exists. Different superimposed harmonic waves of various frequencies in the form of wave packages begin to travel to

* Shorter lines in closed rooms between discharge sources and loads which have been constructed like overhead lines are included in this term.

and fro between capacitor battery and load. Therefore, oscillograms of flashovers at the load end taken at different laboratories always show different wave patterns at flashover. This peculiarity can be explained by the fact that during the few microseconds between beginning of rupture and decay of the high frequency incited on the line, oscillations of great magnitude have occurred. On a line of 30 meters' length one can observe a main frequency of about 5 megacycles. This presents a half-wave within 10^{-7} sec, The actual discharge mechanism can be strongly affected by maxima and minima of this modulation. For quantitative measurements of flashover phenomena, it is therefore advisable to connect the capacitor, including its switching means as described in Chapter B, by extraordinarily short conductors to the load or test sample. The waves developing on the line then have a very high frequency in the range of 10 to 100 megacycles and do not falsify the results of the test. This is because they die out so quickly by natural damping that they cannot interfere with the discharge mechanism of a longer test line.

3. The Coaxial Cable

Coaxial cables in general are cables that have an outer conductor in the form of a tubelike envelope that may be grounded to the body, and a concentric inner conductor accommodated in the outer conductor and carried by a highly insulating substance that preferably has also a low high-frequency loss angle. The inner conductor is utilized to carry the energy impulse insulated from the body. Coaxial cables are manufactured for operating voltages up to 50 kv ac and for breakthrough voltages on the order of 250 kv. Current values of a few amperes can be passed through the concentric inner conductor in continuous operation.

Common coaxial cables of high-frequency antenna application can be used without alteration for impulse discharge circuits. Impedances of these cables lie in the range of 30 to 100 ohms. The statements made with respect to overhead lines also hold true here: When the impedance of the cable is much lower than the load resistance, the cable presents no bottleneck and leads the impulse unhampered to the load. It must, however, be expected that, owing to mismatch between cable impedance and load resistance, stationary waves will form on the cable and, in the case of sparkovers for oscillographic measurements, or in case of other time-dependent resistances, lead to a misrepresentation at the beginning of the discharge phenomenon. But this superimposing of stationary waves lasts only for the very first moments of the discharges. To avoid such disturbances it is advisable to make the cable between capacitor battery and load not longer than about 5 meters. Since the wave expansion velocity in cables is only one-half of light velocity, 5-meter cables for wave expansion are equal to 10-meter overhead lines. The basic wave forming on such a piece of cable has therefore a frequency of about 15 megacycles. With rapidly progressing spark oscillograms in the voltage range up to about 30 kv such frequencies can become disturbing. However, within 1 to 3×10^{-7} sec they have died out so much that they can falsify the main discharge trace very little. In all cases of spark discharges as external loads, one must recognize the fact that exact matching of capacitor battery and spark load is impossible, since the spark resistance varies rapidly with time, and the physical structure of the spark prohibits any kind of matching. Only its average value, based on time and energy content, gives access to matching.

A special case is dielectric drying of cables by stationary waves. This case is of interest insofar as here the cable itself is purposely the energy consumer. The paper used for cable insulation is generally very hygroscopic. In dry air conditions it has a water content that is dependent on the relative humidity of the air during manufacture of the cable. This can be up to 10% of the paper weight. Absorbed water exists in the paper not in its original form but in an intermicellar distribution. The water is bound to the boundary faces of the cellulose and fills the capillary spaces between the fibrils. This fact can also be recognized by the reduction in vapor pressure compared to that of free water; this is considerable, particularly with small water contents, and makes complete drying very difficult. In the manufacture of high-grade cables with low dielectric losses, it is essential to remove any moisture from the cable paper before impregnation with insulating oil. In general, the required drying of the paper insulation for that purpose is effected by prolonged heating of the coiled cable conductors in evacuated and heated containers or by heating of the cable conductors by current heat, Drying, however, is much more effective when the required heat is produced directly in the dielectric. For that purpose the arrangement proposed by Brinkmann and Marx and depicted in Fig. C3-1 serves best. The cable to be dried is placed where the oscillatory circuit for inductive heating otherwise would be connected [16].

In such an arrangement for cable drying, the blast spark gap sends an energy impulse into the cable at every breakdown. In the interval between breakdowns this develops into oscillations with reflections at the open end of the cable. Owing to rapid damping of these oscillations the band width of frequencies is considerable. This fact prevents the formation of selective frequency-dependent patterns at fixed points in the cable that could result in dangerous overheating at wave node or crest points. On the contrary, owing to the broad frequency band, a uniform dielectric heating process develops within the connected cable to be dried.

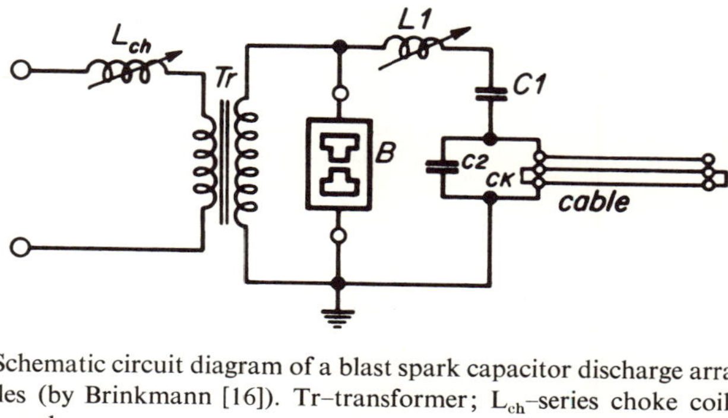

FIG. C3-1. Schematic circuit diagram of a blast spark capacitor discharge arrangement for drying of cables (by Brinkmann [16]). Tr–transformer; L_{ch}–series choke coil; Ll–inductance; B–blast spark gap.

The frequencies of the oscillations utilized for this purpose lie in the range of 100 to 600 kilocycles. The high-voltage energy is fed into a chamber through an especially designed lead-in bushing because it is still necessary to withdraw the dampness of the cable under vacuum. The strongly damped trains of oscillations enter the cable in continual succession like traveling waves at a certain propagation velocity. At the open end of the cable the wave trains are reflected and combine with the forward-traveling waves.

When a reflected wave reaches the beginning of the cable, it has shrunk so much that an uneven heating is practically excluded. It is important to assure continuously uniform operation of the blast spark gap, since sparkovers at overpotentials could lead to excessive strain on the cable dielectric. It is therefore advisable to employ quick-response spark gaps, as described, in a previous chapter, for this purpose.

A special design of the coaxial cable is the so-called coaxial return sheet. Its electrical characteristics are nearly the same as that of sandwich lines (see Chapter C, Section 5).

4. Energy Considerations

Whenever we are considering connecting lines with regard to energy, the line inductance is also of fundamental interest. For the sake of simplicity it may be assumed that the load as well as the line has negligibly low resistance. If that is the case the capacitor energy of $\frac{1}{2}C \times U^2$ is converted into the magnetic energy $\frac{1}{2}L \times I^2$ at actuation of the switching means.

The maximum current

$$I = \sqrt{\frac{C \times U^2}{L}}$$

can be calculated as illustrated in the following example:

Consider a capacitor of 10 μf capacitance at 30 kv charging voltage and a concentric or coaxial cable of 10 meter length having an inductance of 0.5 μh/meter, and thus a total of 5 μh. The current becomes

$$I = 4.25 \times 10^4 = 42{,}500 \text{ amp}$$

Since conductors of opposite polarity repulse each other, the outer cylindrical conductor tends to depart from the inner center conductor. Such forces can lead to rupture of the cable along its periphery. Moreover, the mechanical destruction of the insulation taking place can lead to a spontaneous electrical breakdown of the insulation at this point which reduces the inductance even more. All this may lead to further increased forces. In the case of a dead short circuit at the end of the cable one must assume, as given in the calculation example, that the capacitor energy is

$$\tfrac{1}{2} \times CU^2 = 4.5 \text{ kilojoules}$$

of which the mechanical equivalent is about 450 meter-kg. If this force is exerted by dropping a weight of 450 kg from a height of 1 meter on the cable, there is little doubt of its destructive effects. It is of benefit regularly to visualize in this way the forces acting during capacitor energy discharges, in order to have a sensible impression of possible destruction that could occur when an unpredicted short circuit through breakdown or flashover releases mechanical energy.

5. Sandwich Lines

The sandwich line is a special configuration of the concentric or coaxial cable in which the impedance has been considerably reduced. As we know, the impedance is

$$Z = \sqrt{\frac{L}{C}}$$

and it decreases with increasing C and decreasing L. It is therefore important to make the surface of the forward and return conductor large and the spacing between conductors rather small. The two most common constructions are attempts to meet these requirements. One construction consists of two or three metal strips which are pressed together with interlays of insulating material and are held together by strong outer insulation plates secured by bolts. The bolts serve to resist the repellent forces effective under current flow. This form of sandwich line can be readily improvised with shop facilities.

More suitable for larger-scale production is a form with two concentrically superimposed tubes insulated by an interposed dielectric. The tubes have sufficient cross-sectional area to be capable of absorbing radially acting, uniformly distributed forces.

Let us assume that two copper strips of 1 mm thickness and 100 mm width with an interlay of 1 mm dielectric hardpaper as their spacing are tightly pressed together. We obtain then, with a dielectric constant of 2.5, a capacitance of about 2.5 pf per sq cm. With a 100 cm band of conductor, giving 1000 sq cm of one-sided surface, the total is 2500 pf.

With this in mind, we see that two such grounded strips of 100 cm length with an interposed strip serving as a high-voltage conductor give 5000 pf capacitance per meter. The measured inductance is 0.01 μh/meter.

The impedance becomes thus

$$Z = \sqrt{\frac{10^{-8}}{5 \times 10^{-9}}} = 1.4 \text{ ohm}$$

From this it is apparent that it is possible to realize impedances on the order of a few ohms and even lower by corresponding parallel connection of such sandwich-line systems. Since spark resistances in general have values in the range of 0.6 to 2 ohms (average over the whole duration of the spark discharge), it is possible to operate, without loss in efficiency, a spark gap in which the spark itself is the useful load, through sandwich lines at a few meters distance from the feeding capacitor battery. There is no bottleneck and practically perfect matching of impedances. However, such special sandwich lines are rather costly. They are usually tailored for a particular application and care must be taken to ensure that by embedding in insulation material sufficient pressure is applied over the whole conductor length so that the strips lie actually in close spacing to each other. In the case of sparks with a resistance of about 1 ohm as external load, the peak currents can become extremely large. For instance, at a capacitor voltage of 10,000 volts with current on the order of 10 kamp, the operative dynamic repulsing forces between the strips of a cable are very great (10 to 100 kg per cm band length) so that the supporting insulating plates which are spaced along the cable for clamping are ribbed to increase their strength against cable shock forces. In soft band cables every discharge produces, by its force stroke, mechanical work in moving and deforming the conductors. These strokes represent a real energy loss. If for a small capacitor of 10 kv and 1 μf—thus having an energy of 50 watt-sec or 5 meter-kg—we assume a force of 100 kg exerted over the length of the current conductors at an elasticity of 1 cm/100 kg between the conductors, the mechanical loss through conductor movement would amount to 100 cm-kg = 1 meter-kg; that is, a loss of 20% of the original discharge energy. Therefore, it must be strongly urged that flexible conductors and connections of any kind be avoided in discharge systems whose external resistances have low ohmic values (a few ohms), because every movement in such flexible connections represents a loss in discharge energy by way of electrodynamically effected work. This loss is intolerable in any measuring arrangement. Purely elastic deformation results in a return of energy with hysteresis-like curve deformation. The whole design and construction of capacitor discharge circuits for such large current values must therefore be governed by static considerations just as in bridge construction, where dynamic force exertion leads to destruction of the bridge. Repulsion forces between conductors can be calculated with the help of the basic formulas in electrical engineering, which will now be briefly reviewed:

$$P = \frac{U_0 \times I^2 \times l}{2\pi r}$$

where P (watt-sec/cm) = repulsion force with counterflow current in both conductors, I (amp) = current in conductor, l, r (meters) or (cm) = conductor length and conductor spacing.

A special form of sandwich line is the so-called coaxial return sheet. If two coaxial cylinders are separated by a very small distance, each element of the periphery of the cylinders will have the property of a sandwich line. A special advantage of this design is the compensation of all forces. Like the coaxial cable, such a line is characterized by the absence of an external field and by the lack of susceptibility to external fields from other sources. Especially if one has to connect in parallel circuitry a great number of capacitors, e.g., for highest current plasma discharges, each capacitor has a coaxial return sheet to the common consumer. This type of connection ensures that no feedback occurs. All capacitors are commonly discharged by means of common synchroneous ignition of all discharge spark gaps which are located inside the coaxial return sheets causing a break in the circuit inside the cylinder.

There is also an application to the field of smaller energies. In the so-called Fischer Nanolite [811, 289], a small piece of a coaxial return sheet serves as capacitor and is discharged through a very small Teflon-guided pinching spark, thereby producing nanosecond light flashes.

6. Compensation Conductor Forms

Although the sandwich line conductor systems just described have very low inductance, it is very often necessary to decrease the inductance of a geometrical structure such as a column of capacitor construction components. The possibility of reducing the inductance of a conductor system to practically zero by so-called bifilar winding is a known fact.

When bifilar winding is referred to, one usually thinks only of coils which have just as many turns in the clockwise as in counterclockwise direction. The inductance of the turns is thus compensated. The same principle, however, can be applied in space. Figure C6-1 shows a cross section through an arrangement of single, cylindrical, capacitor building blocks with compensating connections. For the sake of simplicity only two capacitor columns are shown. Each capacitor consists of three windings, 1, pressed upon one another. The capacitance is thus available at the end faces, 2. The problem at hand is to effect a compensation of the inductances of these fixed units. For that purpose a sheet-metal jacket, 3, is pressed by means of contact maker 4 upon the capacitor layers, 2. The capacitor windings themselves are generally wound on small insulating tubes through which is passed a bolt, 5, in order to tighten the whole structure with its compensating outer jacket by nuts, 6. By means of flange portions 7 the outer compensation sleeves are now connected with a plate conductor system, 8, which for its insulation carries a thin plate, 9. It is quite natural to provide for mounting of the sheets, 8-8, against each

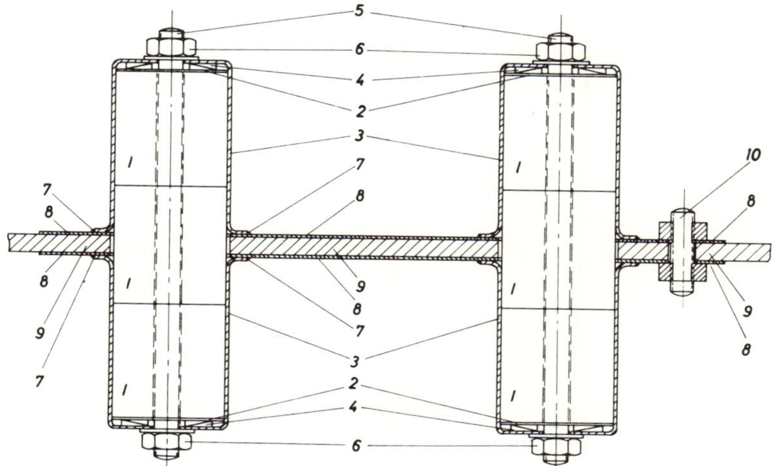

Fig. C6-1. Cross section through compensating arrangement of current paths in a combination of capacitor construction elements (parallel connection). (For explanation of reference numerals see text.)

other to avoid dynamic work losses by vibration. This mounting is done by suitable insulated screws as indicated in Fig. C6-1 at 10. It is clear that it is possible to flange to such plates as many capacitor units as desirable. Each one of these units thus is bifilarly compensated for inductance. It is possible by this means, to join even very large capacitor batteries for low losses, i.e. without inductance losses (see Chapter A), and to draw off the total discharge energy at any point of the plate-conductor system. Total inductances of batteries that have a few μf at a few 10 kv can thus be kept to a value on the order of 0.03 μh. The measures just described are applied today in the field of spark discharges for the realization of high temperatures (see also experiments by Kurchatov [2]).

In low-resistance spark discharge circuits, when it is desirable to achieve current values between 10^5 and 2×10^6 amp, at voltages in the range of 10–40 kv, with power peaks on the order of 40 million kilowatts, it is of primary importance to avoid inductive storing in the discharge circuit. Construction of the capacitors themselves is in this case no longer critical. With relatively low-quality papers, the end faces of the windings are tightly compressed and single inductance values on the order of 0.25 μh are readily obtainable. However, when several hundred such elements are connected together—which corresponds to division of the inductance—it becomes clear that with the increasing number of single elements the single inductance of each capacitor becomes less and less critical and takes practically no part, so that finally the total inductance can be found only in the connecting plate harness. In a construction as illustrated, with care and the use of high quality insulating materials and rigid bracing, the total inductance can be controlled to a large degree. The production of the steepest magnetic fields [17] with very steep rises of the discharge current requires the same connecting technique.

In conclusion consider an example: Two cylindrical capacitors C1 and C2 are connected in series by similar reduction of their inductance to a minimum so that the capacitance will be available on a band conductor system. Figure

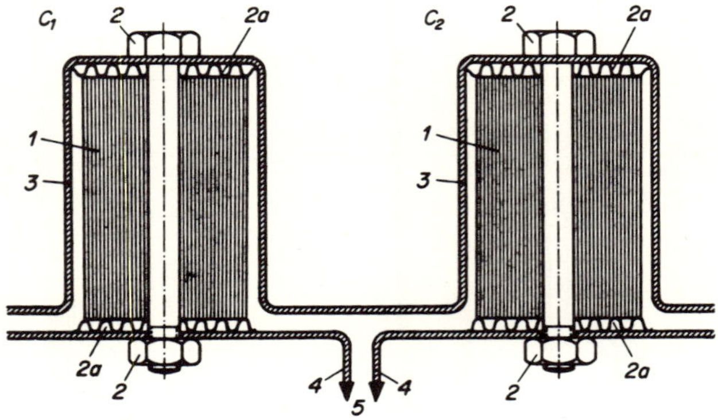

Fig. C6-2. Low-inductance series connection of two capacitors. Tandem connection with bifilar cup conductors, 3, and sandwich lines, 4. (For explanation of reference numerals see text.)

C6-2 shows the solution. Both capacitor windings are illustrated as item 1. They have insulated pressure bolts, 2, with contact discs, 2a. A sheet metal cup system, 3, is used for series connection. The cups are drawn from a thin sheet of metal. The bottom terminals, 2 and 2a, of the capacitors are connected by a band conductor forming the terminals, 4, which, as clearly shown in the figure, are accessible for connection to the circuit. Measurement of the inductance of this construction at the terminals, 5, will indicate that an inductance as low as between 0.003 and 0.01 μh has been realized. Here it is also possible to achieve much by subtle and carefully planned current path arrangement.

D. Conversion of capacitor energy into current impulses

1. Direct Discharge Through a Conductor

The conversion of capacitor discharges into current impulses must generally satisfy one of two requirements. On the one hand, it may be important to obtain a steep buildup in the current impulse, e.g., a rise to 10,000 amp in 1/10 μsec, and thus 10^{11} amp-sec without a pronounced peak. On the other hand, merely an extremely high peak value may be desirable. If a capacitor discharge battery with switching means and conductor is connected to a measuring point at which the steep current impulse will become effective, there exists a prior limitation caused by the impedance of the connecting line. By using band conductors of low impedance it is possible to obtain extremely short rise times because the inductance is very low.

However, when large absolute currents must be produced, an impulse transformer is usually employed. It offers current peaks one to two powers of 10 higher than a direct discharge, because it is possible to dimension the heavy current circuit properly and to reduce thereby the undesirable magnetic field that tends to build up. The engineering limits of impulse generation will now be traced.

The current peak obtainable by direct capacitor discharge is dependent on two factors, namely, the capacitor energy and the inductance of the whole discharge path. The highest current peak will be reached just after the capacitor energy as static energy has first dynamically converted itself into current. The capacitive energy storage is in this case converted into pure inductive energy storage in the whole conductor system. Thereafter, oscillation occurs and, owing to damping, becomes weaker and weaker, so that the maximum current is practically reached after one quarter-cycle of the first oscillation. Based on this principle, what maximum current values can be reached? Assume there is at hand a large capacitor battery of 10,000 watt-sec which is connected to be operated by switching means (as described in Chapter C6) so that its inductance is about 0.01 μh (a value to be strived for but hardly realizable). The discharge circuit proper, which might be constructed as a massive band conductor system, has an assumed inductance of 0.01 μh. The final conductor on which the experiment is made may have again 0.01 μh inductance. Thus the total inductance may be assumed as 0.03 μh.
With these values, the current

$$J = \sqrt{\frac{CU^2}{L}} = \sqrt{\frac{20{,}000}{0.03 \times 10^{-6}}} = 800{,}000 \text{ amp}$$

If the reader finds this current magnitude unbelievable, his doubt is justified; for there are again two factors which are responsible for considerable limitation of this calculation result.

First, because of the rapid discharge, skin effect becomes apparent in the conductor system. Therefore, it is unwise to use connecting plates of more than 1 mm thickness, as only a thin layer will be utilized anyhow. Accordingly, by all construction precautions it is possible to reduce the total resistance to a lowest limit that is still on the order of 10^{-4} ohms. A current of 800,000 amp would create a voltage drop of 80 volts in such a resistance. The second factor is of even graver consequence. At a current magnitude of 800,000 amp such dynamic forces are set up between conductor portions that it is practically impossible to take them up completely by external bracing means. Under these circumstances, conversion of capacitive energy into dynamic deformation work occurs during current rise. The forces that are observed in band

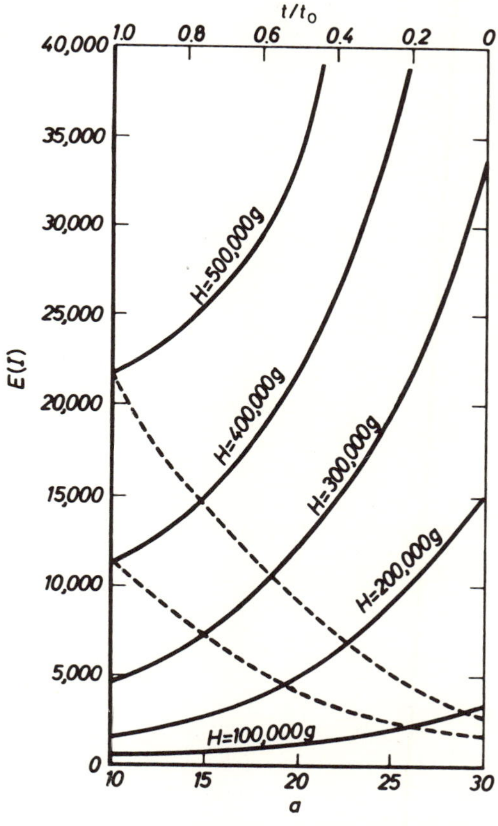

FIG. D1-1. Energy requirement of Helmholtz coils in dependence of field strength, H, for different coil sizes, a (solid line), and relative impulse time, t/t_0 (broken line).

conductors lie between 10 and 10^2 tons. This shows that the elastic limit of the material is reached. The discharge band conductor that makes the connection, as well as the leads in the capacitor battery itself, cause the chief losses by deformation. Therefore, the permissible impulse current limit is dictated by the mechanical construction of the impulse capacitors. In a paper by Furth and Waniek [17] this limit is explained. It lies to a considerable extent also in the switching means. For the so-called "rapid discharge" between 50 and 500 μsec discharge duration, a mechanical switch or a controlled spark gap is therefore most suitable. This switching unit is arranged adjacent to the magnet or combined with the magnet inductor. Since in general, it is desirable to produce directly a magnetic field with such high current values, the values in the following graph are given in gausses* (see Fig. D1-1).

It can be shown that inductances on the order of 100,000 gausses are readily obtainable, but that for 300,000 gausses there are required about as many joules as the numerical value of the coil diameter (5000–30,000 joules).

Taking into consideration mechanical deformation in capacitors and conductor systems, it appears possible to realize 300,000 amps in the direct discharge of very large capacitor batteries with energies between 5000 and 30,000 joules as peak. The main problem here is not the high peak current but the current rise time. A very small capacitor battery of 10 kv and 0.1 μf, corresponding to 5 joules electrostatic energy, can discharge itself through an inductance of 0.03 μh with a current rise time between zero and peak current of less than 10^{-7} sec, because the quarter-cycle time is given by the formula:

$$T = \frac{\pi}{2}\sqrt{L \times C}$$

$$T = 1.6 \times \sqrt{0.03 \times 10^{-6} \times 0.1 \times 10^{-6}}$$

$$T \approx 0.9 \times 10^{-7} \text{ sec}$$

The maximum current value reached after this time would be (without losses)

$$J = \sqrt{\frac{CU^2}{L}} \approx 18{,}000 \text{ amp}$$

The steepness of current rise is then about

$$\frac{\Delta J}{\Delta t} = \frac{1.8 \times 10^4}{0.9 \times 10^{-7}} = 2 \times 10^{11} \text{ amp/sec}$$

The outlay is modest and involves only a small styroflex capacitor battery with compensated inductance and a small ignited spark gap. Although the current values are also considerable, they are so short lived that the inertia inherent in any mass is sufficient to exclude any movement of the single conductor elements in the connecting circuit. This is because within 10^{-7} sec there is no time for considerable acceleration and therewith no time for deformation work so that mechanical work output can be assumed to be zero. With such experimental arrangements it appears that the obtainable limits of current rise time can be almost realized. By increase in size of the construction components very little improvement can be made. With increase in size of the capacitor battery there is an increase in the discharge time as well as the volume and the inductance of the whole arrangement. Probably the absolute current peak value can be increased but not the steepness of current rise. An arrangement with compensated inductances and demountable discharge

* Here the unit gauss is used since it is internationally recognized and has been applied by the authors [17].

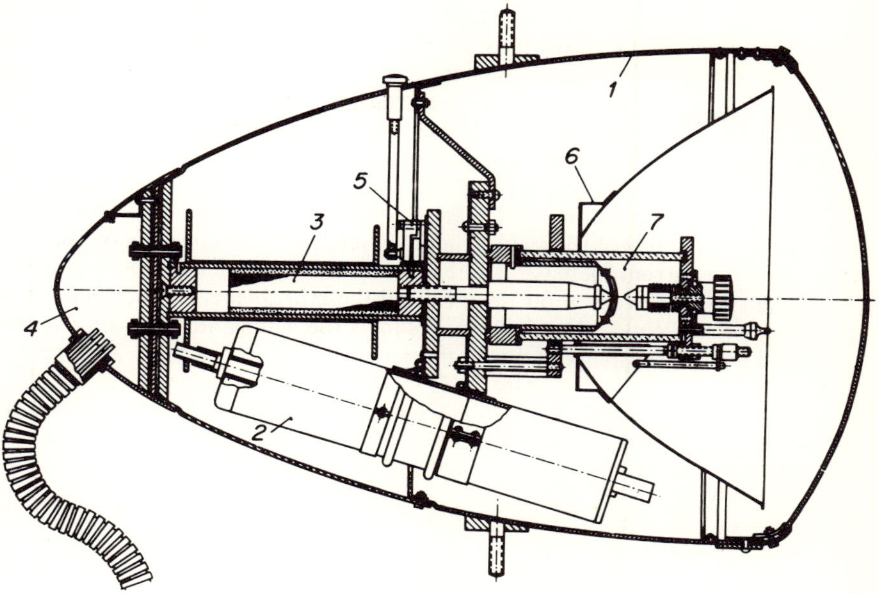

FIG. D1-2. Cross section through superstroboscope lamp, system Früngel, an example of steepest current rise, i.e. shortest rise time. Five capacitor units, 2, consisting of polystyrol windings in series connection whose inductance is bifilarly compensated by two cylindrical partial jackets and which are arranged around the switching spark chamber system, 7, and the ignition impulse transformer, 3, in rear thereof. The capacitor layers are flanged directly to two dished annular collector plates around the axis of the device. The total inductance is about 0.03 μh; the highest capacitance (settable by 5), 0.3 μf; operating voltage 6—9 kv. The steepness of current rise with noninert gas filling in the spark chamber is about 2×10^{11} amp/sec; with argon at high light emission it is about 7×10^{10} amp/sec. Item 4 is the high-voltage supply connection in the form of a cap, 6 is the mirror holder and 1 the lamp housing.

chamber, in itself created for generation of very short-time light flashes, is, without alteration and in accordance with the principles referred to above, fully applicable for producing shortest rise time current impulses. Such a discharge chamber is represented in Fig. D1-2.

The capacitor battery consists here of 3–5 styroflex capacitor units arranged around the axis of the system. Every one of these units is in itself inductance compensated by a surrounding copper jacket. The five capacitor units are joined by a pair of collector plates and connected to the electrodes within the demountable discharge chamber. By igniting the device by means of an ignition electrode, which is controlled by an impulse transformer, one obtains current rise to the peak within $0.09–0.12 \times 10^{-6}$ sec. By selection of proper fill gas the energy loss by radiated optical output can be practically eliminated so that only current rise need be considered. It is possible to dimension the magnetic field or to withdraw the current at several points in the discharge current path—for instance, between the adjustable electrode, seen at the right in the illustration, and the chamber cover—by interposing insulation between these elements and connecting the load between them. To gain full benefit, the load naturally must also have a low inductance. The inductance of the system just illustrated was measured as being 0.03 μh with 0.3 μf capacitor capacitance and 6–9 kv rated impulse operation voltage. Impulse sequence frequencies up to 300 cycles per sec are permissible.

The generation of current in the megampere range comprises a field of physical engineering. It is impossible to completely calibrate a capacitor discharge device owing to the nonlinearity of nearly all parameters (skin effect, decreasing resistance of the switching or switched plasma, heating of conductors while the current increases). The best analog computer in this case is the equipment itself. A very good review of all these problems can be found in a paper by Alfred Vondenbusch [933]. It shows the fields and limits of calculation, and as an example gives a discharge circuit for more than 2000 megawatts.

One should not neglect the results which are obtained from the calculation of the inductive energy storage. In the ideal case of a lossless circuit there is a balance between capacitively produced and inductively stored energy (oscillating circuit) [934].

2. The Transformed Discharge for Highest Current Peaks

While it is possible to obtain very short rise times or steep discharge current rise in a low-inductance discharge circuit, the maximum amplitude of the current value is limited. Increasing capacitor capacitance increases the inductance also by greater spacing. With the considerable higher financial outlay, one gains only small improvement in peak current value. It is, therefore, for production of highest absolute current magnitude, preferable to transform the discharge. Transformers for this particular purpose have to meet special requirements. The true-to-form transformation of a given ac voltage phenomenon is not of interest here, but the conversion of an energy package at the primary side to one (with different characteristics) at the secondary side is. Normally it will be a higher current value at lower voltage—adapted for the secondary circuit with resistance and inductance—that is demanded. The design of the pulse-driving capacitor battery can be executed according to the rules of best economy. With the manufacturing facilities of the capacitor industry today the most favorable price level per kilowatt-sec can be found in capacitors for charging voltages between 5 and 40 kv.

Impulse or pulse transformers for the purpose in question must have extremely low inductance at their secondary sides and must have close coupling between primary and secondary. The secondary winding has generally only one turn and forms a copper block having slots in which the primary winding is embedded. With proper construction (described below) there is no partial coil of the primary winding that can communicate with the other primary coil sections. Fundamentally, only the one-turn secondary coil package is cut by the flux lines induced by the primary coils.

The inductance of the one secondary turn can be readily calculated geometrically by the elementary formula for inductance. It will be remembered that the inductance in centimeters is

$$L_{\text{cm}} = \frac{10 \times D^2 \times n^2 \times K}{l}$$

where D is the diameter of the coil, n the number of turns, and l their length in centimeters. A correction factor, K, is included which, for general purposes, can be found tabulated in pertinent textbooks. This factor can be omitted here since it is approximately 1. If we assume a transformer that has a secondary winding of 50 cm length but only about 15 cm diameter, and only one turn, the inductance of this one-turn winding is then

$$\frac{2250}{50} = 45 \text{ cm} = 0.045 \ \mu\text{h}$$

(Since only dimensional considerations enter here, it is permissible to use such a rough formula that nevertheless gives results within 20% of the correct value.)

The coil selected for the example can now be connected by a band conductor system directly to a load (Helmholtz coil) so that the total inductance becomes 0.055 μh. With a transformer construction such as described here one obtains extraordinary mechanical strength because of the blocklike configuration of this one-turn coil. This strength practically excludes mechanical impulse losses. With such a transformer the capacitor battery to be connected to the primary can be designed for optimal performance: The high transformation ratio possible here means that voltage and capacitance of the capacitor battery can be selected entirely from the economic view point, the inductance in the primary circuit being completely noncritical. If a transformation ratio of 500:1 is selected, the permissible inductances are related as the squares of this ratio. If an additional inductance of 0.01 μ h is assumed as a maximum limit in the secondary circuit, the primary inductance is 250,000 times higher, that is, 2500 μh = 2.5 mh. This is a very convenient value for the designer. The costly wiring of the whole capacitor battery with compensating plates can be discarded and the inexpensive classical method employed. In this method the necessary number of capacitor units is connected in parallel and discharged by way of switching means into the primary coil of the transformer. The switching means in this case can be rather simple. For the selection of the switching means it is here of advantage that a secondary current of, say, 500,000 amp exist (at the ratio mentioned above, there is then only 1000 amp in the primary). A commercial ignitron or a large hydrogen thyratron will serve well as switching means. Simple mechanical impact switches will also do, because a blocking voltage of a few kv and a peak current of 1000 amp lasting for only a few milliseconds present no problem. Since charging of large capacitor batteries usually takes several seconds, one can dispense with a quenching chamber at the switch. There is positively no arcing at circuit interruption because the capacitor battery acts as a spark quencher. This is known from the field of low-voltage applications, where capacitors are used to preserve the life of relay contacts.

A version of an impulse transformer is shown in Fig. D2-1. The laminated iron core, 1, carries, insulated by a thin insulating layer, 2, the longitudinally slotted copper cylinder, 3, whose ends, for mechanical rigidity, are bolted to an insulating bar, 4. Upon copper cylinder, 3, there are pressed tightly copper rings, 5, with interposed strips, 6, which serve to brace the copper rings against one another. Copper cylinder, 3, together with copper rings 5 and 6, forms the secondary winding to which are connected the copper strip conductors 7 and 8 which may lead to the welding electrodes, the impulse current load, the magnetic concentrator, or other loads. Between the copper rings, 5, there lies the primary winding, 9, in the form of disc coils. It may be assumed that each of the seven discs shown has 15 turns, so that the transformation ratio becomes 105:1. The two windings are intimately coupled, as will become apparent from inspection of the figure. This assures lowest stray flux losses in both circuits and consequently the lowest self-inductance.

Experience has taught that in pulse transformers of the kind described, after long continuous operation, dynamic destruction occurs in the windings and in the insulation. The discharge current, flowing in the primary section, 9, induces directly in the surrounding secondary discs, 5, large steep-fronted current pulses. The discs thus have a tendency to expand. The resulting deformations at each pulse are naturally very small. However, after millions of pulses they reach remarkable values and can then lead to destruction. The shock force itself can be figured from the general laws of electrical and magnetic forces. From the peak current, measurable at the secondary, there can be derived a certain number of ampere windings per centimeter when the physical length of the transformer is taken into account. 500,000 amp give, at a transformer length of 30 cm, 17,000 amp-turns/cm.* After this value is known, and assuming a practical radius, the repelling force to which two oppositely lying portions of the secondary winding body are subjected can be calculated by means of the general formulas of electrical engineering.

In the past it was not customary to carry out such calculations in transformer design. However, it is possible to avert a fundamental evil by taking steps to preclude plastic deformation of material in transformers. With this objective in view care must be taken that the mechanical construction comprising the cylindrical part, 6, the slotted discs, 5, and eventually the enclosure of the whole system, is not stressed above its elastic limit by the forces caused by the current; that is to say, not above the limit of Hooke's law for the material used (e.g. hard copper). The coil passed by the heavy secondary current must not be constructed rigidly but be somewhat resilient. However, this resilience must be within the limits determined by exact calculation and experimental results within Hooke's straight line and modified by construction considerations with respect to a safe movement amplitude between secondary output terminals 7 and 8. In order to avoid such movement effecting the primary discs, 9, embedded in the secondary coil, each one of the primary discs, 9, should be designed as a separate integral unit and should be isolated

* A uniform current distribution is hereby assumed. This assumption introduces no great error when the secondary discs are closely stacked.

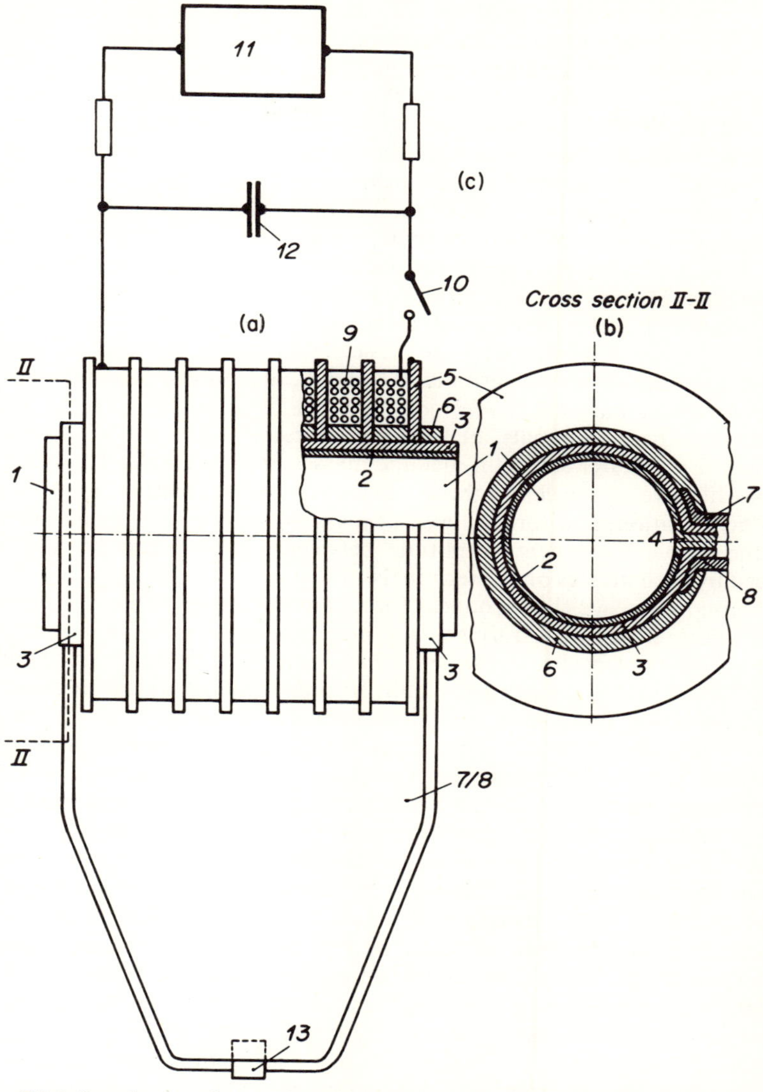

FIG. D2-1. Impulse transformer for high secondary current values represented as a schematic drawing. (For explanation of numerals see text.)

with respect to forces. For instance, isolation may be achieved by insertion of resilient damping material between the two windings. In such a construction, the damping material takes up the movement amplitude difference between primary and secondary and keeps direct force action away from the more sensitive primary winding discs. Such a damping medium must not only have good electrical insulating qualities but also good mechanical damping properties. Particularly its viscosity must be constant over a wide temperature range. Materials with a silicone base have proven to be suitable. Experience has shown that a recurring cause of failure in such transformers is that, by action of the very strong magnetic field created by the heavy currents, fine iron dust is attracted and enters gradually and progressively into the windings and into the crevices between primary and secondary. There it causes friction and abrasion and reduces the life of the transformer. The damping material minimizes such destructive action in that, by its viscosity, it keeps away dust. Under severe operating conditions it should be replenished from time to time. Moreover, it is possible to increase the effectiveness of the damping medium by inserting sealing gaskets between the output electrodes, 7 and 8.

The forces which tend to separate the secondary discs, and which have their highest amplitudes at the terminals 7 and 8, are only momentary shocks. These forces can be effectively counteracted by means of weight masses attached to these terminals, or adjacent thereto, and eventually joined to the discs. For greater strength, metallic masses are preferable. The occurring forces will have to accelerate these masses first. Since force is equal to mass times acceleration, and force as well as mass are now given, the acceleration and the whole movement amplitude between 7 and 8 can be effectively reduced by such masses on the terminals.

A second influence that produces forces among the several components of the transformer is, in transformers with magnetic cores, the magnetic reaction of the core with respect to the several winding elements through which current flows. Transformers are generally built so that windings and iron core form one construction unit. In impulse transformers, however, it is advisable to arrange the iron core, which is usually a bar-shaped open core, to slide axially in the coil system. This arrangement assures that forces between iron core and coils can have no deforming effect, since the iron core has freedom to move somewhat at each shock without transferring this force each time to the coil system. To facilitate such axial movement of the core it can be mounted in lubricated bearing metal slides. In the case of transformer mismatch, the discharge energy from capacitor 12 tends to set up oscillations in the system, particularly if it is unloaded. Oscillations of this kind create increased dynamic stresses, since the forces then act in fast cyclic succession. Such undesirable oscillation can be reduced by making the iron core of the transformer coarsely laminated of material having a high hysteresis loop. When such magnetic material is used, a single discharge impulse, which with proper adaptation of the secondary load is practically nonoscillating, causes only partial circumvention of the hysteresis loop and the energy losses in the iron are negligible. With an unloaded transformer, however, the hysteresis loop is circumvented several times, due to the oscillation then set up. Additionally, there are induced eddy currents in the coarsely laminated core leading to very high energy consumption in the iron. In order to minimize these losses it is advisable to use normal soft steel sheets of 0.5–1 mm thickness as core material, or thick dynamo grade sheets of lower quality such as those commonly used for pole pieces in large dc machines. Such grades of sheet steels are generally not used in power transformer construction on account of their high hysteresis losses in regular ac operation; here, however, they are of benefit, since the requirements are high permeability at high magnetic excitation and large hysteresis losses.

In order to hold the movement amplitude at very high shock currents, i.e. above 100,000 amp, to a minimum, it is of importance to provide the secondary elements of the transformer with a high mechanical section modulus and in order to gain rigidity, to make the secondary discs in a U-shaped section. A practical design consists of saucer-shaped pressed discs fitting pairwise in each other and enclosing the primary discs. Cutouts, which are preferably electrostatically rounded and in which, by chip-free forming, care has been taken that uniform copper section is preserved, provide outlets for the high-voltage leads of each primary disc. In a construction of this kind it is rather simple to utilize the damping medium simultaneously as an effective dust protector.

In order to confine the longitudinal forces (those parallel to the coil axis) in the transformer in their amplitude action, it has been found appropriate to subject the whole package of primary and secondary coils to a certain mechanical prestress, such as by means of axis-parallel bolts which derive their tightening force from the mounting structure. Assume, for instance, that at every dynamic shock the coil package tends to contract itself with a force of 3 tons. This contraction would lead, by variation of this force between zero and three tons, to considerable movement amplitudes. However, with the system under a steady state stress of about 3 tons, the force varies between 3 and 6 tons. In the latter case the coils and discs are pressed originally so tightly upon each other that doubling of the force causes no appreciable movement and, most important, no frictional wear between the component parts.

The primary discs can be made very rigid by encapsulation in synthetic resin, particularly that known under the name Araldite, which has mechanical elastic properties which are somewhat similar to the properties of metals. The damping lubricant of high insulating quality helps automatically to close slits that might occur in the hard resin by entering therein. A premature failure of an otherwise nondefective transformer can thus be much delayed.

The transformer depicted in Fig. D2-2 is built in conformity with the basic principles enumerated above, It is an improved version of that shown in Fig. D2-1.

Figure D2-2a shows the cross section through a primary disc. It is embedded in baked synthetic resin (Araldite). Each single coil consists of two half-coils with about 4 turns each. The Araldite itself is shown at 1, and 2 represents the rectangular wire turns. At 3 there is, as can be seen in Fig. D2-2b, a projection on the coil through which the leads, 4, extend. Figure D2-2b indicates how the two half-coils are wound, one in opposite direction to the other. Figure D2-2c shows a section through the complete transformer. All coil leads, 4, are connected by cross connections, 5. Between every third and fourth coil there is inserted a watercooled coil in which water tubes, 11, are cast in lead metal, 12. In order to avoid a short-circuited winding being formed by the water tubing, the tubes follow first for about 300 degrees around the coil, then turn backward and return to avoid a closed turn. The secondary winding

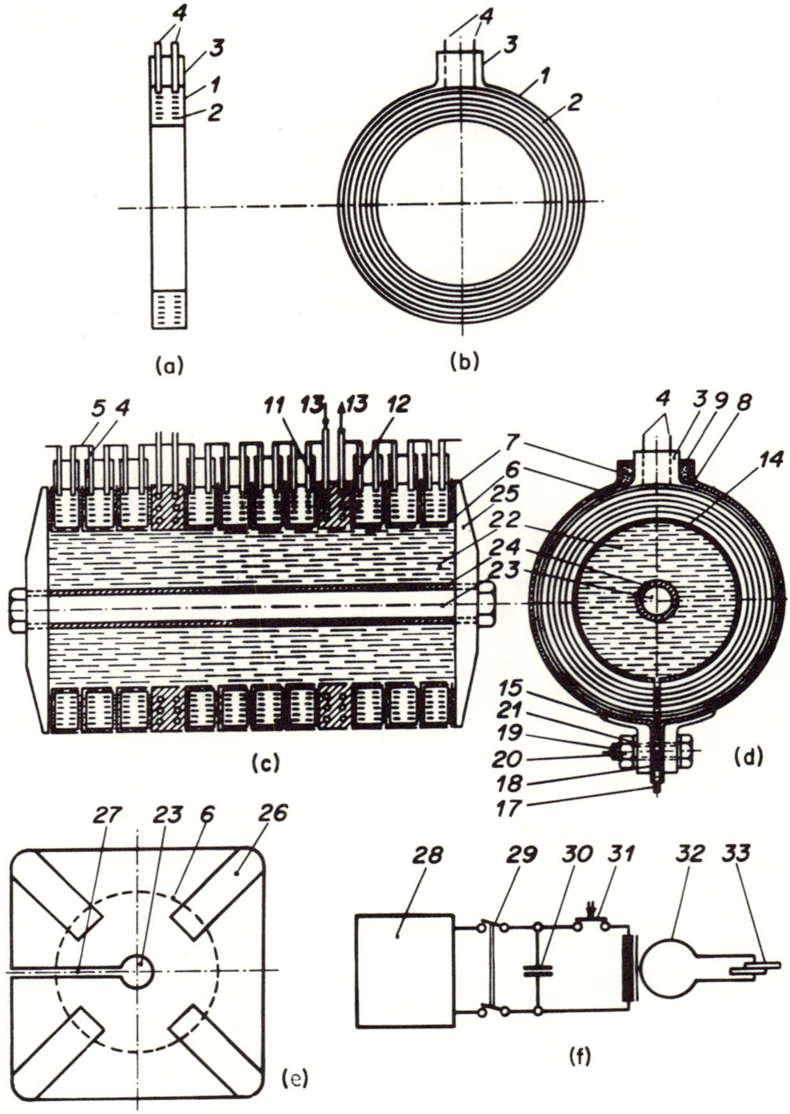

FIG. D2-2. Improved impulse transformer. a–cross section through a primary winding disc embedded in synthetic resin; b–side view of a primary disc, current leads at 4; c–longitudinal section through a long-lived impulse transformer having primary discs embedded in synthetic resin and water cooling (11) (for explanation of reference numerals see text); d–cross section through transformer as in Fig. D2–2c (for explanation of reference numerals see text); e–side plate of transformer as in Figs. D2-2c and D2-2d; f–schematic wiring diagram with secondary load for impulse transformers as in Figs. D2-2a to D2-2e.

discs which together form a winding of low inductance are represented at 6. At 7 there can be seen the damping medium in which the primary discs float enclosed in the secondary coil body (drawn in bold lines). The whole assembly is mounted on an insulation tube, 14, and is tightened by a concentric through bolt, 23, which is insulated from the iron core by a tube 24, and two end plates, 25, which are cast with ribs for greater strength, This arrangement produces an original pressure of about 3 tons. The cross section can be seen in Fig. B2-2d. The secondary terminals are visible at 18. A thin insulation layer, 17, is interposed between these terminals. Moreover, they are secured by a bolt, 19–20, which, insulated by a layer, 21, keeps the whole tightly together. In order to obtain lowest electrical resistance at the point of current outflow, the secondary sleeves, 6, are increased in cross section by welded or brazed reinforcements, 15, at their connecting points as shown at the bottom. The coarsely laminated iron core can be seen at 22.

The side plates are clearly illustrated in Fig. D2-2e. They are also provided with slots, 27, to avoid forming a shorted winding turn. At the center there can be seen the hole, 23, for the through bolt. The ribs, 26, serve to increase the section modulus for greater strength. A schematic diagram of the complete circuit is sketched once more in Fig. D2-2f. A high-voltage generator, as indicated at 28, charges through a switch, 29, which at a desired voltage switches off the capacitor battery, 30, automatically. With the capacitor charged, the impact switch, 31, is actuated to release the current impulse through the transformer primary, wherein, with a transformation ratio of 275:1 and a secondary current of 275,000 amp, flow 1000 amp. To the secondary, 32, there is connected, in accordance with the described example, a welding load, 33.*

An essential feature of this transformer construction is the possibility of mounting as many units as desirable on a common current collecting bus strip (Fig. D2-2). In practice, a value of 1.2 million amp has been reached over a strip width of 3 meters. Unfortunately, because of strip conductor resistance alone, it is impossible to withdraw this current value at a single point, but it is basically available over a greater conductor width. This fact is of benefit in welding problems in which a great number of welding spots have to be supplied simultaneously. It is necessary for feeding a very long tube-like magnetizing conductor of one turn, whereby one million ampere turns can be obtained over a total length of 3 meters, corresponding to 3000 amp-turns/cm over this great length. The transformer depicted in the illustration can dissipate about 2000 watts in loss heat by means of the cooling ribs and water cooling. It has proven its reliability and long life over 10,000 impulse operating hours at 1 discharge per sec.

In connecting Helmholtz concentrators [17] it is possible to produce by means of two transformers flanged to the connecting strips at opposite sides

* In impulse welding there are required at the work piece about 4–6 volts. In double spot welding, two welding spots lie in series, The impulse transformer here described then has to deliver 8–12 volts at its secondary. At the transformation ratio of 275:1 the capacitor battery is designed for an operative voltage range between 2400 and 3600 volts.

of the concentrator about 500,000 amp, or possibly even 700,000 amp, in the concentrator winding.

As a special case there is now described a transformer type of somewhat similar construction which has been designed for transmission of medium impulse energy values of very high frequencies. For example, it might be used in connection with a blast spark gap or quenching spark gap as switching means or for producing an effective output of up to 20,000 watts in continuous operation. With such a water-cooled impulse transformer it is possible to effect low ohmic excitation of oscillatory circuits at the transformer primary and to feed at the secondary side into the load, microsecond impulses with steepnesses of up to 10^9 amp/sec. In Fig. D2-3 a high-voltage transformer, 1, charges capacitor 3 by way of resistor 2. The discharge of 3 takes place through the quenching of blast spark gap 4 into the primary winding, 5, of a transformer. The inductance of the transformer together with the capacitance forms a circuit that is tuned to the desired frequency. In order to reduce damping, the transformer coil, 5, can be extended by an auxiliary coil, 6, and an oscillatory capacitor, 7, can be added, so that with coils 6 and 5 the same frequency arises, when it is desirable, for example, to produce a damped oscillation by the capacitor discharge.

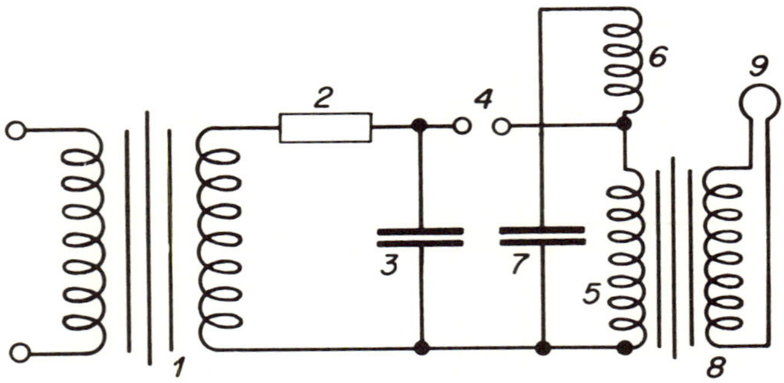

FIG. D2-3. Schematic circuit diagram of a high-frequency generator for high-frequency impulse sequence.

The inclusion of a blast spark gap is of advantage because by switching on the second oscillatory circuit the original impulse energy, which stems from capacitor 3, is momentarily drawn off by way of capacitor 7 and inductance 5 and 6. There is, thereby, developed instantaneously in the spark gaps, 4, a currentless condition which effects the spark's rapid extinction. Such shock current transformer, having primary 5 and secondary 8, and also a small heater conductor, 9, connected to the latter for large currents, has the purpose of consuming by close coupling high outputs of, e.g., 20 to 30 kw in continuous operation at a high rate of impulse sequences. The version shown in Fig. D2-4 is adapted for such service. The primary can be in the form of copper tubes, 5, which carry cooling water at a rate of 1 liter/min. Water passed through a very thin windable copper tubing at this rate is sufficient

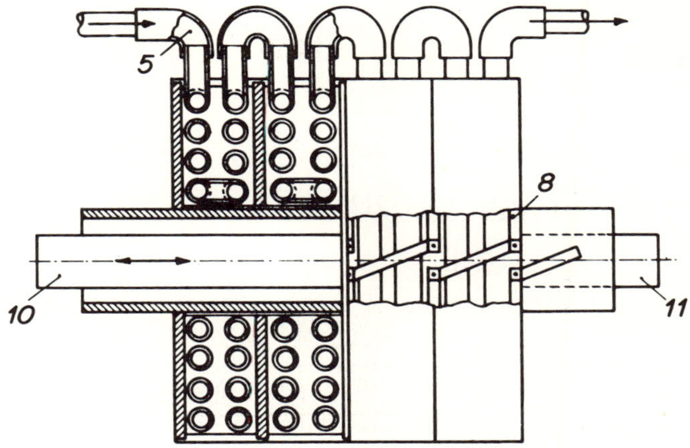

FIG. D2-4. Cross section through the impulse shock transformer 5–8 of Fig. D2-3 with water-cooled primary.

to carry off, at a permissible temperature rise of 60°C, 1 kg-calorie/sec or 4000 watt-sec/sec. If it is assumed that these permissible thermal losses of 4 kw represent 10% of the total energy transformed, the rating will be 40 kw. Higher outputs can hardly be expected from such transformers. Thus it will be obvious that the use of tubing instead of wire imposes no great difficulty with the very small cooling medium volume required, because, at 16 atm pressure within a length of only about 100 meters, tubing of 3 mm outer diameter and 2 mm inner diameter passes the necessary volume of cooling medium.

The slotted secondary discs, which can be seen in the illustration, are somewhat similar to the one in the impulse transformer first described. The damping iron core, 10, within the windings is arranged to slide so that it can produce no reacting force on the coil package that could lead to friction wear with life-shortening effects.

The single secondary coils, 8, are completely enclosed. In case it is desirable to employ the transformer for high-frequency generation it is possible to use ferritic material for the iron core. By this means, fine tuning of resonance between the circuit elements 3, 5, 6, and 7 can be facilitated. However, a ferritic core must be water cooled since the hysteresis losses can easily reach 1000 watts on account of the high magnetization effected. In the field of high-frequency hardening it is customary to cool the ferritic concentrators as well as the heating conductors (item 9 in Fig. D2-3) with high-pressure water.

E. Conversion of capacitor energy into voltage impulses

1. The Cascade Circuit

Fundamentally, it is possible to convert energy from a storage capacitor of given charging voltage and capacitance into a pulse of other voltage, by means of a special impulse transformer. However, when only voltage multiplication is at stake, the impulse voltage circuit published by Marx can serve the purpose. This circuit has the advantage that by insertion of resistors in the discharge path the pulse rise time as well as the decay time can be influenced within wide limits and can be adapted to conform with the standardized rules for impulse voltage tests [18]. These rules specify 1 μsec rise time and 50 μsec decay time. However, it must be recognized that the cost of a circuit with a great number of capacitors, which are connected in parallel for charging and in series for discharging, is considerably higher than the circuit variant employing impulse voltage transformers. Therefore it can generally be said that in all cases where an exact predetermined wave form of the voltage impulse is not important, the impulse transformer is the more economical solution. In the cascade connection by Marx there are charged in parallel, capacitors of similar size, capacitance, and charging voltage, through resistors. These capacitors are discharged in series through spark gaps or switching means. Corresponding circuits are shown in Fig. E1-1, 2, 3, and 4. In the example shown in Fig. E1-1, the three capacitors, 1, are connected by resistors, 2, of similar value, with the charging high-voltage source, 3. Three spark gaps, 4, serve as switching means and must be adjusted so that they do not break down at the charging voltage of source 3, but do break down with certainty at twice that voltage. Such an adjustment is readily obtainable. The load is represented at 5. If a certain spark gap is ignited—for instance, the one at the right in the figure—by ignition electrode 6, at the instant of ignition the capacitor at the right is in series with the first gap and the break-downs of the other spark gaps follow in fast succession because the impulse wave steepens.

As soon as the discharge process is over—that is, when all capacitors are returned to zero potential—the sparkover in all three spark gaps has stopped and the charging process by way of resistors 2 begins anew.

The design calculations for such a circuit are simple. If we assume that a test sample is subjected to an energy of 10,000 joules at a voltage of 10^6 volts, the total capacitance after transposing of the formula $A = \frac{1}{2}CU^2$, becomes

$$C = \frac{2A}{U^2} = \frac{2 \times 10^4}{10^{12}} = 2 \times 10^{-8} \text{ farads}$$

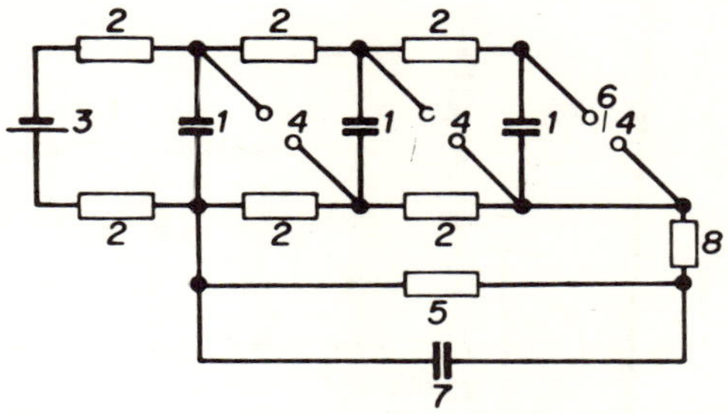

FIG. E1-1. Schematic circuit diagram of an impulse voltage multiplier by Marx, [31] (for reference numerals see text).

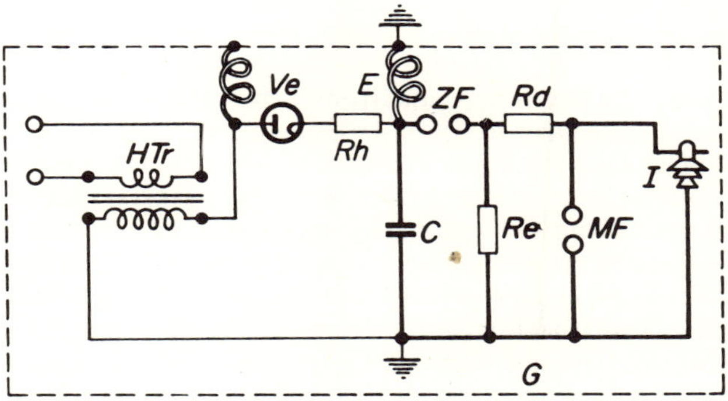

FIG. E1-2. Schematic circuit diagram fo the simplest arrangement for producing impulse voltages (for reference numerals see Fig. E1-3).

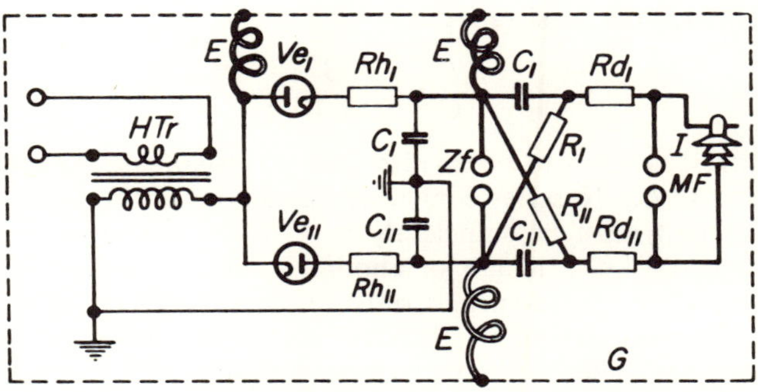

FIG. E1-3. Schematic diagram of a circuit for doubling of the voltage value (Marx [31]). E–movable ground wires (the figure shows protective grounding after finish test); G–protective enclosure; HTr–high-voltage test transformer; I–insulator under test; MF–sphere spark gap for measurement of the peak value; Rd–damping resistor; Re–discharge resistor; Rh–high-potential resistor; Ve–electrical valve; Zf–triggering spark gap.

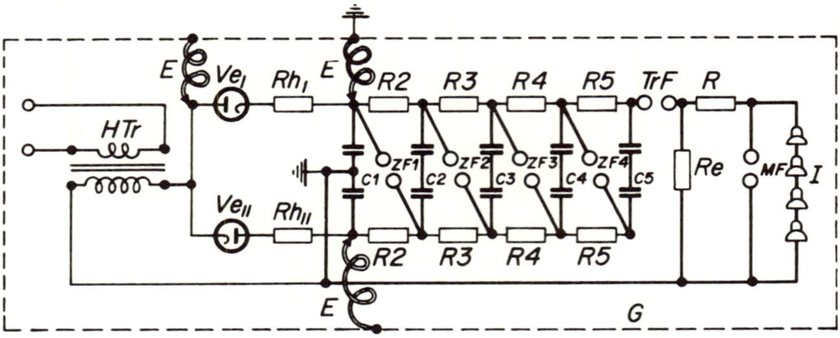

FIG. E1-4. Multiple circuit for producing very high impulse voltage (for reference numerals see Fig. E1–3). C_1–C_5 are the storage capacitors which are charged in parallel connection and are discharged in series connection through ZF 1–4.

It is advisable to obtain this value from a combination of 10 capacitors of which each one must have 10 times the capacitance; that is 2×10^{-7} farads $= 0.2\ \mu f$. The charging voltage of each capacitor must be correspondingly 1/10 of the total voltage, 100,000 volts. The spark gaps are preferably in the form of spheres having 4–5 cm radius and a spacing of about 5 cm. This ensures that there is no breakdown at 100 kv, but a definite breakdown at 200 kv. Such setups can be erected by arranging the capacitor units one above the other. The capacitors are super-imposed in a stack and are charged through resistors accommodated in the stack. Choke coils can be used instead of the resistors. Selection of the resistor size must be made for a proper time constant.

If it is desirable to charge the whole capacitance of 2 μf, consisting of $10 \times 0.2\ \mu f$ in parallel connection, within 10 seconds, the whole effective resistance must be selected for about 1/3 of the given charging time, that is, for 3 sec. This requires a resistance of

$$R = \frac{T}{C} = \frac{3}{2 \times 10^{-6}} = 1.5 \text{ megohm}$$

This total resistance is to be divided into 20 resistors of 75 kilohms each, corresponding to the 10 series connected capacitors of the present example. Each resistor unit must be sufficiently long to prevent an over-the-surface flashover. For 100 kv, a length of about 50 cm should be considered the minimum length for glazed-surface resistors. Since such setups are generally not installed in open air and are therefore not subjected to the weather, it is not necessary to consider any reduction in sparking distance due to precipitation.

In order to assure, in such an impulse voltage installation as that by Marx, the proper rise time and decay time constants, it is necessary to tune the items 7 and 8 in Fig. E1-1 properly to each other. Item 7 may be assumed to be the capacitance of the unit under test. An additional resistance, 8, is inserted during charging. When capacitance 7 and resistance 8 are so selected that the product R × C = 1 μsec a rise time of 1 μsec results. On the other hand, for a decay time of 50 μsec, as, for example, recognized in the German VDE Standard [18], the product of capacitance 7 and resistance 5 must be R × C = 50 μsec. Moreover, capacitance 7 must be rather small relative to the series-connected capacitance of all impulse voltage capacitors combined. In the example they have a combined capacitance of

$$2 \times 10^{-8} \text{ farad} = 0.02\ \mu\text{f}$$

The unit to be tested or the test sample may have a capacitance which is 10% of this value, corresponding to 10% voltage drop, or 2000 pf. Consequently, for a test sample of 2000 pf capacitance, the charging resistance, 8, in order to obtain a rise time of 1 μsec, must be

$$R = \frac{10^{-6}}{2 \times 10^{-9}} = 500 \text{ ohms}$$

For a 50 μsec discharge time constant the discharge resistance, 5, must have a resistance value 50 times greater; thus, 25 kohms.

Capacitor 7 forms a short circuit at the instant of an impinging impulse wave and the whole discharge voltage becomes effective on resistor 8. With regard to voltage strains this resistor must be just as safely dimensioned as resistor 5. (If not, a predetermined test sample of given high resistance value, an insulator for instance, is connected.)

In Marx's impulse circuit the influence of the spark resistance is negligibly small. If it is to be considered, this can be done by making the resistance, 8, somewhat lower than calculated. If at a voltage in accordance with the example under discussion, there are assumed 10 spark gaps each 5 cm long, and thus a total spark length of 50 cm, the spark resistance in accordance with Toepler's formula becomes

$$R = \frac{K \times l}{C \times U} = \frac{0.2 \times 10^{-3} \times 50}{2 \times 10^{-8} \times 10^{6}} = 0.5 \text{ ohms}$$

when Toepler's spark constant is approximated to be 0.2×10^{-3}.

A resistance of this value, however, is negligible even if it should be off by a factor of 2 owing to choice of an inaccurate spark constant.

In every case in which Marx's impulse circuit is used for feeding special impulse tubes at high sequence frequencies it is necessary to replace the resistors by choke coils in order to avoid thermal losses, which in resistance charging amount to 50% (see Chapter A, Section 6). The best procedure in selecting chokes is to adjust the whole circuit for a predetermined cyclic frequency. If this frequency is 50/sec, it is possible to adjust all the chokes replacing resistors so that, in combination with the adjacent storage capacitor, they have a self-frequency of 25 cps. In this case, the charging voltage of the capacitors will have gained after 1/4 cycle (at 25 cps in 10^{-2} sec) a value somewhat above the charging voltage of the feeding source. The sparkover then takes place and again excites the oscillatory circuit. With 50 cps sequence frequency, in the present example having ten series-connected capacitor elements of 0.2 μf, there would be as a total for the 20 partial chokes and $\omega = 2\pi \times 25 \text{ sec}^{-1}$:

$$L = \frac{1}{\omega^2 \times C} \approx \frac{1}{25000 \times 2 \times 10^{-6}} \text{ h} = 20 \text{ h*}$$

From this result it is apparent that at a peak voltage of 100 kv, which becomes effective at the terminals of the choke, the chokes will be rather voluminous, particularly, since 20 h at 100 kv blocking voltage will be available during magnetization by the charging direct current. They will be comparable in size with a potential transformer of about 60 kv terminal voltage. Nevertheless, in large installations such an investment is beneficial, since the efficiency is remarkably increased (from 40% to 70%).

With regard to ignition of the starting spark gap it can be said that any desirable ignition method can be applied. Also the number of stages in such an impulse circuit installation is limited only by the necessary mechanical stability and a sensible number of stages, say, 20 as a maximum.

The development of techniques for the use of highest pulse voltages is described in a very good paper by Helmchen [847]. This paper shows especially the values of all damping resistors in cascade circuits that avoid parasitic oscillations. Very smooth curves must be obtained with cascade discharge circuitry because the objects being tested can show varying degrees of damage, and often only very small deviations of the oscilloscope trace indicate an internal damage, for instance, a spark over an internal turn of a transformer.

A table, given in [847], allows one to compare the designs of cascades equipment made by ten manufacturers. The unit with the highest voltage output is a construction made in Russia. It generates 7.2 megavolts and 400 kilojoules, and operates in 24 steps with a 300-kv charge. Each capacitor has 0.016 μf. In [847] 67 additional references to impulse voltage testing are given.

2. The Nondistorting Pulse Transformer

The elementary theory of pulse transformers and their design methods is given by W. H. Bostick [611].**

Pulse transformers normally have been used:

(1) To transfer or control energy by raising a voltage above a threshold or barrier level.

(2) To invert the sign of voltage.

(3) To effect dc isolation between source and load.

---

* Design is generally such that the main inductance is put immediately between the high-voltage source and the first capacitor. Therefore, 2 times 10 h, and all other 18 chokes, lie between the capacitors and are only low-inductance, single-layer, air cylinder coils having about 1/100 of the inductance of the two main charging chokes, thus, e.g. each case $\frac{1}{10}$ h = 100 mh.

** Chapter E, Section 2, partly reprinted by permission of the publisher from "Pulse Generators" by Glasoe and Lebacqz, pp. 499–541, McGraw-Hill, New York, 1948.

(4) To deliver the correct amount of power to a load of a given resistance by changing the voltage to the proper value.

(5) To effect maximum transfer of energy (or power) from source to load by a transformation of energy to the proper impedance level.

It is possible to conceive of a perfect electromagnetic transformer which would operate successfully over any range of frequency, imepdance, and power levels. For various reasons, such a perfect transformer cannot be built. Any transformer that can be built has certain limitations in its range of operation and should be designed to give optimum performance under the conditions of operation for which it is intended.

Many different types of transformers have been developed to perform the many functions and to cover the various ranges of operation of electromagnetic transformers. For example, there are filament and power transformers, which operate steadily on a sinusoidal input at the various standard power frequencies and voltages. There are audio transformers, which are designed to operate over the audible range of frequencies, some at high power levels and others at low power levels. There are rf transformers, which, when used in the circuits for which they are intended, pass only a narrow frequency band of rf energy.

The development of pulsed radar created a need for a relatively new type of transformer, a transformer that transforms the energy in a pulse having a more or less rectangular shape and a duration of the order of magnitude of 1 μsec.

In general, such pulse transformers are used to perform any or all of the functions listed at the beginning of this chapter. One individual pulse transformer is often used over a fairly wide range of pulse durations, voltages, and impedances.

Pulse transformers are also used extensively (usually at low power) in one type of pulse generating circuit, the regenerative pulse generator or blocking oscillator, in which energy is introduced from the plate to the grid circuit so that the polarity of the voltage is inverted and regeneration occurs. Transformers designed for regenerative pulse generators frequently have extra windings that can be used to transfer power to other circuits.

Pulse transformers are effectively used as interstage coupling and inverting devices in pulse and video amplifiers. These transformers usually operate at medium or low power.

Perhaps the most important use of pulse transformers, however, has been the transformation of the energy in a pulse from a pulse generator to the impedance level of an rf oscillator, which is frequently a magnetron. The pulse power level at which these output transformers operate is usually between 10 kw and 10 Mw.

W. H. Bostick [611] described in the chapters of his handbook the influence of feeding magnetic energy into the core, electrostatic energy into the capacity between core and primary layer, and shows the calculation of leakage inductance.

The calculation of leakage inductance as a lumped parameter follows the process of reasoning wherein a given amount of load current I is assumed to be flowing in the coil and the total magnetic energy resulting from this load current is computed and equated to $\frac{1}{2}L_L I^2$. In computing the leakage inductance of a transformer on this basis the first example to be considered is a transformer with a one-layer primary and a one-layer secondary of equal length. Since the primary and secondary ampere turns are approximately equal in magnitude, and since the currents flow in opposite directions around the core, the field between the two winding layers is very nearly equal to that within a solenoid having the same number of ampere turns. Since the length of the coil is usually large in comparison with the distance between layers, the solenoid may be considered infinite in length, and practically all of the energy in its magnetic field may be considered to be inside the solenoid.

The simplest type of winding consists of a one layer primary (low voltage) winding and a one layer secondary (high voltage) winding. A second type consists of a one layer primary and a secondary broken into two equal layers, both outside the primary (or, the secondary may be broken into three or more layers.) A third general type consists of a primary interleaved between two equal layers of a secondary. Two two layer coils can be put on the two legs of a transformer, and the primaries and secondaries both connected in parallel, the primaries and secondaries both connected in series, or the primaries connected in parallel and the secondaries in series. Two-leg variations on the second and third general type can also be constructed. Each of the above winding arrangements can be constructed as an isolation (or iso) or an auto version (that is, iso or auto for pulse operation).

The magnitude of H in oersteds (and therefore of B in gausses) is given by the following expressions:

In the primary:

$$H_1 = \frac{4\pi N_p I_p}{\mathscr{L}} \frac{x_p}{a_p} = \frac{4\pi N_s I_s}{\mathscr{L}} \frac{x_p}{a_p} \text{ oersteds}$$

where x_p is measured from the side of the primary next to the core, a_p is the thickness of the primary layer in cm, N_p is the number of turns on the primary, and N_s on the secondary. I_p is the primary current in abamperes, I_s is the secondary current, and $\mathscr{L}$ is the winding length in cm.

Between layers:

$$H_2 = \frac{4\pi N_s I_s}{\mathscr{L}}$$

In the secondary:

$$H_3 = \frac{4\pi N_s I_s}{\mathscr{L}} \left(1 - \frac{x_s}{a_s}\right)$$

where x_s is measured from the inside of the secondary.

The energy in a magnetic field in air ($\mu = 1$) is given by

$$W = \frac{H^2}{8\pi} \text{ ergs/cm}^3$$

where H is in oersteds. Therefore, the energy stored in the leakage field yields

$$W = \frac{2\pi N_s^2 I_s^2 \mathscr{U}}{\mathscr{L}} \left(\Delta + \frac{\Sigma a}{3}\right)$$

where $\mathscr{U}$ is the average circumference of the layers in centimeters, Δ is the distance between layers in centimeters, and Σa is the sum of the thicknesses of the layers. Since the energy stored in an inductance is given by $W = \frac{1}{2}LI^2$, the following equation may be solved for the leakage inductance, referred to the secondary:

$$\tfrac{1}{2}L_L I_s^2 = \frac{2\pi N_s^2 I_s^2 \mathscr{U}}{\mathscr{L}} \left(\Delta + \frac{\Sigma a}{3}\right)$$

Table E2-1 shows some design data.

TABLE E2-1

RELATIVE VALUES (IN ARBITRARY UNITS) OF L_L, C_D, AND L_LC_D CALCULATED FOR THE VARIOUS WINDING ARRANGEMENTS

Winding type	Constant length		Constant secondary-wire size		
	L_L	C_D	L_L	C_D	L_LC_D
		Windings of Fig. E2-1			
a	12.00	16.0	12.00	16.0	192
b	6.75	21.3	9.00	16.0	144
c	6.00	32.0	12.00	16.0	192
d	4.22	34.7	11.25	13.0	146
e	4.30	37.3	12.89	12.4	161
f	3.50	40.0	14.00	10.0	140
g	4.00	37.3	8.00	18.7	149
h	2.53	58.6	6.75	22.0	149
		Windings of Fig. E2-2			
a	4.50	36.0	9.00	18.0	162
b	2.81	48.3	7.50	18.1	136
c	6.00	32.0	6.00	32.0	192
d	3.38	42.7	4.50	32.0	144
e	4.00	37.3	8.00	18.7	149
f	2.53	51.0	6.75	19.1	129
g	3.00	64.0	6.00	32.0	192
h	2.11	69.3	5.63	26.0	146

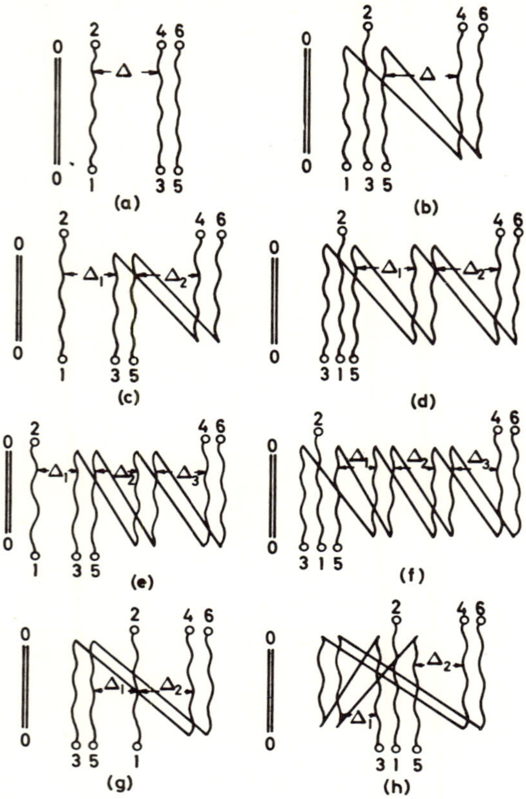

$$(a)\ L_L = \frac{4\pi N_s^2 \mathcal{U}}{d}\left(\Delta + \frac{\Sigma a}{3}\right);\quad (b)\ L_L = \frac{4\pi N_s^2 \mathcal{U}}{d}\left(\frac{n-1}{n}\right)^2\left(\Delta + \frac{\Sigma a}{3}\right);$$

$$(c)\ L_L = \frac{4\pi N_s^2 \mathcal{U}}{d}\left(\Delta_1 + \frac{\Delta_2}{4} + \frac{\Sigma a}{3}\right);\quad (d)\ L_L = \frac{4\pi N_s^2 \mathcal{U}}{d}\left(\frac{n-1}{n}\right)^2\left(\Delta_1 + \frac{\Delta_2}{4} + \frac{\Sigma a}{3}\right);$$

$$(e)\ L_L = \frac{4\pi N_s^2 \mathcal{U}}{d}\left(\Delta_1 + \frac{4\Delta_2}{9} + \frac{\Delta_3}{9} + \frac{\Sigma a}{3}\right);$$

$$(f)\ L_L = \frac{4\pi N_s^2 \mathcal{U}}{d}\left(\frac{n-1}{n}\right)^2\left(\Delta_1 + \frac{4\Delta_2}{9} + \frac{\Delta_3}{9} + \frac{\Sigma a}{3}\right);$$

$$(g)\ L_L = \frac{4\pi N_s^2 \mathcal{U}}{d}\frac{1}{4}\left(\Delta_1 + \Delta_2 + \frac{\Sigma a}{3}\right);\quad (h)\ L_L = \frac{4\pi N_s^2 \mathcal{U}}{d}\frac{1}{4}\left(\Delta_1 + \Delta_2 + \frac{\Sigma a}{3}\right)$$

FIG. E2-1. Various winding schemes that have been used in the construction of pulse transformers for magnetrons. By permission from Glasoe and Lebacqz, "Pulse Generators," McGraw-Hill, 1948.

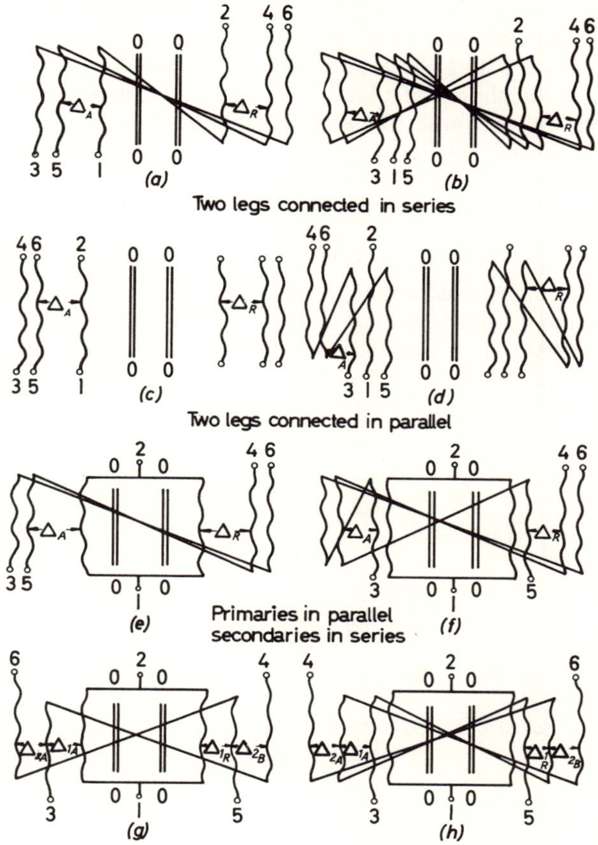

$$(a)\ L_L = \frac{4\pi N_s^2 \mathscr{U}}{\mathscr{L}} \frac{1}{4} \left(\Delta_A + \Delta_B + \frac{\Sigma a}{3} \right); \quad (b)\ L_L = \frac{4\pi N_s^2 \mathscr{U}}{\mathscr{L}} \frac{1}{4} \left(\frac{n-1}{n} \right)^2 \left(\Sigma\Delta + \frac{\Sigma a}{3} \right);$$

$$(c)\ L_L = \frac{4\pi N_s^2 \mathscr{U}}{\mathscr{L}} \frac{1}{4} \left(\Delta_A + \Delta_B + \frac{\Sigma a}{3} \right); \quad (d)\ L_L = \frac{4\pi N_s^2 \mathscr{U}}{\mathscr{L}} \frac{1}{4} \left(\frac{n-1}{n} \right)^2 \left(\Sigma\Delta + \frac{\Sigma a}{3} \right);$$

$$(e)\ L_L = \frac{4\pi N_s^2 \mathscr{U}}{\mathscr{L}} \frac{1}{4} \left(\Delta_A + \Delta_B + \frac{\Sigma a}{3} \right); \quad (f)\ L_L = \frac{4\pi N_s^2 \mathscr{U}}{\mathscr{L}} \frac{1}{4} \left(\frac{n-1}{n} \right)^2 \left(\Sigma\Delta + \frac{\Sigma a}{3} \right);$$

$$(g)\ L_L = \frac{4\pi N_s^2 \mathscr{U}}{\mathscr{L}} \frac{1}{4} \left(\Sigma\Delta_1 + \frac{\Sigma\Delta_2}{4} + \frac{\Sigma a}{3} \right);$$

$$(h)\ L_L = \frac{4\pi N_s^2 \mathscr{U}}{\mathscr{L}} \frac{1}{4} \left(\frac{n-1}{n} \right)^2 \left(\Sigma\Delta_1 + \frac{\Sigma\Delta_2}{4} + \frac{\Sigma a}{3} \right)$$

FIG. E2-2. Various winding schemes that have been used in the construction of pulse transformers for magnetrons. By permission from Glasoe and Lebacqz, "Pulse Generators," McGraw-Hill, 1948.

Like leakage inductance, the effective distributed capacitance of various types of windings can be calculated from general formulas to be found in many different texts and handbooks. The considerations in this section are confined to the calculation of effective distributed capacitance for the various types of power-output pulse-transformer windings that have been developed.

The calculation of distributed capacitance as a lumped parameter follows the process of reasoning wherein a given voltage V is placed across one winding of the transformer and the total energy W stored in the electrostatic field is equated to $\frac{1}{2}C_D V^2$, where C_D is defined as the effective distributed capacitance.

The effective distributed capacitance of a pulse-transformer winding can also be measured by means of an rf bridge.

It is shown that $\sqrt{L_L/C_D}$ may be thought of as a characteristic impedance of a winding for a transformer. For a simple coil with a one-layer primary and a one-layer secondary winding of equal length, and with negligible winding-layer thickness,

$$L_L = \frac{4\pi N_s^2 \Delta \mathcal{U}_i}{\mathcal{L}} \times 10^{-9} \text{ henrys}$$

and

$$C_D = \frac{0.0885\, \epsilon \mathcal{U} \mathcal{L} f_1}{\Delta} \times 10^{-12} \text{ farads}$$

f_1 is a fraction depending upon the voltage distribution between the primary and the secondary.

Then the characteristic impedance is

$$Z_T = \sqrt{\frac{L_L}{C_D}} = \frac{377 N_s \Delta}{\mathcal{L}\sqrt{\epsilon f_1}} \text{ ohms}$$

The concept of characteristic impedance is, however, more meaningful when applied to circuits with distributed rather than lumped parameters. A simple example of a circuit having distributed inductance and capacitance and a characteristic impedance Z_0 is one composed of two long parallel strips of conducting material, 1 meter wide and 1 meter apart, which are part of two conducting sheets infinite in extent [612]. If the permeability and dielectric constant of the material between the sheets are respectively μ_1 and ϵ_1 in mks units, it can be shown that, for a plane electromagnetic wave propagated between the two strips,

$$Z_0 = \frac{E}{H} = \sqrt{\frac{\mu_1}{\epsilon_1}} = \sqrt{\frac{\mu}{\epsilon}} \times \sqrt{\frac{\mu_0}{\epsilon_0}} = \sqrt{\frac{\mu}{\epsilon}}\, 377 \text{ ohms}$$

where μ_0 and ϵ_0 are, respectively, the permeability and the dielectric constant of free space in mks units, and 377 ohms is the characteristic impedance of free space.

If the spacing between the strips is reduced to Δ meters and the width of the strips becomes $\mathcal{L}$ meters, and $\mu = 1$,

$$Z_0 = \frac{\Delta}{\mathscr{L}\sqrt{\epsilon}}\ 377 \text{ ohms}$$

and $\mathscr{L}$ and Δ may be measured in cm or meters.

If these parallel strips are now wrapped into two concentric cylinders whose circumferences are large compared with the distance Δ between the cylinders and whose length is $\mathscr{L}$, the characteristic impedance for a plane electromagnetic wave traveling circumferentially in the space between the two cylinders is still equal to $(\Delta/\mathscr{L}\epsilon^{\frac{1}{2}})$ 377 ohms.

If the outer and inner cylinders are now slit helically so that they become coils of N_s and N_p turns respectively, and the electromagnetic wave is ushered in at one end of the coil and out at the other end, the inductance per unit length of circumference is greater by a factor of N_s^2 (for the secondary) than it is for the unslit cylinders. The capacitance per unit length of circumference is different by a factor of f_1, depending upon the particular type of voltage distribution that is set up beween the primary and secondary. Therefore, the characteristic impedance of the secondary winding may be expressed by

$$Z_T = \frac{N_s\Delta}{\mathscr{L}\sqrt{\epsilon f_1}}\ 377 \text{ ohms}$$

When a pulse transformer is operated into a load of low dynamic impedance (for example, a magnetron), oscillations whose period is usually considerably less than $2\pi(L_LC_D)^{\frac{1}{2}}$ often appear on the top of the current pulse. These oscillations may be considered to be the result of the transmission and reflection in the coil of disturbances that are probably initiated at the end of the secondary to which the oscillator is attached. The impedance of the secondary winding to the transmission of such disturbances is

$$\frac{N_s\Delta}{\mathscr{L}\sqrt{\epsilon f_1}}\ 377 \text{ ohms}$$

The time taken for the disturbance to travel along a length of secondary wire of N_s' turns to the first discontinuity in the impedance is

$$\delta = \frac{N_s'\mathscr{U}}{c}\sqrt{\epsilon} \text{ sec}$$

where $\mathscr{U}$ is the mean length of turns of the secondary winding in centimeters, c is the velocity of light in centimeters per second, and N_s^1 is the number of turns on the secondary winding between the high voltage end and the first sharp discontinuity in impedance (μ, of course, is equal to unity for the space between the coils). The time transpired before this disturbance is reflected back to the source is 2δ, and when, upon successive reflections that set up shock-excited standing waves, the fluctuations of voltage and current at the source take on the approximate semblance of a sine wave, the period T of the fundamental of these oscillations is either 2δ or 4δ, depending upon the nature of the discontinuity in impedance. The effect is somewhat analogous to the two different fundamental periods of oscillation that are set up in closed and open organ pipes.

Many successful pulse transformers have been designed without consideration or cognizance of certain relationships between the elements in the equivalent circuit and the transformer rating and efficiency.

In the early history of pulse transformers it was customary to use the following design procedure: The magnetizing current I_m at the end of the pulse was permitted to be equal to 0.1 of the load current I_l. Then, from the relationships

$$\Delta\bar{B} = \frac{10^8 V\tau}{NA}$$

$$H = \frac{4\pi N I_m}{10 l}$$

and

$$\Delta\bar{B} = \mu_e H$$

the core volume was computed to be

$$Al = \frac{4\pi \cdot 10^7 \mu_e V\tau I_m}{(\Delta\bar{B})^2} = \frac{4\pi \cdot 10^6 \mu_e V\tau I_l}{(\Delta\bar{B})^2}$$

where

$\Delta\bar{B}$ is the increment in gauss (at the end of the pulse) of the average flux density of the core above the remanent value of B;
H is the magnetic field in oersteds created by I_m;
A is the cross sectional area of the core in square centimeters;
l is the mean magnetic path length of the core in centimeters;
μ_e is the effective permeability of the core;
V is the pulse voltage; and
N is the number of turns on the winding to which is applied the pulse voltage V.

A guess was made as to a suitable value of $\Delta\bar{B}$, the core volume Al was computed, and a core having this product, for which the factors A and l were suitably estimated, was then chosen. The value of N was computed, and a simple winding scheme selected. The insulation thickness Δ was chosen so that it would be sufficient to withstand the voltage puncture stress, and sufficient coil margins were left to withstand the voltage creep stress.

Formerly, an operational requirement frequently placed on a pulse transformer was the specification of $t_{r,\,\max}$, the maximum allowable time taken for the pulse to rise to 0.9 of the full amplitude of the output of the pulse transformer on a resistance load R_l. For the transformer and load, the resultant time of rise t_r, usually computed without a consideration of the effective distributed capacitance, was obtained from the expression

$$0.9 = 1 - e^{-(R_l/L_L)t_r}$$

$$t_r = -\frac{L_L}{R_l}\ln 0.1 = \frac{L_L}{R_l}\ln 10$$

If the computed t_r exceeded the specified $t_{r,\,\max}$, L_L had to be reduced by reducing N or A, thereby increasing $\Delta\overline{B}$, or reducing Δ. If the computed t_r was less than $t_{r,\,max}$, the design was considered satisfactory.

There are, of course, complicating factors such as the saturation characteristics of the core, the range of pulse durations that be passed, and the magnitude of the backswing voltage that can be tolerated. A compromise must often be struck between these circumstances. Also, except under unusual circumstances, an approximate core size must be used because, for economic reasons, cores are manufactured in a set of standard sizes, finite in number.

W. H. Bostick [611] shows, that it is helpful at the outset to express the second law of thermodynamics in the following way: The equilibrium or most probable state of an unisolated system is usually achieved when the energy of that system has become a minimum. In a corollary of the second law of thermodynamics, the law of equipartition, it may be stated that the energy of the system tends to be distributed equally to each degree of freedom that enters quadratically into the expression of the energy of the entire system.

In the pulse transformer, the total volume of the coil is assumed to be equal to the space between the primary and secondary windings, and the total volume of the transformer in which energy is stored is assumed to be equal to the sum of the volumes of the core and coil. If this transformer is to be used to transform and transfer a pulse of electromagnetic energy, it is obviously desirable to have a minimum of energy in the transformer at the end of the pulse, that is, to leave the transformer in a minimum energy state. If the energies stored in the core and coil are considered, by way of a thermodynamical analogue, to constitute two degrees of freedom for the transformer, the equilibrium or minimum energy state of the transformer is that, according to the second law of thermodynamics, the equilibrium (and hence the minimum energy) state of the transformer is (among other things) that in which the "electromagnetic temperature" is constant throughout the transformer, and hence that in which the energy densities of core and coil are equal.

Here electromagnetic temperature may be thought of as analogous to ordinary temperature which, in a gas for example, is proportional to the average kinetic energy density.

The procedure given by Bostick bears some formal resemblance to general practice in the design of power transformers. In power transformers operating at a constant frequency, the energy stored in the coil and core is, for the most part, not lost but returned to the circuit. It is the sum of the average power dissipated in the iron of the core and the copper of the coil that is of primary importance, It might be said by way of thermodynamical analogue that a power transformer will remain in its lowest or equilibrium energy state if the rate of dissipation of energy per unit volume is approximately constant throughout its whole volume. Minimum total power will then be dissipated in the transformer when the transformer volume is a minimum (consistent, of course, with the saturation properties of the core and maximum temperature characteristics of the insulation). The transformer volume will be a minimum when the perimeters and volumes of core and coil are approximately equal. The resultant power transformer will thus have total iron losses and copper losses that are approximately equal.

The average energy density of the coil (that is, of the space between the primary and secondary) is

$$\frac{\epsilon f_1 V^2}{9 \times 10^4 \times 8\pi\Delta^2} + \frac{H^2}{8\pi} \text{ ergs/cm}^3$$

where

ϵ is the dielectric constant of the insulation;
V is the pulse potential of the high-voltage winding in volts;
Δ is the spacing between the primary and secondary in cm;
f_1 is a constant depending upon the voltage distribution between the windings that results from a given winding configuration and step-up ratio;
$H = 4\pi NI/10\mathscr{L}$ oersteds;
$N =$ the number of turns on the high-voltage winding;
$I =$ the load current in that winding in amperes;
$\mathscr{L} =$ the length of the windings in cm (the primary and secondary windings are equal in length).

3. High Peak Power Transformers with Differentiating Characteristics

While pulse transformers normally have a very small leakage inductance and a high efficiency, but relatively small pulse power, another type of transformer called a peak power or energy transformer has a high leakage inductance but allows production of peak power amounts up to 2000 megawatts and higher.

Pulse transformers operate in accordance with the circuit shown in Fig. E3-1. Here again, the charging current source is indicated at 3 and can be in the form of a powerful rectifier. However, only one charging resistance or charging choke, 2, is required here and the capacitor, which supplies the total discharge energy, consists of a single battery or a single unit, 1. Moreover, only one spark gap, 4, with an ignition electrode, 6, is necessary. The discharge method also differs from the impulse voltage circuit described above because here the total capacitor energy is passed into the primary of a pulse transformer having the primary 10 and the secondary 11.

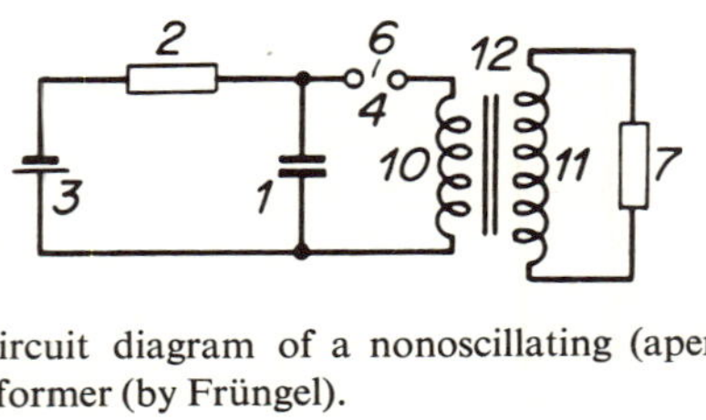

FIG. E3-1. Schematic circuit diagram of a nonoscillating (aperiodically damped) discharge through pulse transformer (by Früngel).

The iron core is indicated at 12; the external load is represented by a resistor at 7. When the discharge is released by means of the spark gap, 4–6, the full capacitor voltage, under the condition that the inductance or impedance of the discharge circuit is sufficiently low, becomes effective in the few turns of primary, 10. Winding turn voltages in the range between 3 and 30 kv have proven to be practical. The value of 3 kv is good for very small pulse transformers having core cross sections of a few square centimeters, and the value of 30 kv is good for large units with core cross sections of more than 100 cm^2.

Figure E3-2 shows the curves of a discharge plotted against time. The current rising in accordance with curve 2 creates in proportion to the magnetizing characteristic a uniformly increasing magnetic flux in iron core 12. Correspondingly the voltage induced in the secondary rises rather steeply to its maximum value. It does not drop until the current curve becomes flatter, and it has its zero value when the current curve is at its maximum. Note that the secondary voltage remains approximately constant for a short duration.

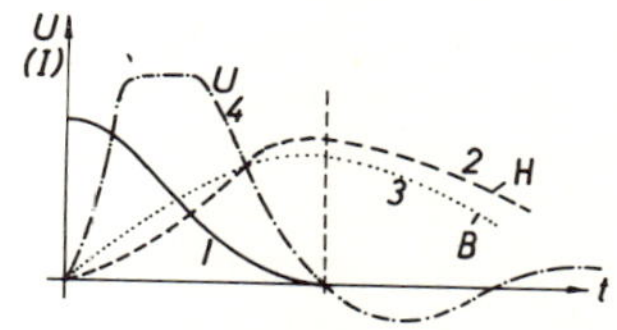

FIG. E3-2. Wave forms created by a discharge. Primary voltage (1); magnetic excitation *H* (2); magnetic induction *B* (3); and secondary voltage (4) in a circuit according to Fig. E3-1.

The further progress of current and magnetic field is not plotted in the graph, since it is not of interest for the discharge mechanism proper. Only the secondary output voltage to be expected has been plotted by continuation of the dot-dash line (on a contracted time axis).

Reversal of secondary voltage direction has the following physical reason. The wave form of *B* (3 in Fig. E3-2) shows first an increasing then a decreasing magnetic flux in the same direction. According to the law of electrical induction there must be induced during flux decrease a voltage that has negative direction with respect to the voltage induced during flux increase. The voltage passes through zero at flux maximum during the instant when there is no change in flux value. The occurrence of positive and negative partial oscillations of a similar current sum but of different voltage, that is, of different energy, is physically necessary because in the secondary of each transformer the positive and the negative charge values resulting from current × time, must be equal to each other. From this consideration it will be clear that it is impossible to suppress the undesirable negative half-waves completely; but it is possible to reduce the peak values to a harmless level by retardation of the zero passages with suitable means whereby the areas above and below the zero line in the diagram remain equal.

Since in voltage impulse transformers, not the energy efficiency but the realizable peak value of the secondary impulse voltage and its peak energy are of importance, it is possible to achieve this aim by selection of a magnetic core material that is highly magnetizable but produces high losses during demagnetization. The peak voltage value will thereby not be adversely affected, but the high losses slow down the drop-off of the magnetic field with the result that only a small induced secondary emf is produced. When reduced to a value of 10% of the main impulse, this emf does not materially disturb test results. A prognosis as to the voltage progression after passing through zero, however, can not be practically made. The worse the core material in respect to hysteresis losses (here all the considerations of Chapter D, Section 2 are in a proper sense applicable), the slower and less harmful is the progression of the negative voltage half wave. Low quality, relatively thick, soft steel sheets are well suited for this purpose.

The closer the coupling between secondary 11 and primary 10, the more the excitation of iron core 12 is equalized by counterexcitation through the load current flowing in the secondary. The iron core will then be less excited at increased secondary load and the hysteresis losses will be lower too. The efficiency of the arrangement (energy of the output impulse/discharge energy) increases at first with rising secondary current flow and drops later to zero at zero voltage of the secondary, that is, with short-circuited secondary. Impulse current transformers having close coupling are suitable to impress the primary energy on the load even under mismatch conditions. With loose coupling they assure a repeatedly constant current wave form and with resistance loads a good form of the voltage wave [$U = (dI/dt) \times R$].

The discharge will thereby never tend to oscillate even with the secondary unloaded because of the great hysteresis loop caused by the low-grade core material.

The following considerations will prove worthwhile in selecting the core material and the thickness of the core sheets:

During the initial upswing of the magnetic energy in the core, 12, the magnetic field first penetrates the whole iron core. A skin effect cannot occur, since growing fields, when originating from primeval condition, penetrate a conductor or a magnetic material completely. Skin effect or other surface phenomena develop only during some alternating action or during a change in rate of growth of fields. Owing to these fundamental properties of swing-in phenomena during the instant of switching on, it is possible to use almost any arbitrary ferrous sheet material for the core, 12, in impulse transformers of this type, as it is certain to be penetrated by the field. Thin sheets will serve as well as thick sheets of 0.2 mm. Only in heavier sheets of 0.35 or 0.5 mm thickness can there be observed a drop in voltage. This drop can be attributed to eddy currents induced after the first magnetizing rise.

It is very important that the single core laminations are well insulated from one another. The voltage per winding turn, as already mentioned, can be 3–30 kv and a similar voltage will be induced around the periphery of the core. This means that the core laminations must be insulated to safely withstand such potentials in service.

Another consideration might be of benefit. When, for instance, the size of the capacitor, 1, is about 1 μf and the inductance of the coil 10 is 1 μh, the resulting discharge times will be of the order of 1 μsec. The ferromagnetism in the iron core can spread only to a small extent during such a short impulse duration. One finds, therefore, a steep gradient of the magnetic field in cores of such impulse transformers at only a few centimeters' distance from the origin of excitation, the surrounding primary winding. Where excitation through the coil itself is not given, a magnetic flux is practically not present. In such transformers it is therefore useless to employ closed magnetic cores, since during its short life the impulse proper is not able to utilize the lower magnetic reluctance created by a closed magnetic circuit. As long as during the short impulse duration there is sufficient core volume for magnetic field expansion or for orientation of sufficient magnetic domains, it is without consequence whether the magnetic circuit outside the coil field is closed or not. For rough approximation it can be said that the iron core should project at each end of the coil for about 1/4 of the coil length.

The design of such impulse transformers involves no great problems. Since an open core effects a reduction of magnetic reluctance in the coil by the factor of permeability, and since the permeability of iron, being 50–100, is high with respect to air, it is permissible to assume the iron core as a magnetic short circuit within the coil. In accordance with the elementary laws of solenoids, 9/10 of the magnetic energy lies within the coil and only 1/10 in the field outside of the coil; consequently there is also 9/10 of the reluctance within and 1/10 outside the coil.

$$E_m = \tfrac{1}{2}LI^2 = \tfrac{1}{2}n^2\Lambda I^2$$

$$n \times I = \Theta; \quad \Lambda = \frac{1}{R}; \quad R = R_i + R_a; \quad R_a = \frac{1}{9}R_i \rightarrow R = \frac{10}{9}R_i$$

Therefore

$$E_m = \frac{1}{2}\frac{\Theta^2}{(10/9)R_i} \quad \text{and} \quad \frac{9}{10}E_m = \frac{1}{2}\frac{\Theta^2}{R_i}$$

where

E_m	= magnetic energy	Λ	= permeance
L	= inductance	R	= reluctance
n	= number of turns	R_i	= reluctance within coil
I	= amperes in coil	R_a	= reluctance outside coil
Θ	= flux		

By insertion of a magnetic core, the reluctance within the coil is reduced to 1/10 of the value that would be present with only air in the coil. The inductance of the primary becomes thereby 10 times as high as would be derived from its physical dimensions as an air solenoid. A complete design example of an impulse transformer setup follows.

Problem: An impulse transformer shall subject a test sample having 5000 pf capacitance to a shock impulse potential of 500 kv. The charging voltage of the capacitor is limited by construction considerations to 25 kv. To meet this requirement the impulse transformer must have a no-load ratio of 1:20, and taking some load into consideration, it is advisable to add 10%; thus a ratio 1:22 will be chosen. The energy input to the test sample of 5000 pf becomes $\frac{1}{2}C \times U^2 = 625$ watt-sec. Since the test sample should not weaken the impulse voltage source proper more than 10%, which is the same percentage referred to in Marx's impulse voltage circuits, it is necessary to charge the primary capacitor to a 10-fold energy content, that is, 6.25 kw-sec at a charging voltage of 25 kv. This results in a capacitance value of

$$C = \frac{2 \times 10 \times 625}{U^2} = 20\ \mu\text{f}$$

at a charging voltage of 25 kv. In order to give the impulse a sufficient steepness or a short rise time to the peak of, e.g., 2 μsec at an impulse duration of a few additional μsec, it will be assumed that the primary winding of the impulse transformer has 5 turns, and thus 5 kv/turn. The magnetic core might have a length of 800 mm and a square cross section of 150·150 mm². Since the core is to project at each end of the coil for 1/4 of the coil length, the primary coil will have a length of about 550 mm. Thus, the inductance of the primary coil, when figured by the elementary formulas for solenoids, becomes

$$L = \frac{10 \times d^2 \times n^2}{L}$$

$$L_{\text{iron}} = 10 \times L_{\text{air}}$$

(10 = effective permeability factor of iron filled internal field of coil)

$$L = \frac{10 \times 10 \times 225 \times 25}{55} \times 10^{-9}\ \text{h}$$

$$L = 10^{-5}\ \text{h}$$

The duration of 1/4 cycle, approximately equivalent to the duration of the secondary impulse voltage maximum, can be derived thus

$$T = \frac{\pi}{2} \times \sqrt{L \times C}$$

$$T = 1.6 \times \sqrt{10^{-5} \times 2 \times 10^{-5}}$$

$$T = 25 \times 10^{-6}\ \text{sec}$$

Consequently, the test sample capacitor will be subjected to a shock voltage impulse of 25 μsec duration. After this the shock decays slowly and nearly in accordance with the decay time of 50 μsec as generally standardized. The rise time lies in the neighborhood of 2 μsec. By insertion of a series spark gap between winding and test sample the rise time can be reduced to 1 μsec. It then meets the standard wave form specification of 1/50 of the decay time. From this discussion it is apparent that one can reach the goal with small physical dimensions and elementary construction forms, whereby secondary voltage of any desired magnitude can be tapped off and the primary voltage can be made continuously variable without change of a spark gap.

Should it be desirable to check the performance on an ohmic resistance, it can be assumed from calorimetric measurements that the charging energy stored in the primary capacitor, 6.25 kw-sec in the present case, is available on the secondary side at an efficiency of about 70%. This would result in a usefully available energy of 4500 watt-sec. Since the impulse duration has been set at 25 μsec, there can be withdrawn as average output

$$\frac{4500 \text{ watt-sec}}{25 \times 10^{-6}} = 180 \text{ megawatts}$$

In accordance with

$$R = \frac{U^2}{N}$$

the permissible load resistance becomes

$$R = \frac{25 \times 10^{10}}{180 \times 10^6} = 1400 \text{ ohms}$$

It is also possible to subject resistors with a resistance as low as 1400 ohms to voltage surges up to 500 kv with the circuit just described.

With the ratio of 1:22 as selected and a primary of 5 turns the secondary has 110 turns. The secondary can be provided with taps and one then has the choice of lower voltages and higher loading possibilities. Care, however, must be taken that the close coupling between primary and secondary is not impaired by such tapping, because otherwise energy will be lost, that is, simply wasted in the core.

Practical circuit arrangements of this type provide for impulse voltages up to 2 megavolts at a voltage of 10 kv per turn. They have only 200 secondary turns and can be accommodated in a space of about 2 meters length. The housing is preferably an oil tank of synthetic resin (Trovidur) or plastic in order to avoid eddy currents.

In practice, furthermore, mechanical impact switches are preferred instead of the spark gap indicated in Fig. E3-1 and thereby a voltage-independent switching means (in contrast to the Marx circuit) is gained. Such switches are usually electromagnetically released and manually reset.

Figures E3-3 and 4 show a complete schematic circuit diagram and a diagrammatic view of the impulse transformer. The discussion herein is in part a description of this installation. It has the capacity to deliver about 180 to 200 megawatts to external resistors or, with capacitive load of 5000 pf, 500 kv impulse voltage. Stepless voltage control is another feature. For that purpose the voltage source takes the form of a variac with subsequently arranged high-voltage transformer and rectifier, so that at the impulse capacitor of 20 μf there are available all voltages in the range between 0 and 25 kv.

It is worth mentioning that for measuring purposes a voltage divider is arranged parallel to the secondary coil, which has about 50 kilohms resistance and consumes only a negligible trifle of the impulse energy. At the lower 5% tap of this resistance the output voltage can be measured by means of a precision measuring spark gap with ultraviolet irradiation. At 600-kv no-load voltage, which can generally be attained at 25-kv capacitor voltage, there are

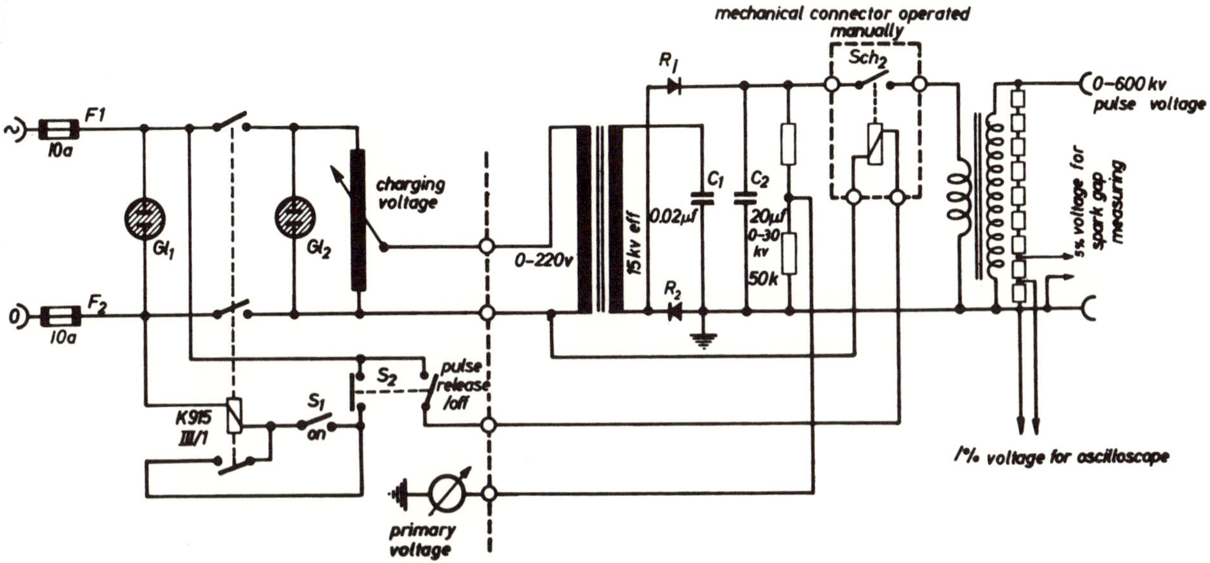

FIG. E3-3. Complete schematic circuit diagram of a 600 kv impulse voltage installation with impulse transformer for stepless voltage control. Release by Sch 2. The measuring voltage divider introduces in spite of its small resistance of 50 kilohms only a low loss. At 5000 pf, 500 kv, with resistance load 180 megawatts, impulse duration 25–50 μsec.

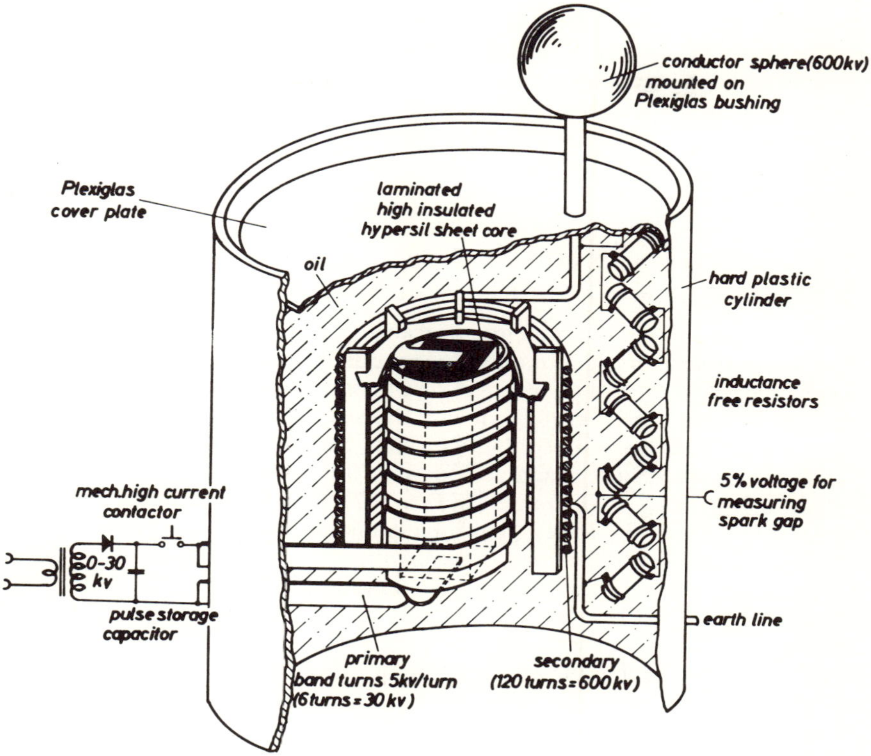

FIG. E3-4. Impulse transformer assembly for the 600 kv installation shown in Fig. E3-3 with stepless adjustment for voltage values from 0 to 600 kv, shock load output 180 megawatts, permissible capacitive load 5000 pf.

effective at the 5% tap, 30 kv. Consequently, the measuring spark gap is constructed as a 1-cm sphere spark gap having micrometer calibration so that very accurate readings can be taken within the measuring range from 3 kv to 30 kv. For measurements within the μsec range, a short spark measuring length is mandatory in every case. Besides, with 1% resistance values an oscilloscope can be connected which permits tracing the voltage wave at the sample under test.

The great advantage of the setup just described over the impulse voltage circuit with series-connected capacitors is that with a short-circuited discharge switch the capacitor remains automatically empty and cannot recharge itself. The negligible resistance of the transformer coil forms an almost perfect short circuit. Only after pressing the push button S_1 is recharge possible. Another advantage of the transformer method is that, by way of the secondary winding, there is a basically galvanic connection between the transformer body and the high-voltage terminal. A residual charge can never remain there. The operation of such impulse voltage transformers requires, therefore, less attention and is more convenient for the operators since there is practically no danger of electrical shock. At the start of a test there is a conscious action necessary, namely winding and setting of the switch mechanism; then return to the control desk; adjustment of charging voltage relative to the desired impulse voltage, and thereafter push button actuation with both hands, so that the operator is locally bound and safe from dangerous high voltage.

4. Impulse Setups for Cyclic Operation and Very Short Impulse Duration

In accordance with the principle of transformed discharges just discussed, it is also possible to produce impulses of considerably shorter duration for such uses as radar service and the production of corpuscular radiation. In utilizing the compensation method for reducing the self-inductance of the capacitor (see Chapter C, Section 6), so that capacitor batteries of only about 0.02 μh effective inductance are available, one can use inductances of the impulse transformer primary in the range between 0.2 to 1 μh. If we accept an inductance of 0.2 μh as the lowest realizable value and assume a nonoscillatory condition in the discharge circuit, then, with an impulse capacitor of 0.1 μf, the voltage impulse duration, equal to 1/4 cycle, will become 2.5×10^{-7} sec. Thus one can expect impulse durations of 0.3 μsec as an average. The small inductance of 0.2 μh here necessary can be realized with a one-turn primary of a transformer having the physical size of a commercial 200-watt core-type transformer. Also since here it is senseless to have the core project much beyond the coil but, on the other hand, it is desirable to surround the coil by a steel jacket in order to avoid external stray flux, it has been found very practical to use an E-shaped core of E-shaped laminations. The bridging yoke can be omitted. The E core surrounds the coil on two sides and projects somewhat on its open end.

If a voltage of about 10 kv is applied to the one winding turn by means of a quick-response spark gap, a three-electrode spark gap, or a thyratron, a six-turn secondary will produce an output voltage of 60 kv. Although the sparking distance, according to Binder [19], sags with fast voltage impulses considerably more than with a static voltage in oil, it is possible, even with the small dimensions given by the commercial 200-watt core, to produce in this normal winding space a winding system which has the 6-turn secondary and sufficient insulation for 60 kv in oil. Thus one obtains an impulse generator that has an output capacity on the order of 2–10 megawatts and renders impulses of 0.3 μsec duration with half sinusoidal wave form. The secondary voltage is thereby variable between about 30 and 60 kv, depending on the limitation of the controlling spark gap or thyratron. However, for obtaining such outputs it is important to provide sufficient cooling means, because in periodic operation, at 500 impulses/sec, considerable losses in the iron core arise. Effective cooling can be had by mounting the transformer in a circulating oil tank. A typical design for such construction will now be described.

The assembly drawing presented in Fig. E4-1 shows two examples. In the shock (impulse) transformer of Fig. E4-1a, the cube-shaped iron core, 1, of commercial transformer sheets carries on insulation sleeve 2 the primary winding, 3, of only 2 or 3 turns. This primary lies (as in Fig. E3-1) in series with

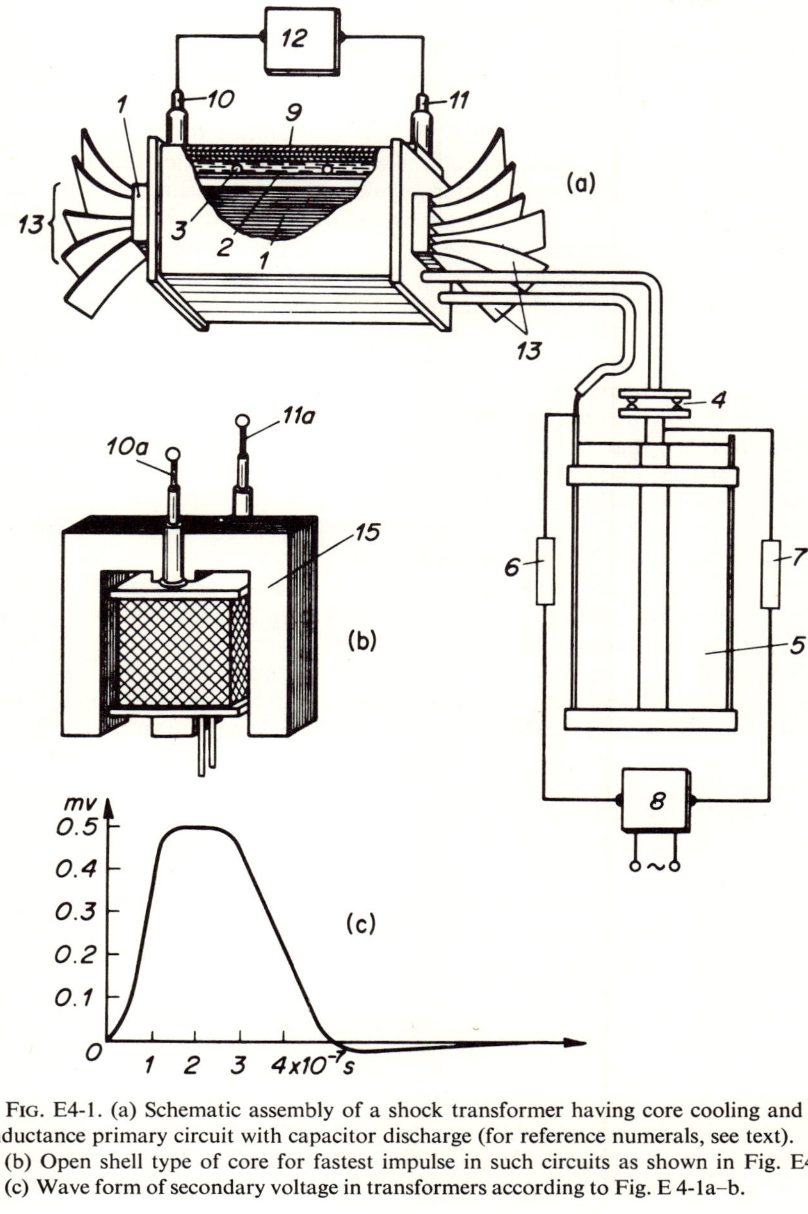

FIG. E4-1. (a) Schematic assembly of a shock transformer having core cooling and low inductance primary circuit with capacitor discharge (for reference numerals, see text).
(b) Open shell type of core for fastest impulse in such circuits as shown in Fig. E4-1a
(c) Wave form of secondary voltage in transformers according to Fig. E 4-1a–b.

the spark gap, 4, on the terminals of the capacitor, 5, which is charged through resistors 6 and 7 from the voltage source, 8.

The secondary, 9, of about 6–12 turns is wound outside the primary with an interposed heavy insulation layer. The load, 12, is connected at the ends, 10 and 11, of the secondary.

By means of proper dimensioning and laminating of the iron core, 1, the degree of coupling between coils 3 and 9 is brought closer to unity, so that the resistance of the load, 12, is fully integrated in the damping of oscillatory circuits, 3, 4, and 5. The energy not consumed in the main impulse is destroyed through iron losses in the core, 1, which is essential for achieving non-oscillatory (approaching the aperiodic marginal) condition by damping. Size and subdivision of the core are preferably determined experimentally for each apparatus type with the guidance of oscillograms taken of the voltage impulses at the load. For fine adjustment there serve the copper laminations, 13, which are interposed in the iron core, 1, for improving heat conduction toward the core surface. For this additional purpose the number of such laminations can be readily varied and for fine adjustment, cuts can be made in their outwardly spreading external projections, if, for instance, the leading edge of the impulse waveform is critical. Copper or aluminum sheets are inserted in the core in such quantity as appears necessary to prevent overheating of the core by loss heat that could lead to damage of the interlaminar insulation.

Figure E4-1b shows a transformer having an E-shaped core, 15, whose stem portion carries the winding. Transformers of this type have proven suitable for Fourier main frequencies of over 0.8×10^6cps. With their assistance it is possible to obtain impulse voltages having a wave form as presented in Fig. E4-1c. The herein illustrated principle of shock transformers makes it clear that in reality they are not pulse or impulse transformers. A pulse transformer in present-day terminology is a transformer for true-to-form transformation of an impulse introduced into its primary. Only the voltage is changed in accordance with the transformation ratio. In contrast, in a shock transformer the secondary voltage pulse is produced in the transformer itself, and without this action it would be an unformed capacitor discharge.

The following example shows that the inductance of the whole discharge circuit outside the transformer can be greatly reduced by means described in Chapter C of this book. Figure E4-2a shows a schematic arrangement for the formation of shock voltages having impulse durations as low as about 10^{-6} sec, and Fig. E4-2b shows an improved version of this type for shock voltages having an impulse duration of only about 3×10^{-7} sec. Figure E4-2c shows a sketch of the windings for two series-connected capacitors, and Fig. E4-2d shows the same for two parallel-connected capacitors.

As can be seen in Fig. E4-2a, the layers of the capacitor, 1, shown in cross section, project at each end and are soldered to the sheet metal end discs, 2 and 3. Disc 2 is joined to cage 4 which consists either of a shell surrounding the capacitor or of three rods or bars arranged at equal spacing around the capacitor. End disc 3 is joined by way of the low-inductance resistor, 5, directly to one terminal of spark gap 6. The cable, 7, connected to cage 4 is arranged as closely as possible along the spark gap, 6, and leads to the primary, 8, of the

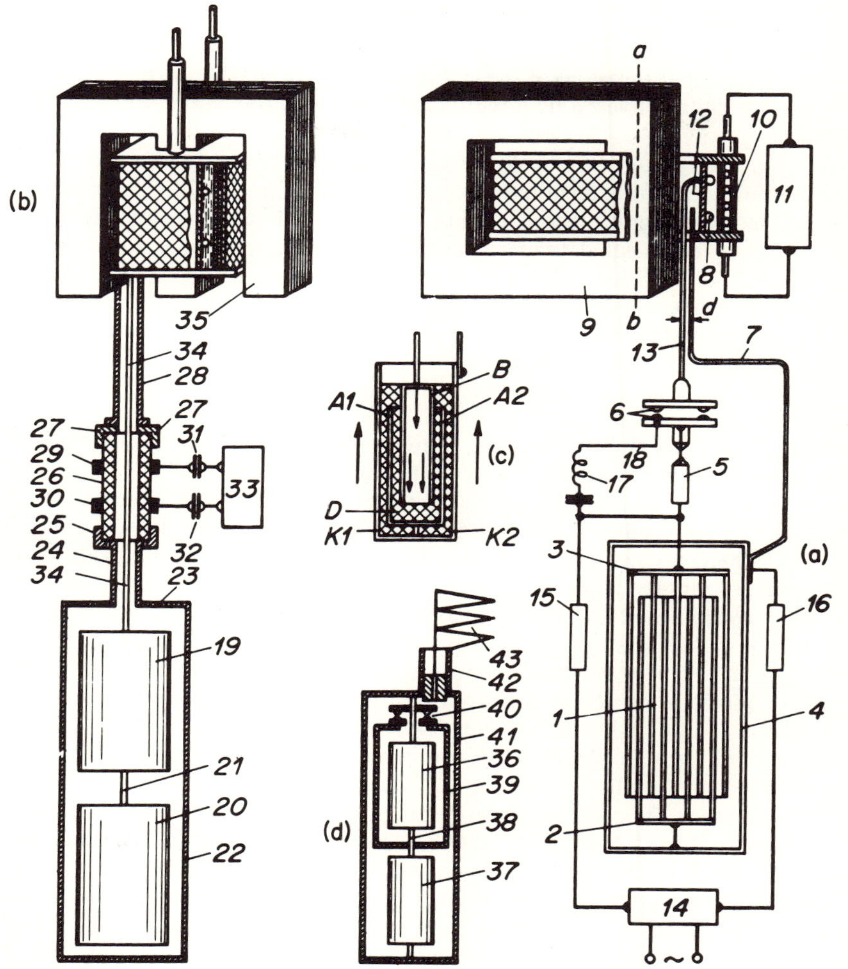

FIG. E4-2. (a) Schematic total arrangement for low-inductance input circuit of a shock transformer.

(b) Axially symmetrical arrangement of the primary capacitor circuit, shock transformer having open E core for impulse durations too short driving the magentic flux around a closed ferromagnetic circuit.

(c) Assembly scheme of a low-inductance shock capacitor.

(d) Schematic diagram of low-inductance wiring for two parallel-connected windings of shock capacitors.

transformer. This is equipped with laminated iron core 9 and its secondary 10, feeds the load, 11. The primary 8, has only two turns wound on an insulating tube, 12. Cable 7 passes through this insulating tube in parallel to tube axis a–b. Also the return cable, 13, lies parallel to the tube axis a–b in an internal jacket line of this insulating tube, 12, which has the necessary distance d from cable 7 to prevent breakthroughs. The return cable, 13, is connected to the other terminal of the symmetrical plate spark gap, 6. The supply voltage, which is delivered by a rectifier, 14, flows to the capacitor, 1, by way of traveling wave blocking resistors 15 and 16. The single discharges of the spark gap are ignited in a well-known manner by an auxiliary spark gap, which is formed by one terminal of spark gap 6 and an ignition electrode, 18, which is fed by an ignition coil, 17. The durations between the single ignitions must thereby be selected long enough to assure intermediate complete recharge of capacitor 1 from the voltage source, 14.

The peak current values one can obtain during discharge of capacitor 1 in such a nonoscillatory (approaching the aperiodic limit case) circuit 1, 6, and 8, are obviously higher the more the damping resistance, 5, can be reduced. In practical installations it has become apparent that small deviations from the parallel run of the two cables, 7 and 13, can cause such an increase in damping resistance 5 that a marked reduction in initial energy output results. The closer, however, cables 7 and 13 lie to each other and the more care taken to prevent magnetic stray fields outside the iron core, 9, the smaller the damping resistance, 5, can be made. It is even possible to dispense with it entirely, when, for instance, it is possible to transfer the required damping, by correspondingly coarse lamination of the core, 9, into the transformer itself (a procedure which is beneficial to close coupling between coils 8 and 10). It is possible to suppress almost completely the residual magnetic stray fields by concentric or partly concentric arrangement of the circuit conductors.

Such considerations have been followed with two series-connected capacitors in the circuit indicated in Fig. E4-2b. The capacitors, 19 and 20, are connected with each other through pin 21 and are enclosed by cage 22 which is connected by annulus 23 to tube 24. The tube, 24, ends in a tungsten ring, 25, into which is pressed a ceramic cylinder, 26. Over the top of this cylinder there is pressed tungsten ring 27 which is connected to tube 28. The tungsten rings, 25 and 27, form in conjunction with ceramic cylinder 26 a low-inductance surface spark gap. The ignition of this gap is effected by ring-like ignition electrodes 29 and 30 surrounding cylinder 26, whenever a corresponding voltage impulse is applied to them by way of capacitors 31 and 32 from a control device, 33.

To the terminal of the capacitor, 19, lying opposite connecting pin 21, the cable, 34, is connected. It leads through tube 24, the ceramic cylinder, 26, and tube 28 to the primary of shock transformer 35. The return path from the primary to the spark gap is through tube 28. The leads of the primary are also brought out of the coil body in parallel relation to the coil axis.

In an arrangement of this kind the value

$$R_d = \sqrt{\frac{4L}{C}}$$

becomes so small that with sufficient energy consumption in the load it is possible to omit an extra damping resistance. Consequently one can obtain, with capacitances of 0.01 to 0.2 μf and primary voltages of 10 to 20 kv in only approximately nonoscillatory (aperiodic limit) circuit condition, peak current values on the order of from a few hundred up to a few thousand amperes. The whole discharge procedure occurs thereby within about 1×10^{-7} sec to 3×10^{-7} sec. Since durations of less than 10^{-6} sec lie within the range of magnetic expansion in iron, it is advisable to use for circuits in which the natural frequency of the oscillatory circuit lies above 0.8×10^{6} cps, an E-shaped iron core of the kind illustrated. The circuit construction as here described in detail is extensively used today in the wiring of capacitor discharge circuits for high-temperature research on gaseous discharges. It was first developed in 1942–43 for pulse generators by the author.

It is known that at 10,000 volts operating voltage a capacitor of 0.1 μf capacitance stores energy of 5 watt-sec. If this energy becomes effective within 2×10^{-7} sec, there are released during this instant 25 megawatts, of which one-half, i.e. about 12 to 15 megawatts, might be consumed by the load. This shows that with such modest means loads of high voltage and small current requirements such as an x-ray flash tube can be supplied with impulse voltages in the range between 30 and 200 kv and with relatively high peak current values. This results in favorable aspects for radar technique, since it makes possible extremely simple impulse triggering devices for large outputs. However, one limit is set with such simple form of impulse transformers in the decimeter range, because they must be used in conjunction with triodes having low impulse wave form fidelity. For magnetrons and klystrons there exists, however, the possibility of correction of the impulse form.

The importance of the cage arrangement for capacitor 1 or the capacitor chain 19 to 20 will become apparent from Fig. E4-2c which presents a schematic section through a capacitor having a middle layer B, a dielectric layer D and the outer layers A_1 and A_2 as well as the return conductor cage bars K_1 and K_2. The current paths, which are closed by the displacement current in the dielectric during the discharge process, are passed by the current in the direction of the arrows and produce stray magnetic fields in opposite direction to one another, which, however, practically compensate each other toward the outside as in the case of a bifilar winding. In case the capacitor has the shape of a cylindrical winding, there is created within the capacitor space surrounded by the cage an enclosed magnetic ring field. Somewhat similar conditions exist in the concentric conductor system of Fig. E4-2b.

In practical application it is often desirable to construct the oscillatory circuit with parallel-connected capacitors. In order to achieve the end of an equivalent suppression of all external magnetic disturbance fields in this case, it becomes necessary to employ a double cage of the kind shown in Fig. E4-2d. Capacitor 36 is joined to capacitor 37 by a pin, 38. Connected to this pin, 38, is the inner cage, 39, from which a conductor leads to one terminal of spark gap 40. Both other layers of capacitors 36 and 37 are connected with each other by the outer cage, 41, to which in turn is joined tube 42 leading to the primary, 43, of the impulse transformer. The return line from the transformer is through a cable arranged concentrically within tube 42 and leading to the other terminal of spark gap 40.

In the examples, the production of steep-fronted wave impulse voltages by transforming a discharge current is made possible in that, first, the discharge proceeds against very little inertia and, second, magnetic stray fields in the discharge circuit are effectively suppressed. The first requirement is met by the switching spark gap, the second by ferromagnetic guidance of the transformer field and arrangement of all circuit elements and conductors for extremely low inductance.

The maximum that can be reached with impulse transformers such as described here is a base width of the secondary impulse wave of 0.1 μsec having an available output of 2 megawatts in a resistive load. Contrary to expectation, there could also be used for the cores of such transformers rather fine laminated dynamo grade sheet steel. (Measurements in a circuit with a transformer of 60 kv rating, having a core of 0.1 mm sheets, 9 cm^2 cross section, 2 megawatts secondary energy, and 0.1 μsec impulse half time value, revealed an efficiency of 40% in the resistive load.)

In shock impulse devices of this kind, it becomes necessary to pay attention to dielectric losses in the capacitor battery. The dielectric losses grow rapidly with the frequencies in the steep-fronted impulses. Some dielectrics show decreasing loss angles with increasing frequencies, others increasing ones. It is advisable to use dielectrics of ceramic or of polystyrol base, which have a sufficiently small loss angle at high frequencies. With such materials, cooling of the capacitor battery itself is superfluous; however, it is necessary to remove the unavoidable heat created by the resistance of the discharge circuit by, for instance, a cooling fan.

Another fact worth mentioning in conclusion is that, in accordance with Fig. E4-3, it is possible to produce pulses of corrected wave form. The feeding high-voltage transformer, 1, in this case delivers, e.g. in small installations for operation of radar tubes, 10 kv at 30 kv secondary voltage. A selenium rectifier, 2, serves as full-wave rectifier; the capacitor, 3, serves as a filtering means; the charging resistor, 4, limits the charging rate during periodic recharging of shock capacitor 5, which preferably is a ceramic battery or a polystyrol capacitor. The capacitor discharges through the controlled spark gap, 6, whereby the ignition electrode, 7, receives its energy periodically in the direction of arrow 9 by way of capacitor 8, e.g., from a hydrogen thyratron. The counterelectrode, 10, preferably the cathode in accordance with what has been said in Chapter B, is connected with the primary of the shock impulse transformer. The primary consists of two portions, 11 and 12. A tap, 13, is connected to a correction capacitor, 14. The E-shaped core, 15, of the transformer also carries the secondary, 16, connected to the radar device or another impulse-consuming load, 17.

For the purpose in view, section 11 of the primary and capacitor 14, are tuned to 3 times the frequency of capacitor 5 along with coil 11–12. One obtains thereby, besides the fundamental wave marked *a*, an additional wave marked *b* which in combination results in the dotted wave *c*. The third harmonic smooths out the fundamental wave by addition and subtraction so that one obtains a strictly trapezoidal wave form. With the assistance of an oscilloscope and variation of capacitor 14 it is readily possible to maintain the

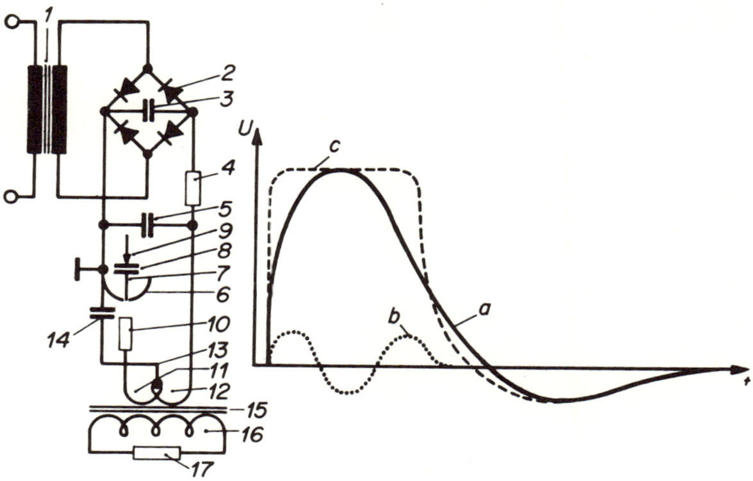

FIG. E4-3. Schematic diagram of a blast spark operated impulse transformer with wave form correction means (for reference numerals see text) (by Früngel and Marx).

upper amplitude of wave c constant to within 1% over a period of about 0.3 μsec. Such an installation can readily be designed for trapezoidal, high-voltage impulse waves within a range of about 20–60 kv.

An example will be given in passing. Assuming that the primary of the shock pulse transformer has 3 turns, and a tap at one turn, then the correction capacitor, 14, will have the same capacitance as the shock pulse capacitor itself, because the relationship between the one-turn inductance and the three-turn inductance is 1:9; that is, threefold frequency with the same capacitance is assured. Surprisingly, there is hardly any sacrifice in amplitude, because no leak of energy is caused by the modification, but only a change in wave form.

Basically, the load resistances must be in conformity with the nonoscillatory, critically damped (aperiodic limit case) condition in the primary discharge circuit. The wave form is consequently load dependent.

Design example: Given a radar or spark-measuring tube of a voltage of 30 kv during the impulse and an average current of 5 amp, that is, 150-kw impulse output, the ohmic resistance is 6000 ohms. If we select a 6-kv voltage as dimensioning base for the shock capacitor, the impulse transformer will have a transformation ratio of 1:5—or 1:6, to provide for voltage drop. Then the primary effective resistance should be 1/36 of the secondary load resistance, and thus roughly 150 ohms. With an average efficiency of 60% for the installation a further reduction to about 100 ohms is justified, since the eddy current and the hysteresis losses in the core manifest themselves as additional resistances during the magnetizing upswing. Therefore, the resistance for nonoscillatory conditions becomes

$$\sqrt{\frac{4L}{C}} = 100 \text{ ohms}$$

and the primary transformer inductance is

$$L = 2500 \text{ ohms}^2 \times C$$

The capacitor capacitance can be deduced from the energy requirement. The secondary side should have, in accordance with this example, an impulse voltage of 30 kv during an impulse duration of 0.3 μsec. The required current flow is assumed to be 5 amp. This results in 150 kw $\times$ 0.3 $\times$ 10^{-6} sec = 45 milliwatt-sec per impulse. To have some reserve, the primary capacitor should store 70 milliwatt-sec energy at 6 kv, the voltage originally planned. Its capacitance will be

$$\frac{140 \times 10^{-3}}{36 \times 10^{6}} \text{ farads} = 4000 \text{ pf}$$

The inductance is then 100 μh. With a winding diameter of 5 cm and a winding length also of 5 cm, the number of turns will be

$$n = \sqrt{\frac{L \times l}{10 \times 10 \times D^2}}$$

where the inductance L is inserted in cm,* and in the present example 100,000 cm = 10^{-4} h. The result is 14 turns for the winding. The secondary turns would correspondingly be sixfold, that is 84 turns. The size of the iron core is again equivalent to the core of a 100 to 200 watt commercial transformer; therefore an E-core is chosen. At an impulse sequence frequency of 1000/sec and 70 milliwatt-sec converted energy per impulse, the average output becomes 70 watts. The charging current source must be designed for about 200 watts when resistance charging is applied, in order to have driving overpotential for extremely simple construction of the circuit and to be able to operate with simple resistance charging. If correction of the wave form is desirable, a tap would have to be provided at 1/3 of the coil, that is at the fourth or fifth turn. The exact tap point would have to be located with the help of an oscilloscope, and a capacitor of about 4000 pf connected at this tap to ground. This example is valid for normal decimeter disc triodes.

As will be understood from the examples, it is possible to obtain impulse outputs of many megawatts and fully satisfy powerful klystrons and disc tubes with very simple generators. The impulse generators can be tubeless. Owing to their convenient voltage and output regulation they are particularly suitable for test-stand use.

However, when impulse emitters of only such low impulse output as 150 kw are required it is not advisable to reach this goal by way of a transformed impulse. It is simpler, as shown in Fig. E4-4, to connect the feeding capacitor, 1, by way of a hydrogen thyratron, 2, directly to a discharge resistance, 3. In the figure the charging resistor is indicated at 4 and the charging high-voltage source at 5. In this circuit, the oscillatory circuit proper or the magnetron with its anode, 7, and cathode, 6, which are of conventional construc-

* In calculating the number of turns from geometrical data, the use of the old cgs system with the unit cm for 10^{-9} h is convenient; therefore, the cm unit is still used in a few places in this book.

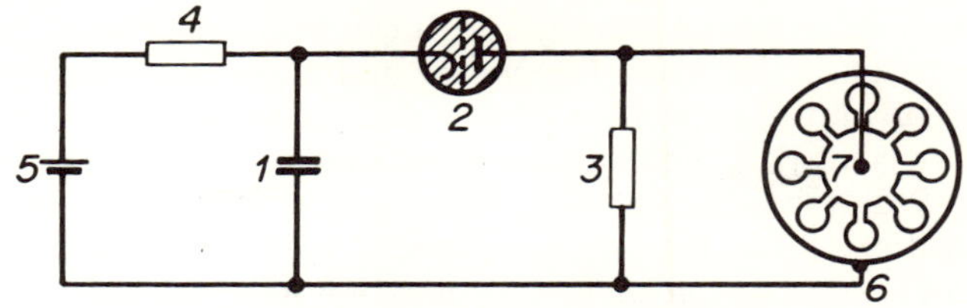

FIG. E4-4. Schematic circuit of a magnetron with switch tube.

tion, is connected in parallel to resistor 3. The latter is able to absorb superfluous energy in case of fluctuating emission of cathode 7. A hydrogen thyratron for the corresponding impulse sequence frequency having a high inverse voltage rating is mandatory for such circuits. Since the electrical data of industrial magnetrons vary within wide limits and also klystron tubes are employed for ultrahigh-frequency generation to a great extent, it is necessary to take the data for a particular case from the tube data sheets. No generally applicable dimensioning rules can be given here. It is important to know the permissible impulse peak current and the voltage to be applied. For smoothing of pointed waves the correction member of threefold frequency can again be employed here. The connections therefore, are somewhat similar to those in the impulse transformer application.

5. Pulse-Forming Network

It is also possible to use pulse-forming networks for stabilizing of the impulse, as represented in Fig. E5-1. Such stabilization results in a constant impulse voltage.

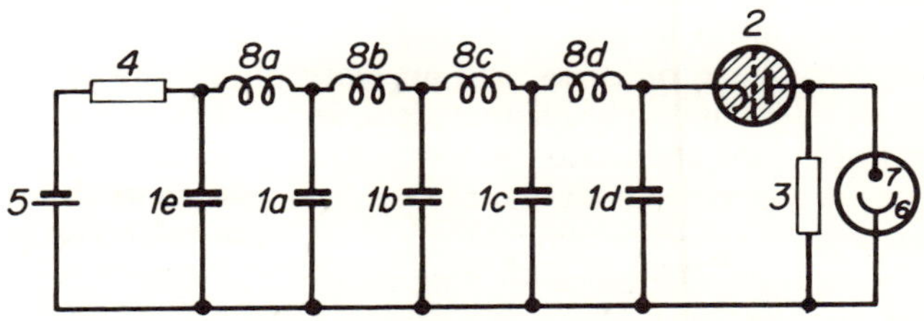

FIG. E5-1. Magnetron circuit with pulse-forming network for trapezoidal or rectangular waves.

Since magnetron tubes fundamentally need a constant impulse voltage for the whole duration of the pulse, the insertion of such a pulse-forming network is recommended. The capacitor in this case is subdivided into partial capacitances 1_a, 1_b, 1_c, and 1_d and these single components are connected to inductances 8_a, 8_b, 8_c, and 8_d. At the instant of ignition of tube 2 capacitor 1_d discharges first; it is followed through inductance 8_d, by capacitor 1_c and then through inductance 8_c by capacitor 1_b, and so forth. In this way, by proper adaptation of this pulse-forming network for the desired impulse duration, one obtains an impulse wave, as represented in Fig. E5-2. It has a slight ripple which is not as pronounced as shown in the graph. It can be considered as a rectangular impulse for practical purposes. Depending on the permissible ripple content, the number of partial capacitors and partial inductances in such a circuit can be rather great [20]. The impulse durations lie in the range between 0.1 and 1 μsec. Progress today, as far as tubes are concerned, is toward shorter impulse duration. Since the swing-in time of magnetrons today still is 0.1 μsec the limit is at present given by the tubes. Klystron tubes as well as the so-called disc triodes are less critical. They have, however, the drawback of not radiating an accurate frequency. Therefore, for spark or radar measuring purposes it is the magnetron that is still extensively used. Each impulse emitter tube in a customary radar installation is pulsed by means of a pulse-forming network and a hydrogen thyratron. The total number of capacitors 1_a to 1_d must have the same energy value as capacitor 1 in Fig. E4-4.

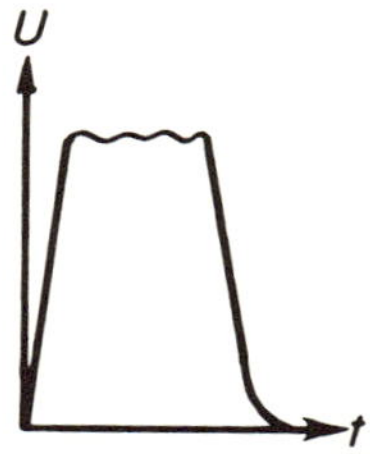

FIG. E5-2. Output pulse of the pulse-forming network shown in Fig. E5-1.

In order to assure proper shielding against the voltage source, an auxilliary capacitor 1_e is necessary, which serves as ground capacitor of the pulse-forming network. It is worth mentioning that, for dimensioning of the impedance, the inductance and capacitance of 8_a and 1_a must be equal to the effective resistance of the connected magnetron. When, for instance, the tube has a triggering voltage of 10 kv and an emission current of 2 amp, this corresponds to 5000 ohms. The characteristic impedance of the whole pulse-forming network must then also be 5000 ohms, in order that undesirable reflections of the capacitor discharge energy do not occur at the magnetron. These, if occurring, would become apparent in artificially produced "parasites" or false echoes. The pulse-forming network is thus, from the standpoint of impulse form, just as critical as the impulse transformer. Assuming that a single member of the pulse-forming chain suffices to supply the energy demand of the magnetron for 10^{-7} sec, and this energy demand is set at 10 kv and 2 amp, the energy content of this capacitor element must then be 2×10^{-3} watt-sec, that is, at a charging voltage of 10 kv equal to a capacitance of 40 pf. Since the inductance must be tuned to an impedance of 5000 ohms there results from

$$Z = \sqrt{\frac{L}{C}}$$

the inductance

$$L = Z^2 \times C = 10^{-3}\,\text{h} = 1\,\text{mh}$$

For an impulse duration of 0.4 μsec there have to be provided four capacitors of 40 pf each and four inductances of 1 mh each.

The basic network derived by the Guillemin* design procedure consists of a series of resonant LC elements connected in parallel. Design parameters

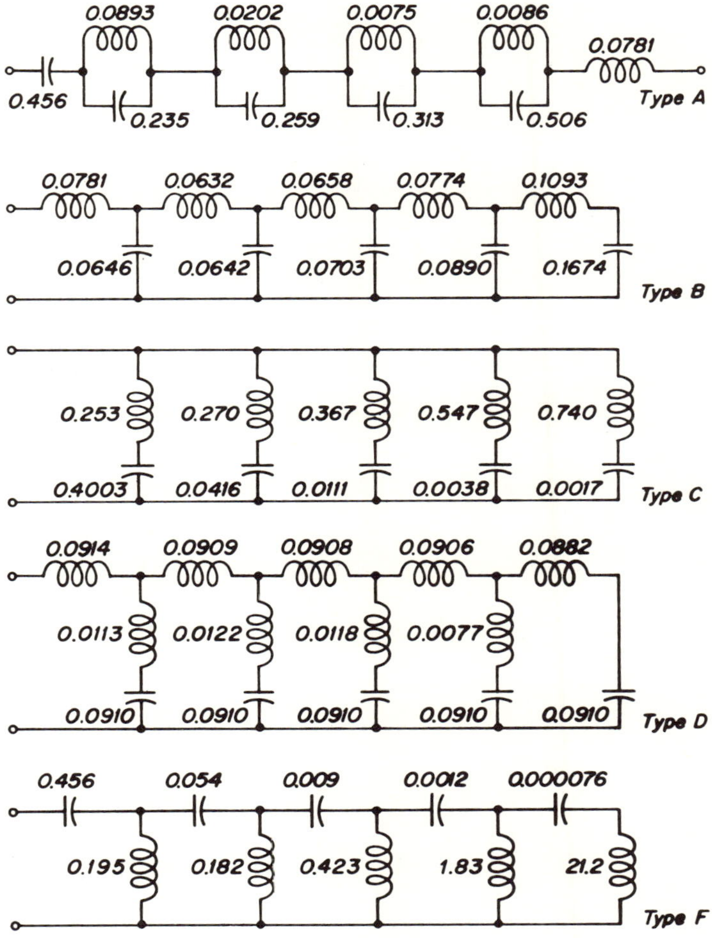

FIG. E5-3. Equivalent forms for five-section, Guillemin voltage-fed network. Multiply the values of the inductances by $Z_N\tau$ and the values of the capacitances by τ/Z_N. The inductances are in henrys and the capacitances in farads if pulse duration τ is expressed in seconds and network impedance Z_N in ohms. By permission from Glasoe and Lebacqz, "Pulse Generators," McGraw-Hill, 1948.

* The remainder of Chapter E, Section 5, is reprinted by permission from Glasoe and Lebacqz, "Pulse Generators," pp. 201–213, McGraw-Hill, New York, 1948.

TABLE E5-1

VALUES OF INDUCTANCES AND CAPACITANCES FOR FIVE-SECTION PULSE-FORMING NETWORK[a]

Wave form	Fourier coefficients						Inductance					Capacitance				
	a	b_1	b_3	b_5	b_7	b_9	L_1	L_3	L_5	L_7	L_9	C_1	C_3	C_5	C_7	C_9
Rectangular	0	1.2732	0.4244	0.2547	0.1819	0.1415	0.2500	0.2500	0.2500	0.2500	0.2500	0.4053	0.04503	0.01621	0.00827	0.00500
Trapezoidal	0.05	1.2679	0.4089	0.2293	0.1474	0.0988	0.2510	0.2595	0.2777	0.3578	0.3578	0.4036	0.04318	0.01459	0.00670	0.00349
Trapezoidal	0.08	1.2601	0.3854	0.1927	0.1015	0.0482	0.2526	0.2753	0.3303	0.4478	0.7340	0.4011	0.04089	0.01227	0.00462	0.00170
Trapezoidal	0.10	1.2524	0.3643	0.1621	0.0669	0.0155	0.2542	0.2912	0.3927	0.6796	2.2875	0.3987	0.03865	0.01032	0.00304	0.00055
Trapezoidal	0.20	1.1911	0.2141	0	−0.0393	−0.0147	0.2672	0.4455	∞	−1.1561	−2.4052	0.3791	0.02272	0	0.00179	−0.00052
Parabolic rise	0.05	1.2699	0.4166	0.2418	0.1640	0.1194	0.2507	0.2547	0.2632	0.2773	0.2961	0.4042	0.04420	0.01539	0.00745	0.00422
Parabolic rise	0.10	1.2627	0.3939	0.2064	0.1194	0.0691	0.2521	0.2694	0.3084	0.3808	0.5122	0.4019	0.04179	0.01314	0.00543	0.00244
Parabolic rise	0.20	1.2319	0.3127	0.1032	0.0246	0.0017	0.2584	0.3393	0.6168	1.8472	20.94	0.3921	0.03318	0.00657	0.00112	0.00006
Parabolic rise	0.25	1.2092	0.2610	0.0564	0.00353	0.0017	0.2632	0.4065	1.1292	12.887	21.37	0.3849	0.02769	0.00359	0.00016	0.00006
Parabolic rise	0.33	1.1609	0.1720	0.00930	0.00338	0.0064	0.2742	0.6168	6.8493	13.44	5.5556	0.3695	0.01825	0.00059	0.00015	0.00023
Parabolic rise	0.40	1.1142	0.1080	0	0.0085	0.0015	0.2857	0.9821	∞	5.346	23.15	0.3547	0.01146	0	0.00039	0.00005
Parabolic rise	0.50	1.0319	0.0382	0.00825	0.00300	0.0014	0.3085	2.7747	7.7160	15.15	25.00	0.3285	0.00406	0.00053	0.00014	0.00005

[a] Multiply the inductances by $Z_N\tau$ and the capacitances by τ/Z_N. The inductances are given in henrys and the capacitances in farads if the pulse duration is expressed in seconds and the network impedance is in ohms. a is fractional rise time of wave.

$$i(t) = \frac{V_N}{Z_N} \sum_{\nu=1,3,\ldots}^{\infty} b_\nu \sin \frac{\nu \pi t}{\tau}.$$

for the first five sections of a network of this type, as well as the corresponding Fourier series coefficients, are given in Table E5-1 for a pulse having both linear and parabolic rise and fall, and for a range of values of the rise time parameter *a*.

It may be noted that a few negative values of inductance and capacitance are listed for a trapezoidal pulse with a = 0.20. These negative values are, of course, not realizable in a physical network.

Equivalent networks can be derived from the values listed in Table E5-1 by the mathematical procedures outlined above. A set of equivalent networks for the five section network producing a trapezoidal pulse with a rise time of about 8% is shown in Fig. E5-3.

Examples of one-, two-, and three-section networks, all for pulses with parabolic rise and for various values of the rise time factor *a*, are shown in Fig. E5-4. The two- and three-section examples are of the type known as type *A* and consist of one or more antiresonant sections in series with an inductance and a storage capacitor. Calculated pulse shapes produced on a matched resistance load by these three networks are shown in Fig. E5-5, E5-6a, and E5-7a, respectively.

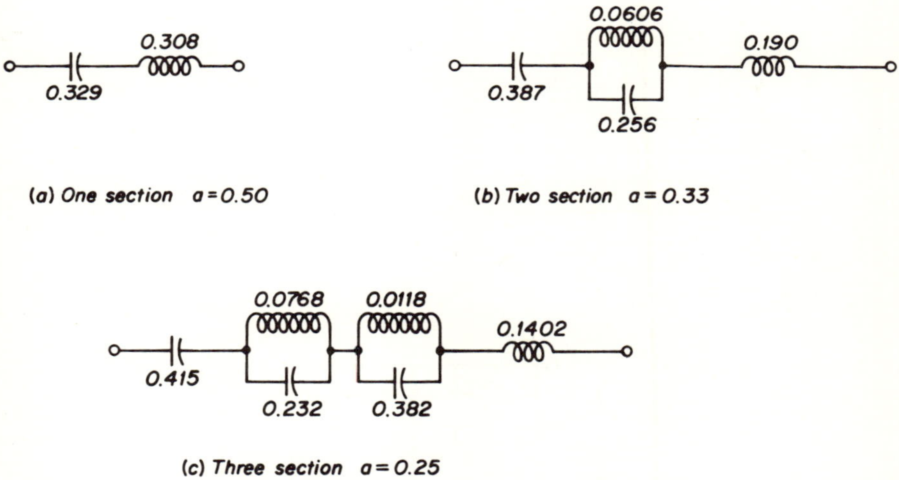

FIG. E5-4. Theoretical parameters of one-, two-, and three-section type A voltage-fed networks for parabolic rise. Multiply the inductances by $Z_N\tau$ and the capacitances by τ/Z_N. The inductances are in henrys and the capacitances in farads if pulse duration τ is expressed in seconds and network impedance Z_N in ohms. By permission from Glasoe and Lebacqz, "Pulse Generators," McGraw-Hill, 1948.

Considerable improvement in the pulse shapes from networks of more than one section can be produced by slight departures from the theoretical values of the parameters. Figure E5-6b shows the improvement over Fig. E5-6a resulting from a 15% increase in the series inductance. Figure E5-6c shows the further improvement resulting from a 5% increase in the capacitance of the storage capacitor. Figure E5-7b shows the improvement over Fig. E5-7a resulting from similar increases in inductance and capacitance in the three-section network. The fact that these improvements can be made is perhaps an indication that the correspondence between the steady state wave form and the transient pulse, which was assumed to be exact, is in reality only approximate.

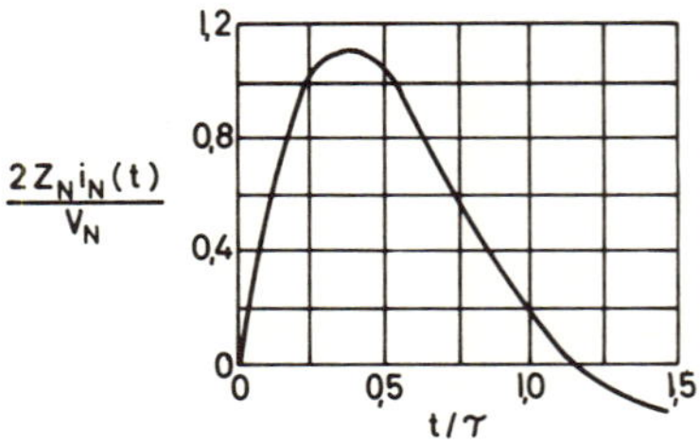

FIG. E5-5. Calculated pulse shape for the one-section network. By permission from Glasoe and Lebacqz, "Pulse Generators," McGraw-Hill, 1948.

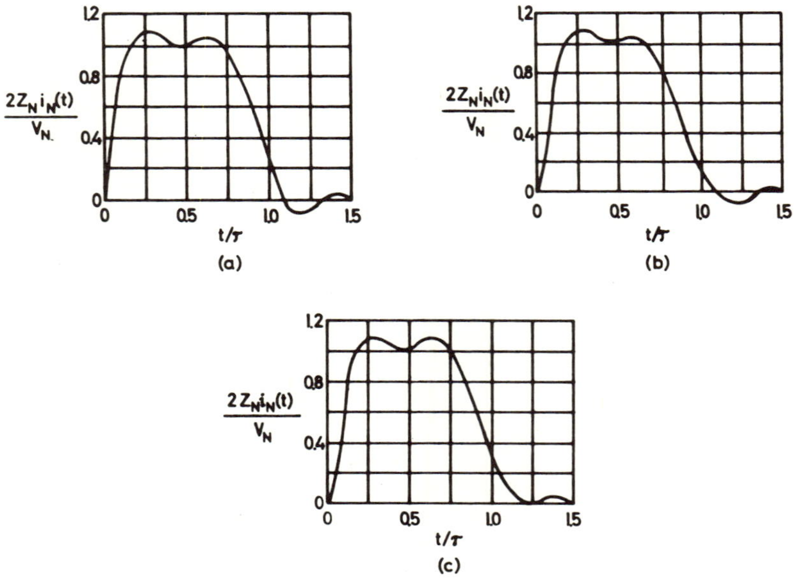

FIG. E5-6. Calculated pulse shape for the two-section network. (a) Theoretical parameters; (b) L_4 increased 16% from theoretical value; (c) L_4 increased 16%, C_N increased 5% from theoretical value. By permission from Glasoe and Lebacqz, "Pulse Generators," McGraw-Hill, 1948.

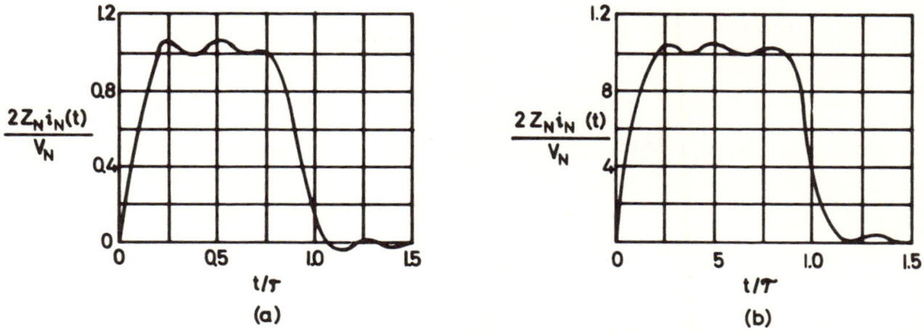

FIG. E5-7. Calculated pulse shapes for the three-section network. (a) Theoretical parameters; (*b*) L_6 increased 21%, C_N increased 4% from theoretical value. By permission from Glasoe and Lebacqz, "Pulse Generators," McGraw-Hill, 1948.

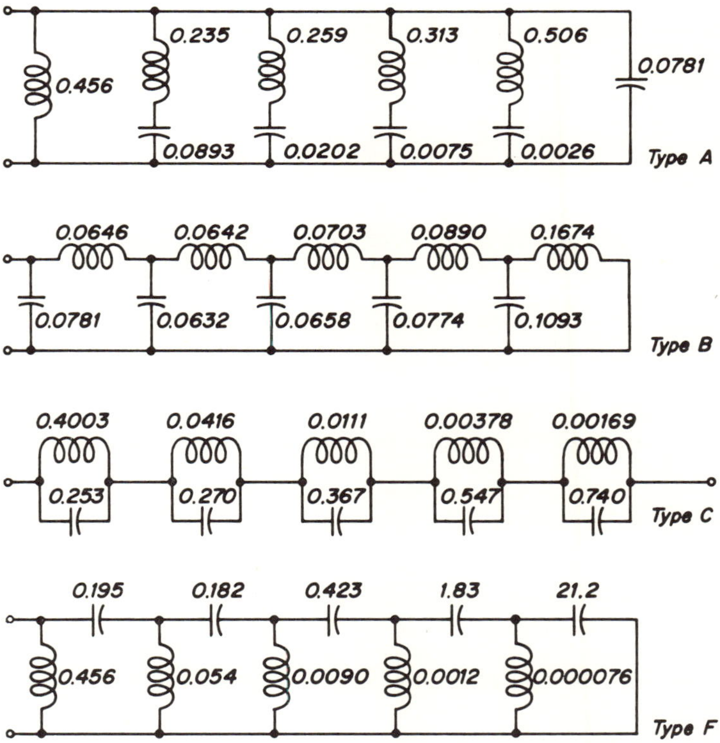

FIG. E5-8. Equivalent forms for five-section Guillemin current-fed networks. Multiply the inductances by $Z_N\tau$ and the capacitances by τ/Z_N. Inductances are given in henrys and capacitances in farads if the pulse duration τ is in seconds and the network impedance Z_N is in ohms. By permission from Glasoe and Lebacqz, "Pulse Generators," McGraw-Hill, 1948.

Some examples of current-fed network circuit dimensions are shown in Figs. E5-8 and E5-9, using the formula

$$L_\nu = \frac{4Z_N\tau}{\nu^2\pi^2}, \qquad C_\nu = \frac{\tau}{4Z_N}$$

The current-fed networks thus derived are shown in Figs. E5-8 and E5-9. The five-section networks are correspondingly designated in Figs. E5-3 and E5-8.

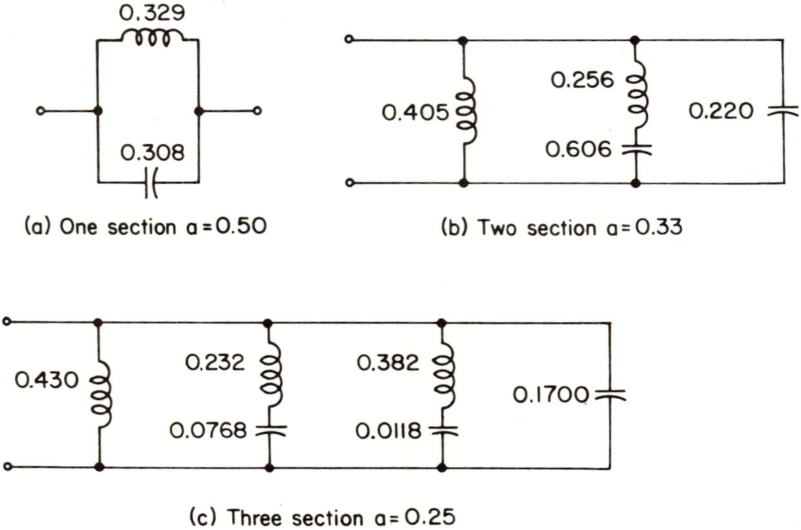

FIG. E5-9. One-, two-, and three-section, type A, current-fed networks. Multiply the inductances by $Z_N\tau$ and the capacitances by τ/Z_N. Inductances are given in henrys and capacitances in farads if the pulse duration τ is in seconds and the network impedance Z_N is in ohms. By permission from Glasoe and Lebacqz, "Pulse Generators," McGraw-Hill, 1948.

6. CABLE DISCHARGE LINES AND PULSE GENERATORS [680a]*

An essential piece of equipment for testing high speed circuits is the pulse generator. A rectangular output pulse having as short a rise time as possible is usually required, and the amplitude and duration should both be controllable and of known value. The various types of circuits readily fall into one of two classes. In the first group the pulses are produced at instants of time determined solely by the generator itself. Usually they occur more or less regularly at a known controllable repetition rate but in some applications a random source of pulses is employed [680b]. In generators of the second class

* Chapter E, Section 6, reprinted by permission from I.A.D. Lewis and F. H. Wells, "Millimicrosecond Pulse Techniques" (2nd edition), pp. 115–132, Pergamon, London, 1959.

the circuit is triggered by some external agency and we require that the pulse be produced with a minimum of delay, usually, and such delay as there is should not be subject to random variation, i.e. should be jitter free. Circuits falling into the latter class can be employed as primary generators by the addition of a relaxation or other oscillator arranged to produce triggering pulses at the desired recurrence frequency. The rise time requirements on the triggering pulses may be several orders of magnitude less stringent than the requirements which the final output pulse must meet. Conventional microsecond techniques may therefore be employed in this part of the unit.

A circuit described by Espley [681, 682] is of interest in illustrating how far the conventional methods of employing several stages of pulse squaring, followed by amplification, may be extended. The repetition rate is accurately controlled by a 166.7 kc/sec quartz crystal and a 35 volt, 20 nanosec pulse is obtained across a 35 ohm load. The arrangement, though elaborate, has the advantages of a high recurrence rate and stability in pulse timing to the order of 1 nanosec.

The simplest type of pulse generator [684, 685, 686] is illustrated in Fig. E6-1a. An open ended length of transmission line (or lumped delay line, or even a single condenser) is charged through a high resistance to a dc potential V. The line is discharged into a load resistance Z by means of a switching device. This opens after each closure to allow the line to recharge, thus enabling repetitive pulses to be produced. The load Z will normally be equal to the characteristic impedance Z_0 of the line but it is instructive to consider the case in which the two are different. The effect of closing the switch is the same as that due to a generator connected as shown in Fig. E6-1b which applies a step function emf of magnitude V.

This emf is written as a function of time in the form $[V] = V[1]_0$, the switch being operated at time $t = 0$.

The Laplace transform of the emf is then $[\bar{V}] = V/p$. On looking into the line at A a certain impedance Z_{in} is seen. The signal voltage $\bar{V}_B$ which appears across the load Z is given by the simple potential divider formula as

$$\bar{V}_B = \frac{Z}{Z + Z_{\text{in}}} \times [\bar{V}]$$

Since the line is open circuited at the end remote from the generator, it follows:

$$\bar{V}_B = \frac{\dfrac{Z}{Z + Z_0} \times \left(1 - e^{-2pTl}\right)}{1 - \dfrac{Z - Z_0}{Z + Z_0} \times e^{-2pTl}} \times [\bar{V}] \qquad \text{(E6-1)}$$

For the case $Z = Z_0$ this reduces to

$$\bar{V}_B = (1 - e^{-2pTl}) \times [\bar{V}]/2$$

and when the inverse Laplace transform is taken it is found that the output pulse is rectangular in shape, has an amplitude $V/2$, and lasts for a time $2Tl$, i.e. the double delay down the line.

In the case when Z differs slightly from Z_0 we may put $Z/Z_0 = 1 + \epsilon$ where ϵ is a small quantity. On expanding the denominator of relation (E6-1) it is found that

$$\bar{V}_B = \left[\left(1 + \frac{\epsilon}{2}\right) - e^{-2pTl} - \frac{\epsilon}{2} \times e^{-4pTl}\right] \times \frac{[V]}{2} \tag{E6-2}$$

where terms in ϵ^2 are neglected*

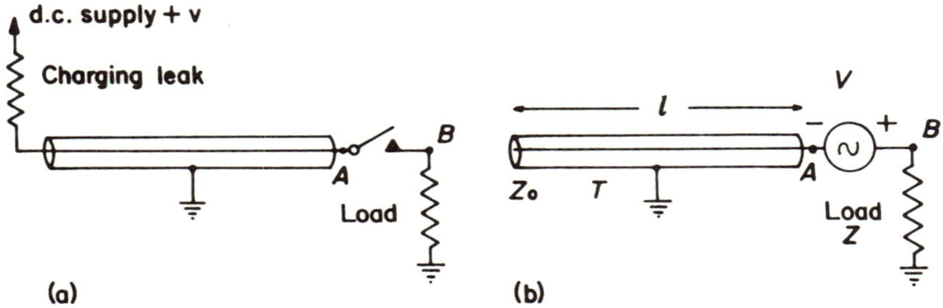

FIG. E6-1. (a) Basic circuit of discharge line pulse generator. (b) Equivalent circuit for analysis. By permission from I. A. D. Lewis and F. H. Wells, "Millimicroecond Pulses Techniques," Pergamon, 1959.

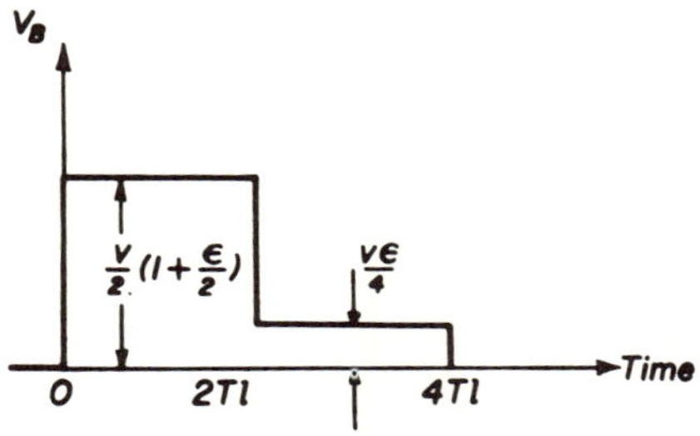

FIG. E6-2. Output pulse when $Z = Z_0(1 + \epsilon)$, $\epsilon \ll 1$. By permission from I. A. D. Lewis and F. H. Wells, "Millimicrosecond Pulse Techniques," Pergamon, 1959.

---

* The reader may question the rigor of the present analysis on two points. First, the switching theorem applies to a system containing emf, not necessarily charged capacitors; and, second, in our Laplace transform analysis we have implicitly assumed that the system is unenergized up to time $t = 0$. The following procedure may be adopted if preferred.

First we note that the Laplace transform of df/dt is $p\bar{f}-f(0)$ where $f(0)$ is the value of the function at $t = 0$. We then have

$$\frac{d\bar{I}_x}{dx} = -pC\bar{V}_x + CV$$

where V is the dc potential to which the line is charged everywhere initially. The wave solutions and the current values are unaffected but a term V/p must be added. The same relation as before applies at the open end, where no current flows, but the additional term must be included when applying Ohm's law at the load end. Exactly the same final result is obtained.

The output pulse now has the form shown in Fig. E6-2. The wave form is drawn for the case $Z > Z_0$, i.e. $\epsilon > 0$, and further subsequent steps, being of negligible amplitude, are not shown

In practice the switch may have a finite, possible variable, resistance ϵZ_0 which may be included in the resistance Z for purposes of analysis. The actual output obtained across a load Z_0 will then be equal to $V_B/(1 + \epsilon)$ where $\bar{V}_B$ is as given in Eq. (E6-2) above. An observed experimental wave form, like that shown in Fig. E6-2, indicates the presence of switch resistance, and any jitter in amplitude, particularly in the small secondary pulse, is attributable to resistance variations.

It is now well established that certain types of electromagnetic relay satisfactorily perform the switching function. Generators employing relay switches naturally belong to the first class mentioned above.

Simple metal to metal contacts are not usually satisfactory because of contact bounce which may lead to the production of multiple pulses. A technique has been developed by Brown and Pollard [686, 687] in which solid platinum contact surfaces are continuously wetted with mercury by means of a capillary connection to a reservoir below the contacts. The mercury film prevents the contact from breaking when the armature bounces slightly on making contact. Only one pulse is accordingly produced for each relay operation. The whole assembly is sealed in an atmosphere of hydrogen at some 10 atm. This enables a high voltage gradient to be withstood right up to the instant of contact and the high gas mobility assists in the dissipation of heat. Such relays, which are of the small high speed type, are considered by Brown and Pollard to operate satisfactorily if no spark is visible at make, i.e. provided the line voltage does not exceed 50 volts, or, for higher voltages, provided the energy flowing is not greater than 40 μjoules. It is recommended that the current build-up rate should not exceed 25 amp/μsec; otherwise the mercury bridge may be vaporized, giving a discontinuity which may last for a few microseconds.

Circuits are described by Moody *et al.* [688] and by Garwin [689], using Western Electric relays types 275 B, 276 B, and D-168479, and also by Schultz and Smith [690]. These relays are not coaxial types and require mounting in a coaxial housing (Fish [691]). Pulse rise (and decay) times of less than 0.2 nanosec are claimed by Garwin [689] at repetition rates of 60–120 cps. It is interesting to note that appreciable distortion is suffered by a pulse on passing through 10 meters of solid dielectric shaping cable.

It appears that such mercury relays are reliable, for the present application, at line potentials of less than 10 volts (when no spark occurs at make) or at potentials greater than 100 volts when a definite gas discharge occurs prior to mechanical make. Two types of contact mechanism may thus be called into play. At intermediate voltages uncertain operation may be encountered in which the main pulse may be reduced in amplitude, being followed, throughout an interval of a few microseconds, by irregular secondary pulses. If multiple pulses occur, for one reason or another, and are troublesome, a second relay may be incorporated and arranged to short-circuit the output from the first after a suitable time delay.

As already mentioned, the above relays were designed as high speed types, with no view to any rf application. Their success as nanosecond pulse generators is due to the small size which enables them to be mounted as part of a coaxial system in which little reflection occurs. The design is not ideal, however, from the rf point of view and Whitby [692] has constructed a vibrating relay in which attention has been paid to preserving the coaxial nature throughout. Simple platinum contacts are employed and the device, as here described, tends to produce multiple pulses.*

The arrangement, which is comparatively large in size and therefore rugged, easy to construct, and capable of handling higher powers than the mercury relays, is shown in Fig. E6-3. The inner conductors, A, B, and C, are made of steel, silver plated to reduce skin effect losses, and the vibrating member B is hinged by means of a piece of steel spring strip D, 0.02 inches thick. The magnetic field due to ac flowing in the coil, E, (5000 turns No. 38 S. W. G. enamel) induces opposite magnetic poles across the contact gap F and attraction provides the driving force (in the equilibrium position the member B makes a small angle with the axis and the contacts do not touch). The outer conductor G is split longitudinally into two halves and the whole inner assembly is clamped rigidly against the split perspex insulating supports H by pairs of saddles I (1/8 in. thick brass) on the outside of the outer conductor. The longitudinal gap is not quite closed so that the alternating magnetic field shall not be diminished by a solid shorted turn effect. The longitudinal gap does not affect the wave propagation in the principal mode. The ratio of the conductor diameters is chosen to give an impedance of 72 ohms and the effect of the dielectric supports H is corrected for, to a first approximation, by the appropriate change in diameter.

The inner and outer conductors are stepped near the ends in order to facilitate plug-in connection into the rest of the circuit. A conical adapter is required to provide a gradual change from relay dimensions to coaxial cable diameters. A pulse rise time of the order of 1 nanosec was obtained (the limitation lying in the display equipment used rather than in the relay itself) at a repetition frequency of the order of 100 cps. At line voltages exceeding 100 volts a spark is clearly visible before make. Rise time performance is not impaired and the pulse edges are equally sharp at all voltages.

A typical relay type pulse generator circuit, for general test purposes, is shown in Fig. E6-4. The dc charging potential can be easily varied and its value read on a voltmeter. Pulses of either polarity can be obtained at will (a very attractive feature possessed by the relay type of circuit). The relay may be driven from the ac mains or from a low frequency oscillator.** The frequency of the latter should be variable so that the relay may be pulsed at the frequency at which it vibrates most smoothly. Assorted lengths of cable or plug-in sections of coaxial line should be available to provide various pulse lengths. The number of plug and socket connections should be kept to a minimum to avoid unnecessary reflections.

---

* A very short platinum wire dipping into a minute mercury cup might be employed.

** When only a very low repetition rate is required the unit may be simplified by using one of the all-relay oscillators described by Ives [693].

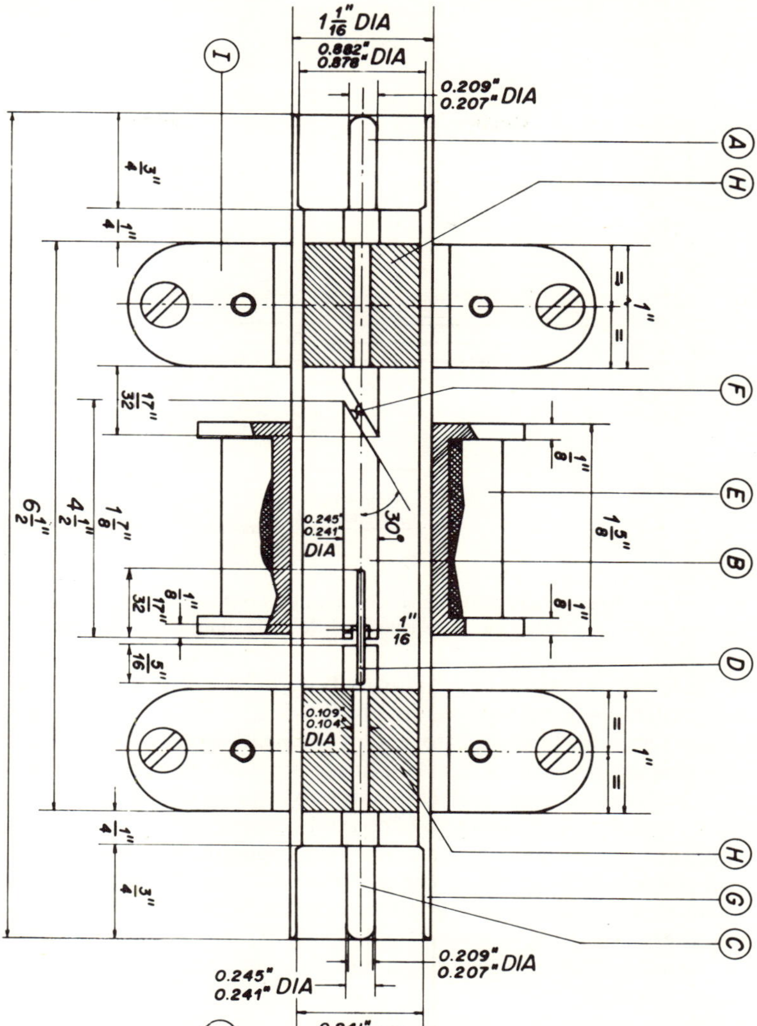

FIG E6-3. Section through 72 ohm coaxial vibrating switch. By permission from I. A. D. Lewis and F. H. Wells, "Millimicrosecond Pulse Techniques," Pergamon, 1959.

It is important that the charging leak resistor be mounted as closely as possible to the end of the line in order to avoid stray capacitance which would modify the reflection coefficient at the nominally open end. If the relay is prone to give rise to multiple pulses their effect can be much reduced by suitably choosing the cable charging time constant (equal to the product of the leak resistance times the total capacity of the line). The time constant should be as long as possible to that negligible partial recharging can occur between bounces after the first make, but short enough to allow complete charging during the off part of the cycle.

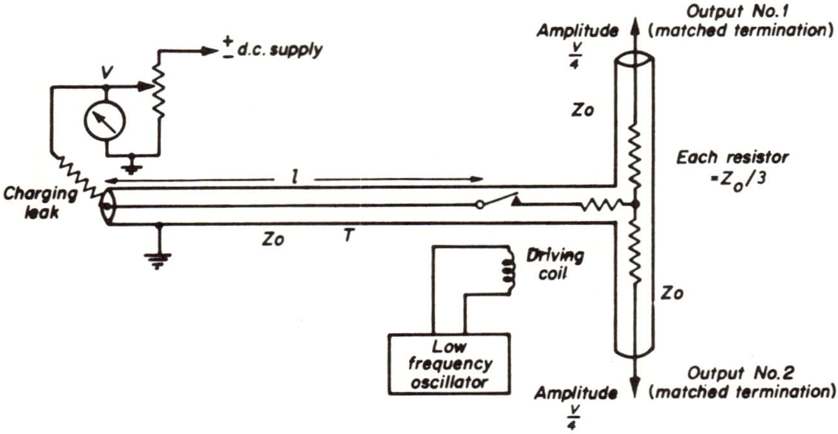

FIG. E6-4. Typical relay type pulse generator giving two outputs. By permission from I. A. D. Lewis and F. H. Wells, "Millimicrosecond Pulse Techniques," Pergamon, 1959.

More than one output channel is usually required, and the two-way and three-way matching pads depicted in Fig. E6-5 may be used. These provide a resistance match Z_0 looking into any of the arms when all the other arms are terminated by resistances Z_0. Amplitude attenuation figures are quoted. Care must be taken to avoid reflections due to stray reactance, and the resistors must be mounted in some form of coaxial housing.

One output channel is usually required for triggering an oscillograph and the signal pulse will have to be delayed to allow for the time base circuit start-up time. This may be accomplished by including a suitable length of cable anywhere in the signal circuit under test. When small signal pulses are required, together with a large trigger pulse, an attenuator may be inserted in the signal channel.

Such a pulse generator is attractive from the points of view of both simplicity and accuracy in pulse amplitude and duration. The latter depends on the propagation characteristics of the line and it has been found by Brown [694] that the electrical length of a typical dielectric filled cable varies by only 0.1% over a period of one year including temperature changes of 10°C.

Relay-type pulse generators are marketed by E.M.I. Ltd., Electrical and Physical Instruments Corp., Spencer-Kennedy Laboratories Inc., and Teletronics Laboratory Inc. These instruments produce pulses of rise time about 1 nanosec and of amplitude variable up to ± 100 volts. Several outputs, with variable relative delay between pulses are available on some models, and the repetition frequencies are variable between 50 and 200 cps.

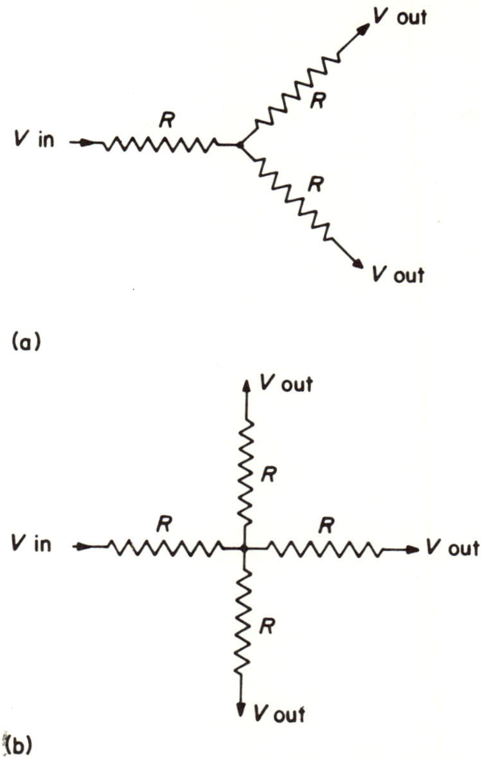

FIG. E6-5. (a). Two way matching pad $R = Z_0/3$; $V_{out}/V_{in} = 1/2$. (b) Three way matching pad $R = Z_0/2$; $V_{out}/V_{in} = 1/3$. By permission from I. A. D. Lewis and F. H. Wells, "Millimicrosecond Pulse Techniques," Pergamon, 1959.

A disadvantage of relay-type generators is the low repetition rate usually encountered and the difficulty of obtaining jitter-free triggering from external sources. Finch [695], however, has used a modified form of the reed type mercury-fed relay, described by Brown and Pollard [686], which operates at 480 cps. A model was later constructed to function at 1000 cps. These relays are incorporated in an instrument designed primarily for testing networks that can be arranged to store a dc charge. (A potential of about 50 volts is employed.) The network is discharged by the relay, and produces a pulse characteristic of itself. The wave form is displayed together with one produced simultaneously by a standard similar network and direct visual comparison can then be made.

Dorn [696] has constructed a pulse generator which gives pulse rise times of less than 2 nanosec at repetition rates in excess of 10 kcps with a peak pulse power of 50 watts. The basis of the device is a Deltaswitch, made by Detroit Controls Corporation, which is a single-pole multiposition switch utilizing a rotating jet of mercury which impinges in turn on a series of 81 contact elements arranged in a circle. The switch may be driven up to a speed of 150 rpm indicating that a pulse repetition rate of 12,150 pulses/sec should be achievable. Some simple modifications to the connections are required, in order to improve the rf properties. Trains of pulses, of differing amplitudes, polarities, and durations, could readily be generated by suitably grouping the contact elements together, and connecting the several common points to pulse-forming lines of different lengths and charged to different potentials. The maximum current handling capacity quoted by the manufacturers is 75 milliamp but the switch has functioned satisfactorily for over 8 hr at currents of 1 amp. This was presumably due to the fact that the pulse length was determined by the shaping line and thus the mercury jet was only called upon to make and pass a current of 1 amp, and not to break a circuit carrying a current of this magnitude.

While on the topic of relays we may digress to mention certain types, which, although not suitable for pulse generation, have been designed for wide band rf switching purposes. They may be applied to the remote switching of pulse channels but may also be used to advantage in place of conventional wafer or toggle switches even when distant operation is not involved.

Coaxial relays which function up to several hundred megacycles per second are available from Londex Ltd. and Besson and Robinson Ltd. Advance Components Ltd. lists a manually operated, single pole, six way switch which is usable up to 300 Mcps. Several types of relays are reviewed by Rockett [697] and a rapid acting mechanical switch for coupling and decoupling a pair of coaxial lines is described by Boothby [698]. Other relays and switches, some of which function up to as high as 10,000 Mcps., are listed by Barker and Williamson Inc., Danbury-Knudsen Inc., Thomson Products Inc., and Transco Products Inc. Mention may also be made of the high-speed switch type 223, described by Volz [699], comprising a mercury relay mounted in appropriate coaxial housing.

A gas discharge device, or thyratron valve, may be used to perform the switching function in a line type pulse generator. Such tubes can be triggered from an external pulse source. The jitter in triggering delay may be too large for some applications, however, and the generator should rather be treated as belonging to the primary class. Thyratron circuits have one advantage over relay pulse generators in respect of the considerably higher recurrence rate at which they may be operated.

Much work has been done on thyratron design for use in low and high power modulators in the microsecond region. Information may be sought in papers by Birnbaum [700], Chance [701], Knight [702, 703], Mullin [704], Romanowitz and Dow [705], Webster [706], Wittenberg [707], Appel and Fuenfer [708], Chantereau and Leduc [709], Mullard Ltd. [710], Charles and Warnecke [711], and in the book by Glasoe and Lebacqz [683]. The application of thyratrons to the nanosecond range is discussed by Yu *et al.* [712], general conclusions being that a jitter of down to 2 nanosec can be achieved, but with difficulty, and maximum recurrence frequencies af 5–10 kcps are obtainable. When low jitter is desired the heater should be run from a dc supply. The presence of a permanent magnetic field has a pronounced effect on the firing delay and it has also been found that jitter may be reduced by suitable positioning of the magnet.

The miniature thyratron type 2D21 (CV797) has been used with success. The valve is a xenon-filled tetrode capable of passing 500 milliamp peak current. Fowler [713] described its use in a noise source application in which the thyratron discharges a single condenser to produce a saw-tooth shaped pulse which is then differentiated. The ionization time is about 300 nanosec when the screening grid is connected to cathode and a 60 volt positive drive pulse is applied to the control grid. The deionization time, which determines the time interval which must elapse before the full anode voltage may be restored, can be reduced by holding the control grid highly negative between pulses. Steady operation at a recurrence frequency of 7 kcps is obtained when the grid is biased to − 95 volts.

The thyratron type 2050 and the hydrogen filled thyratrons types 3C45 and BT79 (see Knight and Hooker [714], Heins [715]) have also been employed by other workers with satisfactory results. The hydrogen tubes are suitable for use at recurrence frequencies well in excess of 10 kcps. Factors affecting the rise time, particularly in the case of the hydrogen thyratron type 5C22, are discussed by Woodford and Williams [716]. A rise time of a few nanoseconds is obtainable with this valve.

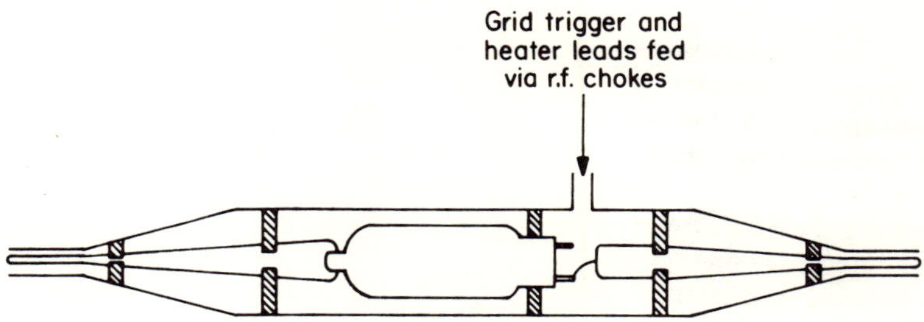

FIG. E6-6. Sketch of coaxial housing for hydrogen thyratron. By permission from I. A. D. Lewis and F. H. Wells, "Millimicrosecond Pulse Techniques," Pergamon, 1959.

The output pulse rise time may be limited by circuit inductance rather than by the thyratron itself.* Switching times of down to 2 nanosec may be obtained when the valve is mounted in a coaxial housing such as that shown in Fig. E6-6. Heard [717] has employed two thyratrons type 5C22 connected in series in a coaxial arrangement. When they were triggered with a 200–500 volt pulse, of 5 nanosec rise time, output pulses of up to 25 kv were obtained at 50 ohms impedance with a rise time of 1.75 nanosec. Two thyratrons were used in order to give the required output amplitude and it was found that no deterioration in rise time occurred when two thyratrons, rather than one, were employed. The triggering of the first tube causes a large voltage to appear across

---

* The maximum permissible rate of rise of current quoted by the manufacturer should not be exceeded or damage may result.

the second sufficient to cause it to fire within a time of the order of 1 nanosec. It is suggested that any number of discharge devices (not necessarily all thyratrons) could be used in series to obtain an arbitrarily large output.

The arrangement of Fig. E6-6, as applied to a conventional thyratron (even with a top cap anode), is by no means perfect from the rf point of view. In an ideal design the anode together with its lead out should be a straight circular rod and the cathode output should be in the form of a hollow tube of the same diameter. The emitting surface might be placed across the end of the cathode tube, the whole thus forming a coaxial line when enclosed in a cylindrical case. Heater leads and the grid trigger lead could run inside the cathode tube without affecting the wave propagation in the main line. Commercial production of such a thyratron or a triggered spark gap, with high permissible rate of rise but only moderate final current, would be desirable.

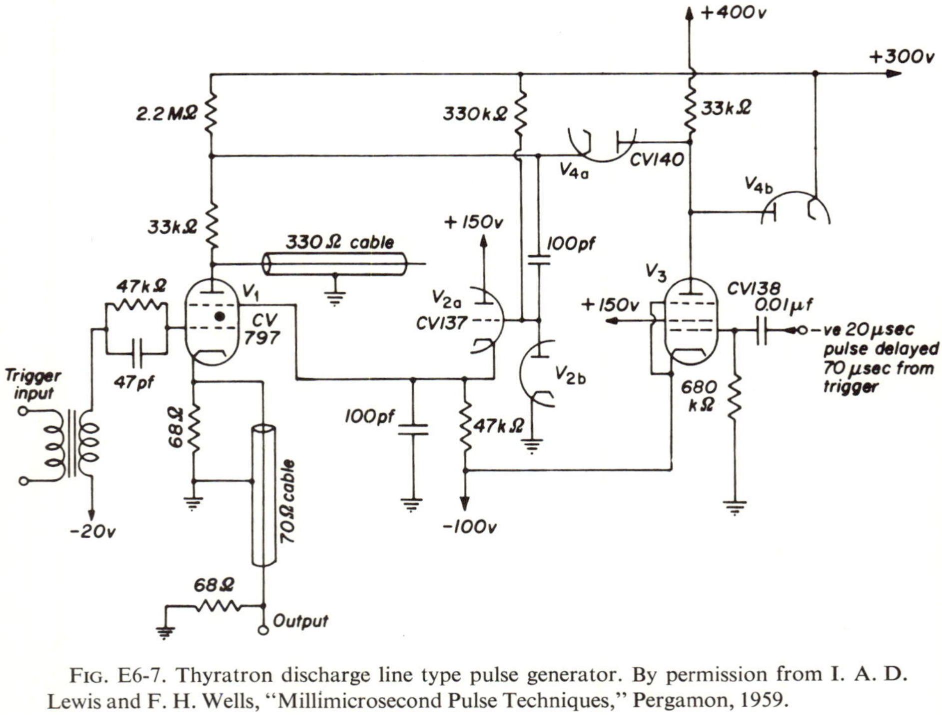

FIG. E6-7. Thyratron discharge line type pulse generator. By permission from I. A. D. Lewis and F. H. Wells, "Millimicrosecond Pulse Techniques," Pergamon, 1959.

A practical circuit described by Wells [680] is shown in Fig. E6-7. The 330 ohm open circuited cable in the anode circuit of the thyratron V_1, which determines the pulse length, is discharged into the 70 ohm output connecting cable. The latter is matched at the near end to absorb any pulses which would be reflected from the far end should the load not provide the correct termination. In the circuit as shown the 330 ohm cable is terminated by an impedance equal to that of the thyratron in series with about 35 ohms (the impedance of the output cable in parallel with the 68 ohm resistor). The pulse shaping line is thus not properly matched and multiple pulses are produced by reflection in this line, the secondary pulses which reach the output are of negligible amplitude, however.

In order that a well-shaped output pulse may be obtained care must be taken in the first instance to reduce as far as possible the stray capacitance to earth at the anode of the valve. The value may then be adjusted to give optimum pulse shape by the addition of a small, 5 pf, variable capacitor. In the case of the thyratron type 2D21 the screening grid must be taken to earth through a resistance of at least 10 kohms* by-passed by a small capacitance.

The thyratron grid is biased beyond cut-off by a potential of $-$ 20 volts and the circuit is triggered by applying a large positive pulse. The amplitude of this should be at least 100 volts in order to keep the ionization time below 100 nanosec. This triggering delay time was found to be remarkably constant for thyratrons type 2D21 and 2050; random variations did not exceed 1 nanosec provided the power supplies were stable.

A positive output pulse of 30 volts amplitude is delivered into the 70 ohm cable. The amplitude is determined by the supply voltage, the impedance of the pulse shaping cable, and the impedance (35 ohms) at the thyratron cathode plus the series impedance of the valve itself. The latter is small and the output amplitude is almost independent of the thyratron used. Good stability is achieved.

The rise time** of the front edge of the output pulse is about 10 nanosec for the 2D21 and about 5 nanosec in the case of the 2050.

After the thyratron has been triggered the anode voltage must be restored slowly to avoid premature restriking. With the normal anode resistance of about 1 megohm the recurrence frequency is limited to about 500 cps. One method by which this figure may be raised has been mentioned. An alternative arrangement utilizes the fact that the high voltage may be reapplied quickly to the anode of the valve, without causing restriking, provided the anode voltage is kept very low, after each pulse, for a time sufficient to allow the ions to disappear. Recurrence frequencies of up to 10 kcps are possible with the circuit of Fig. E6-7. After the thyratron has fired the anode potential remains low because of the high, 2.2 megohm resistor. The valve V_3 then drives the anode of the thyratron, through V_{4a}, back to the 300 volt supply potential after a delay of 70 μsec. The anode potential is then held at 300 volts by the 2.2 megohm resistor since the diode V_{4a} ceases to conduct after the drive pulse (of duration 20 μsec) is over.

---

* In the circuit of Fig. E6-7 the valve V_2 is cut off by the pulse which appears at the thyratron anode; the input impedance at the cathode accordingly has the high value necessary.

** Dixon and Neher (unpublished) have found that high voltages of the order of 1–5 kv reduce the conduction time of thyratrons 2D21 and 2050. Immediately after triggering, the current rises to about half the final value in about 10 nanosec and then increases to the full value in a further short interval of 0.3 nanosec (in a low inductance circuit). In the case of 2D21 a peak current of 8 amp was obtained in a 50 ohm load.

A positive output pulse, with a shorter rise time than that given by the circuit just described, can be obtained by allowing the negative pulse which is available at the thyratron anode to switch off a pentode which is normally taking current. A length of coaxial cable, short circuited at the far end, is included in the anode circuit of the valve to shape the output pulse. Two CV 2127 pentodes, connected in parallel, each taking about 100 milliamp of anode current have been used and an output pulse rise time of about 2 nanosec obtained. In order that the wattage rating of the valves be not exceeded, and to economize on current, the pentodes are slowly brought into the heavily conducting state a microsecond or so before the output pulse is due, by applying an independent pulse to the grids, derived from the trigger pulse. Anode current is then cut off very rapidly by the pulse from the thyratron,* which is applied to the control grids.

Again, if a given positive going pulse is of sufficiently large amplitude, but has comparatively slow rates of rise and fall, it may be sharpened by feeding into a distributed amplifier which is biased beyond cut-off. The amplifier exercises a clipping effect when overloaded, combined with amplification.

Pearce and Smith [718] describe and analyze a pulse generator using a gas tetrode. Pulse durations of down to 10 nanosec were obtained at repetition rates of up to 10,000 pulses/sec.

While not of the simple discharge line type, a number of other pulse generators employing thyratrons may be mentioned. Naray [719] uses two thyratrons arranged such that one initiates the pulse and the other controls the trailing edge. The pulse duration is changed by varying the relative time delay between the triggering pulses applied to the thyratrons. Reversal of the order of firing causes a change in the sign of the output pulse. Pulses of 80 volts amplitude and duration as low as 6 nanosec are obtained.

Edwards [720] described a similar pulse generator employing two 3C45 hydrogen thyratrons in series. The external load is in parallel with the second thyratron and the combination constitutes the cathode load of the first. The output pulse is again initiated by the triggering of the first thyratron and clipped by triggering the second after a variable delay which determines the pulse length. The output is taken from a hard valve type 5D21 and pulses of duration 0.05–80 μsec are available at amplitudes between 0 and 20 kv.

Querzoli [721] has used two tetrode thyratrons arranged such that the cathode of the first drives the grid of the second, the anode of the latter being coupled back to the screen of the first. Pulses of 300 volts amplitude and duration 0.2–10 μsec are generated with rise and fall times not exceeding 10% of the duration.

---

* A similar sort of pedestal pulse technique, used for example by Espley, can often be employed with advantage in cases when a valve stage is required to accept a short negative going input pulse. The screen voltage, rather than the control grid voltage, may be raised comparatively slowly from a low value to the working level a few microseconds before the arrival of the main pulse and afterwards slowly reduced again. When the stage is required to accept positive signal pulses no dissipation difficulties arise since the valve is cut off for most of the repetition period.

The spark gap appears to offer the means whereby pulses of large amplitude and the shortest possible rise time may be generated. It has been observed (Whitby [692]) with the mechanical switch that at potentials exceeding a few hundred volts, and with a critical amplitude of vibration such that the contacts never touched each other mechanically, the output pulse was as sharp as ever. Under this condition the spark mechanism alone is operative and line potentials of up to 600 volts have been used satisfactorily with this particular switch. A surge generator, employing spark gaps giving 80 kv 1700 amp pulses of duration 0.8 μsec and rise time less than 10 nanosec has been used by Goodman *et al.* [722], and Fletcher [723, 724] described in detail how electrically triggered gaps can be employed for generating high voltage impulses having rise times on the order of 1 nanosec. The technique involves the use of highly overvolted pressurized gaps for pulse sharpening and Heard [717] has produced a pulse with rise time 0.6 nanosec, from one rising in about 2 nanosec, by this means. He estimates that pulses of rise time on the order of 0.1 nanosec may be attainable. Further information on spark gaps, the mechanism of the discharge itself, and on associated circuits is contained in papers by Craggs *et al.* [725], and Wilson [726], and in the books by the Bell Laboratory Staff [727], Craggs and Meek [684], Glasoe and Lebacqz [683], Reintjes and Coate [728], and Ridenour [729].

A successful experimental spark gap pulse generator has been constructed by J. B. Gunn (R.R.E.) which delivers an output of up to 8 kv with a rise time of less than 0.5 nanosec. The anode potential for a 4C35 hydrogen thyratron (see Fig. E6-8) is derived from an external variable e.h.t. supply via a resonant choke charging system.

The thyratron is triggered (A) by a 1 μsec pulse of 300–400 volts amplitude and discharges the 500 pf capacitor into the cable system. The comparatively long pulse so produced (B) causes the gap to breakdown after a short delay thus providing a pulse with very short rise time (C). This pulse is then clipped by a variable length of short-circuited transmission line. The last section of the system is composed of strip line for ease in varying the position of the short circuit (G) and of the output tapping point (F).

The same output pulse was used in a very simple and ingenious manner to provide the time base wave form for the associated cathode ray oscilloscope. A 5–10 yard length (CD) of UR31 attenuating cable is incorporated. Owing to skin effect the attenuation varies with frequency. The higher frequency components in the pulse are attenuated to a greater extent than are the lower frequencies and accordingly a pulse with a considerably longer time of rise results. This slow rising edge provides the time base (D). Plenty of amplitude is available and there is no jitter problem since both the signal and time base are derived from one and the same output pulse. The linearity is not very good but is adequate for many applications.

The pulse from the gap is also arranged to provide a brightening wave form for the cathode ray tube (E). The various wave forms are depicted in Fig. E6-8b which also illustrates the relative timing of events; the quantity T_{CG}, for example, is the time delay of propagation in the line between points C and G.

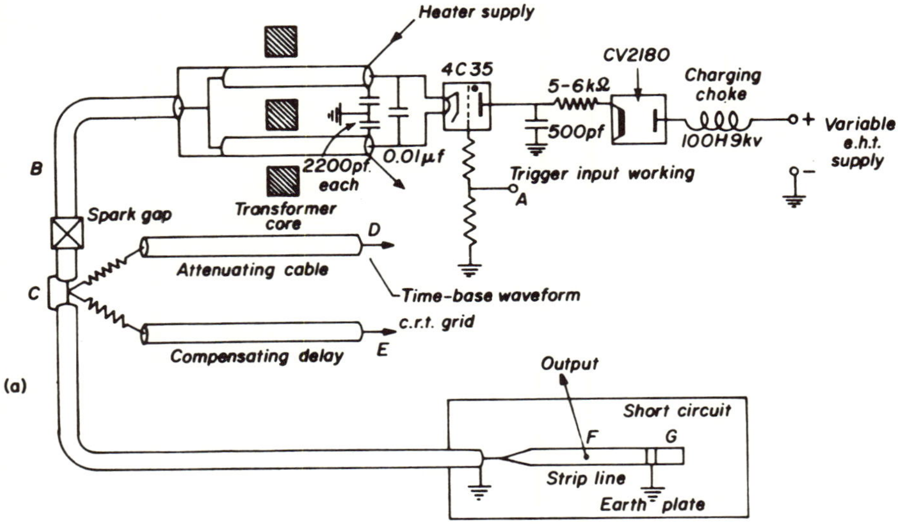

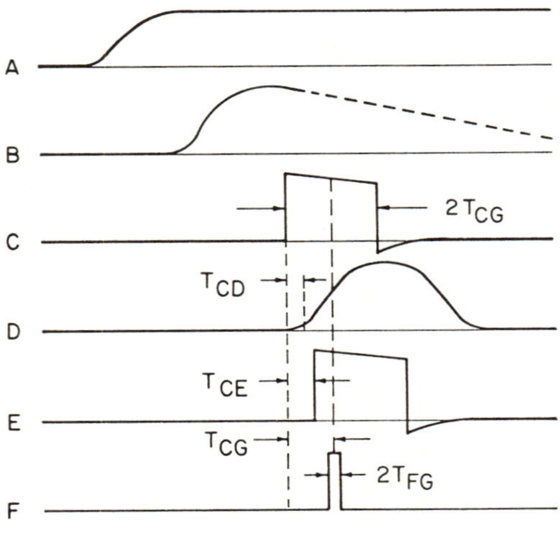

FIG. E6-8. Spark gap pulse generator (a) circuit diagram; (b) wave forms. By permission from I. A. D. Lewis and F. H. Wells, "Millimicrosecond Pulse Techniques," Pergamon, 1959.

Ordinary interval combustion engine type of sparking plugs form the basis of the spark gap itself which is illustrated in Fig. E6-9. The negative electrode is fitted with a short piece of 0.01 inch diameter tungsten wire and the positive electrode is filed flat. The gap is surrounded by a quartz tube to prevent flash-over to the mounting block which must enclose the gap as closely as possible in order to minimize series inductance. The gap is filled with argon at a pressure of 200 lb/inch^2 and means are provided for adjusting the electrode separation.

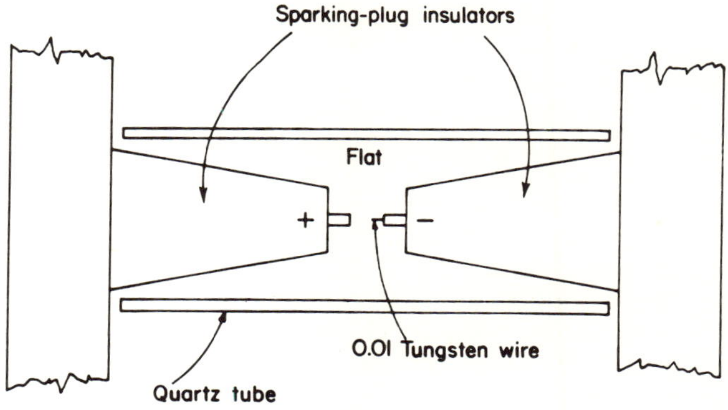

FIG. E6-9. Spark gap assembly. By permission from I. A. D. Lewis, and F. H. Wells, "Millimicrosecond Pulse Techniques," Pergamon, 1959.

The pulse generator as described runs at a repetition frequency of 200 pulses/sec but the gap appears to be capable of functioning at much higher rates. After a few hours running the positive electrode burns concave and the gap becomes uncertain in operation. A flat tungsten electrode might be used here with advantage.

7. Spark Gap Triggering Device* by Marx

The purpose of spark gap triggering devices is to produce approximately rectangular high-voltage impulses for triggering of transmission tubes. The requirements of the particular magnetron (LMS 101) considered at this point are as follows: voltage about 30 kv, current value up to 300 amp, impulse output 9 megawatts, impulse duration 1 μsec or shorter, number of impulses about 1500/sec, initiation deviations not more than about 10^{-7} sec, and amplitude fluctuation of less than 1%. These requirements necessitate particular circuit measures, on the one hand, and a suitable switching means, on the other hand.

The particular circuit requirements will be considered only briefly here. A main current circuit and an ignition circuit. For the main current circuit there are available chiefly the following selections.

1. Direct production of the high voltage from a capacitor discharge without a transformer or shock impulse multiplier circuit. When this arrangement is selected the triggering device must handle the whole voltage. Moreover, a

* Report of a session on Oct. 20, 1944, concerning pulse-forming networks for radar equipment. Since that time no development work has been done in this field.

direct current source for high voltage is necessary, and this in turn needs large rectifier tubes. To meet these requirements, the direct current source must have, e.g., six tubes for inverse blocking voltage of about 40 kv and an emission capacity of 0.5 amp (e.g. U.S.A. Type 705a).

2. Application of an impulse transformer. The industry today builds such transformers* with ratios of from 1:2 to 1:4. However, with such commercial transformers, small deviations from the rectangular wave must be tolerated.

3. Application of shock-impulse multiplier circuits. This possibility appears to be the most favorable for high voltages and large outputs. The direct current source can be rather small and the remaining circuit presents no problems as far as insulating is concerned. With a shock-impulse multiplier circuit, the necessary voltage of 30 kv can be produced by four members of 7.5 kv each. These four members are arranged similarly to one another. The capacitors of the single members are charged in parallel by way of resistances or, better, choke coils. They are then connected periodically in series by switching spark gaps. One of these switching spark gaps must be in the form if a triggering spark gap with external ignition means.

The triggering device has to perform only the function of regular impulse initiation. The rectangular wave form must be assured by proper circuit means under observation of the specified maximum permissible deviation from accurate timing. Such circuit means are the superposition of various sinusoidal or exponential voltages or the application of pulse-timing chains. A steep drop of voltage at the end of the impulse cannot be achieved by the triggering device, since an arc breaks only when, in the discharge circuit, the current is forced to pass through zero. Such action can be obtained by the use of a second switching device that cuts off the discharge voltage periodically at the proper instant.

A consistently constant impulse can also be achieved by series insertion of a tube with saturation characteristic.

In order to assure reliable operation of the triggering device, a slow rise of the capacitor voltage after each impulse is necessary so that the trigger device is not ignited during deionization by the returning voltage. Such slow return and rise of the voltage is assured when a swing-in choke is applied. The voltage at the trigger device must have substantially the same magnitude for each ignition.

Figure E7-1 represents a circuit which meets the requirements mentioned. The direct voltage source is to be connected at the points marked plus and minus at the upper left in the figure. This source charges, by way of swing-in choke L_1 the capacitors of the pulse-timing chain. The voltage between electrodes A and B of the trigger spark gap rises simultaneously therewith. Each time the charge of the capacitors of the pulse-timing chain has reached the potential of the direct current source by way of L_1, the trigger spark gap is ignited by means of the ignition circuit depicted at the right side of the

* The transformers here referred to are for wave form–true transformation of voltage impulses.

diagram. A short circuit occurs between A and B which connects the pulse-timing chain directly to the magnetron. The impedance of the pulse-timing chain is selected to be equal to the resistance of the magnetron. Therefore, at the magnetron there exists one-half of the voltage up to which the capacitors of the pulse-timing chain have been charged. The duration of the voltage impulse depends on the dimensioning of the pulse-timing chain. After a given time the voltage at the magnetron drops suddenly. The arc between A and B extinguishes and charging of the pulse-timing chain begins anew.

A version of an ignition circuit which has proven very successful is depicted in Fig. E7-1. To the screen of the 715B, there is applied a periodic rectangular impulse of about 600 volts by means of a multivibrator ignition device. When this impulse, which is negative with respect to the cathode of the tube, occurs the tube is instantly blocked and the direct current flowing through this tube interrupted.

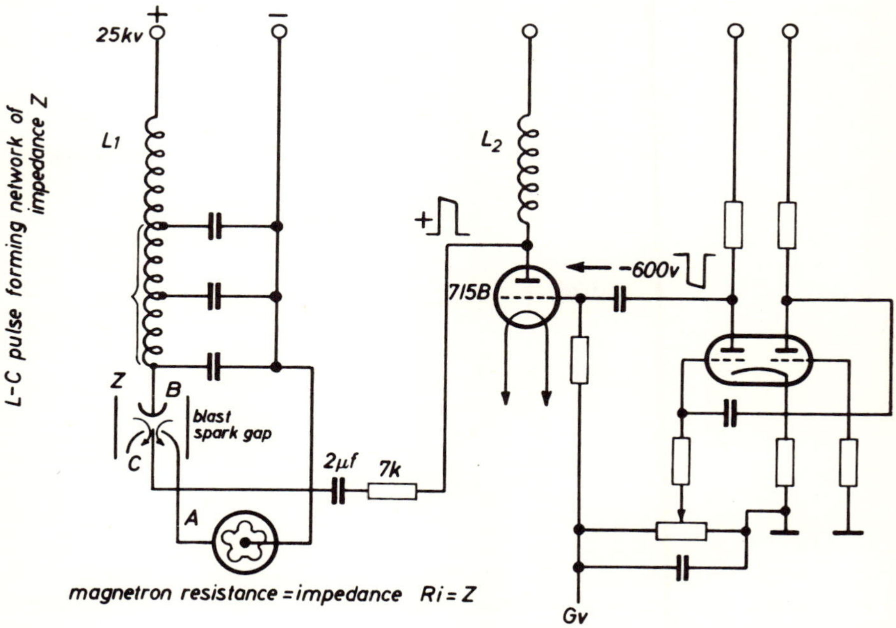

FIG. E7-1. Schematic circuit diagram of a magnetron with spark gap trigger device and tube control of the sequence frequency. A multivibrator periodically blocks the emission strong trigger tube (715B); the stored energy of choke L_2 delivers the ignition sparks for the blast spark gap (by Marx, Kreft, and Helmchen).

The current in L_2 cannot change suddenly; therefore electrode C receives a positive charge by way of the resistor of 7 kohms and the capacitance of 2 μf shown in the diagram. An ignition spark is caused between electrodes C and B, which disappears after a short interval.

The most important portion of the installation for impulse generation is the switching device. It can operate either with stationary electrodes and with external ignition (as in the circuit of Fig. E7-1) or with slidable electrodes. Devices known today operate as high-vacuum tubes or as tubes under overpressure (see Fig. E7-2). Thyratrons have also been tried. At this point a description will be given of a new device that operates under atmospheric pressure or in an overpressure air flow.

The arrangement shown in Fig. E7-2 has gained acceptance as an impulse spark gap.* Such a device was developed years ago as a valve for high dc voltages and large currents in the field of dc power transmission.

As mentioned above, the ignition pulse is effected between electrodes B and C. During occurrence of the ignition pulse there exists between electrodes A and B a high dc voltage, the magnitude of which can lie just below the pure sparkover potential between electrodes A and B. The polarities which have proved most reliable are electrode B, ground; electrode A, negative; and ignition pin C, during ignition, positive.

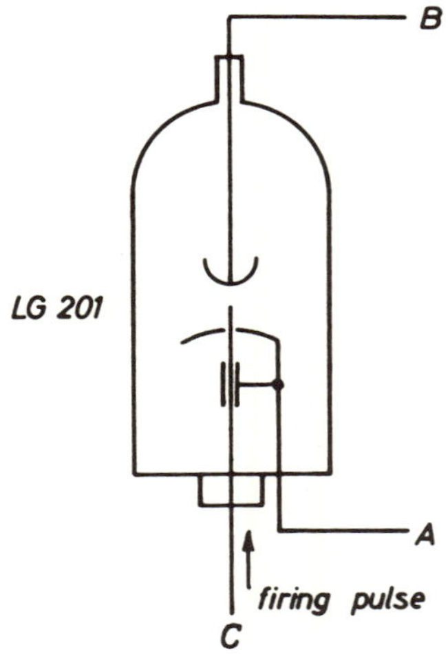

FIG. E7-2. Glass sealed spark gap type LG 201 having concentric control electrode. Attention is called to the small cylindrical capacitor inserted between C and A directly below the electrode to assist fast rising ionization.

It is in conformity with theory that these polarities must be particularly favorable for proper operation of the impulse spark gap. Between a positive pointed electrode and a large well-rounded counterelectrode there develop, at relatively low voltages, predischarges in the form of discharge channels. The electric field between the electrodes, A on one side and B and C on the

* The development of this trigger spark gap was conducted by Helmchen in Braunschweig. In taking measurements Donaty gave valuable assistance.

other side, is at first practically homogeneous (as long as B and C are at the same potential) because the point electrode, C, projects very little beyond the ring electrode, A. However, at the instant at which, through the ignition circuit, the point electrode becomes suddenly positive, the field strength at this point increases considerably and very fast, and instantly causes in turn the mentioned positive discharge channels at the pin electrode. Simultaneously therewith, the breakdown between the pin electrode and the surrounding ring electrode occurs. This breakdown spark irradiates the space between electrodes A and B (Fig. E7-1 and E7-2) and supplies thereby a sufficient quantity of electrons. The well-known succession in gaseous discharges, (1) formation of positive discharge channels, (2) creation of intense positive space charges in the shock ionization range, (3) increase of voltage drop in front of the cathode, and (4) shock ionization of the positive electrodes in front of the cathode, can thus elapse within a very short time interval. For a complete breakthrough only movement of the electrons is necessary, not movement of ions. This fact must be recognized for judgment of the breakdown time. The velocity of electrons is several times higher than the velocity of positive ions and can approach the order of light velocity. Breakthrough durations on the order of 10^{-9} sec have been measured by a cathode ray oscilloscope in the research laboratory of Rogowski. Experiments in the high-voltage Institute of the Technische Hochschule at Braunschweig with exceptionally steep shock impulses in short conductor loops have shown that, with voltages of 60 kv, breakthrough durations on the order of 10^{-9} sec can be achieved. Proof of these findings can be derived from the fact that, with steepest shock impulse voltages in a conductor loop of 30 cm length, a potential difference of 60 kv can be produced and measured with a sphere spark gap. (This result is at the same time very significant for generation of ultrahigh frequency with spark gaps; but then it is possible to obtain powerful short-wave oscillations with shock impulse voltages, when the steepness of these voltage pulses is equal to the voltage steepness of the generated high-frequency oscillations).

If both of the following conditions are met, namely, a blowlike intense field strength increase at the pin electrode, and a simultaneous ionization of the entire breakthrough space, there are attainable extremely short breakthrough durations and, as a consequence thereof, exact initiations of ignition pulses and accurate pulse timing with a minimum of deviation. One of the requirements for trigger devices of magnetrons is a timing accuracy with deviation of not more than 10^{-7} sec. This requirement has been fully met by the trigger spark gap just described, as tests at the firm of Telefunken have proved. The deviations from accurate timing, according to these tests, were considerably less than 10^{-7} sec.

An air flow, as indicated in Fig. E7-1 by arrows, is applied in pulse spark gaps. The velocity of air between pin electrode and ring electrode is about 100 meters/sec. This flow of air has no influence upon timing accuracy, ignition quality and voltage progression, since the travel distance of the air during 1 μsec is only 0.1 mm. Thus the moving air has only two purposes, to effect cooling of the electrodes in the interim between pulses and to assist in deionization and heat dissipation in the space between the electrodes. Every ignition pulse must find the electrodes as well as the space between the electrodes in a cool and deionized condition, if reliable uniform operation with higher current values is to be expected. All four arc bases must be effectively reached by the flow of air. The air entering from the outside must furthermore prevent the accumulation of burn-off residues within the envelope.

The pulse spark gap functioning is trouble free with respect to ignition, deionization, and cooling in continuous operation over long periods. The arc itself is located in the space of strong air flow and is kept by such flow in a straight line. During operation the spark travels uniformly around the pin electrode and the rim of the air outflow opening of electrode A. From the fact that the arc does not adhere to a single point it can be realized that there are no after-effects of one breakthrough spark on the next following one.

To appreciate the difficulties encountered in developing spark gap trigger devices it should be remembered that, without proper measuring, the deviation from accurate breakthrough timing of a spark gap in air fed from sinusoidal ac voltage of 50 cps would be rather great.

To be successful, extensive experiments with impulse spark gaps have to be conducted to arrive at the most suitable dimensions for the electrodes, the most favorable polarities, the necessary air pressure, and the required air quantities for the given electrical values in each case. It has been found that the blocking voltage at the spark gap can be raised during operation up to 25 kv without signs of trouble in operation.

Higher voltage values could not be reached with the direct current source available during the tests but from previous ac/dc and dc/ac high power converter experiments it can be taken that a blast spark gap without appreciable overpressure can be used up to a blocking voltage of about 50 kv. When a few atmospheres of overpressure are applied, the blocking voltage can be raised to about 100 kv. Current values up to 550 amp have been handled in the experiments and with this approach a high degree of usefulness can be attained. High pressure arc current rectifiers have been built up to an average continuous rating of 150 megawatts. The following additional experiment results might be valuable and of general interest.

The pin electrode C may project by 1 to 4 mm from the ring electrode without adversly affecting the operating behavior of the spark gap. This fact is of benefit as it increases the life expectancy of the spark gap. During operation there occurs natural burn-off of the electrodes and a certain quantity of material must be provided for this. Without readjustment of the electrodes continuous faultless operation of 90 hours has been achieved, for example, with an impulse output of 1 megawatt, 16 kv, 0.7 μsec impulse duration, and 1500 impulses/sec. After about 100 hours of operation it becomes necessary to advance the pin electrode, C, by a few mm. This adjustment can be made without interruption of operation. The ring electrode, B, has a useful life of about 500 hours at a load of 1 megawatt. After such a period an annular insert has to be replaced in this electrode, which can be done quickly and easily. Pulse spark gaps for heavy outputs and extended periods of operation can be built but such heavier units need pin electrodes of increased diameter, and for cooling and arc stabilization, a greater quantity of air.

The air quantity for normal spark gaps of this type is about 20 to 30 cubic meters per hour. The pressure drop at the spark gap is about 0.1 atm. A relatively small blower will supply these needs. Nevertheless, the necessity of a blower deserves attention, because these spark gaps are likely to be used in airplanes at high altitudes. Operation would there be impossible without modification, owing to the low air pressure encountered, or the required air would have to be drawn from the supercharger of the aircraft engines.

Further research work concerning the most favorable electrode dimensions, air requirements at very heavy outputs, operation with closed gas circulation, and electrode spacing with different materials and gases, resulted in dimensioning directives for triggers having ratings up to several hundred kilovolts at outputs up to about 20 megawatts, impulse frequencies from 50 to 5000/sec, and impulse duration from 0.1 to 1 μsec.*

In conclusion a short report about experiments with a rotary spark gap may be of value. In 1942, these experiments in the U.S.A. led to impulse forming for radar instruments. A rotary disc, that carries four pin electrodes, faces a stationary disc provided for selective operation with one or four pin electrodes. When it is driven by a motor at 6000 rpm, there will thus be produced either 400 or 1600 impulses/sec. The revolving electrodes are arranged on a circle of 134 mm diameter. A preionization needle is arranged adjacent to each stationary electrode. The distance of this needle from the revolving electrodes must be adjusted so closely that at each passing of a moving electrode in front of a stationary electrode, a predischarge occurs at the needle point which ionizes the breakdown path. Experiments with a reconstructed spark gap of this kind proved that such preionization is very effective.

Proper functioning with little timing diversion is only possible with proper adjustment of these needle electrodes. The timing accuracy of the breakdown sparks in atmospheric air and at 3000 rpm was still within about 5 μsec. This rotary spark gap is designed for mounting in a pressure chamber which is kept under a pressure of about 2–3 atm by means of a hand air pump. In operation under pressure it can be assumed that timing accuracy can be further improved. It is, however, very doubtful whether it can be minimized to a value of 10^{-7} sec as is expected from other impulse spark gaps. The purpose of the rotary impulse spark gaps is application in position finding and scanning radar devices where lower timing accuracy can be tolerated.

In summarizing it can be said that with the air flow impulse spark gaps by Marx all the exacting requirements of transmitter operation can be met up to the highest outputs.

With rotary spark gaps, there can also be produced perfect high-voltage impulses. However, consideration must hereby be given to the timing inaccuracy of 5 μsec.

---

* The arrangement of Fig. E7-1 is somewhat similar to the current gate LG201 (Fig. E7-2) built after an English model. The valve development for high dc voltage was known in England from publications and patents of Marx. The General Electric Co. Ltd., London, supplied valves of this kind before World War II. It is therefore likely that the English development of the impulse spark gap has been influenced by the earlier German arc current rectifier development.

8. Ignition Transformer for Internal Combustion Engines

A special field for shock impulse transformers is their adaptation for producing ignition sparks in internal combustion engines [20, 522]. By the nature of this application, an accurate adaptation is not possible. It should be visualized that, on the one hand, the voltage with badly sooted spark plugs, equal to an ohmic resistance burden of about 1000 ohms, may not be decreased considerably; and, on the other hand, it should be ensured that the thermal energy in the spark is a considerable portion of the energy previously stored in a capacitor.

The basic circuit for engine ignition application is shown in Fig. E8-1. A current source of, say, 300 volts can also be replaced by low-voltage equipment giving induction pulses when a few trick connections are utilized but, for the sake of simplicity, is just indicated at 1. A contact maker, 2, replaces the usual breaker and is actuated in synchronism with the engine, as otherwise the breaker is actuated. The capacitor, 3, furnishes the energy for the ignition spark and is fed from the current source, 1, through a resistor or choke, 4, so rapidly that with a combustion cycle of about 200/sec in a 6-cylinder engine sufficient charging energy is supplied.

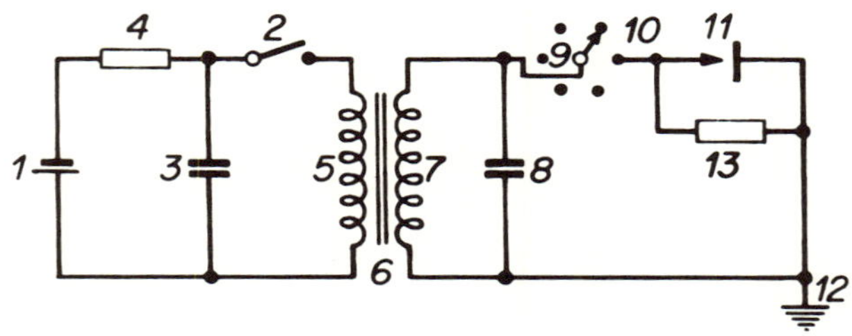

Fig. E8-1. Schematic circuit diagram of an ignition system for an internal combustion engine with transformed capacitor discharges (by Altrogge and Früngel [114]).

With each contact made at 2 the capacitor, 3, discharges through the primary, 5, of a special small pulse transformer having an iron core, 6, in the form of a bar of sheet iron laminations. (Ferritic cores will also serve for this purpose.) With a chosen transformation ratio, in which about 300 volts primary voltage are applied to 4 turns, corresponding to 75 volts per turn, there are induced in the 200-turn secondary, 7, 15 kv as peak voltage. An auxiliary capacitor, 8, is charged by this voltage. The relationship between the size of this auxiliary capacitor and that of capacitor 3 is so chosen that capacitor 8 has only 50% of the energy capacity of capacitor 3. The conventional distributor of an automobile is indicated at 9. The different contact elements, 10, are connected to the several spark plugs, one of which is indicated at 11. The engine body or mass is indicated at 12 and is given in a spark plug automatically by its outer thread. Since in a conventional distributor the contacts never touch each other, but only pass each other at close spacing, a short circuit at capacitor 8 is never possible.

After only a few kilovolts of charging voltage have been reached, the sparks will begin to jump between the moving arm, 9, and the stationary contacts, 10. In the case of a parasite resistance, 13, lying in parallel to the spark gap, 11, this resistance is subjected to the full voltage if it is higher than the resistance of the discharge circuit consisting of capacitor 8 and the inductance of the circuit conductors, if this circuit is damped for the critical case. Only when this resistance is extremely low, will the voltage break down. An example follows.

Capacitor 3 is assumed to have a capacitance of 1 μf at 300 volts charging voltage, and thus an energy capacity of 45 milliwatt-sec. Then, with a relationship as previously mentioned, capacitor 8 has, at 15 kv charging voltage and one-half energy capacity, 100 pf. The inductance of the circuit conductors is assumed to be 1 μh. Then the resistance necessary to establish critically damped or nonoscillatory conditions in the circuit becomes

$$R = \sqrt{\frac{4L}{C}} = 200 \text{ ohms}$$

This shows that resistances above 200 ohms—for instance, 1000 ohms, which may be the resistance of a heavily sooted spark plug—are entirely without influence upon the peak voltage. The full voltage value will always be available at the spark plug, no matter how badly sooted it may be. On the other hand, the breakthrough of the spark leads to a critically damped discharge process, because the spark resistance is about 500 ohms. (In accordance with Toepler's spark equation, which can here by applied only with reservations, there would result with a spark length of 0.5 mm and 8 atm pressure, a spark resistance of

$$R_F = \frac{K \times l \times p}{C \times U} = \frac{0.2 \times 10^{-3} \times 0.5 \times 8}{10^{-10} \times 15 \times 10^3} \approx 500 \text{ ohms}).$$

A spark resistance of 500 ohms does not differ seriously from the resistance of 200 ohms required in the circuit to reach critical case, nonoscillatory damped condition. With such fair resistance adaptation the spark resistance will safely be the major load resistance. This fact assures the high efficiency and reliability of such ignition systems. They have in fact, as Altrogge [522] and the author [20] have demonstrated, a reliability that tolerates badly contaminated engine fuels and spark plugs of particularly low temperatures without ill effects. The system operates under all conditions with absolutely certain sparking, whereby, owing to the high energy investment in the spark and the high peak energy (15 kv on 500 ohms, $N_{peak} = U^2/R = 225 \times 10^6/500 = 450$ kw!); a noticeable ultrasonic effect is added that appears to accelerate catalytically the initiation of the combustion.

9. Voltage Impulses in Extensive Networks

In conclusion a further application of capacitor discharges for the formation of voltage pulses in extended circuit paths will be briefly described, namely, utilization in so-called fence chargers. The purpose of these devices is to frighten persons or animals when they touch a fence, by electrical shocks that are physiologically effective but not injurious or harmful to health. Most of these devices utilize the energy stored in an inductance coil coupled with a self-interrupter. At each interruption of the current, a quantity of energy is released that is equal to $\frac{1}{2} LI^2$, where L is the self-inductance of the coil and I the current flowing through it. In most cases the voltage pulse that is induced at current interruption is transformed to a higher potential by a kind of Ruhmkorff's inductor and is then applied to the fence wire. Another species of these devices operates with capacitor discharges. Both systems are hampered by one disadvantage, namely, that it is necessary to make the pulses strong enough to provide sufficient voltage and remain effective under conditions of considerable leakage, such as with wet insulators. To provide such strong impulses takes considerable energy, and it becomes difficult to supply—for instance, from a dry cell battery—a fence throughout the summer. To improve the fences in that respect several constructions have been offered that give pointed peak pulses but do not take into account the fact that only a certain pulse duration can create an optimum physiological effect. However, it is possible to obtain good results with small energy quantities per impulse, while dimensioning the single current pulses in such a way that they achieve optimum physiological effect. The average energy requirement can be further reduced, by not continuously applying the impulse energy necessary in wet weather but adjusting the same to the weather-dependent loading of the fence. To achieve this end it is necessary to change the energy content of the capacitor feeding the pulse transformer automatically in conformity with fence loading. In the simplest case such energy variation is obtained by switching on an additional capacitor when loading increases. A corresponding circuit diagram is shown in Fig. E9-1. The battery, 1, supplying the energy for the device is connected by way of resistor 2 to the capacitor, 3, for charging the same. A controlled contact mechanism, an example of which is described below, short circuits contact 4 periodically, say, once every second. During each contact closing, capacitor 3 discharges by way of contact 4 into the primary of transformer 5.

A closely coupled transformer has its secondary, 6, preferably provided with taps, connected through disturbance protectors, 7 and 8, with the terminals, 9 and 10, of the fence. Close coupling of the transformer, 5–6, assures that the wave forms in primary and secondary are practically similar. For an optimum physiological effect the auxiliary capacitor and the main capacitor are so dimensioned that in cooperation with the transformer primary the circuit is tuned to a discharge frequency of about 250 cps, a frequency very effective in animals as well as in human beings. Experiments have shown that in all mammals of comparatively similar size, the same frequency has opti-

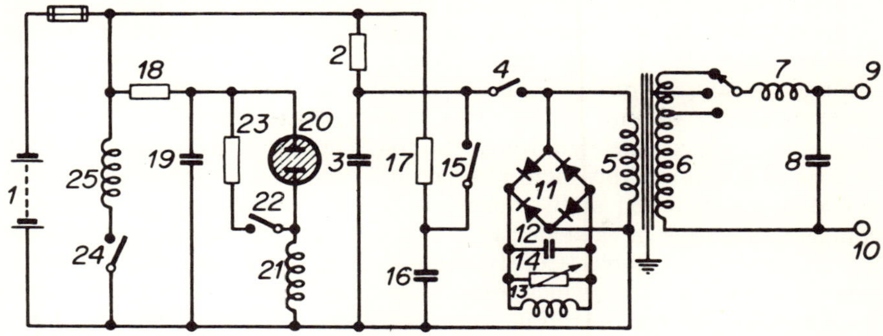

FIG. E9-1. Shock impulse circuit for slowly rising millisec impulses in extensive conductor lines (pasture fence protection) using transformed capacitor discharges and automatic *RC* burden dependent control (switching) means.

mum effect because the sensory nervous systems of men and other mammals are almost identical. A small portion of the voltage effective at the primary, 5, is drawn off by the rectifier, 11, and fed to a relay, preferably in such manner that rectifier, 11, feeds capacitor 12 which in turn is connected to the energizing coil, 13, of a switching relay that is shunted by a rheostat, 14. Now, if the current value that becomes effective with every pulse at the primary, 5, is not impaired, the relay with coil 13 will respond. It will be kept closed by the successive pulses continuously or periodically and thereby keeps a normally closed contact open. This action can be explained by the missing external load (or a well-insulated fence). The oscillation that is created by discharge of capacitor 3 through winding 5 has a periodic characteristic so that by rectification, in case of a heavy oscillation, a large current volume will be measurable at capacitor 12. Since the current volume $I \times t$ is identical with the capacitor charge $C \times U$, U becomes sufficiently high to supply coil 13 with enough energy to keep the relay closed. With increasing load, however, the oscillation becomes more and more damped and even aperiodically damped so that the current volume $I \times t$ is insufficient to supply the energizing coil, 13, with the necessary voltage, U. In this case the relay drops out and actuates its normally closed contact, 15. By way of this contact, that is, with insufficiently energized relay, the capacitor 16, which is also charged from the current source through resistor 17, becomes connected in parallel with capacitor 3. The capacitor 16 can have 10 times the capacitance of capacitor 3 so that the current value of the pulses occurring is 10 times higher. By proper proportioning, this capacitor size, owing to considerably higher magnetization of the iron core of transformer 5–6 and consequently lower inductance, gives approximately the same frequency as capacitor 3 with the primary 5. The latter capacitor delivers only weak pulses. Consequently it is possible, if necessary with capacitor 16 added, to operate near the frequency that has the optimum physiological effect.

An auxiliary circuit for periodic control of the device will now be described. The capacitor, 19, is charged through resistor 18 in such manner that the charging time constant is about 1 sec or another desirable time. As soon as the voltage at the condenser, 19, has reached the ignition voltage of a glow discharge gap, 20, the latter ignites and a portion of the electricity flows through the coil, 21, of an auxiliary relay. If a contact of relay 21 is used to short circuit the gaseous discharge gap by way of resistor 23, capacitor 19 discharges itself completely through the coil. The quantity of energy flowing to the coil is considerably increased, because otherwise the discharge gap 20 would be extinguished at the quenching voltage and would not permit a complete discharge of capacitor 19. From experience it has been found that the factor 10 is suitable. Correspondingly only one-tenth of the size of capacitor 19 is necessary; which results in saving of current, a fact particularly favorable for battery operation. A second contact, 24, of relay 21 closes the circuit of coil, 25, actuating a magnetic switch, such as a mercury switch, periodically. The contact 4 belongs to the relay having coil 25.

In order to avoid radio interference caused by the long wire lines, which act like antennas it is advisable to design the transformer with proper arrangement of primary and secondary—for instance, by elimination of capacitive coupling—so that hard discharge pulses, which cannot be prevented in the primary, are carried over to the secondary in such a manner that frequency portions above 10 kilocycles are avoided. An arrangement conducive to that end is the separation of primary and secondary adjacent to each other so that a voltage pulse is induced in the secondary only by detour through the iron of the transformer core. A core of normal dynamo grade sheets, especially the low grades, cannot carry over interfering high frequencies owing to its high iron losses. It is, however, also possible to arrange additional inductances and capacitances in a manner similar to that used for low pass filters in communication, whereby only frequency portions below 10 kilocycles are passed. Instead of relays, cathode glow relays or glow lamps are frequently used. Multivibrator circuits with transistors are used nowadays, since the latter need only voltages below 10 volts for operation. These new devices open the way for the design of extremely simple, battery, fence charging devices.

10. Automatic Overhead Line Protection by Means of Periodic Capacitor Discharges

The pertinent industry has conducted developmental work in the transmission of steep voltage impulses, produced by impulse transformers, through overhead power lines. For such transmission, the impulse width is between 1 and 4 μsec and the voltage between 20 and 30 kv; these parameters must be produced in an impedance of about 300 ohms which is an average value for overhead lines. It is necessary to use sequence frequencies that diverge completely from the line frequency of 50 cps (in U.S.A. and Canada, 60 cps) so that no interaction phenomena can occur. A completely irregular sequence frequency between 127 and 154 cps has proved successful. However, harmonics, such as 150 cycles, must be strictly guarded against.

It is possible to introduce at distribution points in each line a scanning frequency so that the single impulses never coincide. After introduction, the single impulses travel along the line. They cannot be absorbed in the transformers at the end of the line because in all high-voltage lines there is the customary feeding choke, which prohibits passage owing to its high inductance. The impulse travels therefore to the other end of the overhead line where it encounters a similar choke forcing it to return as a reflection. Simultaneously with introduction of the impulse, a small portion of it is branched off as a signal into a delay line of similar time length. Whenever the overhead line is in sound condition, the returning impulse coincides in timing with the signal sent through the delay line. Both signals are introduced into a so-called coincidence hexode 6CS6, i.e., applied to the first and third grid thereof. If both pulses arrive simultaneously and with predetermined amplitudes, the anode current of this tube is released for this instant and subsequently actuates a multivibrator, which energizes a relay in the anode circuit [21]. When, however, because of a disturbance in the line, the impulse does not return, the signal at the coincidence hexode is missing. The multivibrator receives no signal and consequently drops the relay. This, in turn, causes switching off of the overhead line.

It is possible to release into one of the three overhead lines an auxiliary discharge by way of a spark gap from a large reserve capacitor. This discharge travels along the line to its other end and there passes through a low pass filter and an inserted rectifier to actuate a properly preset relay. This relay is thus actuated by the energy content of the traveling wave package and needs no auxiliary conductor. It in turn throws out, at the other overhead line distribution point, the corresponding station breaker and disconnects the faulty line at both ends. A coincidence disturbance occurs, then, for instance, when an arc burns between two conductors of the line. This is a pronounced disturbance of the impedance, particularly since arc resistances lie in the order of 2 to 30 ohms as against 300 ohms natural impedance of the line, and leads immediately to a disturbance of the received image and release of the coincidence hexode. Also fractures of insulators and deeply sagging line portions lead to a massive reduction of impedance and return signal strength so that considerable geometrical deformation appears to assure drop-out of the defective line. Line breaks or line short circuits will cause a sure drop-out. Further experience will teach whether such transit time controlled, line scanning arrangements with transformed capacitor discharges of high energy and accurately controlled sequence frequency will reach such perfection that they will be considered for replacing in certain cases the extensively employed conductance emergency circuits.*

The scanning impulses must be in the megawatt range in order to oversound strong white noises of corona, also at extended transit times and under unfavorable conditions (such as hoar frost). The evaluating circuit with coincidence hexode is represented in Fig. E10-1. At the top there can be seen as a phantom the pulse-timing cable given as an image of the sound condition of the overhead line. The circuit represented appears to be suitable for 110 kv

---

* Development of Hamburgische Elektrizitätswerke in cooperation with FF Impulsphysics Corporation/Impulsphysik A.G.

and 220 kv networks in which the transmitted power volume is important enough to justify economically the cost of installing additional protective hookups.

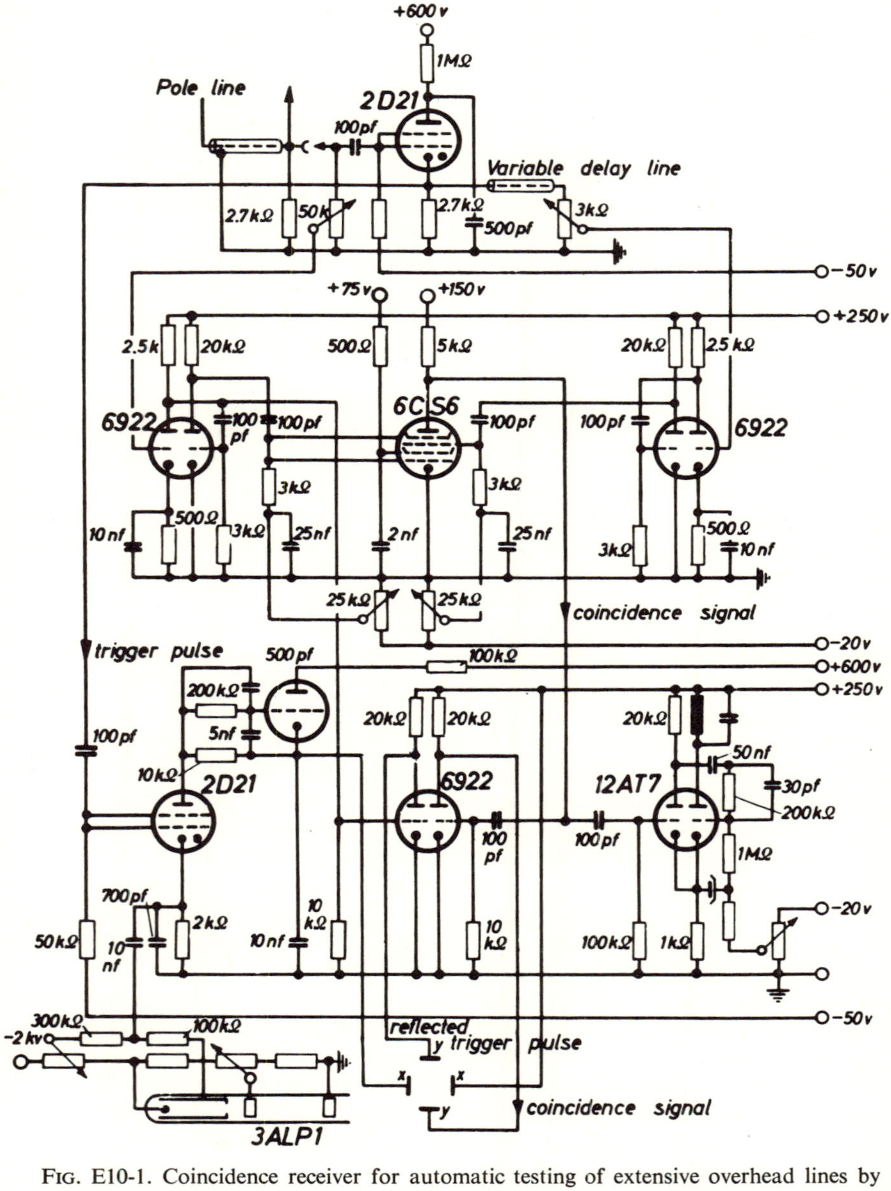

FIG. E10-1. Coincidence receiver for automatic testing of extensive overhead lines by traveling waves. Cooperation of Hamburger Elektrizitäts Werke and FF Impulsphysics Corp.

11. Lightning Flashes in Nature as Capacitor Discharges

The following data are taken from a summary report of Baatz [22]. Measurement of shock current values in lightning strokes is experimentally difficult, since the occurrence cannot be released at will. One approach is to attach small magnetizable bars at several spots on overhead lines and poles. Lightning stroke currents flowing through the overhead systems cause magnetization that can be compared with similar magnetization produced by permanent currents or artificial capacitor discharges. Most of the lightning strokes have shock current values below 40,000 amp. A few reach current values up to 100,000 amp and a few cases of 400,000 amp have been observed.

From the sum of the currents flowing off in the current distribution pattern of each lightning stroke, the presumable current magnitude in the lightning channel can be computed. From the direction of the currents, the polarity of the cloud discharge can be determined. The total number of lightning discharges considered is then tabulated in groups of 10 kv increments. Such a record is shown in Table E11-1. Of the total number of lightning strokes, 87% originated in negatively and 13% in positively charged clouds. Attention is called to the fact that the number of lightning strokes from positively charged cloud portions decreases continually after the first group, whereas strokes from negatively charged cloud portions increase in number up to currents between 10 and 20 kamp and then begin to drop off gradually. Lightning strokes having current values up to 10 kamp occur perhaps more frequently than measured, because the low limit response of the magnetizable bars on poles lies at about 2.5 kamp and on ground wire ropes, at about 1 kamp.

Since these low values are not recorded it is possible that strokes on the order of 3 kamp and lower which strike poles with lines having ground wires are not registered by the magnetizable bar measuring method. In case the ground wire is not provided with test bars, it is also possible that strokes up to nearly 5 kamp which strike ground wire ropes escape observation. Also, in measurements with probes, with response limits of about 0.5 kamp, there is reason to believe that some strokes with low current values occur but do not affect the line as long as they do not hit the line conductors themselves.

When the number of lightning currents is subdivided by the number of ground wires, there results, as presented in Fig. E11-1, a certain dependency. Since the number of results from transmission lines having 3 ground wires is relatively small, these have been grouped with those having two ground wires, since no basic differences could be recognized. The total evaluation shows 77% of the currents to be up to 40 kamp, 94% up to 70 kamp, and 97.4% up to 100 kamp. An explanation for the distribution of the current values in dependence of the number of ground wires has not yet been found. The reason is perhaps the low number of measuring results in lines with zero or with two or three ground wires as against those with one ground wire.

TABLE E11-1

RECURRENCE OF CURRENT VALUES IN A LIGHTNING CHANNEL AND POLARITY OF CLOUD CHARGES

Current value (kamp)	Number of lightning currents		
	Cloud charge		
	Positive	Negative	Total
Up to 10	65	172	237
10 to 20	24	194	218
20 to 30	9	177	186
30 to 40	6	115	121
40 to 50	5	84	89
50 to 60	5	41	46
60 to 70	3	22	25
70 to 80	2	14	16
80 to 90	1	13	14
90 to 100	—	5	5
100 to 110	1	8	9
110 to 120	—	2	2
120 to 130	—	4	4
130 to 140	1	2	3
140 to 150	—	3	3
150 to 160	—	2	2
160 to 170	—	—	—
170 to 180	1	1	2
400	1	—	1
Total number	124	859	983
Total %	13	87	100

Perhaps the numerically more extensive inclusion of strokes with lower current values in the presence of a reduced number of ground wires is the explanation. But it could also be that other, as yet unknown, phenomena are effective in producing the resultant relationship. It is surprising that the number of lightning strokes from positively charged clouds increases considerably with smaller current values. With currents up to 10 kamp the share of positive strokes is 27%, more than twice the average value. It is surprising that with an increasing number of ground wires, proportionally fewer lightning strokes from positively charged cloud portions are observed.

Lightning strokes with currents over 100 kamp have been determined in 25 cases (2.6%). The stroke with the highest current value 400 kamp, was measured with a relatively high certainty of accuracy in 1937 as a positive cloud charge. This lightning stroke was followed by a second one with 100 kamp and negative polarity. Such high lightning currents, however, appear to be very rare.

Previously there had been observed only once, in 1936, a lightning current of 400 kamp and negative polarity which was followed by a second one with

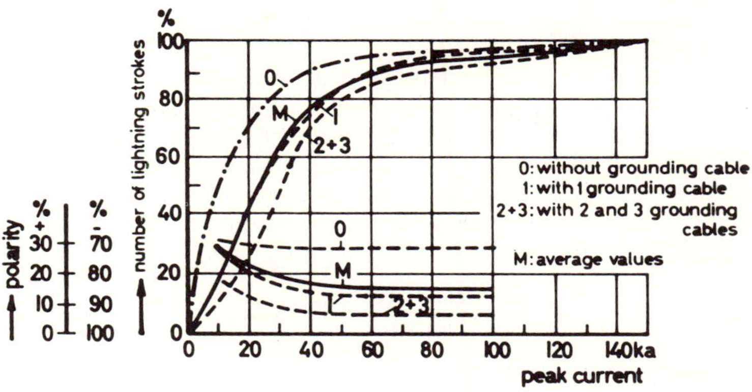

FIG. E11-1. Current magnitudes in a lightning stroke channel and polarities of cloud charges striking lines having none, 1, 2, and 3 ground wires.

more than 200 kamp and positive polarity. The remaining 24 lightning strokes lie in the range between 100 and 180 kamp.

These high lightning stroke current values can be explained by the fact that clouds often can have an enormous capacitance against earth. A cloud of, say, one square kilometer area at 1000 meter height has a calculated capacitance of 10,000 pf. Since the charging potential can be assumed to be 50 to 100 million volts maximum, there results an electricity quantity of

$$Q = C \times U = 0.5 \text{ to } 1 \text{ coulomb}$$

At 1 coulomb and 1/10 millisec duration of the crest current value, 10,000 amp must be transported. Since clouds can have an extension considerably larger than one square kilometer and several lightning flashes usually combine into a massive discharge, the measured heavy lightning stroke currents are plausible. In comparison with cloud capacitance and measurable lightning stroke discharge duration there can be drawn indirect conclusions as to the charging voltage of the cloud. It is only with sufficient charging voltage that the measured quantity of electricity in a stroke can be physically realized. A suggestion at this point is to take electricity quantity measurements of lightning discharges and at the same time estimate the cloud capacitance in order to be able to figure backward the static cloud potential. It appears necessary here to clarify some controversy. According to the table, and based on the actually measured highest current of a lightning stroke of 400,000 amp, there is, assuming a cloud capacitance of 0.1 μf and a discharge duration of 1 millisec, a cloud potential of 4×10^9 volts; and, assuming a discharge duration of 100 μsec, 400×10^6 volts.

The utilization of cloud capacitance and lightning discharge for impulses of extraordinary voltage and current values for x-ray flash tube operation has been tried by Brasch [41]. He achieved more than 40 megavolts on an overhead line strung between mountain tops at Monte Ceneri.

12. Steepness of Voltage Shock Pulses

Very steep voltage pulses can be realized with shock voltage setups as developed by Marx and described in Section 1 of this chapter. If the total number of spark gaps responds within 0.2 μsec (this can be achieved with certainty) it is possible to obtain at a total shock pulse voltage of 10^6 volts a rise steepness of 5×10^{12} volts/sec. This value becomes more impressive when it is computed in light velocity. At a propagation velocity of 3×10^{10} cm/sec the voltage rise per centimeter wire length becomes

$$\frac{5 \times 10^{12} \text{ volts/sec}}{3 \times 10^{10} \text{ cm/sec}} = 170 \text{ volts/cm}$$

Even considerably steeper current rises in conjunction with correspondingly steep leading edges of the voltage waves are obtainable with a steepening spark gap by Marx as shown in Fig. E12-1.

In this device the discharge voltage pulse of capacitor 1 is passed first through a normal spark gap, 2, with ignition electrode 3 to a resistance, 4, so that at the customary velocity of the spark breakdown in spark gap 2 the capacitor voltage of 1 becomes effective at resistance 4. Inserted at the other side of resistance 4 there is a pulse steepening spark gap having two spheres, 5, surrounded by a closely fitting hard glass tube, 6. This spark gap now is apparently well stabilized by the hard glass tube so that breakdown between the spheres, 5, does not occur until a very high overpotential has been reached. By connecting a conductor system, 7, as a load, it is possible to measure the steepness of the pulse by means of a measuring spark gap, 9. The maximum that could be measured in this way up to now were voltage steepnesses of about 1 kv/cm. With such steep leading edges, or short rise times, it is possible to excite centimeter waves in antennas.

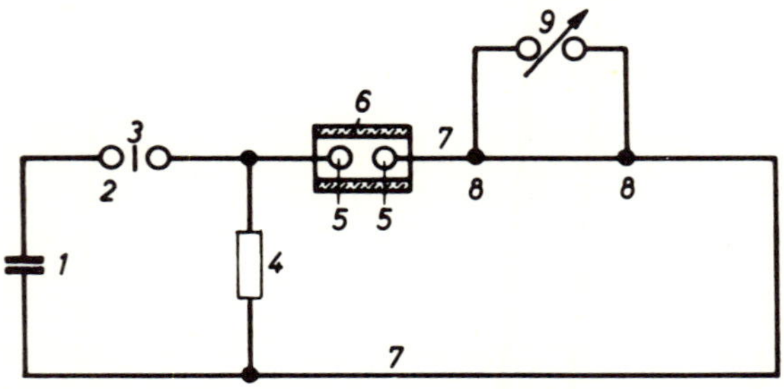

Fig. E12-1. Measuring arrangement for steepness of voltage sparks (by Marx) (for reference numerals see text).

It is most likely that Heinrich Hertz in his early experiments operated with similarly steep voltage waves which excited the dipole with which he conducted his first transmission tests sufficiently to enable him to draw sparks from a receiving antenna several meters away. The steepness that Hertz must have applied lies in the order of 0.1 to 1 kv/cm. It would be of benefit to check Hertz's original papers with today's knowledge, since many of the results obtained at that time have never been repeated and they contain, from the viewpoint of high-voltage engineering, interesting concepts.

Fletcher [23] has measured the breakthrough of spark gaps in the order of 10^{-9} sec, thus the thousandth part of a μsec, under normal atmospheric pressure. For his experiments he used, in a manner similar to that just described for the Marx spark gap, a spark gap arrangement as illustrated in Figs. E12-2 and E12-3.

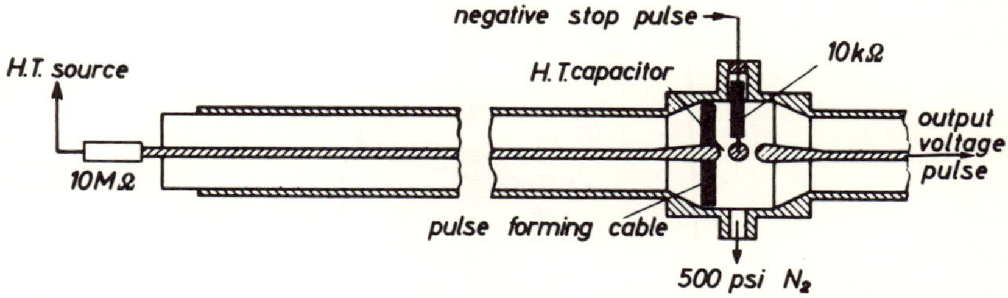

FIG. E12-2. Schematic section of an impulse generator for extremely great steepness, which can deliver 20 kv impulses within 4×10^{-10} sec rise time. The coaxial capacitor which is the energy carrier serves to maintain the full sharpness of the impulse peak from (Fletcher [23]).

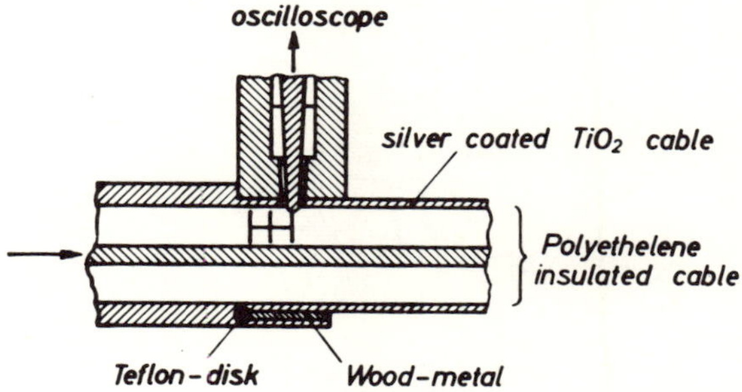

FIG. E12-3. Connecting unit of voltage divider for extremely steep voltage pulses, consisting of two concentric coaxial transmission cables whose characteristic impedances form a voltage divider relationship 100:1 (from Fletcher [23]).

Figure E12-4 shows a high-voltage impulse spark gap created in the USSR [128a–f] which, somewhat similar to the Fletcher spark gap, is capable of producing steep discharges, but, owing to its particular design, permits high thermal loading of about 250 watts. Because this construction form represents, from the engineering viewpoint, an elegant solution of a spark gap for steep discharges, it will be described in more detail. Basically it is a variant of a lamp design introduced by the author [6] and previously mentioned herein.

The perspective view of the illustration shows some design detail. The lamp body, 1, appears like a glass cylinder (160 mm dia.) but is made of brass. It has ribbed outer surfaces for cooling. At the front portion of the lamp body a window, 2, of quartz is mounted, assuring light radiation within a radiating angle of 120 degrees. A special ribbed tube, 20, of hard ceramic is secured to the rear portion of the lamp body. This hard ceramic is a material for heat compensation and the conductors, 19, which lead to the tungsten electrode, 4, are sealed pressure tight therein. The other tungsten electrode, 3, of the spark gap is mounted to the bottom of the lamp body and is independent of other supporting means. Also the stem, 5, through which the ignition impulses for initiation of the discharges are conducted to electrode 6, extends through the lamp body.

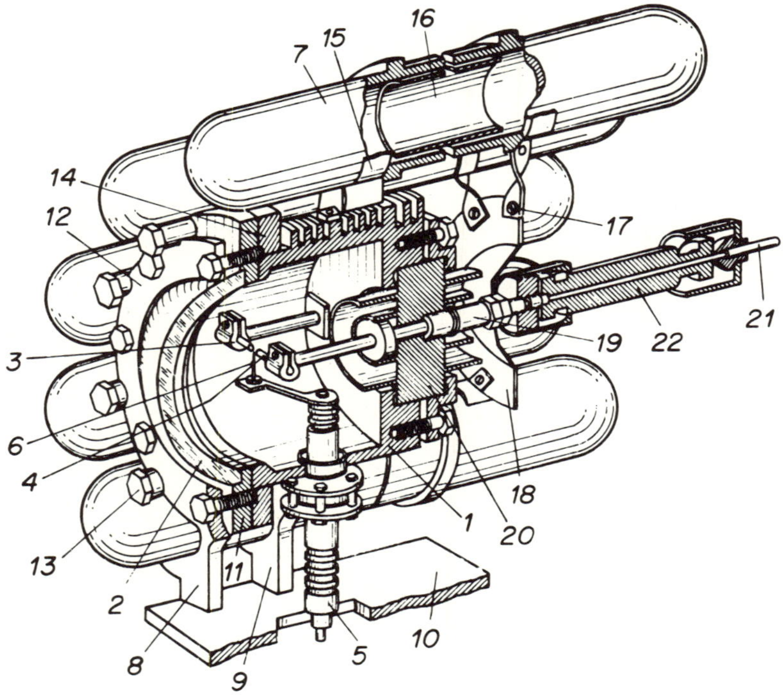

FIG. E12-4. Spark gap system with ceramic construction components and quartz half sphere for outlet of the radiation (for explanation see text) (from Wanjukow and Marshak [128]).

A short and broad strip, 17, of brass with ceramic cup capacitor, 7, is arranged independently of the other structure. The inner layers of each capacitor pair are connected with each other by a brass tube, 16; the outer layers of a capacitor group are connected to terminals 15 and lead to lamp electrode 3. The other group is connected through a brass disc, 1, and a conductor channel, 19, to electrode 4. In the external discharge circuit two series-connected capacitors are provided for self-triggering operation. Each capacitor has a capacitance of 4700 pf. There are 20 capacitors with a total capacitance of 0.022 μf. The inductance of the discharge circuit is 0.6 μh.

Two brackets, 8 and 9, secure the lamp body to base plate, 10. The outlet window, 2, forming a half-sphere of quartz is secured to the lamp body by a bracket, 11. A copper foil of 0.2 mm serves for sealing. Quartz half-sphere and bracket are pressed together by invar ring 12, asbestos lining, and invar bolts. At the other side, the bracket is mounted to the lamp body by steel bolts 13 and ring 14.

Argon is introduced into the lamp from a supply flask through the inner bore of the concentric conductor bolt. A tube, 22, serves to insulate the gas line, 21, from the concentric conductor line, 19. The lamp is good for experiments at up to 10 atm pressure and a maximum voltage of 4 kv. In continuous operation 2 atm pressure were applied. The electrodes were subjected to 28 kv at a spacing (spark length) of 7 to 10 mm.

Continuous operation is permissible for 0.5–2 hours with 1–4 kv load and forced air cooling. The permissible duration of operation is dependent on blackening of the quartz window by tungsten particles from the electrodes. The light flash duration, measured as half-value time against the light peak, was found to be below 1 μsec in every flash and the light density is about 10^8 cd/cm^2.

13. The Operation of Kerr Cells by High-Voltage Impulses

a. *The Kerr Cell and Its Application*

A number of liquids like nitrobenzene become birefringent in an electric field. A beam of light, plane polarized at 45° with respect to the field and directed normal to it, emerges from the medium in general in a state of elliptic polarization. If the delay between the two components is half a wavelength, the emerging light will be linearly polarized again, but in a plane normal to the original plane. Therefore, the light passes completely through the analyzer.

The voltage necessary to produce this phase difference is about 20 kv for a Kerr cell of 2 cm in length and electrodes 1 cm apart. Such a cell has a capacity of 20–30 pf. The time constant of this capacity is about 1×10^{-9} sec. The Kerr cell is therefore the ideal shutter for exposure times of 10^{-5} to 10^{-9} sec. The optical losses in the polarizing prisms are the same for all electro-optical or magneto-optical light valves.

The Kerr cell is a vessel normally filled with nitrobenzene. It utilizes the fact that the plane of light polarization is turned in the space between two electrodes if a high field strength grows. The turning of the plane of polarization follows the changes in field strength almost instantaneously (less than 10^{-12} sec) as a cosine function.

In order to achieve rectangular opening and closing of the optical transparency of such a Kerr cell which lies between two polarization filters or crystals, a step growth as well as a step decrease is desired.

The physics and some applications of Kerr cells are published in [318–323, 532–541]. W. Müller [319] showed the use of Kerr cell shutters in schlieren photography. Owing to their extremely short exposure times, such shutters permit the schlieren photography of projectiles in full daylight.

The field for application of spark schlieren photography as developed by Toepler, Mach, and Cranz and introduced into the practice of measurement technique can be extended by adding an electro-optical shutter. Only the Kerr cell shutter and the image converter tube need be taken into consideration for applications in which extremely short exposure times are wanted.

By such a shutter, schlieren photography of self-luminous events in full daylight becomes possible for any testing station. The exposure time of the photographic plate is no longer determined by the luminescence of the spark but by the opening time of the shutter. Thus, requirements for sparks are quite different from those for ordinary spark photography. Schlieren pictures using a Kerr cell shutter with ultra short exposure times of 50 nanosec are shown as examples.

Today, schlieren photographs rank among the most important measuring methods, extensively employed at ballistic ranges, in wind tunnel work, shock wave tubes, or similar applications. Unfortunately, many workers still operate largely without optical shutters, i.e. fast action shutters, and cling to the method of darkening the measuring room and illuminating the event for a sufficiently brief period to give the impression of frozen movement.

Owing to the lack of a suitable optical shutter the applications of this method are rather limited and it fails the very moment the event as such has luminosity of its own. Detonations and the phenomena at and close to the muzzles of weapons are examples in point. Even air currents at high Mach number offer difficulties as soon as the shock waves become luminous. Furthermore, the above-described method proves impracticable when the event under observation cannot be set off in a darkened measuring room, e.g., explosions and detonations. Other examples to prove this point are photographs of large and fast-flying models. Finally, the method is bound to fail when the lighting time of the illuminating spark is no longer small compared to the duration of the event. With most of today's supersonic or superballistic events the illuminating spark's limits have already been reached if not exceeded.

The difficulties outlined above can, however, be overcome by equipping the photographic gear with a suitable fast action shutter. Thereby the field of application of spark photography is considerably widened. Such a shutter eliminates the influence of detrimental foreign light and even makes possible the photography of self-luminous events. Thus it no longer is necessary to move to a special darkened measuring room but the photographs can be made in full daylight at any testing ground. Finally, the exposure time is no longer fixed by the lighting-up time of the illuminating spark. It is solely determined by the effective exposure time of the shutter.

This, of course, makes a considerable difference in selecting a light source, as will be illustrated by Fig. E13a-1.

Besides the time t the diagrams show the intensity J radiated from the light source. In ordinary spark photography, i.e. without the use of a Kerr

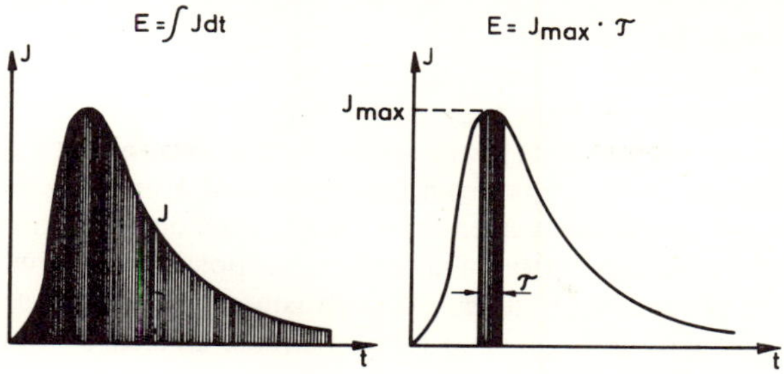

FIG. 13a-1. Exposure time and light sum with and without Kerr cell shutter.

cell shutter, the total light emitted from the spark becomes effective on the photographic plate. The blackening of the plate will then be proportional to the logarithm of the light sum.

$$E = \int_0^\infty J dt$$

As is generally known E is unfortunately determined not merely by the width at half maximum intensity of the light pulse but can be considerably increased by afterglow. When a fast shutter is used, however, only the brightness maximum J_{max} of the light source and the effective opening time t of the shutter count, assuming, of course, that the shutter is opened at precisely the moment of maximum brightness. Consequently, the proportionate blackening of the photographic plate is $\log E = \log J_{max} \times t$. It is evident that in this way extremely short exposure times with sufficient blackening can be realized. It certainly will not be easy to give exact figures for the above statements. There are still lacking systematic measurements on the peak brightness of electric sparks as a function of discharge circuit data such as voltage, capacitance, spark resistance, internal induction, gas type, and gas pressure. In many optical setups the size of the luminous source as such is a determining factor too, so that, strictly speaking, the luminous density as such should count (not to speak of the fact that in the end the photographic effectiveness is decidedly determined by the spectral composition of the light). Finally it should also be pointed out that with such a fast action shutter not only the electric spark but also any other suitable bright light source can be used. When one is designing light sources for such purposes, concepts and considerations quite different from those ordinarily applied in spark photography will have to be followed.

Because of the extremely short exposure times required, only electro-optical shutters will qualify. Among these the most suitable will be the Kerr cell shutter and the image converter tube. The faraday shutter will not work as it hardly permits exposure times shorter than a microsecond.

The Kerr cell shutter has the disadvantage of absorbing 75–80% of the available light but compensates for this by offering high resolution and very short exposure times. Despite the large-size Kerr cell used in the experiments (diameter 35 mm), photographs with 50 nanosec exposure time could be taken without any difficulty.

Contrary to any other shutter type known, image converter tubes will cause an intensification of light rather than reduce its intensity. With suitably designed image converter tubes spark schlieren photographs at exposure times of 10 nanosec and less should be feasible. Up to now Kerr cells were generally preferred because image converter tubes have considerably lower resoluton and because the above-mentioned ultrashort exposure times will hardly be achievable with commercial tubes. (At ultrashort shutter times the picture becomes very poor owing to space charge effects caused by clouds of electrons.)

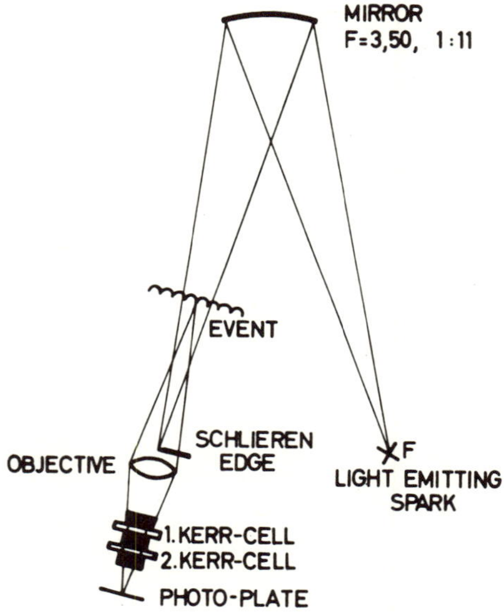

FIG. E13a-2. Optical setup.

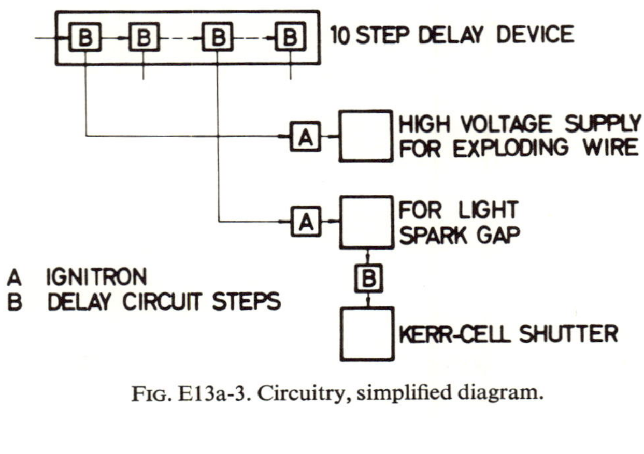

FIG. E13a-3. Circuitry, simplified diagram.

W. Müller [320] has reported on Kerr cell schlieren photographs taken of exploding wires. In these experiments two Kerr cell shutters were used behind the photographing lens because the blocking factor of the polarizing filters was not sufficient for the extreme brightness of the event. Figure E13a-2 is a schematic diagram of the optical setup which follows the Toepler schlieren method using a concave mirror as schlieren head. A spark flashlamp was used for a light source, having a capillary discharge gap so that merely one schlieren edge was required in front of the camera lens. The two Kerr cell shutters were arranged behind the lens. Figure E13a-3 shows a schematic diagram of the electrical circuitry. The event and illuminating spark are triggered at preset intervals. The opening of the Kerr cell shutter follows the firing of the illuminating spark with such a delay that the shutter opens at exactly the moment of maximum spark brightness.

The same setup, with merely one Kerr cell, was also used in full sunlight at a ballistic range for photography of flying projectiles.

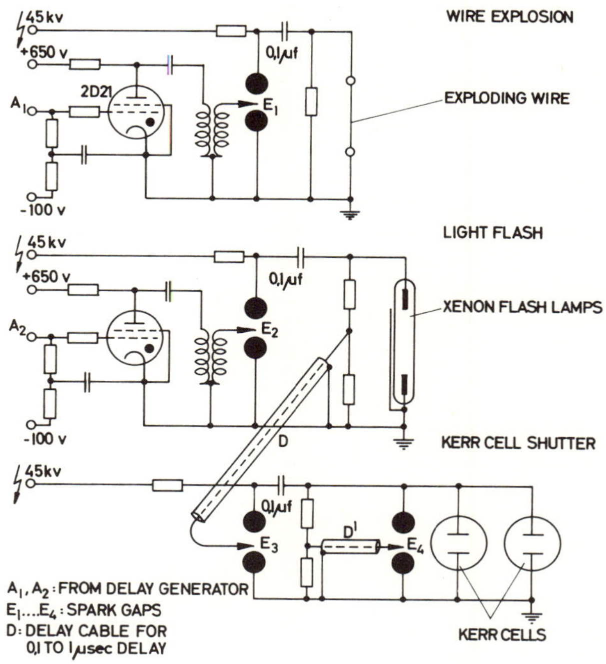

FIG. E13a-4. Arrangement of the high voltage and delay circuit for single picture Kerr cell photography of exploding wires.

Often it is sufficient to open a Kerr cell shutter in combination with a xenon flash lamp only once at a predetermined time. Figure E13a-4 shows the circuit diagram [188] of an arrangement corresponding to Fig. E13a-3 in which first the xenon flash lamp is started for a long duration flash, second the wire gets its discharge to start the explosion process, and last the Kerr cell shutter opens to shoot a photograph at a certain phase of the explosion. Owing to its simplicity this circuit seems useful.

In 1950, Quinn *et al.* [534] delivered a paper in the field of ballistic photography about a Kerr cell camera combined with flash illumination. The earliest paper may be that of Beams [935], who used the Kerr cell for a study of the successive appearance of spectral lines in spark discharges down to some nanoseconds, and this was accomplished without the aid of modern electronics.

A more popular description of Kerr cell function is given in [885]. For the molecular polarizability of different liquids, especially sydnones, see [837]. In 1958, Zarem *et al.* designed a nanosecond Kerr cell shutter [936], which is commercially available from Electro-Optical Instruments, Inc. This nearly foolproof equipment allows automatic synchronization between the event, the synchronizing pulser tube, the pulse-forming network and the Kerr cell camera. The shortest shutter time provided for is about 5 nanosec. A recent paper by Hauser and Quan [937] describes the further development to get low time jitter, low rf noise, and high rates of current rise. The pulser operates with the hydrogen thyratron 6587. This tube uses a separate heated filament of the hydrogen reservoir.

b. *Multiple Kerr Cell System with Square Shuttering Characteristic* [315]

For extremely short exposure times, Liebing and Früngel developed a multiple Kerr cell system with a square shuttering characteristic. This Kerr cell shutter is designed entirely in cable form. The operating principle of this shutter is as follows.

A 15-ohm cable, charged to a voltage of 80 kv, is connected via a coaxially connected spark gap to another 15-ohm flexible cable, the latter being fed into a band transmitter of like wave resistance. This band transmitter contains nitrobenzene as dielectric and forms the Kerr cell. This arrangement is closed by a terminal resistor of 15 ohms.

If the spark gap is fired by a triggering pulse, an electromagnetic wave of 40 kv passes through the cell, the passing time being double the transit time of the wave in the 15-ohm cable. The electric field generated in the Kerr cell is directed from the outer electrodes to the middle electrode and its field strength is calculated when the cell is completely open.

The rise of the voltage pulse is flattened by disturbance spots in the wave resistance by reflecting an oncoming wave gradually through the disturbance spot so that the output as step form Fig. E13b-1. The length of one step is equal to double the transit time in the spot of disturbance. The amplitude of the $-n$th step is

$$U_n = U_0\left[1 - \left(\frac{Z - z}{Z + z}\right)\right]^{2n}$$

Such a disturbance spot is, for instance, the spark gap. According to the above reasoning one could safely assume that on realization of this Kerr cell shutter one could expect a rise time of only a few nanoseconds.

The technical design of this Kerr cell shutter will be explained according to Fig. E 13b-2.

(a) The impulse forming cable is designed as a delay cable for fixed impulse durations of 10–50 nanosec.

(b) The spark gap is designed as a pressure spark gap for 10 atm gas pressure nitrogen in order to keep the disturbance spot in the wave resistance as short as possible. By a special circuitry it can be triggered with an accuracy of 10 nanosec.

(c) The Kerr cell as such is made of solid brass with plane windows and carefully tuned wave resistance.

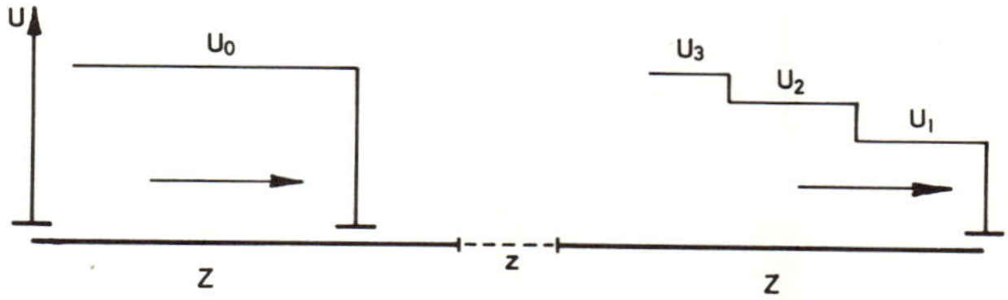

FIG. E13b-1. Change of a steep impulse at a place of disturbance.

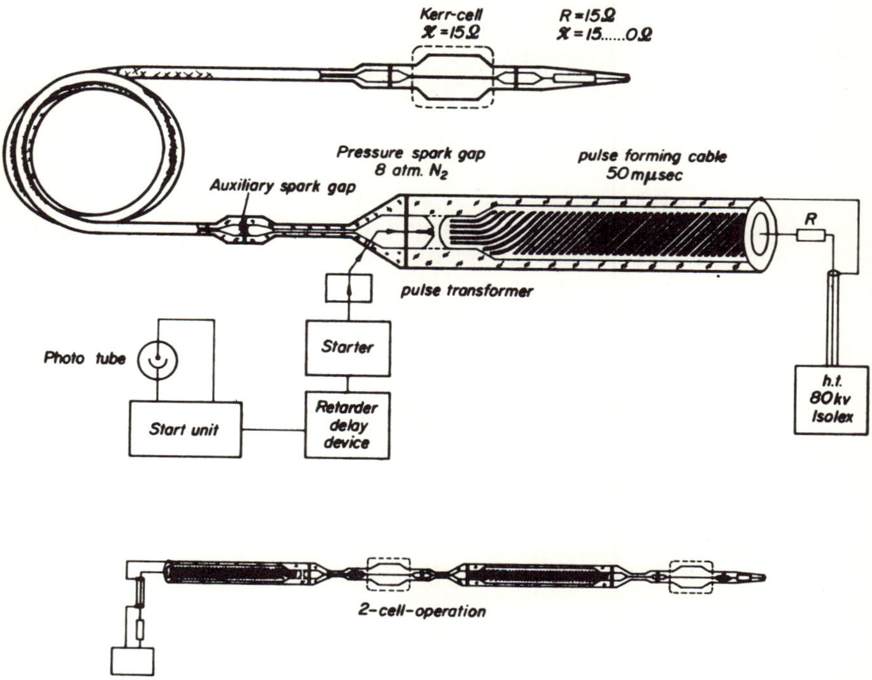

Fig. E13b-2. Block diagram of the Kerr cell shutter.

(d) The terminal resistor is a special carbon resistor, araldite sealed. The outer transmitters are connected in wedge fashion to the resistor so that no reflexes can occur.

The spark gap, flexible cable, Kerr cell, and terminal resistor are designed as plug-in units.

Liebing [315] shows the time characteristic of the current curve during one opening period of the shutter. The measuring is effected by means of a probe at the end of the terminal resistor.

The above described method of a traveling wave Kerr cell offers the following advantages:

(a) The cell capacitance has no influence on the pulse form. Merely the wave resistance of the system, including any number of cells, must be constant.

(b) In photographing through two cells, the cells are opened by a single pulse practically simultaneously. This arrangement is preferred if an increase of the blocking factor is desired.

(c) If two cells are interconnected by means of a delay cable, the cells can be opened at exactly defined intervals. This arrangement is preferred for measuring velocities.

(d) If the pulse is guided through the cell along a meandering path, it becomes possible to construct large-size cells without the need for modification of any of the other parts of the system.

A commercial multiple Kerr cell camera for several independent Kerr cells and exposure times between 7 nanosec and 7 μsec is described by Zarem and Marshall [532]. It has been developed by Electro-Optical Instruments Inc. Stuart M. Hauser [938] describes a Kerr cell camera with 6 Kerr cells and exposure times of 3 times 5 nanosec and 3 times 20 nanosec. By use of Polaroid films in combination with a high-speed, high-resolution lens, this camera system is a powerful tool for plasma research. This system, KFC 600, also provides a wide range of recording capabilities. For example, recording can be accomplished on standard 4 $\times$ 5 cut film or on cut or roll Polaroid Land film. Special film holder adapters avoid the need for refocusing of the lens in switching from one type of film to another. There is also a slow acting electromechanical capping shutter which effectively isolates the individual film planes before and after the even period of interest even when the dark slides are not immediately reinserted.

The complete control circuit for the Kerr cell framing camera includes a framing or delay generator chassis (for initiation of the framing action), a high-voltage power supply, a thyratron power supply chassis, and a central control unit chassis from which all operational and functional adjustments can be made. All cables between the camera assembly and the control unit are connected to the face of the control chassis, rather than at the rear, to allow for greater ease and versatility in handling.

c. *A High-Frequency, Spark Light, Kerr Cell Combination*

An unexpectedly versatile application of capacitor discharges with high sparking rate has resulted from the combination of spark light illumination with Kerr cell shutters. Here the basic idea is to cut off the leading and trailing edges of 0.5 μsec sparks by means of a Kerr cell, thus retaining for effective exposure merely the very intense light peak, for ultrashort exposures at ultrarapid exposure rates. In this way one succeeds in considerably increasing the quality of still pictures and films of very rapid events. Owing to the Kerr cell's extremely fast shuttering action, even self-luminous events, or even those extremely difficult phenomena which start with a dark phase and progress to a luminous phase, can easily be photographed with ordinary drum cameras, e.g. the deformation or the melting of fuses.

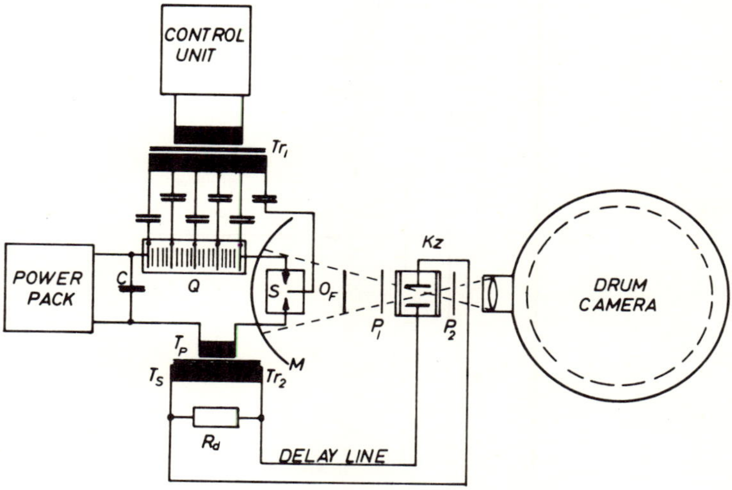

FIG. E13c-1. High frequency spark light Kerr cell combination.

The following is a description of the circuitry of this rather original arrangement based on the latest work of the author. Figure E13c-1 shows a spark chamber S into which are fed capacitor discharges from C. The discharges of C have to pass over the quenching spark gap, Q, so that spark bursts of any desired frequency up to approximately 300,000/sec can be fired. The sparking rate is enforced upon the pulse transformer, Tr_1, by means of a control unit. In the discharge circuit of the capacitor C there is an auxiliary turn, T_p, of the pulse transformer, Tr_2, which is a transformer with differential characteristic and on the secondary of which there is an attenuator, R_D. The value of this resistor corresponds to the critical damping case. During the simultaneous current and light rise as in Fig. E13c-2a, there occurs at the secondary of Tr_2 a voltage pulse according to the first differential power of the primary current as shown in Fig. E13c-2b. Thus the voltage passes through zero when the light is at its maximum. By means of a correctly adapted delay line (usually a uhf antenna cable will suffice) the pulse is now moved so that its voltage after the transit time coincides with the light peak (Fig. E13c-2c). This voltage pulse now opens the Kerr cell KZ in front of the drum camera. Since the opening and closing of the Kerr cell proceeds according to a cosine² function, there results the shuttering action which is represented by the curve in Fig. E13c-2d. One thus cuts exactly into the light peak of the illuminating spark. The light of the spark gap S is projected onto the object O_f by the elliptic or parabolic reflector M. This arrangement is particularly valuable when the switching-off process of a melting fuse wire or a similar phenomenon is to be examined.

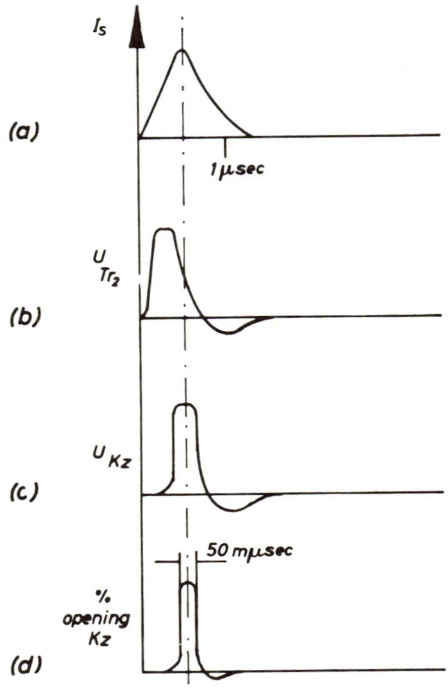

FIG. E13c-2. Primary current and secondary voltage of differential pulse transformer and Kerr cell.

Figure E13c-3 shows a series of pictures produced in this way. In the beginning stage, one first sees a shadowgraph of the fuse wires under examination with back lighting from the spark. Thereupon the voltage load is applied. The wires are first deformed, and then they burn through, producing a brilliant light arc between the wire ends. Owing to the short opening time of the Kerr cell shutter this self-luminous phase of the event is, despite the brilliancy of the emitted light, nevertheless relatively dark compared to the coincidental spark back lighting and thus does not become effective for the exposure. Because of this suppression of the event's self-luminosity the photographs obtained have a certain likeness to x-ray photographs. This novel technique offers the desired result, namely to observe the fuse mechanism and any irregularities occurring without disturbance from the melting wire's luminosity. This method is already commercially available and known as the Strobokin-Strobodrum-Strobokerr combination. Due to its assembly kit design this equipment permits photographing both dark and self-luminous rapid events, and even events proceeding from a dark phase to brilliant luminosity.

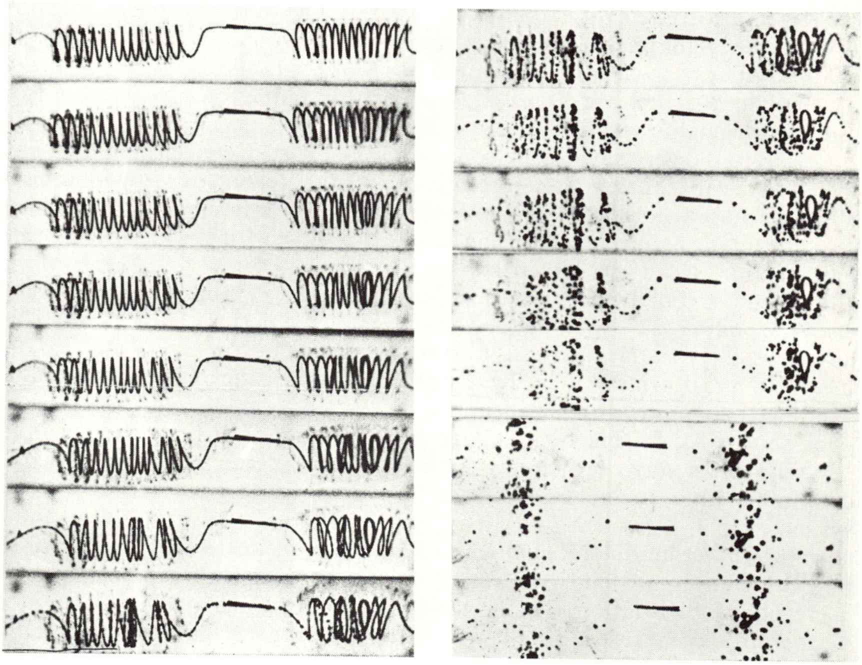

FIG. E13c-3. A blowing fuse photographed by Strobokerr, corresponding to Fig. E13c-1.

W. Thorwart [809] showed how the knowledge of the application and the valuable information obtainable by means of high speed photographic methods increasingly makes them a most valuable measuring device in research and engineering projects. W. Thorwart describes the current techniques of high speed motion analysis methods and the fundamental principles involved, as well as the most important equipment available, such as light sources, cameras, shutters, and synchronization devices. Some special methods and techniques chosen from the wide field of high speed motion analysis may be an indication of some of the most advanced fields of research into which speed motion analysis photography has now entered. The combination of high repetition rate Kerr cell cinematography with synchronous spark illumination has already led to important results in the field of electrical arc welding. The spark light radiates through the arc plasma and allows the cinematography of the forms and speed of the metal droplets which leave the welding electrode. The above-mentioned circuit is described by Früngel [845] and Früngel *et al.* [895]. A review of the technique given at the 6th International Symposium of High Speed Photography in 1962 at Den Haag is given by R. Schall [852].

d. *Kerr Cells for Special Applications* (*Ultraviolet, Laser-Q-Spoiling*)

Certain cases of cutting light in the invisible spectrum, especially in the short ultraviolet, do not allow the application of nitrobenzene filled Kerr cells. L. Frommhold [813] found that a water filled Kerr cell can also be used. He combined a Kerr cell with a controlled spark light source for the production of 5 nanosec light pulses. The quartz ultraviolet region down to a wavelength of 1900 Å may be transmitted using water as Kerr cell liquid. Some properties of a water filled Kerr cell are discussed. The Kerr constant is found to be 10×10^{-7} esu at wavelengths between 2100 and 3000 Å.

Owing to the very small Kerr constant, water filled Kerr cells are long and have a high capacitance. Therefore, a powerful nanosecond circuit with unusually high pulse voltage is needed to feed them. L. Frommhold and H. Raether describe this technique in [896] using a cable capacitor with 30 kv and a nitrogen high pressure spark gap for switching.

Recently, a broad field of Kerr cell application was found in the so-called laser-Q-spoiling. The spontaneous discharge of the accumulated energy of a laser crystal is possible by means of rapidly rotating mirrors or, controlled, by means of the Kerr cell. The Kerr cell must have either extremely parallel windows or windows arranged under the Brewster angle and, of course, a good shutter factor. The high-voltage pulse generator must have a very short rise time and a very exact timing circuit to start the laser operation at a predetermined time. References [834] and [935] describe Kerr cell laser-Q-spoiling systems.

In basic laser modulators the Kerr cell normally is unstressed, and is pulsed into the quarter- or half-wave retardation condition. Half-wave retardation corresponds to a rotation of an input, linearly polarized ray of 90°. This method of operation comprises a simple shutter or optical gate when the Kerr cell is used between crossed polarizers. Pulsing briefly into the quarter-wave retardation condition will correspond to more sophisticated laser modulation technique in conjunction with quarter-wave plates. Pulse durations are determined by means of interchangeable pulse-forming networks. Kerr cells incorporate parallel plate electrodes and are normally filled with nitrobenzene. The windows are flat to 1/4 fringe over the useful diameter.

Normally, the system has a bias network substituted for the pulse-forming network, and the Kerr cell is filled with specially purified nitrobenzene. This system is used for the "Q-spoiling" technique by use of the "pulsed reflector" mode. In this operation, the Kerr cell is optically biased by the application of a dc voltage into the quarterwave retardation condition. Under these conditions there is no coupling between the incident and reflected rays, and lasing action is held off. A good bias network system shorts the Kerr cell in 5–6 nanosec producing very high peak laser powers.

In order to use this bias mode Q-spoiling technique it is essential that the Kerr cell be supplied with specially purified nitrobenzene to hold off the high dc voltage.

Another system (Electro-Optical Instruments Inc., LPM 60/B) provides the combined capability for laser operation in the pulsed open and the pulsed reflector modes. It contains the usual control electronics, module, and Kerr cell, and is provided with both a pulse-forming network and a bias network.

The minimum delay from initiation to firing is approximately 100 nanosec, and the maximum jitter approximately 15 nanosec.

e. *Solid State Voltage-Pulsed Electrooptical Shutter*

The plane of polarization can be rotated not only by means of a Kerr cell but also by applying electrical fields to special crystals. Such a solid state modulator [814] operates by virtue of the so-called Pockels effect which is analogous to the Kerr effect. The solid state modulator offers the advantages of smaller size, shorter path length, longer life, less power consumption, and elimination of toxic liquids such as nitrobenzene.

Such crystals consist of either a flat, polished plate of ammonium dihydrogen phosphate (ADP), potassium dihydrogen phosphate (KDP), or potassium dideuterium phosphate (KD*P) whose faces are perpendicular to the crystallographic Z axis. The crystal is placed between two transparent electrodes allowing light to pass in the same direction as the applied electric field. The crystal-electrode configuration is potted with silicone rubber into a phenolic case.

Crystal modulators offer great advantages where high driving voltages must be avoided. Another advantage is the high transparency which already exists in the medium ultraviolet range. KD*P can be used between 2500 Å and 1.4 μ wavelength and KDP between 2500 Å and 1.2 μ. ADP is similar to KDP.

If one substitutes for usual glass windows quartz windows and for the foil polarizer a Glan-Thomson prism, a low loss, white light, broad-band modulation can be effected.

The limit of the solid state light modulators is given by the physics of non-linear optics. If the light amplitude reaches the level of megawatts e.g., in a Q-spoiled laser system, a KDP-crystal begins to double the light frequency.

14. Biological Applications of Voltage Impulses in Electrical Fishing [323]*

a. *Reaction of Fish to Direct Current*

When direct current is passed into a fish tank, it can be observed that nearly all fish swim toward the anode, but, before reaching it, are usually stupefied and turn upside down. When the current is switched off, they recover and swim away.

There are several phases of reaction with the gradually rising current density. The first visible reaction of the fish is that, being in parallel position to the current direction, their bodies are vibrated by the current. But if the fish are in transverse position to the current, they turn their heads toward the anode.

The second reaction occurs when the fish, if in parallel position to the current, swim toward the anode. If the fish are in a transverse position to the current, then the body axis of the fish turns parallel to the direction of the current when the latter is turned on. But the fish do not in all cases swim toward the anode. That will only occur when a new pulse is given. This second reaction is called galvanotaxis.

---

* Chapter E, Section 14, reprinted by permission from P. F. Meyer-Waarden, "Electrical Fishing," pp. 3–51, Food and Agriculture Organization of the United Nations, Rome (1957).

The third reaction occurs when the fish turn on their sides, toward the anode, and are no longer able to move of their own accord. This reaction is called galvanonarcosis.

Table E14-1 shows the pulse values needed to induce the above-mentioned reactions. The threshold values of the pulses are constant and reproducible for fish of the same species and size. No reaction can be observed from pulses below that given for the first reaction. It is significant that these pulse threshold values decrease with the increasing size of the fish, although the proportion among the pulse threshold values remains approximately constant (Table E14-2). Holzer [342] calls the voltage tension between head and tail which is required to produce one of these reactions, "body voltage." It amounts, in the case of trout for instance, to 0.4 volt for the first reaction, to 1.2 volts for second reaction, and to 2 volts for the third reaction. Since large fish receive a greater voltage in the water than smaller ones, they can be influenced more quickly and by relatively smaller voltages (Table E14-3).

Fish which receive a lower voltage than that required for galvanotaxis can escape from the field.

TABLE E14-1

VALUES (δ) OF THE CURRENT DENSITY FOR FRESH-WATER AND SEA FISH ACCORDING TO HALSBAND

Species of fish	Length (cm)	Weight (gm)	First reaction δ	Second reaction δ	Third reaction δ
Stickleback	5–6	3–4	0.48	1.88	3.94
Loach	10–12	7–9	0.08	1.55	2.43
Plaice	20–22	95–100	39.91	44.34	70.95
Steinpicker	8–10	14–16	18.18	50.99	115.00

TABLE E14-2

RELATION OF BODY LENGTH TO PULSE THRESHOLD VALUES FOR THE FIRST GALVANOTAXIS AND GALVANONARCOSIS WITH THE MINNOW (*Phoxinus laevis*)[a]

Length group (mm)	Number of fishes	Medium length (mm)	Current density thresholds δ		
			First reaction	Galvano-taxis	Galvano-narcosis
15–20	5	19	0.19	3.30	4.74
21–30	17	24	0.18	3.13	4.74
31–40	10	36	0.13	3.32	3.94
51–60	7	57	0.11	1.43	2.50
61–75	14	66	0.13	1.28	2.15

[a] According to Scheminzky.

TABLE E14-3

RELATION BETWEEN FISH LENGTH AND CURRENT DENSITY[a]

Length of fish (cm)	Current intensity μA_{eff}	Current density in water δ	Voltage at the electrodes V_{eff}	Voltage in the electric field V_{eff}/cm	Body voltage between head and tail V_{eff}
3.0	123.0	2.49	11.2	0.476	1.429
3.2	114.0	2.31	10.4	0.425	1.360
4.1	87.8	1.78	8.0	0.342	1.399
4.7	75.6	1.53	6.9	0.293	1.380
5.0	77.4	1.57	7.0	0.301	1.500
5.6	64.6	1.31	5.9	0.251	1.410
5.9	61.4	1.24	5.6	0.238	1.405
6.0	60.2	1.22	5.5	0.234	1.402
6.3	58.1	1.18	5.3	0.225	1.420
6.4	55.9	1.13	5.1	0.217	1.390
7.4	49.3	1.00	4.5	0.192	1.420
7.8	47.1	0.955	4.3	0.183	1.430
8.0	43.8	0.890	4.0	0.170	1.360
9.0	39.5	1.800	3.6	0.153	1.385
10.5	32.9	0.666	3.0	0.128	1.340

[a] Constancy of body voltage for roach (*Leuciscus rutilis*), Crucian carp (*Carassius vulgaris*), and bleak (*Alburnus lucidus*), with alternating current at 50 cps, according to Holzer.

b. *Reaction of Fish to Alternating Current*

When alternating current is used, the fish do not swim toward one of the two electrodes, as they do in the case of direct and interrupted direct current (which will be explained later). They take a transverse position to the direction of the current between the two electrodes, namely in such a way that they tap off a minimum voltage. This phenomenon is called oscillotaxis.

The colors of the body of a fish narcotized by alternating current fade owing to a pigment contraction. Moreover, most of the fish (for instance, trout, carp, tench, catfish, etc.) undergo, according to Scheminzky [356], a kind of "hypnosis" after the current is switched off. They do not return immediately to normal swimming position, but stay for some minutes in a lateral or dorsal position. When "hypnosis" vanishes, the fish swim away as they do in the case of direct current.

c. *Reaction of Fish to Interrupted Current*

When using direct current pulses of rectangular form (Fig. E14-1A) or pulsating current in the form of a condenser discharge (Fig. E14-1D), or in the form of a quarter sinus wave (Fig. E14-1C), more or less heavy vibrations occur in the fish body, depending on the number of pulses used per time unit. As soon as the pulse threshold for the galvanotaxis is reached, the fish turns and swims toward the anode. This happens with all above-mentioned forms of pulsating currents, except when the current is increased slowly and then decreased steeply. In the latter case, the fish turn toward the cathode, and narcosis occurs in the same way as with direct current.

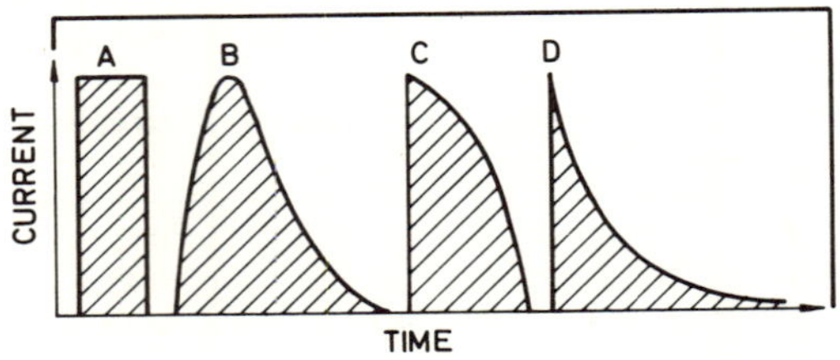

FIG. E14-1. Various current impulses (according to Kreutzer [343]). A–rectangular pulse, B–unperiodic pulse, C–part of a sinus-curved pulse, D–capacitor discharge. By permission from P. F. Meyer-Waarden, "Electrical Fishing," Food and Agriculture Organization of the United Nations, 1957.

By simple algebraic derivation from the formula for the sphere surface one obtains the equation developed by Kreutzer [343, 344, 345] for catching fish in sea water (see Fig. E14-2). The extension of the equipotential lines can be spherical in sea water as well as in conducting fresh water with corresponding electrode shapes and application.

d. *Main Types of Pulses Used in Electrical Fishing*

Owing to the conditions of conductivity in sea water, it is quite obvious that one cannot work with alternating and direct currents. According to investigations already made, it would be necessary to produce up to approximately 10,000 kw to obtain an electrical fishing range (spherical range) of a radius of only 10 meters. In such a case it is only possible to operate with short pulses such as those of electric fishes. Only 200 to 300 kw would then be required for the same effect. Leduc operated with a current of 100 periods, namely with impulses of 1 millisec duration, followed by current interruptions of 9 millisec each. The body voltage of the fish in case of such currents requires an effective value substantially smaller than when normal direct current is used. The peak value of voltage used by Leduc amounts to from 1.5 to 2.5 times the body voltage required in the case of normal direct current.

However, as these rectangular impulses with greater pulses of 10 to 20 kamp and 1.5 to 3.5 kv cannot be produced economically, Kreutzer began [343], in the first experiments he made in electrical fishing in 1947, to use currents produced by capacitor discharges with a steep rise and a gradual decrease. It is to his credit that he recognized then that the shape of the individual impulse is essential. Figure E14-1 shows a survey of the several types of impulses. The curve forms A and B are less suited to electrical fishing than the curves C and D. The curve C is specially used to block fish or to guide them in a desired direction by means of movable gear. It is similar to the type of pulse produced by the electrical eel (Fig. E14-3).

Curve D is the most effective form of pulse for attaining galvanotaxis. Most fish react to these pulses anodically.

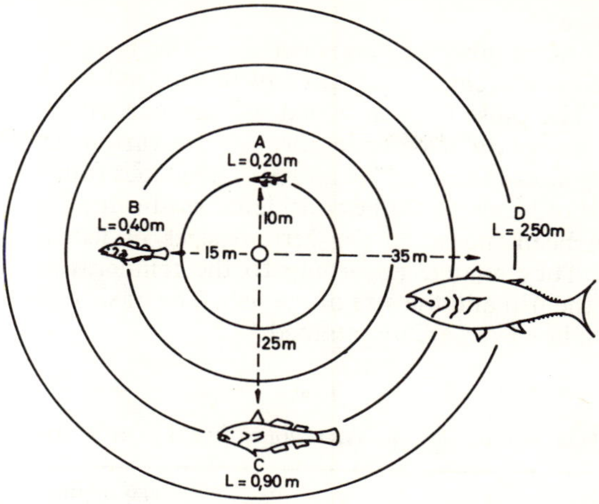

FIG. E14-2. The formula for computing the electrical catching area in sea water (Kreutzer) is:

$$R = \sqrt{\frac{ILK_1}{G4\pi}}$$

R = catching area (distance from the electrode in meters), I = current potential in water in amp, L = length of the fish in meters, G = body voltage in volts, K_1 = specific resistance of sea water (0.3 ohm × meter).

The electrical energy expended for the ranges in this diagram was about 250 kw. A–herring, B, C–cod, D–tuna. By permission from P. F. Meyer-Waarden, "Electrical Fishing," Food and Agriculture Organization of the United Nations, 1957.

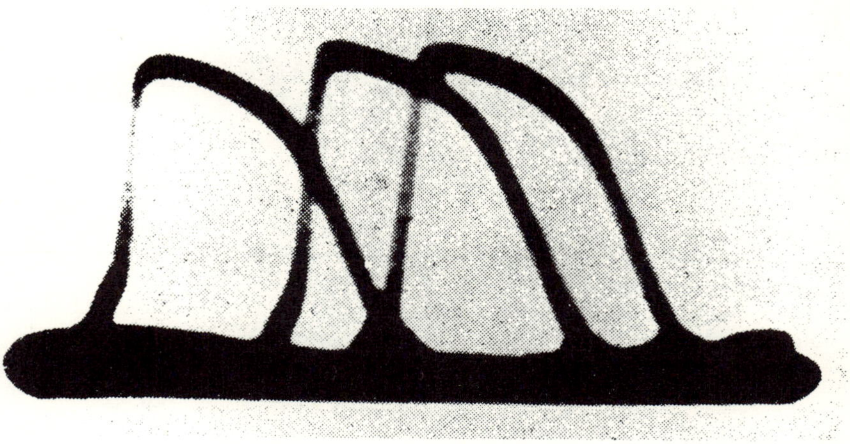

FIG. E14-3. Curve shapes of impulses transmitted by the electrical eel (*Electrophorus electricus*) (300 to 600 volts). By permission from P. F. Meyer-Waarden, "Electrical Fishing," Food and Agriculture Organization of the United Nations, 1957.

e. *Rate of Pulse*

Moreover, the number of pulses per time unit is decisive for the efficiency of the pulsating current, a fact already recognized by Kreutzer in his first experiments. The pulse rates at which fish are narcotized with the minimum of electrical energy differ considerably for the various species as well as for fish of varying sizes (Table E14-4 and 5). The values found for the narcotizing impulse thresholds are not to be considered as absolute, but as relative values, as they have been measured for certain sizes of fish and a certain water temperature. They change according to the temperature, the physiological condition of the fish and the size of the fish. The basic differences between the species of fish, however, are maintained.

TABLE E14-4

NARCOTIZING PULSE THRESHOLDS FOR FRESH-WATER FISH[a, b]

Species of fish	Average length (cm)	Voltage of the individual pulse required for galvanonarcosis[c] (volts)	Narcotizing pulse threshold values[d] (impulse rate/sec)
Rainbow trout (*Salmo gairdnerii*)	15–17	6.5	80
Carp (*Cyprinus carpio*)	12–15	7.5	50
Catfish (*Ameiurus nebulosus*)	14–16	11.5	40
Eel (*Anguilla anguilla*)	20–22	13.5	50
Goldorfe (*Idus melanotus*)	13–15	7.5	30
Stone Perch (*Acerina cernua*)	14–16	7.5	50
White bream (*Blicca björkna*)	12–15	11.5	40
Minnow (*Phoxinus loevis*)	7–9	8.5	90
Stickleback (*Gasterosteus oculeatus*)	6–7	13.5	100
Perch (*Perca fluviatilis*)	12–14	9.5	70
Schlammpeitzger (*Misgurnus fossilis*)	24–27	9.5	40
Tench (*Tinca tinca*)	16–18	7.5	40

[a]According to Halsband.

[b]Note: Size of the experimental basin, 35 × 22 × 22 cm; size of electrode, 22 × 22 cm; water temperature, + 15°C.

[c]Measured between the electrodes.

[d]Capacitor discharge with form D: see Fig. E14-1.

f. *Narcotizing Impulse Threshold*

Fish within a specific area remain stunned when they are affected by impulse rates which are above the threshold values required for narcotizing them. But if the impulse rates are kept below the limit which narcotizes them, then the fish swim from a larger area toward the anode and, when they are close to it, become narcotized. With a low pulse rate the required electrical energy is reduced by half as compared with the purely stupefying method. That fact is of great importance to electrical fishing. And because different species of fish react differently to varying pulse rates, and as the threshold values of the pulse rates are known, it is possible to select the fish to be caught by electricity within a certain range according to size and kind.

The term "narcotizing pulse limit" was formulated by Kreutzer. He was also the first to recognize the importance of this idea for electrical fishing.

TABLE E14-5

NARCOTIZING PULSE THRESHOLDS FOR SEA-FISH AT WATER TEMPERATURE OF + 15°C

Species of fish	Average length (cm)	Voltage of the individual pulse required for galvanonarcosis	Narcotizing pulse threshold hold values (impulse rate/sec)
A. According to Halsband[a]			
Fatherlasher (*Myxocephalus scorpius*)	15–19	6.2	40
Eelpout (*Zoarces viviparus*)	17–21	10.0	60
Plaice (*Pleuronectes platessa*)	23–26	10.0	30
Sea burbots (*Ciliata mustela*)	16–17	11.5	60
Smelt (*Osmerus eperlanus*)	18–22	12.0	50
Flounder (*Platichthys Pleuronectes flesus*)	16–20	11.5	40
B. According to Kreutzer			
Herring of medium size	—	—	45
Cod of medium size	—	—	25
Red tuna (200–300 kg)	—	—	7–10

[a]Measured in the experimental basin 35 × 22 × 22 cm; size of electrode 22 × 22 cm.

The importance of the size of the fish in relation to the narcotizing pulse limit can be explained as follows (Fig. E14-4). Each pulse causes a vibration of the muscles in the fish. If the next pulse occurs before the mechanical movement caused by the preceding pulse ends, the muscles are continuously stimulated and cramp sets in. This phenomenon is slower in developing in larger fish because larger masses of muscle have to be moved, and big fish do not, therefore, require such a quick sequence of pulses to produce cramp as small fish. Of course, a small fish can also be narcotized electrically at low pulse rates, but the length of the pulses must be substantially greater or the pulse voltage must be greater, and that would mean deviating from the desired minimum of electrical energy, at which the necessary effect would just be reached.

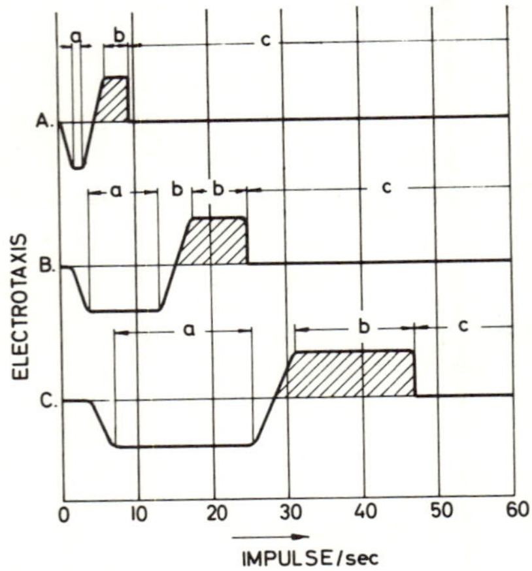

FIG. E14-4. Influence of impulses per second on various fish (according to Kreutzer [343]). A–tuna, B–cod, C–herring. (a) Possibility of escaping, (b) Swimming towards the anode, (c) Electro narcosis. By permission from P. F. Meyer-Waarden, "Electrical Fishing," Food and Agriculture Organization of the United Nations, 1957.

The battery gear developed by the Institut für Küsten-und Binnenfischerei, Hamburg, does not operate with a vibrator, transformer or converter, but with current taken from a 12 volt battery which is passed over a switch shaft specially designed for the purpose (Fig. E14-5). The switch shaft is driven by a small motor. When it turns on around 360°, the capacitors are switched twice in parallel and twice in series. Forty capacitors loaded in parallel and discharged in series thus produce a peak voltage of 480 volts. By turning the switch shaft either quickly or slowly the impulse rate (in contrast to the previously mentioned pulse gear) can be altered in any desired way and the gear adjusted to a certain extent to catching the various species and size of fish. By switching to several sets of capacitors, the impulse period can be adjusted and thus the gear can be adjusted to varying conductivities of water. This impulse gear can also be used in larger waters. For that purpose it must, depending on the area of the water, have correspondingly greater energy. This can be attained by (a) a larger size of the gear, or (b) connecting the gear to gasoline generators of 0.6 to 5.0 kw.

The blocking or driving effect is now being used in fresh water for the following purposes:

(a) As barriers for controlling the migration of Chinese crabs.
(b) To prevent fish from being driven into turbines and pumps of hydroelectric plants, dams, etc.
(c) For guiding fish to fishways and into new river systems.
(d) For fencing off water areas which are to be managed (for anglers, etc).

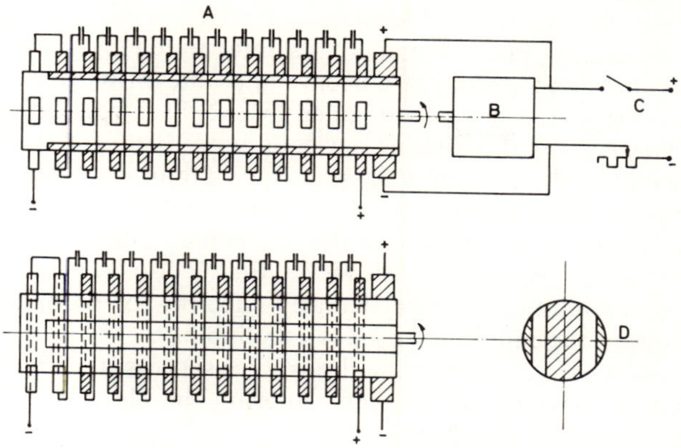

FIG. E14-5. Diagram showing connection of an impulse gear designed by Halsband and Plöger [337]. According to the number and size of capacitors the gear can be used for fishing in fresh waters (with conductivity values of 1000 to 15,000 ohms × cm, and also for killing fish in sea water. However, the range in sea water is only 1.5 to 2 meters radius. The gear operates with 12 and 24 volt batteries respectively. A–switch, B–motor, C–battery, D–axis of the switch. By permission from P. F. Meyer-Waarden, "Electrical Fishing," Food and Agriculture Organization of the United Nations, 1957.

As the Chinese crab (*Eriocheir sinensis*), originally from China, is considered a great pest to shore installations and gear, electrical experiments were carried out in Germany to kill or damage the young Chinese crabs which are born in the North Sea and ascend the rivers in Spring in immense swarms. A similar attack was made on the old crabs descending to the sea for spawning.

Fences were designed similar to the electrical screens in rivers, bearing in mind that it is the particular characteristic of Chinese crabs to wander along the river bottoms, and the electrical fence was installed on the bottom across the river. Two wires were used, about 40 cm (16 inches) distant from each other. Later it proved practical to install a second pair of electrodes between the first, so that each of the interior electrodes was 16 cm ($6\frac{1}{4}$ inches) from the nearest outer electrode. A galvanized 8 sq mm iron cable consisting of 11 wires, strengthened by a 3 sq mm bright copper wire, was used to prevent a big drop in potential. As the river Elbe was 330 meters (1800 ft) in width at the place of experiment and the unavoidable drop in potential threatened the efficiency of the fence, a second current supply was installed in the middle of the electrical fence by using an underwater copper cable of 2 × 35 sq mm in cross section. The electrodes were fastened with clasps to oakwood blocks of 400 × 80 × 30 mm. The chain was anchored by means of large stones at about 10 meter (33 ft) intervals to prevent it being driven away by the relatively strong current of the river. In this way three electrical fields were produced. Crabs, which entered the electrical field during an interruption of the current, got no further, at best, than the interior electrical field, where they usually grabbed, in a state of cramp, the two electrodes situated at 8 cm ($3\frac{1}{4}$ inches). Numerous electrical shocks finally killed them.

Three methods were investigated, all of which worked successfully. The first used alternating current which was interrupted by a cam switch; the second operated with pulsating current; and the third used, like the first, alternating current, but it was controlled by a 2-thyratron ignition system.

For the latter method a similar device is used as for a fish barrier. From the lighting circuit the current is passed over an ignitron, which is controlled by a thyratron, with adjustable frequencies, and sent into the chains of electrodes. The ignitron potential of the ignitron amounts to 300 to 500 volts. The diagram of connections is shown in Fig. E14-6. It was found that 30 to 40 pulses per minute were particularly effective for damaging the crabs. The advantage of the equipment used is its reliability.

As mentioned above, all three methods were successful. While all the crabs were not killed outright, all of them entering the electrical field lost their limbs and usually died in 24 hours.

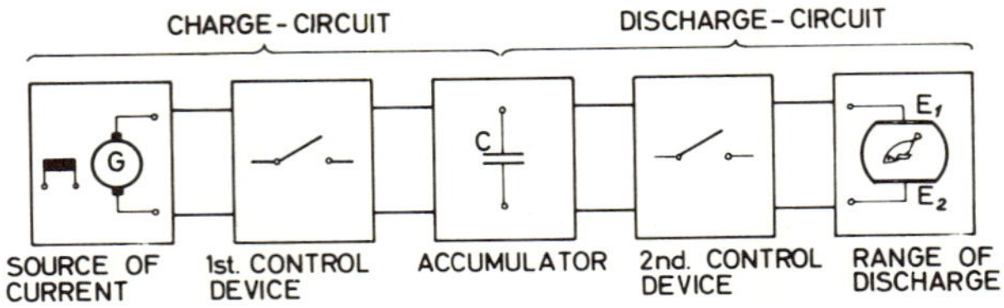

FIG. E14-6a. Basic principle of producing impulse current (according to Haier [334]). By permission from P. F. Meyer-Waarden, "Electrical Fishing," Food and Agriculture Organization of the United Nations, 1957.

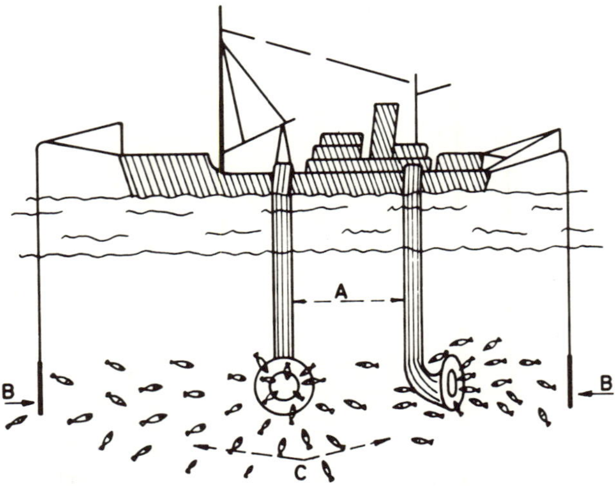

FIG. E14-6b. Diagram of electrical fishing (according to Skoskiewicz [358]). A–Fish section pump; B–Electrode; C–Electrical field. By permission from P. F. Meyer-Waarden, "Electrical Fishing," Food and Agriculture Organization of the United Nations, 1957.

In larger rivers with stronger currents, such as the river Elbe, these installations meet some difficulties. The fencing wires get easily covered with sand, thus reducing their efficiency. The wires must be lifted from time to time to shift the sand. A comparison between a German and an American control device is made in Fig. E14-7.

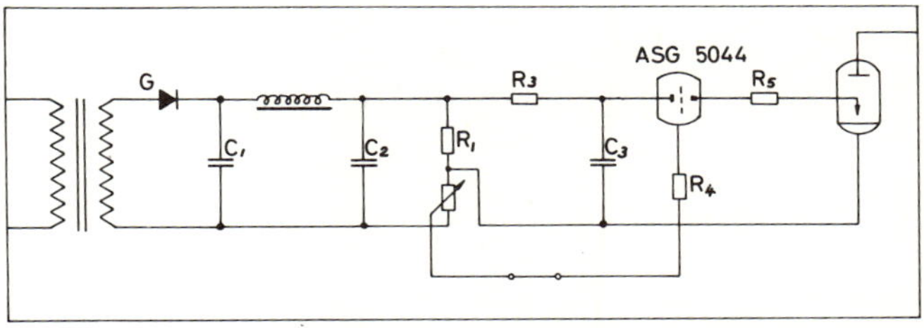

FIG. E14-7a. Diagram of a controlling device for Chinese crab or fish barriers. C_1, C_2 and C_3–capacitors; G–rectifier; R_1, R_3, R_4 and R_5–rheostats; ASG–ignitron. By permission from P. F. Meyer-Waarden, "Electrical Fishing," Food and Agriculture Organization of the United Nations, 1957.

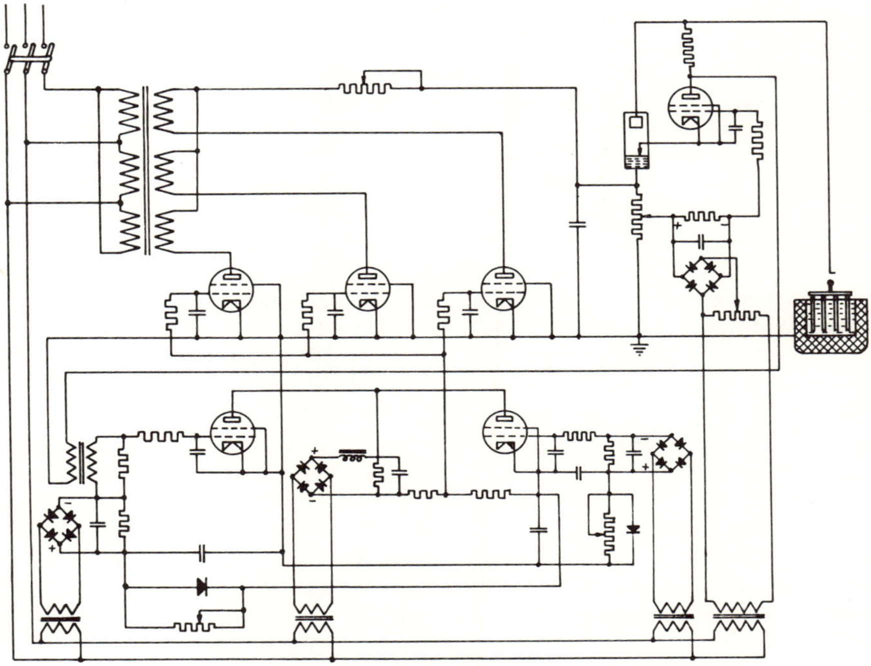

FIG. E14-7b. Diagram of an American controlling gear. The gear produced impulse current by means of a tube system. Several governing tubes (american thyratron) and a discharge tube (american ignitron) are built in. The current produced is quadratic. Compare the diagram of a german barrier in Fig. E14-7a. By permission from P. F. Meyer-Waarden, "Electrical Fishing," Food and Agriculture Organization of the United Nations, 1957.

F. Conversion of capacitor energy into x-ray flashes and neutrons

1. Physics, Design, and Circuitry in x-Ray Flashing

a. *The Production of Short x-Ray Flashes*

In the conversion of capacitor energy into x-ray radiation one has to reckon with efficiencies similar to those encountered in the stationary production of the rays. The efficiency is about proportional to $U^{3/2}$. With voltages above about 200 kv, in addition to efficiency we must consider the Compton effect, that is, the emission of x-radiation in a preferred direction instead of radiation with a spherical characteristic. In every capacitor discharge through an x-ray tube one has to consider good efficiency in conjunction with the Compton effect for ray concentration. The efficiency drops quickly with decreasing voltage. Therefore, the emitted x-ray impulse appears to be considerably shorter than the electrical discharge progression. Radiation at low tube voltage is, moreover, soft and is not capable of penetrating the tube wall easily.

By the term "x-ray flashes" we understand x-radiation that is emitted from tubes in which strong discharges of flashlike duration occur [24]. M. Steenbeck has reported a discharge tube and a method for producing intense x-ray light flashes. The gas-filled x-ray flash tube with mercury anode and striction capillary disclosed by Steenbeck has been further developed by J. Mühlenpfordt with more favorable arrangement of the construction elements.

From his first tube he developed a series of other types on which he made observations which led him finally to complete abandonment of the capillary and to the development of the high-vacuum x-ray flash tubes.

x-Ray flash arrangements utilizing capacitor discharges can be used in single flash operation as well as in periodic operation (to an extent limited by the thermal capacity of the anode). The temperature of the anode block must not be driven so high that evaporation or melting of the tungsten sets in. On the other hand, the intensity of a single flash, during and owing to whose extremely short duration no heat conduction can occur in the anode, should not be so high that a single flash is able directly to evaporate larger portions of the anode surface. If such evaporation should take place, the intensity of x-radiation would be markedly reduced during a single flash series by vapor condensation of tungsten on the window of the tube.

The flash tube constructed by Mühlenpfordt, before the first disclosure by Slack and Ehrke [25] on the same subject, consists of a steel jacket that is mirror finished on the inside and carries at one end a porcelain bushing for an anode rod which has a massive tungsten cone as a point. At the other end is a molybdenum cathode in the form of a hollow cone provided with an aluminum window for outlet of x-radiation. By means of a flanged on diffusion pump the interior of the tube is continuously kept under high vacuum during operation.

The anode of this tube, in accordance with Fig. F1b-3, is connected to a charged high voltage capacitor by way of an interposed spark gap. Ignition is effected when an overpotential is applied to the spark gap by means of an ignition transformer. The potential at the gap breaks down and the capacitor voltage is applied to the flash tube. During a discharge between the anode and cathode of the tube, a very intense x-ray flash is radiated from the point of the anode. This flash, at a voltage of 100 kv, is able to irradiate a steel plate of 4.5 mm thickness at a distance of 20 cm from the window.

Since the spark gap releasing the x-ray flash lies in series with the tube, a certain portion of the capacitor energy is lost in the spark gap. It is furthermore very annoying that ignition in this circuit, even with considerable overpotential, is not certain and not very accurate in timing. This creates a problem when it is desirable to release the x-ray flash automatically in proper dependence on a phenomenon to be investigated. In order to improve this flash tube in this respect, Slack and Ehrke arranged a special ignition electrode above the cathode and applied the voltage of the ignition transformer to the ignition electrode. The idea of forming the anode as a cone in order to give the burning spot the shape of a conical jacket and arrange the cathode annularly around the anode originated as early as 1934 (R. Thaller's *Kestralröhre*). The only difference is that the cathode of this early tube was a glow cathode.

In continuation of this development work, the x-ray flash tube represented in Fig. F1a-1 was perfected. This tube first served for study of the sound problem during electric sparkovers. Facing the hollow cone cathode, K, is the anode, A, in the form of a pointed solid tungsten cone carried by a porcelain insulator. The cathode, K, carries a glass ring with the disc-shaped ignition electrode, Z. Between the sharp inner edge of cathode K and ignition electrode Z there is a small gap which, for the purpose of sound wave investigation, has the width of 1/10 mm.

All vacuum sealing is effected by rubber gaskets. A 3-stage mercury diffusion pump (Leybold, model E) with a cooling trap of liquid air serves as vacuum pump. Since owing to the heavy discharges, on the order of 10,000 amp, considerable gas quantities are sometimes released from the anode, the tube must remain continuously connected to the diffusion pump.

The electric circuit. The operating voltages range up to 100 kv. The inductance-free, high-voltage capacitor, C, is permanently connected to the anode of the flash tube, RB. Capacitor C is charged from a transformer, Tr, through a valve tube, Vi, and a high ohmic resistance, R. Adjustment of capacitor voltage is effected at the primary of transformer Tr and can be measured by a sphere spark gap, F.

Ignition of the x-ray flash tube is effected by applying a voltage to the ignition electrode, Z. This voltage is taken from a current gate circuit. The process to be investigated must cause, by mechanical or electrical interlinkage, lowering of the screen bias voltage in the current gate, S, such that a capacitor, C_s, which has been charged from the line through a device, N, discharges through current gate S and the primary of a shock impulse transformer, T_s. In the secondary of this transformer a voltage of 10–20 kv is induced. This voltage breaks down F_v and delivers the ignition spark.

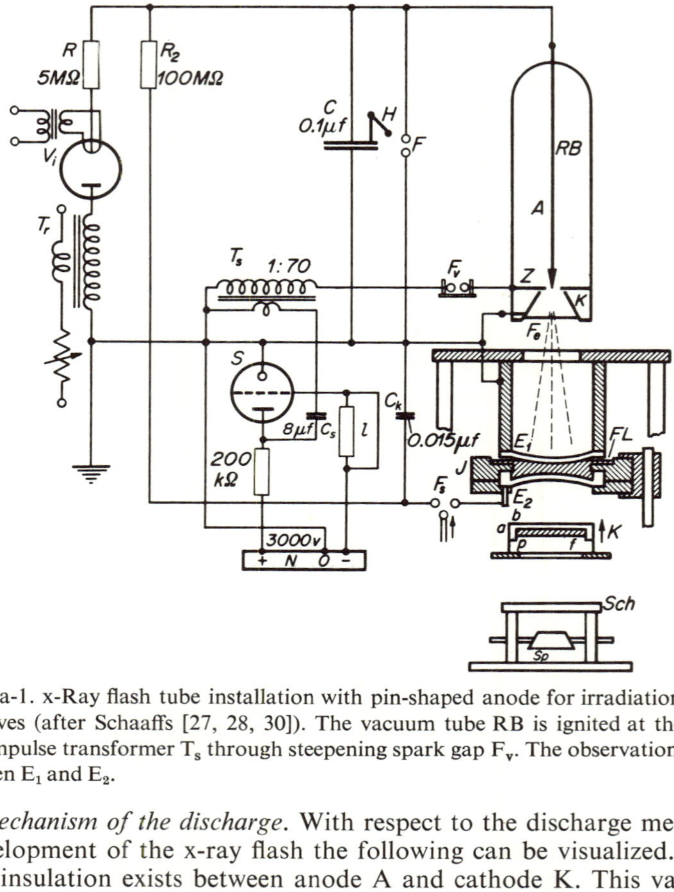

FIG. F1a-1. x-Ray flash tube installation with pin-shaped anode for irradiation of spark sound waves (after Schaaffs [27, 28, 30]). The vacuum tube RB is ignited at the cathode from an impulse transformer T_s through steepening spark gap F_v. The observation chamber lies between E_1 and E_2.

The mechanism of the discharge. With respect to the discharge mechanism and development of the x-ray flash the following can be visualized. At first vacuum insulation exists between anode A and cathode K. This vacuum is broken down at a given instant determined by an external occurrence. With voltages of 50 to 100 kv between anode A and cathode K and an electrode spacing of 10 mm, as presently assumed, the cathode gradient is too small to effect a field emission of electrons [26]. As soon as electrons enter the gap, they will extract ions from the anode. These ions will travel to the cathode and cause the release of additional electrons.

Furthermore, it is likely that photons also assist in loosening more electrons. When this exchange reaches an upper limit, the sparkover occurs in high vacuum. By the tremendous impact of electrons upon the anode at current values of about 100,000 amp the anode radiates x-rays and a portion of the anode material evaporates [27, 28].

In order to obtain the primary electrons necessary for the breakthrough, an ignition voltage of 10–20 kv is imposed on the gap between the sharp edges of ignition electrode Z and cathode K, in accordance with the teachings of Mühlenpfordt referred to herein. The particular shape of these electrodes and the spacing of 1/10 mm result in a cathode gradient of about 2000 kv/cm. This causes field emission and sparkover in this auxiliary gap and creates the electricity carriers necessary for the main discharge.

The focus. The shortest paths that the electrons travel toward the anode terminate at the anode tip. It is true that the whole cone is hit by electrons; however, the hardest components of the ray mixture emerge from the anode point. The intensity of the hard rays is limited to an area of about 1 mm in diameter.

Output and Current Loading. The output of an x-ray flash tube can best be judged and specified by the thickness of a steel plate that can be irradiated by a single flash. With an anode voltage of 100 kv it is possible to irradiate steel plates of 3 mm thickness at a distance of 20 cm. Plates 10 mm thick can be irradiated in the immediate vicinity of the tube. However, attention must be called to the fact that, for unknown reasons, the intensity of x-radiation is subject to variations so that the output indicated above is often not reached, although it is sometimes surpassed.

The average current values prevailing during the discharge can be estimated. The total duration of an x-ray flash is less than 2×10^{-7} sec. During this time interval the magnitude of current as well as voltage must be relatively great. The capacitor C(0.1 μf) discharges from 100 kv to about 20 kv. Then, according to an established relationship, the quantity of electricity.

$$Q = \int_0^t i \, dt = C \Delta V$$

has been discharged. For an average current value i_m, this is approximately $i_m = C\Delta V/\Delta t$.

As a safety factor, it is assumed that the actual discharge duration is about 2–3 times as long as the observed x-ray flash duration. Inserting values in these formulas, 100 kv results in an average current value of $i_m \approx 15{,}000$ amp; and 50 kv, in current $i_m \approx 7500$ amp. This current maximum, however, is not reached at the full voltage, but presumably at below one-half of it. In spite of such heavy loading, the constructions shown are capable of withstanding several thousand strokes at voltages between 60 and 100 kv.

b. *x-Ray Flash Duration and Circuitry*

In order to gain some insight into the above estimated duration of x-ray flashes, an aluminum disc of 20-cm diameter was secured to the shaft of a high speed motor. At two diametrically opposite points close to its periphery the disc was provided with two windows in which, in radial direction, a 0.1 mm tantalum thread was strung. A comparison of the sharpness or blur of this thread in exposed film images, one with the motor at standstill, the other with the motor running, indicated an x-ray flash duration of 2×10^{-7} sec. Such measurement, however, is infected by a deficiency that cannot be traced to the upper speed limitation of the motor. A measurement of this kind can give at the best only an indication of the total duration of the radiation of all wavelengths of the x-ray flash. The actually effective x-ray flash duration in irradiation of an object is dependent on the absorption of the hardest radiation. In an irradiation of fast sound waves at 50–60 kv, which will be explained below, the effective flash duration was

$$t_{\text{eff}} \approx 5 \times 10^{-8} \text{ sec}$$

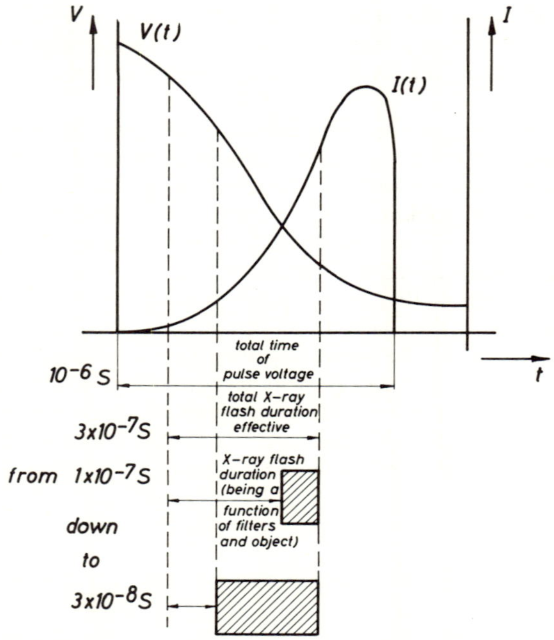

FIG. F1b-1. Difference between discharge duration, total x-ray flash duration, and effective x-ray flash duration.

Figure F1b-1 shows the difference among discharge duration, total x-ray flash duration, and effective x-ray flash duration (by W. Schaaffs [27]).

Figure F1b-2 shows schematically the construction of a high vacuum x-ray flash tube for a 200 kv shock impulse voltage.* In an evacuated chamber, which is formed by a metal cup, 1, and a porcelain insulator, 2, the hollow cone-shaped cathode, 3, a cone-shaped anode, 4, and an annular ignition electrode, 5, are arranged. In order to obtain a very intense but extremely short x-ray impulse, a capacitor-fed high vacuum spark is released between anode and cathode. A spark, jumping through the gap between

* Construction of Siemens, partly excerpted from a leaflet of Siemens & Halske. The construction is here described in detail, since it can be logically converted for higher voltages and thus represents a physical expansion of the tube construction discussed in Chapter F1c.

ignition electrode and cathode, initiates the impulse. This ignition spark is caused by a high voltage impulse of weak energy which is initiated in synchronism with the occurrence to be recorded on film. The x-radiation, originating at the anode, passes to the outside through the window, 6, of the tube. The duration of the x-ray emission is about 5×10^{-7} sec.

An electrical circuit for an x-ray flash tube for voltages above 100 kv is shown in principle in Fig. F1b-3. A switching spark gap is interposed between the tube and the charged capacitor. The ignition impulse, which causes the

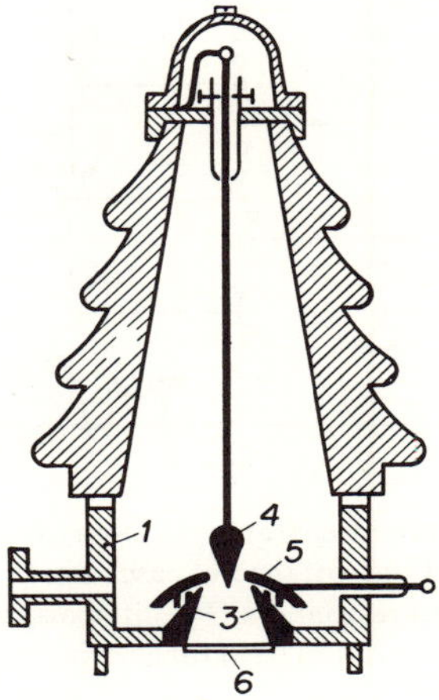

FIG. F1b-2. Longitudinal section through a high vacuum, x-ray flash tube with conical anode (design of Siemens) (100 kv).

breakdown flash or x-ray flash at the desired instant, originates in the shock impulse transformer. The capacitor discharge then, assisted by the ignition spark simultaneously occurring at the ignition electrode, occurs within about 10^{-6} sec through the flash tube.

The shock impulse transformer is connected with a thyratron impulse circuit, not shown in Fig. F1b-3, which permits an accurate synchronization of the x-ray flashes to within 10^{-6} sec with the occurrence to be recorded.

A Marx shock impulse generator [31, 32] (Fig. F1b-4) serves to supply

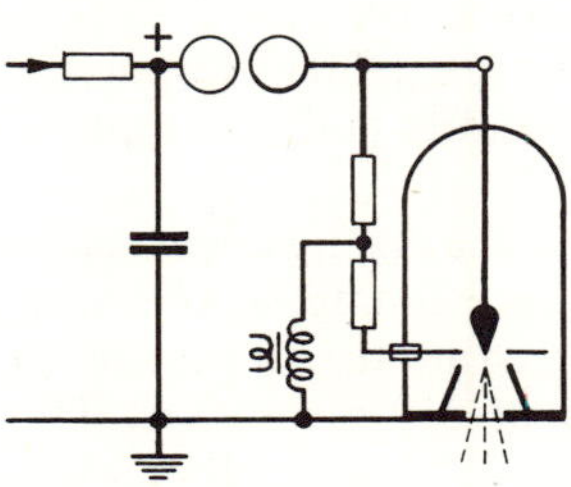

FIG. F1b-3. Schematic circuit of an x-ray flash tube for more than 100 kv operating voltage with series spark gap.

the energy to the x-ray tube. Ignition of the shock impulse is effected by a small thyratron device acting at the slit spark gap, F_z. The number of main capacitors, C_1 to C_5, can be varied between one and seven, so that it is possible to apply shock impulse voltages of between 20 kv and 140 kv at the x-ray flash tube, depending on the desired hardness of radiation.

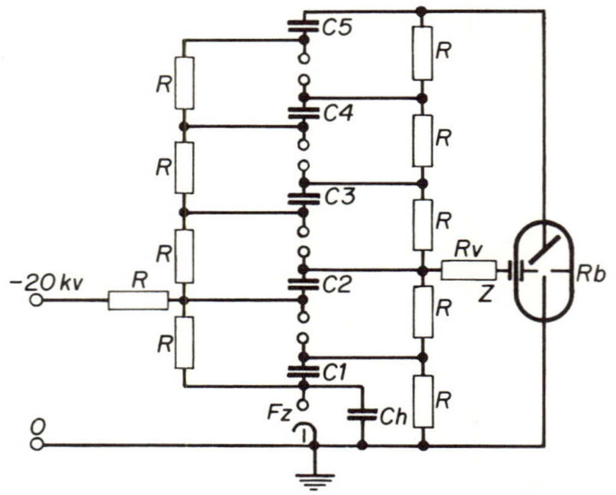

FIG. F1b-4. Schematic shock impulse circuit with x-ray flash tube (100 kv). The ignition voltage is tapped off at the second stage of the five cascades.

C_1—C_5–capacitors, R–charging resistances, F_z–slit spark gap for ignition, Ch–auxiliary capacitor, R_v–retarding resistor, Rb–x-ray flash tube, Z–ignition electrode.

If the vacuum within the x-ray flash tube is not good, the shock impulse voltage will not suffice to initiate a discharge by field emission. However, the discharge can be initiated by means of an ignition electrode, Z, when for that purpose a voltage of 40 or 20 kv is tapped off from the impulse circuit. The resistance R_v in series with the ignition electrode of the x-ray tube assures that with little delay, and thus shortly after the main voltage is applied to the main electrode of the x-ray tube, an ignition spark occurs between the ignition electrode, Z, and the cathode of the x-ray tube, which initiates the main discharge. The circuit presents a convenient combination of an external ignition of the impulse circuit and an internal ignition of the x-ray tube. Because in this circuit the conventional large series spark gap is not present, the circuit conductors can be very short so that the inductance for the shock impulse circuit with one main capacitor, C_1, (20 kv impulse voltage) is only 0.5×10^{-6} h, and with seven main capacitors (140 kv impulse voltage) it is only 1.2×10^{-6} h.

The x-ray flash duration of the arrangement referred to above could only be estimated. To obtain such an estimate, a steel wire of 0.2 mm thickness was swung around by a high-speed motor so that the peripheral end of the wire reached a peripheral speed of 210 meters/sec. This revolving wire end was then photographed by an x-ray flash together with a wire of the same thickness but at rest. The corresponding shadow pictures could be compared. A motion sharpness of 0.2 mm could then be estimated without difficulty. This corresponds to an accuracy of 10^{-6} sec. Estimates with the assistance of such photographs gave as a result an effective x-ray flash duration of less than 10^{-6} sec.

Flash x-rays can elucidate high-speed phenomena. Recent increases in peak intensity (to 10^8 rad/sec) and decreases in pulse length (to 30 nsec) have given x-ray sources advantages in many photographic assignments. When the phenomena under study occur in an opaque medium (e.g., shock waves in solids), or when the phenomenon is itself an intense source of light (e.g., exploding wires), x-rays may be the only convenient method by which a sufficient amount of information can be acquired at reasonable cost.

New field emission x-ray tubes [850] have made multiple-tube radiography practical. Small size (1 inch × 6 inches at 100 kv) allows several tubes to be closely grouped; low impedance permits the tube to be operated remote from its voltage source on a coaxial cable; low cost has made multiple-tube systems economical.

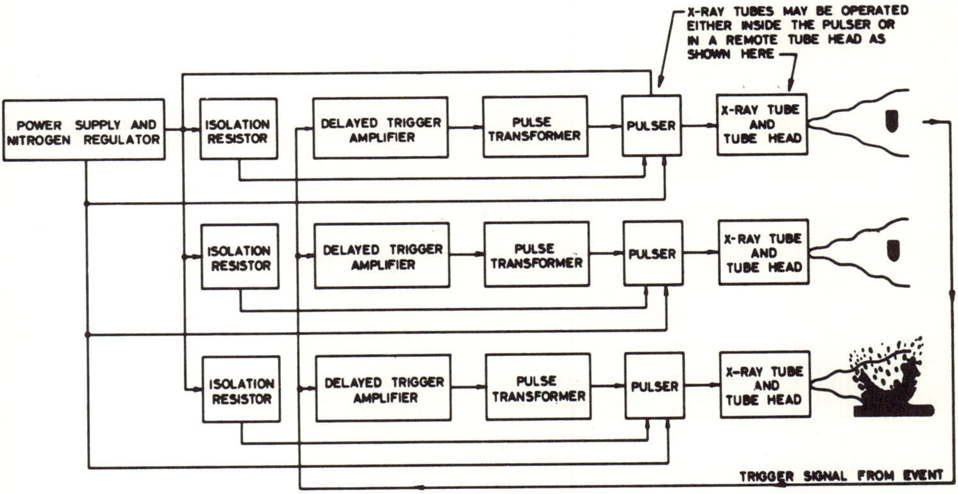

FIG. F1b-5. Block diagram of a three-channel system for three independent x-ray flashtubes (Fexitron, Type 730).

Components and a schematic for such multiple-tube systems are shown in Fig. F1b-5. All high voltage pulsers are compatible with the same dc supply and trigger-amplifier delay units. Thus a single system may include a variety of pulsers and tubes. The latter may be optimized for each phase of the event under study. For example, 400 kv may be needed to penetrate a gun barrel and to observe a bullet's acceleration, when the bullet emerges into air. 100 kv may be adequate for chronographs; a 30-nsec pulse length may be necessary to "stop" terminal ballistic fragments, particularly at hypervelocities; all of these different pulsers and tubes may be operated in the same x-ray system. As another example, studies of radiation effects may thus benefit from a wide range of dose rates, pulse length, frame rates, and spectra.

High signal-to-noise ratio is provided by enclosing all components (including pulser and tube) in grounded metallic containers. This enhances the detection of radiation effects and avoids in many cases the screen rooms required with earlier equipment. It also minimizes the coupling between channels and provides a high degree of integrity in the delay and timing circuits.

c. *Arrangement of the Anode; Stereoscopic Operation*

In order to obtain sharp shadow pictures with x-radiation, it is always desirable to have a small focal spot. It is difficult to meet this requirement, because of the extraordinary loading of the x-ray flash tube anode.

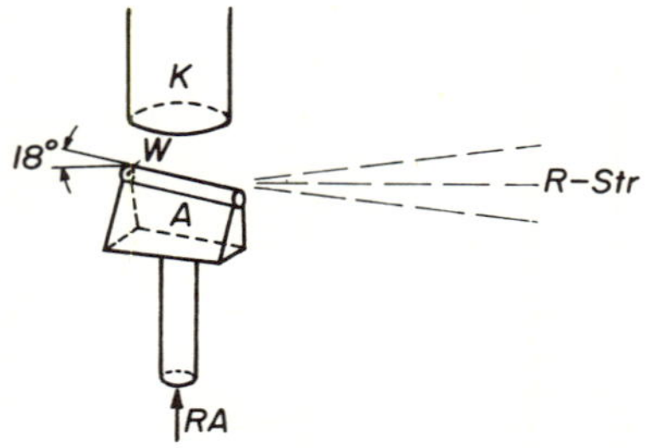

FIG. F1c-1. Line focus anode, as in the tube of Fig. F1d-1. A–anode, K–cathode, W–tungsten pin, RA–x-ray tube axis, R-Str–x-ray radiation (beam).

The arrangement of the Goetze focus presents the possibility of reducing the expansion of the apparent burning spot, at least in one direction. The holder of the anode in this arrangement is rooflike, as depicted in Fig. F1c-1, but such that the ridge is inclined by an angle of 18 degrees. The anode, in the form of a tungsten pin, W, of 1.5 mm diameter and 16 mm length, is welded along this ridge. The x-radiation (RStr) utilized in this construction is in the direction in which the ridge points.

The anode, inclined under an angle of 72 degrees with respect to the axis, RA, of the x-ray tube, forms a Goetze focus in the direction of the x-radiation. Thus one obtains a burning spot of about 2 mm diameter.

The advent of extremely short-duration x-ray flashes made accessible for research study various fast progressions which otherwise are hidden from direct visual observation or direct flash photography. Attention is here called to the possibility of applying the capacitor impulse technique to x-ray flash production in which two flashes occur simultaneously. Such an effect can be obtained with a double anode tube, [29].

The rooflike arrangement of the anode can be doubled so that the spacing, *a* (Fig. F1c-2), of the two tungsten pins, that is, the spacing between the two burning spots created, is 20 mm.

Symmetrically facing this double anode is a cylindrical cathode. During the discharge, the current passes to both tungsten pins. The timing difference in the initiation of the two burning spots is smaller than the x-ray flash duration and in most cases irrelevant.

The two burning spots of a double anode arrangement offer the possibility of taking two shadow pictures of a small object practically simultaneously. From the distance between the shadow images on the film and the distance between the burning spots in the x-ray tube, it is possible to reconstruct afterward the location of the object at the instant of exposure, as is customary in evaluating stereo pictures.

FIG. F1c-2. Double anode for line focus radiation (from Elle [29]). W–tungsten pin, a–spacing between the two burning spots, RA–x-ray tube axis.

With proper arrangement, two burning spots, spaced 20 mm apart, can be produced simultaneously. The spacing of the two shadow images of an object thereby produced reveals afterward a measure of its position in space at the instant of exposure. The double arrangement offers furthermore the possibility of comparing different anode materials. From the results of such an investigation it can be found that evaporation of anode metals plays an important role in the discharge mechanism of an x-ray flash tube.

d. *The structure of an x-Ray Flash Tube for a Broad Impulse Voltage Range*

Figure F1d-1 represents the structural arrangement of an x-ray flash tube. It consists of a steel chamber, E, which is in communication with an oil diffusion pump by way of a tube stub and a flange. The front of the chamber is sealed vacuum tight by a hard plastic insulator, J, and an interposed rubber gasket, G. The rear side is sealed by a disc, S, of the same plastic. The cathode, K, is a cylinder of 20 mm diameter and is conductively joined to chamber E and to ground. The ignition electrode, Z, is located at the center of the cathode and is insulated from it. The field emission initiating the ignition of the tube is started in the annular gap, Sp, of about 1/10 mm spacing. The ignition electrode is capacitively coupled to the outside [30]. The dielectric of the coupling capacitor, C_K, is the plastic of the closure disc, S. Facing the cathode and connected to the high-voltage side there is arranged the anode, A, whose face is inclined against the tube axis by an angle of 72 degrees. In this manner, when viewed in the main radiation direction vertical with respect to the tube axis, the anode represents a Goetze focus. The casting resin* insulator has in this direction a recess of 30 mm diameter, so that the remaining plastic layer forms the outlet window, F, for the x-radiation. The penetrability of this

* Leguval, Araldite, or other two-component casting resins.

window is relatively good, since the plastic consists primarily of elements of low ordinal number. A lead baffle at the outside of the window serves for intercepting scattered radiation originating in the insulator. The axis of the x-ray tubes is journaled horizontally. The insulator, J, with the anode can be conveniently turned about this axis without opening the tube. Thus it is possible to direct the x-radiation in any desired direction, depending on the subject to be investigated.

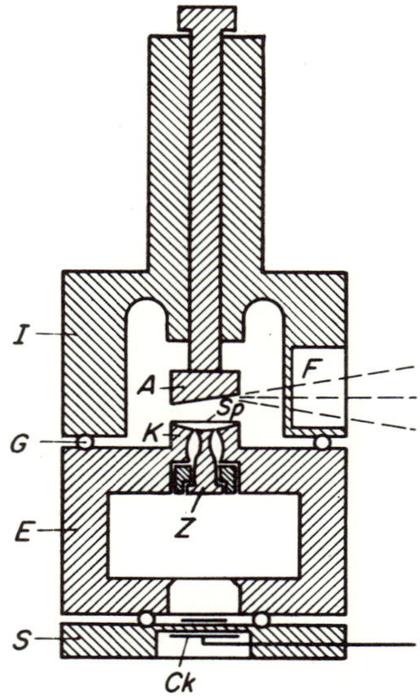

FIG. F1d-1. x-Ray flash tube with line focus and lateral ray outlet. The anode has great heat capacity. E–steel chamber, J–insulator of synthetic resin, G–rubber sealing gasket, S–disc of synthetic resin, K–cathode, Z–ignition electrode, A–anode, Sp–annular gap, Ck–coupling capacitor, F–x-ray beam window.

The dimensions of the x-ray flash tube are relatively small; the tube length is 20 cm and the outer diameter is 10 cm. This fact offers the opportunity to make the discharge path very short. that is, to hold its inductance, L, very low. The discharge space has been transferred from the steel chamber, E, into the insulator, J, in order to avoid undesirable discharge from the walls of the chamber and therewith the creation of stray rays. Wide channels have been provided in the mounting plate of the cathode. They form the connection between discharge space and vacuum pump. One disadvantage of this arrangement is that protective lead walls have to be installed outside the tube for absorption of dangerous radiation.

x-Rays can cause currents to flow in insulators. At a dose rate of 1×10^8r/sec, currents of several milliamperes can be induced in small, dielectric components. The Fexitron Model 515 x-ray tube encased within the metal-shielded Model 233 pulser provides a small, economical pulsed radiation source [850]. This maximizes the x-ray signal-to-noise ratio.

Owing to the rapid development of heavy duty x-ray flash equipment, definitions of the performance specifications may be given:

Pulse width tolerance: ±10% as measured from 80% of the leading edge to 80% of the trailing edge of wave.

Pulse rise time: 20% maximum of pulse width measured from 10% to 80% of leading edge of wave.

Pulse fall time: 50% maximum of pulse width measured from 80% to 10% of trailing edge of wave.

Pulse overshoot: less than 5% measured at nominal voltage with matched impedance load.

Pulse undershoot: less than 10% measured at nominal voltage with matched impedance load.

The Fexitron pulser, Model 233, has an output voltage of 150–300 kv, an output current of 1400 amp, an output impedance of 215 ohms, and a pulse width of 0.1 μsec.

The equivalent x-ray tube, Model 515, then gets a peak power of 420 megawatts and delivers a dose rate at the tube surface with the unbelievable amount of 1×10^8 r/sec. One flash has a penetrability of 6.0 inches of aluminum. Some applications of high power nanosecond x-ray flashes are shown in [849], e.g., the radiography of shaped charges.

e. *x-Ray Flash Device for* 200 *kv and Application*

For the 200 kv x-ray flash unit, a design has been selected that assures universal applicability. The circuit of the x-ray tube for voltages up to 100 kv is shown in Fig. F1e-1. In this case the charged capacitor is connected directly to the flash tube and thus the tube itself must block the total operating voltage. The x-ray flash is here released by the ignition spark between cathode and ignition electrode.

The x-ray flash tube, in principle similar to the one shown in Fig. F1b-2, communicates with an oil diffusion pump. All joints of the connection are sealed by rubber gaskets. The assembly can be readily taken apart for cleaning and for replacement of component parts.

This possibility of component replacement makes tube life independent of the number of recordings taken and permits, moreover, adaptation of anode setting and anode material, of radiation outlet window, and of the electrode

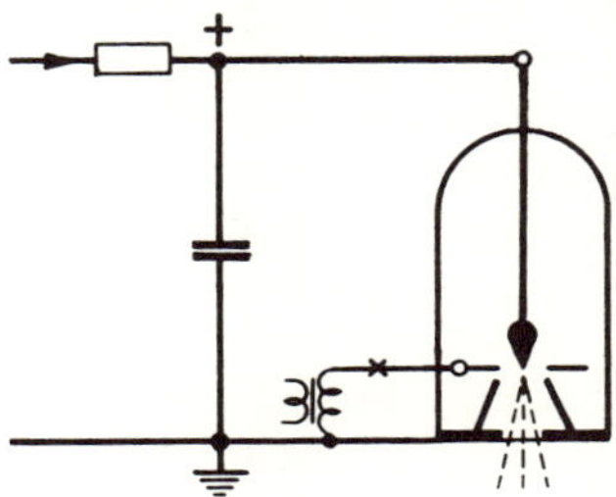

FIG. F1e-1. Schematic circuit of an x-ray flash tube for voltage below 100 kv. The tube itself must block the total operating voltage.

system, to the requirements of a certain problem to be x-ray recorded. The tube is mounted on the crossbeam of a jacketed tripod with its ray outlet window inclined toward the bottom. It can be adjusted as to height and can be tilted without affecting the vacuum condition. Owing to its rugged construction, the apparatus can withstand mechanical abuse to a large degree. This fact is of importance for explosion and detonation research work.

The pump arrangement and the ignition device are accommodated within the jacket of the tripod. At the top, the tripod carries the capacitor buildup and the switching spark gap, in accordance with Fig. F1b-3. An automatically controlled, high-voltage switching arm forms the connection between capacitor battery and cable. The latter connects to the high-voltage source at its other end. When operation is exclusively with voltages below 100 kv or with an impulse transformer, the switching spark gap and the upper half of the capacitor buildup can be dispensed with (Fig. F1e-1).

The necessary switching elements, control lamps, and voltage divider are accommodated in a mobile control console. Thus the setup can be operated at a safe distance from the occurrence to be recorded and from a radiation-protected point.

The electrical tube data are:

$$\text{Voltage } U_{max} = 200 \text{ kv}$$
$$\text{Capacitance } C = 0.08\ \mu\text{f}$$

At a focal distance of 30 cm, an irradiation strength (i.e. the layer thickness behind which a 1 mm stage can still be recognized) of 1.2 cm steel can be achieved.

The effective duration of the x-ray flash is about 5×10^{-7} sec. It can drop below 10^{-7} sec when the radiation is filtered, an effect which can in some cases be caused by the film recorded object itself.

The actuation of the ignition device for release of flashes can to a great extent be adapted to the peculiarities of the test occurrence. Selective switching on the face of the apparatus can be set for: contact closing or opening, positive or negative low-voltage impulse, and synchronization with sine-wave or sweep frequency.

The ability of x-ray flashes to irradiate rapidly progressing occurrences within opaque bodies to make them accessible to film recording has led to their widespread application in detonation, explosion, and ballistics research. Furthermore, they have made possible the investigation of extreme dynamic loading problems, of fracture occurrences, of projectile penetration into solid bodies, and of a few problems in rocket research.

Moreover, the study of the behavior of shock waves in solid bodies, liquids, and gases is made possible by x-ray flashes. x-Ray flash recordings in many cases allow determination of compression of the medium during the occurrences enumerated above. As an example, the acoustic occurrences in sparkovers through insulating liquids have been investigated with the assistance of x-ray flashes.

Some occurrences manifest themselves by such strong illumination that light optical recordings can give no results. Being independent of light condition, the x-ray flash is able to present a picture of the density structure of the object. In the investigation of light arcs, which have only a short burning duration or which are so luminous that the film can be brought close to the arc for only a very short time, the x-ray flash tube has rendered valuable service.

In the medical field also, the x-ray flash tube is likely to gain importance for research into high-speed occurrences. As an example, the actions in the human larynx in uttering sound can be studied with x-ray flashes.

The investigation of microstructures by means of x-ray flash interferences is still in its infancy. With exposure methods using wide apertures (e.g. by Seeman's edge method) it is possible to produce interference recordings on monocrystals with a single flash, that is, within less than 10^{-6} sec.

Investigations of the biological effectiveness of x-ray flashes have also been conducted. Since high-speed production of extremely high dose outputs (10^8 r/sec) is possible, it appears that interesting problems are still to be solved in this field.

f. *Megavolt x-Ray Experiments* [175–181]

x-Ray flashes wherein the energy quantity to be discharged is of very high potential were produced by Brasch [41]. Initially, Brasch proposed, in order to advance into the range of many megavolts, an antenna-like expanded lightning reception rope on Monte Generoso (Switzerland). He insulated the rope first by porcelain insulators, later by hemp ropes. During thunderstorms he obtained 12–15 megavolts. The construction of his x-ray flash tube was relatively simple; he used a laminated tube with an anticathode and established the voltage symmetry along the tube by metal and hard paper discs in spaced relation somewhat similar to a quenching spark gap. The vaccuum difficulty was avoided by the tube being joined to a very effective pump together with a cooling trap. The tube was thus subjected to the high-voltage only during the impulse itself. This action was assured by the function of a sphere spark gap. In such an arrangement the breakthrough voltage of the tube can lie considerably below the operating voltage.

After having completed his investigations with "atmospheric capacitors" in the form of lightning discharges from clouds, Brasch diverted his efforts to a 2.4 megavolt Marx shock impulse voltage generator. The x-ray flashes obtained therewith permitted irradiation of 20 cm thickness of lead. The shock impulse generator used had 12 capacitors of 48,000 pf each, with a total of 4000 pf at 2.4 megavolts; thus about 12,000 joules. The discharge duration was between 10^{-4} and 10^{-6} sec. In place of the anticathode Brasch now employed an outlet window for the projection of his electrodes into the air. Crater effects in metals struck by fast electrons as well as fluorescence phenomena have been studied. Some metals, e.g. brass, lean toward bubble formation. Cellon is very often penetrated by the fast electrons due to the formation of static charges. The utilization of atmospheric discharges has been abandoned because the uncertainty of thunderstorm occurrence results in much waste of time.

A particularly simple x-ray flash tube for use as a diode for high voltages has been developed in the USSR. Figure F1f-1a shows this tube in principle. A single-layer coil, 6, is wound on an insulation tube, 1, that at the same time serves as a vacuum chamber. Spacing and angular inclination of the turns determine the impulse potential distribution along the tube, 1. The anode is the tungsten wire needle, 5, which is secured in the metal holder, 4. The cover plate, 2, is sealed to the insulating tube and carries the terminal conductor, 3. The cathode is formed by the metal sleeve, 7, and the bottom portion, 8. The latter is adjustably mounted. The shock impulse voltage is supplied through connecting lines, 9 (grounded at 10) and 11, by way of a spark gap, 12, from the charging capacitor, 13.

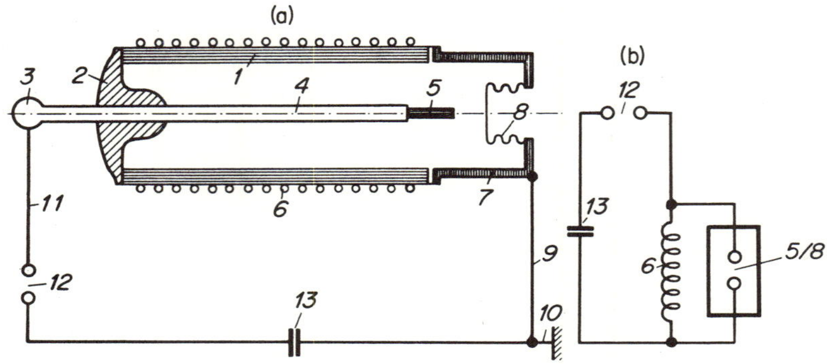

FIG. F1f-1a. x-Ray flash diode (from Zuckermann and Bogulew [175-181]) with voltage stabilization and simultaneous electron focusing by enveloping cylindrical coil.

FIG. F1f-1b. Schematic circuit for x-ray flash diode with parallel-connected inductance (for reference numerals see text).

Normally, capacitor 13 is in the form of a shock impulse capacitor battery in cascade connection. Figure F1f-1b shows the electrical equivalent circuit diagram. The inductance, 6, must be so chosen that along with capacitor 13 the duration of a half-wave is about twice the duration of the x-ray flash duration of 0.3 μsec. The proper frequency is about 800 kilocycles. If such a relationship is attained it is not necessary to maintain an especially high vacuum in the tube. Voltage distribution and duration are then fixed. Immediately after breakdown of spark gap 12, the coil, 6, is passed by a current of rising magnitude ($U = L \times dI/dt$) and focuses the electron flow for an instant in a point on the anode, 5.

Based on the principle shown, reliable tubes up to 1.5 megawatt shock impulse voltage have been built, which have withstood a few thousand flashes and then could be readily regenerated.

The arrangement of the anode point in the cathode cup ensures that only a small portion of the tube becomes metallized and must be cleaned.

Figure F1f-2 shows in principle an arrangement by which the circuit of Fig. F1f-1a can be rendered operative with fast sequence flashes. The capacitors, C, are charged from the voltage source, U, by way of charging resistors,

R. Their discharge is controlled at the ignition spark gaps, St 1–3, by the control device.

This circuit was developed as a true Cranz-Schardin circuit having 16 different tubes and capacitors rated at 1.5 megavolts. Successful x-ray flash recordings were made with such a setup including lead-mask hole cameras adapted for that purpose.* Figure F1f-2 shows in diagrammatic form 3 flash tubes and 3 lead-mask hole cameras which, in the plane of an object such as an armor plate penetration, take the film recordings in the point of intersection of the rays. At the high voltage of 1.5 megavolts, the radiation, owing to the Compton effect, is concentrated in a narrow beam and has a high efficiency (η increases in proportion with $U^{3/2}$).

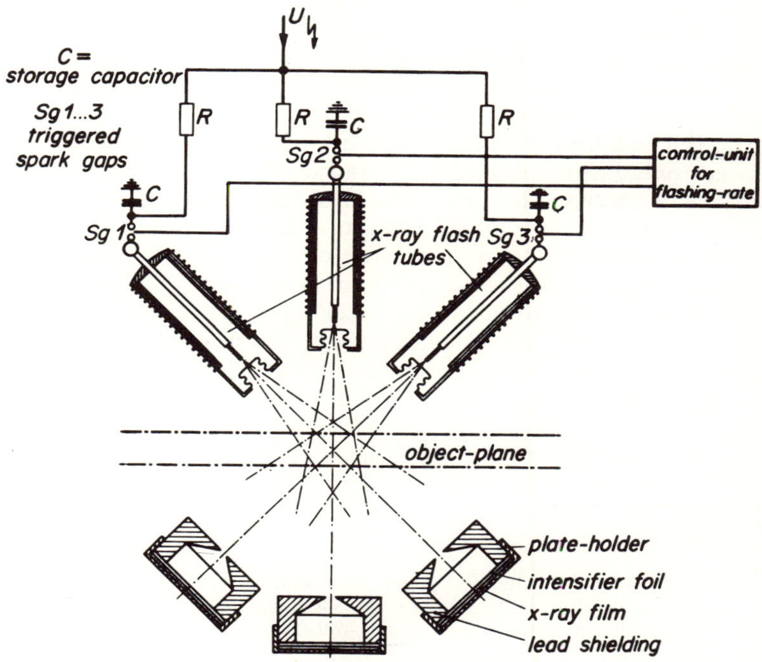

FIG. F1f-2. Arrangement of setup for x-ray flash series by method of Cranz and Schardin with cooperating lead mask cameras [179, 181].

g. *Application of Image Intensifier in Flash Radiography*

G. Thomer and R. Schall [258] have measured the possibility of amplifying the x-ray flash image intensity.

---

* A film recording of the penetration of an armor plate by a hollow charge taken with this apparatus was shown at the 4th International Congress of High Speed Photography, September 1958, in Cologne, Germany.

For certain applications of flash radiography, particularly diffraction studies, the x-ray intensity of a single flash is not sufficient for direct photographic registration. A further increase of flash specific intensity is not to be expected because of the limited current density at the anode and the required short flash duration, so an improvement in sensitivity of registration might be expected to open new fields of application to flash radiography. One possibility for realizing this is given by the electronic x-ray image intensifier. The above authors have studied the possibilities of application of such a device in flash technique.

A slight loss in definition must be accepted, but this is nearly the same as in the stationary case. The image blur due to development of space charge in the electron optics by high impulse loads is negligible for dose rates up to 10^5 r/sec. If an objective with a very large aperture and highly sensitive film is used the gain in sensitivity is by about a factor of 50, compared with direct registration using a highly sensitive film. The method will be of interest not only for single-flash photography but also for cinematography with x-ray flashes.

As in optical high-speed photography, the development of flash radiography depends not only on the production of shorter intense flashes, but also on improvement of the recording medium. Flash radiography has benefited in the past two decades by the development of intensifier screens and of emulsions of higher speed. For certain applications, however, the intensity of a single flash is not sufficient to blacken a film. For example, diffraction studies with x-ray flashes, undertaken by Schall [266] and Schaaffs [267], have, for intensity reasons, hardly found practical applications until now although there are interesting problems that await solution, particularly in dynamic strength phenomena. Another field where it is desired to increase the recording sensitivity is the submicrosecond exposure range. Today x-ray flash sources with a flash duration of 10^{-7} sec are available. Just as for visible light photography efforts are being made to reduce this time to a few millimicroseconds. The development of the source needs to be paralleled by an improvement of the recording system.

One possibility of increasing the recording sensitivity in radiography is created by the image intensifiers which have been available for several years. These intensifiers were originally built for static use, but the research work of Thomer and Schall [258] was concerned with the examination of their applicability in flash techniques.

Tests were run with a Philips Image Intensifier for material testing built by C. H. F. Müller, Hamburg, Type MB 12. Its input screen has a diameter of 135 mm, and the output screen is a yellow-green phosphor with a diameter of 14 mm. For visual observations, a Zeiss binocular microscope was used. For the photographic records, a Contaflex camera (Tessar $f/2.8$) or a Zeiss-R objective lens ($f/0.85$) adapted to a Foca camera was used. As a static x-ray source, medical equipment (80 kv, 5–20 milliamp) was available. As a flash source, a 1-megavolt impulse generator (10 stages) with a sealed-off glass tube was used. For comparison purposes, the flash installation was run at 100-300 kv.

To study the influence of flash illumination, flash radiographs were taken at 120 kv and at different distances of the intensifier from the x-ray source. Figure F1g-1 shows the experimental arrangement. For these records the image converter is used with its normal static acceleration voltage. By variation of the distance from the flash tube, a convenient mean film blackening is obtained which allows a comparison between the flash record and the static record. It seems that there is no essential difference between the pulsed and the static record. Flash duration is 0.5 μsec. x-Ray dose rate falling on the photosensitive converter screen corresponds to about 10^3r/sec. Even for still higher dose rates, i.e., for smaller distances from the source, no essential loss in definition could be noted if the aperture of the lens was chosen to obtain a suitable film blackening.

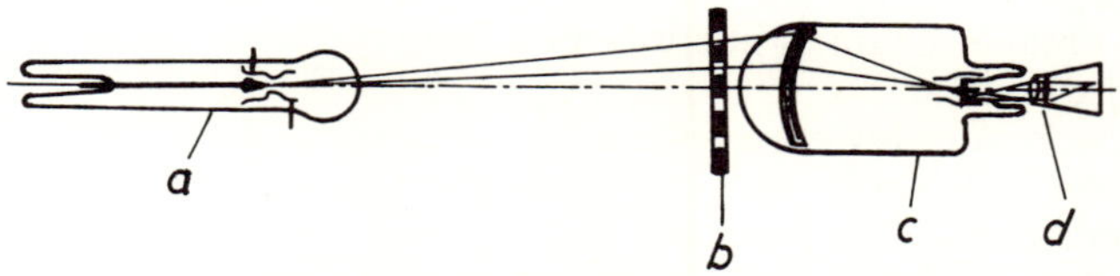

FIG. F1g-1. Experimental arrangement. a–x-ray flash tube, b–test grid, c–image intensifier, d–camera.

Manufacturers of x-ray intensifiers indicate intensification factors on the order of 10^3. This factor indicates the gain in brightness by the fluorescent screen of the intensifier compared with that of a screen illuminated directly by x-rays. This includes the increase in luminosity due to the electron optical size reduction, and the intensity increase by electron acceleration.

For high-speed measurement this radioscopic intensification factor is less interesting than the radiographic factor resulting from the comparison of a direct x-ray record to a record of the intensified image. It is obvious that this comparison depends not only on the qualities of the image intensifier, but also on those of the films, the intensifier screens, and the camera used. In fact, the intensification factor not only is an index of the performance of the image intensifier, but it also reflects the actual relative state of visible light photography and radiography.

The camera aperture is of course of great influence. An R-Biotar $f/0.85$, which was particularly designed for fluorescent screen photography was used. Because of the geometry of the apparatus, the images of the screen on the film had to be reduced 3:1. The final images were only 5 mm in diameter. This means that, owing to the photographic grain size, losses in definition arise, depending on the film used.

The intensification factor of the system used was determined as follows: The blackening of high-speed x-ray film between intensifier screens (CGR Fulgurix) was compared with the blackening of a Scopix film in a camera with the Biotar lens focused on the intensifier screen.

This factor, F, is obtained from a series of exposures. $F = (r_1/r_2)^2$, where r_1 is the distance, source to film, giving an absolute blackening, $s = 0.8$, and r_2 is that distance, source to image intensifier, giving the same blackening, s.

In these series, flash conditions were of course held constant, and flash intensity dispersion was eliminated by taking average values. Reduction of x-ray intensity was achieved by varying the distance of the recording device from the x-ray source. Under these conditions F values of between 40 and 64, depending somewhat on processing conditions were obtained. Thus there is a considerable gain in intensity making possible many new applications. The gain is accompanied, however, by a small loss in definition. As in visible light flash technique, a choice must be made between highest sensitivity and best definition.

x-Ray image intensifiers are convenient for recording fast events, provided that highest definition is not necessary. For impulse use, definition is not remarkably worse than for static or continuous radiation. Using an $f/0.85$ lens, one obtains a gain in intensity of a factor of 50 compared with a direct record on x-ray film.

The gain in intensity enlarges the field of application of flash radiography, particularly for records with extremely short exposure times, and also for diffraction records. The transformation of the x-ray image to an optical one is especially worthwhile in flash cineradiography. In the direct recording of cineradiographs, with presently attainable flash frequencies on the order of 10^4/sec, and modern drum cameras, the image height can hardly be greater than 1 cm. Thus normally image separation could be achieved at higher repetition rates only by reduction of frame size.

The image converter effects a size reduction, easily giving an intensification at the same time. Flash cinematographic series could therefore easily be obtained by means of the image intensifier by filming the converter screen with a high-speed camera of the Fastax type, flash triggering being controlled by the camera.

If one wishes to combine an x-ray image intensifier with a high-speed camera it is necessary to use phosphors with a short decay time. The P2-phosphor shows the best light output, but has a very long decay time. The P11-phosphor has $\frac{1}{3}$ the light but a submicrosecond decay time. To obtain the shortest decay time, the application of a P15-phosphor would be preferable, however, the loss of light compared with the P11-phosphor is in the range 3:1.

2. x-Ray Irradiation of Subjects in Mechanical Motion

a. *Application of Short Bursts of High Energy x-Ray Flashes*

The application possibilities of x-ray flash tubes, as visualized early by Mühlenpfordt, include structural investigations of rapidly moving dynamically loaded machine elements, particularly wear and fracture occurrences in highly loaded gear drives, deformation and loosening of blades in high-speed turbines (machines), and explosion of windings in electrical machines.* One could consider investigations of tool materials in operation, for example,

* Such an investigation on windings has been conducted and gave different results from those theoretically expected.

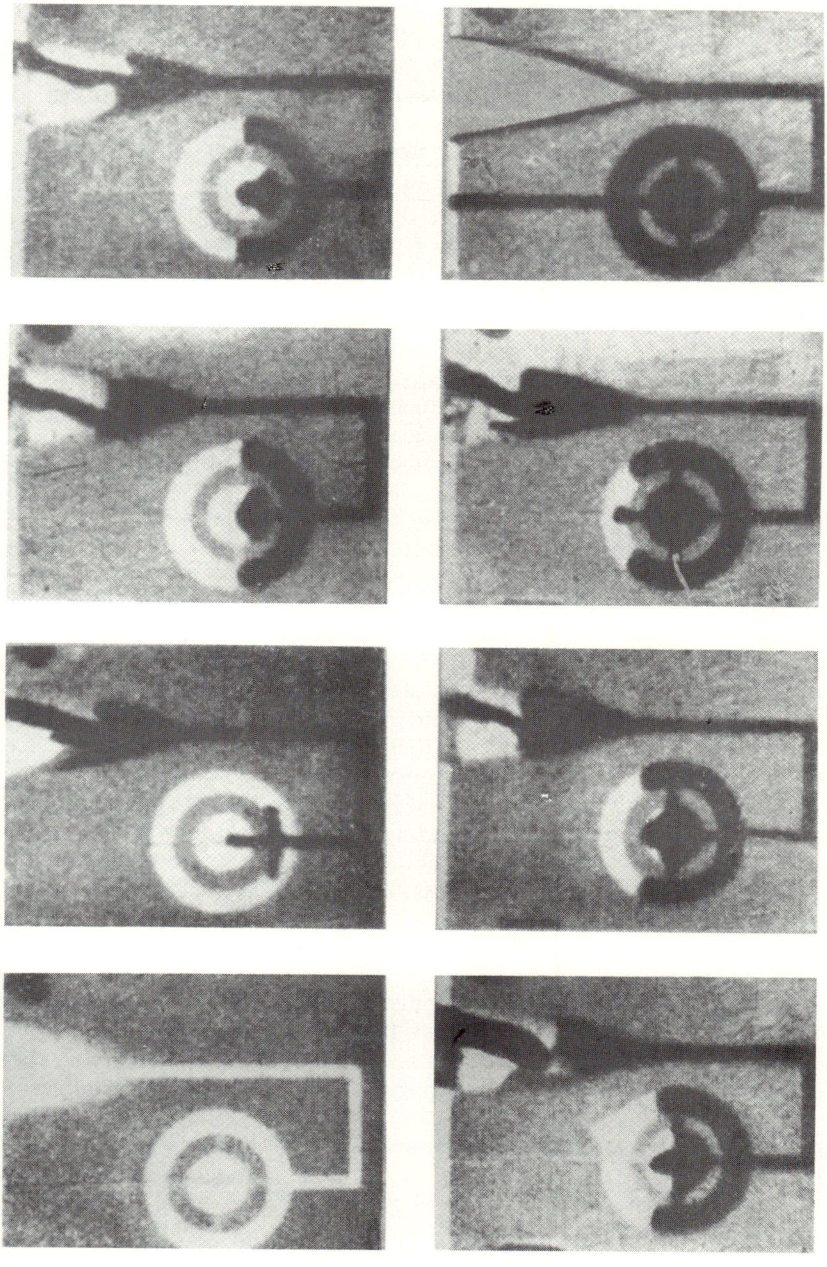

FIG. F2a-1. x-Ray flash photos taken during casting of a lead wheel in an aluminum mold, showing eight phases.

during drilling of soft metals and rocks, of wear in brakes when they are overstrained, and of heavy shock loading (fracture strength tests) (examples of ballistic investigations have been given by Steenbeck).

Of further interest might be the study of the chronological progress of casting processes, such as die casting in plaster or metal molds and centrifugal casting. Proper judgment of later fractures or fracture possibilities may depend on the knowledge of casting procedure steps. Figure F2a-1 illustrates the chronological steps in the casting of a lead wheel in an aluminum mold in eight successive phases, a reproduction of photographs taken by Mühlenpfordt.

In the field of explosive physics, the x-ray flash technique is indispensable for the study of hollow charges.

In case the objects for observation are very dense and composed of heavy metals, or if they are extensive and it is necessary, in order to obtain clear shots, to operate at greater distance from the focus, it is possible, as shown by Steenbeck, to synchronize periodic movements with the x-ray flashes, release the flashes by the action of such movements, and complete the exposure of the photographic plate after a certain number of x-ray flashes [28, 493].

b. *Periodic or Stroboscopic x-Ray Flashing with Low Energy* [493].

Often it is necessary to obtain a long burst of periodic x-ray flashes. Such cases are the observation of optical screened periodically moved subjects, e.g. steel gears in aluminum boxes, tracing the process of burning of fuses, and x-ray high-speed filming by means of drum cameras.

A subject of a paper by Früngel [250] is a method for operating x-ray apparatus and x-ray tubes in the form of electric impulses with the result of higher efficiency in the tube and higher yield in x-radiation.

Generally, x-ray tubes are operated either by pulsating or by continuous high-voltage direct current. The maximum safe working voltage of tubes for such operation depends on their design and is limited by their rating and safe temperature rise. Surpassing the rated voltage results in gaseous discharges, usually accompanied by arclike discharges, and finally leads to destruction of the tube.

It has been an object of development to provide a method of operation for x-ray tubes whereby the peak voltage applied to a given tube can be multiplied without shortening the life of the tube. To achieve the purpose, a given x-ray tube is fed by a train of electric impulses of short duration instead of a continuous current supply. The duration of the impulses is so short that they cannot create a detrimental current avalanche within the tube. Thus it is possible to considerably increase the safe operating voltage of a particular x-ray tube without adverse effects. However, with impulses of such very short duration to obtain x-radiation of the same average magnitude as with continuous current supply, it is necessary to adjust the current of each impulse to a correspondingly higher value. Assuming for illustration an impulse duration of 10^{-6} sec and input to the tube of 10^3 such impulses per second, it becomes necessary, for a similar loading of the tube, that the current of each impulse carry a current 10^3 times greater than in continuous current operation. Thus, it is advisable to reduce the duration of the voltage impulses so much that they cannot cause gaseous discharges within or around the tube but to increase the current so that the root-mean-square value of a train of impulses is equivalent in order of magnitude to the maximum continuous current rating of the tube.

An advantage of the impulse operation of x-ray tubes is the fact that in case the voltage amplitude of the impulses is not greater than the voltage rating for continuous current operation of a tube, the safety of operation is materially increased owing to reduced danger of breakdown.

Another advantage is gained by the fact that, without creating dangerous operating conditions, it is possible to operate with peak impulses voltages that are several times higher than the permissible rated voltage of the tube. Since the efficiency of x-ray production increases approximately proportionally to the square of the operating voltage, it is thus possible to achieve with e.g., an impulse voltage of twice the rated operating voltage, four times the yield in x-radiation. In cases where increase in radiation is neither necessary nor desirable, the use of smaller and lower priced tubes can be considered if the present impulse method is to be employed.

It becomes possible for very intensive x-radiation normally requiring, for instance, a peak voltage of 500 kv, to employ commercial tubes rated at e.g., 200 kv. Figure F2b-1 is a graphic representation of the voltage-time relationship of the impulses for operating an x-ray tube. Figure F2b-2 shows diagrammatically an arrangement of an impulse voltage source connected to an x-ray tube.

The graphic representation of Fig. F2b-1 will help to visualize the time relation between impulse durations and intervals. The duration of the impulses t_1 is very short, on the order of 1 μsec, and the intervals t_2 between successive impulses are relatively long, as a rule more than 1000 times the duration of t_1. Figure F2b-1 serves for illustration only and the relation t_1 to t_2 is not drawn to scale. It should be understood that t_1 must be very short, in fact so short that no harmful breakdown within the tube can be caused by the high potential of the impulse, which, as aforementioned, is much higher than the kilovolt rating of a particular tube.

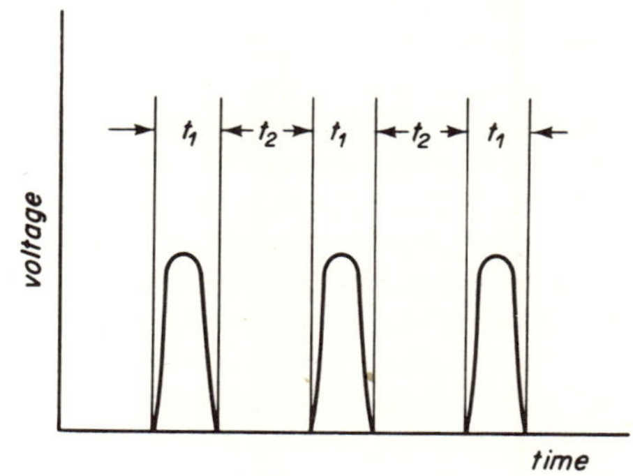

FIG. F2b-1. Voltage-time relationship of impulses operating an x-ray tube.

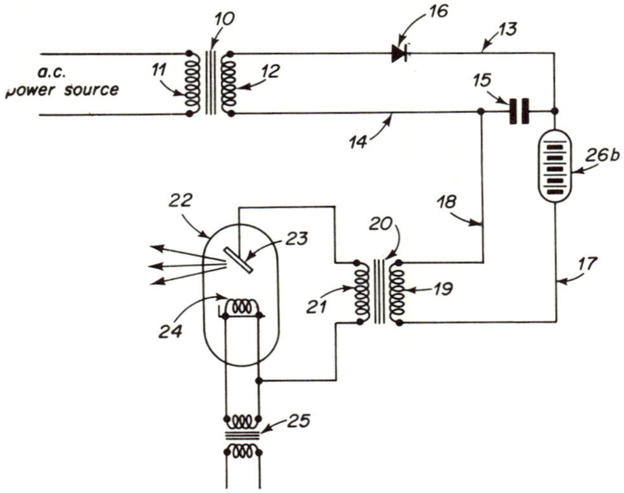

FIG. F2b-2. Diagram of an impulse voltage source connected to an x-ray tube.

In order to take full advantage of the energy delivered to the x-ray tube by the short but strong impulses it is necessary that the hot filament cathode of the tube be designed for higher electron emission than for conventional operation. Considering a time relationship $t_1/t_2 = 1/1000$, the emission must be 1000-fold in order to get the whole benefit of impulse operation and must not exceed the safe temperature rise of the tube as given for continuous operation.

Such high emission as mentioned above could not be achieved in previous x-ray tube design because it was impossible to employ cathode material of high emissivity, such as thoriated tungsten or pill material, on account of the fact that such material can not endure the continuous ion bombardment due to traces of gas in the tube. However, in operation by the impulse method the instant during which the impulse voltage is effective is too short for developing a detrimental ion avalanche and the time intervals between impulses are long enough for de-ionizing the traces of gas in the tube before a subsequent impulse occurs. Therefore, highly emissive cathode materials can be incorporated in tubes designed particularly for this purpose [752].

If low cost of the tubes is a primary factor, good results in the production of x-radiation by the present impulse method can be obtained with tubes of the getter type in which the getter pill can be attached at a protected spot. Other relatively inexpensive tubes applicable in this new method are metal tubes somewhat similar to metal radio tubes, wherein the metal envelope is designed in accordance with electron optics to concentrate the electron stream in a focal spot for x-radiation at its inside; the lead-ins to a hot filament cathode of high emissivity project through an insulator secured in the metal wall; the vacuum within the tube is maintained by getter means attached to the inside of the envelope; and the metal envelope, serving simultaneously as anode and target, can be cooled from the outside.

Such special x-ray tubes as described above, or tubes of conventional construction operated by the impulse method, are used in conjunction with a circuit as diagrammatically shown in Fig. F2b-2. The primary winding, 11, of the high-tension transformer, 10, is connected to an alternating power source not shown. The high-tension secondary, 12, of this transformer is connected by means of conductors 13 and 14 to a storage capacitor, 15. Since capacitor 15 is to be charged by high-potential direct current, a rectifier, 16, is shown inserted in conductor 13. In this connection the capacitor is continually charged as long as the primary of transformer 10 is in circuit with the power source. Capacitor 15 is designed for impulse discharges, i.e. it has extremely low inductance and low losses. In parallel connection with capacitor 15 is shown a discharge circuit consisting of conductors 17 and 18 and the primary winding, 19, of impulse transformer, 20. Shown inserted in line 17 is a periodically operating switching means, for high discharge sequences, a quenching spark gap, 26b, triggered up to 50,000/sec, which successively breaks down whenever the capacitor potential surpasses the breakdown voltage of the gap. It will of course be understood that other periodic switching means can be employed and that they may be inserted anywhere in the discharge circuit. The secondary winding, 21, of impulse transformer 20 is connected directly to the x-ray tube, 22, which is shown diagrammatically as having an anode or target, 23, and a hot-filament cathode, 24, being heated by electrical energy derived from a power source (not shown) through transformer 25. The x-ray beam emanating from the tube is indicated by arrows.

In operating, assuming the secondary winding of transformer 10 to be designed for charging the storage capacitor, 15, with 10 kv, and assuming the primary winding, 19, of impulse transformer 20 to have 10 turns and the secondary winding 200 turns, the anode, 23, of the x-ray tube will receive impulses at a peak voltage of 200 kv. These impulses carry relatively large currents, e.g., on the order of 5 amp, and the hot filament cathode, 24, must therefore be capable of emitting current of that order which is considerably higher than the average in conventional x-ray tubes. Thus, in order to take full advantage of the benefits rendered possible by the present impulse method, it is advisable to use thoriated tungsten as the construction material for the hot filament cathode of the x-ray tube employed, and to design the cathode to have an emissivity of more than 1 amp. The connections with respect to rectifier 16 must of course be made so that the secondary impulse wave of impulse transformer 20 is positive at the anode of the x-ray tube.

It should be understood that the operation of an x-ray tube in the manner described herein is not intended for uses where each electrical impulse must furnish sufficient x-radiation for a particular exposure, but is provided for general x-ray practice in which the period for passing the train of impulses is essentially of the same duration as the length of time usual in continuous current operation.

An example of a design of a combined system with x-ray tube and pulse transformer, cooled by H_2 or Helium, is shown in Fig. F2b-3.

At an operative voltage of about 200 kv and more, and relatively low peak

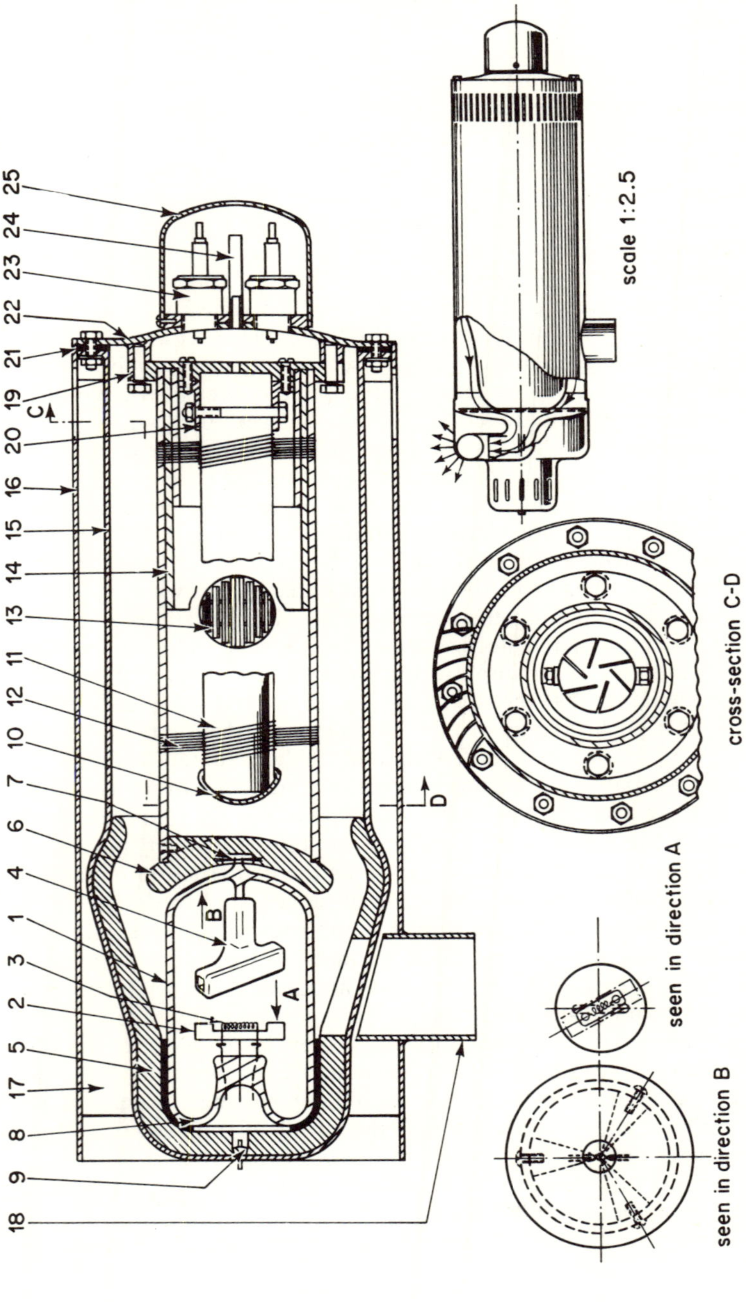
25
24
23
22
21
19
C
20
16
15
14
13
11
12
10
7
D
6
4
B
1
3
A
2
5
17
8
9
18
scale 1:2.5
cross-section C-D
seen in direction A
seen in direction B

currents of about 10 amp, the equivalent ohmic resistance must be assumed to be on the order of 20,000 ohms. Therefore, the impulse transformer must be designed for a burden of such magnitude at its secondary that the impulse duration will preferably be not longer than 1 μsec, and the product

$$200 \text{ kv} \times 10 \text{ amp} = 2 \text{ megawatts}$$

can be assumed, when the impulse duration is 1 μsec; thus 2 watt-sec/impulse. In operation with 1000 impulses/sec, the thermal anode output would be 2000 watts, which would have to be dissipated by the tube. Then the energy to be supplied to the impulse transformer must be 3000 watts, that is, 3 watt-sec/impulse, on account of an efficiency of about 70%. This results furthermore, at 10 kv capacitor voltage, in a capacitance of 60,000 pf which is provided by a polystyrol capacitor with small loss angle. The primary inductance of the shock impulse transformer having a ratio 1:20 to 1:25 lies in the order of $L = 4t^2/10C$ (impulse duration taken as one quarter cycle)* and in the present example becomes

$$\frac{4 \times 10^{-12}}{6 \times 10^{-7}} = 0.7 \times 10^{-5} \text{ h} = 7 \ \mu\text{h}$$

This result shows the possibility of variance by change in winding turns.

In normal operation of x-ray tubes for hard radiation, a deflection of the filament can generally be observed that is due to static attraction of the filament toward the anode. Deflection represents work for which, in addition to force, space is necessary, namely for motion of the filament from its state of rest toward the deflected state. The mechanical work for this filament movement is the integral of the force over the deflection distance, and the latter depends on the duration of the force exertion.

When the anode is not operated by dc voltage, but with short voltage impulses which have large time spacing from each other, it is possible to

FIG. F2b-3. Design of an x-ray tube for impulse operation. 1–x-ray tube hard glass body; 2–cathode screening; 3–glow cathode, thorium coated; 4–anode plate with line focus, tungsten; 5, 6–lead shields; 7–anode clip; 8–tube support, e.g. teflon; 9–current lead for cathode heating; 10–slotted calotte, Rogowski profile, for impulse transformer electrostatic screening of the conductor; 11–primary; 12–secondary; 13–laminated hyperm or permalloy metal racks, (the space between packs serves for channeling the cooling gas); 14–Mylar insulation, primary-secondary (also teflon may be used); 15–inner metal housing, copper; 16–outer metal sleeve, aluminum; 17–space for cooling agent, e.g. oil; 18–filling nozzle for cooling agent; 19–pulse transformer base; 20–iron core support with primary; 21–packing; 22–cover plate with connectors; 23–current leads for primary pulse; 24–filling nozzle for H_2, He, or cooling agent; 25–cover for current connectors.

* It is assumed that after a quarter cycle, referred to the periodic discharge, the primary voltage passes through zero and the current has reached its maximum value. At this instant also the secondary voltage and – since the x-ray tube represents an ohmic load – the secondary current reached zero value, so that the effective pulse can be considered finished at this instant.

reduce considerably the duration in which the electrostatic forces are effective and thus to keep any deformation that occurs within the elastic limit. To achieve this end it is, however, necessary to increase the emission of the filament, e.g., by use of thoriated wire. In order to reduce the danger of filament fracture in a tube of 200 kv rated voltage, it can be operated at a rate of 1000 impulses/sec and an impulse duration of 1 μsec, so that each impulse is separated from the next one by a time interval that is 1000 times longer than its own duration. Thus, for an intensity of x-radiation equal to that in continuous dc voltage operation, each impulse emission of the cathode must be 1000-fold. Cathodes which did not appear useful for normal operation on account of their brittleness or softness (close to their melting temperature), for instance, thoriated cathode, are perfectly suitable for impulse operation.

In every periodic sequence of high voltage impulses there exists the danger that mechanical oscillations will occur at the filament, since its mechanical natural frequency, if not counteracted by some design measures, can come readily into resonance with the impulse sequence frequency or one of its harmonics. The system of the filament and its supporting means must therefore be designed so that its natural frequency is not excited by the im sequence frequency or its even or odd harmonics. In filament construc available for test it is readily possible to determine the resonance freqı of the filament when it is excited by lightly tapping the tube and measuring the frequency by a body-sound microphone. The frequency thus determined must be guarded against and not be used as the impulse sequence frequency.

In such a circuit, the primary voltage and frequency of the impulses as well as their duration and energy can be arbitrarily changed by simple adjustments in the apparatus. An advantage of this setup is that connection between the x-ray tube and the control device can be made by a commercial, coaxial, flexible, medium-voltage cable.

c. *x-Ray Flash High-Speed Cinematography of Small Objects*

Technical research encounters many problems where comparatively small objects must be filmed by means of x-ray flashes at high flashing rates and where also a great number of frames must be obtained. Problems of this kind, include the melting of electrical fuses, operation and performance of electrical contacts in sealed relay systems, the snapping of drilled cables, and the stapling process for periodicals, in which the staples must penetrate the paper at high velocity. Particularly, in this stapling process, the number of waste and rejects is still very high.

A further problem is the distribution of mass particles and their respective velocities in the combustion process of rocket propellants, where the results obtained by high-speed photography with visual light do not yet permit conclusions as to the respective mass values involved.

While until recently x-ray flash cinematography has only been performed by the Cranz-Schardin method in which several x-ray flash tubes werd fired in succession, Thomer [260] can now show film strips of a melting wire taken with a drum camera.

For reasons intrinsic in the subject matter, high-speed x-ray flash cinematography is mainly restricted to smaller objects. When a drum camera with direct film exposure is used, the frame height is determined by the rotating speed of the film and the flashing rate. At a given object size of maximum 35 × 35 mm the framing rate is approximately 3000 fps; at a frame height of 35 × 10 mm the maximum frame rate will be 10,000 fps. Larger objects up to approximately 130 mm diameter could be filmed by means of a commercial x-ray image intensifier whereby the intensity gain would be 40- to 100-fold [258].

As the quality of materials examined with x-ray flash cinematography is rather varied, a commercial x-ray flasher must have a wide range of radiation quality, i.e. 15–150 kv. It is therefore desirable to have, for the examination of the above-mentioned problems, equipment which is reliable in operation and yet is easy to handle, if necessary, even semiskilled personnel should be able to operate it.

When a discharge of 100 kv and 0.5 μf, that is, an energy of 2500 joules, can be released in an x-ray tube, it is also permissible to release in fast sequence 100 single discharge pulses each having 1% of this energy, or 5000 pf at 100 kv. In this case, however, it is advisable to use a differentiating impulse transformer with periodic control of the primary, as presented in Chapter E, Section 3. It is also possible to release the impulses by means of a quenching spark gap or a blast spark gap and reach thereby extremely fast flash series in the range between 1000 and 5000 cps sequence frequency within the limits of thermal tube capacity. When one considers that a 16-mm film for normal projection needs 16 frames/sec and that a good scene length that can be evaluated by the eye lasts for 3 sec, that is, for 50 frames, it can be realized that it is of interest to take x-ray flash series of 50 flashes at arbitrary speeds for observation of rapidly proceeding phenomena and, therefore, to revolve the film which is to be exposed to the x-ray flashes by means of a drum-type camera. Drum-type cameras with the film around the periphery of the drum, which permit direct exposure of the film material to the x-radiation through a plastic material window, operate at film speeds up to 100 meters/sec.*

If the picture height of each single exposure were 10 mm at the normal negative film material width of 35 mm (nonperforated 35 mm; only 24 mm is useful if it is perforated) then at 10,000 images/sec and 1.5 meter film length, it would still be possible to utilize 150 x-ray flashes successfully in a rational manner. Whereas static x-ray irradiation is more and more replaced by other processes such as neutron irradiation or gamma radiation, it is impossible to replace x-ray stroboscopy and x-ray flash series, as well as single x-ray flash film exposures, by isotopic preparations or the like. Some progress has recently been made in the development of x-ray exploration methods. Figure F2c-1 shows a drum-type camera having provisions for taking a film at its periphery and an exchangeable lead mask that can thus be adapted to suit x-ray flash series as to their speed and number of images.

The following discussion is intended to give some inspiration for constructive developments, particularly in the field of x-ray stroboscopy. In this field

---

* "Strobodrum," FF Impulsphysics Corporation/Impulsphysik A. G.

the impulse transformer is excited by a spark chamber or ignition similar to that used for control and production of light flashes in light-flash stroboscopy.

Figure F2c-1 shows an operative circuit for x-ray cinematography having an x-ray tube operating at an increased short time heater output for flash trains of short duration with control at the impulse transformer. Such flash trains can be photographed by a drum-type camera and can be synchronized with the perforation of the film just as in the case of spark cinematography.

The methods for control of high frequency flash devices, which will be described herein are, in their basic principle, also applicable to x-ray flash devices [40].

Commercially manufactured apparatus delivers a rated total energy of about 50,000 joules maximum with impulses in the range between 16 and 50,000/sec and energies per impulse adjustable from a maximum of 10 joules/

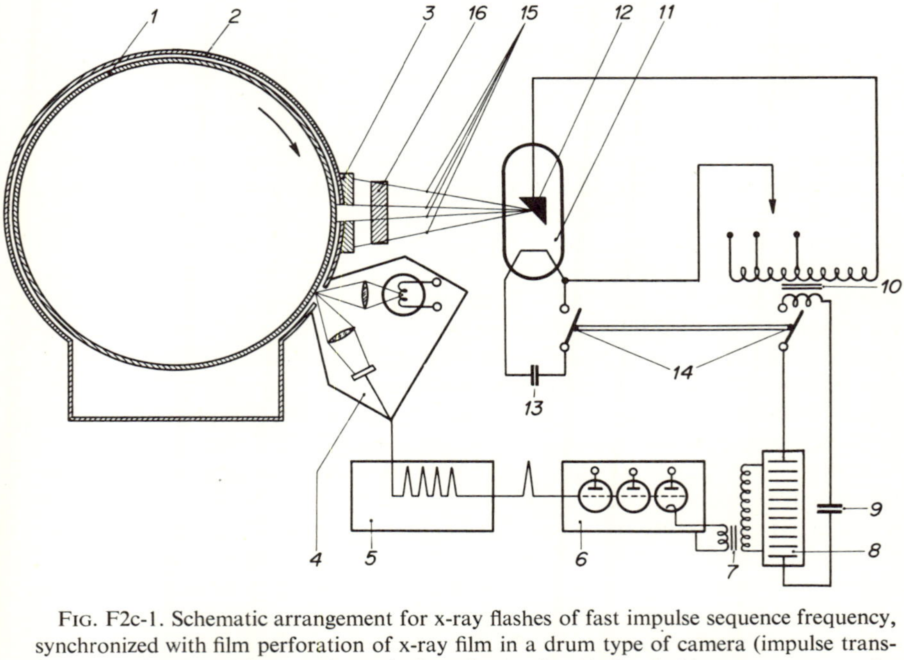

FIG. F2c-1. Schematic arrangement for x-ray flashes of fast impulse sequence frequency, synchronized with film perforation of x-ray film in a drum type of camera (impulse transformer having quenching spark gap in its primary circuit) (Strobokin x-ray accessory by FF Impulsphysics Corp.). Synchronized flashes up to 5000/sec with standard film size. Quartz controlled flashes up to 50,000/sec with smaller film mask.

1–rotary film drum (Strobodrum); 2–housing; 3–slidable lead mask for film framing; 4–infrared pickup for sensing film perforation; 5–frequency divider 4:1 (Synchrophot); 6–control device with H_2-thyratron; 7–ignition impulse transformer; 8–quenching spark gap; 9–polystyrol capacitor battery; 10–impulse transformer 100 to 500 kv; 11–x-ray tube; 12–anode of high thermal storage capacity; 13–impulse capacitor for spontaneous heating of the cathode for about 1/100 sec; 14–interlocking switch for simultaneous release of the flash series and impulse tube heating; 15–radiation beam; 16–object to be photographed.

impulse to a minimum of 0.5 joules/impulse. The impulses are transformed by an impulse transformer into high-voltage impulses and fed into the x-ray tube at such a rate that the energy of the flash series is adapted to the thermal anode capacity (solid tungsten anode). At 200-kv impulse voltage and 2-μsec impulse duration, with 10 joules anode energy, there can be tolerated

$$I = \frac{10}{2 \times 10^5 \times 2 \times 10^{-6}} \text{ amp} = 25 \text{ amp emission}$$

The permissible emission at the highest allowable burning voltage of irradiation tubes in position for exposure (about 20 kw anode output at 50 kv) amounts to barely 1 amp. For that reason a separate impulse heating device, 13, is shown in Fig. F2c-1.

Since, in the application of impulse operation shown here, the duration of the whole flash series is very short, the tungsten filament is heated impulse-like to just below its melting temperature by a small capacitor charge simultaneously with initiation of the flash series. Because radiation of heat losses is very low in such a rapid heating process, the necessary energy of the capacitor charge has to be figured on the basis of tungsten filament mass, tungsten melting temperature, and specific heat in order to avoid overheating of the filament.

Assumption:

Specific heat of tungsten: 0.036 cal/gram $\times$ degree
Tungsten mass of filament: 1 gram
1 calorie = 4.19 joules
Melting temperature: 3380°C.
Melting heat: 46 calories/gram (considered as reserve safety against burnout of the filament, so that the capacitor energy is less than necessary to supply the melting heat that would destroy the filament)
Capacitor energy:

$$A_{\text{joule}} = 4.19 \times 1 \times 3380 \times 0.036 \text{ joules}$$
$$= 510 \text{ joules}$$

Capacitor voltage: at least 2 to 3 times the rated maximum voltage of the filament, e.g., 300 volts (this makes possible the use of standard photo-flash capacitors)

$$\text{Capacitance } C = \frac{2A}{U^2} = \frac{1020}{90{,}000} \times 10^6 \ \mu\text{f} = 11{,}400 \ \mu\text{f, 300 volts.}$$

It is necessary to determine the required 510 joules by calorimetric pretesting owing to great manufacturing tolerances in photo-flash electrolytic capacitors. The accurate filament dimensions of commercial x-ray tubes can, of course, be had from the manufacturers.

Since the emission of tungsten increases sharply just below its melting point, a fine adjustment of the charging voltage for the heater impulse capacitor with respect to the "stiffness" of the high-voltage impulse transformer is readily possible. Its voltage should drop by about 20% against the no-load voltage.

The flashes of the Strobokin device and here also the x-ray flashes can be synchronized with the film perforation by an infrared sensing device, 4, (Fig. F2c-1). Since standard film has four perforation holes for each 18 × 24 mm frame, a frequency divider, 5, is interposed. In operating with a film velocity of 90 meters/sec, it is possible to obtain 5000 film images of normal size. With a total film length of 1.5 meter, which length is predetermined by camera design, there is space for 80 film images. From this explanation some of the technical data can be derived:

Frame frequency: 5000/sec

Number of frames: 80

Flash energy: 10 joules/electrical impulse 0.1 joule/impulse at 1% efficiency as x-ray

Tube voltage (example): 200 kv

Emission current: 25 amp

Anode impulse output: 5 megawatts

Impulse duration: 2 μsec

Wave form: bell-shaped curve. Increased hardness of radiation results practically in shortening of effective flash duration.)

Operating time (duration of flash series): 80/5000 sec = 16 millisec

Thermal requirements of anode: 80 × 10 joules = 800 joules, about 200 calories which are liberated in 16 millisec (nearly the same time as taken by a 50 cps cycle).

It is quite obvious that all common x-ray flash tubes can be operated in a circuit as shown in Fig. F2c-1, with the exception that in some other cases the circuitry for impulse heating can be eliminated.

Emission and efficiency become more favorable with increasing voltage at the impulse transformer, but smaller objects under observation become too contrastless in the exposure, owing to extreme hardness of the radiation at higher voltages. The optical synchronization with film perforation can be replaced in original equipment by a quartz crystal-stabilized control frequency if desirable. With a low total number of flashes it is also possible to change to operation without motion, if the frequency is predetermined as to time limitation of the series by means of a Retarder device and phonograph recording, and when the object under observation (for instance, a pressure wave) itself moves.

The equipment now described [264] uses the known technique of direct capacitor discharges, utilizing high-voltage pulses by which a demountable x-ray flash tube to be described here is operated. A differential pulse transformer is used as a high-voltage source, the primary of which is fed with a low voltage and high current of any desired frequency.

These transformers [261, 255, 262] were originally designed for aerodynamic research work. The secondary of these transformers permits tapping off the plate voltages required for any desired radiation quality. If extremely low voltages, e.g., down to approximately 15 kv, are required, it is also possible to replace the high vacuum by a higher pressure in the tube, e.g. 10^{-3}

Torr. The electrons arriving at the plate are then slower than they would be according to the pulse voltage fed in.

In order to ensure economical operation in combination with universal adaptation the x-ray flash tube was designed as a demountable tube. Electrodes and emanation window can thus be exchanged. Similar to the form described in ref. [260], the tube comprises a replaceable conical tungsten anode and a cathode shaped as a hollow cone so that the x-rays penetrate in the direction of the tube axis. For firing the x-ray flash, a triggering electrode surrounds the cathode concentrically. In order to adapt the tube to different pulse voltages the gap between anode and cathode is made variable by means of a spring-loaded adjusting device.

The x-ray flash tube as such, however, permits operation at voltage pulses up to 250 kv. A hostaphane foil serves as an emanation window which, however, also permits penetration of radiations below 30 kv. The x-ray flash tube is evacuated by means of a two-step rotation pump and a subsequent airproof oil diffusion pump. As the x-ray flash tube does not operate unless it is properly evacuated, wrong handling or wrong dismantling and reassembly cannot lead to bodily harm of the operator. The vacuum meters are likewise designed in such a way that any attempt to operate the x-ray flash tube with residual traces of air still present does not result in a destruction of the meters.

Photographic recording is effected by means of a modified drum camera similar to the model used with the ordinary light flash Strobokin [263]. The only difference is that the lens has been replaced by lead masks for limitation of frame height. The photographic material used is ordinary x-ray film with double coating; an intenser foil can be used in order to increase sensitivity.

By using a cone-shaped anode one achieves a comparatively small focal spot. Thus the distance between anode and film plane need not be great. As the width of the film is given the object as such must of course be small. Normally the distance of the focus to the image plane will be less than 200 mm. Consequently the flash energy required at the x-ray tube is extremely small. Experiments have shown that with a pulse peak voltage of 60 kv an energy of 1.5 joule per pulse is still sufficient.

Figure F2c-2 shows the Strobokin with adapter for x-ray flash, high-speed cinematography. At the right-hand side is shown the power supply and control unit Strobokin; next to it the impulse transformer with the pulse-generating Strobokin lamp in the bottom section; to the left the x-ray flash tube with its circuit and vacuum pumps; and to the far left the drum camera with tripod.

Figure F2c-3 shows a block diagram of the arrangement. The capacitive voltage divider C3 and C4 supplies over R2 the triggering pulse for the x-ray flash tube and furthermore serves as a storage capacitor for the individual x-ray flash. In the anode circuit of the x-ray flash tube is a quenching spark gap which prevents arc formation during flash intervals and which does not trigger unless a certain voltage has been built up again at C3 and C4.

The primary of the pulse transformer is controlled by a quenching spark gap which in turn is triggered by a control unit and, if required, quartz triggering can also be provided.

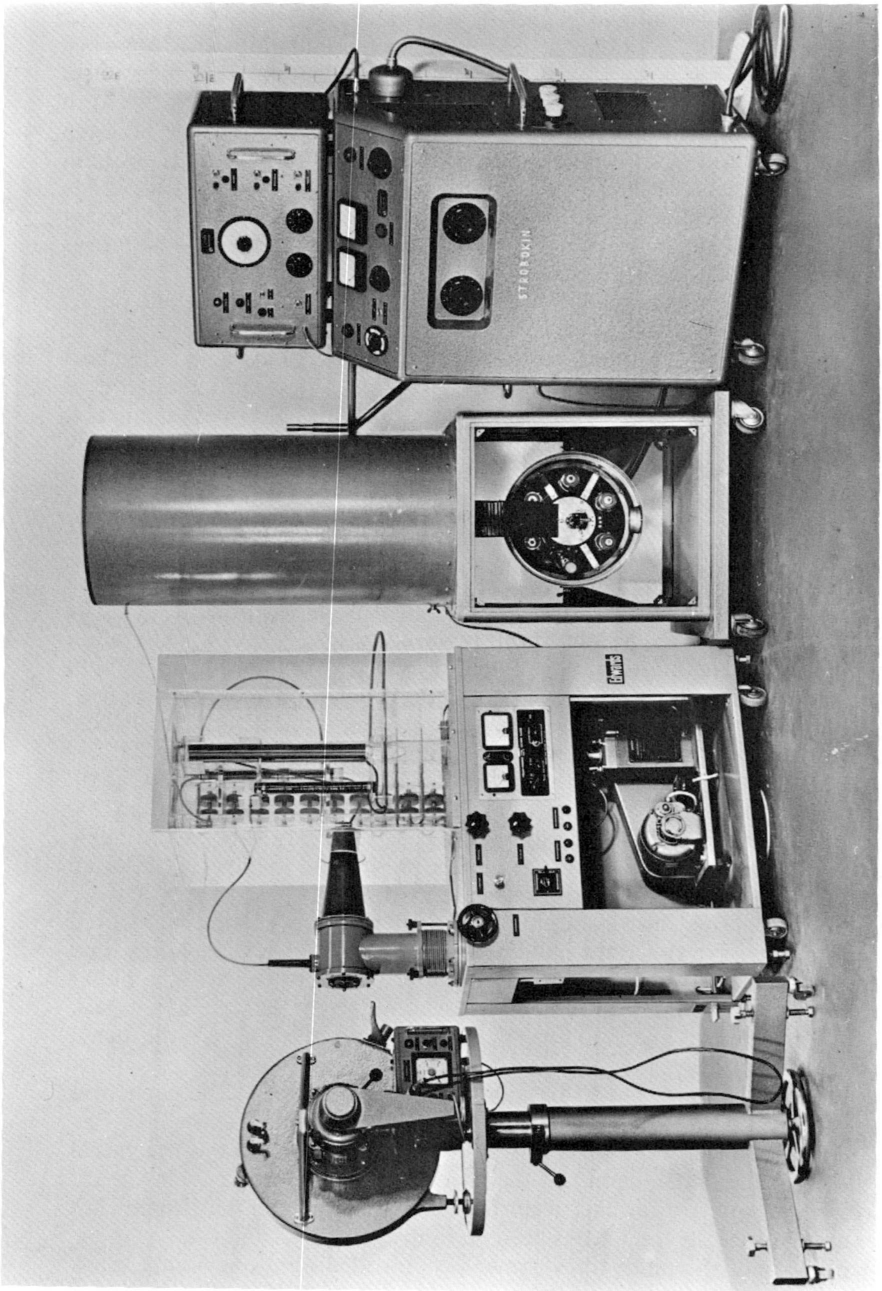

FIG. F2c-2. Arrangement of a Strobokin unit for x-ray flash, high-speed cinematography.

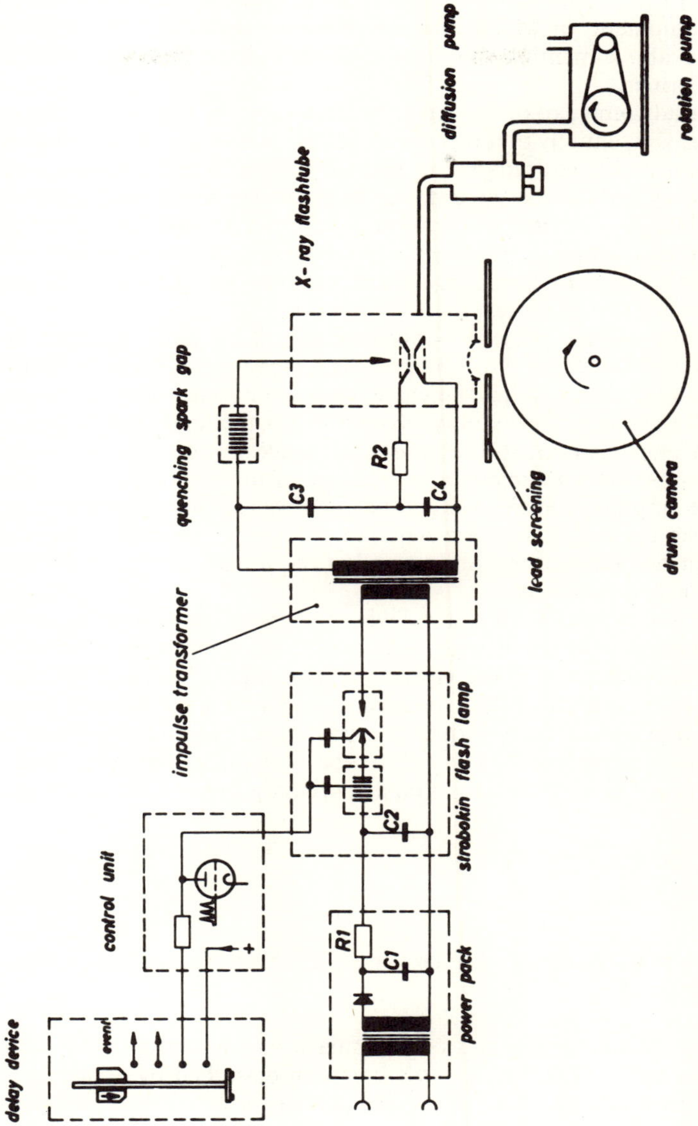

FIG. F2c-3. Diagram of the Strobokin x-ray flash equipment.

The discharge capacitor C2 is a polysterol capacitor of variable capacity. The energy can be varied from 1 to 8 joules. The complete system is designed in such a way that it can be operated as an attachment to the so-called Strobokin equipment, in which case the already existing light-emitting spark gap of the above-mentioned unit serves as a control device for the impulse transformer.

The advantage of operating x-ray flash tubes by means of an impulse transformer is essentially that there is no voltage present in the impulse transformer between individual pulses and that all control functions, as e.g. the pulsing frequency and the energy feed, can be effected to the primary at comparatively low voltages. Generally operation is at 8 to 9 kv primary pulse voltage. For this reason ordinary hydrogen thyratrons, as they are sometimes also used for the control of quenching spark gaps up to 40 megawatts, can also be applied for this purpose.

At low voltage pulses of approximately 8 kv, peak currents up to 5,000 amps may flow for a short interval. The proven system of the hydrogen filled quenching spark gap has therefore, been maintained, though nowadays there are also available heavy duty thyratrons for the loads just mentioned. These thyratrons, however, still require much space so that low inductance adaptation to the low characteristic impedance of the sandwich line system leading to the impulse transformer cannot yet be realized. Generally impedances between 0.5 and 1 ohm are required, which no longer can be realized with concentric hollow tube conductors but require an extremely flat design of all construction elements.

The above x-ray flash method was made possible by limiting its application to comparatively small objects as previously mentioned. Owing to the low energy transduction in the x-ray flash tube, e.g. at bursts of 200 flashes of 1.5 joule each, a total of 300 watt-sec, it is possible to shoot a comparatively large number of high-frequency x-ray flash films, approximately 50–100, without demounting the x-ray flash tube or changing the electrodes.

Figures F2c-4 a,b show two film strips photographed with the above-described unit.

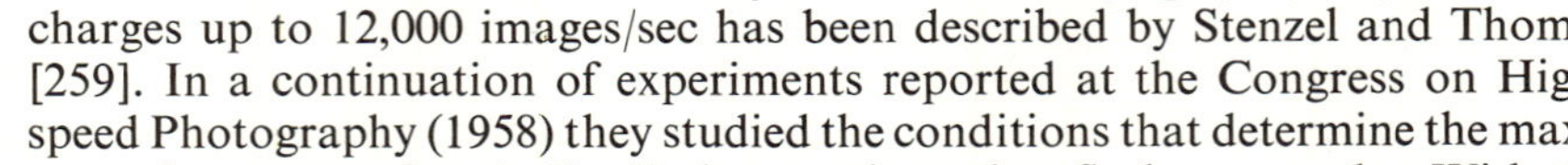

FIG. F2c-4a. Melting of wire, 2000 pulses/sec, 12 volts wet cell battery. Velocity of film: 20 meters/sec.

Another similar method of x-ray flash cinematography by capacitor discharges up to 12,000 images/sec has been described by Stenzel and Thomer [259]. In a continuation of experiments reported at the Congress on High-speed Photography (1958) they studied the conditions that determine the maximum frequency of periodic discharges through a flash x-ray tube. With the earlier device, using direct triggering of the tube and simple *RC* or *RLC* recharging, the limit frequency is about 5000/sec. To get a higher rate, it is necessary to isolate the anode from the discharge capacitor during the pauses.

FIG. F2c-4b. Explosion of wire, 10,000 pulses/sec, 600 volts, 200 μf discharge. Velocity of film: 50 meters/sec.

In the new circuit this controlled separation is attained by a quenched spark gap, which is periodically triggered by an electronic pulse generator. The device allows a precise control of frequency and total frame number. A capacitive voltage divider assures the synchronized pulses to the tube trigger electrode. With an apparatus following this principle, periodic x-ray flashes up to 12,000/sec and a total number of 60 were produced. Image separation is accomplished by means of a drum camera running at 80 meters/sec. As applications, Stenzel and Thomer show frame series of the firing of a squib and of the liquid metal transition in arc welding.

To record a fast event by cineradiography at high frequencies, several individual flash sources, or one tube with repeated flashes, may be used. These two techniques have been reported in several papers, and each of them has found its field of application. Although equipment having several tubes with individual pulse circuits is rather expensive, especially at higher voltages, nevertheless, it has sometimes been used [251]. For this technique the upper frequency limit is obviously given only by flash duration and could therefore be pushed up to 10^7/sec. A disadvantage of this method is the limited number of frames. No apparatus described to date has more than eight flash sources. Moreover, parallax is sometimes a handicap. The repeated flash technique obviates these disadvantages, but entails difficulties with respect to obtainable frequency and image separation.

One of the earlier reported circuits [252] has been modified as shown in Fig. F2c-5. Its crucial part is a special quenched spark gap working in a hydrogen atmosphere as proposed in refs. [253–255]. This gap consists of 80 carefully machined and cleaned copper electrodes with central holes. They are insulated from each other by plastic rings to give a gap width of 0.15 mm

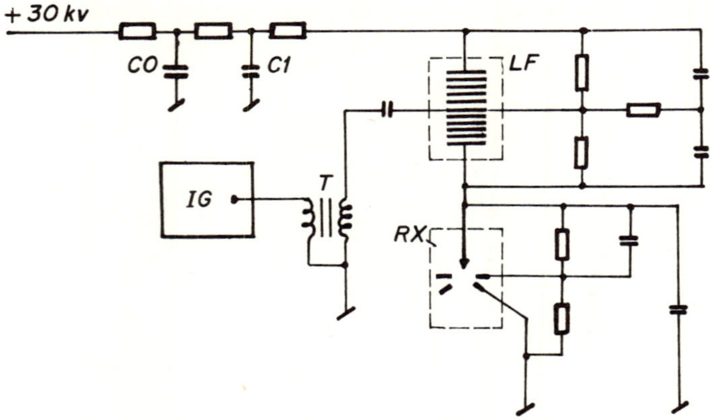

FIG. F2c-5. Circuit for high-frequency x-ray flash cinematography at 30 kv (from Stenzel and Thomer [259]). C_0–storage capacitor; C_1–discharge capacitor; LF–quenched spark gap in hydrogen; IG–pulse generator; T–pulse transformer; RX–x-ray flash tube.

and an active electrode surface of 1.2 cm^2. The complete pile is mounted in a glass tube with brass end plates and sealed by O-rings. A mercury manometer indicates gas pressure, maintained at 80 mm Hg above atmospheric, to prevent the entrance of air. One of the middle electrodes is connected with a wire, which is sealed in the glass, for frequency control. An *RC* network parallel to the spark gap ensures a symmetrical potential reincrease, avoiding unwanted firings.

As described in earlier papers, a continuously evacuated three-electrode flash tube with conical copper anode and cellophane window was used. The reliability of the tube trigger circuit was improved by adding capacitors to the resistance chain. The discharges are controlled by pulses to the previously mentioned trigger electrode of the spark gap. They are produced by an amplitude-controlled pulse generator. A special device limits the number of frames and allows synchronization with the phenomenon. With this installation, a series of 60 flashes at frequencies up to 12,000/sec could be obtained. Higher frequencies seem to be possible. No efforts were made to increase the frequency further, since image separtion set a practical upper recording limit of about 10^4/sec. As in earlier tests, a Leybold drum camera, running at a maximum speed of 90 meters/sec, was used for image separation.

The circuit of Fig. F2c-5 may also be used for higher voltages, up to 120 kv, provided that capacitors and resistances are suitably chosen. Stenzel and Thomer preferred, however, a voltage-multiplying device as shown in Fig. F2c-6, consisting of four stages of capacitors charged in parallel and discharged in series. All spark gaps are of the quenching type, but only the first is hydrogen filled. This is sufficient for the present frequencies as the later stages are fired far over the static breakdown voltage. Only the first gap is working under critical conditions. In the future, for increased frequencies, hydrogen gaps will generally be used.

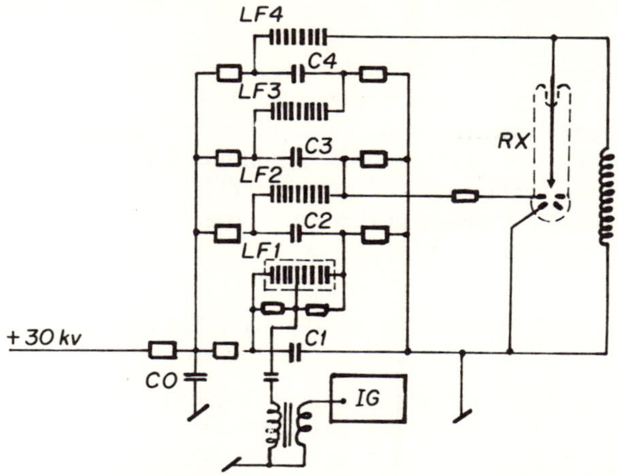

FIG. F2c-6. Circuit for x-ray flash cinematography at 120 kv (from Stenzel and Thomer [259]). C_0–storage capacitor; C_1, C_2, C_3, C_4–discharge capacitors; LF_1–quenched spark gap in hydrogen; LF_2, LF_3, LF_4–quenched spark gap in air, IG–pulse generator; RX–sealed-off x-ray flash tube.

The x-ray flash source used for the 120-kv experiments was a sealed-off Pyrex glass tube with tungsten anode, developed and manufactured in their Institute's laboratory [256]. Image separation was performed in the same way as for the 30-kv tests. Film blackening, however, was remarkably increased by cementing a flexible intensifier screen (Auer HV) on the drum surface.

An interesting field of application of cineradiography is in electric arc welding, where material transport phenomena can be visualized. As reported in a detailed monograph by Conn [257], optical cinematographic methods have already been extensively applied in this field. Optical records, however, are less able to give information about material transport because of the very bright self-luminescence of the arc. Furthermore, it is of course not possible to use optical recordings for studying the techniques of welding under powder nor is it possible to look through the cover when covered electrodes are used. In such cases x-ray cinematography gives valuable supplementary information.

Film transport speed ($\sim$ 100 meters/sec for an exterior film on a drum) limits the product of image width and frequency. x-Ray flash frequencies achieved today (10,000–12,000/sec) make it necessary therefore to reduce the image width to 8–10 mm. Further work in this field will therefore require other optical and electronic arrangements, e.g. one might photograph a fluorescent screen. It would seem that a particularly valuable technique would involve the application of an image intensifier [258] (See Chapter F section 1g).

When one uses image intensifiers the size of the radiated object can be the same as the diameter of the cathode. Intensifier tubes with diameters of up to 10 inches are available.

A review of modern developments is given by H. Schardin and R. Schall in [881].

The application of x-rays to nondestructive testing is described in a paper [884] that was presented at a conference organized by the Institution of Electrical Engineers, which was held in London on Nov. 8–10, 1961.

Früngel and Alberti [900] describe a commercial high-speed x-ray cinematography equipment capable of up to 20,000 fps. It takes 10 min to change the anode-cathode electrodes in this tube system, which includes pumping necessitated by vacuum.

3. Capacitor Discharges in Feeding Ion and Neutron Sources

a. *A Simple Pulse Neutron Source Based on Crossed Field Trapping*

A simple pulsed neutron source has been constructed by J. D. Gow and L. Ruby [269] consists of an anode formed by a cylindrical shell, permanent magnet, and two disc cathodes. One of the cathodes forms the neutron producing target and the device is exhausted adjacent to the other. In operation, the anode is given a positive pulse at 140 kv for 1 μsec at up to 200 pulses/sec. Continuous operation has also been achieved in a slightly different configuration. Ions are produced adjacent to the anode in a toroidal trapping region which exists as a result of the crossed electric and magnetic fields. The ionization is enhanced by the action of electrons reflecting through the inside of the magnetic shell. As a consequence of this trapping, large fluxes of soft x-rays are created which may be injurious to personnel.

On the basis of the experimental observation by J. D. Gow *et al.*[609] that certain configurations of crossed electric and magnetic fields efficiently trap and accelerate charged particles, a particularly simple pulsed neutron source has been constructed. The device consists only of a permanent magnet encased in a stainless steel shell which constitutes the anode, and two disc electrodes which constitute the cathodes. The anode is given a positive pulse and the cathodes are maintained at ground potential. The cathode has fastened on it a neutron producing target containing occluded deuterium or tritium. The other cathode is similar except that it has no target. It has, however, several holes spaced around the periphery for the purpose of exhausting the system, and one small opening connected to a supply of deuterium gas. It has a 3 inch industrial glass elbow which joins this cathode with a metal mercury diffusion pump. Two pieces of 3 inch industrial glass pipe, each 6 inches long, form the insulating walls of the system. They will stand off, in dry air, up to 140 kv peak. The glass pipes make O-ring seals onto brass flanges at each end, and the whole structure is bolted together with Bakelite studs. Recently, these have been replaced with threaded nylon rods, which have less tendency to shear when tightened.

A sketch of the magnet and magnet case is shown in Fig. F3a-1. The presence of soft-solder joints exposed to the plasma causes no difficulty. The projections of the stainless steel at each end constitute shields which minimize heavy electron bombardment of the glass walls. Indox V ceramic magnets in the form of face magnetized rings were chosen, as they are capable of a particularly high residual field. The resultant field distribution has been described elsewhere [610] in another connection.

The anode is pulsed by means of a high-voltage transformer which is driven by a pulse-forming network. The lower limit on the pulse width is about 1 μsec, as this is the characteristic buildup time of the discharge. The upper limit is determined for this system by the external high-voltage standoff, but a modified configuration has been run both with long pulses and with dc. The upper limit of the pulse repetition rate is limited by heat dissipation in the device. With a 1 μsec pulse, satisfactory operation has been achieved over a long period at up to 200 pulses/sec.

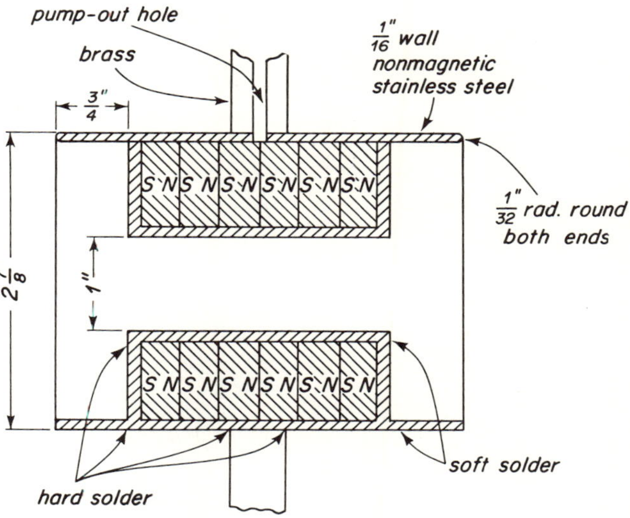

FIG. F3a-1. Sketch of anode assembly.

As a circuit element, the device presents an impedance that is a function of both voltage and deuterium pressure (see Fig. F3a-2). Below about 9 μ of deuterium pressure, essentially no current is drawn as a result of pulsing the anode. However, at this value an abrupt change occurs, resulting in 1 amp or more of tube current, which increases either with additional voltage or with additional pressure. The upper pressure limit is determined by internal arc breakdown, which sets in at progressively lower pressures as the high voltage is raised. Representative operating parameters are 125 kv peak, 15 amp tube current, and 12 μ of deuterium. A significant experimental observation is that, irrespective of the type of gas used, the operating range as measured by an ion gauge is the same. That is, the operation of the device depends only on the degree of ionization produced and not on the absolute pressure.

Table F3a-1 lists some neutron yields for the $H^2(d, n)He^3$ reaction. Recently the device has been situated at the Livermore Laboratory, where neutrons from the $H^3(d, n)He^4$ reaction have been employed in critical assembly experiments.

In order to obtain stable operation on a steady state dc basis, it was found necessary to increase the anode-cathode spacing, and to install water cooling on both electrodes. Such a device has been constructed by using a 2 ft length of 4 inch i.d. industrial glass pipe as an insulator. Yields in excess of 10^7 neutrons/sec have been obtained from the reaction $H^2(d, n)He^3$ at 100 kv and 3 to 5 milliamp.

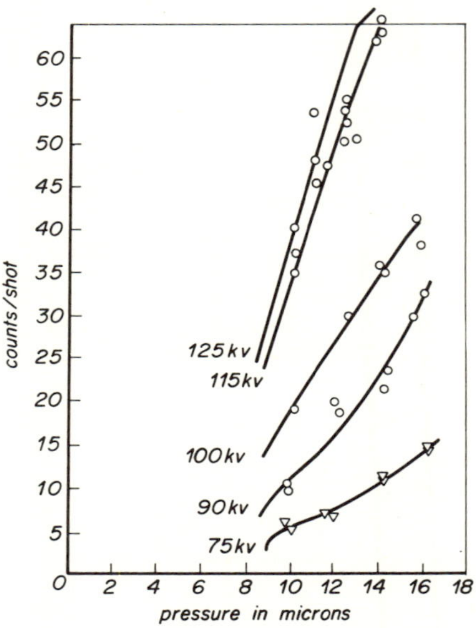

FIG. F3a-2. Yields vs. pressure at various voltages.

The theory of operation of this neutron source evolved from the study of several other crossed field, ion trapping devices. Consider the region adjacent to the side of the anode corresponding, for example, to the north magnetic pole. An electron born in this region is accelerated toward the anode by the electric field. The radial component of the magnetic field, acting on the axial velocity thus acquired, deflects the motion toward the positive azimuthal direction, finally returning the particle to the axial plane from which it originally came. The force due to the axial component of the magnetic field on the azimuthal component of velocity is directed toward the axis, and results in a composite motion in the form of cycloids which encircle the axis. During these excursions an ionizing collision can occur which releases another low-energy electron and moves the average axial position of the first electron slightly toward the anode. Thus electrons can be trapped and multiplied in such a configuration of fields.

Actually, the foregoing explanation is highly oversimplified. Integration of the equations of motion with an analog computer has shown certain conditions to be required for trapped orbits. In particular, from the values of electric fields required, it can be said that the above mechanism is principally effective in first establishing a region of intense ionization, which results in a radical alteration of the electric potential distribution. Thus, after a microsecond or two, when the discharge has become stable, practically all the potential drop occurs within about 1 cm of the cathode. Electrons on the axis of the device are not trapped by the mechanism described but are accelerated into the throat of the magnet. Emerging on the other side, they are reflected by the decelerating electric field, so that the magnet throat region becomes another source of intense ionization. This second region merely augments the ion current reaching the targets, and, in fact, if the throat is plugged, both sections of the device operate independently. Other advantages of the double-ended design over the simpler configuration are the avoidance of the intense bombardment of the anode by an axially collimated electron beam, and the possibility of maintaining the pump-out flange as well as the target flange at ground potential but at opposite ends of the device.

Because of the existence close to the anode of a region containing trapped electrons, i.e. about $10^{10}/cm^3$, with average energies in the kilovolt range, this type of device is an intense source of soft x-rays. Within a meter of the apparatus, the flux is dangerous to personnel at repetition rates as low as 1 pulse/sec. When the device is operating with 100 kv potential difference at 3 to 5 milliamp, radiation levels of about 80 r/hr are observed at 8 inches from the glass tube. Electron bombardment also produces radiation damage to the glass insulators, as evidenced by discoloration and finally by crazing of the inside surface, especially in the area adjacent to the anode.

One has found that it is not possible to use this type of neutron source for energy calibration purposes when it is desired to obtain 14 Mev monoenergetic neutrons from the reaction $H^3(d, n)He^4$. Because the tritium target is bathed in a deuterium atmosphere at least a thousand times as dense as is

TABLE F3a-1

SOME NEUTRON YIELDS FOR THE $H^2(d, n) He^3$

Yields from $H^2(d, n) He^3$ (neutrons/μsec)	Peak voltage (kilovolts)
0.8×10^3	70
2.2×10^3	85
4.8×10^3	95
8.1×10^3	110
16.6×10^3	125
23.7×10^3	140

common in deuteron accelerators, progressive exchange of the tritium has been found to take place. The magnitude of the effect is easily observable as a gradual degradation in yield and as a dilution of the energy spectrum with neutrons of 2.5 Mev obtained from the reaction $H^2(d, n)He^3$. This effect should not be confused with the usual gradual loss in yield associated with use of tritium targets in accelerators, which arises from the buildup of an oil film on the surface of the target.

b. *A High Output Ion Source* (*Duoplasmatron*) [752]

The Duoplasmatron is a high output ion source (with approximately 1 amp peak ion current) with auxiliary magnetic field developed by von Ardenne [752].

This source is capable of delivering proton currents of nearly 100 milliamp in continuous operation, at an acceleration voltage of 60 kv and electrostatic concentration of the beam, and is suitable, for example, for installation in accelerators constructed for ions entering in linear or tangential direction. The outstanding advantages of this source are its high efficiency of ionization, which approaches 1, and its relatively low discharge current.

At discharge currents of more than 7 amp in pulsed operation the source delivers ion emission currents of 1 amp max. Extraction and focusing of the beam require special ion optical systems, which have to be designed in accordance with the unit in which the source is intended to be used.

In the system an acceleration voltage of 35 kv max is employed at a pumping speed of approximately 400 liters/sec (at the ion-optical system). Data resulting from such operation are presented in the first column of Table F3b-1.

TABLE F3b-1

TECHNICAL DATA OF DUOPLASMATRON ION SOURCE

Gas discharge voltage	100 volts	approx. 100 volts
Gas discharge current	1.5 amp	0.6–3 amp
Magnetic excitation	500 amp-turns	1250 amp-turns
Diameter of the extraction opening	0.7 mm	1.2 mm
Cathode heater power	approx. 500 watts	approx. 500 watts
Ion acceleration voltage	35 kv	approx. 60 kv
Ion beam current	approx. 12 milliamp	approx. 100 milliamp
Efficiency of ionization	—	50–80%
Hydrogen consumption	approx. 20 cu cm $\times$ h^{-1}	20–50 cu cm $\times$ h^{-1}
Pressure near the discharge chamber	3.5×10^{-2} mm Hg	$2\text{–}5 \times 10^{-2}$ mm Hg
Pumping speed of the vacuum system (at the ion-optical system)	approx. 400 liters $\times$ sec^{-1}	approx. 2000 liters $\times$ sec^{-1}

The second column is data taken from "Tables of electron physics, ion physics and electron microscopy" Vol. I, by von Ardenne, pp. 544–557 [753]. These were obtained by using a vacuum system having a pumping speed of approximately 2000 liters/sec (in the ion-optical system) and enlarging the ion extraction aperture, and by increasing the magnetic excitation and the current of the gas discharge as well as the acceleration voltage.

c. *Neutron Production by Pinching Discharges* [2]

One of the most interesting effects observed in intense pulsed discharge in light gases is the appearance of penetrating radiations. In 1952, soon after experiments with pulsed discharges were started, it was found that at sufficiently high currents the discharge in deuterium becomes a source of neutrons.

The first experiments performed with the aim of studying this phenomenon showed that neutrons appear when the peak discharge current is 400–500 kamp and the initial deuterium pressure is about 0.1 mm Hg. The neutron emission was observed in a relatively narrow pressure range and its intensity rapidly increased with increase of the applied voltage, i.e. with increase of the peak current. In these first experiments the radioactivity induced in a silver target embedded in a paraffin block placed near the discharge tube served as the neutron detector. A possible explanation of this phenomenon was that the neutron emission resulted exclusively from collisions between the accelerated deuterons and deuterium absorbed by the electrodes on tube walls; however, control experiments did not confirm this explanation.

At the early stages of investigation it was quite natural to assume that the neutrons resulted from thermonuclear reactions in the plasma heated to a high temperature. This was exactly what was expected from the beginning and the fact that the effect was detected under conditions that completely corresponded to the theoretical predictions seemed to speak in favor of this viewpoint. The behavior of the neutron radiation (its dependence on pressure and current) observed in the first experiments qualitatively concorded with the assumption that the phenomenon was due to thermonuclear mechanism.

However, very soon serious doubt concerning the correctness of this assumption began to appear. This happened after it was found that neutrons can be observed at comparatively small discharge currents, on the order of 150 kamp. According to the initial calculations the intensity of the thermonuclear reaction at currents of 150 kamp should be practically zero.

In subsequent experiments the neutrons were recorded with a scintillation counter fed to an oscilloscope. This technique made it possible to find the relation between the discharge properties and the moment of appearance of neutrons. It was found that the neutrons were always emitted when the second kink appeared on the current oscillogram, i.e. at the moment when the plasma was subjected to the second contraction. No neutrons were produced during the first contraction.

The neutrons were always emitted as short pulses with a steep front. The rise time of the pulses was several tenths of a microsecond. The chief results of the oscillographic investigations were not consistent with the initial assumption that the neutron emission is the result of quasi-stationary heating of the plasma during which the temperature increases proportionally to the square of the current.

Further investigation yielded new interesting facts pertaining to the plasma neutron radiation. In particular, it was established that in specially designed discharge tubes the neutrons could appear at fairly high deuterium densities, as much as several tenths of a millimeter of initial pressure. This fact signified that the neutron emission was certainly not a trivial effect.

It was found that not only neutrons but also hard x-rays were produced in pulsed discharges. Penetrating x-rays were found when large currents were passed through deuterium by short spurts of discharge. The pulses due to the neutrons and x-ray quanta always appeared simultaneously on the oscillograms. The energy of the x-ray quanta produced in pulsed electrical processes in hydrogen and deuterium reached 300–400 kev. It is noteworthy that at the time of emission of such high energy quanta the voltage applied to the discharge tube was only about 10 kv.

Theoretical analysis of the complex phenomena that occur in the plasma of a pulsed discharge oscillating under the action of electrodynamic forces is still at such a stage that quite a number of facts remain to be explained. However, the general picture of the process is gradually becoming clear and some of the peculiarities of the phenomenon seem to have been sufficiently elucidated.

It is now clear that contraction and expansion of the plasma are not quasi-stationary processes characterized by equilibrium between the external and internal pressures.

In the equations describing the dynamics of a pulsed discharge the main term is that which accounts for momentum changes in the ionized gas due to magnetic pressure. Thus, the kinetic energy of the ordered motion should at some stages of the process greatly exceed the thermal energy concentrated in the plasma.

At the initial stage of the discharge the internal pressure in the plasma is very small; thus the electrodynamic forces produce acceleration along the radius toward the discharge tube axis. Thus, the work of the electrodynamic forces is not expended in raising the temperature but in imparting kinetic energy to the converging plasma layer. At this stage the discharge tube operates as a peculiar type of accelerator in which the particles are driven by the magnetic field. Since charged particles, irrespective of sign, will move with the same velocity, the kinetic energy acquired by the ions will be quite large. But the kinetic energy of the electrons practically will not change in virtue of the small mass of these particles.

From the viewpoint of gas dynamics, the contraction process should be considered as a phenomenon in which a cylindrical shock wave converging towards the axis is produced in the plasma. At first, the gas located before the inner wave front is not ionized. When the wave begins to move the gas is carried along together with the charged particles of the plasma and its atoms simultaneously become ionized. The amount of matter that begins to move gradually increases and the total amount of ions and electrons in the plasma increases rapidly. The duration of the contraction phase can be determined by calculating the velocity acquired by the contracting gas. It was found to be approximately proportional to $(M/V_0^2)^{1/4}$, where M is the mass of gas per unit length of discharge tube and V_0 is the initial voltage. This is exactly what one finds experimentally for the interval between breakdown and appearance of the first kink in the current oscillogram.

The final stage of cumulative contraction sets in when the plasma accelerated by the magnetic field reaches the axis. At this moment a great part of the energy of ordered motion changes into heat and the pressure and plasma temperature sharply increase. During maximum contraction, the plasma temperature may be on the order of 10^{6}°C.

The nature of the processes occurring during maximum contraction are not very clear, but after maximum cumulation a diverging shock wave should appear to drive the plasma toward the walls. Inside the outgoing wave there should be a rarified zone. Under the action of electrodynamic forces that tend to compress the current, the outgoing wave should be decelerated abruptly and a new phase of contraction should ensue. This stage differs from the first in that the density in the inner region of the discharge is small and the gas in this region is probably almost completely ionized.

As a result, during the second contraction, conditions are produced that are favorable for accelerating in the longitudinal electric field a certain group of ions and electrons located near the discharge axis, i.e. in the region in which the magnetic field is small.

One may note here a certain analogy with the accelerating mechanism proposed by Fermi in his theory of origin of cosmic rays. A plasma of high conductivity will move together with its magnetic field, and with respect to particles located in the inner zone it will be equivalent to a converging magnetic wall from which the enclosed electrons and ions repeatedly will be reflected, their energy increasing after each reflection.

Acceleration of ions and electrons in the longitudinal electrical field near the discharge axis is possibly the explanation of the appearance of neutrons and penetrating x-rays. The electric field strength during the second contraction may be very high. It can exceed the instantaneous external voltage applied to the discharge tube by many times.

However, it must be mentioned that not all in this acceleration mechanism is yet clear. Under certain conditions acceleration of ions in a longitudinal electric field may also be possible outside the central zone of the discharge owing to the presence of space charges. Some types of stability that are peculiar to the column may play an important role in accelerating particles in the plasma. In particular, one type of instability observed experimentally may be of importance for acceleration of electrons. It consists of spontaneous creation of a longitudinal magnetic field in the plasma as a result of spiraling of the plasma column.

If the second contraction is followed by a few more oscillations of the plasma column, the acceleration of the particles may be repeated several times. Experimentally not more than three successive oscillations have been observed. A possible explanation of this is that the plasma may begin to interact with the discharge tube walls with the result that the wall material begins to evaporate and appreciable amounts of foreign gases appear in the volume.

Here have been considered some features of the phenomena that accompany the passage of intense pulsed discharges through rarefied gases. The success of further work in this direction will greatly depend on the possibility of creating conditions under which the plasma column will experience multiple oscillations during buildup of the current without coming into contact with the walls. However, there are serious reasons to believe that this cannot be achieved.

G. Conversion of capacitively stored energy into heat

1. IMPULSE WELDING, DIRECT CAPACITOR DISCHARGE

The simplest form of a capacitive welding device is that for welding by direct capacitor discharge. The capacitor, 1, of Fig. G1-1 is being charged through resistance, 2, to a certain dc potential and can be discharged through a damping resistance, 3, inserted for energy control, and the electrodes, 4-4, so that the discharge energy is available at the electrodes. At the faces of the electrodes there are placed small metal parts, 5-5, which are to be welded together. The electrodes are kept apart by a lock pin, 8, but are urged into contact by a spring, 6, held by cap, 7. When lock pin, 8, is disengaged, the electrodes speed toward each other, and, shortly before they make contact, a spark of the capacitor discharge crosses between their faces. The time constant of this discharge occurrence can be variably set at 3.

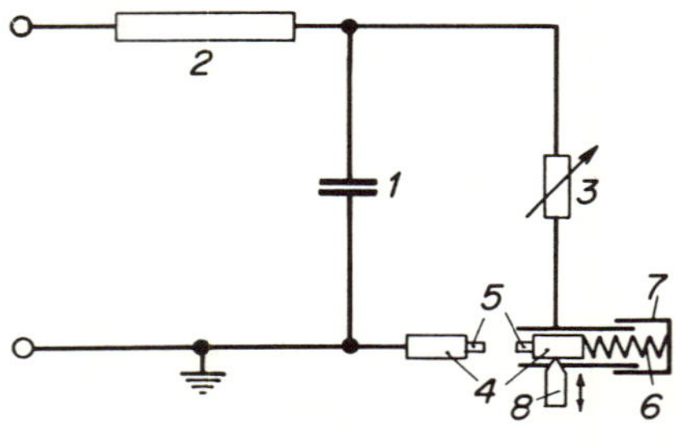

FIG. G1-1. Arrangement for impulse welding with direct capacitor discharge showing the weldment parts to be "shot" together. 1–energy capacitor; 2–charging resistance; 3–Current limiting resistance; 4–work piece holder; 5–work piece to be welded; 6–compression spring which speeds the masses 4–5 together faster than the time constant of 1–3; 7–spring cap; 8–electrode locking and release pin.

In this action, the cathode at least of the welding spark gap is extremely heated and even partly evaporated, and thus through rapid approach of the weld pieces, 5–5, toward each other, welding takes place at the point of their contact. This very simple method is, at an operating voltage on the order of a few hundred volts, suitable for welding of very small parts difficult to weld by other methods, such as welding together molybdenum wire and thinnest tungsten or hard metal sheets, foils and so forth. A grave disadvantage of this method is that the work pieces cannot be in contact before actual welding takes place, so accurate relative positioning is made difficult. In some special cases, such as spot welding of thin wires to aluminum foils, it is possible to insert in the discharge path an ignitron that initiates the current pulse. The efficiency of such direct capacitor discharge welding is very low. The resistance of the welding pieces between the electrodes can be measured in fractions of 1 milliohm. The peak current magnitude that can be reached in the discharge circuit with ignitron switching is, depending on the rating of this tube, only a few hundred amperes. At 1 milliohm and 1000 amp the voltage drop across the welding spot is only 1 volt. The capacitors, however, are good for several hundred volts, because the capacitor storage of energy, as is well known, requires more and more outlay the lower the voltage. The efficiency, owing to the voltage division, is from the outset less than 1%. In electrolytic capacitors one electrode is formed by the electrolyte and this enters in the current path as a considerable series resistance, which accounts for an even lower efficiency with electrolytic capacitors than with paper capacitors.

Such efficiency limitation applies fundamentally in every circuit in which a switching means is inserted in the discharge path in order to obtain a really controllable welding procedure. When, however, as in the diagram of Fig. G1-1, a switching means is dispensed with and the parts to be welded are "shot" together, the spark resistance between the metals is practically the only substantial resistance in the circuit. This resistance, depending on circuit dimensioning, can be in a range between 1 and 50 ohm, so that high spark resistance here becomes a thermally useful resistance leading to improved efficiency. However, it is regrettable that even in this case only a small portion of the spark heat becomes effective in weld heating, whereas the greater part is converted into useless ionization work, sound, and radiation, so that the thermal efficiency measurable at the electrodes even under this improved condition is not more than 5%. However, in special small weldments the efficiency may become of minor importance, because the advantage of reaching practically unlimited temperatures at the spark base might become a necessity for a successful weld not obtainable by other methods. For several variants of this method attention is called to references [44, 48].

Very extensive investigations about direct capacitor impulse welding have been conducted by Hofmann and Schumacher [44]. They studied, oscillographically, a particular case with a setup having 107 μf, 1500 volts charging voltage, 15 μh inductances, and a discharge circuit resistance of 0.1 ohm. In order to vary the duration of discharge an additional serial inductance of 0–250 μh was inserted in the circuit. The work pieces, chucked in a manner similar to a watchmaker's lathe and presenting a moving mass of 96 grams, were speeded together by springs having spring constants of 57, 172, and 850 ponds/mm. Hard metal wires up to 3 mm in diameter could be welded perfectly. The 172 ponds/mm spring was set under initial tension for 25 mm and work pieces were spaced 10 mm apart before the welding movement. During the sparkover that occurs shortly before actual touching of the work pieces, 3000 amp was measured in the discharge circuit. The volt-ampere load was thus 1500 volts $\times$ 3000 amp = 4.5 megavolt-amp. The pressure at the electrodes during the welding process is not of great consequence.

A totally different use of the spontaneous heat produced by direct capacitor discharge was found by the United States Army Signal Corps in a research project, the results of which were published in 1961, [901, 932]. The capacitor discharge is engaged to produce the extremely high pressure necessary for the synthesis of diamonds. A well-defined piece of graphite is heated to 3000°C., then the capacitor battery is discharged in about 1 msec., producing by heat a pressure of 125,000 atm. As a result, the graphite is converted into a fine crystalline mass of diamonds.

2. The Transformed Capacitor Discharge in Welding Engineering

The discovery that concentrated heating can be derived from stored electrical energy dates back to the last decade of the 19th century. E. Thomson [241] read a paper before a meeting of the Franklin Institute in 1887, according to which he discharged a battery of Leyden jars into a coil with the result of considerable heating; the ends of the coil winding, when brought in contact, melted together. Van Nostrand [242] reports practically similar results, and from E. de Fodor [243] it seems that the first practical applications took place in the works of the Thomson-Houston Company and of Alex. Siemens a short time later. These were actually the first experiments toward electrical resistance welding by stored electrical energy, that is, by capacitor discharges.

In this method, the welding energy as such is fed from a special energy storage capacitor battery and not, as in normal resistance welding devices, directly from the power line. The principle of separating the actual welding phase from the energy supply offers considerable advantages, provided that design and construction of the equipment are of highest reliability. Some of the advantages of the stored energy welding method are the prevention of heavy current surges on the power line, independence of line voltage fluctuation, heavy welding current if it is desired, fast welding time, and welds of consistently uniform quality. The equipment and its qualities desirable for practicing this method will now be briefly discussed.

For coordinating between the very low resistance of welding spots at adjusted and compressed metal surfaces, lying in the range between 10 μohms and 1 mohm, and for the most economical operating voltages for electrical energy storage means, lying in the order of a few kilovolts, an impulse transformer is advantageously employed (schematic diagram Fig. G2-1). In the early days [49] the performance of this transformer was considered of little consequence and normal power transformers were used despite their shortcomings in oscillatory behavior and high dc magnetization when employed

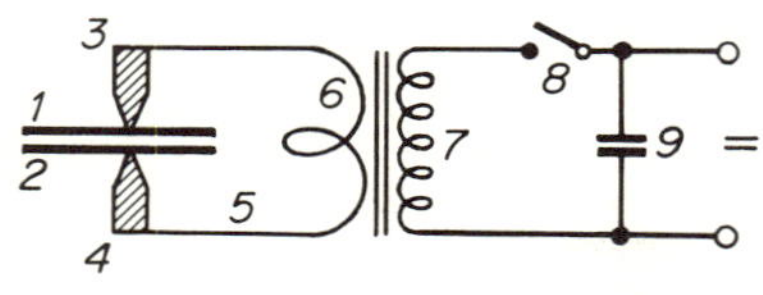

Fig. G2-1. Schematic circuit diagram for impulse welding by transformed capacitor discharges.

for transforming capacitor discharges. A particular machine on the market operated with capacitive energy storage in the range between 10,000 and 30,000 μf with a charging voltage of 200–500 volts,* that is, with an energy between 200 and 3750 joules. With this machine it was possible to weld sheets up to 2 mm at a rather slow discharge time. The low efficiency and mediocre welding results obtained from these first capacitor welding machines denied them general acceptance.

On the other hand, a line of famous pulse welders became very important in the field of high reliability welding in the range of small energies between 0.1 and 500 joules, e.g. for welding of micromoduls.**, †

The method which we will term "capacitor impulse welding" [50] is distinguished from the other methods referred to by the application of a closely coupled transformer having an open nonsaturable core as described in Chapter D, or even an air core. With such a transformer it is assured that the primary and secondary current and voltage values are so closely interlinked that the transformer in its action can be considered like a lever in mechanics, that is, only with respect to its transformation ratio. The construction details of such transformers have been discussed in Chapter D.

Whereas normally the welding energy is controlled by welding voltage and welding time, in capacitor impulse welding such control is obtained by the selection of charging voltage and capacitance of the storage capacitor. In this difference lies a considerable advantage of capacitor impulse welding over common resistance welding: Joule's heat produced by resistance welding in the weld is dependent on the product $U \times I \times t$, or $I^2R \times t$, where U is the welding voltage, I is the welding current, R is the resistance of the welding spot, and t is the welding time. Thus, the particular resistance of the welding spot participates in fixing the welding heat and can, if it varies from weld to weld—owing to resistive differences of surface structure, for instance—cause serious variations in weld quality. Such variation in quality is impossible in capacitor impulse welding, because the welding energy taken from the power line—that is, the charging energy of the storage capacitor—is constant. Also, the product $U \times I \times t$ is primarily constant on account of the practically uniform discharge condition in the discharge circuit including impulse transformer. Varying the resistance of the weld material in the welding spot effects a change only in the adaptation of the discharge circuit, or, in other words, the degree of approach to the critically damped condition in this circuit. (See Figs. G2-2, 3.) However, even with slight variation in that respect, the welding energy supplied to the weld remains substantially the same. Practical experience has shown that variations in welding spot resistances of the order of 1:5 still render practically uniform weld quality in impulse welding, whereas in resistance welding with welding time control, resistance variations of the order of 1:2 cause trouble.

* Philips, Eindhoven.

** Weldmatic, Monrovia, California.

† Micromodules are complete electronic circuits of very small size specifically designed for airborne equipment.

Besides uniform welding results, the impulse welding method offers further advantages: In contrast to other welding methods, the welding energy is taken from the power line during the intervals between welding actions, since the storage capacitor is charged during these intervals. This charging, when compared with the welding impulse duration, takes place rather slowly and without causing disturbing current surges in the power supply (relationship of welding duration to charging duration, e.g., 1:100). In heavy equipment installations it is moreover possible to connect the power supply to a 3-phase line to obtain completely symmetrical phase loading and a power factor of practically 1. Because the storage capacitor charges itself each time up to a preset voltage, uniform welding energy is always assured in impulse welding in spite of power line voltage fluctuations; only the charging time is influenced by line voltage conditions.

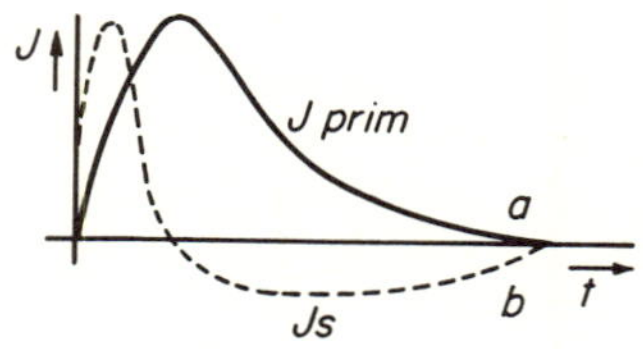

FIG. G2-2. Primary and secondary impulse current wave under ideal critically damped circuit adaptation ($R^2 = 4L/C$).

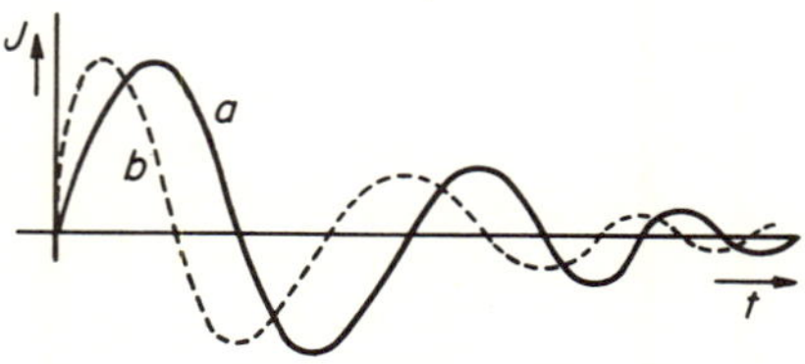

FIG. G2-3. Primary and secondary impulse current wave under bad adaptation (oscillation). (a) primary current; (b) secondary current.

Another advantage is the very short welding time of 1/50 to 1/500 sec at very high current magnitudes. During such short welding time, heat dissipation by heat conduction is of no avail and the heating zone is locally bound. In most cases, no tempering colors show even directly adjacent to the welding zone. Within the welding zone itself the grain structure is generally more finely crystalline than in the unwelded structure. This crystalline structure lends great strength to the impulse welded spots. The fact that the welding spot is locally closely limited offers the opportunity to weld sheets of different thicknesses together, for example, a sheet of 0.5 mm to a sheet of 10 mm, or the head of a bolt to a thin sheet, without noticeable changes at the back of the sheet.

Short welding times become of decisive importance in welding of good electrical and heat conductors, e.g. silver, copper, and aluminum. In the product $I^2 \times R \times t$, which determines heating, t must be kept small in order to avoid heat loss. Besides, in these metals the resistance R is low. Thus, to obtain sufficient welding temperature, higher current values must be employed, which fortunately enter with their second power into the product of heating. Since aluminum particularly has a narrow plastic state range, proper dosing of the welding energy is mandatory for achievement of good quality weldments in this material. This requirement can be met to a high degree by the impulse welding method in which an adjustable and successively repeatable, uniform welding energy is available and becomes effective in shortest welding times and at extremely high welding currents. Impulse welding is exceptionally suitable for welding of non-ferrous metals, in particular.

The highest efficiency in impulse welding is assured when the discharge wave form of the storage capacitor reaches the critically damped condition, i.e., when the oscillatory circuit formed by the primary of the impulse transformer and the storage capacitor is damped correspondingly by the loading of the secondary circuit (weldment).

Under such ideal adaptation conditions the welding spot resistance, as it becomes effective through the ratio of the transformer at the primary thereof, is the resistance necessary for establishing critically damped condition in the discharge circuit. The relationship

$$R\ddot{u}^2 = \sqrt{4L/C}$$

holds true for such damping, where $\ddot{u}$ is the transformation ratio, L is the resultant stray inductance of the transformer, and C is the capacitance. Thus, if the resistance of the welding spot is known, it is possible, by proper

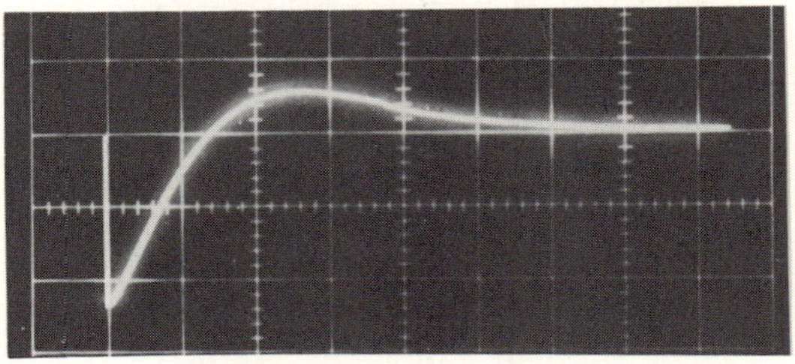

(a)

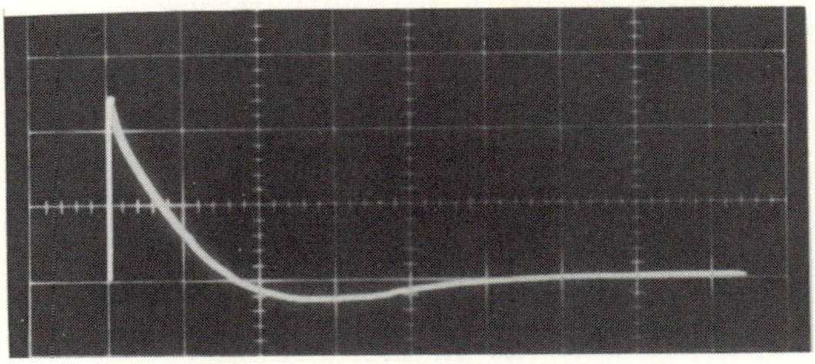

(b)

FIG. G2-4. Oscillograms of the primary and secondary voltage wave at the impulse transformer. (a) Primary discharge voltage, capacitance 240 μf. $U = 1$ kv/div, $t = 4$ msec/div (b) Secondary discharge voltage of the welding electrodes, $U = 1$ volt/div, $t = 4$ msec/div

selection of *ü* and *C* (and to a certain extent also of *L*), to obtain a favorable adaptation. The oscillograms in Fig. G2-4 show the primary and the secondary voltage wave. Whether the fact that in impulse welding the welding current rises rapidly to its maximum value and then decays in accordance with an exponential function (whereas in normal resistance welding it fluctuates about a uniform average value) is beneficial to the weldment remains to be investigated more thoroughly, but from inspection of the welding results a favorable influence can be presumed. The critically damped condition of the capacitor discharge not only results in the highest efficiency of conversion from electrical energy into heat, but also permits a less bulky and expensive design of the high-voltage insulation system in the storage capacitor. Since the capacitor charge does not change polarity during a critically damped discharge, the high-voltage insulation in the capacitors can have a much lower breakdown value than in capacitors for ac voltage.

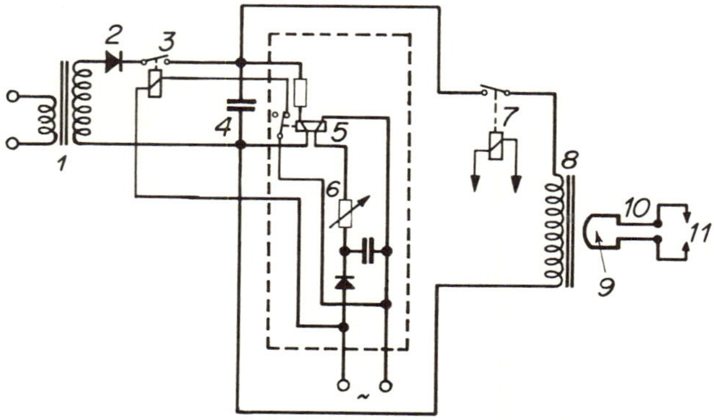

FIG. G2-5. Schematic circuit diagram of an impulse welding device operating with transformed capacitor discharges.

An impulse welding device consists of an automatic charging device and the impulse welding transformer. The charging device can be a mobile or a stationary unit and the relatively light and small impulse transformer can be connected by an extension cable to the charging device and can be located directly at the welding operator's stand. Fundamentally it is possible to supply several welding transformers from one charging device, but it is then practicable for each transformer to have its own storage capacitor and discharge switch.

The operation of the device can best be explained with reference to Fig. G2-5. The large capacitor battery, 4, of from a few 100 to 1000 μf capacitance is charged from a high-voltage source that includes the high-voltage transformer, 1, the rectifier, 2, and the magnetic contactor, 3. The charging voltage can be adjusted by a small control device, 5, that includes a sensitive relay and an auxiliary rectifier.

The voltage is adjusted once for a particular job by a rheostat, 6, arranged in control device, 5, for such predetermined value as is suitable for the particular welding job conditions. The magnetic contactor, 3, permits charging of the capacitor battery, 4, until the voltage set at, 6, is reached. After this, contactor, 3, switches off the capacitor, 4, which is charged to the desired potential and is ready for action. For the welding action, contactor, 7, is now closed to establish the path from the storage capacitor, 4, to the primary of impulse transformer, 8, through which the capacitor then discharges. Contactor 7, is of special construction and can switch on heavy loads at high voltage. In order that no energy can be consumed by a switching-on spark, the contactor is actuated rapidly by a capacitor discharge through its energizing coil, whereby copper contacts effect an ignition delay at closing of the contacts. Switching off causes no difficulty, because then the storage capacitor is already empty and voltageless. Thus switching off of heavy currents does not occur here. From the secondary, 9, of the impulse transformer there leads a sandwich line, 10, to the pair of electrodes, 11, between which the welding load is located.

The rapidly changing current flow occurring during capacitor discharge in the primary circuit can, of course, be transformed by a correspondingly rated impulse transformer, whereby the secondary becomes the continuation of the primary current. Figures G2-2, and G2-3 represent the current wave at good adaptation (critically damped) and inferior adaptation (e.g. wrong transformation ratio or excessive circuit inductance), respectively. In principle it is possible to operate with impulse transformers having no iron core. Because of the rapid current rise at the beginning of the discharge, an iron core becomes oversaturated and then, of course, the transformer performs more or less like an air core transformer anyhow. Besides the light weight of an air core transformer, there are other advantages that in some cases might enter into consideration.

Devices for practical use, however, are usually built with an open or even closed iron core having an air gap, because with such transformers a better adaptation such as set forth above can be obtained.*

Very essential for a good impulse welding transformer is an extremely close coupling between its primary and secondary winding, which is achieved, preferably, by making the secondary consist of a single compact winding turn that encloses the several sections of the primary winding completely on all sides. The insulation between the primary and the secondary must have thereby an insulation value of about 10 kv, since it is necessary that the secondary be grounded for safe operation of the welding circuit. Provisions are generally made to change the transformation ratio of the values 600:1, 300:1, and 150:1.

In designing impulse transformers, particular pains must be taken to

* The machines of Raytheon Mfg. Co. Waltham, Mass., and Weldmatic, Monrovia, Calif., use closed iron cores of highly magnetizable material (e.g. Hypersil at 18,000 gausses); the larger types of FF Impulsphysics Corp., Impulsa and Vitessa, use open iron cores, because the transformers are working as air core types while the iron damps only the first steep current rise.

reduce the inductance of the secondary circuit as much as possible so that this circuit does not act as a choke against the high rate of change of impulse current. Inductances below 1 μh must be reached if the efficiency is not to suffer markedly. For this reason the conductors for supplying the current to the welding electrodes are usually in the form of relatively thin-walled, closely combined band conductors which are separated just as much as needed for the necessary insulation and depart from each other only adjacent to the electrodes in order to make the connections thereto. A more favorable conductor and electrode arrangement can be achieved by double spot welding, in which the counterelectrode is formed by a passive conductor without lead connection (Fig. G2-6).

Table G2-1 shows technical data of various impulse welding machines. Of particular interest are welding times, welding energies, and peak currents, as well as the rate of welds, which are typical design properties.

Of chief importance in impulse welding is the electrode pressure, which generally must be higher than in conventional resistance welding. During the welding process the material becomes liquefied very rapidly. Under the electrode pressure, this causes compacting of the two pieces to be welded together. The pressure on the electrodes must be high enough to make the electrodes follow elastically and fast enough such a change in thickness as it causes the two work pieces to move closer together. Such a follow-through effect can

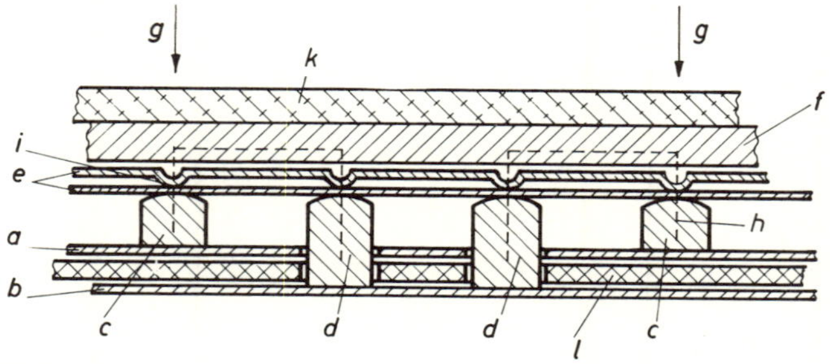

FIG. G2-6. Electrode arrangement in double spot welding. a, b–low-induction sandwich lines coming from the impulse transformer; c, d–welding electrodes mounted on the bands a, b; e–weld load, here two sheets of metal to be welded together; f–passive counterelectrode for current passing between c and d at the rear of the weld load; g–direction of pressure force which acts against rubber pad k and forms, together with it, a system of high mechanical natural frequency, which should be higher than the highest Fourier component of the impulse welding current; h–current paths; i–projections in the weld work piece to limit the area of current passage through the weld and to confine the weld spots later; k–rubber pad; l–insulation.

be achieved by a spring which has a high natural frequency and is under initial tension. Arranged directly above a welding electrode, it urges the electrodes to follow at least at the rate at which the material is being compacted by the progressing liquefaction. For the welding electrodes, any materials

TABLE G2-1

SPECIFICATIONS OF SOME INDUSTRIAL PULSE WELDING MACHINES

Manufacturer	Raytheon Mfg. Co., Waltham, Mass.						Peco, Munich		Weldmatic, Monrovia, Calif.						Impulsphysics, Dr. Früngel GmbH, Hamburg		
Type	18A	60A	225B	1100	1101	1103	FP 1K	FP 2K	1039 Dual range	1016	1026	1034	1027	1024	Impulsa	Vitessa Cabinet	Vitessa Table
Height (cm)	39	46	61	127	127	127	107	46	12	15	32	48	63	85	130	1010	80/130
Width (cm)	20	28	28	43	43	43	90	29	30	20	23	25	58	58	125	70	70
Depth (cm)	51	36	36	81	81	81	55	56	23	33	30	38	40	40	75	60	60
Capacitor energy (watts-sec)	18	63	225	1080	2700	8100	300	1000	3/15	40	80	160	250	500	2000	6000 (with aux. cap. 12000)	
Peak current (amp)	1700	3000	8000	16000	—	—	10000	30000	1500	2000	3500	5000	8000	10000	35000	70000	
Discharge-time welding (millisec)	3.3–10.8	3–17	3–17	80–32	—	—	2.5–5	2	0.5	1	1.5	2	2.5	4	2–20	ca. 2–20	
Capacitor capacitance (μf)	28	14, 28, 56	50, 100, 200	60, 120, 240	120	120	336, 500	80	200	600	1200	2400	3200	6400	30, 80, 250	90, 240, 750	
Charging voltage, adjustable (volts)	425–1150	650–1500	650–1500	1000–3000	—	—	300–1100	1000–5000	20–365	20–365	20–365	20–365	20–394	20–394	1000–4000	1000–4000	
Rate of welds per minute at max. energy	50	60	60	60	35	8	60	20	120	50	55	60	45	30	20	20	
at min. energy	250	120	120	300	300	300	3000	100	250	150	130	150	90	80	60	60	
Welding head Type	I	ISS	J	ISS	ILSS	H	—	—	1032MB	1032	1040 Butt-Welder	1037	1029	1038 Dual head	Impulsa head	Vitessa head	
Working pressure (kg)	0.5–5	0.25–7	0.5–5	0.25–7	0.25–1	20–250	20–250	20–250	2.5	7.5	5.0	25, 50	25	25	25–250	50–1000	
The machine equals in its output about a half-cycle machine of kilo-voltamp	2.5	5.8	15	60	350	470	60–120	500	—	—	—	—	—	—	100–1000	300–3000	
Welding voltage (volts)	10–40	10–40	10–40	10–40	10–40	10–40	5–20	5–20	1–18	1-18	1–18	1–18	1–18	1–18	3–12	6–24	

can be used that have proved to be suitable in conventional welding methods. Very little Joule's heat loss is caused in the transformer windings, welding electrode leads, and the electrodes themselves, so the efficiency is high and only minimal heating becomes apparent at the electrodes. Thus, in most cases cooling is either not necessary or (in heavily loaded installations) can be effected sufficiently by blowing air.

In impulse welding it is important to have an accurately limited contact area between the two pieces to be welded together, because only in physical point or line contact areas can a sufficient pressure per unit area and the necessary current density be reached, the latter by contracting the current path in such limited contact areas. A perfect weld is possible only with sufficient current density and sufficient pressure.

Furthermore, it is necessary to have the electrode contact area larger than the welding spot area in order that the current density in the electrode contact area does not become so high that there is danger of welding the electrode to the material. A relationship between electrode contact area and welding spot area of 3:1 has been found practicable. In a practical example of two crossing wires (e.g., each wire of 3 mm dia and an electrode diameter of 6 mm) this relationship has been well met. The contact area between the wires is practically a point and the contact between wire and electrode is on a line; however, as soon as a pressure is exerted on the electrode the point contact increases to an area of about 1 mm^2 and the electrode contact area, to about 0.5 mm $\times$ 6 mm $=$ 3 mm^2. When two sheets of metal have to be welded, it is advisable to provide one sheet with point or line projections (projection welding) which define and locate the welding spot or welding seam. For welding of stranded copper wires on metal sheets, the wire ends are first joined in a metal bead by drawing an arc, and this copper bead can then be welded by a welding impulse upon any sublayer. In butt welding (i.e. a rivet on a plate) the plate can be provided with a projection or the end of the piece to be welded on can be made conical so that a small contact area and therewith the necessary local heating, is assured. As the elements melt into each other during the welding cycle, the joint (e.g., between rivet and sublayer) expands again to the full cross section of the shank.

The energy stored in the storage capacitor is $E = \frac{1}{2}CU^2$ measured in watt-seconds, where C is the capacitance of the storage capacitor in microfarads and U is the charging voltage of the same in kilovolts. For example, 1000 watt-sec are required for melting 1 gram of steel. With an obtainable efficiency of 50%, it would be necessary to store twice this energy, that is, 2 kilowatt-sec, to obtain a weld nugget of 1 gram weight or 1/8 cm^3 volume. For welding of nonferrous metals, the welding energy must be even higher owing to the lower Joule's heat generated in the material of the weld, but this disadvantage is offset to a certain degree by the lower melting points of these metals. The welding energy and the electrode force necessary for a certain welding process can be determined most quickly by experiment and can then be used for all subsequent welds of the same kind. Depending on the required welding energy and output rating of the charging device, the rate of welds can be 20–60 per minute, but in seam welding substantially lower.*

As early as in 1943** the impulse welding process was successfully employed for high vacuum tight welding of electronic tubes and metal stabilizers in one operation and with a reject rate of about only 1%. Successful vacuum-tight welding of a seam 100 mm long with one welding impulse gives evidence of effectiveness, reliability, and quality of impulse welding. More recently, impulse welding has also been employed for mass production of steel wire mats for reinforcing concrete.†

Worth mentioning are the good welding results that have been achieved in precious metal and light metal welding, particularly in welding together different materials. One of the many industrial applications is welding of switching contacts upon their underlays. This is of importance from an economic viewpoint as it replaces the slow conventional riveting practice and improves the contact cooling condition in later practical use of the contacts.

The welding examples cited are performed between electrodes of normal copper. The polished sections show, in observation under the microscope, a very homogeneous, uniform structure in similar metal welding; but also with different metals joined in the weld, there can be observed a flowless contact line in the microsection and melting in each other of both metals (Fig. G2-7). Tensile rupture test samples proved the strength to be that of a real weld. In contrast to so-called cold-press welding, which also allows for joining two different metals by welding, no pretreatment of the two metals is necessary here. Thin oxide layers on the surface of aluminum and copper do not hamper the welding process, but rust spots do hinder the quality of the weld.

It is a fact that impulse welding machines require a higher investment and are more complicated than conventional spot welders, but it is often likely that higher initial cost can be compensated by reduction of rejects and savings in material, particularly in precious metal welding. Impulse welding is therefore chiefly to be considered for special welding problems, particularly for mass production or for welding results that cannot be obtained with other welding means.

The advantages that can be gained by the capacitor discharge impulse welding method can be summarized as follows: Low power line demand; possibility of uniform load distribution by 3-phase supply with good power factor (often the saving in comparison with the kva installation otherwise required, are higher than the first cost of an impulse welding machine); independence of the charging voltage from the line voltage by comparing and

* In the impulse welding machines of Peco, charging of the capacitor takes place during one half-cycle of line frequency, and discharge during the second half-cycle. In this manner, one welding spot can be produced during each cycle (i.e., 50/sec or 60/sec, depending on line frequency). Such a welding rate is necessary, for instance, in seam welding of thin spring material.

** Osram GmbH, Berlin, impulse welding machines of Geräteentwicklung Danzig GmbH, a former company of Impulsphysics Ass.

† Bau-Stahlgewebe GmbH, Düsseldorf.

(a)

(b)

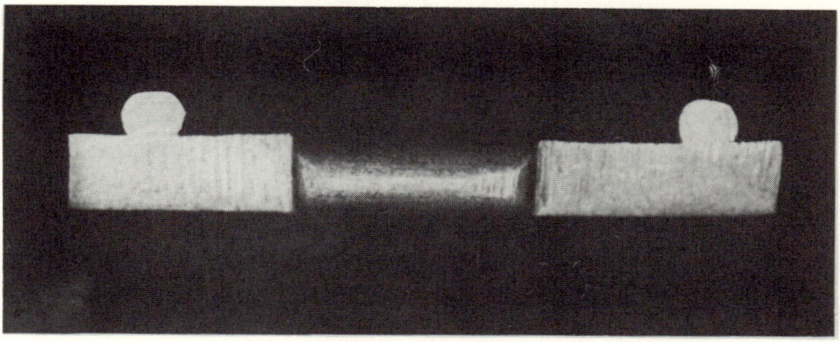

(c)

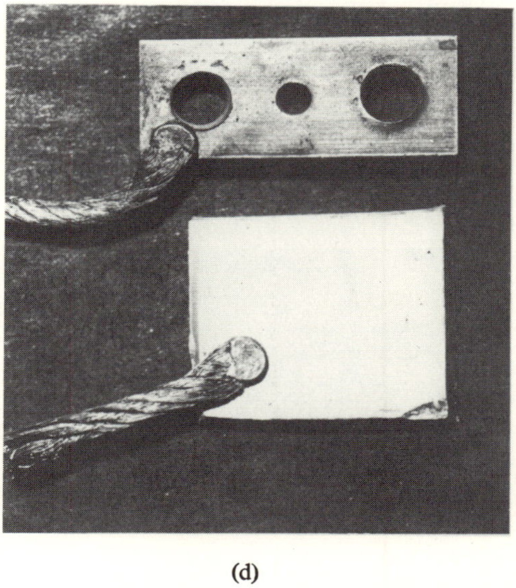

(d)

FIG. G2-7. Photographic reproductions of views and polished microsections of impulse weldments. (a) Al-Al weldment of two wires each having 3 mm dia. (b) Welding of a copper rivet upon bimetal. (c) Welding of two silver wires on a silver sublayer. (d) Welding of a 100-strand copper wire on copper (top) on aluminum (bottom).

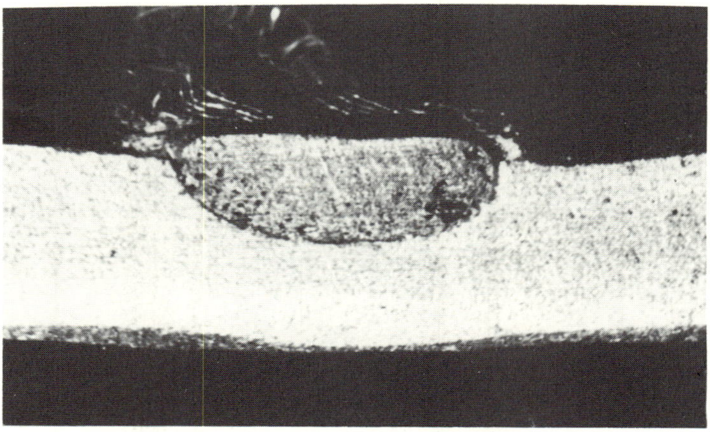

(e)

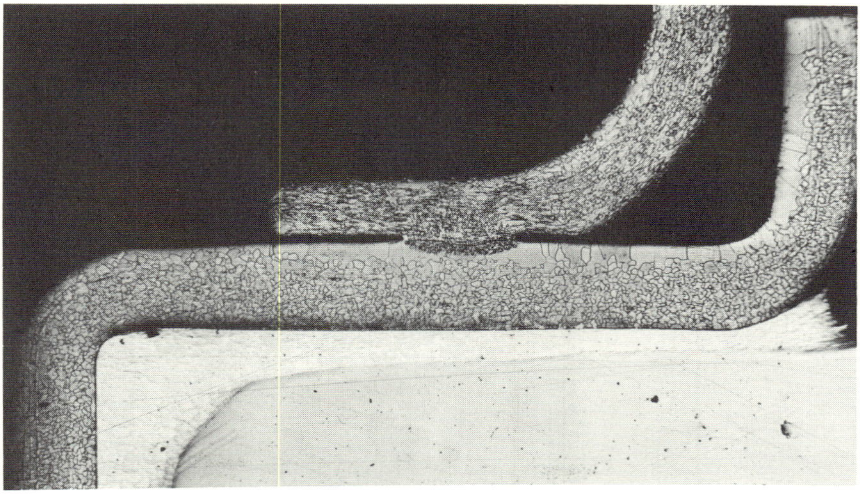

(f)

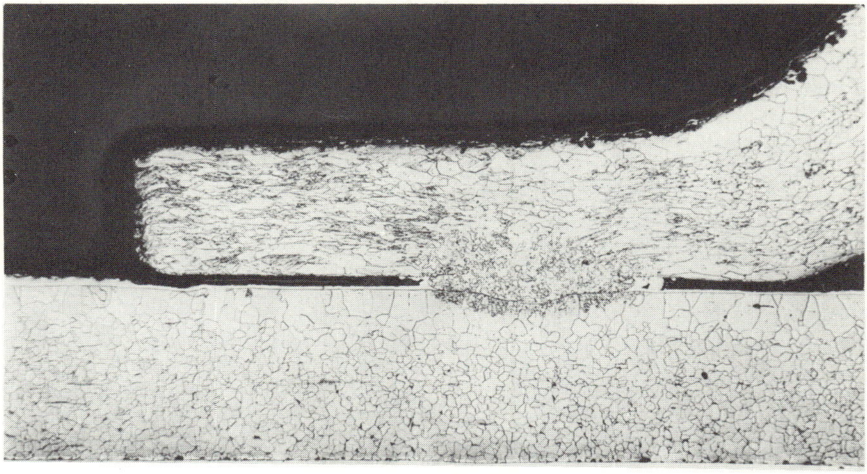

(g)

(h)

(e) Polished microsection of stranded copper wire welded to sheet aluminum. (f) Polished microsection of a vacuum tight welded ring zone of a transistor housing (kovar-Fe). (g) Enlarged zone of (f). (h) Heavy duty impulse butt welding of metal wings to an iron cup of an electromotor vent.

controlling the former with a stabilized voltage; and low total heating of the weldment, thereby sustaining fine crystalline microstructure therein, and making welding close to heat-sensitive materials possible. The surfaces around the weldments remain metallic since, within the extremely short welding duration, they have no time to oxidize. Work pieces of different cross sections and of different metals can be welded together if no metallurgical questions arise. Also good conducting metals, particularly contact metals, can be welded to a sublayer; however, proper cross section relationships must be strictly observed, viz., the contact areas of the electrodes must be larger than the contact areas of the welding spots. Still further advantages are rapid rate of weld production owing to low electrode heating and simple operation on account of the possibility—indeed necessity—automation. The net results are weld uniformity in mass production and a small percentage of rejects.

Figure G2-8 shows a schematic circuit diagram of small capacitor impulse welding machines rated at 1 to 500 watt-sec. A continuously adjustable toroidal control transformer, Tr1, having a very close coupling to assure rapid recharge, is used to set the desired voltage on the high-voltage transformer, Tr2. The full-wave rectifier, Gr1, charges the capacitor battery, C2-4, which can be selectively connected for three different values of capacitance. The impulse welding transformer proper can be switched by relay, M3 for two transformation ratios, e.g., 250:1 and 500:1, depending on the material of the welding stock. For release of a welding pulse, the relays, M1 and M2 are actuated simultaneously. relay, M1 disconnects the capacitor from the high-voltage rectifier for the duration of the welding action and, after interruption of charging, a normally closed contact of this relay closes the energizing coil circuit of relay, M2, which in turn closes its contacts and opens the path for the capacitor discharge to the primary of impulse transformer, Tr 3. By propeı subdivision of the capacitor battery in the capacitance relationship 1:3:10 and variation of the charging voltage up to

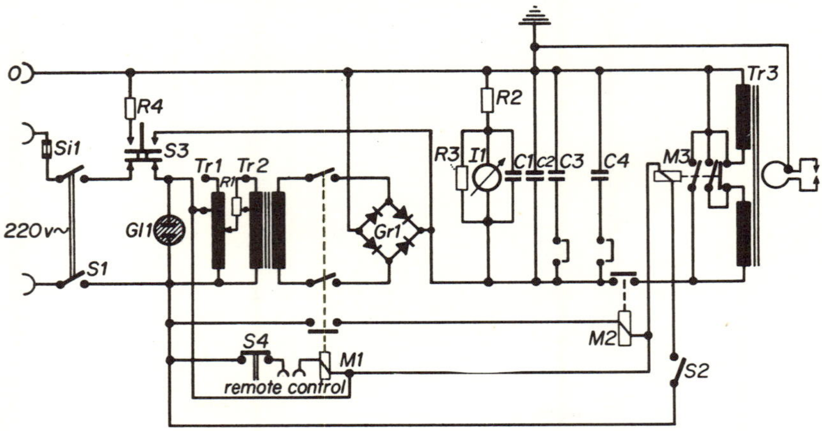

FIG. G2-8. Schematic circuit diagram of small capacitor impulse welding machines Impulse energy 1–500 watt-sec.

1:3, an energy control range of 1:90 is obtained. For extremely small welding objects, say, thermal elements of thin platinum-rhodium wires, it is possible to reduce C2 even more.

Another somewhat heavier machine on the market operating on the stored energy principle makes available impulse welding energies that are comparable to outputs of conventional resistance spot welders of more than 1000 kilovolt-amp rating. The energy per weld spot is variable between 140 and 6000 joules. Setting for desired weld energy is effected by change of storage capacitor connections (rough setting) and a precision control of the charging voltage (fine setting). Voltage stabilization means are included to be independent of line voltage fluctuations. Provisions are made to insert an additional capacitor for increasing the welding energy to 12,000 joules. About 30 welds/min can be made with the capacitors in the machine, and 10 welds/min with the external capacitor added.

The basic machine consists of two units, the mobile cabinet power pack containing line transformer, rectifier, storage capacitor, and switching means; and the work bench with impulse transformer, transformer control box, and, welding head. The connection between the two units is made by a coaxial cable of 10 meters maximum length. The maximum welding current is about 100,000 amp and the maximum welding voltage 12 volts. The charging rectifier is rated at maximum 20 kilovolt-amp, 3-phase.

A special form of this machine uses a so-called coaxial electrode system as described by Früngel [930].

The principle of pulse welding with transformed capacitor discharge found an expanded application in the welding of plastic-plated steel [869, 882]. If the rise time of the pulse current is shorter than 1 msec it produces so small a quantity of thermal energy on the electrodes on the metal surface of 0,5 mm thick sheets that on the plastic-covered back the temperature is too low to cause any damage in the plastics. Expanded sheets for covering houses or equipment can be connected in this way with suitable impulse welders. A review of all these applications is given by Alberti and Früngel in [787a].

In welding different types of metal, Sosoka [872] found that the polarity of the secondary pulse voltage affects the weld strength. Positive or negative polarity can vary the strength by a ratio of 1:2.

Another important application is the production of steel wire meshes for heating the ground-covering of airports [883]. By means of heavy current pulse welding, stainless steel wires become welded crosswise with standard thicker steel wires. The stainless steel wires have a high ohmic resistance and allow the application of reasonable ac voltages and currents, without any insulation, to heat the ground.

Last but not least, pulse welding has resulted in the highest reliability in the interconnections [897] of military guided missile systems.

The rapid advance of modern circuitry has focused increasing attention on the design problems of interconnections. At present, the weakest link in the chain of circuit reliability lies specifically within this area. The electronics industry is now at a crossroad in selecting an interconnection system which will fit the requirements of a rapidly developing new generation of sophisticated electronic equipment. Urgently required is a more efficient harnessing system for modules and miniature components in high density packaging through: (1) a substantial increase in system reliability; (2) strict limitation of mass and space allowances; and (3) the development of a technique for simple low cost production.

Worthy of consideration in fulfilling these needs is the newly developed system for module interconnections called the Universal Transmission Line.

With less space and smaller mass required for harnessing this application of impulse welding, it is ideally suited for application in airborne electronics. The small mass of the harness significantly reduces the usual mechanical resonance problems associated with heavy mechanical environments.

3. Electrodynamic "Cooling" by Particle Ejection in Capacitor Impulse Welding [246, 247]

In very heavy capacitor impulse welding machines, which convert up to 50,000 joules per impulse, impulse durations between 1/50 sec and 1/10 sec are encountered. At such loading the thermal burden becomes so high that a relatively coarse crystalline structure is created in the weldments, and thus their strength suffers.

In order to obviate such inferior weld quality, a heavy current of short duration is utilized in a manner mentioned previously. The magnetic field of this heavy current causes molten metal droplets to be blown off at high velocity. In this way, even with considerable quantities of molten welding stock or deep contraction of the joined pieces into each other, and with consequently large welding spot areas, a clean weldment can be produced. In addition to the magnetic blowing action of the current impulse, the molten welding stock is also pressed out of the welding zone by action of the electrode force. This action is modified by providing at least one of the electrodes with an elastic or plastic-elastic pressure-transmitting means arranged directly behind the electrode. By such means the mechanical natural frequency is raised to above 100 cps and mechanical inertia becomes so small that the electrode pressure continues to act steadily on the work pieces during the entire period in which molten metal particles are ejected from the welding zone. The welding circuit is preferably formed in the double-spot welding arrangement with current feed-in and return at the same weldment side and is constructed for extremely low electrode circuit inductance (below 0.1 μh) so that the welding current rises at the rate of about 10,000 amp within 1 millisec. The welding current path at the opposite side of the work piece in the double- or multiple-spot method is formed by a bridging electrode having no connection to the transformer but being in contact only with the welding stock. It is this electrode that is cushioned by the elastic or plastic-elastic material (rubber), whereby the electrode together with this cushion forms a mechanical oscillating system having a natural frequency above 100 cps. For operation at more than one welding spot with just as many welding transformers, one for each spot, all being fed from a common current source by a common current impulse, the electrode system with its feeding conductors is preferably designed so that the electromagnetic force field of the current pulse throws out the welding beads of all the welding spots in the same direction.

The improvement gained by this method lies, first, in the perfect appearance of the weldments with their smooth joints and, second, in that the two sections welded together, owing to the large melting or softening area, produce more extensive welding spots which in turn increase the strength of each welded joint.

Since the electrodes used in welding of wire mesh have relatively large masses and are good conductors, the heat taken on by the electrodes can be readily dissipated by radiation, conduction, and convection alone. Forced cooling by artificial cooling means is thus rendered completely unnecessary, mostly owing to the fact that molten welding stock is ejected in heat-carrying liquid droplets.

This method can in principle also be employed for the production of hammer scale or other metal oxides that are extraordinarily clean chemically, since molten metal is ejected in larger quantity than in other welding methods and is subject to very intense oxidation owing to the small droplets.

Some further insight into the principle of ejection cooling will be given by some schematic illustrations. Figure G3-1 depicts the welding transformer with iron core, 1, primary, 2, to be connected to a suitable power source, and secondary, 3, closely coupled to the primary and consisting generally of only one winding turn. The secondary is connected by conductors, 4 and, 5, to the

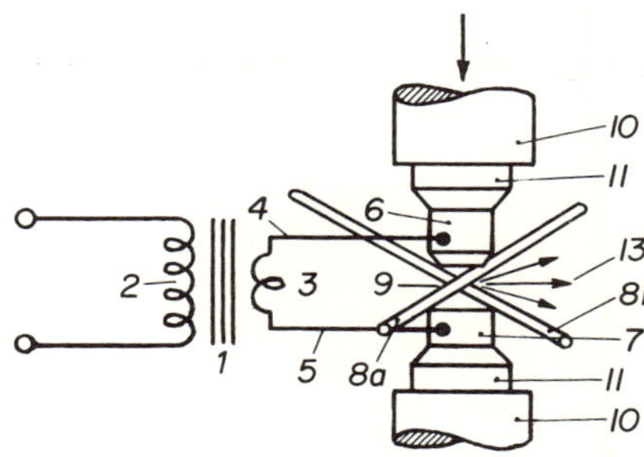

FIG. G3-1. Functional diagram depicting electrodynamic ejection of molten beads of welding stock from the weld in impulse welding.

electrode system which includes the electrodes, 6 and 7. The electrodes touch the welding load, 8a and 8b, shown as two wires crossing each other welded together at 9. The pressure between the parts to be welded and the electrodes is produced by a welding head, and the elastic system of the device is designated in Fig. G3-1 by numerals 10 and 11, 10 being the pressure jaws and 11, resilient cushioning material. Whenever a welding current impulse is passed through the secondary of transformer, 1, and thus through the welding circuit, the current traversing the work load melts the material in the welding zone at 9, and some of the molten material, subjected to the force created by electromagnetic action in the circuit, is ejected in the direction indicated by arrows, 13, in the form of welding beads. The quantity of ejected material depends thus to a large extent on the current condition in the welding circuit. Since such ejection was not considered beneficial in prior welding devices, only a small portion of the total quantity of molten material, stemming mostly only from the rim areas of the weld, was ejected; whereas the largest part would remain in the weld and, after interruption of current flow, solidify again. In order to achieve substantial ejection of molten material from the welding zone and thus considerable removal of heat from the weld it is necessary to increase the magnitude of welding current by means about to be described.

A welding circuit more suitable for obtaining large instantaneous welding currents is depicted in Fig. G3-2. In this arrangement, the secondary, 3, of transformer, 1, is connected to the main electrodes, 6 and 7, by the flat conductors, 4 and 5, which are closely superposed upon each other to reduce their inductance as much as possible. This arrangement is for operation according to the double-spot welding method, whereby the welding current flows successively through two welding zones formed between the work pieces, 8a and 8b, and an underlying work piece, 8c, to which the two former pieces are to be welded. The work pieces are shown as crossing wires, but, of course, parts of other shape also can be welded. In such an arrangement, in which positioning of the welding current conductors in the immediate vicinity of the welding zone is possible, the inductance of the secondary or welding circuit can be made very small, for instance, of an order lower than 0.1 μh. Experience has shown that low inductance is essential for success in a welding operation where rise to very large instantaneous welding currents for intensive electromagnetic force action is of primary importance. In comparison with the system depicted in Fig. G3-1—where the welding current must be fed to opposite sides of the work load and superposition of the conductors, 4 and 5, close to the welding zone is therefore impossible—the electromagnetic force achievable here is a multiple, usually several powers of ten higher. The favorable circuit condition here gives rise to ejection of molten material in the direction of arrows, 13, at a velocity generally of the order of more than 5 meters/sec. It can be seen that the welding circuit at the opposite side of the work load is completed through passive electrodes, 6a and 7a (electrodes not connected to the welding transformer), and a bridging member, 12. The jaws of the welding machine producing the welding pressure are indicated at 10, and the interposed resilient material at 11. Both the main electrodes, 6 and 7, and the passive electrodes, 6a and 7a, are shown resiliently supported, but in some cases it will suffice to have resilient mounting only on one side of the work load. The resilient members, 11, are indicated as a mass of rubber or the like.

In order that no current interruption is caused by the jerking action of ejection of molten material from the weld—for this would lead to a complete failure of the intended joint—it is important that the speed at which the work pieces are pressed together is higher than the velocity at which, by the amount of ejected material, a vacuum is created in the welding zone; or, in other words, the electrodes must follow fast enough to keep the metal surfaces of the parts

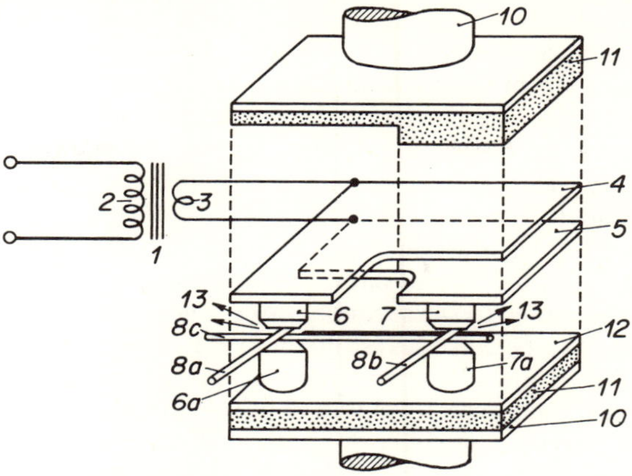

FIG. G3-2. Exploded view of low inductance arrangement of welding electrodes in impulse welding of wire mesh.

to be welded in close contact at every instant of the welding process irrespective of the quantity of metal ejected. Vacuum formation can be prevented only when the electrode system—comprising the main electrodes, 6 and 7, and the counterelectrodes, 6a and 7a, including bridging member, 12, in conjunction with the resilient mounting means, 11, which are preferably arranged directly adjacent to the electrodes—is designed for a relatively high natural frequency. It has been found, for instance, that 100 cps is sufficiently high in most practical cases. For best results, this natural frequency must be strongly damped. In achieving the natural, damped frequency of the desired

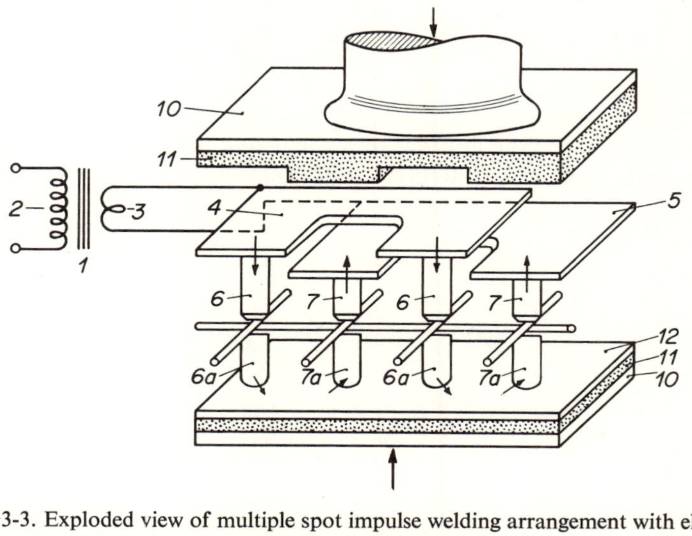

FIG. G3-3. Exploded view of multiple spot impulse welding arrangement with electrodes and low inertia welding pressure means, including rubber pad.

character, the resilient members, 11, play an essential role. An electrode system having a natural frequency of the order referred to above in conjunction with resilient pressure means has hitherto been unknown in welding technique, but it offers the particular advantage that, in addition to the ejection realized by electromagnetic forces, sudden squeezing-out of molten material from the welding zone takes place so that the liquid particles leave at even higher velocity.

The welding circuit shown in Fig. G3-2 includes only one pair of main electrodes in the welding current path, but a plurality of electrodes in parallel connection can be joined to the secondary of transformer, 1. Such an arrangement with two pairs of electrodes is shown in Fig. G3-3. The passive counter-electrodes, 6a and 7a, are all joined by the bridging member, 12, and the flow of current through the several welding zones of the work load will be in the direction of the arrows. This, of course, is true if we assume a certain polarity which can be changed without affecting the results. In Fig. G3-3 are shown notches in the flat conducting members, 4 and 5, and corresponding cut-outs in the cushioning means, 11, to assure that uniform pressure is applied to all electrodes alike. A modification of the conductor arrangement is indicated in Fig. G3-4, where the electrodes, 6, in electrical contact with conductor, 4, pass through clearance holes in the lower conductor, 5, and where the conductors are separated from each other by an insulating layer, 15. In this case the resilient member, 11, can have a plane face. The upper welding pressure and cushioning means are indicated in exploded relationship to show more detail of the electrode system.

The current source to be connected to the primary, 2, of transformer, 1, can be of any high-potential type, but it is advisable to utilize a capacitor discharge circuit as diagrammatically indicated in Fig. G3-5. Here a large storage capacitor, 16, is connected to be charged from a high-voltage power source of more than 2000 volts. High charging voltage is essential for gaining considerable advantage over normal power supply operation. The capacitor discharge circuit shown includes the conductors, 18 and 19, a switch, 20, interposed in conductor, 18, and the primary, 2, of transformer, 1. Switch, 20, is of the quick make type in order to prevent spark formation at its contacts. There occur only make but no break sparks at these contacts, since capacitor,

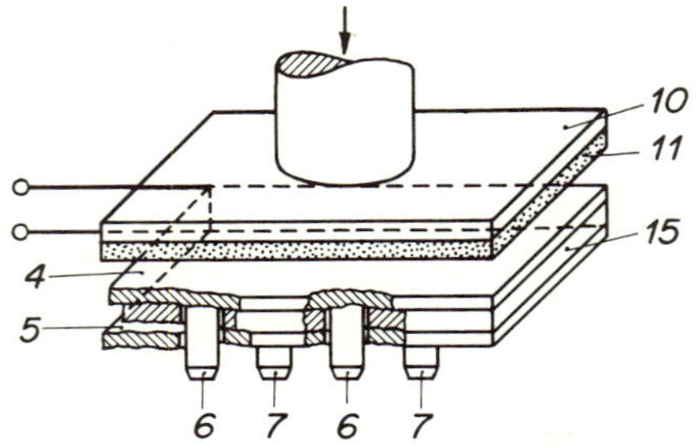

FIG. G3-4. Low inductance electrode arrangement for an even number of welding spots with one-sided current feed to the work pieces.

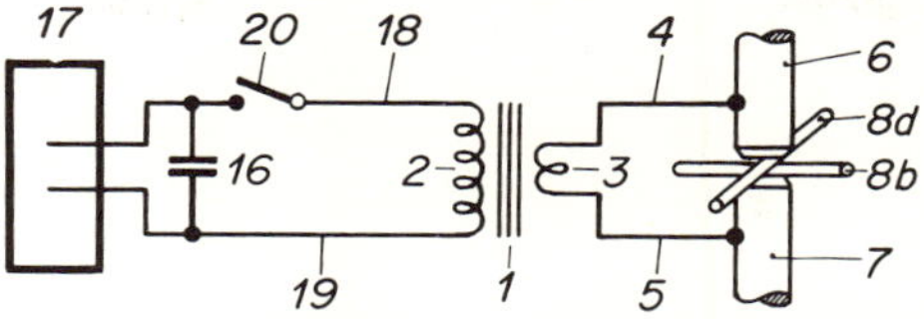

FIG. G3-5. Elementary principal diagram for capacitor discharge impulse welding.

16, is in completely discharged condition whenever the switch opens. The high-potential power source, 17, from which capacitor, 16, is charged preferably is provided with a voltage stabilizing device. Stabilized charging voltage is necessary for gaining the full benefit of an absolutely uniform effect by molten-metal ejection.

For uniform results of successive welding operations on similar parts it is furthermore of importance that the welding load conductivities have substantially similar values before initiation of the welding discharge pulses. An adjustment for a predetermined proper conductivity can be had by an arrangement illustrated in Fig. G3-6 where a motion- or pressure-dependent switch, 21, is interposed in a circuit including a power source, 22, and an energizing winding, 23, for actuating the capacitor discharge switch, 20, which in this case is magnetically actuated. As schematically shown in Fig. G3-6 switch, 21, will be closed when the pressure on the upper pressure jaw, 10, exerts a certain pressure on resilient member, 11, which compresses in proper relation to the pressure. By proper setting of switch member, 21, relative to jaw, 10, the predetermined pressure can be adjusted. Thus the welding load resistance can be reduced by sufficient electrode pressure, for example, to a value of 1/100,000 ohm, before capacitor discharge starts. Low welding load resistances of the order mentioned above assure that the welding current rises immediately after closing of switch, 20, to a magnitude necessary for effective forceful ejection of molten material and thus effective ejection cooling. By combined

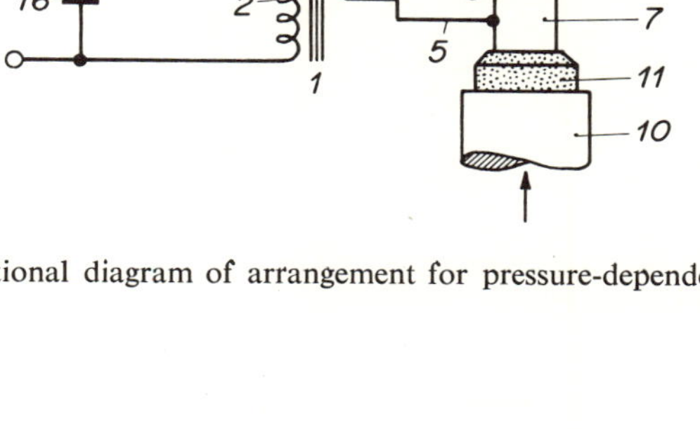

FIG. G3-6. Functional diagram of arrangement for pressure-dependent release of the discharge.

action of a properly designed welding circuit, an electrode system of proper natural frequency, a stabilized capacitor charging source, and a switching system set for proper welding load resistance, it is possible to obtain absolutely homogeneous, uniformly welded products, independent of supply line voltage fluctuation.

In a heavy impulse welding machine in operation, the uniform velocity of the electromagnetically ejected molten particles is surprising. Photographs have shown that they travel a distance of about $\frac{1}{2}$ meter during 1/25 sec, i.e., at a velocity of 12.5 meters/sec.

The strongest magnetic force fields of such impulse transformer welding machines occur at no load. They are noticeable throughout the whole vicinity and thus heat steel construction parts of the machine. Sensitive permanent magnets, such as those in measuring instruments, must be shielded when used close to such machines. In impulse welding on hardened objects, such as ball bearing races with inserted balls, it may become necessary to include a demagnetization step after the welding process; for instance, passing the object through a strong ac field.

In closing, attention is called to a certain analogy between impulse welding and impulse hardening [3]: For both purposes the energy needed for fast, extreme heating is stored in a charged capacitor. During the discharge, the electrical energy is converted in the work piece at high efficiency into heat energy for welding or hardening. Small impulse welding setups can be provided with an accessory device containing a high-frequency generator and can then be used for impulse hardening too.

4. Microinduction Hardening—Application of Instant Heating by a Capacitive Energy Impulse [244, 245]

The microinduction hardening process presents another utilization of capacitor impulse energy. The term implies hardening of thin surface layers or regions on surfaces of hardenable carbon steel construction elements that are to be subjected to friction. For such problems in surface hardening of steel, high-frequency generators are used as power sources. They are first accurately adjusted for energy quantity and duration, and are connected to a heater coil, the so-called inductor, of a configuration adapted to fit closely around the surface to be hardened. It is now possible to supply, particularly to very small work pieces, a predetermined quantity of energy for heating to a desired degree.

The difference with respect to the previously described impulse welding method, in which the ohmic heat is generated in a welding spot, is that here the energy impulse is converted by a high-frequency generator of high peak output into a high-frequency oscillation in the meter wavelength range. It is entirely feasible to employ several portions of an impulse welding machine, e.g., the whole capacitor battery with charging mechanism and switching means, in common for welding and hardening projects.

With the method described here it is possible to produce very thin hardened surface layers on delicate parts with all the advantages of the impulse welding method, such as line voltage independence, low reject percentage, and so forth. For induction hardening by means of high-frequency impulses, an accurately adjustable and measured quantity of energy is stored in a capacitor. The charged storage capacitor feeds—during the short duration of the discharge—a small UHF emitter which converts the electrical energy of the capacitor into induction heat for hardening of the work piece. Such a hardening impulse is very short but intensive. By means of such high-frequency impulses it is possible to harden, for instance, small gears or shafts for watches and other similar small parts after final finishing, without troublesome hardening distortions. Also odd-shaped surfaces of construction elements, often inaccessible for other methods, can be hardened by fast high-frequency impulses, previous to or after machine finishing or grinding, in spots where intense friction or impact wear is to be expected. The application limit of this method is reached when the cold masses of the work piece, which here actually effect quenching, become too small in comparison to the mass to be hardened. Impulse hardening can be applied on work pieces which in volume are smaller by one order of ten, than any that could previously be hardened by high frequency.

Inductance heat treatment of steel, particularly for surface hardening, has gained considerable acceptance in recent years. Surface hardening is proper in all cases in which construction element surfaces are subjected to hard wear. In view of previous surface hardening methods, which depend partly on chemical and partly on physical principles, inductive hardening in a high-frequency magnetic field with its exact energy dosing and its fast action as well—and, not least, owing to the fact that distortion-free work pieces are obtained—gains more and more favor. For a comparison between the micro-induction hardening method and known induction hardening, the principle of high-frequency hardening will be discussed briefly.

In inductive heating, the heating energy is imposed on the work piece through an alternating magnetic field around the output coil, which induces high-frequency eddy currents in the conducting work piece. In ferromagnetic materials the heat produced by hysteresis losses is added. Inductive heating increases, of course, with the increase of induced eddy currents. The rate of

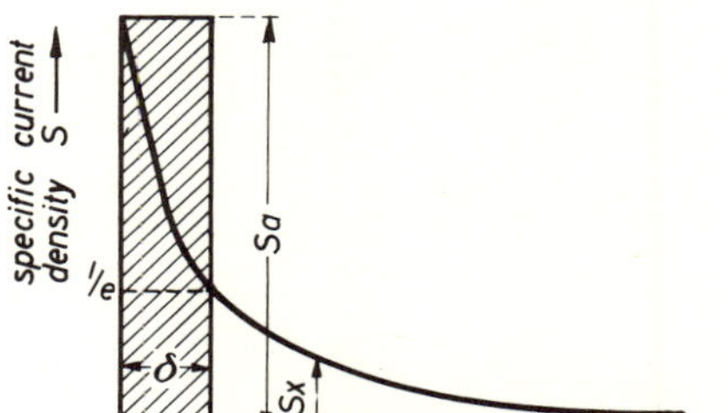

FIG. G4-1. Penetration of alternating current into metals. δ—Depth of penetration; S_a—current density at the work piece surface; S_x—current density within the work piece.

heating depends furthermore on the metal properties and space extension of the body to be heated. Advantage is taken of the facts that, due to the skin effect, the body heats only at the surface and that heat reaches its inner portions only by heat conduction. The eddy current density S is highest directly at the surface of the work piece and diminishes toward the interior in accordance with an exponential function (Fig. G4-1).

The eddy current density S_x at a certain depth measured from the surface is

$$S_x = S_a \times e^{-x/\delta}$$

where

$$\delta = 503\sqrt{\frac{\rho}{\mu f}}$$

In the formulas, S_a is the initial current density, ρ is the specific resistance (in ohm mm²/meter), μ is the relative magnetic permeability, f is the emitter frequency (in cps) and δ is the penetration depth (in mm) of the high-frequency current measured from the surface of the work piece. The penetration depth is directly proportional to the specific resistance and inversely proportional to the magnetic permeability; furthermore, the value δ decreases with increase of frequency. The values ρ and μ change with temperature. The value ρ rises with temperature; the permeability μ, which in steel at saturation assumes values between 100 and 300, becomes 1 above the Curie point (steel at 760°C), because then the magnetic properties disappear. The penetration depth in steel thus changes abruptly to higher values above the Curie point. Figure G4-2 shows the penetration depth of high-frequency current in dependence on the frequency for various materials, and Fig. G4-3, for steel. In view of the dependence of penetration depth on the frequency of the high-frequency current and the spreading of the induced heat by heat conduction, it can generally be said: If it is desirable to obtain a very thin hardened surface

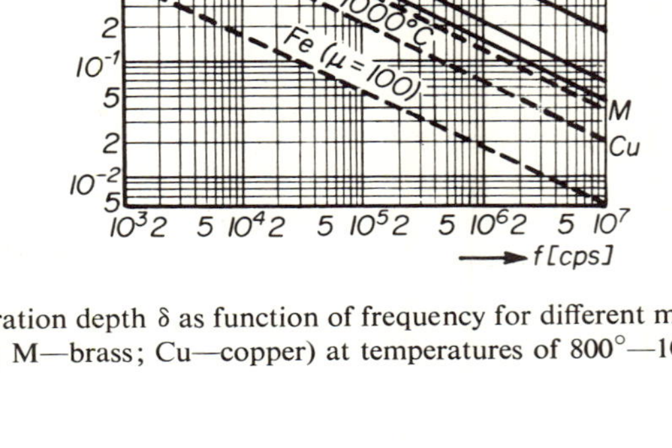

FIG. G4-2. Penetration depth δ as function of frequency for different materials (graph—graphite; Fe—iron; M—brass; Cu—copper) at temperatures of 800°—1000°C.

layer, it is necessary to employ higher frequencies and shortest hardening durations. This takes emitters of high output, which can give high energy values practically instantly.

In hardening processes, quenching is necessary immediately after heating. Distinction is made between external quenching (e.g. by a cold water spray) and self-quenching. In self-quenching the work piece cools itself by action of the internal masses that are not affected by heating and thus remain cold. From this explanation it will be realized that thin hardening surface layers are the more difficult to produce—particularly owing to the lower self-quenching action—the smaller the work piece to be hardened. As soon as the

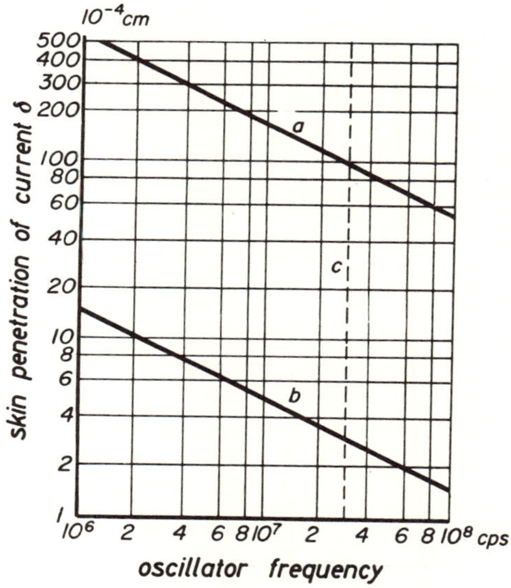

Fig. G4-3. Penetration depth δ of high-frequency currents in steel versus oscillator frequency f. (a) Steel at 1000°C; μ—1; ρ—1.2 ohm-mm²/meter. (b) Steel at 20°C; $\mu = 100$. (c) Applied oscillator frequency in impulse hardening; frequency—30 megacycles/sec.

volume of the work piece to be hardened becomes equal to the volume not to be hardened, there exists the danger of hardening throughout by heat conduction during the hardening process, because the self-quenching effect of the cold inner masses becomes too small. Attempts have been made to improve such surface hardening with high-frequency energy by extremely short and accurately set hardening durations. But to be able to adjust accurately durations of about 1/100 to 1/1000 sec—as far as this is at all possible—one must face considerable complications. High-frequency hardening of very small parts, e.g., pinions and shafts of watches ranging in size within fractions of millimeters, was thus discouraged up to the present by such difficulties.

In impulse hardening, in contrast to high-frequency hardening, it is not necessary to control the hardening durations by circuit means, for the necessary duration is automatically obtained by a capacitor discharge. The energy required for a certain heat treatment process is stored for that purpose in a capacitor which during its discharge feeds a high-frequency generator that in turn converts the energy into a high-frequency impulse which becomes finally effective in heating the work piece. The energy E stored in a charged capacitor is

$$E = \tfrac{1}{2}CU^2 \text{ in watt-sec or joules}$$

where C is the capacitance of the storage capacitor in microfarads and U is the charging voltage in kilovolts.

To achiëve the desired result, it is possible, e.g., to change the capacitance in steps and the charging voltage continuously and thus have any desired energy content available in the capacitor. Assuming a consistent efficiency of the emitter, the quantity of heat generated in the work piece is given by the charge and capacitance of the storage capacitor and can be repeatedly reproduced. The consumption of the capacitor charge in the high-frequency generator occurs very fast. Durations ranging from 10^{-2} to 10^{-4} sec for the whole heating cycle are readily obtainable.

The quantity of heat Q required in the work piece for hardening can be calculated from

$$Q = Gc_p\Delta t$$

In this formula G is the weight of the work piece in grams, c_p is the specific heat in calories/gram°C (for steel $c_p \approx 0.16$ as an average for the temperature interval), and Δt is the temperature differential in degrees C.

The heat induced in the work piece equals the product of efficiency and discharged energy from the storage capacitor. The efficiency can be determined by tests and depends considerably on the so-called adaptation, i.e., a favorable relation of work piece resistance and operating resistance of the emitter. The heat quantity Q and energy that must be fed into the capacitor (e.g., for a certain capacitor capacitance, the charging voltage) can thus be computed approximately, and by this result an adjustment of the emitter is possible. The exact dose can then be found by a few practical hardening tests. The circuit of the emitter is depicted in Fig. G4-4. The storage capacitor, C, is charged to the energy content necessary for a certain heat treatment process through a high-frequency transformer, A, and glow valve (rectifier). This energy reservoir feeds—through a high-voltage contactor, D, during the capacitor discharge time—the high-frequency oscillatory circuit with anode current. The high-frequency emitter, E, oscillates during a short interval that depends in duration on the energy content of the storage capacitor and the effective discharge resistance. The coupling coil, K, that is closely coupled with oscillatory coil, G, extracts the energy from the high-frequency oscillatory circuit and imposes it through heater coil, M (induction coil), on the work piece, N. The heater coil is adapted in form to fit a certain work piece.

It is the purpose of circuit adaptation to relate the work piece resistance favorably to the emitter because the emitter tube can only deliver its highest output when its operating resistance, that is, the resistance of the oscillatory

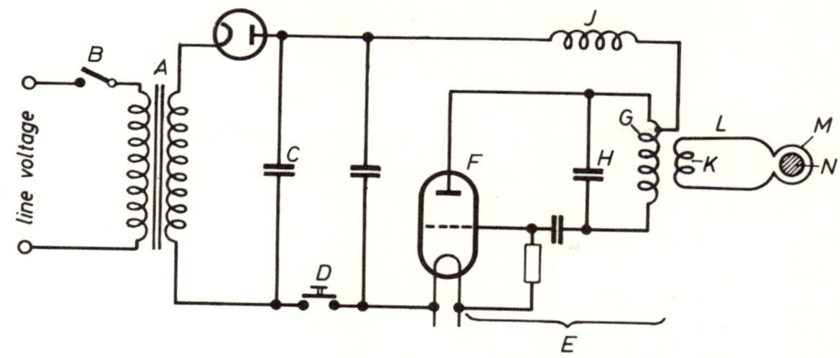

FIG. G4-4. Schematic circuit diagram of a high-frequency generator for induction hardening impulses. A—high-voltage transformer; B—charging contact; C—storage capacitor; D—high-voltage contactor or ignitron; E—UHF generator; F—oscillator tube; G—oscillatory circuit with capacitor, H; dc feeding through choke, J; secondary coil, K, with band-type conductors, L, to heater coil, M, surrounding work piece, N.

circuit, is coordinated with the resistance of the work piece. Thus the work piece resistance, transformed to a suitable value by means of the coupling transformer, is included in the effective oscillatory circuit. The work piece resistance effective in the heater coil cannot be readily calculated, since it is dependent on the emitter frequency, the degree of coupling, the specific resistance, and the permeability of the work piece material. The two last-named values are, moreover, temperature dependent. In hardening minute work pieces, circuit adaptation can only be approximately computed and skillful practical experimenting in that respect usually leads to success more efficiently [52].

For reasons of space and safety—the coil might carry up to 10 kv potential—the heater coil must be at some distance from the emitter. The connection between coupling coil and heater coil is preferably formed by a low inductance band-type conductor (inflicting low losses) that can be in the form of a mobile extension cord. For concentration of the field in the heater coil, high-frequency iron has been found suitable, especially when it is impossible to heat inside the heater coil, yet necessary to approach the surface to be heated with the outside of the heater loop. Fitting and mechanical adaptation of the heater loop to the work piece is a task even in normal heat treating and is even more exacting in hardening of minute or micro parts; extraordinary care has to be exercised in properly forming the heater unit. It is obvious that for electrical reasons alone (high potential, heat) the heater cannot be made as small as desirable, and with difficult work pieces it may be necessary to be satisfied with low efficiency on account of poor heater approach; but at the very low energies involved, this fact is not of serious economic consequence.

A major aim is to avoid further finishing steps on the work piece after hardening. If the thin oxide film (tempering colors) that forms during hardening in open air is objectionable, the process can be performed in protective gases, for instance, by interlocking the emitter with a pneumatic valve by means of which, during the instant of heat treatment, a blast of argon is blown upon the work piece. Especially in minute parts (e.g. watch parts) a metallic, untarnished surface will be demanded. The tempering colors occurring without protective atmosphere, however, do not coincide with the hardening zone. In order to make the hardening zones visible, it is necessary to polish and cauterize the surfaces in the hardening zone. When so pretreated, the hardening zone will become a dark spot in a brighter surrounding. One difficulty should be mentioned that arises when the hardening cycle is of extremely short duration: The conversion of pearlite carbon or austenite into martensite, which takes place during hardening, requires a certain time just as the dissolution of graphite in the ferrite crystals. At the usual hardening temperatures, between 800 and 900°C, this conversion time is longer than the impulse hardening time, but the conversion is speeded up at higher temperatures, say 1100°C. With longer hardening duration such higher temperatures would lead to rupture strains and so-called quenching cracks, but they can cause no harmful results in such thin surface skins as are heated by the impulse hardening method. Thus higher hardening temperatures can be employed to advantage here.

That it is possible to harden minute parts satisfactorily by means of the impulse method, and this in spite of difficulties encountered in microhardening, will now be illustrated by a few examples and several microphotographs.

In a small wind-up pinion of a watch, the teeth were hardened in one operation for about 2/3 of their height, whereby the tooth bases were left soft to avoid breaking of the teeth (Fig. G4-5).

In large gears usually only the teeth flanks are hardened. With such extremely small gears this may be theoretically possible, but practically it is too cumbersome and costly and, in fact, unnecessary, since here it is more the wearing quality than the breaking strength that counts. Figure G4-6 shows the impressions of a Rockwell diamond in the hardened and in the soft zone of the

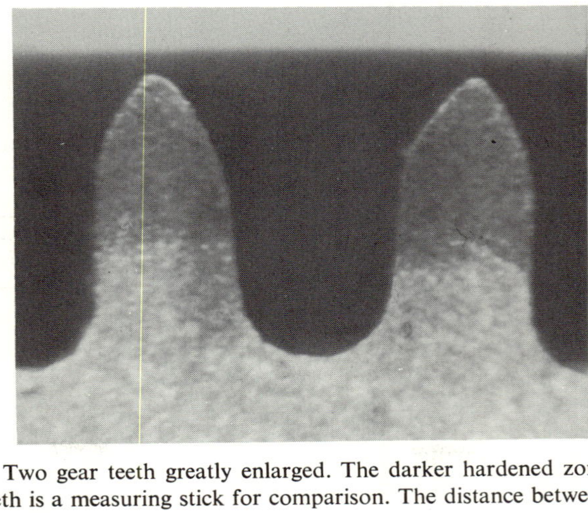

FIG. G4-5. Two gear teeth greatly enlarged. The darker hardened zone is well visible. Above the teeth is a measuring stick for comparison. The distance between the somewhat blurred lines is ½ mm.

gear shown in Fig. G4-5. It will be noticed that the tooth base remained soft with a higher breaking strength.

Another problem was to harden the row of teeth in a gear rack without distortion in order to prevent refinishing after hardening. It was possible to

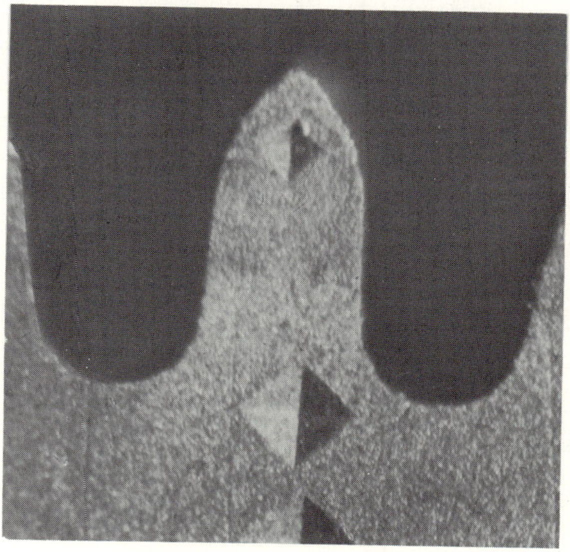

FIG. G4-6. Hardness test on a tooth by the microhardness tester: a pyramid-shaped polished diamond presses with 5 kg weight against the surface to be tested. From the average value of the two diagonals of the impression pyramid the Rockwell hardness can be determined by a table. The impression in the hardened tooth tip resulted in 68 Rockwell, both impressions in the soft rear portion, 35 Rockwell.

harden the 23 teeth uniformly in one operation. The result of this hardening is represented in Fig. G4-7. The penetration depth of the hardening effect could be adjusted as desired by changing the spacing between heater loop and work piece. Steel wires less than 1 mm in diameter can be surface hardened, so that the remaining soft tough core prevents breaking of the wire. Figure G4-8 shows two steel wires of 2 mm dia in cross section; one is unhardened and the other surface hardened. The annular hardened zone can be easily recognized, and, under the microscope, also the finer crystalline structure of the hardened skin. Even in a much finer wire (0.8 mm dia) it was possible to produce a hardened skin layer of 80 microns. The width of the hardening zone in axial direction of the wire can be changed by altering the width of the heater coil.

Besides hardening of minute parts, the microhardening method can be successfully employed for hardening of larger work pieces in special cases. Single hardened zones can be formed on finished plane or curved surfaces—particularly in spots where hard wear is to be expected—by means of the impulse hardening method. For such hardening procedure, a heater coil—

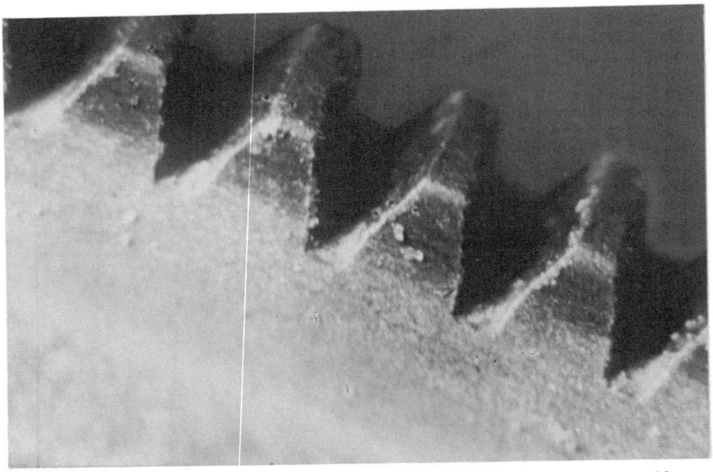

FIG. G4-7. Section through the centre of a hardened row of teeth. The uniform penetration depth of hardening is clearly visible (darker than the sublayer, only the edges are lightened by reflex illumination). The apparent contamination is caused by fine dust particles. Enlarged 45:1.

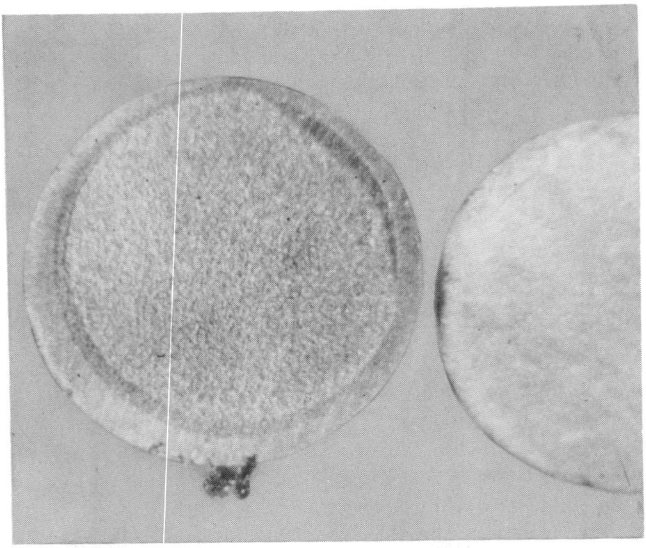

FIG. G4-8. Section through a surface hardened (at the left) and an unhardened (at the right) steel wire of 2 mm dia. The penetration depth of the hardened zone can be readily discerned at the left, as well as the finer crystalline structure of the hardened layer.

properly formed for either a point- or line-like hardened zone—is brought close to the desired zone on the work piece and then a short hardening induction impulse is released. The cold masses surrounding the heating zone provide for sufficient quenching, and, if desirable, a protective gas stream can prevent oxidation and tarnishing of the treated surface. Such hardened spots can be produced in any number and arrangement on the surface of any work piece. For friction wear, thin, force transmitting, hardened spots are formed, whereby the unhardened softer intermediate surfaces carry the necessary lubricant.

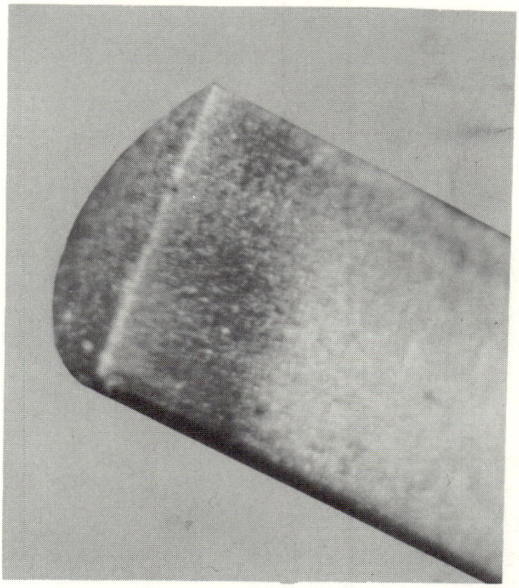

FIG. G4-9. Hardening of a wire end (axial section). A complete through-hardening of the end, as is practical for pivot bearings, has been effected. Diameter of wire 0.8 mm, penetration depth in axial direction 0.5 mm.

Because in this application of microhardening the cooling conditions are exceptionally favorable, the penetration depth of hardening agrees very closely with the theoretical penetration depth of the high-frequency energy. On the other hand, the extension of each single hardened zone is limited. In attempts to heat larger surface areas, heat dissipation becomes irregular and some partial tempering (annealing) results (Fig. G4-9). The upper limit of a hardenable surface area appears to be in the neighbourhood of 10 to 20 mm^2. Figure G4-10 shows the results of such hardening.

Since very small outputs (between about 50 to 1000 watts, depending on the size of the work piece) are sufficient for microhardening, impulse hardening devices can be rather small, simple, and inexpensive. It is possible to solve various hardening problems with one emitter by shaping a particular heater coil for each type of work piece, and by making the intensity of the hardening

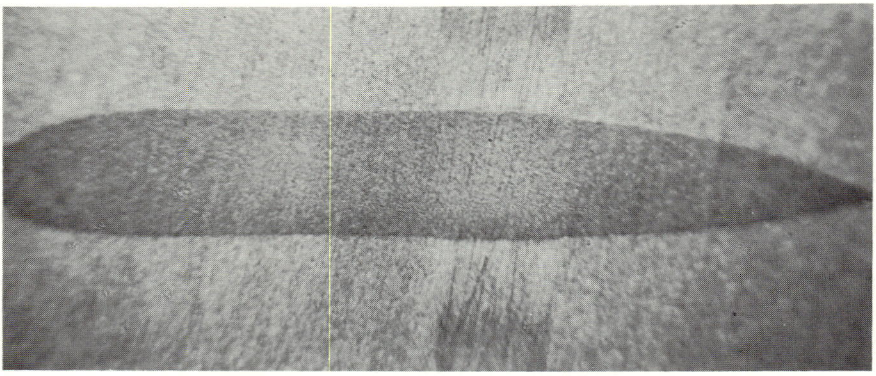

FIG. G4-10. Hardened zone in the form of a cross. The vertical bar has been hardened first. Hardening of the horizontal bar effects tempering of the vertical at the crossing point (tempering color).

impulse variable within wide limits with changeable storage capacitor capacitance and variable charging voltage. Also, in automatic, mass production hardening of parts (it is possible to harden at a repetition rate of up to 5/sec), the surrounding air or the stream of protective gas (if such is applied) suffices for cooling of the heater coil in the intervals between successive hardenings.

Figure G4-11 represents an impulse hardening device.

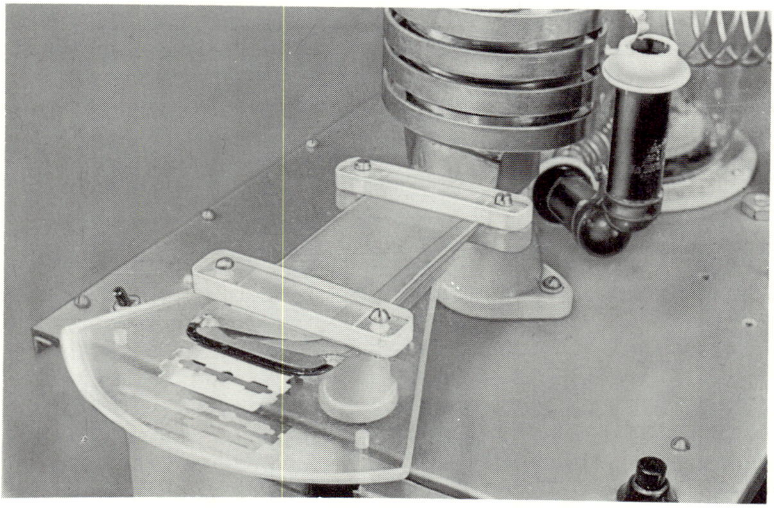

FIG. G4-11. Attached heater coil (inductor) for experimental work in a trial for hardening razor blades only at the surface of the cutting edge. It was possible to harden razor blades only at their rims if the blades were 0.1 mm thick (Works-photo, FF Impulsphysics Inc.).

5. High-Frequency Heating by Rapid Impulse Sequences—the Blast Spark High-Frequency Generator by Marx*

In impulse induction hardening a single capacitively stored energy quantity is converted by a conventional high-frequency tube generator into a single high-frequency wave train of energy for output. The heating method about to be described employs a spark to directly excite a damped oscillatory circuit, and, after decay of the oscillation (see also Chapter B, Section 6c), repeats excitation by a subsequent energy impulse through a following sparkover, and so forth.

Thus it is possible to produce continuously high-frequency energy at a considerable rate by cyclic charging and discharging of a capacitor battery, and the rate can be made adjustable by the sparkover sequence frequency, (operating range between 10 and 35 kw). This type of high-frequency generator operates practically free of radio interference, since no harmonics are excited; there is only a broad band frequency spectrum at both sides of the operating frequency in accordance with the damping decrement; e.g., with a decrement of 0.1, that is, at 300 kilocycles operating frequency, there is a band width of ± 30 kilocycles. Band widths of that order can be rendered completely free of radio interference by choke coils at the transformer, as the author has proved in practice.

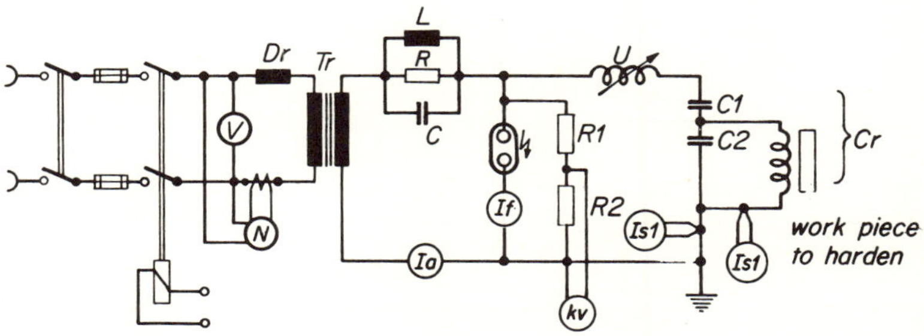

Fig. G5-1. Circuit of high frequency heating setup having a 35 kw blast spark gap (from Marx [15]). The capacitor C_1 is charged by Tr and discharged at high frequency through the blast spark gap (for explanations see text).

The blast spark high-frequency generator is, undeservedly, seldom used in practice although it has real advantages, particularly in the iron and steel processing industries, because of its ruggedness, low dependence on line voltage fluctuations (no tube heating) and simple radio interference protection. Because it is a simple tool for solving induction heating problems on regular work pieces without the use of sensitive electronic tube devices, it appears

* Excerpts from a test report of the Institute for High-Voltage Engineering at the Technische Hochschule Braunschweig, Germany.

proper to describe a complete generator in more detail here. In the generator in question, capacitor charging is effected directly by alternating current. Whether the transformer is provided with a rectifier or not is inconsequential, since a very high number (up to 200) of sparkovers occurs during each half-wave. It is proper to employ stray transformers for reduction of effective losses in the same manner as in the charging units of impulse welding machines and cooled capacitor batteries (Fig. G5-1).

The number of charges is inversely proportional to $\sqrt{L_{Dr}C_r}$ and furthermore controlled by U_1 and U_2. The largest number of impulses occurs at the middle of the half-wave. The feeder circuit and the oscillating heater circuit with the damping work piece are capacitively coupled by C_1–C_2. The circuits I and II can be tuned under load by U. Owing to close coupling, two different frequencies between feeder and heater prevail. The spark extinguishes after 1/2 interference oscillation time.

Owing to the blast in the spark gap, there is—in contrast to Wien's quenching spark gap—also sufficient deionizing action, with larger electrode spacing (3 mm), so that even at a high rate of impulses high ignition voltages can be controlled with one pair of electrodes.

Transformer: 400 volts 15,000 volts, 250 kv-amp
Choke: 3-layer coil of bare aluminum rope, 150 mm² on wooden frame
Cylinder: Outer layer, 119 cm dia, 50 turns, l = 180 mm
Intermediate layer, 113 cm dia, 50 turns, continuously tappable up to 11.25 mh
Inner layer, 109 cm dia
C_1: 792 ceramic capacitors, totally 10,000–20,000 pf, 30 kv, 1580 kv-amp
C_2: 1895 ceramic capacitors, totally 8,000–210,000 pf, 6 kv, 1895 kv-amp
Fusing of C_1 and C_2 by spark gaps
Variometer U: Hard-paper body, 8 turns D = 465 cm dia, stranded wire of 25 mm²
Hard-paper body, 10 turns D = 320 cm dia, stranded wire of 25 mm²
continuously adjustable from 11 to 97 μh
Induction coil: Copper tubing, 1.5 mm walls, D_i = 35 mm dia, l = 130 mm, 17 turns
Water cooling: Measuring cup at inlet and outlet with toluene thermometers
Blocking circuit: Coil, D = 26 mm dia, l = 300 mm, L = 430 μh, 52 turns, C = 1320 pf, R = 1000 ohms

The blast spark gap electrodes have an outer diameter D and inner diameter d. Spacing a is adjustable by turning the electrodes. The air enters through a connecting stub and leaves the chamber through an electrode bore. Tempered cooling water is provided to avoid water-vapor condensation in the spark housing.

The blower for the gap delivers 2.5 atm controllable by bypass* up to 10 meters³/min, power consumption 12 kw at 8 mm bores in the electrodes.

Output measurements: A through-flow bolt of steel inductively heated by the equipment is passed by the cooling water. Inlet and outlet temperature are measured by toluene thermometers, and quantity of water by in-flow time in calibrated cup.

Measured output: 35 kw at 200 kc, P_k = 2.5 atm
Tuning data: $C_1 = 3.14 \times 10^4$ pf $a = 3.5$ mm
$C_2 = 2.86 \times 10^5$ pf $d = 11$ mm (bore)
$L_{Dr} = 3.5$ mh $D = 30$ mm (electrode dia.)

* In designs by the author, 0.6 atm was found to be sufficient and such pressures can be delivered, e.g., by revolving piston blowers of Maschinen Fabrik Ärzen, W. Germany.

The number of impulses drops with rising impulse voltage.
Output rises with increasing pressure.
Output rises with dropping L_{Dr} to max L_{Dr} for max depending on C_r.
Output rises with increasing a (until sparking does not function properly).
Output rises with increasing C_r.
Output rises with increasing C_2 to $\frac{C_1}{C_2} \approx 1/10$.
Maximum pressure given by installed blower.
Maximum distance given by transformer and capacitance of C_1 and C_2.
N can only be improved thereafter by enlarging C_r and corresponding reduction of L_{Dr}.
N reaches an optimum with an electrode bore $d = 6$–8 mm.
Output limit not yet reached.
Electrodes: $d/D = 8/30$ mm dia most favorable with respect to N, η and burn-off. With variation of L_{Dr}, N and η change inversely.

N and η are directly dependent on P_k and a. η is very little dependent on a and just as little on P_k.

An efficiency of 61–67% was measured at a blower pressure between 1.5–2.5 atm. The over-all phase angle was: cos $\varphi \approx 0.26$–0.54, normally 0.42 in permanent operation.

Dates of operation:		
	$C_1 = 1.47 \times 10^4$ pf	$a = 2.6$ mm
	$C_2 = 1.5 \times 10^5$ pf	$P_k = 2.5$ atm
	$L_{Dr} = 8.7$ mh	$d/D = 8/30$ mm

Operation characteristic: At 11.5 kw starting output $\pm$ 2.5%. This output drops after 30 sec by about 10%. In operation over longer periods, the output increases by from 0.9% to 0.4% per hour. The lower value applies for longer operating times (due to increase of a through burn-off until crackling and irregular operation start, then again decrease in output).

Spark gap: Weight loss at 12 kw after 25 hours, 3.5 grams, $d/D = 8/30$
Copper electrodes: 23 grams/1000 kw-hour; 5.5 grams/kamp-hour
Reduction in electrode length after 8 hours, 0.5 mm or 2 mm/1000 kw-hour

Very favorable burn-off conditions (correspond to high pressure arc-valve operation with only eutectic AgCu at certain temperatures and loadings). AgCu (70%–30%) electrodes result in only 19 grams/1000 kw-hour for both electrodes together, and in 4.6 grams/kamp-hour for both electrodes together but impulse voltages vary.*

Output variation 3% at 10% line voltage fluctuation.
In induction hardening of 0.5 kw-hour/kg steel:

Current cost, 1 kw-hour	\$ 0.01/kg steel
Cooling water cost, 0.025 cu meter	\$ 0.001/kg steel
Operator	\$ 0.02/kg steel
Amortization	\$ 0.01/kg steel
Total costs	\$ 0.041/kg steel

Summary: Operation with no extreme accuracy
35 kw, 200 kc, $\eta = 68\%$ (without blower)
Blower output independent of power line output. $\eta = 48\%$ at 20 kw; $\eta = 55\%$ at 35 kw
Owing to electrode heating: Output drops within the first 30 sec of operation by about 10%
Output variation $\pm$ 2.5%
Output rise in continuous operation, at first, 1% per hour; after 24 hours, 0.5% per hour.
Controllable spark gap for compensation of burn-off. Automatic control is possible.

---

* The low-inertia or quick response spark gaps described in Chapter B, Section 7a can here be employed to advantage.

Uniform burn-off. 700 hour operation at 20 kw possible.

Practical application: Hardening emitter, stationary crucible heater or through-flow melting heater.*

The high-frequency heater in blast spark operation represents an analogy to high-frequency cinematography with high sequence frequency of sparks.

6. Annealing of Wires by Means of Capacitor Discharge Pulses

A special application of the capacitor discharge for spontaneous heat generation has been described by the author [754], who used transformed capacitor discharges for annealing copper wires in wire manufacturing. Since wire leaving the drawing mill has been hardened by the drawing process, its original softness and ductility must be restored by a heating and subsequent quenching process. The heating can be accomplished by an electric current.

However, the process of recrystallizing is dependent on a minimum temperature, which must be kept very accurately since the annealing process would be ineffective below this temperature. For the manufacture of cables this means one or two additional working operations, which have to be performed suitably after the drawing process and prior to the coating or sheathing process. In former times the wire was annealed in special annealing ovens. This method was uneconomical, because apart from requiring much space it was practically impossible to achieve a continuous manufacturing process, not to speak of the difficulties in maintaining the rather critical annealing temperature. This disadvantage was largely overcome with modern machines, in which the wire is annealed between contact rollers by resistance heating. This method usually operates in three steps. A pre-heating phase is followed by the actual heating to re-crystallization temperature. Then follows a tempering step, and finally the quenching and a liquid cooling bath. Though this method permits continuous operation, results can only be obtained when the continuity of the whole process is absolutely guaranteed. Since the drawing process cannot be performed at a continuous and even rate, automatic current controls are required, and the difficulty here is that these controls do not always respond quickly enough. Every maladjustment causes an interruption of the recrystallization process.

These disadvantages can be avoided if the wire in a cooling medium such as protective gas, water, oil, or trichlorethylene, is heated in overlapping lengths, by means of short current pulses, the duration of which is very short compared to convectional heating and cooling mechanism. On the other hand these pulses are long enough to avoid the occurrence of any skin effects in the wire. In this process the pulsing rate is controlled by the wire feed in such a way that every partial length is annealed by just one pulse of sufficient amplitude. The required pulse energy of, for instance, 2000 watts in 10^{-4} sec is generated by capacitive storage of tension from 1 up to 5 kv and then fed into the length of wire to be annealed over a pulse transformer with open iron core, great

* For such high-frequency heat producers the so-called zone melting process is an actual application field for cultivation of extremely clean monocrystals.

external field, and close coupled windings such as is used in the pulse welding technique, or by a sufficiently dimensioned transformer of conventional design. For the work described here the power supply of a commercial pulse welding machine (Impulsa, FF Impulsphysics Inc.) was used [50]. The feeding of the voltage pulse into the wire length is effected by means of contact rollers, through which the wire passes. These rollers are arranged in such a way that by symmetrical feeding of rectified opposite currents the respective magnetic fields cancel each other (Fig. G6-1).

In order to avoid magnetic movement of the wire due to the high pulse amplitude of 10^4 amp/mm^2, special current feeding bars are arranged alongside the length of wire under process so that the wire is centered on an electromagnetically determined path. Lengthening of the wire during the annealing process is avoided by the wire being slack between the contact rollers. The firing of the pulses is effected by the ignitron of the power pack of the pulse welding machine and controlled by the contact rollers by means of a microswitch.

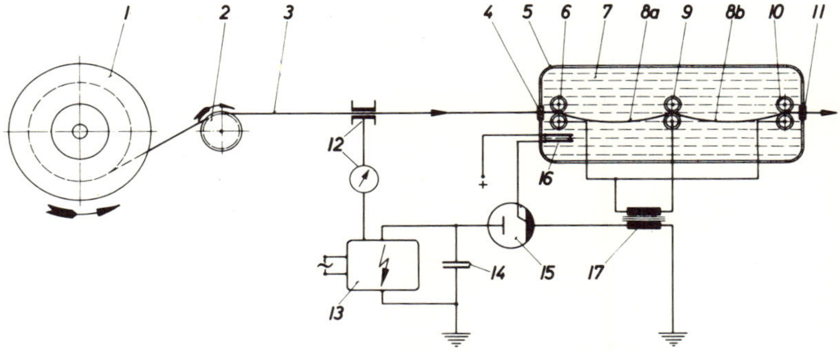

FIG. G6-1. Schematic diagram of the method of annealing of wires by means of capacitor discharge pulses.

From the feeding roll, 1, the wire, 3, is fed over the guiding roller, 2, through lock, 4, of the container, 5, filled with cooling agent, 7, into the contact rollers, 6, whereupon the wire passes within this cooling container through a further wire roller, 9, and after passing through the roller, 10, and the lock, 11, leaves the cooling container, 5. The lengths of the wire between the rollers 6, 9, and 10 are marked 8a and 8b. Both these wire lengths (8a and 8b) are under annealing. The roller pair 6, 10 and roller 9 are connected to the secondary of the pulse transformer, 17, the primary of which is fed from the built-in pulse capacitor, 14. The discharge pulse is fired by the ignitron, 15, which in turn is controlled by switch, 16. Item 13 is the power supply with charger and rectifier and 12 is an electric meter for automatic adjustment of the charging voltage in relation to the wire's diameter. Compared to the passing time of the wire, the annealing pulse must be very short in order not to interrupt the continuity of the working process, but it must according to its Fourier spectrum be long enough so that no skin effect can occur in the wire, because in the latter case only the surface of the wire would be annealed. The pulse duration required can easily be calculated. The penetration depth in copper is $\delta = 2.12/\sqrt{f}$ mm, where f stands for the frequency in kcps. If $\delta = 1$ mm, the frequency f is 4.5 kcps. This means that at this frequency a wire with a diameter of 2 mm can be annealed without occurrence of any skin effect. Now in first approximation the pulse duration must be assumed to be equal to a half-wave of this frequency, that is to say, 1/9000 sec and thus the limit for the pulse duration is roughly 10^{-4} sec.

The pulse energy results from the product of the length and diameter of the wire to be heated up by one pulse, the specific heat of the copper (0.093) and the annealing temperature (600°C). If, for instance, the annealing length is 100 cm then there results for a wire cross section of 1 mm^2 and with copper of specific weight 8.9: $Q = (100 \times 0.01 \times 8.9) \times 0.093 \times 600$, that is ≈ 500 cal or 2000 watt-sec. These 2000 watt-sec must be fed into the wire length within a period of 10^{-4} sec, corresponding to a peak energy of $2 \times 10^3/10^{-4} =$ 20 megawatt.

The circuitry requires certain changes, since there is only a very low input impedance. The energy storage is effected by the capacitor, 14, for the charging of which voltages between 1 and 5 kv are employed. The pulse transformer, 17, is determined on its primary by the dc component of the capacitor; and at its secondary by the resistance of the wire. To achieve a short pulse duration an open iron core with great outer field is used. Further more, the windings must be very closely coupled [249].

One may also use a transformer of sufficient dimensions and conventional design, though difficulties may arise with regard to the required pulse duration. The control of the annealing process adjusted by 16 and 15 is independent of the wire's speed. If the latter varies the pulsing rate also varies, because of the triggering control, which is coupled to the annealing length. In this way every wire length receives the same energy, which heats the wire just to recrystallization temperature.

It is essential that the wire is annealed in the cooling medium. For this purpose all contact rollers are arranged in the vessel where the quenching occurs, as can be seen from the drawing. For quenching there may be used protective gases and liquids such as water, oil, or alcohol since the cooling means should at the same time prevent oxidation. Owing to the sudden quenching of the wire its structure becomes softer than in the conventional annealing methods, in which the wire on its way from the annealing place to the cooling liquid has already lost temperature by radiation and heat dissipation. If a longer wire length is pulse annealed it may happen that lengthening of the wire occurs. In most cases this lengthening will be infinitesimal, so that it can be ignored. However, for precision wire, where the diameter requires close tolerances, such a lengthening through pulse annealing may not be permissible. It can be avoided; for instance, by slackening the wire between the contact rollers. The amount of slackening must be suitably limited. If input and output rollers are coupled in such a way that the output roller feeds off at a slightly slower rate than that of the feeding input roller, this objective may already have been achieved.

7. Exploding Wire Discharges

a. *Electrically Exploded Wires**

Exploding wires are an increasingly important experimental means [274] of high-temperature research.

Compilations of the rather extensive literature on the electrical explosion of metallic wires are found in papers by Conn [275] and Chace [276]. They studied phenomena taking place inside of the vapor column formed upon explosion of the wire.

Figure G7a-1 shows, according to Conn [606], the wiring diagram for work in vacuum or certain gases. Voltages used in the Kansas City setup amount to a maximum of 25 kv at a maximum of 2 μf, while equipment made available at the Airforce Cambridge Research Center, Maynard, Mass., Test Station could be operated successfully up to 72 μf and 4 to 7 kv.

The general setup of equipment for this type of work is normally a high voltage transformer, A, such as for medical x-ray work, for charging a battery of condensers, D, by way of one or two H.T. rectifiers, B. Upon triggering a spark gap, E, or by pre-setting the breakdown voltage, the condensers are discharged over the spark gap and the wire F, to be exploded. The wire explodes, emitting a very intense flash of light and a considerable amount of noise; one or more strong shock fronts are generated during the same time interval. A second spark, gap G, for safety and for cutting down oscillations is provided.

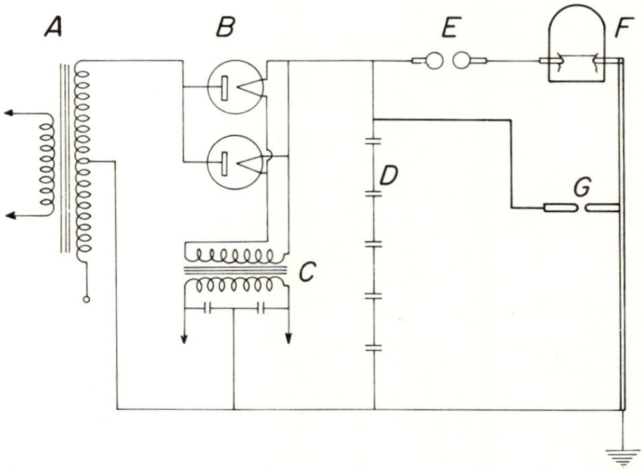

Fig. G7a-1. Wiring diagram of exploding wires in vacuum or certain gases. By permission from Combination of electrically exploded wires and electric arc, W. M. Conn, in "Conference on Extremely High Temperatures," *Wiley, New York*, 1958.

---

* Chapter G, Section 7a, is reprinted by permission from W. M. Conn, Combination of electrically exploded wires and electric arc, *in* "Conference on Extremely High Temperatures," pp. 29–41, *Wiley, New York*, 1958.

The mechanism of the exploding wire is believed at this time to operate in 6 steps:

(a) Heating of the wire by the heavy current passing through it.

(b) Formation of a liquid column of metal replacing the wire.

(c) Development of instability of the wire which forms unduloids (figures of rotation) and causes the appearance of striations in the metal vapor. . . Unduloids were first studied by Plateau [607], and correlated to the exploding wires by Kleen [608].

(d) Upon disruption of the arcs formed between the remaining unduloids, the so-called "dark interval" is obtained during which no current flows through the wire, the voltage remaining constant.

(e) A sudden flash of light is next observed, the spectrum of which is continuous and independent of the material used. Thereafter emission and absorption lines are observed depending on the materials used.

(f) One or several shock fronts are observed during the same interval.

The duration of the entire sequence can be controlled for a given material by the energy supplied, the adjustment of the capacitance and inductance of the system, and the diameter and length of the wire. Other factors entering are the pressure of the tests (atmospheric pressure or vacuum). The total duration of an "explosion" extends from 4 μsec to 15 msec. The duration of the intensive flash amounts to from 1 to 2 μsec and occurs about 2 to 20 μsec after triggering or starting the discharge.

In order to find out what is happening inside the vapor column surrounding the wire or its residue during stages b through d, two methods were developed in Conn's experiments:

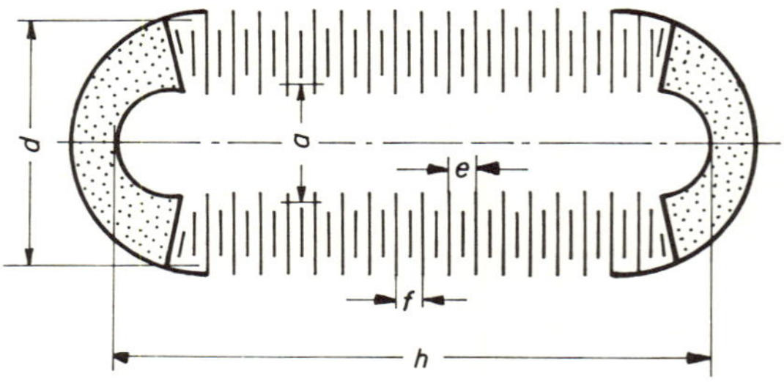

FIG. G7a-2. Schematic sections through the metal vapor surrounding the wire along the axis. By permission from Combination of electrically exploded wires and electric arc, W. M. Conn, in "Conference on Extremely High Temperatures," *Wiley*, *New York*, 1958.

Figures G7a-2 and 3 show, according to Conn [275], schematic sections through the metal vapor surrounding the wire along the axis and perpendicular to the axis (wire), respectively. Glass slides or plastic strips E_4 through E_1 are placed, one at a time, in the distances from the wire indicated. The exploding wire produces characteristic patterns due to the metallic vapor. They are composed of striations perpendicular to the axis of the wire alternating with regions of less density of the material. If the glass slide is sufficiently close to the wire, a pattern appears. This pattern shows metal deposits in the

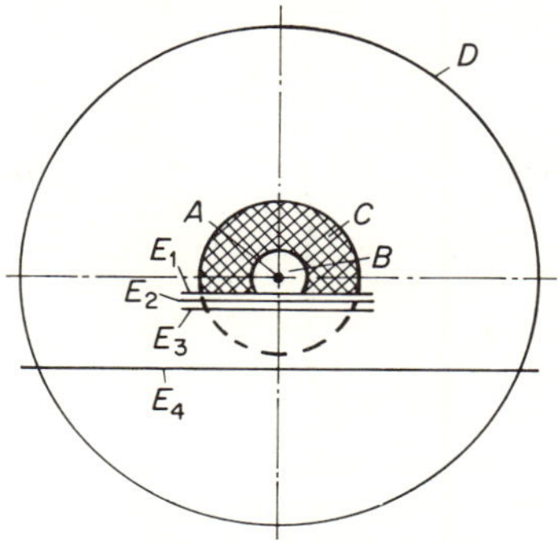

FIG. G7a-3. Schematic sections through the metal vapor perpendicular to the axis. By permission from Combination of electrically exploded wires and electric arc, W. M. Conn, in "Conference on Extremely High Tempeartures," *Wiley*, *New York*, 1958.

shape of striations which correspond to primary, secondary, and, in part, to third-order unduloids. The central part of the pattern remains practically free of metallic deposits. Additional experiments were carried out in which a glass plate or sheet of Mylar (plastic) ,0.007 inch thick, was placed at a short distance, for example, $\frac{1}{8}$ inch or 3 mm, from the wire to be exploded. The most important results were found in the case of glass slides broken into small sections (which could be reassembled to obtain the complete pattern) which exhibited three types of metallic deposits:

(1) A continuous, uniform deposit.
(2) The characteristic pattern.
(3) Individual droplets which slid along the glass or Mylar leaving traces of their movement in the continuous deposit.

All three types of deposit were evident on the broken pieces, even if they were widely scattered. The conclusion was drawn that the shock or shocks causing the breakage of the glass occurred only *after* formation of the three types of deposit.

Strips of Mylar were not shattered as were the glass slides, but were forced, in many cases, for a fraction of 1 mm, into the supporting pieces of Bakelite. Experiments were carried out with the following metals and alloys: Ag, Cu, Fe, Pb, and steel. The tests were carried out between 3 and 11 kv, but the number of tests at 11 kv was rather limited.

The significant result was that in practically all cases striations appeared in the patterns. The striations seemed to occur at regular intervals. Whether an increase in the energy supply changes the dimensions of the characteristic pattern could not be ascertained, due to the lack of a suitable measuring microscope. From a visual comparison, the differences in pattern dimensions did not seem very great.

A "Rapatronic" camera, according to Edgerton and Germeshausen [310], equipped with a magneto-optic shutter (Faraday effect) and permitting exposures of the order of 1 μsec was used. A "window" was placed, according to Conn [312], as near to the wire as desired for the exposure according to one of the following methods:

(1) A transparent circular window, for example, $\frac{1}{2}$ inch in diameter, is placed at the distance to be studied from the wire. A rectangular window may also be used. The windows are protected against lateral disturbances by a cylindrically shaped tube or by side walls forming an open box, its closed side being nearest the wire.

(2) A second embodiment of this method includes placing a solid piece of Plexiglas at the level to be studied.

(3) Finally, a thin piece of Mylar about $\frac{1}{2}$ to 1 inch wide is placed at the desired distance from the wire.

Most experiments were carried out with the window placed right against the wire. This was found permissible after two types of preliminary tests:

(1) Wires were exploded under identical operating conditions and at an identical optical setting of the camera using either a solid Plexiglas window or a thin Mylar window. These tests gave practically identical results, the picture through the Mylar window being only slightly inferior in definition to the picture obtained through the Plexiglas. The difference could be readily explained by the lower transmissivity in integrated light of the Mylar in comparison to the Plexiglas.

(2) The characteristic pattern, consisting of striations, etc., was formed on the side of the window adjacent to the wire. This showed that flattening or destruction of the Mylar window by shock fronts took place only *after* deposit of the pattern.

The most promising method for observations of this type is by means of x-ray flash exposure. However, an attempt was made to approximate this method by window tests which permit photographs of the *inside of the vapor column*. Owing to the lack of a modern type of sweep camera or similar device, it took a great deal of work and time to ascertain the time interval at which formation of unduloids and successive stages take place for a given material and given conditions.

Tests were made for example, with sequences of iron wires and steel wires. The camera was set to take a photograph of about 1 μsec duration. The photograph of the exploding wire shows in the central region, which was made accessible by the window, a sequence of steps which are characteristic for the wire itself during the initial stages of the "explosion." The regions to the left and right of the window show the usual pattern for an exploding wire similar to that already obtained by Anderson and many others who only photographed the outside of the vapor column.

A typical sequence shows the following steps:

(1) Formation of unduloids in the wire and formation of a tubular, uniform column or metallic vapor which is rapidly expanding.

(2) Formation of arcs or sparks at regular intervals along the wire, in part overlapping with the next step.

(3) Appearance of striation-forming regions in regular intervals, but still within the continuous vapor cylinder.

(4) Breaking through of the striation-forming circular or semicircular regions through the continuous vapor cylinder.

The windows were blown down or shattered during the explosion. However, the small pieces which were recovered showed the characteristic pattern of the material. The relatively small amount of metallic vapor directed during the explosion toward the window seems to be diverted to the edges of the window eliminating a possible interference of the window with the sequence of events. Very thin, long wires may be affected by the window, but not the rather short wires of considerable diameter used here.

These tests were initiated by an accidental observation. It was found occasionally that a piece of wire was broken out of the wire before starting the experiment, the wire being stretched between two electrodes. Under ordinary circumstances, such a broken wire was replaced, after discharge of the condensers, by a new piece of wire. Occasionally, the wire from which a piece was missing exploded, as could be seen from the pattern obtained. It was then ascertained that the explosion of the remainder of the wire took place as usual, giving the characteristic pattern, provided, however, that the "gap" did not exceed the length of the spark gap used in the wire circuit. These results were very surprising, because they showed that somehow the circuit must have been closed, although, of course, the inductance differed from the standard conditions with uninterrupted wire.

The patterns obtained in experiments with a gap showed at the location of the gap a break in the pattern on the glass slide. The break in the pattern was often found to be of greater length than the length of the gap between wires.

The next step was to bring the gap to the end of the wire by clamping the wire only at one electrode. Investigation was made only of what would happen if the wire clamped at one end was opposite an extended target of a material different from the wire, the target being placed perpendicularly to the axis of the wire. This set up is similar to arc welding with a consumable electrode opposite extended plates to be welded. In both cases, nonhomogeneous fields are present due to the differences in size of the electrodes and in their arrangement.

Pictures of the sequence of phases during an explosion of steel wires of 0.025 inch = 0.6 mm diameter were obtained. The length of wire was $1\frac{15}{16}$ inch and the gap between wire and target was $\frac{1}{16}$ inch wide; 72 μf and 4 kv were used. Such pictures show that an electric arc is formed between the wire and the target. The arc occurs before the wire starts to disintegrate and remains in existence even on formation of a vapor cylinder of large diameter.

b. *Rapid Wire Feed for Exploding Wire Experiments*

There is a region between "exploding" and "glowing" wires that is worth mentioning. The spontaneous heating of wire can be utilized also for light flash illumination in cases where long flash durations of 100–300 μsec are required with a normal glow bulb color temperature. But owing to the expense of burning glow bulbs with too high tension, Marschak [671] has devised a practical way to charge tungsten and other metal wires in air. Figure G7b-1 shows the method.

Between two contact nozzles, 1 and 2, a 60-mm piece of tungsten wire of 1 mm diameter passes by means of a transport mechanism, 5–8. The capacitor, 9, is charged to such an energy amount that the piece of wire does not burn off completely. The remaining nonvaporized core of wire must be only stiff enough to tie the wire following from the roll, 7. Principally it is possible to flash the wire in intervals of 1–5 sec. Marschak uses a capacitor of 5.1 μf, charging voltage 30 kv, and gets a half-value time of 125 μsec. Because of the skin effect of the steep discharge current, a core of the wire remains undestroyed each time and allows a highly reliable operation.

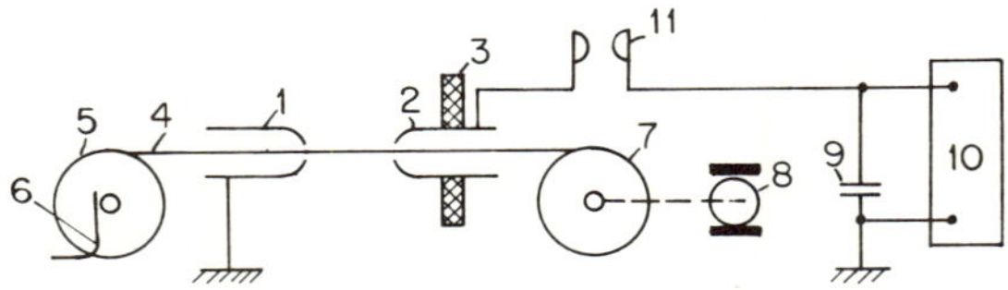

FIG. G7b-1. Schemes for brightly glowing wires with a feeding device for rapid charging with new wire pieces (from Marshak [671]. 1, 2–wire nozzles; 3–insulated holder; 4–wire; 5–transport roll for moving the wire; 6–contactor; 8–moving device for roll 7; 9–discharge capacitor for spontaneous heating of the wire between 1, 2; 10–H.T. power supply; 11–switching spark gap.

In the above paper an exact calculation of emitted radiation also is given, corresponding to a paper of A. Anderson and S. Smith [672]. The papers [673, (2–16)] may also be useful for further study of illumination of objects by glowing and exploding wires.

c. *High-Speed Cinemicrographic Results of Electrically Exploded Wires*

High-speed cinemicrography of electrically exploded tungsten and molybdenum wires was achieved by Zernow, Woffinden and Kreyenhagen [314] (Aerojet-General Corp., Downey, Calif.).

Cinemicrographic observations of electrically exploded tungsten and molybdenum wires, having diameters of 1 to $3\frac{1}{2}$ mils, have been carried out at framing rates up to 1.2×10^6 frames/sec and magnifications of about $25\times$. Both undulant and striated breakup modes were observed. The growth of vapor bubbles inside the molten wire is an important factor in the striated breakup mode.

Cinemicrography is the name commonly applied to the technique combining motion picture photography (cinematography) and microscopy. Although it is not a new technique, the development of modern ultrahigh-speed framing cameras has made possible the useful application of cinemicrography to high-speed, short duration events which were beyond the limits of slower framing cameras. The exploding wire phenomenon which is observed when an electrically charged condenser is discharged through a fine metallic wire provides the kind of event which is usefully studied by means of high-speed cinemicrography. It is surprising that relatively little use of this technique for the study of exploding wires has been reported. Some incidental observations of melting wires were made by Yoffe in connection with studies of the ignition of explosives. These were reported by Courtney-Pratt who devised the lenticular lens cinemicroscope used by Yoffe. Somewhat more extensive work has been reported by Zernow and Hauver who first applied the ultrahigh-speed framing camera, capable of more than 10^6 frames/sec, to the cinemicroscopic observation of exploding wires. More recently Zernow and Woffinden have reported further observations made with this technique. It is the purpose of this section to comment briefly on the advantages and limitations of the technique, and to discuss the observations in the light of theoretical and experimental data obtained by other observers with other optical observation techniques.

Cinemicrography can normally be carried out either with a framing camera or with a repetitively pulsed light source. In systems of low framing rate for frame separation a rotating prism is normally used. In ultrahigh-speed framing cameras, a rotating mirror normally sweeps the image over stationary film. A very clever method of achieving effective frame separation with stationary film and without a rotating mirror was described by Froome. He observed "cathode spots" in an electrical discharge by means of a Kerr cell shutter, but used the mobility of the cathode spots under the influence of an intense external magnetic field to obtain the equivalent of frame separation.

The magnification obtained from the microscope provides some advantages as well as some disadvantages. The major advantage, obviously, is the possibility of observing microscopic details of the event not normally seen without magnification.

However, the penalties that must be paid for increasing magnification are fairly severe. As the magnification increases, the field of view becomes smaller. This tends to restrict the application of the technique to events whose spatial extent is consistent with the field of view. This restriction makes it advantageous to use the least magnification consistent with the nature of the fine detail desired.

The depth of field decreases as the magnification increases, and hence the event must not move very far in the direction of the optical axis if blurring due to defocusing is to be avoided.

Finally, geometrical magnification carries with it the effect of velocity magnification, since the image velocity on the film is the product of the true velocity of the event in the image plane and the magnification. Thus a shock wave moving at 10,000 ft/sec in the image plane would cast an image moving at 250,000 ft/sec when viewed at 25 $\times$ magnification. It is thus easy to see why an ultrahigh-speed framing camera is a necessity when one is studying high-velocity phenomena.

Self-luminous phenomena, such as sparks and exploding wires, simplify the observation problem considerably. However, for nonself-luminous events, external light sources can readily be synchronized to provide either reflective illumination or back illumination for shadowgraphic observation. The exploding wire itself is often used as a light source for backlighting.

Where the event is accompanied by relatively violent consequences, it is sometimes necessary to protect the microscope objective from harm, since its separation from the event normally decreases as the magnification increases. This can be handled in several ways:

(1) Special lens designs can be used to increase the separation between front lens surface and object.

(2) Covering the objective lens with a transparent protective shield will often be satisfactory.

(3) An auxiliary relay lens can be used to cast an image of the event. The image, rather than the event itself, is then viewed by the microscope objective.

Methods (1) and (2) are suitable for events entailing moderate violence only. For very violent events such as those involving sizable amounts of explosives it is clearly necessary to use method (3). In this case the relay lens quality is of major importance.

In the case of self-luminous events, the luminosity may continue long enough to cause rewrite difficulties as the rotating mirror on the framing camera sweeps through successive turns. This is definitely a problem in the study of exploding wires. A capping shutter of the type first proposed by Edgerton and Strabala has been found to be very convenient, since it serves also as a protective shield in front of the objective lens. The capping shutter contains a 5-mil lead wire which, by means of a condenser discharge, is vaporized to form an opaque coating on a glass plate. This shutter goes from full open to full closed in about 30 μsec and is therefore very easy to synchronize. Other capping shutters which can be used include the so-called blast shutters in which an explosive charge is used to introduce opacity into a glass or a plastic plate by fracturing it. The lead wire shutter has been found to be much more convenient, safer, and much less noisy than the blast shutter. Proper ventilation for the lead fumes from the lead wire shutter is of course necessary.

In a recent paper on the liquid behavior of exploding wires, Chace has contended on the basis of his observations with a Faraday shutter that the unduloid theory of Kleen [608] is not applicable to the exploding wire problem. In fact, Chace states that "it will be shown that the formation of unduloids is highly improbable if not impossible."

Kleen proposed his unduloid theory as a basis for explaining the striated pattern obtained when observing an exploding wire. According to Kleen's theory, the molten wire is unstable and in accordance with the work of Plateau [607] and Rayleigh [652] should first form an unduloid shape and finally break up into individual droplets. Kleen proposed that the unduloid shape resulted in higher resistance in the constricted portions. The heating rate was therefore expected to be greater in the constricted area. Kleen suggested that the resulting higher evaporation rate in the constricted area was responsible for the formation of the striations.

The exploding wire observations which will be described indicate that neither Chace nor Kleen are entirely correct. Although Kleen's unduloid hypothesis does appear to be verified under some conditions of the experiments by [314], his proposed mechanism for striation formation does not appear to be correct for either tungsten or molybdenum wire under the experimental conditions used by [314]. Instead, a mechanism involving the interaction of adjacent bubble-like formations is clearly seen to be responsible for the formation of striations under the experimental conditions by [314]. It is important to stress the experimental conditions because these, unfortunately, vary rather widely among the various experimenters.

The experiments were carried out with tungsten and molybdenum wires. The tungsten wires were 0.001, 0.002 or 0.003 inch in diameter. The molybdenum wires were 0.0035 inch in diameter. The wires were mounted in a coaxial holder which in turn was mounted on the movable stage of a microscope. The microscope was substituted for the objective lens in a rotating mirror framing camera (Beckman & Whitley Model 189) so that the magnified image was cast on the face of the mirror. A 2 μf condenser charged to various voltages between 1000 and 5000 volts was discharged through the wire by means of a thyratron switching circuit. The magnification was adjusted to about 25 ×. Camera framing rates were varied between 125,000 frames/sec and 1.2×10^6 frames/sec. The lower and intermediate framing rates appeared most useful in the analysis of the particular aspects discussed in [314].

The best illustration of the formation of unduloids and subsequent breakup into droplets is found in the observations made with 0.001 inch diameter tungsten wires. The behavior of this experiment is clearly in accordance with the general requirements of Kleen's theory. In fact, careful examination indicates the formation of an initial, widely spaced unduloid system followed by a build-up of secondary unduloids between the primary ones. It is interesting to note that the primary unduloid system appears to be more widely spaced than those predicted by Kleen on the basis of his analysis of other wire materials. Kleen, however, did predict the formation of both primary and secondary unduloids, and, since the spacing of the unduloids is dependent on the wire material, this deviation is not properly considered to indicate a defect in Kleen's theory.

Experiments with a 0.003 inch diameter and a 0.002 inch diameter tungsten wire again show the formation of a primary unduloid system with very little evidence of secondary unduloid formation. The process proceeds more slowly, and at increased voltage, for the larger wire diameter.

It is evident then, that unduloid formation and breakup can do and occur under experimental conditions, but as the energy increases a striated system is formed. The interaction process between two bubble-like formations forms a central striated layer almost exactly perpendicular to the wire axis. The interaction of many bubble-like formations and the subsequent generation of many striations in molybdenum wire are shown in [314]. The energy levels for the corresponding behavior of the molybdenum are clearly reduced. This would be expected from the lower melting point, compared with tungsten.

The origin of the bubble-like formations on the tungsten and molybdenum wire is by no means clear at this time, and, although several hypotheses have been proposed, none of them has been verified.

For additional reference, see Liebing and Früngel [315]. Figure G7c-1 shows 6 Kerr cell pictures with 50 mμsec exposure time. Figure G7c-2 shows the same event, taken by x-ray flash high-speed photography, a sequence with 10,000 fps and 0.3 μsec exposure time [264].

Recently, Clark *et al.* [899] took high-speed photography pictures of exploding wires with image converter cameras.

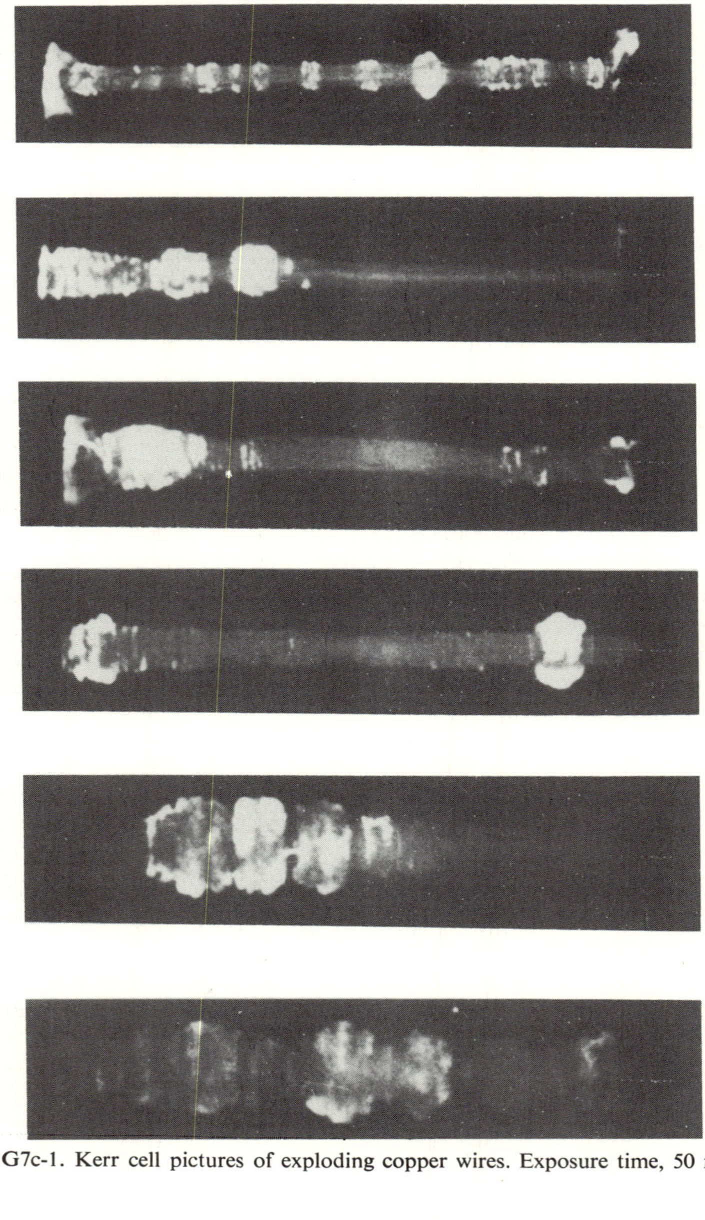

FIG. G7c-1. Kerr cell pictures of exploding copper wires. Exposure time, 50 mμsec.

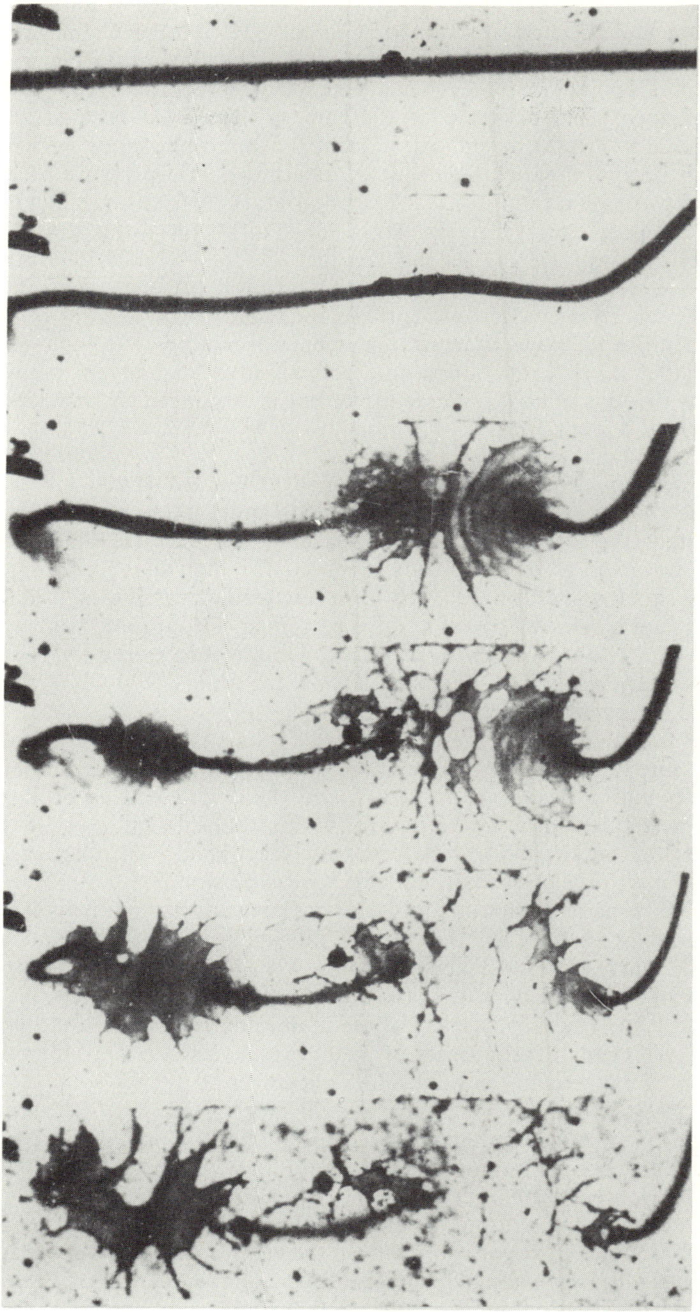

FIG. G7c-2. x-Ray photograph of an exploded constantan wire showing hyperbolic power fields surrounding the centers of primary explosion.

The explosion of fine silver wires by the fast discharge of a low inductance capacitor has been photographed with an STL Model C image converter camera operated as a streak camera. All of the previously observed shock waves and contact surfaces have been clearly recorded. In addition, the large effective *f*/0.5 aperture of the camera (necessary because of the fast event), 50 × 1 light amplification, and 0.5 msec phosphor persistence has allowed the initial shock wave in air at atmospheric pressure to be photographed by its own luminosity. Also, photographs of both radially or circumferentially propagating shock waves, depending on dwell duration, have been recorded during the second conduction. In reference [899] are shown impressive pictures in smear and streak operation.

The observations indicate that there are two distinct regimes of exploding wire phenomena, depending on the capacitor voltage. At high voltages, a discharge channel forms almost immediately after the current break. In this case, the dwell is almost imperceptible, but is evidenced by a sudden change in the current wave form.

d. *Metal Forming by Means of Electrically Exploded Wires*

Chace [787b] says: "Exploding wires will quite likely one day be used to guide interplanetary vehicles through space. This, to be sure, is far in the future. Today, however, wire explosions are used for high-speed photography, to produce shock waves for study, to form intricate metal shapes, as a source of plasmas (hot electrified gases) and for certain spectrographic studies."

The "exploding-wire" phenomenon, a subject of intensive study in laboratories in many countries, has been known for almost 200 years. It was first reported on by Edward Nairne in 1774.

If the rise in current in the wire provided is very slow, so that the discharge time is more than 100 μsec, the molten wire will break up into droplets, owing to gravity and surface tension forces, and the circuit will be broken by this purely mechanical process. Little arcs may establish themselves between the droplets, or the circuit may be completely broken. This is the way a fuse behaves, but this is not usually called a wire explosion.

But if the time of discharge (until the wire has vaporized) is between 5 and 20 μsec, too short a time for the wire to fall apart, mechanical and magnetic forces will have time to distort the wire, with the result that, before vaporization can occur, the wire will take the form known as unduloid (see Chapter G, Section 7c). The thin "necks" of the unduloids vaporize more quickly than the thicker "beads" and a series of metal bridges and arcs are formed. Eventually, most of the metal in the bridges vaporizes.

It is possible to deliver the energy still more quickly to the wire, so that vaporization time is less than 2 μsec. Here the mechanical distortions do not have time to become significant and the liquid wire remains essentially cylindrical. The temperature of the cylinder rises and the liquid superheats. Superheating occurs when a liquid is heated in the absence of nuclei foreign matter on which bubbles of vapor can form. In the case of a molten wire there are no such nuclei—the wire hangs in space—so the temperature rises far above the metal's normal boiling point, until sufficient free energy has built up throughout the wire to induce bubbles to appear spontaneously in the molten wire. The result is a violent detonation.

Very recently, exploding wires have been found useful for shaping intricate parts of very tough metals. The principle of the technique is illustrated in Fig. G7d-1. The metal to be shaped is placed over a die or mold and a liquid—usually water—is poured in to a considerable depth. The exploding wire is immersed in this fluid. The space into which the metal is to be forced is evacuated. When the explosion occurs, the metal is forced into the mold in milliseconds. Many metals which can be die-formed in no other manner behave very well under these explosive-forming conditions.

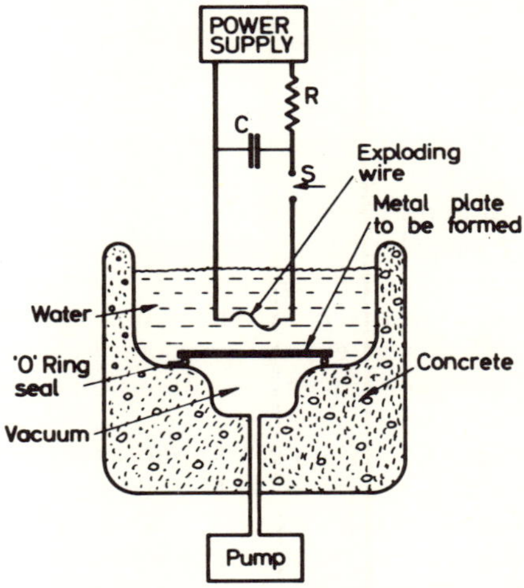

FIG. G7d-1. Diagram of an exploding wire metal-forming setup, using a concrete mold.

Capacitors used to feed exploding wires must have extremely low ohmic resistance, and should have a self inductance less than that of the discharge system. Normally, one uses 30-kv, 1-μf units circuited parallel in the number necessary. For energies up to 12,000 joules, special heavy current ignitrons are available, but for higher energies the old 2- or 3-electrode spark gap systems are preferable.

8. HIGH-TEMPERATURE PLASMA GENERATION BY HIGH-ENERGY CAPACITOR DISCHARGES

a. *The Aim of High Temperature Discharges* [2, 755]

Of foremost significance among the more important problems of modern engineering science is the utilization of energy of thermonuclear reactions. Physicists over the whole world are attracted by the extraordinarily interesting and very difficult task of controlling thermonuclear reactions. Investigations in this field are being carried out by the Academician Artsimovich. A leading role in the theoretical investigations was played by Academician M. A. Leontovich.

As is well known, thermonuclear reactions can arise if the temperature of matter is sufficiently high for atomic nuclei to surmount the forces of the Coulomb barrier with appreciable probability during thermal collisions. The excitation of thermonuclear reactions in deuterium or in a mixture of deuterium and tritium is especially interesting as in this case a noticeable effect should be obtainable at relatively low temperatures.

Physics is indebted to the founder of nuclear physics, Ernest Rutherford, for information regarding the interaction of deuterons. In one of his last investigations Rutherford studied the nuclear reactions that occur when two deuterons collide. It was difficult to suspect at that time that the new facts discovered by him would make more realistic our hope of mastering the energy sources that up to now have existed only in the hot remote parts of the sun and distant stars that shine above us.

The fusion of light nuclei releases energy. If one would get substantial numbers of light nuclei to fuse in a controlled fashion a new source of power would be available to man. However, the fusion reaction is reasonably probable only when the particles have sufficient kinetic energy to overcome the repulsive electric force acting between them. For many particles to have such energies the assemblage must be at extremely high temperatures, hundreds of millions of degrees. However, even at temperatures as "low" as 200,000°C one expects some energy-producing fusion—about 1 d-d reaction/sec/gm of deuterium.

At these temperatures deuterium is an almost totally ionized plasma. The thermal conductivity of a plasma is so high that heat losses would make it impossible to reach temperatures much above several tens of thousands of degrees unless some way were found to insulate the plasma. No material container can be used because these all are vaporized at the temperatures of interest. Fortunately magnetic fields can be used to constrain the plasma, thereby lowering its thermal conductivity.

In particular, if intense currents are passed through the plasma a magnetic field will be generated that will serve to pinch the plasma onto the axis of current flow. Simultaneously, the current will heat the plasma. If the whole thing were to hold together long enough for an appreciable fraction of the deuterons to fuse, this could be the makings of a thermonuclear reactor.

Suitable heavy current discharges have been studied mostly in deuterium at low pressures, 0.005–10 mm. Discharge tubes were cylinders of various sizes. Discharge currents were 0.1–2 million amp at 10–40 kv. Peak power input was 40 million kw.

Typically the discharge begins at the outside wall of the cylinder and spreads toward the axis. The self-magnetic field of the very high current causes the discharge to collapse onto the axis, but it re-expands after reaching a minimum size. This is followed by a second contraction as the magnetic forces again draw the discharge toward the axis. It is during this second contraction that strong electric fields are established along the axis and accelerate the ions there, leading to the production of neutrons and x-rays. A third contraction and expansion phase is sometimes observed but never more. Thus neutrons are not produced as the result of reaching truly thermonuclear temperatures.

The transient nature of the whole process makes it less promising than continuous processes, which Kurchatov refers to but never specifies.

A notion regarding the conditions under which thermonuclear reactions should be experimentally observable can be obtained by considering some concrete cases.

In deuterium whose density equals that of a solid body under normal conditions a temperature of 2×10^5°C would be required to obtain 1 neutron/sec/gm of deuterium. If the deuterium were a highly rarified gas with a concentration of about 10^{13} atoms/cm^3 a temperature of about 5×10^5°C would be required to produce some effect in a gram of deuterium, which would occupy a volume of 30,000 cubic meters. Thus, even to approach the threshold for production of thermonuclear processes the temperature of matter must be raised to a very high level. At such temperatures and under stationary conditions the deuterium should be an almost totally ionized plasma.

The amount of energy that must be concentrated in the plasma to raise its temperature to a level sufficient for the production of intense thermonuclear reactions should be comparatively small. Thus, the amount of thermal energy necessary to raise the temperature of 1 gm of deuterium to 10^6°C equals only a few kilowatt-hours.

Therefore, if one were able to invent a method of heating the plasma with practically no thermal losses, even a low-power energy source could be used to induce intense thermonuclear processes. The main problem is to exclude heat losses, which rapidly increase with the temperature since the thermal conductivity of the plasma is proportional to $T^{5/2}$. If matter is heated to a temperature of only several tens of thousands of degrees in the absence of thermal insulation, the losses will be so great that further increase of temperature will be practically impossible.

There is another obstacle that arises when dense substances are heated, one must overcome the enormous mechanical forces that result from increase of the pressure with temperature. On heating initially solid or liquid deuterium one finds that already at $T = 10^5$°C the pressure exceeds 10 atm. Therefore, thermonuclear reactions can be induced only during a very short period of time in dense substances and such a process will always resemble an explosion or a brief pulse.

On considering the various ways of generating intense controllable thermonuclear processes one finds that there are quite a number of directions that can be followed in attempting to solve the problem.

On the one hand, there are the approaches that lead to stationary thermonuclear reactions, and, on the other, those that are based on the idea of utilizing an instantaneous temperature rise in transient processes of very brief duration. However, irrespective of the way the investigation is carried out, there is one problem that is inevitably encountered, namely, the isolation of the plasma heated to a high temperature from the walls of the vessel in which it is confined. In other words, a means must be found to keep the fast particles within the plasma over a period sufficient for the particles to have a good chance to react with each other.

In a sufficiently strong magnetic field, electrons and ions can move freely only along the lines of magnetic force. In a plane normal to these lines of force the particles will move along circles of small radius. The positions of the centers of these circles can vary only as a result of collisions, each collision displacing the center by a distance of the same order of magnitude as the radius of curvature of the particle trajectory. If the radius of curvature of the trajectory is small compared to the mean free path, diffusion of the particles and thermal conductivity of the plasma in the plane normal to the magnetic field will be greatly diminished.

The theory of processes taking place in completely ionized plasma indicates that at high field strengths H and high temperatures the transverse thermal conductivity coefficient is inversely proportional to H^2 and decreases by many orders of magnitude with respect to the value found in the absence of a magnetic field. However, under these conditions radiation losses must be considered.

The magnetic field required to provide the thermal insulation can be produced by passing a sufficiently intense current through the plasma. The current will also heat the plasma as a result of joule losses and of the work of electrodynamic forces. These considerations were the basis for carrying out theoretical and experimental studies of the physical processes that occur in a plasma on passage of intense currents.

b. *The Pinch Effect*

When a current is passed through a plasma the latter should contract under the action of electrodynamic forces (attraction of parallel currents). An increase of plasma temperature should follow. If a contracted column of the plasma is detached from the vessel wall as a result of electrodynamic contraction (pinch effect) its temperature may be estimated from the condition that the pressure of the ionized gas is balanced by the electrodynamic forces.

A simple computation shows that in such quasi-stationary contraction processes the plasma temperature should be proportional to the square of the current. As is well known, if the electrons and ions are in thermal equilibrium the plasma temperature can be expressed by the equation:

$$T = I^2/4Nk$$

Here I is the current expressed in electromagnetic units, N is the number of particles of a given sign per centimeter length of discharge tube, and k is the Boltzmann constant, 1.38×10^{-16} erg/deg. C. Investigation of the conditions of thermal equilibrium showed that for $N \sim 10^{17}$ the electron and ion temperature should be practically identical. At appreciably lower values of N only the electron temperature will increase.

A contracted plasma column detached from the walls can exist only as long as the current is building up. If the current is constant the column will disintegrate and come in contact with the walls.

It is evident that a thermonuclear reaction with a constant yield over an appreciable period of time cannot be produced by passing a current through a plasma. In principle, it should be feasible to periodically heat the plasma and induce thermonuclear reactions in phase with the peak current.

Calculations of the expected thermonuclear effect led to the following result, which at first glance may seem paradoxical. It was found that during a single heating cycle the total number of elementary nuclear interactions pertaining to a given value of the peak current should be independent of the duration of this cycle. Thus, it seemed that it should be possible to excite very intense thermonuclear reactions by sending pulsed discharges of very short duration through deuterium if the current was sufficiently large. Theoretical calculations indicated that even a current of 300 kamp should be sufficient to produce an appreciable emission of neutrons of thermonuclear origin. At currents of several million amperes the emission should be very intense. Such were the theoretical predictions that preceded the experimental work.

Further development of concepts regarding the nature of the processes occurring in a plasma during the passage of an intense current was profoundly influenced by the new facts discovered during experimental investigation of powerful pulsed discharges. These results completely altered the landscape and colors of the picture created by the first spurts of theoretical effort.

Experimental investigation of intense pulsed discharges was carried out in a broad range of the parameters characterizing the initial discharge conditions.

Discharges through hydrogen, deuterium, helium, argon, xenon, and gas mixtures (deuterium-helium, deuterium-argon, deuterium-xenon) of various relative content were studied. The measurements were carried out at gas pressures ranging from 0.005 mm Hg to 1 atm. Most of the experiments were performed with straight discharge tubes. The length of the discharge gap varied from several centimeters up to 2 meters and the diameter from 5 to 60 cm. The discharge was produced by a voltage of several tens of kilovolts. The peak current varied from 100 kamp up to 2×10^6 amp, the rate of buildup of the current lying between 10^{10} amp/sec and 10^{12} amp/sec. The maximum instantaneous power released in the plasma in these experiments was as much as 40×10^6 kw.

Banks of high-voltage capacitors were used to produce the discharges. The leads that carried the current from the capacitor to the discharge gap were designed to keep the parasitic inductance of the electric circuit down to minimum since this factor restricted the magnitude of the current and its rate of growth. For a voltage of 50 kv and total capacity of the capacitor bank of several hundred μf, the parasitic inductance of the circuit and switch was only 0.02–0.03 μh (in those cases when the current growth was maximal).

Oscillographic methods of measurement of the main parameters characterizing the state of the plasma during passage of a current were developed and these were used to study the intense pulsed discharges. Besides oscillography, ultrahigh-speed moving picture cameras (up to 2×10^6 frames/sec) were used, as well as photography by aid of Kerr cells supplied with special electroexplosive types of shutters.

c. *Experimental Research on High-Temperature Plasmas* [407]*

The general theme of this section is "the interpretation of plasma dynamics information on the basis of experimental observations". Not only is this a very broad topic in the field of plasma dynamics but also it appears to be a topic which is perhaps the Achilles' heel of the whole plasma physics field. Although there has been a great deal of excellent theoretical work done in the field of plasma dynamics and a great deal of observational information has been accumulated through astrophysical observations, nevertheless in the area of real, controlled experimentation in the laboratory there are not a great many checks with theory. Thus in discussing the interpretation of plasma dynamics information on the basis of experimental observations, we are starting off, in a sense, with the weakest part of the whole field of plasma dynamics.

There appear to be several areas which should be fruitful for such a discussion. The first in obvious importance is to compare experimental work and theory as it now stands, to assess what is known. At the same time, it will be important to examine our knowledge critically in the light of whether the word "known" is a correct appellation for such experimental information. Also of importance is the deduction of physical principles or scaling laws by which one can relate easily accessible experimental conditions to those of interest under more extreme conditions. For example, it is possible to create low density, low temperature plasmas rather easily in weak magnetic fields and from such experiments to draw conclusions about areas which one cannot now reach, namely, areas of very much higher temperatures and pressures. As an example, at Livermore** we have explored the role of magnetic pressure *vs.* particle pressure by lowering the magnetic pressure in a confining magnetic bottle to very low values. In such a case, the particle pressure can be correspondingly low and still make a major perturbation in the fields. Some of the information which we can deduce from these experiments, under conditions which we can reach right now, can be used to deduce phenomena that are likely to occur in regions of temperature and pressure which we cannot yet achieve.

Another important topic is the conception of some single experiments to check specific theoretical predictions. Usually experiments in the plasma field have many kinds of complexities built into them. But perhaps we can devise particularly simple, "clean" experiments which will check particular theoretical points.

A powerful tool familiar to physicists in other experimental fields is the deduction of new information from analysis of the effect of parametric variations. In the field of plasma dynamics, this method has not been adequately explored.

An item which holds a key to progress in experimental plasma dynamics

---

* Chapter G, Section 8c, is reprinted here by permission from R. F. Post, Experimental research on high temperature plasmas, "Symposium of Plasma Dynamics," pp. 1–7 Pergamon Press, London (1960).

** University of California, Livermore, California.

is the question of the validity of the diagnostic techniques which are now being used. For example, there has been some criticism of the magnetic probe as a diagnostic tool. A magnetic probe is, of course, simply a small pickup loop inserted within a plasma in order to deduce the behavior of the magnetic field in the vicinity of the loop. What is not yet clear is what constitutes a valid interpretation of such data.

In another area, there has been criticism of the usual spectroscopic observations made on hot plasma and of their interpretation in terms of a "plasma temperature." These questions illustrate why experimental results constitute the Achilles' heel of the field. It is not only very difficult to make reliable measurements, it is also difficult to interpret them after they have been made.

Two last items: we need a better understanding of the role of symmetry in plasma experimental physics. Astrophysicists deal with situations that are seldom characterized by a high degree of symmetry. Plasmas in the universe are usually amorphous blobs, the magnetic fields which exist in interstellar space are seldom symmetric. But usually we perform laboratory experiments which are characterized by a very high degree of symmetry. What can we deduce from such experiments which is really of general significance? Lastly, are there any really novel diagnostic methods not yet developed in this field which should be studied carefully for the information that they can give? If we accept the premise that the rate of progress in plasma physics is geared to the rate of progress in physical measurements, obviously the development of new and more refined physical measurements is of prime importance.

To summarize: because of the extreme complexity of the field of plasma dynamics and because of our inability to treat real plasma dynamic situations exactly by any theory which we now know, it is of great importance that experiments and their interpretation provide a guide to understanding and a standard for deciding which of many possible theoretical calculations most closely represents the true physical situation.

The problem of controlled nuclear fusion: As is now fairly well known, the central problem of research on controlled thermonuclear fusion is to achieve a confined plasma having a density of 10^{14} to 10^{18} particles/cm^3 and a temperature of 10,000 to 50,000 ev (1 ev $= 1.16 \times 10^{4}$°K). It is hardly necessary to say that we have not as yet reached this goal. To make a high temperature plasma, it is usual to attempt to start with a fully ionized cool plasma. Even such a starting point is very hard to achieve. The difficulty is associated with the intense line radiation occurring during the transition from the normal condition, in which 0.1% to 1.0% of the particles are ionized, to the condition in which the atoms are fully stripped of their electrons. Once this latter state has been reached, the radiation becomes the ordinary free-free or brake radiation, and the rate of energy loss is much reduced. The intermediate condition, in which some of the electrons are still bound to the atoms, constitutes a formidable barrier, and our experience has indicated that it requires enormous power to break through this barrier. Most plasmas that have achieved the fully stripped state have been produced by pulse techniques, and their lifetimes have been relatively short.

At this point, let us consider the results that have been obtained from four experimental configurations. The first of these is the now familiar pinch. This, of course, is a selfconstricted linear discharge of circular cross section. The fundamental equation (due to Bennett) that expresses the condition necessary for pinching to occur is $I^2 = 2NkT$, where I is the current in emu, N is the number of particles/cm of length of the pinch, T is the temperature in ev, and $k = 1.6 \times 10^{-12}$. The pinches we use typically have a tube diameter of 10 cm or more and the current rises to perhaps half a million amperes at the peak of the pulse, which lasts about 10 μsec. The pinch may be endless, produced by magnetic induction in a torus. Such a configuration is known to us as the Perhapsatron, and is illustrated in Fig. G8c-1. Or, for short duration experiments where cooling at the ends is unimportant, the pinch may be in a straight tube. We call this the Columbus geometry. A typical example is shown in Fig. G8c-2.

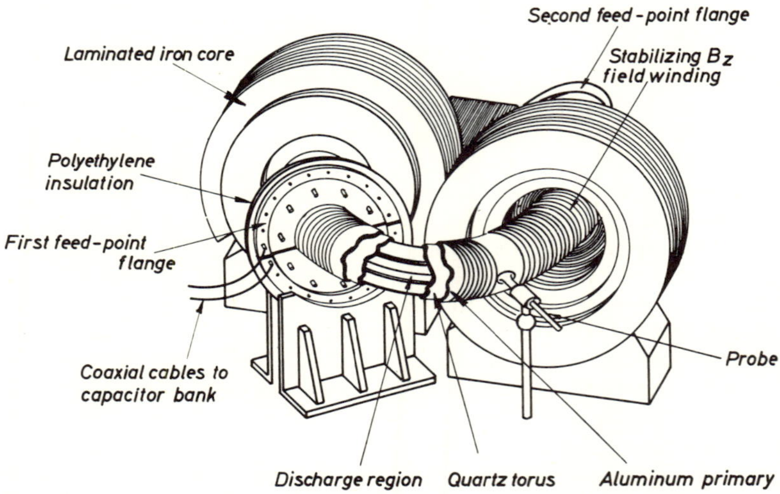

FIG. G8c-1. Perhapsatron apparatus. By permission from Experimental research on high temperature plasmas, R. F. Post, in "Symposium of Plasma Dynamics," Pergamon Press, 1960.

It should be recognized that each of the geometries that we shall discuss has its counterpart in various parts of the world and everybody has his own particular name for each. The next device to be discussed here is the Ixion, which has an axially symmetric magnetic field, B, which is greater at the ends, i.e. it has "magnetic mirrors" which confine the plasma as shown in Fig. G8c-3. In addition, there is a crossed electric field, E. Under equilibrium conditions, the plasma rotates with a tangential speed of cE/B such that, relative to the plasma, the electric field disappears. Now one might ask about the purpose of such an electric field, which simply produces a tangential drift motion. Such a macroscopic motion does not produce collisions and, after

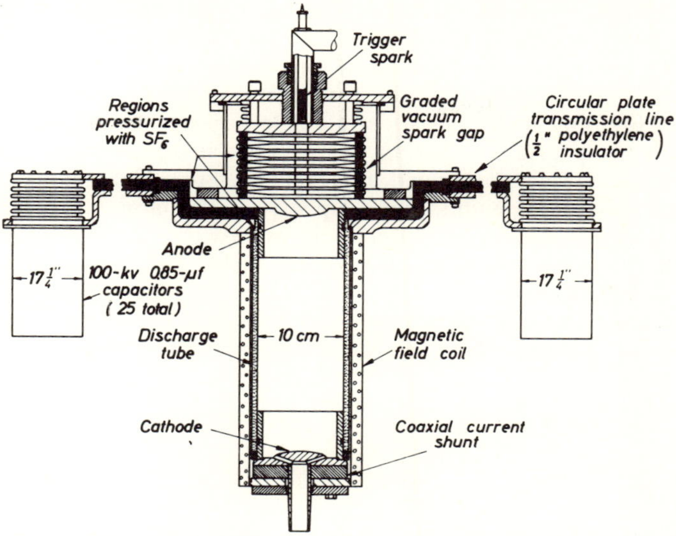

FIG. G8c-2. Columbus apparatus. By permission from Experimental research on high temperature plasmas, R. F. Post, in "Symposium of Plasma Dynamics," Pergamon Press, 1960.

all, thermonuclear devices aim at producing energetic collisions in order to obtain reactions. In systems of this kind, two types of particle energy can be identified. First, there is the energy of the macroscopic drift motion, and second, there is the Larmor energy associated with the sudden, nonadiabatic birth of particles in crossed electric and magnetic fields. These two energies

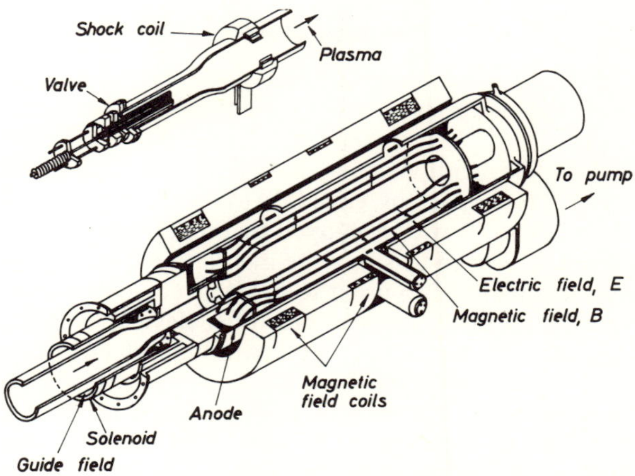

FIG. G8c-3. Ixion apparatus. By permission from Experimental research on high temperature plasmas, R. F. Post, in "Symposium of Plasma Dynamics," Pergamon Press, 1960.

are comparable in size, and since the Larmor orbits do intersect, the desired collisions are produced. The interesting point is that a colliding process is obtained which is not initiated by collisions.

The third configuration in which we are interested is called Scylla. It is shown in Fig. G8c-4. It consists of a single-turn coil placed around a tube containing low-pressure deuterium gas. The coil is shaped so as to produce a mirror magnetic field at each end. It is connected to a high voltage condenser by a transmission line with a spark gap switch. The large induced tangential field which is produced inside the coil ionizes the gas, and the resulting plasma is then compressed by the rising magnetic field. This has produced many interesting effects.

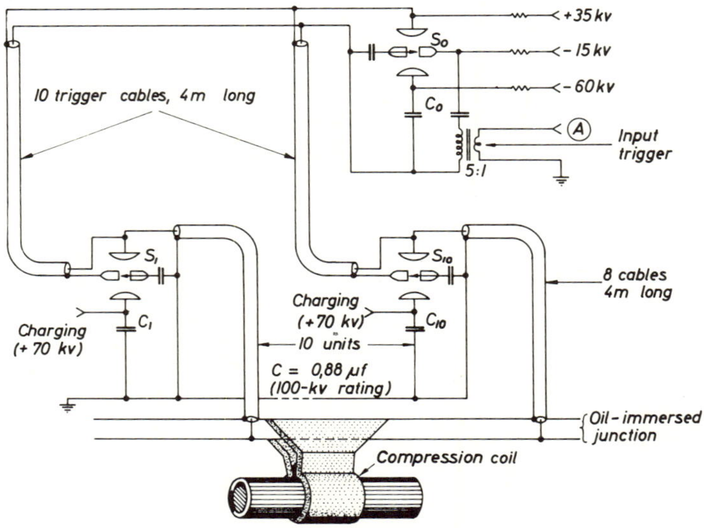

FIG. G8c-4. Scylla circuit. By permission from Experimental research on high temperature plasmas, R. F. Post, in "Symposium of Plasma Dynamics," Pergamon Press, 1960.

The last device to be mentioned may have interesting applications in several fields of research. This apparatus, called PAM (plasma accelerating machine), is shown in Fig. G8c-5. It was developed by one of our experimenters, John Marshall [407], who wished to study the macroscopic motion of plasma in a vacuum. Later, it is planned to use it to produce colliding slugs of plasma. It has also been used to inject plasma into other devices. The apparatus consists of a long tube having a solenoidal winding which produces an axial magnetic field. At one end of the tube is a valve of special design. The tube is initially evacuated, and the valve is briefly opened to let in a puff of gas. This gas is then ionized by any one of a number of methods: for example, by making a small pinch discharge or by using a method analogous to Scylla. The partially ionized plasma is then accelerated magnetically down the tube, leaving the neutral gas behind. With this system it is possible to produce

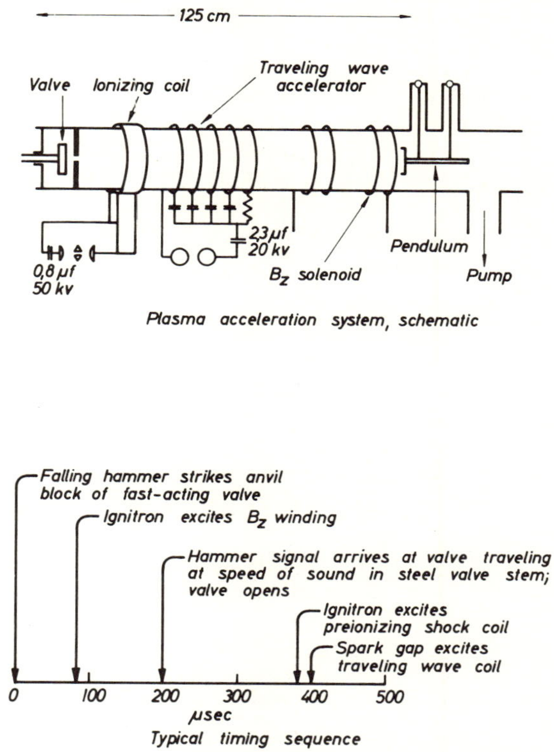

FIG. G8c-5. PAM apparatus and timing sequence. By permission from Experimental research on high temperature plasmas, R. F. Post, in "Symposium of Plasma Dynamics," Pergamon Press, 1960.

quite marked mechanical effects. For example, a microgram puff of gas moving down the tube at 40 km/sec will deflect a pendulum by its impact.

Let us now return to a discussion of the pinch effect. Our most refined experimental work has been associated with studies of this phenomenon. The magnetic probe has proved to be a powerful experimental technique for such work. Almost as soon as we started using such probes, they revealed a fundamental difficulty. The resistance of the fully ionized plasma of a pinch does not seem to be at all related to the values given by the Spitzer-Cowling formula. It is too early as yet to say that the conductivity formula is wrong, because the high apparent resistivity of the pinch column may arise from a masking of the true resistivity by some other process.

H. Conversion of capacitively stored energy into magnetic fields

1. The Creation of Very Intensive Magnetic Fields for Short Duration

For many years, scientific interest in the investigation of phenomena related to the Zeemann effect was centered about permanent magnetic fields. But after discovery of the hyperfine structure by nuclear quanta, an interest was aroused in creating extremely intensive magnetic fields for toric experiments. Stationary fields for that purpose are practically not realizable owing to immense outlay in equipment, etc.

Fürth and Waniek [17] were leaders in utilizing capacitor discharges for the creation of magnetic fields. Many experiments involving the use of high magnetic fields may be conducted on a transient basis. The electrical requirement then appears in terms of energy rather than mean power dissipation, with the stored energy of the magnetic field as an irreducible minimum loss per experiment. If considerations of pulse duration and operating temperature allow this minimum to be approached, and if the pulse repetition rate is not critical, it proves possible, by using relatively modest electrical equipment, to realize magnetic fields which are limited only by the structural strength of the system. The problems of magnet cooling, which are of a large order in the case of dc high field magnets, are greatly eased by intermittent operation.

The present development has been aimed at the exploitation of practical modern capacitor-discharge equipment, and at the construction of special impact-resistant solenoids. By supplanting the use of wire with the use of massive single layer helices, it has been found possible to raise the safe level of high field experimentation beyond the half-megagauss mark in the pulse range 50 to 10,000 μsec. Multiple turn coils of the present type are expected to be operable up to a megagauss, pending some improvements in insulation technique and in the energy source. Low-inductance discharge methods permitting operation of single turn coils in the multi-megagauss, 10 μsec range will be described later.

Pulsed magnetic fields in the megagauss region may well find important applications in the containment of high temperature ionized materials, the acceleration of particles in small orbits, and the alignment of magnetic dipoles.

While Kapitza made successful use of a mechanically shorted generator-flywheel arrangement, this type of equipment suffers from many disadvantages, notably expense. A generator is required which is either very large in terms of mean power consumption, or very strongly reinforced, with special high current brushes. A shockproof foundation may be needed, if the energy of the machine is large. It is estimated that only in the million joule range will the generator begin to compete with the capacitor bank in terms of economical energy storage, and the same is true of a generator-inductance system. Capacitor storage is least costly in terms of joules per dollar in the range from 3000 to 4500 volts. These voltages have the additional merit that they are neither so high as to constitute an extreme safety hazard nor so low as to create unmanageable lead inductance problems. Most industrial equipment of the high-energy discharge type operates in this range, and thus furnishes a useful base for adaptation. Fürth and Waniek [17] devised a cross between the Raytheon 1100 weld control unit and the 8100 impulse magnetizer, which has worked out most satisfactorily. The maximum repetition rate of the unit for a 1725 μf, 3000 volt capacitor bank is 4.5 per second. In Fig. H1-1, two alternate modes of discharge are indicated: "slow discharge" (500 to 10,000 μsec) through a system of ignitrons, which permits accurate timing and prevents voltage backswing, into a step-down transformer; and "fast discharge" (50 to 500 μsec) through the ignitrons, or through a mechanical switch, or a triggered spark gap directly into the magnet. The standard Class D, 3000 volt ignitrons used are not designed to pass much more than 10,000 amp. It is desirable to prevent voltage backswing, in order to increase the lifetime of the capacitors and minimize heating of the magnet.

If the operating voltage and the total energy storage have been determined

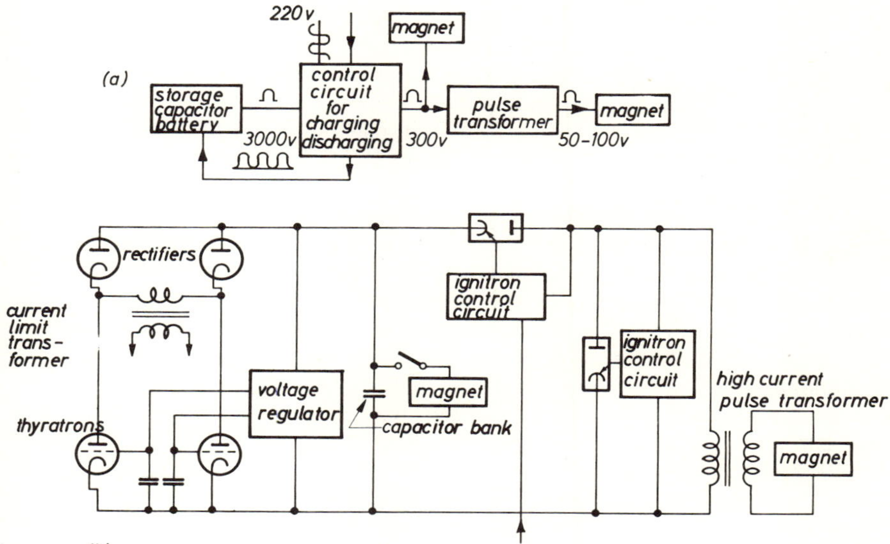

FIG. H1-1. Block diagram and partial schematic of discharge system showing alternate modes of operation. (a) 40 to 500 μsec. (b) 500 to 10,000 μsec discharge times. From Fürth and Waniek [17].

on the basis of economic considerations, then the source capacity is also fixed, and the desired discharge time of the system must be obtained through adjustment of the coil impedance. The size of the coil is set by the availability of energy and the expectation of field strength. The structure, the material, and the operating temperature are determined from considerations of efficiency. There remains, however, the arbitrary subdivision of the conducting volume of the coil to give, within limits, whatever time constant is needed. The lower limit to subdivision of the coil is naturally attained with the use of one heavy single turn. Such structures are very sturdy, but tend to be impractical, owing to severe energy loss problems in the leads and source. The upper limit to subdivision is set by considerations of mechanical simplicity and ruggedness, which vary, of course, with the desired field strength, but in general favor the use of single layer solenoids made of small numbers of thick turns over solenoids which have numerous thin windings. A coil of the single layer type is also of advantage from the point of view of cooling in situations where repetition rate is a consideration. Few of the coils made by Fürth and Waniek had more than 200 turns/inch, and the convenient level was found to be 50 turns/inch and less. Variation of subdivision alone does not, in general, give satisfactory flexibility in the choice of time constant.

An extension in flexibility can be achieved through the use of pulse transformers. In the case of pulse times in the 50 μsec range, air core transformers can be made and operated efficiently. If the problem is to obtain long pulse times (10 msec range) from an economical high-voltage capacitor bank, it is advisable to resort to industrial equipment. A similar problem which arises in the construction of impulse magnetizers has been solved through the development of large steel pulse transformers. The cost of energy passage through such a transformer is entirely comparable with that of energy storage in a capacitor bank; nonetheless, the combination may be advantageous. Fürth and Waniek had a variable 2200 lb transformer made by Raytheon, and have thus been able to work in a continuous range from 50 μsec, 3000 volts to 10,000 μsec, 50 volts.

The production of extreme magnetic fields is largely a matter of the availability of a large and low impedance energy source, which will permit the operation of electrically inefficient coils of simple construction. The ultimate in high field solenoids is evidently a single turn of a very hard metal. There is, however, a secondary limitation in that a minimum temperature rise of several hundred degrees/megagauss must be associated with the pulsing of an unlaminated coil, independently of size, material, or pulse time. In the conduct of most actual experiments, it is practical to devise moderately strong but efficient coils having reasonable self-inductance.

While the magnet current flows in such a manner as to minimize the coil impedance, and is thus driven to the inside surface of the coil, the conducting matter tends to move in the opposite sense, so as to maximize the coil inductance. The radial explosive force in the coil is

$$F_r = \tfrac{1}{2} I^2 \frac{dL}{dr}$$

Similarly, the axial contractive force is

$$F_z = \tfrac{1}{2}I^2\frac{dL}{dz}$$

One also knows

$$\tfrac{1}{2}I^2L = \frac{1}{8\pi}\int H^2dV$$

from which it follows that the coil pressure increases as the square of the field strength and is independent of the size of the coil. For $H = 500{,}000$ gausses, pressures on the order of 7 tons/cm^2 are encountered.

The nature of the magnetic impact may be visualized to some extent by imagining the solenoid compacted between two centerpunches. In the case of mechanically marginal shots, the inside diameter of a coil is sometimes enlarged by a factor of two or more. In such a case, the mean impact force of a 10,000 joules discharge will have been equivalent to that of a 1 ton centerpunch dropping through a meter into the coil.

J. D. Cockroft has given a detailed analysis of the distribution of pressures in a solenoid. The radial outward pressures are cumulative, being zero on the inside surface of the coil, and generally reaching a maximum in the interior. Similarly, the axial inward forces vanish at the top and bottom of the coil and become maximal at half-height. The basic design problem is to make the magnetic forces relatively isotropic, and thus allow transgression of the normal elastic limit of the coil material, without serious deformations.

The design choice of Kapitza was a wire-wound coil, reinforced peripherally by a steel band and shaped to an hour-glass inside contour, so as to conform with the surfaces of constant pressure. Such a coil could be contained without side slipping of turns, but some cumulative longitudinal slippage was encountered on first application of high fields. This effect, which often led to the rupture of the magnet wire, was finally allowed for by means of sliding contacts.

Figure H1-2 shows the basic design of the present high-field coils. Following the example of Bitter, wire windings have been supplanted by a thick-walled conductor insulator double helix composed of individual split rings. This design shares the virtue of efficient cooling, which is important in dc magnets, and in addition has several others. While the magnet current flows mostly on a narrow inside strip, this strip is fixed integrally to the much larger and sturdier strip which constitutes the main body of the helix. The forces on the periphery of the helix are reduced almost in the ratio of inside to outside diameters from their maximum value, which is near the inside wall. Thus the radial containment of the coil becomes an easy matter. A second advantage is that one may precompress the helical coils axially with forces which are comparable to the magnetic forces, thus improving in convenience and efficiency of design on the method of conformal coil shapes. Lastly, the use of split rings permits nonviolent slippage, instead of rupture, when the applied magnetic field exceeds the holding capacity of the coil, an important advantage in any actual experimental context. It is well known that dangerous high voltage may be produced by interrupting the current of an electromagnet.

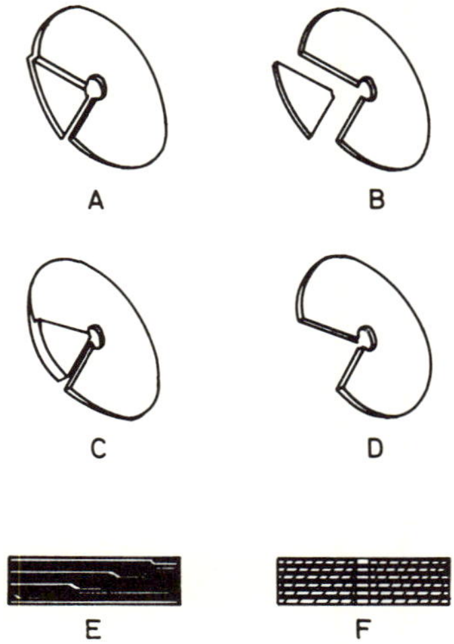

FIG. H1-2. Coil components: (A) stamped conductor disk with insulator disk (B); (C) milled conductor disk with insulator disk (D); (E) assembly, side view; (F) assembly, cross section. From Fürth and Waniek [17].

The modes of deterioration which troubled Kapitza are thus largely eliminated. The principal remaining problem is that of durable insulation. The axial attraction of the conducting disks for one another, combined with the tendency of the inside edges of the disks to thicken under the impact of the radial force, produces a punching effect on the intervening insulator disks. This effect is aggravated by strong precompression, which augments the axial force, as well as by inadequate precompression, which increases the tendency of the conducting edge to thicken and buckle under the radial force. Fortunately these troubles can be eliminated to a large extent, at the expense of the efficiency of the coil, by using hard conducting metals, thick conducting disks, and a large ratio of insulator to conductor thickness. Speaking generally, pure copper coils cannot be operated above 350,000 gausses without appreciable aging, and deteriorate rapidly above 500,000 gausses. Coils made of hardened Berylco 25 have shown only minor mechanical deformation at 650,000 gausses, the coil deterioration being chiefly due to insulator failure. Immersion in liquid nitrogen tends to improve the mechanical properties of most cuprous conductors, but to impair those of most insulators.

The properties of insulators in various configurations and at various operating temperatures have been studied by driving a large number (about 150) of coils of similar construction to the mechanical limit and examining the resultant deterioration. It is convenient in this work that coil components can be stamped in bulk and inserted like cartridges into the containing chamber of the magnet. The ratio of insulator to conductor thickness was generally 1/4 to 1/8, conductor thickness being between 5 and 125 mils. Of the materials tested, high quality mica was found to have the best over-all properties in the range of 5 mils and up. Deterioration occurred because of cracking and splintering of the inside edges of the mica disks, which allowed the material of the conducting disks to ooze or spark together. Mylar was found useful, especially in the range 0.5 to 5 mils, but since it melts easily, a small flaw of insulation will lead to a cumulative discharge. Teflon is resistant to cracking and does not melt easily, but it extrudes under pressure. Mica, Mylar, and Teflon all perform reasonably well at liquid nitrogen temperature, in contrast to rubber-like and cellophane-like materials. Teflon-Fiberglas loses its tensile strength when crushed by the axial pressures of the coils. Papers and cloths of various kinds shear on the coil edges. Varnishes and enamels crack when the conductor flows. Thus far, the most promising results have been obtained with a mica-Teflon combination under oil, and with Okun bonding agents (metal and epoxy resins) which are currently being tested.

The coil disks are inserted into a plastic collar, which is in turn contained in a Dural or brass frame. Dural is more easily worked with and can be made acceptable from the point of view of contact resistance by means of silver-plating or oxide-removers such as No. 2 Alcoa Jointing Compound. Figure H1-3 illustrates the typical construction of a high-field magnet. The leads are perforated to reduce heat conduction and kept close together, in order to minimize stray inductance. The magnet frame may be split radially from the center hole, as needed, since, in general, eddy currents are unwanted. If low temperature cooling is necessary, the magnets are generally immersed directly into liquid nitrogen, unless close access is a consideration, in which case, as in Fig. H1-3, an indirect cooling system may be used.

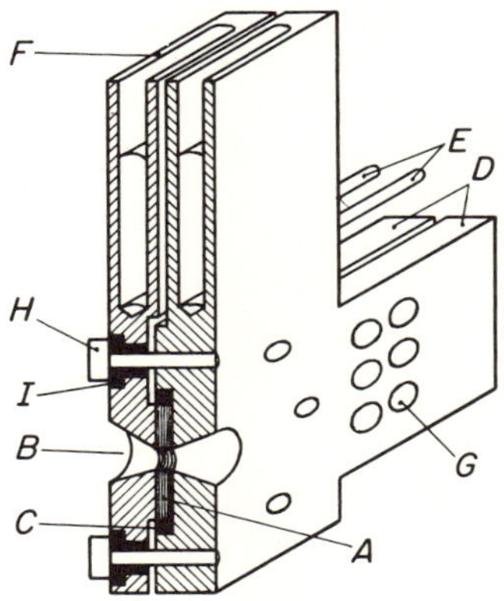

FIG. H1-3. 500,000-gauss magnet for optical experiments. From Fürth and Waniek [17]. (A) Magnet coil; (B) optical channel; (C) Formica ring; (D) Dural leads to transformer; (E) liquid nitrogen inlet; (F) nitrogen vent filled with baffling material; (G) perforations for thermal isolation; (H) steel bolts; (I) Formica bushings.

Most of the high-field experiments were done with coils having an inside diameter of 3/16 inch and a height of 3/8 inch. In that case, the field is uniform to 5% in a cylinder of 3/32 inch diameter and 1/8 inch height. The theory of magnetic field variation in thick solenoids has been treated by H. W. Koch.

Three principal modes of operation were used. The 7800 joules capacitor bank was discharged in 1 to 3 msec through the pulse transformer, set for 35:1 stepdown, into an uncooled copper coil. This arrangement was extremely inefficient. For 3 msec a maximum field of 400,000 gausses was attained, giving a magnetic energy on the inside of the coil of only 200 joules. The magnetic energy stored in the whole coil and its leads was estimated at 1500 joules, and the thermal dissipation before maximum at 2500 joules. The remaining 4000 joules were lost chiefly in the transformer (which is designed for 85% efficiency at 10 msec pulse time). At liquid nitrogen temperature, it proved possible to attain 600,000 gausses, the total magnetic storage now being 3000 joules and the thermal dissipation before maximum only 1000 joules. About 20 ml of liquid nitrogen was vaporized per shot, which was usually negligible compared to conduction and convection losses of about 6 liters/hour in the low temperature system itself. The third method of operation consisted of direct discharge in 50 to 200 μsec from the capacitor bank into the coil. For such short pulses thermal dissipation was sufficiently small, even at room temperature, to permit the use of hard conductors, generally Berylco 10 and 25. Another helpful aspect of short pulse operation is that inertial and elastic effects begin to play a role in limiting coil deterioration. These advantages are partly offset by the need for higher driving voltages. In the direct discharge operation, about 7000 joules can actually be converted to magnetic energy in the coil, so that 850,000 gausses are attainable. Only 600,000 gausses have been reached thus far, due to source impedance troubles, which are being remedied,

Slightly higher fields were produced in experiments with a 62,500 joules, 25,000 volts bank made accessible to Fürth and Waniek by Dr. M. O'Day of the Cambridge Air Force Research Center. Since it was found extremely difficult to insulate the magnet coils against voltages in excess of 8000 volts, an air core stepdown pulse transformer has been designed, which will make possible the release of energies in the 40,000 joule range at safe operating voltages.

The magnetic field strength is measured by means of a pickup coil feeding through an RC integrator into an oscilloscope, according to the equation,

$$H = 10^8 \frac{VRC}{An}$$

where V is the measured potential in volts, An is the effective area of the pick-up coil, and H is the field in gausses. An, which is generally about 1 cm^2, is determined geometrically. The integrator is an ordinary multiple choice RC (from 2000 to 200,000 μsec) made of a number of low inductance resistors and a calibrated linear capacitor. The oscilloscope used is a Tektronix 513 D with preamplifier 121, which is externally calibrated. The largest single error in the measurement is due to the width of the oscilloscope trace, and total error is estimated as $\pm 3\%$.

High-field pulsed electromagnets have also been developed by Advanced Kinetics, Inc. [949]. These magnets were developed to provide transient magnetic fields up to 500,000 oersteds. The transient pressures produced (up to 150,000 psi) require a careful selection of coil material and housing. The coils are constructed of the most appropriate type of beryllium-copper alloy which has been subjected to special heat treatment to optimize tensile strength and electrical conductivity. The shells are of massive stainless steel, and incorporate a vacuum-cast epoxy resin which has been impregnated with glass for insulation purposes. The bore has a 1 inch diameter, and a working length of 24 inches.

2. Application of Strong Magnetic Shock Fields in Nuclear Physics

For the creation of free-burning toric discharges in production of highest velocities or temperatures in so-called plasmoids there are required for supply of energy to the toric discharges uniformly rising magnetic field gradients. For this purpose there are employed capacitor energies up to 200,000 watt-sec with discharge durations of a few hundred μsec to a few msec. In conversion of such large discharge energies into strong magnetic fields there arise problems similar to those discussed in Chapter A, Section 3, B, Section 2, and Section 6, and also C, Section 5. Particularly reliable ignitrons designed as demountable units have been described in reference [140]. With impulse currents in the order of 20–90 kamp the discharge in an ignitron starts not from a single burning spot but from numerous burning spots of which each one emits perhaps 10,000 amp. Each of these spots hurls powerful mercury corpucular rays against the walls of the discharge vessel and they destroy the wall surface in the course of a few thousand discharges. In the case of hard-glass vessels care must therefore be taken that under no circumstances the mercury rays strike the glass walls directly. In reference [140] there is disclosed, besides others, a flat tube type having a disk-like graphite anode which at a spacing of a few centimeters faces an equally large mercury level. With such construction all mercury rays are thrown against the graphite disk.

For such and even heavier currents the mechanical impact switch has gained acceptance, particularly in the field of nuclear fusion. This impact switch is designed on the basis of the same principles considered in Chapter B, Sections 6 and 7. An impact switch with pneumatic actuator, rated 20–100 kamp at 20 kv, is described in [141]. The switching mechanism of this switch is actuated by 12 atm compressed air. This switch operates satisfactorily for 350 discharges at 25 kv in the range between 10 and 100 kamp whereafter a dressing of the contacts becomes necessary. The contacts themselves are resiliently mounted. In the Zeta device, developed in Great Britain, there were possible, with 20 kamp peak current and impulse durations of 500 μsec, 3500 switchings within 10 hours, and recent results gave 80,000 trouble-free magnet field switching operations.

The energy balance in converting capacitive into magnetic energy is explained in [142]. In this paper there can also be found cost comparisons of capacitors having impregnated paper, titanium-dioxide, and barium-titanite dielectrics, according to which the cost of paper capacitors is 40% lower than the cost of those with titanite or titanium-dioxide dielectric. Also flywheel generators can be employed economically instead of capacitors for the generation of very extensive powerful magnetic field impulses. When very large energy quantities are involved it becomes more economical not to store the energy electrically but in the flywheel mass of a large generator which is braked during switching. The requirements for such flywheel generators are also described in [142], where also the possibility of taking impulse currents from storage batteries is discussed. A large project including mechanical storage of the required magnetic energy has been realized in Canberra with a rotating disk of 20 tons weight. An energy of 50 watt-sec is dynamically stored in each cubic centimeter of this disk—thus more than it would ever be possible to store in a capacitor of equal volume. Disadvantageously, however, it is impossible to convert such huge mechanical energies into magnetic field energies spontaneously as can be done from a capacitor battery. The supply of energy from storage in revolving masses will therefore remain reserved for current impulses of at least a few seconds duration.

Design directives for discharge tubes in strong magnetic field application are given in [143, 144], illustrating also the measuring technique for high-current discharges and describing test probes which can be entered into the high-current discharge channel. Since in plasmoid discharges the heavy current does not flow in metallic conductors, the generally applied measuring methods are not applicable, and only an indirect conclusion can be made as to current flow, by observation or photography of the plasma itself.

An interesting compilation of all materialized construction features in the British project Zeta is contained in [145]. This technical paper offers valuable stimulation as to space arrangement of the various construction elements, such as capacitor units, switching cabinets, and impulse transformers. The impulse transformers proper with their disk-type windings (see also Chapter D, Section 2) are represented in [147].

Reference [148] compiles data on high-current impulse apparatus that has been built throughout the world for producing ultrastrong magnetic fields in nuclear fusion research.

In discharge circuits of low inductance, whenever steeply rising impulse currents for creation of magnetic fields are required, it is expedient to employ parallel-connected spark gaps that are ignited simultaneously. Design and calculation rules for multiple spark gap switching are given in [146] (see also Chapter B, Sections 5d and 6d). Marx's circuit (see Chapter E, Section 1) is used for producing steep-fronted ignition impulses to be imparted on the switching spark gaps for release of the main discharge.

Extremely effective cooling is necessary because the participation of Joule's heat is exceptionally high with such heavy currents and, with insufficient cooling, would unavoidably lead very shortly to melting of the material. Conventional cooling means such as high-pressure water are sufficient to keep the material within safe temperature limits, but cooling with liquid nitrogen results in the tremendous advantage that the ohmic resistance of the material, due to the low temperature, drops to about 20% of its value at room temperature. Consequent reduction of ohmic losses leads to considerable savings in capacitor energy and reduces initial cost of the installation so much that cooling by liquid nitrogen or liquid air becomes an attractive method.

The creation of magnetic inductions in the order of 800,000 gausses, corresponding to about just as many ampere turns, appears at the present standing of the art to be the limit that can be realized for current magnitudes in metallic conductors. Whether Ohm's law is still valid with such extremely high current densities is questionable.

For the transformer feeding such a Helmholz concentrator, there apply the fundamental directives explained in Chapter C, Section 5 with respect to the acting mechanical forces, dust protection, a highly magnetizable but high-loss bar-type iron core, and eventually cooling.

Advanced Kinetics, Inc. [950]. has developed high-energy, high-current pulse transformers. The transformers are designed to transfer stored energy at 80% efficiency into inductive loads as small as 0.015 μh. These specifications call for (secondary) leakage inductances as low as 0.0015 μh, equivalent to about a centimeter of ordinary coaxial cable. The extremely restrictive condition on the leakage inductance dominates the transformer design and limits its physical size. The small size, together with the requirement of operability at the high power level, makes the use of an iron core uneconomical. The secondary shunt inductance of the transformer with an effective one-turn secondary connection is required to be about 0.15 μh, which is met in the present air-core transformer.

Some typical parameters are summarized in Table H2-1 for a 10 kv, 1000 μf, 50 kilojoule source, and a 0.015 μh optimum load. The input current and discharge time (quarter-cycle time) depend on the transformer-turn ratio. The efficiency tends to drop off for the longer pulse times due to ohmic losses.

TABLE H2-1

Transformer ratio	Input current (kamp)	Discharge time (μsec)	Output current (megamp)	Output voltage (volts)
4 : 1	640	25	2.3	2200
8 : 1	320	50	2.3	1100
5 : 1	500	30	2.3	1800
10 : 1	250	60	2.2	900
10 : 1	250	60	2.2	900
20 : 1	130	120	2.1	400

3. CAPACITOR DISCHARGES IN MAGNETIC PLASMA GENERATION AND ACCELERATION

a. *Plasma Acceleration Experiments*

There are many areas of both physics and engineering where it is helpful to understand the basic properties of a high-density, high-temperature plasma. Such a plasma can be produced in the laboratory behind a strong shock wave, and the technique for creating strong shocks utilizes shock tubes, high explosives, or spark discharges. It is naturally of interest to explore methods by which the shock strength can be increased, since this would not only permit extending the range of densities and temperatures but would also lead to study of some entirely new phenomena.

Speculations on phenomena of interaction between the flow and the magnetic field have been presented also by Elsasser [429] and Hoyle, who have also pointed out that an exact mathematical treatment is at present impossible.

Lin, Resler, and Kantrowitz [436] have shown that strongly shocked gases are good conductors, which presents the attractive possibility of driving shock waves magnetically. In one group of experiments [437, 438, 439] referring to the pinch effect discussed by Rosenbluth, the magnetic field was generated by the current in the plasma itself. Other experiments make use of magnetic field generated by currents in external metallic conductors.

The production of high-energy plasmas by magnetically driven shock waves is also the object of papers of A. C. Kolb [402, 403]. Fowler *et al.* [442, 443, 444] have studied shock waves produced by the ohmic heating of various gases in a T-shaped tube. In these experiments a discharge was struck between a pair of electrodes in a quartz tube and shock waves were observed to propagate down a side arm perpendicular to the current path. With this apparatus plasma temperatures of the order 30,000°K were obtained in hydrogen for discharge voltages in the 1–10 kv range.

Because the electrical resistivity of a completely ionized gas depends inversely on the three-halves power of the temperature, the transfer of electrical energy to a plasma by resistive losses becomes inefficient with increasing temperature.

The above experiments were extended by the use of external magnetic fields to accelerate the plasma as well as by utilizing higher voltages and lower inductance in the discharge circuit.

Deuterium plasma with ion energies > 100 ev/ion in the ordered motion at densities $> 5 \times 10^{16}$ ions/cm^3 have been produced behind strong shock waves generated by high current discharges in a transverse magnetic field [440, 441, 402]. After the initial acceleration by the electromagnetic driving forces, the shock velocity and thermal energy decay rapidly a few centimeters from the discharge due to heat conduction to the tube walls. The term shock wave is used here advisedly to describe the high velocity luminous fronts observed in these experiments.

A rising longitudinal magnetic field along the expansion chamber has been used to insulate the plasma from the walls of the shock tube and to further accelerate the shock wave by compressing the plasma behind the shock front. The shock wave is produced initially by a discharge between two electrodes (50–100 kv, 0.66 μf, 500 kc/sec) at one end of a T-shaped, 3 cm diameter quartz tube. The return lead to one of the electrodes is placed at the base of the T and parallel to the discharge current [440, 402]. The current through this lead generates a magnetic field which drives the plasma into the expansion chamber. The longitudinal field was generated by a four turn coil with each turn connected in parallel to a capacitor bank (10–40 kv, 5 μf, 120 kc/sec). The coil array was 7 cm long with a volume of 50 cu cm. Temperatures (Rankine-Hugoniot) between 70,000 and 700,000°K were measured.

By connecting the coils in parallel, the high-conductivity shock-heated plasma successively reduces the inductance and increases the current in each turn as the front moves down the tube. Circular currents are then induced in the plasma, causing the magnetic energy density at the surface of the plasma behind the advancing front to be large compared to the magnetic energy density ahead of the shock wave. Kolb [402, 403] suggested that this mechanism is responsible for the compression and acceleration of the shock-heated plasma. With these techniques it is possible to eliminate the experimental difficulties encountered if each coil is connected to a separate capacitor bank and then switched on just after the shock passes.

Korneff, Nadig, and Bohn [405]* tried further methods of acceleration experiments. The plasma accelerations have been studied by three different methods, viz., an image converter, a rotating mirror camera, and biased gaps. The image converter and rotating mirror studies permit direct acceleration measurements while a biased gap placed a known distance from the wire corroborates these measurements by giving an average acceleration.

An electric gun effect arises whenever a geometrical asymmetry exists in a circuit. In this particular case, the magnetic field is stronger behind the wire and this produces a magnetic pressure differential which accelerates the plasma in such a direction as to reduce the pressure gradient.

Velocity measurements were made by using a cathode ray oscilloscope triggered by the initial pulse of current which vaporized the wire. At the end of a measured length, a gap biased by a 300 volt battery is used to detect the arrival of the ionized particles. The time taken to traverse the prescribed distance is therefore measured from the beginning of the oscilloscope trace to the beginning of the 300 volt pulse which occurs when the biased gap breaks down. The measurement does not take into account the time required to produce the ions so that this time must be subtracted from the total time measured.

Figure H3a-1 is a diagram of the experimental setup. The capacitor bank used is 590 μf at 2000 volts dc. The switch is an air gap adjusted in spacing to break down at a voltage 10% higher than the applied voltage. Breakdown, and consequent closing of the "switch," is accomplished by a 9000 volt pulse within 0.2 μsec of the initiation of the pulse.

A coaxial shunt with a 930 μohm calibrated resistance is also included to measure the current by presenting the voltage drop across the shunt on an oscilloscope. A glass tank 28 inches long, 12 inches in diameter encloses the electric gun. To reduce induced image forces, the end plate is 3/4 inch Plexiglas, which incidentally facilitates the boring of necessary holes for measuring devices. A mechanical fore pump evacuates the system. Figure H3a-1 corresponds to the actual physical setup and clearly shows the asymmetrical design involved.

---

*The remainder of Chapter H, Section 3a (except for the last two paragraphs), is reprinted by permission from T. Korneff, F. H. Nadig, and L. Bohn, Plasma acceleration experiments, in "Conference on Extremely High Temperatures." Wiley, New York, pp. 197–207 (1958).

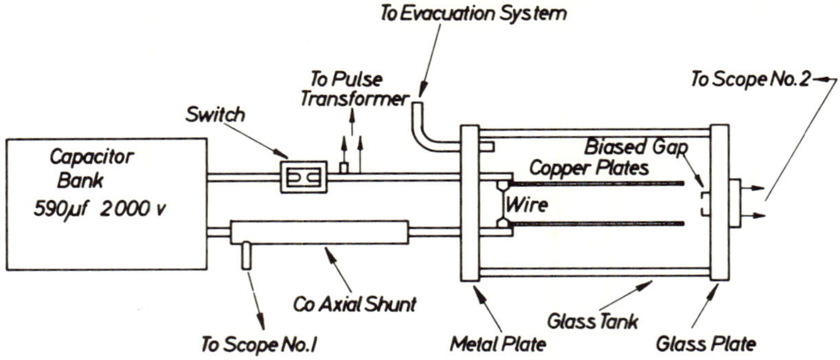

FIG. H3a-1. Diagram of experimental setup for plasma acceleration. By permission from Plasma acceleration experiments, T. Korneff, F. H. Nadig, and L. Bohn, *in* "Conference on Extremely High Temperatures," Wiley, New York, 1958.

As to the question of whether it is the ions injected by the wire explosion that travel down the tracks, rather than a shock wave, a simple demonstration easily shows that the shock wave at this reduced pressure is negligible. Of two cards mounted on either side of the wire, the one on the left is propelled to the left with considerable speed and in fact has a hole ruptured in it, while the one on the right barely vibrates when the wire is exploded (Fig. H3a-2).

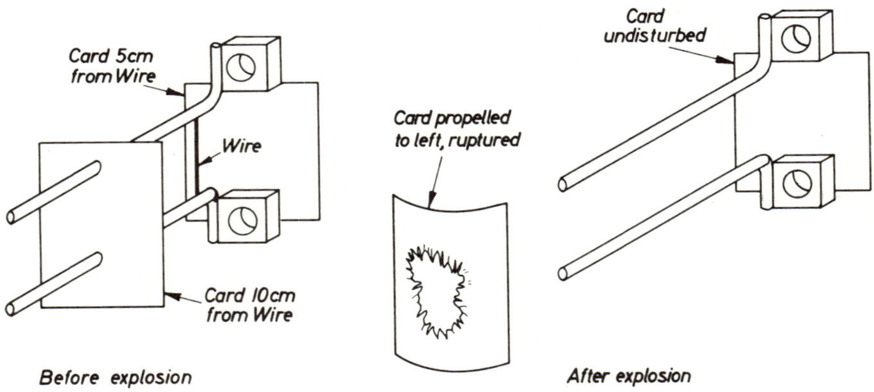

FIG. H3a-2. Illustrating propulsion of card in one direction only by plasma energy. By permission from Plasma acceleration experiments, T. Korneff, F. H. Nadig, and L. Bohn, *in* "Conference on Extremely High Temperatures," Wiley, New York, 1958.

As to the ions being composed of copper, fiber plates, selected because of the ease of obtaining deposits, situated at varying distances from the wire give ample proof of the presence of copper atoms in a preferred direction as well as a pattern indicative of the path followed by the ions. The presence of these atoms showed up unexpectedly in another way. Initially, the end plate was a 1 inch glass plate and after approximately 30 shots, definitely abraded spots appeared due to the cumulative bombardment of the accelerated plasma. Also, the glass plate soon became coated with a deposit of copper.

Two types of gun (Fig. H3a-3) were used in the experiments. In the first case, the tracks consisted simply of two parallel, round copper rods with the wire soldered between them. Pictures show that the plasma tends to crowd between the rods and expand laterally as it travels down the rods. There is a definite leading edge but no well defined packet.

An interesting effect is noticed at the ends of the rods. After the plasma has reached the ends of the rods, a thin beam extends from each rod in an end fire configuration. At first the thought was that this was an arc of some

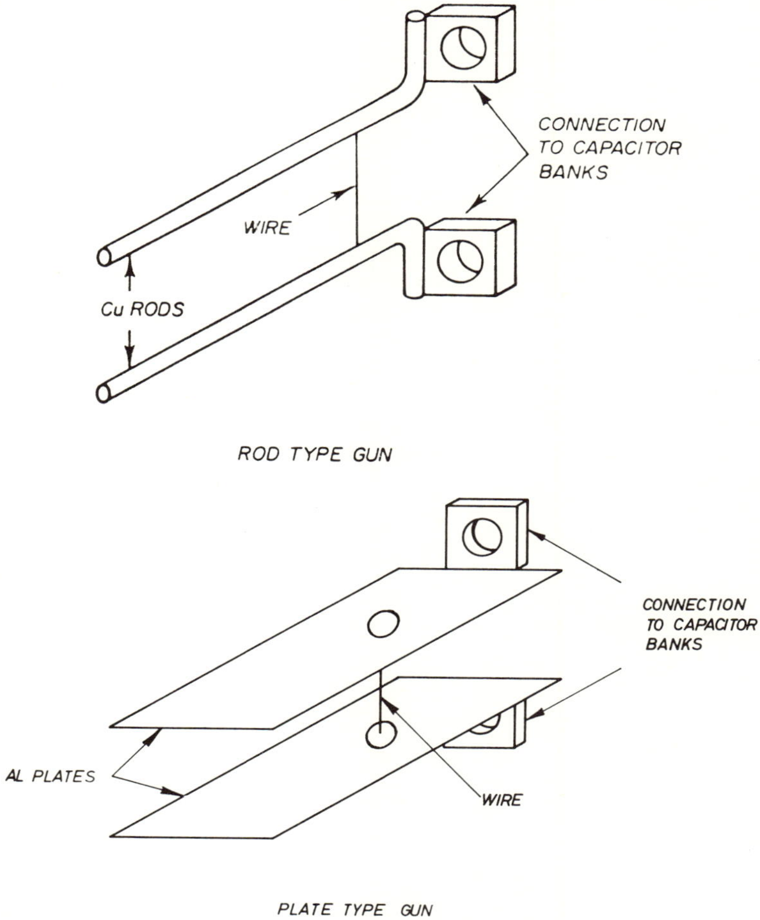

FIG. H3a-3. Two types of plasma guns. By permission from Plasma acceleration experiments, T. Korneff, F. H. Nadig, and L. Bohn, *in* "Conference on Extremely High Temperatures," Wiley, New York, 1958.

sort, but it has been shown to be dependent on the plasma reaching the rod ends. It is this same end fire that produced the abrasive pattern in the glass end plates. The pictures show that the brightest part of the plasma is near the rods but a good portion of the ions travel down in a close pattern between the rods. There is considerable side loss in this configuration and the fact that the magnetic field is divergent should produce a separation of charge and loss of momentum change in the preferred direction.

Mostly to counteract the side losses, a new configuration was tried. In place of the rods, flat aluminum plates were substituted, one foot long, and 4 inches wide. Results were immediately evident in the image converter picture [405].

With the explosion of the wire, the plasma moves down the plates in a well defined packet which spreads with time. The acceleration is quite noticeable in these pictures and agrees with results obtained using average velocity methods. The two conclusions arrived at from these measurements are that the acceleration increases with the current and is in effect while the plasma is between the plates. The plates were of such length that the acceleration took place during the first quarter cycle of the damped oscillatory current waveform. The current incidentally reached a peak value of 50 kamp during the discharge. Since all the energy available in the bank was not used to best advantage, longer plates should result in higher final velocities. There is some dispersion due to the initial radial velocities of the wire explosion, but in general focusing is enhanced by using plates.

The dependence of velocity on ambient pressure is clearly shown here. In each case, for the rods and the plates, the initial voltage of the capacitor bank was 2000 volts. It should be noted that the ordinate is the average velocity; the final velocity can be calculated as twice this figure. (See Fig. H3a-4.) In conclusion, on the basis of these velocities, the average acceleration becomes of the order of 9×10^{10} meters/sec.

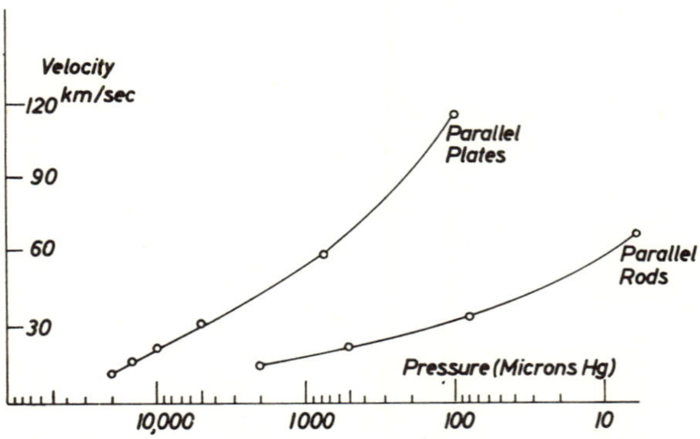

FIG. H3a-4. Drift velocity as a function of ambient pressure. By permission from Plasma acceleration experiments, T. Korneff, F. H. Nadig, and L. Bohn, *in* "Conference on Extremely High Temperatures," Wiley, New York, 1958.

Because of the geometry, so long as a plasma is present to sustain current, there will always be a magnetic pressure gradient in one direction, even though the current may oscillate; and therefore a force accelerating the plasma. This is a different result from that which is obtained using the simple assumption of uniform crossed electric and magnetic fields which are not distorted by the introduced ions; here a limiting drift velocity given by E/B is obtained.

It is of interest to calculate the kinetic reaction for the acceleration involved. Assuming that two milligrams of the wire ionize, the thrust becomes 40 lbs of force for the electric gun.

Also the magnetic driving field of the strap behind a T-tube can be enhanced by using external field coils to increase the shock velocity and the internal energy of the plasma. Two different configurations of the external field coils have been tried, with the coils being pulsed independently of the primary gas discharge in the shock tube; this causes the field to rise slowly compared to the rise time of the main discharge. Figure H3a-5 shows the experimental arrangement. Field coils can be placed directly under the tube

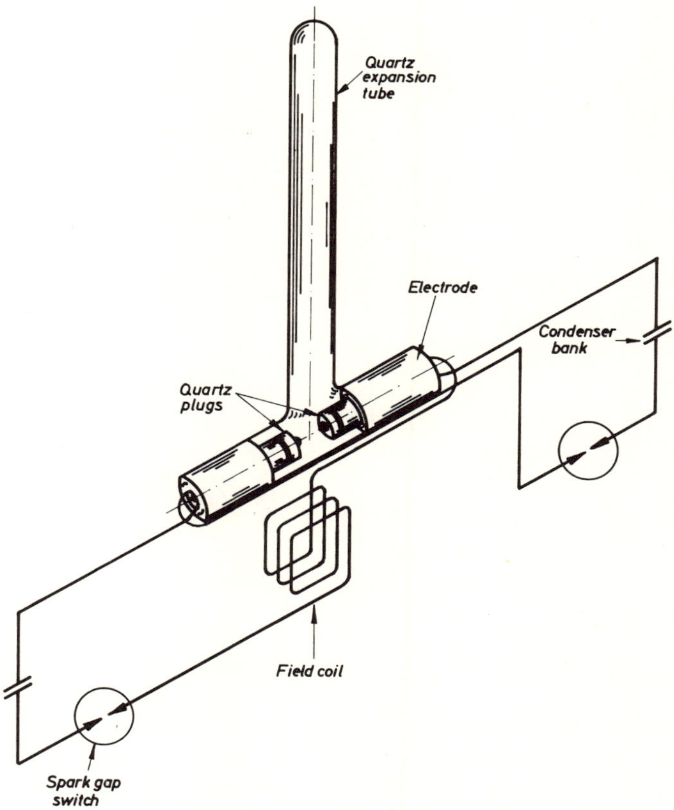

FIG. H3a-5. Schematic diagram of a shock tube with field coil. From Kolb [402, 403].

at the base of the T. There is a magnetic field perpendicular to both the discharges.

Recently, schlieren photography has also become applicable in the field of magnetically driven plasma. Ascoli-Bartoli and Martelluci [812] have studied the possibility of utilizing the schlieren method for detecting and measuring the density gradients of a plasma produced by a pulsed discharge, in an orthogonal pinch arrangement. Due to the lack of light in earlier phases the observation of these phases necessitates the application of an image intensifier.

b. *Magnetically Driven Shock Waves*

The following experiments, which were performed at Lockheed Missile Systems Division [401], have attempted the use of a hydromagnetic shock tube similar to that originally used by Fowler [442–444]. Activities so far have been confined to observation of the effects of magnetic propulsion on shock velocity, the development of instrumentation for measuring shock velocities, and an examination of the visible radiation by the luminous shocks produced.

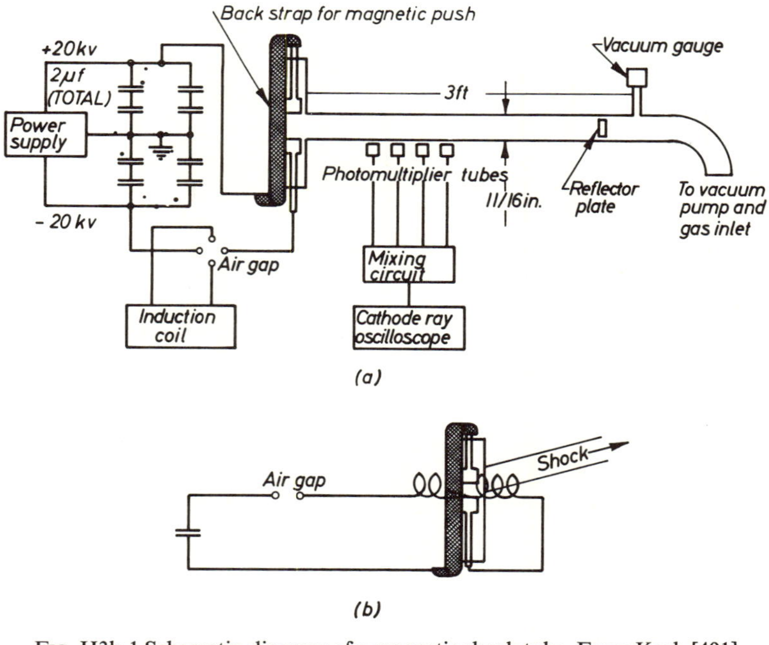

FIG. H3b-1 Schematic diagram of a magnetic shock tube. From Kash [401].

Kash's shock tube (see Fig. H3b-1) was a 3-ft long, 11/16-inch inside diameter, T-shaped pyrex tube with nickel electrodes sealed in the T. A discharge is produced by current flow from a bank of capacitors triggered by the breakdown of an external series spark gap. The bank has a total capacitance of 2μf, and can be charged to 40 kv. The total inductance of the capacitors and the discharge circuit is about 1 μh, most of which is in the capacitors, giving the discharge current a ringing frequency of about 100 kc. From the damping of the discharge oscillations, both with normal electrode spacing and with the electrodes shorted, it is estimated that only about 1% of the available electrical energy is delivered to the gas in the shock tube. This amount can be increased by applying auxiliary magnetic fields in the vicinity of the discharge. The tube differs from Fowler's and from Kolb's principally in its greater length, which enables observation of the progress of a single shock pulse over a greater range of velocity. Air has been used as a working gas, and attention has been given mainly to velocities below Mach 50.

Two techniques have been employed in measuring shock wave velocity. The passage of the highly luminous, ionized gas past fixed positions along the shock tube was recorded with photomultipliers, and its advance down the length of the tube was observed through the use of microwave reflections. The average velocity between stations is readily obtained from the position and time data. These preliminary measurements assume repeatability of a given shock situation, and accordingly limit the velocity determinations to averages between widely spaced stations.

The effect of magnetic field on the discharge current was observed through the use of a backstrap similar to that employed by Kolb [403], in addition to a 9-turn, 2-inch diameter coil. As shown by Fig. H3b-1 the backstrap and the coil were located to provide at the discharge a magnetic field perpendicular to both the discharge current and the axis of the shock tube.

At the 1958 Conference on Extremely High Temperatures in Boston, Janes and Patrick [404]* of the AVCO Research Laboratories read a paper showing new experimental apparatus for higher plasma speeds.

Figure H3b-2 is a schematic diagram of the electrodeless gas accelerator shown in two views. There is on the outside a single-turn copper coil whose leads are connected to a 66-kv, 1.4-μf capacitor through a pressurized spark gap and a nonlinear transmission line. Inside of this coil a pill-box-shaped pyrex chamber contains hydrogen or deuterium at a low pressure. This chamber is surrounded by a radio frequency excited preionization coil which is used to provide the initial gas conductivity. When a voltage is applied to the main coil leads, the induced gas currents are acted upon by the rising magnetic field in such a manner as to produce a radial inward accelerating force. Thus a moving slug of hot gas is created while the magnetic field lines tend to be excluded from the center by the electrical gas conductivity. Cold gas ahead of this moving slug should then be heated by collisions which occur during the acceleration process.

This device eliminates electrode surface problems, but has other serious limitations. One of these is the necessity of preionization for reproducible breakdown. Without electrodes, this becomes intrinsically difficult below 50

*The remainder of Section 3b (except the last paragraph) is reprinted by permission of G. S. Janes and R. M. Patrick, the production of high temperature gases by magnetic acceleration, *in* "Conference on Extremely High Temperatures." Wiley, New York, pp. 3–11 (1958).

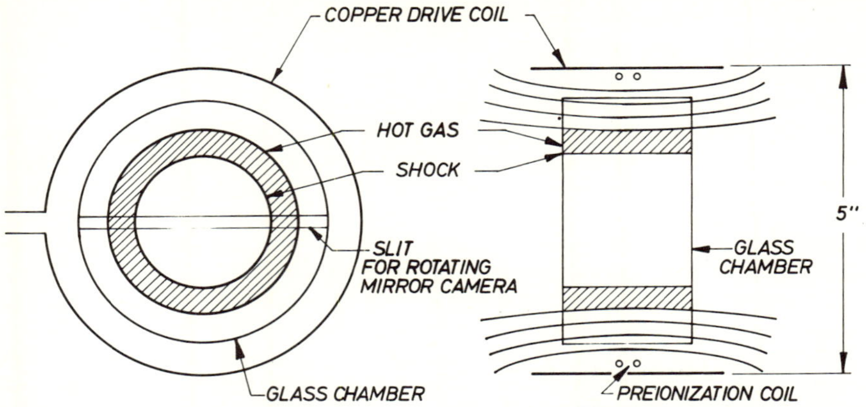

FIG. H3b-2. Schematic diagram of a gas accelerator. By permission from The production of high temperature gases by magnetic acceleration, G. S. Janes and R. M. Patrick, *in* "Conference on Extremely High Temperatures," Wiley, New York, 1958.

microns gas pressure. Another difficulty which is basic to all single-coil inductive coupling mechanisms has to do with the short-range characteristics of the dipole magnetic field. In order to attain the conditions of interest to us, it is necessary to produce an accelerating field of 12,000 gausses in a time comparable with the maximum gas velocity divided by the chamber radius. This time is about 1/4 μsec.

The latter problem is analogous to that involved in attempting to drive a chemical shock tube with a big valve rather than a diaphragm. Figure H3b-3 is a schematic diagram of one stage of the inductive energy storage system. This is actually a nonlinear transmission line where the magnetic

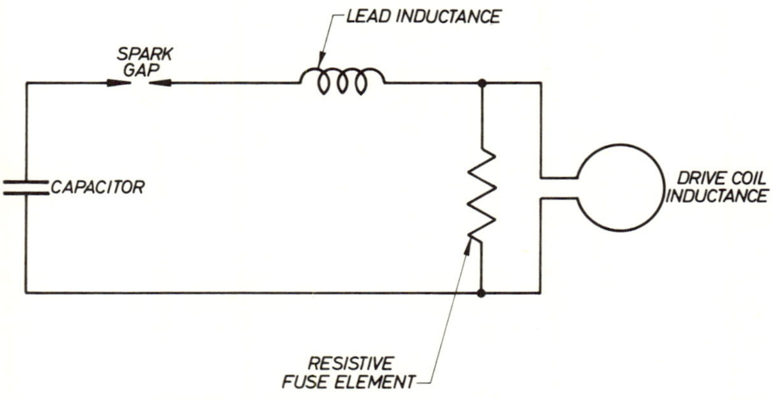

FIG. H3b-3. Schematic diagram of exploding wire diaphragms. By permission from The production of high temperature gases by magnetic acceleration, G. S. Janes and R. M. Patrick, *in* "Conference on Extremely High Temperatures," Wiley, New York, 1958.

energy is stored in the lead inductances and rapidly transferred into (or, more correctly, shared with) succeeding stages by the vaporization of exploding wire resistive fuse elements. In this case, these resistive fuse elements consist of 20 to 50 1-mil copper wires held in place across a 2 inch gap with pressure sensitive tape. By this device one has increased the maximum rate of current rise from 300,000 amp/μsec to 800,000 amp/μsec. The magnetic field is thus built up in 0.15 μsec. This also provides excellent damping of the capacitor bank.

If a moderate field is provided before initiation of breakdown, then for an initial field of 1320 gausses in hydrogen at 80 microns initial pressure, the time at which the gas motion reverses corresponds to the time of maximum field at the center.

All plasma high-speed processes can also be referred to as processes of high temperature. Temperature and gaseous speeds are directly proportional. The best bibliography on high-temperature chemistry and physics is given by Diamond [795, 796], and includes more than 500 references.

c. *Plasma Motors and Guns* [406]

Winston H. Bostick has called a special method of plasma accelerators "plasma motors."*

It has been demonstrated that plasma consisting of titanium ions, deuterium ions, and electrons can be propelled by a small button plasma gun at speeds up to 2×10^7 cm/sec. Plasma motors employing rails should be able to obtain speeds of 10^8 cm/sec without any difficulty. The corresponding specific impulse is 10^5 sec. There are several types of plasma motors.

Button-Type Plasma Gun: It has been demonstrated [423, 424, 425] that

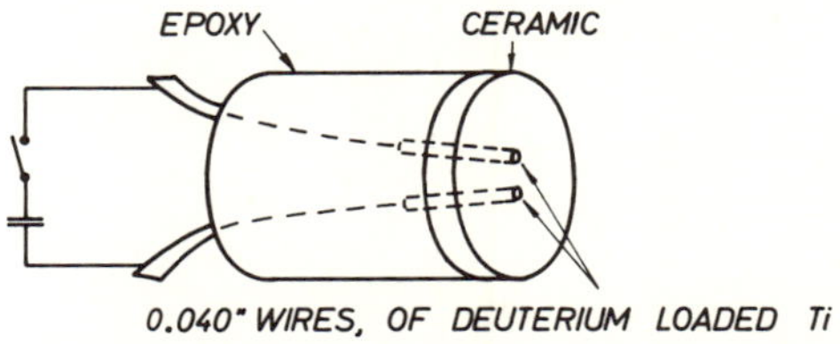

FIG. H3c-1. Button plasma gun or source. By permission from Plasma motors, W. H. Bostick, *in* "Conference on Extremely High Temperatures," Wiley, New York, 1958.

a small button probe (Fig. H3c–1) can project plasma consisting of metallic ions, deuterium ions, and electrons at speeds up to 2×10^7 cm/sec. These speeds are measured in a vacuum chamber by time of flight by means of an oscilloscope. The first-arriving plasma signal corresponds to this high speed of 2×10^7 cm/sec. There are later signals corresponding to other portions of the plasma which are traveling more slowly. However, from the predominantly

* The remainder of Chapter H, Section 3c, is reprinted by permission from W. H. Bostick, Plasma motors, *in* "Conference on Extremely High Temperatures." Wiley, New York, pp. 169–178 (1958).

positive sign of these slower signals, it can be inferred that much of this slower plasma corresponds to plasma which encountered the walls of the cylindrical vacuum chamber and thereby was slowed down.

It is necessary to devise, therefore, some method which measures the total momentum of the plasma and also the mass of the plasma. Crude ballistic pendulum measurements of the plasma momentum were made at the University of California Radiation Laboratory in Livermore in 1955 by David Finkelstein and W. H. Bostick. More refined measurements are now planned. The method is diagrammed in Fig. H3c-2. The total momentum of the plasma will be measured by the swing of the ballistic pendulum bob which is a ping pong ball with a $\frac{1}{2}$ inch hole cut in it to receive the plasma. The gun, for the initial measurements, will consist of a two wire, button-type source where the wires are made of copper. The capacitor will be discharged through the spark gap switch and permitted to ring until its energy has been dissipated, and the momentum mv of the plasma measured by the swing of the ballistic pendulum. The plasma gun and ping pong ball will each be weighed accurately. The gun will then be fired into the ping pong ball n number of times. The ping pong ball and the plasma gun will then be weighed again. The difference in the weight of the ping pong ball before and after gives the total weight of copper plasma ejected. The difference in the weight of the plasma gun before and after will serve as a check on this value. The mass m of plasma ejected per discharge of the capacitor can then be computed and since mv is already known from the ballistic pendulum method, the effective v can be calculated.

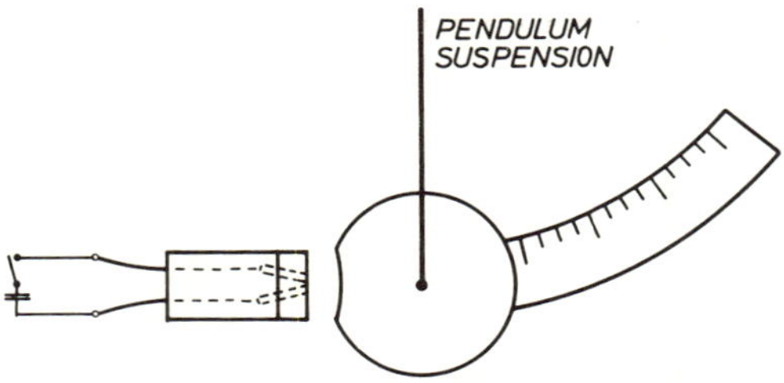

FIG. H3c-2. Ballistic pendulum method for measuring momentum of the plasma. By permission from Plasma motors, W. H. Bostick, *in* "Conference on Extremely High Temperatures," Wiley, New York, 1958.

Although oscilloscope measurements show speeds of 2×10^7 cm/sec. for the first arriving plasma from the button-type gun shown in Fig. H3c-1 and also the coaxial (Finkelstein) gun shown in Fig. H3c-3, from the practical point of view these guns suffer from the following weaknesses:

(a). The back emf (the gun is essentially a linear motor, whose armature delivers a back emf) delivered by the guns is in general small compared with the voltages which are suitable for capacitors. Consequently, the discharges of the capacitor are not anywhere near critically damped, and the capacitors

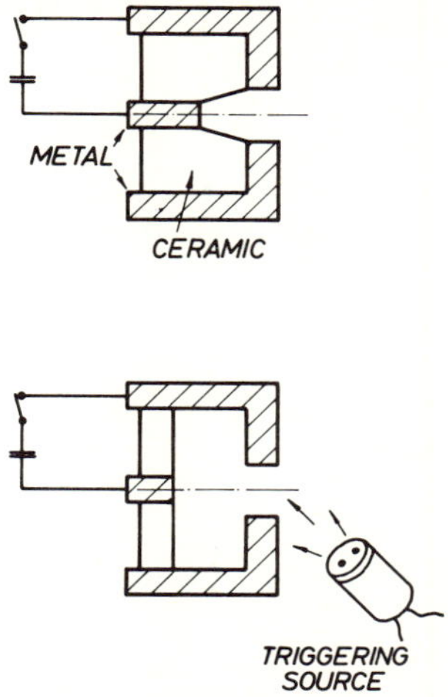

FIG. H3c-3. Coaxial plasma guns. By permission from Plasma motors, W. H. Bostick, *in* "Conference on Extremely High Temperatures," Wiley, New York, 1958.

ring for many cycles, thus dissipating much of their energy in circulating currents rather than storing that energy as kinetic energy of the plasma. The use of pulse transformer coupling between the capacitor and the plasma gun can improve the matching of impedance and bring about a critically damped discharge, but then the discharge time of the capacitor is correspondingly lengthened.

(b) The region where the magnetic forces are concentrated and hence effective is within only a few millimeters of the gun. The time taken for a fast moving plasma ($\sim 10^7$ cm/sec) to travel this short distance is a small fraction of a microsecond. It is impractical to try to have capacitors which store reasonable amounts of energy ($\sim 10^3$ joules) discharge in this short time. Slower capacitors cannot discharge their energy efficiently into the kinetic energy of the plasma with these kinds of guns.

(c) Crude measurements indicate and theory predicts that the button source, at least, is not at all unidirectional in its plasma pattern, but fires over a fairly wide angle. For thrust purposes, only the forward momentum and energy can be utilized. This large-angle pattern of fire therefore represents a reduction in efficiency.

A new system is that of accelerating a sample of plasma by passing a current through the plasma as it rides on the rails, as indicated in Fig. H3c-4 This scheme for accelerating plasma by the current in the rails is essentially the electromagnetic gun, except that the bullet in this case is a mass of plasma or ionized gas. The analysis of the acceleration of a bullet by a rail system has been dealt with elsewhere [426]. Russian investigators [427] have made a theoretical analysis of the acceleration of plasma on a rail system where an externally excited magnetic field is applied to the plasma. However, they have not considered the effect of the magnetic field due to the current in the rails.

Russian investigators [428] have also conducted experiments involving the acceleration of a plasma sample, which begins as a metallic wire, by means of rails, and have achieved speeds of 10^7 cm/sec.

The simple analysis that follows gives a relatively easy way of assessing the effect of various parameters in a rail-type plasma motor without the laborious task of numerical integration of the equations of motion.

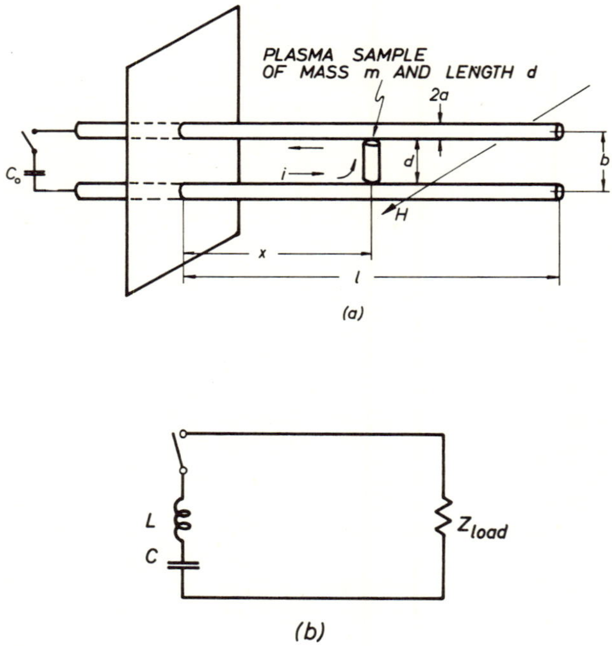

FIG. H3c-4 (a) Rail-type plasma accelerator or gun, and (b) an approximate equivalent circuit. By permission from Plasma motors, W. H. Bostick, *in* "Conference on Extremely High Temperatures," Wiley, New York, 1958.

The Series Plasma Motor: If, as is indicated in Fig. H3c-4 m is the mass of the plasma sample which is placed between two rails of radius a and spacing b, and if a current i flows down one rail, through the sample, and back the other rail, the plasma sample is immersed in a magnetic field of value

$$H_{\text{av}} = \frac{2i}{10b} \ln b/a \cong \frac{2i}{10d} \ln b/a \qquad b \cong d$$

The force on this sample is then

$$f = \frac{2i^2}{10^2} \ln b/a$$

The acceleration of the plasma sample is given by $d^2x/dt^2 = f/m = (2i^2/10^2m) \cdot \ln b/a$. The velocity acquired by the plasma sample is

$$v = \int_0^t (d^2x/dt^2)\, dt = (2/10^2m)\,[\ln b/a] \int_0^t i^2\, dt$$

$$= (4.6/10^2m)\,[\log_{10} b/a] \int_0^t i^2\, dt$$

The impedance Z_{load} may be thought of as an approximation to the impedance of the load which consists of the two wire transmission line and the plasma. Fairly efficient transfer (about 50%) of energy to the load occurs when $Z_{\text{load}} \cong Z_0$. This energy is shared between the kinetic energy of plasma and the inductive energy stored in the two-wire transmission line. Almost complete transfer of energy to the plasma occurs when $Z_{\text{load}} \gg Z_0$. Under these circumstances the back emf due to the plasma traveling in H_{av} is sufficient to reduce the current almost to zero, and there is then very little magnetic energy left in the transmission line. Table H3c-1 shows some practical numbers. It can be seen that with a pair of rails 50 cm long, when $b/a = 10$, and a current I of 10^4 amp, a 10^{-7} gm sample can be given a speed of 4×10^7 cm/sec. The effective plasma impedance $Z_{\text{load}} = 0.42$ ohm is fairly high and it is easy to obtain a capacitor with $C_0 = 2.4\ \mu$f, and L_0 low enough so that $Z_0 = (L_0/C_0)^{\frac{1}{2}} \leqq 0.42$ ohm.

TABLE H3c-1

ELECTRICAL DATA FOR PLASMA MOTORS

$I = 10^4$ amp, $m = 6 \times 10^{16} \times 1.6 \times 10^{-24} = 10^{-7}$ gm, $b/a = 10$, $t = 10^{-6}$ sec

	$H = 0$ (series motor)	$H = 10^4$ (series shunt motor)
v	4.6×10^7 cm/sec	1.46×10^8 cm/sec
l	23 cm	73 cm
$V_x = 0$	4.24×10^3 volts	2.8×10^4 volts
E_{in}	22 joules	140 joules
C_0 from $E_{\text{in}} = \frac{1}{2} C_0 V_0^2$	2.4 μf	0.36 μf
L_0 from $\sqrt{L_0 C_0} = t$	0.42×10^{-6}h	2.76×10^{-6}h
Z_{load}	0.42 ohm	2.8 ohms
Z_0	0.42 ohm	2.8 ohms

An obvious step is to add an externally excited magnetic field H, as indicated in Fig. H3c-4a, in order to obtain the analog of a series shunt wound motor. The effect of such additional shunt field H is to increase the field in which the current i in the sample is flowing (The calculation is given in [406].)

Presumably the most efficient way to operate the motor is to adjust the parameters so that the back emf reduces the current to zero, and hence leaves no energy stored in the transmission line, just as the plasma leaves the end of the rails. Under these circumstances all of the energy stored in the capacitor is transformed to kinetic energy of motion of the plasma during the first half-cycle of current. Moreover, no arc will be drawn at the end of the rails as the plasma leaves because no current will be flowing.

The series and series shunt motors diagrammed in Fig. H3c-4 put their energy predominantly in the forward direction. They also are capable of developing adequate back emf's. It can thus be seen that they do not suffer from the same difficulties as the button sources.

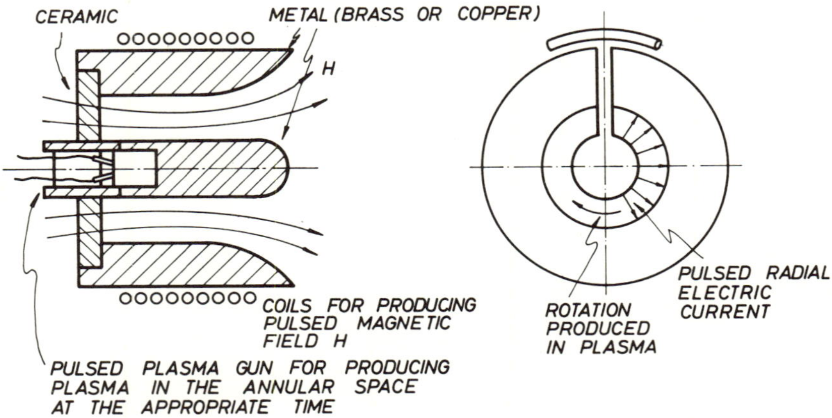

FIG. H3c-5. Arrangement for a rotary, shunt plasma motor. By permission from Plasma motors, W. H. Bostick, *in* "Conference on Extremely High Temperatures," Wiley, New York, 1958.

A variation on the shunt plasma motor is to arrange the magnetic field as shown in Fig. H3c-5. Here the plasma will pick up rotational kinetic energy and the duration of the application of the current can be chosen to be as long as one pleases. Hence, the plasma can be accelerated in principle to speeds which are limited only by the mechanical strength of the materials used in the apparatus. The rotational kinetic energy will be transformed to translational kinetic energy as the plasma ring is propelled to the right by the gradient in the magnetic field.

The matching of the impedance of the power source can presumably be accomplished by making an array of button sources and connecting them all in series. Although such an arrangement would seem *a priori* to be inferior to the rail-type plasma motors described in Fig. H3c-4, it will nevertheless present some interesting phenomena involving the manner in which the individual pieces of plasma ejected from the individual guns interact with one another.

It has been shown that, in principle, it is possible to transfer most of the energy stored in a capacitor to the plasma in the form of kinetic energy. The energy expended in ionizing the gas is a small fraction of the kinetic energy stored in these ions, the latter being 5000 ev for hydrogen ions at $v = 10^8$ cm/sec. The kinetic energy of metallic ions at these speeds is correspondingly larger.

The specific impulse is given by v/g. Thus for speeds of 10^8 cm/sec the specific impulse is 10^5 sec.

Fig. H3c-5 shows an arrangement for a rotary, shunt plasma motor where the rotational velocity of the plasma ring becomes transformed to linear velocity v as the ring is ejected from the motor. The copper electrodes also serve as flux concentrators so that the magnetic field H can be made as high as 10^5 gausses without any great difficulties.

d. *An Electrodeless Magnetic Shock Tube* [408]

H. Blackman and B. Niblett described the different methods for producing high-velocity plasmas in a low-density gas by producing currents which can then lead to an interaction with an external magnetic field and resultant forces on the gas so as to induce an electrodeless discharge in the gas [420]. An attractive advantage of this technique is the possibility of obtaining plasma without electrode contamination. In order for this approach to work, the electric field induced by a rapidly varying magnetic field must be strong enough to cause a breakdown in the gas, and the induced current loops must be closed in the gas, thus eliminating the need for electrodes. To produce such a field, the geometry indicated in Fig. H3d-1 was used at Princeton University.

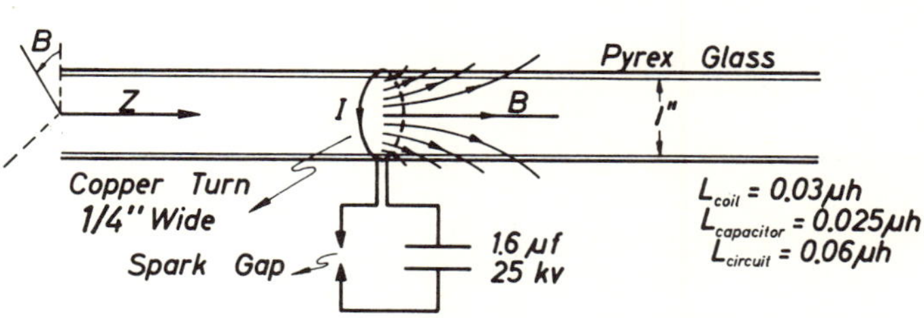

FIG. H3d-1. Schematic diagram of apparatus for producing high velocity plasmas. $\nu = 380$ kc; $I = 75{,}000$ amp; max $dI/dt = 2 \times 10^{11}$ amp/sec; max $B_{axis} = 36$ kgauss; max $dB_{axis}/dt = 90$ kgauss/μsec; max $dB/dt = 350$ kgauss/μsec; max $E_\theta = 900$ volts/cm. From Blackman and Niblett [408].

A low-inductance capacitor is discharged through a single-turn copper coil wrapped about a glass tube. The induced E field is in the tangential (Θ) direction and, as shown by Fig. H3d-2, has its maximum value at the surface of the glass tube. The electric field (about 900 volts/cm maximum for most of the work reported here) is sufficient to cause an electrodeless breakdown in

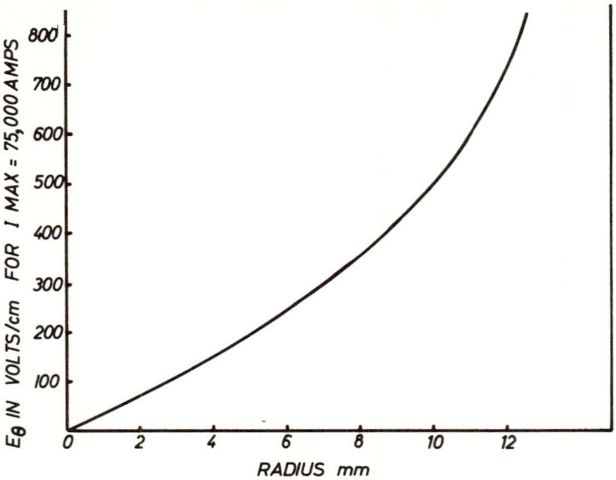

FIG. H3d-2. Dependence of induced electric field on radial position in the plane of a circular loop of 1.5 cm radius at a ringing frequency of 380 kc/sec. From Blackman and Niblett [408].

argon for pressures from the micron region up to 50 mm, while in air the upper limit for breakdown is about 10 mm. The gas breaks down initially at the walls of the tube and implodes inward; then high-velocity plasma is observed to propagate outward in both directions from the plane of the coil. Rotating drum photographs have been used to measure the velocity and attenuation of the plasma as a function of distance from the coil. Photographs taken with a slit perpendicular to the direction of the shock propagation show that the shock becomes plane within about three tube diameters [408].

A series of measurements was carried out to study the breakdown process and the relation between electric field strength and gas pressure at the onset of breakdown. The electric field induced in the gas was varied by changing the charging voltage on the capacitor while the inductance remained constant. A number of tests were conducted for each pressure and the reproducibility of breakdown was reasonably good. The results for air are shown in Fig. H3d-3. It can be seen that the minimum breakdown field for a frequency of 380 kcps occurs at a pressure of about 0.3 mm Hg pressure and at a field strength of about 380 volts/cm. The experimental error in both field and pressure readings was about 12% for each point. For comparison, the results for breakdown in air for dc fields between metal electrodes as given in Meek and Craggs have also been included. It is possible to make an order of magnitude comparison between these graphs only because the Townsend theory, rather than the diffusion criterion, determines the breakdown pressure limit of a gas between metal electrodes. Some points were also taken in argon and helium at higher pressures and these measurements agreed fairly well with the curves of Meek and Craggs and also Schumann.

The delay between the initiation of the external current and the gas break-

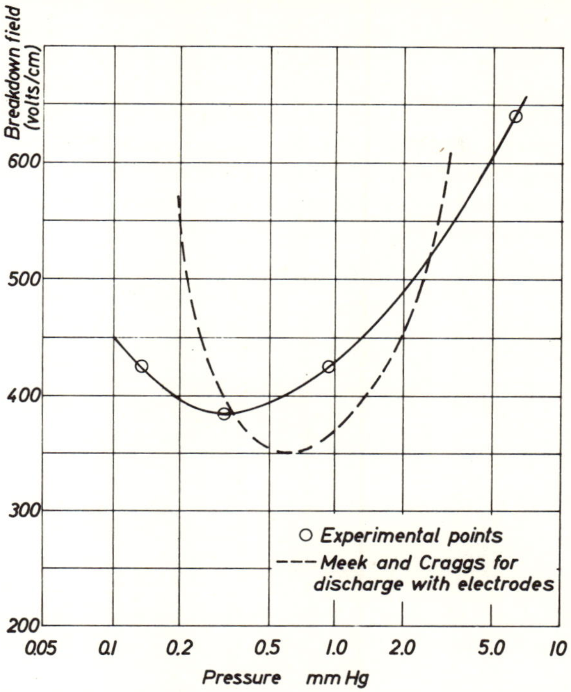

FIG. H3d-3. Dependence of breakdown field strength on gas pressure at a frequency of 380 kc/sec. From Blackman and Niblett [408].

down in the tube was also determined. The light from the spark gap was viewed with a photocell which triggered an oscilloscope and the light from the tube discharge was picked up with another photocell and viewed on the trace. The time lag for light to appear in the tube after the capacitor discharge started was 1.3 μsec, just half of the 2.6 μsec period of a full oscillation. Thus, most evidence showed that in this setup, for the given discharge conditions, the gas breaks down on the second half of the first cycle. When the same experiments were repeated at near-critical pressure the tube took a full cycle to break down.

Since this method can produce relatively plane shock waves of moderate strength ($M = 10$–30) at a distance of 10 to 20 cm from the coil, studies were undertaken to determine the ionization times of various gases. Photographs taken with the streak camera record the motion of the luminosity behind the shock wave since the shock front itself is not visible on the photographs, and this luminosity is observed experimentally to have a sharp front. For the noble gases the luminous front is taken as coincident with the point at which ionization equilibrium is attained [421]. The ionization time is very short for strong shocks and the luminous front draws progressively further ahead of the luminosity; the intermediate dark space is the region in which the gas "relaxes" to its equilibrium condition.

If a series of pictures are taken with the same gas pressure and the same capacitor voltage the streak records can be accurately superimposed and a master diagram constructed showing the locus of the luminous front and the shock front. Such a master diagram is shown in Fig. H3d-4. The Mach number of the shock can be obtained from the angle α and the film speed, and the distance τ along the time axis is a measure of the ionization time. This measured time must be multiplied by the mean density ratio across the shock to give the true ionization time.

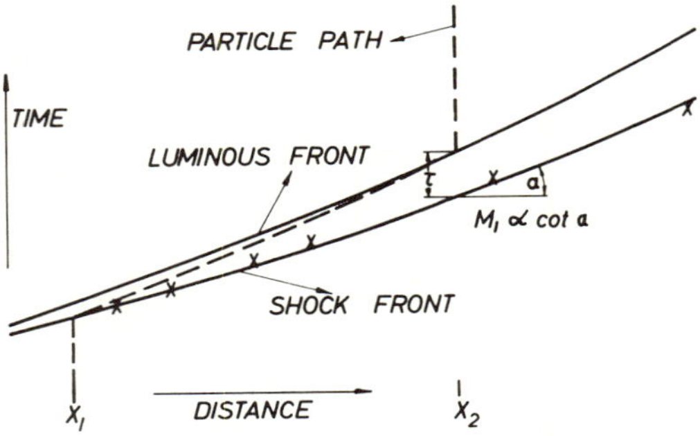

FIG. H3d-4. Diagram indicating the method of evaluating ionization times from streak photographs. From Blackman and Niblett [408].

The time lag of the luminous front in air has also been measured, for the range of ionization times over which there is reasonably good agreement between the measurements for argon reported here and the previous results of Petschek and Byron [421]. The results are plotted in Fig. H3d-5 as log $p_1\tau$ versus Mach number, where once again the Mach number given represents a minimum value. The experimental points fall roughly on a straight line when plotted in this manner.

It is believed that the radiation is produced mainly by processes which involve free electrons. This assumption seems reasonable since in the range of photographic sensitivity the radiation is produced principally by oxygen free-free and oxygen free-bound transitions together with nitrogen band spectra which are likely to be excited by free electrons [422]. Thus the luminous front is taken to be coincident with the ionization front and the times given in Fig. H3d-5 represent approximate values for the ionization time in air.

This method of producing high-velocity plasmas, utilizing an electrodeless discharge, lends itself immediately to the study of shock interactions and the interaction of high-velocity plasmas with magnetic fields. Additional coils can be put about the tube and fired in various time sequences to study such effects.

A more recent method for the determination of plasma parameters a theta-pinch plasma has been developed by Fünfer *et al.* [787c]. They have imaged the radiation from a high-power ruby laser onto a theta-pinch plasma of high electron density. The light scattered and emitted

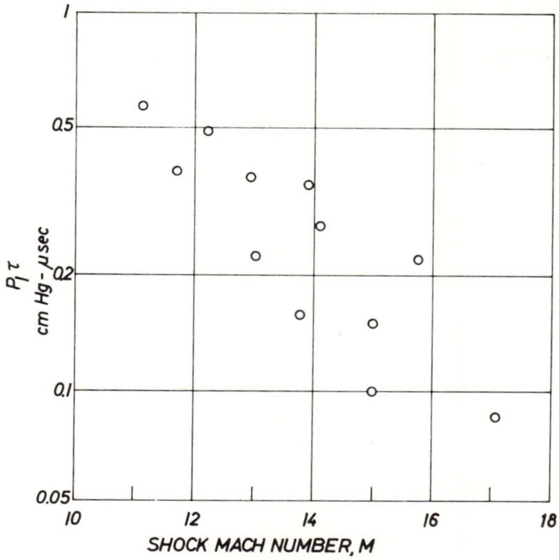

FIG. H3d-5. Results of ionization time measurements for air. From Blackman and Niblett [408].

perpendicular to the incident laser beam is limited by a diaphragm by means of which a certain plasma volume is selected. The light is analyzed by a monochromator followed by photomultipliers and oscilloscopes. In order to suppress the large *bremsstrahlung* from the plasma a differential method is used. The time resolution is limited from 0.1 to 1 μsec for the scattering signal. Excluding many possible sources of error, the experimental results are presented and discussed.

Another recent method for measuring, e.g., the plasmoid rotation in an electrodeless theta-pinch, has been described by Clark and Wuerker [903]. By means of high-resolution STL image-converter cameras, ultra high-speed photographic observations have been made of the behavior of plasma rings or plasmoids due to trapping of magnetic fields within the plasma. The nonrelativistic cyclotron radius of the individual particle orbits has been calculated.

e. *Magnetic Channeling of a Strong Shock*

In the study of plane high-velocity shocks, one serious deterrent to a quantitative understanding of the conditions behind the propagating disturbance has been the strong interaction with the containing walls. The following experiment was initiated to minimize this effect and also to examine a specific interaction of a shock-produced plasma and a magnetic field. Although this is by no means a complete study, it does include interesting macroscopic measurements. Also, it shows that a strongly ionizing shock produced outside an axial magnetic field will channel along the axis using the magnetic field as a "wall." In the configuration used, measurements indicate that the magnetic interaction does not contribute any noticeable heating of the shock-produced plasma as it channels.

For these experiments Scott, Basman, Little and Thomson [410] produced the shock by a low-inductance linear discharge tube with a hollow electrode. Figure H3e-1 is a schematic diagram of the system. The shock propagates along the axis into the cylinder attached to the hollow electrode. Because this hollow electrode is a solid ring conductor, there is good isolation between any alternating magnetic field in the tapered shock tube and any alternating field in the cylinder. Any mixing will decrease the strength of the plane shock by allowing axial flow along lines of force between the two fields. Also, mixing makes any measurements of the interaction of hydrodynamic shock with a magnetic field ambiguous.

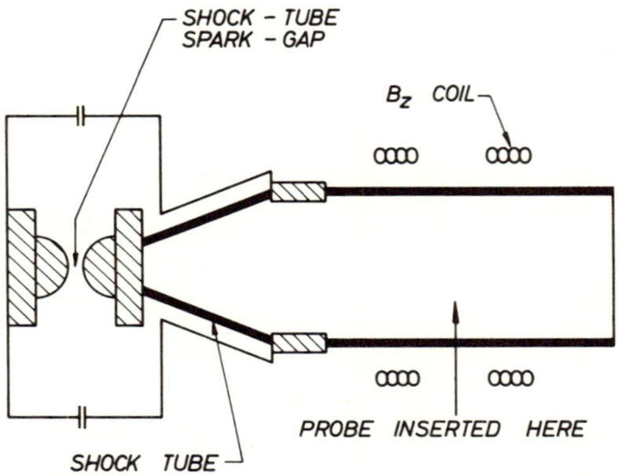

FIG. H3e-1. Schematic diagram of the system producing the shock. From Scott *et al.* [410].

The electrical system for the production of the shock consisted of twelve 1-μf capacitors, rated at 25 kv. Each capacitor was connected in parallel to a coaxial spark gap through a separate length of RG 8/U cable. The spark gap was coaxially connected to the tube. This system had a ringing frequency of 160 kc and a peak current of 200,000 amp at 20 kv. To study the effects of the resulting shock on a magnetic field, a separate low-frequency, low-voltage bank of eleven 100 μf capacitors supplied current to a solenoid wound around the cylinder. The half-cycle period of this system was 6000 μsec, and axial fields up to 6000 gausses were produced in the cylinder.

For these experiments, the slow axial field was triggered first. Then when the field had reached a maximum B_{z0}, the shock tube was discharged. For a typical shock velocity of 8 cm/μsec the time for the shock to traverse the length of the axial coil was less than 3 μsec, a time small compared to the period of the field. During the time of interest the magnetic field was essentially constant.

The magnetic channeling of plasma shocks and the speeding of the ions seems to permit the construction of plasma motors for space vehicles. Wichmann [861] describes some of these motors. Figure H3e-2 shows an advanced model. There is a plurality of magnet field coils, 1, fed by the successively discharged capacitors, 2, by means of spark gaps, 3.

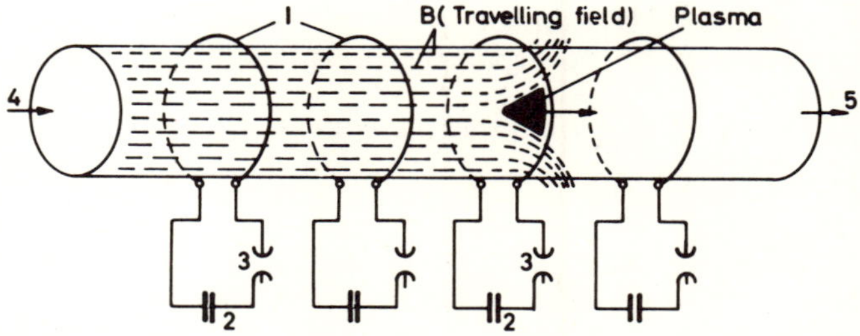

FIG. H3e-2. Schematic of a traveling ave accelerator for space propulsion.

The plasma enters with low speed at 4 and leaves highly accelerated at 5, creating the propulsion. The production of the low-speed plasma can be made by contact ionization of alkali or halogen gases, delivering a stream of positive or negative ions. Granet and Guman [948] show the application of the pinch process for space propulsion which would allow essentially an increase in the efficiency by boosting the ion speed.

A summary of about 20 designs of space propulsion devices is given by Löb [942] (see 30 additional references in [942]). At last a vision of the future

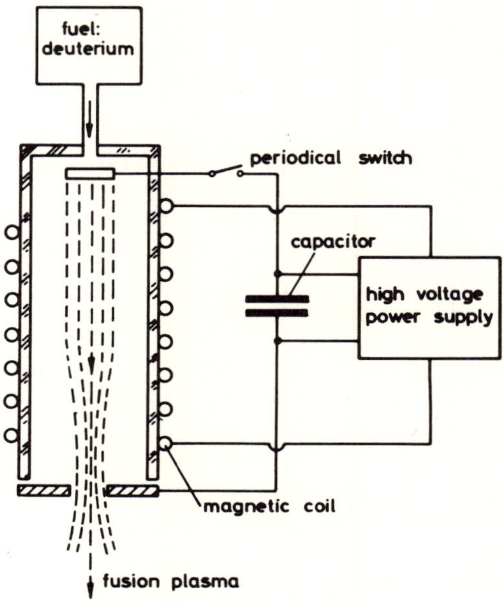

FIG. H3e-3. Schematic of a D_2-fusion space propulsion device.

may be seen in Fig. H3e-3. The ion fuel enters above, a capacitor charged and discharged periodically feeds the electric field for acceleration of the ions, and in a second current path a magnetic coil causes the plasma to pinch. There is hope of acquiring fusion processes and of using the thermonuclear energy produced by fusion of D_2 atoms into He in the form of the extremely high temperature or speed of the departing He atoms for propulsion purposes.

4. Capacitor Discharge Magnetizers

a. *Circuitry and Basic Considerations*

In the manufacture of permanent magnets it is necessary to magnetize the magnets after hardening. The conventional method for that purpose is to subject the magnet to the field of a strong electromagnet, but this practice is becoming more and more obsolete owing to the high cost of strong field electromagnets, and is being replaced by the impulse magnetizing method. Basically it is possible to store the impulse energy either inductively or capacitively. Inductive storage was for more than 40 years the preferred selection, because for a similar energy content the dc magnetic choke coil was less costly than a capacitor. With the advent of economical capacitor construction the picture has changed and, for similar energy-storing capacity, the storage capacitor has become more economical than the storage inductance, particularly since improvements in inductance design have not been made and for practical reasons, do not seem likely. The energy of a choke, as well known, resides chiefly in the field of its air gap and this field as well as the magnet coil of copper have not been the subject of fundamentally new discoveries. This is the reason for the fact that magnetizers today generally utilize the capacitive energy storing principle. The simplest devices of this kind employ electrolytic capacitors such as those used in photoflash devices and as fabricated in mass production. For the purpose in view such a capacitor is discharged through the magnetizing coil of the device. When one utilizes, for example, a coil having an inductance of 100 μh (about 100 turns around 1 cm dia for small bar magnets) and discharging a capacitor of 500 μf at 500 volts, the duration of one half-cycle (oscillation is impossible with the polarized electrolytic capacitor, since the capacitor would become conducting in the opposite direction) is

$$t = \pi\sqrt{LC} = \pi\sqrt{500 \times 10^{-6} \times 10^{-4}}\ (\text{sec}) \approx 0.7 \text{ millisec}$$

and, owing to strong damping, somewhat slower, say, about 1 millisec.

The current magnitude, in accordance with the capacitor charge of

$$Q = C \times U = 5 \times 10^{-4} \times 5 \times 10^{2} = 0.25 \text{ amp-sec}$$

released during 1 millisec, becomes thus $I = 250$ amp. With the 100 turns referred to, this gives 25,000 amp-turns. In practice this value will not become effective but must be reduced to about 15,000 amp-turns owing to ohmic losses in the coil and displacement losses in the electrolyte of the capacitor. This example shows that it is possible to magnetize effectively with exceptionally simple means. However, a decisive disadvantage of this method is that, in continuous operation at a high rate of magnetizations, disintegration effects will soon occur in the electrolyte of the capacitor, and this rules out the use of this simple device in mass production. When the electrolytic capacitor is replaced by a metal-paper capacitor and switching is accomplished by an ignitron, the device becomes very rugged but also more expensive and larger.

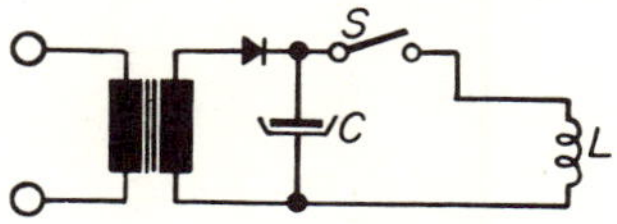

FIG. H4a-1. Schematic circuit for magnetizing by discharge of an electrolytic capacitor C through coil L and switch S. Oscillation impossible owing to electrolyte.

In Figs. H4a-1, 2, and 3, schematic principal diagrams of the simplest device, of a device with ignitron switching, and of a device with transformed capacitor discharges are represented for comparison.

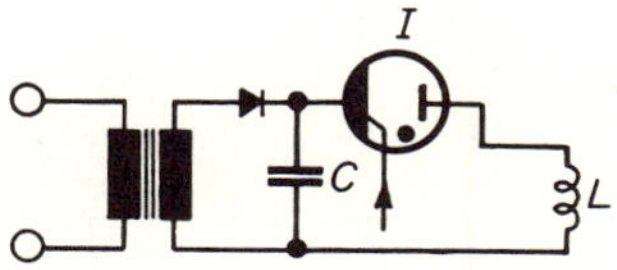

FIG. H4a-2. Schematic circuit for magnetizing by discharge of a high quality, metal-paper capacitor C through an ignitron I, which also prevents oscillation, into coil L.

The device of Fig. H4a-3 can be designed, just as the conventional device, with a choke coil. Figure H4a-4 is a single conductor unit, that is, the magnetizing current flows through only one bar-shaped current conductor, so that, for watt-hour meter magnets and other mass-production magnets that must be very precisely calibrated, the magnet can be magnetized by simple time-saving insertion.*

Ceramic magnets which have a coercivity almost three times greater than alloy magnets may require currents higher than 45,000 amp. To avoid the use

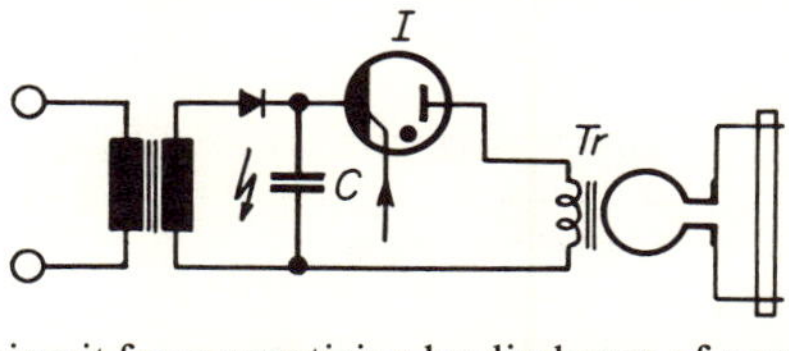

FIG. H4a-3. Schematic circuit for magnetizing by discharge of a great energy, high voltage capacitor C through an ignitron I into impulse transformer Tr. A bar conductor is connected to the one-turn secondary in which are induced 10^4–10^5 amp.

* Single conductor magnetizing makes it possible to excite the magnet in its so-called neutral zone. The poles, which are critical with respect to the emerging direction of the lines of force, are not directly affected by the exciting magnetic field. Thus the change in the direction of magnetic force lines at the poles that is responsible for magnet aging is prevented.

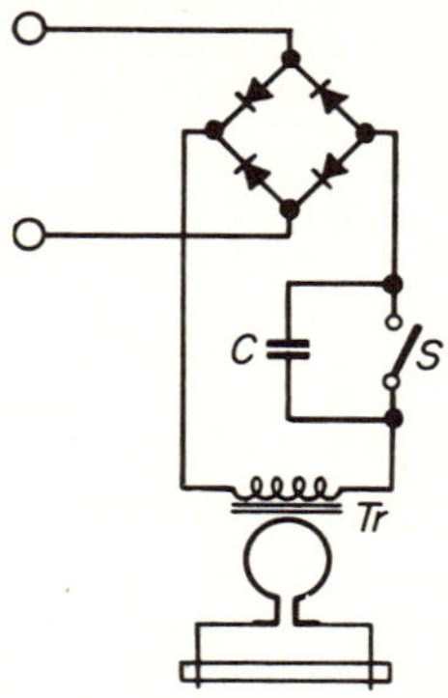

FIG. H4a-4. Schematic circuit diagram of a conventional magnet charger with inductive storage. The primary of transformer Tr is traversed by a dc value saturating the core, then switch S is opened and the spark is quenched by capacitor C. A nonoscillating current impulse will occur in the single-turn secondary.

of larger transformers, reversal of current in the secondary can also be reduced or almost entirely eliminated by various other devices [780].

One method is to bias the transformer to saturation magnetically by means of a dc winding. The transformer is then energized in the opposite direction and the whole flux range of the core becomes available. Output may be doubled for a given core size, and there is no secondary current reversal. A novel method of biasing the core of the transformer is to fill the usual air gap with ceramic magnet material as this material can function satisfactorily in the form of thin plates magnetized through their thickness.

A second ignitron, as shown in Fig. H4a-5, can be used as an inverse shunt tube across the transformer primary, and fired when the primary voltage becomes negative; thus the primary voltage is held to a value near zero and current flow in the primary coil is maintained in the same direction. The secondary current therefore also flows in the same direction and dies away exponentially.

The energy in the secondary can be prevented from oscillating back into the primary circuit if a shunt tube is fired at a time when the capacitor bank

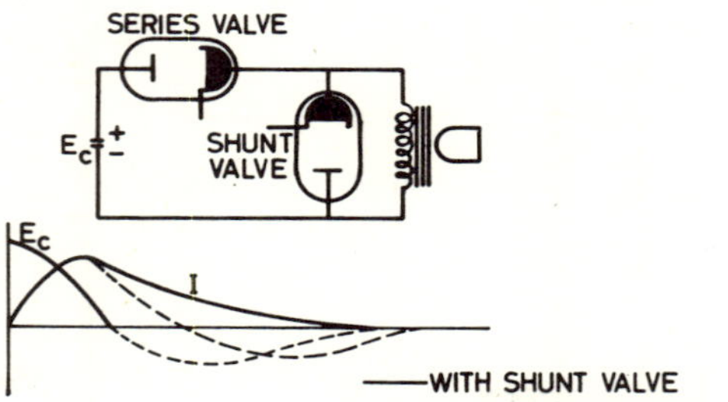

FIG. H4a-5. Ignitron capacitor discharge magnetizer with shunt ignitron clamp.

is fully discharging. The current in the secondary winding decays exponentially and ceases abruptly when the current value will no longer support a cathode hotspot in the shunt ignitron.

When a permanent magnet material is subjected to an excitation field and then care is taken that the field does not decay too rapidly, there remains in the magnetic material a residual magnetism. The magnitude of this remanent magnetism grows with increasing field strength toward a peak which is reached at a definite excitation. With a further increase of excitation above this definite value no further gain in residual magnetism can be had. In order to reach the peak of this residual value, it is not necessary to drive the excitation up to magnetic saturation. There is a somewhat lower value at which the magnetizing progression that leads to the peak residual condition ceases.

In a ferromagnetic material there are present certain domains, the so-called Weiss's domains, which are magnetized up to saturation in one direction. The change from one domain to the other is not abrupt but is spread over a zone in which the electron spines alter from one direction to the other. This transition layer, called the Bloch wall, has a finite half-value thickness. In cases where the magnetizing vectors of adjacent domains lie in diametrically opposite directions, reference is made to a 180° wall; otherwise, to a 90° wall.

In a pressure- and strain-free monocrystal there exist several crystallographically conditioned directions which are magnetically equally important. When the magnetic elementary domain is heated above the Curie point and cooled again, it magnetizes iteslf up to saturation in one of the equally important preferred directions. The crystallographically conditioned preferred directions of magnetization are strongly restricted by compression and tension stresses, and also the geometrical form of the elementary domains influences the orientation of the magnetization factor.

The magnetization factors, which in a demagnetized body are distributed so that the resulting magnetic moment is zero, can be influenced by an external excitation field. With a low excitation field, the volume of an elementary domain whose magnetizing factor lies at a small angle to the direction of the excitation field grows at the expense of the adjacent domain so that the transition layer is shifted toward the less fortunately located magnetizing factor. When the excitation field disappears, the wall returns to its original position. Such change of magnetization is termed "reversible wall displacement." If the transition layer passes over the whole volume of the unfavorably oriented elementary domain it remains in its new position. Then reference is made to an irreversible wall displacement or an irreversible transition. If the transition of a magnetizing vector amounts to a reversal into a diametrically opposite direction and it remains in this position after decay of the excitation field, this is called an irreversible 180° transition. Whenever the change in direction is less than 180°, it is called a 90° transition. No further mechanical deformation in the ferromagnetic material and runoff is caused by 180° transitions, whereas 90° transitions, generally at a low excitation field strength, often cause shortening or elongation of the test sample.

A further magnetizing phenomenon is the rotation process. This phenomenon does not contribute to the remanence value and is mentioned here for completeness only. In case the magnetizing vector is shifted by a 180° or 90° transition into its crystallographically preferred direction as determined by crystallographic structure conditions, which possibly is at a small angle to the direction of the excitation field, a rotation of the magnetizing factor from its preferred position can effect an increase of magnetization in the direction of the excitation field. Upon decrease of excitation the magnetizing factor turns back to its original position. The rotation process is always a reversible magnetization phenomenon.

The progression of magnetization in a sample is, roughly as follows: In demagnetized condition the single magnetizing vectors of the magnetic elementary domains are distributed so that the resulting moment in a sufficiently large volume is zero. The number of magnetically equally valued directions, which are given by the crystallographic condition of the elementary cells within the volume under observation, is large but finite. Under the influence of an external excitation field, a reorientation occurs; the magnetizing vectors move into an energetically favorable position. It can be assumed with fair safety that at first all possible 180° and 90° reversible and irreversible transitions occur. Since, however, the number of 180° and 90° transitions is finite, the magnetization increases up to a value $I = \lambda \times I_s$, whereby $\lambda < 1$. Further increase of magnetization at rising excitation is effected by the rotation process. Only the irreversible magnetizing phenomena contribute to remanence. In order to obtain the maximum remanence value in a permanent magnet material it is necessary to raise the excitation only to the point at which all possible 180° and 90° transitions have occurred. The excitation to reach that point is considerably lower than that at which saturation is obtained.

Figure H4a-6 represents the magnetization curve of a permanent magnet. The saturation value of this material is reached at an excitation of 2000 to 3000 amp/cm. The irreversible occurrences, however, have already ceased at about 800 amp/cm. Thus, in order to magnetize this permanent magnet up to its maximum remanence value, it is necessary to raise the excitation only up to 800 amp/cm and then to lower it so slowly that the forced orientation is not disturbed.

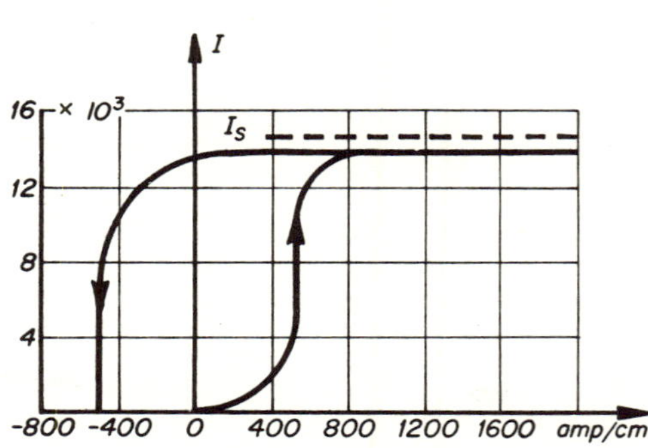

FIG. H4a-6. Magnetizing curve of a permanent magnet material (H-B curve).

The value of the excitation field at which all irreversible phenomena have occurred can be taken from the static magnetizing curve. The conditions in question have been reached at the point where the dropping branch enters the rising branch of the hysteresis loop. In the following discussion this point will be termed the "entering point."

For magnetizing a permanent magnet, an excitation field is required that increases up to the entering point and then drops off at a reasonable speed. Solenoids or impulse methods are employed for producing an excitation field. In case the entering point lies at H_E amp/cm and the magnet has a length of l cm, the maximum value of remanence is reached when the relationship $H_E = nI/l$ is met. In this, formula n is the number of turns in the coil and I is the current value. In the solenoid method, n is selected high and I small; in the impulse method, n low and I large. In the case where $n = 1$, as is usual in the impulse method, the current becomes $H_E \times 1$ cm-amp, a value high enough to magnetize the permanent magnet up to its maximum remanence. In the impulse method it is necessary to give consideration to the rate of change of the excitation field dH/dt particularly during the dropoff. When this rate of change surpasses a critical value, strong eddy currents develop in the work piece, causing nonhomogeneous magnetization.

The basic formulas are given by Speadbury [889] and Chanas [871].

In magnetizers operating on the basis of inductive storage and with direct current, the energy is released at interruption of the dc circuit and becomes available at the secondary in a current pulse. The velocity of the field breakdown cannot be accurately determined, since this breakdown is dependent on the opening speed of the primary switch, the temperature of the core, and the type of secondary conductor, all in combination.

b. *Industrial Models of Magnetizers*

In an impulse magnetizer operating on the capacitive principle the energy is stored in a high-voltage capacitor. For magnetization the discharge is delivered to the primary of a closely coupled transformer through a rugged impact switch or an ignitron. An adjustable damping resistance is inserted in the discharge circuit. The magnetizing conductor is connected to the single turn secondary of the transformer.

The energy equation for such arrangement is

$$\tfrac{1}{2}CU^2 = \int_0^t I_p{}^2 R_D dt + \tfrac{1}{2}(L_{TS} + L_I)\, I_S{}^2 + \int_0^t I_S{}^2 R_S dt + A_m$$

where I_p = primary current,
R_D = primary damping resistance,
I_S = secondary magnetizing current,
L_{TS} = inductance of the single-turn secondary,
L_I = inductance of the magnetizing conductor
R_S = total resistance of secondary conductor,
A_m = energy magnetically invested in the work piece.

The following data were given in the data sheet for one industiial device.

Magneto Vitessa
$L_I = 0.2\ \mu h$
$L_{TS} = 0.7\ \mu h$
$I_p = 700\text{–}1500$ amp
$R_D = 2\text{–}6$ ohm
$\frac{1}{2}CU^2 = 500\text{–}2000$ watt-sec

At maximum excitation the device puts out about 55,000 amp (in short circuit about 110,000 amp). Numerically (that is, by measurement) an energy

TABLE H4b-1

EFFICIENCY AND LOSSES OF A MAGNET CHARGER: MAGNETO VITESSA

Charge	2000 watt-sec
Thermal losses in damping resistance	400 watt-sec
Ohmic losses in transformer and magnet conductor	100 watt-sec
Magnetizing input at the work piece, arbitrarily chosen value	100 watt-sec
Available magnetizing energy	900 watt-sec
Efficiency	70%

balance as illustrated in Table H4b-1 will result with $R_D = 2$ ohm. The following relations exist then in the secondary magnetizing circuit:

$$\tfrac{1}{2}(L_{TS} + L_I) \times I_S^2 = 1400 \text{ watt-sec}, \qquad L_{TS} + L_I = 0.9\ \mu h$$

$$I_S = \sqrt{\frac{1400 \times 2}{0.9 \times 10^{-6}}} = 56{,}000 \text{ amp}$$

A complete circuit diagram of the magnet charger is shown in Fig. H4b-1.

Larger devices, even larger than the Magneto Vitessa discussed above, have capacitor batteries up to 8000 joules. They deliver up to 400,000 amp at the conductor of the impulse transformer under short circuit and under a burden (inductance increase by the work piece to be magnetized), between 150,000 and 250,000 amp.*

The heavy duty magnetizer Magneto Vitessa has an output peak current of about 56,000 amp, and heavy magnets, several kilograms in weight, such as the magnets for the so-called magnetron tubes, can be completely magnetized. The minimum energy that a work piece needs for complete magnetization is determined by the necessary current and the inductance of the magnetizing loop corrected as to the contribution by the permeability of the work piece and is thus basically:

$$\tfrac{1}{2}CU^2 = \tfrac{1}{2}L_m I_m^{\,2}$$

An essential advantage [120] of the impulse magnetizing method is that one

---

* Large magnetizer of Raytheon Mfg. Corp. Waltham, Massachusetts.

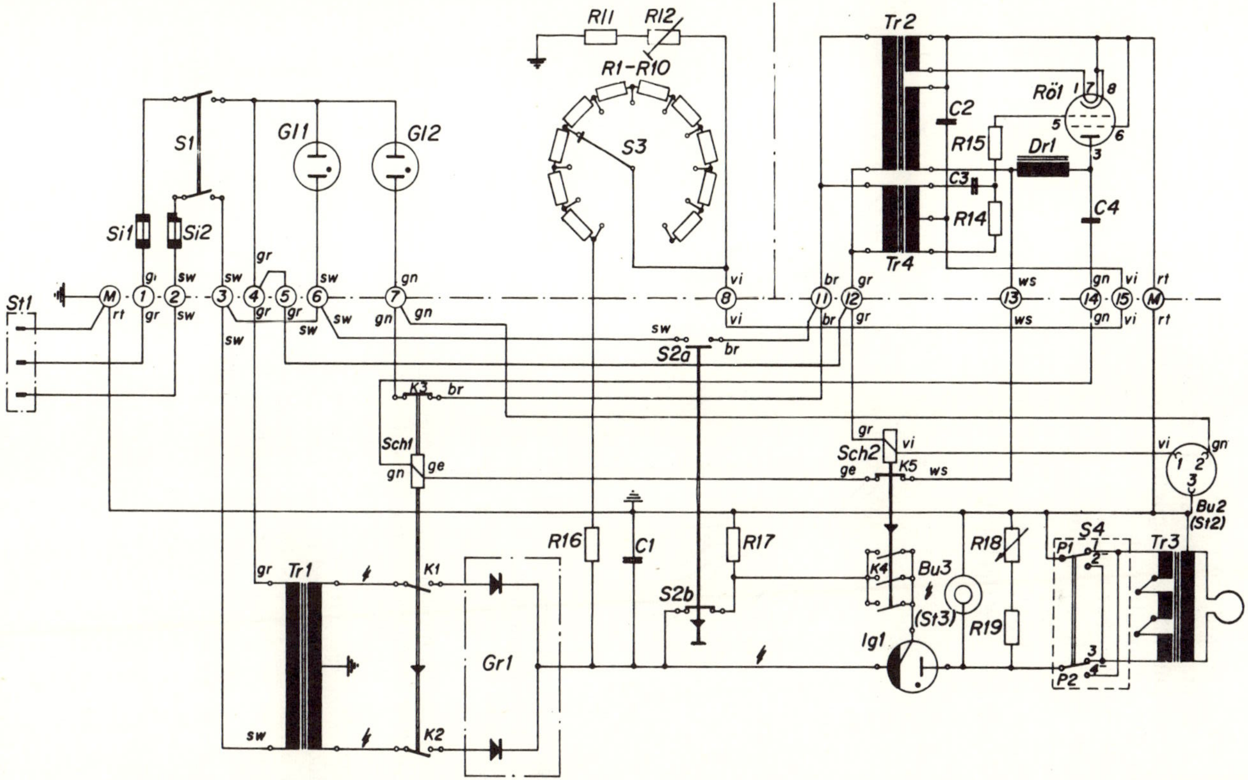

FIG. H4b-1. Schematic circuit diagram of a magnet charger. The capacitor battery C_1 is being discharged by way of Ignitron Ig 1, through the primary of transformer Tr 3. Up to 56,000 amp flow then in the secondary magnetizing conductor. (Works diagram, FF Impulsphysics Corp.)

can magnetize magnets of shapes such that the exact introduction of the magnetizing flux into a yoke of the magnet is impossible. This is true, for instance, of ring magnets for moving coil instruments and watt-hour meters. Older types of magnetic materials of relatively high permeability and small coercive force can still be magnetized by conventional methods. However, the higher alloyed cobalt steels cause difficulties, because a flux distribution in accordance with the shape of the magnet is not possible; the magnet becomes inconsistent and thus useless for devices of highly rated accuracy. In impulse magnetizing, on the contrary, the lines of force form concentrically around the conductor which is inserted in the ring magnet and the magnet is thus uniformly penetrated by the flux (Fig. H4b-2).

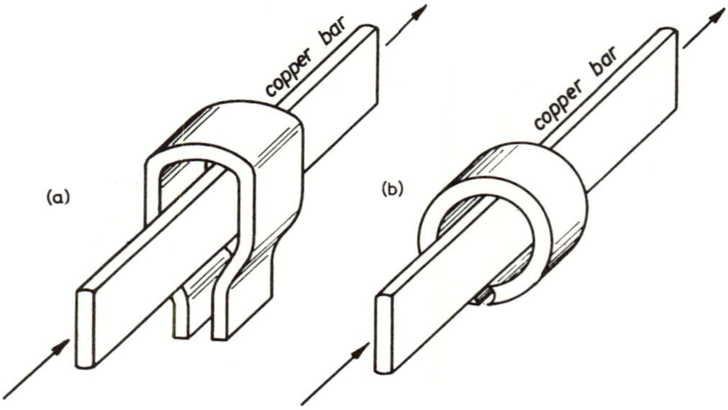

FIG. H 4b-2. (a) Copper bar of an impulse magnet charger with ferromagnetic guide for the magnetic field around the bar for magnetizing of prismatic magnets. (b) Copper conductor bar inserted in a ring magnet to be magnetized by a current impulse.

The open gap or mouth of the magnet must be closed during magnetizing by an iron slug or disk; also the air gap provided for its later use should be tightly closed by a suitable piece of brass. In taking the magnetized part from the magnetizer, care should be taken that the circuit of magnetic lines of force, which will be assured in use, is not even temporarily interrupted. It is therefore necessary always to provide keepers or magnetic bridges during shipment and for any length of storage time.

It is obvious that magnetization of straight magnets and magnet systems also is possible by the impulse method. For that purpose the magnetizing current must be properly shaped to conform to the magnet. In certain cases a sufficiently heavy cable can be used as conductor (observe adaptation resistance). However, a properly formed magnetizing yoke, around which the current bar is to be arranged, is more effective. (See example in Fig. H4b-3.)

For magnetizing of already completely assembled ferritic loudspeaker magnets a setup similar to that depicted in Fig. H4b-3a is used by the magnetizer Magneto Vitessa. In such magnets, the closing of the field flux line

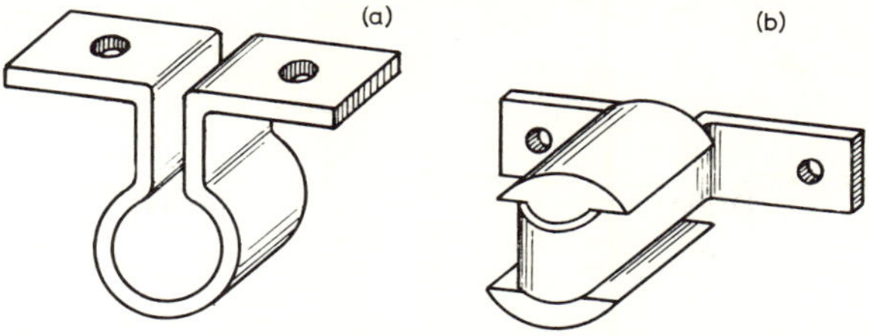

FIG. H4b-3. (a) Omega-shaped magnetizing conductor for magnetization of bar magnets and loudspeaker magnets of high coercive force. (b) Double-T armature of highly magnetizable dynamo sheets with surrounding magnetizing conductor for magnetization of internal fields in ring or stator magnets.

circuit by a keeper is not necessary, since the cup magnet itself later produces the flux in the air gap in which the oscillatory loudspeaker coil is to operate. Figure H4b-4 shows an oscillogram of the magnetizing current flowing during magnetizing with such a fixture. A peak current of about 70 kamp is indicated. By application of the mono-bar conductor fixture it is possible to protect the air gap of the loudspeaker magnet against any iron particles that might enter during magnetization. This is very important for reducing reject percentage, because the finest dust formation in the gap adversely affects later acoustical speaker diaphragm vibration.

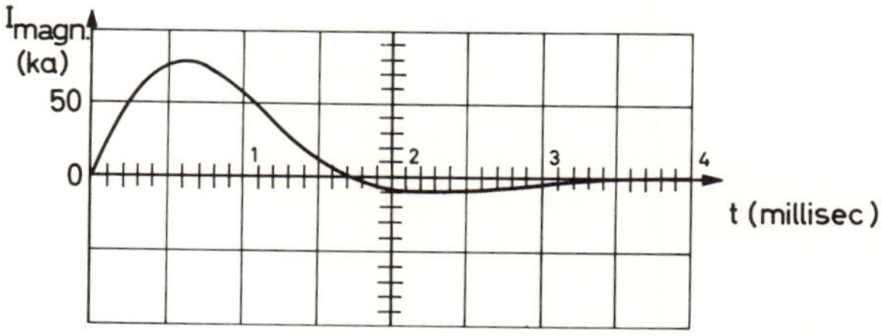

FIG. H4b-4. Oscillogram of a magnetizing impulse in the magnetizing conductor of Fig. H4b-3a.

Modern magnetizing equipment [789] must have the following characteristics:

1. High magnetizing field strength, to assure a magnetization to saturation for all technologically common materials and all possible magnet shapes.
2. High production speed. The prerequisite for an economical magnetizing process for mass production is that the magnetizing energy can be applied in adequately short time intervals.
3. Simple operation by trained personnel.
4. Possible fully automatic operation, e.g., by photocell control.
5. Applicability of formed parts in any shape for the magnetization (bar-, ring-, horseshoe-, pot-shaped magnets, etc.).
6. Reproducibility of the maximum magnetizing condition with identical samples.

The time-amplitude characteristics of the decaying magnetizing current is of decisive importance for the attainment of the saturation. As is well known, the magnetic field of a changing current induces a current of opposite direction in every adjacent conductor, which, in turn, again produces a magnetic field opposite to the original one. For example, if the magnetizing current, after it has reached its maximum amplitude, suddenly drops to zero then the just-magnetized sample becomes re-magnetized in the reverse sense and therefore can in no case possess the largest possible remanence. The inducing current and the current induced in the sample are timewise comparatively equal when the current decay follows an exponential function. It took several years of extensive developmental work until this relationship could be realized completely with high-current equipment.

Principally, every permanent magnet, initially, should be magnetized to saturation, regardless of whether the maximum or a lower work-induction is desired as a final value, if a stable permanent magnetic condition is to be obtained. If the final value has to be lower than the maximum then it is advantageous to obtain this condition by a demagnetization of the magnet from the maximum to the desired value. It would be wrong to produce a weak magnetizing condition by a correspondingly low magnetization; instability and nonuniformity of the magnetic condition would be the result.

5. The Magneto-Optical Shutter—Application of the Faraday Effect

a. *A Microsecond Faraday Shutter and Its Application*

A light shutter using the Faraday effect consists of a short cylinder of a suitable medium such as heavy flint glass arranged in a solenoid of few turns. The plane of polarization of plane-polarized light traveling through the medium is rotated under the influence of a current in the solenoid. With no current in the solenoid and with crossed polarizers, a negligible amount of light passes through the optical system. The rotation of the plane of polarization under the action of the magnetic field produces a light flux through the Faraday cell, increasing with the square of the sine of the rotation angle. There exists no measurable time lag between the variation of the magnetic field and the light flux. According to the theory of the Faraday effect the time lag is 10^{-10} or 10^{-11} sec, this inertia being of the same order for the Kerr effect.

The solenoid has an inductance of about 1 μh. In order to produce a sufficient rotation of the plane of polarization a current of more than 1000 amp must exist during the operation of the shutter. Such a current surge is obtained by discharging a capacitor into the coil. The inductance of the coil together with the inertia of the switching device for the current, a spark gap or a thyratron, limits the duration of the opening cycle of the shutter to about 1 μsec.

A Faraday shutter with an effective open time of about 1 μsec is used in a camera described by Edgerton and Germeshausen [310b]. This camera is specially designed to photograph high explosives during detonation. Precision adjustment of the exposure instant by a time delay circuit triggered by the explosion light is used for synchronization. Optical systems of focal lengths of 6 inches to 6 ft have been employed. Examples given include pentolite and TNT explosions.

Experiments with high explosives using a previously described magneto-optic shutter indicated that a shorter exposure time would be advantageous in studying high-velocity shock waves and flame fronts. Accordingly the equipment described here was designed with every effort to obtain a simple, rugged field instrument with a 1-μsec shutter open time.

The microsecond shutter is located in a square-shaped aluminum casting. On the back of this casting is a mounting position to accept a 4 × 5 inch Eastman view camera, although almost any camera can be used with slight modification of the base. Provision is made on the back side of the magneto-optic shutter to fit the lens ring of a Wollensak shutter containing a 163-mm focal length lens. The "X" synchronizing contacts on the Wollensak shutter enable the operator to fire his explosive charge without a long open time, which might fog the film owing to light leakage through the closed polarizers. An image of a subject illuminated by direct sunlight will be dimly exposed in 10 sec with fast film even if the polarizers are accurately crossed.

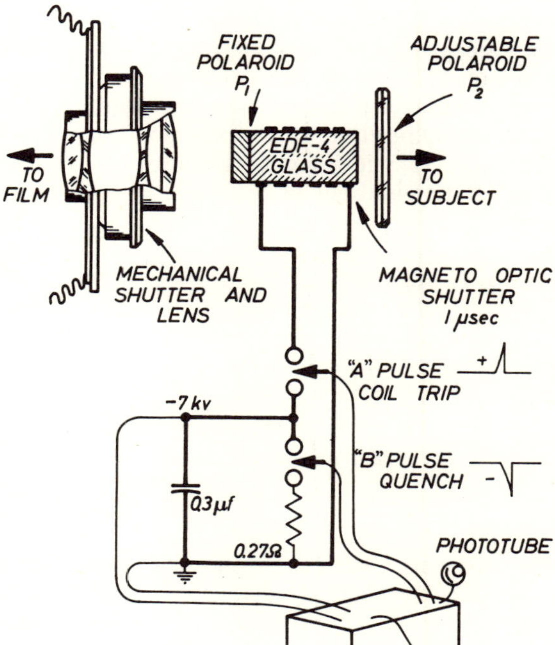

FIG. H5a-1. Cross-sectional view of magneto-optic shutter and driving circuit. From Edgerton [310b].

The magneto-optic shutter described here was the result of a redesign of the previously designed 4 μsec model in the following ways:

1. The aperture was reduced from an inch in diameter to 1 cm.
2. One pair of polarizers instead of two crossed pairs was used.
3. The capacity was decreased from 4 to 0.3 μf.
4. A spark-gap and capacitor assembly was designed to eliminate as much circuit inductance as possible.

The main capacitor circuit consists of ten 0.03 μf capacitors in parallel, arranged to have a low interconnecting inductance.

Figure H5a-1 shows a cross section of the magneto-optic shutter as well as details of the electrical circuit that pulses the 5-turn coil around the extra dense flint glass magneto-optic element. The glass is constructed of Bausch & Lomb Type EDF-4 annealed glass in the form of a cylinder 1 cm in diameter and 2 cm long. The two gap circuit is used to excite the shutter coil for a half-cycle as has been described in the reference given above. These two gaps are shown in the diagram, Fig. H5a-1, together with the pulse forms that trigger them. Pulse A initiates the discharge of the 0.3 μf capacitor into the coil around the glass element. Pulse B triggers the quench air gap which short circuits the main capacitor into a damping resistor after a half-cycle of operation. In this manner, the energy in the circuit is removed so that the capacitance, C, and inductance, L will not oscillate.

The light time transmission of the shutter under normal operating conditions is shown in Fig. H5a-2 as sketched from oscillographic observations. The 100% light transmission refers to the transmission with the polarizers (Type HN 23) in a parallel position which corresponds to a density of about 1. The transient open transmission density is close to that of the uncrossed condition since the rotation is about 90°.

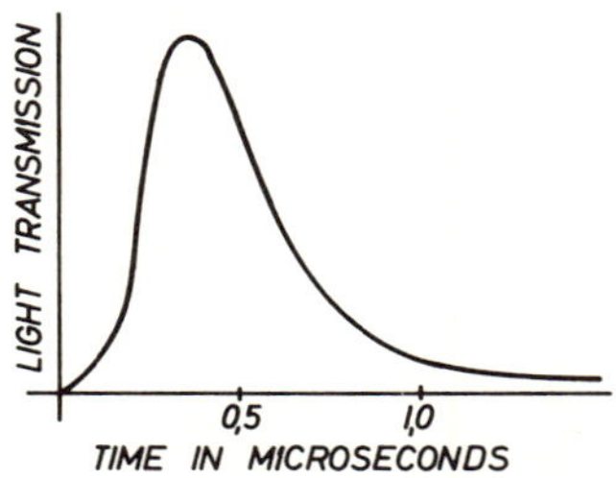

FIG. H5a-2. Transmission of the magneto-optic shutter as a function of time. From Edgerton [310b].

Electrical cables connect the camera and shutter portion to the power supply and control unit. Details of the delay and trigger circuit in the control unit are given in Fig. H5a-3. The trigger portion of the circuit is usually a photoelectric tube, marked 929 on Fig. H5a-3, although a "make" circuit or a positive voltage pulse is equally effective. The flash of light from a subject illuminates the photo tube creating a voltage pulse which trips the thyratron (V 102) and the delay *RC* network. A dial on the unit controls the resistance (50 kohms variable) of the *RC* coupling portion of the circuit. The pick-off thyratron (V 103) triggers after the time delay as determined by the pick-off voltage on the adjustable resistor (10 kohms). The thyratron V 104, which excites coil A, triggers instantaneously with V 103, followed in about $\frac{3}{4}$ μsec by V 105 which triggers coil B and the quench gap.

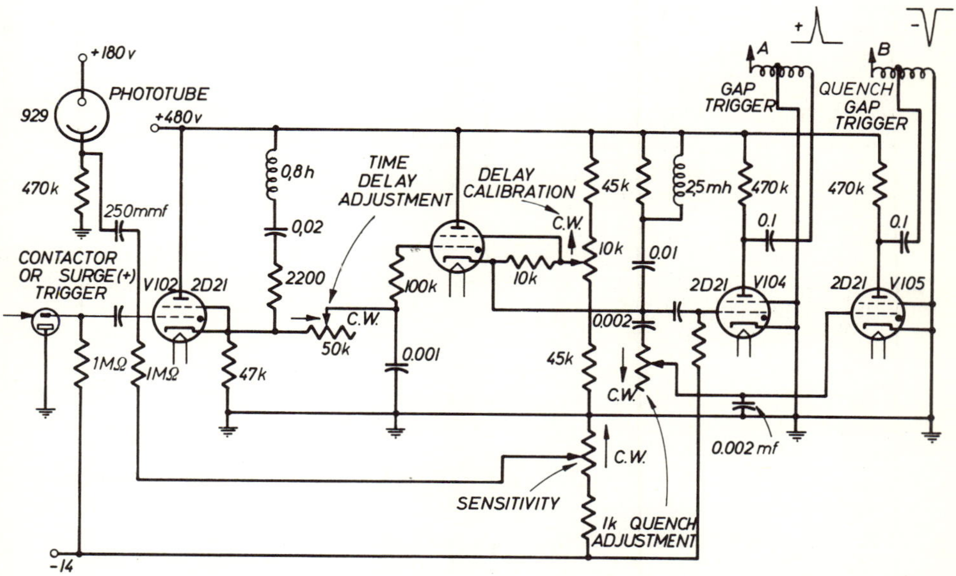

FIG. H5a-3. Control and time delay circuits for magneto-optic shutter. From Edgerton [310b].

An interesting and often useful effect results when the quenching gap is prevented from firing. One method of accomplishing this is to remove the thyratron V 105 from its socket. If the quench gap does not operate, then the current through the shutter coil will oscillate at the natural frequency of the circuit consisting of the capacitance, C, and coil inductance, L, as given by

$$f = 1/(2\pi\sqrt{LC}) \text{ cycles/sec}$$

The shutter does not depend upon the polarity of the current; therefore the shutter will open twice per cycle. The frequency is about one million cycles per second.

b. *A Daylight, Time Accelerator System with Faraday Shutter*

A new method to use the Faraday shutter for eliminating daylight is given by Rieck [265, 321].

Unlike high-speed, slow motion cinematography, time acceleration cinematography does not require extremely short exposure times. However, there is a special application for time acceleration cinematography with short exposure times, involving a problem which hitherto could be solved only with extreme difficulties; namely, motion picture studies outdoors or in lighted rooms which must be carried on day and night without any fluctuation in exposure time. These requirements are very often encountered in biological applications, such as studying the process of plant growth.

Such time accelerated motion studies require a room which is completely darkened during the exposure of the single frames, so that the illumination of the subject will not be disturbed. If such photographs are to be made, e.g., of a plant, which requires daylight for its process of growth, then devices are required to completely darken the photographed area. For this purpose, a glass house was built, which can automatically be darkened for time accelerated studies. The front and the rear walls of this glass house are blackened and 2/3 of the glass walls of the sides and the roof are also blackened. The daylight can only enter through the middle part of walls and roof. To darken these, there is, inside the glass house, a movable screening apparatus, which is coupled by a timing mechanism with relay and motor. This mechanism is expensive, does not operate very reliably, and requires much maintainance.

If one desires to photograph in the field, such devices bring still more difficulties. Rieck once used a darkened house which was movable on rails. An automatic timer control moved the house over the area under photographic observation, closed the door, automatically switched on the lights, released the camera for a single frame, switched off the lights, re-opened the doors and moved the house back to its original position. Merely this listing of all the phases which had to be released automatically gives an idea of the difficulties involved. It is clear that there is an urgent requirement for some other suitable method to dispense with such elaborate screening devices.

The influence of daylight additional to the desired artificial illumination is greater, the longer the exposure time per single frame. In order to eliminate this detrimental daylight influence one therefore must use the shortest exposure possible, for which purpose a very great luminous density of artificial illumination is required. The solution to this problem will be found in a combination of a fast shutter with synchronized flash illumination. For the required fast shutter for the "daylight time accelerator" Rieck utilized the Faraday effect and made a shutter with an effective opening time of roughly 1/5000 sec [321]. A cylindrical flint glass block 5 cm in length with a diameter of 4 cm was surrounded by a magnetizing coil with 60 turns. Polarizing filters were on the front and end surface blocking off all light in axial direction. This shutter device was put in front of the camera lens and care was taken to properly align the shutter cylinder's axis with the optical axis of the camera. When a current is sent through the magnetizing coil a magnetic field is produced in the glass cylinder which turns the polarizing plane of the light. This means that the shutter is now open and the light can pass through as long as the magnetic field is present. Both magnetizing coil and illuminating flash are in the discharge circuit of an electronic flash unit (see Fig. H5b-1) so that shutter opening and flash firing are automatically synchronized by the discharge which, in turn, is set free by a cam switch on the camera. The charging of the capacitors takes approximately 10 seconds, thus determining the shortest time acceleration intervals at 0.1 frames per second. The light absorption

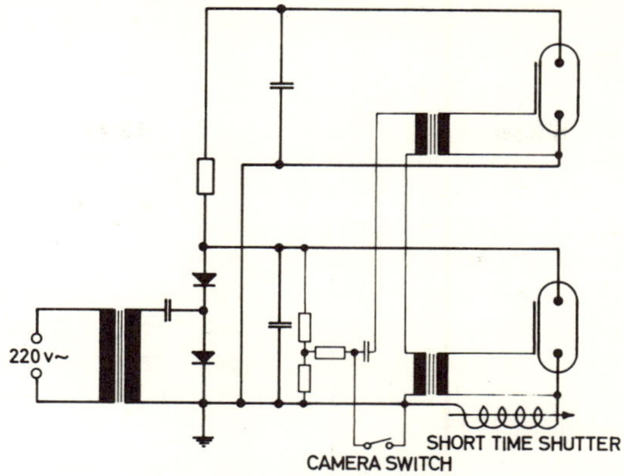

FIG. H5b-1. Circuit of a combination of a big Faraday shutter and capillary flash lamp. From Rieck [322].

of the polar filters is considerable. The transparency of the shutter was merely 5% in open condition.

There still remains the problem of the effective disturbing influence of varying additional daylight illumination over long periods of time accelerated, motion picture studies. The following considerations will apply to the arrangement under consideration, which has a shutter time of 1/5000 sec. The flash unit must be capable of ensuring normal blackening of the negative. With a medium speed film of 325 ASA and a lens stop of 2 this would mean an effective exposure of 20 lux-sec (200 foot candle seconds). If such time accelerated motion picture studies are to be made out-of-doors, then the influence of additional daylight illumination on the film negative, even at its maximum, must be kept low enough so that density differences of single frames do not make themselves seen in projection of the film.

In case of sunshine from a clear sky the maximum daylight illumination level is about 80,000 lux. At an exposure time of 1/5000 sec additional daylight illumination would amount to 16 lux-sec (160 foot candle seconds). At the above-mentioned 5% transparency of the shutter, merely 0.8 lux-sec would remain from the 16 lux-sec. In order to further suppress the undesired daylight illumination and also for greater depth of field the lens was stopped down to *f*/8. Practical experience showed that by this method one could reduce the remaining effective proportion of unwanted additional daylight illumination to a value of 0.65%. Thus, daylight illumination was suppressed sufficiently so that it could no longer cause irregular density of single frames, no matter at what daylight illumination level the flash exposures were made.

The daylight time accelerator arrangement described above [322] has proved its value in extended field tests. Time accelerated motion picture studies have been made outside in fields, in glass houses, at varying temperatures and humidity conditions. Several remarkable research films have resulted and in one of the latest experiments the total shooting period was 4 weeks with single frame exposure round the clock at 10 minute intervals.

A total number of 4000 frames were exposed with no detectable differences in negative densities.

c. *The Collapsing Foil Capping Shutter*

With modern equipment an early state of a phenomenon is often, while emitting a very small amount of light, the object of study. In a later state intense light can be emitted but must not enter the camera. Then, it is necessary to shut the camera before the light greatly increases. The purposes of the so-called collapsing foil capping shutter are (1) to prevent retake in the case of rotating mirror cameras, (2) to avoid optical "feed through" in the case of Kerr cell cameras exposed to ultrahigh-speed recording work and (3) to protect users of optical devices (telescopes, range finders, etc.) from the effects of nuclear flash [943].

Essentially, the capping shutter consists of a cylinder of aluminium foil which fits snugly within a short coil. When energy from associated storage capacitors is transported into the coils, the foil is completely collapsed by the repulsion of the eddy currents induced in the foil by the primary currents in the shutter-coil magnetic field pressure.

Such a shutter needs a capacitor bank of between 1–15 μf charged at 20 kv with an energy of 200–3000 joules. It depends on the diameter which can be between 1.25 inches and 5.00 inches. The smaller models have dosing times of 100 μsec; the greater ones, 400 μsec. The applied inductance of the shutter coil is about 0.15 μh. A specially constructed sandwich line serves to feed the capacitor bank into this low inductance consumer.

The main advantage of the collapsing foil shutter is a completely unobscured aperture before firing. Due to the lack of glass or any window material, it can even be used in the region of soft x-rays.

6. Metal Forming with Pulsed Magnetic Fields

Strong, short-duration magnetic fields may be used for direct forming of metal parts. Since the energy for forming is imparted to the work piece solely through the magnetic field, many unusual capabilities and applications exist for the process. Brower [939–941] gave a survey of this technique which was conceived by the General Atomic Division of General Dynamics Corporation.

By discharging a capacitor through a coil over a period of microseconds, one may readily produce magnetic fields having flux densities of the order of hundreds of kilogausses. By this technique, one is able to impart to a metallic work piece a powerful, uniform impulse solely through the medium of the magnetic field. A pressure of 50,000 psi for a period of many microseconds is typical of the magnitude of impulses that are used. Since the magnetic field does not interact appreciably with insulators, the magnetic impulse can be transmitted to the work-piece through the walls of a nonconducting sheath which may maintain the work piece in a sterile, inert, or other special environment.

Magnetic impulses have been used to swage and expand tubular forms as well as to coin, shear, and form flat sheets, To date, magnetic pulse forming has found most of its applications in the forming of excellent conductors, such as aluminium, copper, and brass. It has been used effectively to form with the use of highly conductive coatings poor conductors such as stainless steel. Equipment for magnetic pulse forming has been available to industrial users only since the beginning 1963. Since then, the technique has had a very wide variety of applications, from the assembly of electromechanical devices to the manufacture of pressure vessels.

Figure H6-1 shows schematically the circuit consisting of the charging circuit, a capacitor, C, of several thousand joules of storage energy, a switch, S, a coil, a field shaper, and the work piece.

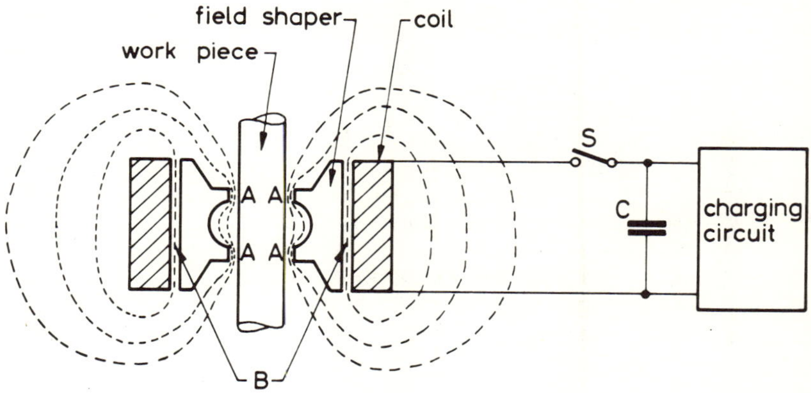

FIG. H6-1. Basic magnetic pulse metal-forming circuit with field shaper to concentrate magnetic energy on a cylindrical work piece. A–high pressure-high current density; B–low pressure-low current density.

In magnetic pulse-forming equipment important components are the energy storage capacitors, the switching equipment, and the work coils. Capacitors must be capable of storing energy measured in thousands of foot pounds in order to create the power needed for practical metal working applications. The switches must be capable of handling voltages of the order of 10 kv and currents of many thousands of amperes. Ignitrons have been found to be reliable and long-lived for this purpose. Again, with the proper selection of maximum operating levels, they can be operated reliably for many millions of operations without failure. Coils must be efficient electrically and, at the same time, must withstand the same impulse that they impart to the work piece for years of operation.

The coils must be designed in such a way that stray inductance is minimized and that current concentrations are avoided. Since high permeability materials cannot be used in this application because of the high flux densities required, fields may be shaped only through the use of current-carrying conductors.

These current-carrying conductors need not be directly connected to the basic coil but may be used in inductively coupled configurations similar to the flux concentrator devices used in induction heating. Figure H6-1 illustrates the use of a massive conducting structure utilizing this technique to concentrate the current, hence the force, in certain regions of the work piece, thus relieving the basic coil from the task of carrying high current. Carefully designed coils operated within their design limits have the capability of lasting an indefinite length of time. There is no theoretical limit to the maximum energy which can be built into magnetic pulse-forming equipment.

The concentration of the magnetic energy transformed into acceleration of the particles of the work piece and their stop on the wall of the tool produce unbelievable forces. A work piece of aluminum accelerated in 10 μsec to a speed of 40,000 cm/sec has available 4,000 inch pounds per square inch of energy for forming. If it comes to rest in 0.02 inch, an average force of more than 2 million psi would have been exerted.

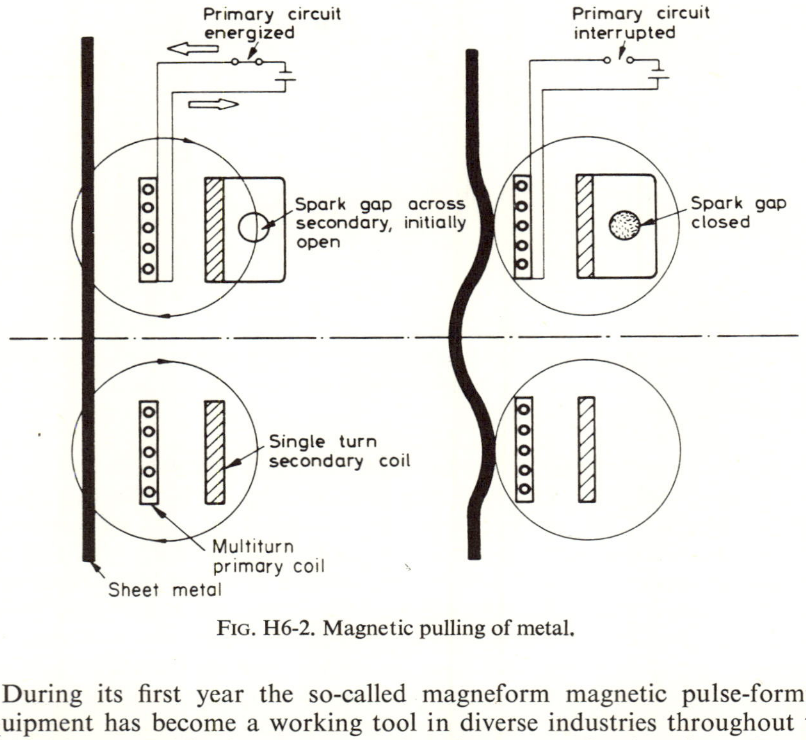

FIG. H6-2. Magnetic pulling of metal.

During its first year the so-called magneform magnetic pulse-forming equipment has become a working tool in diverse industries throughout the United States. While the process has found many applications in performing operations previously done by conventional techniques, its greatest potential will be found in the fabrication of products designed specifically to exploit its unique capabilities (see also patent literature [902]).

New ideas on magnetic forming are published by Furth and Waniek [944].

They constructed a magnetic hammer fed from a capacitor bank and attached to a drill press. A dipole hammer fed through the transmission sandwich line exerts its maximum pressure at the center and develops (5×6 inch size) 100 kilogausses at the center.

Figure H6-2 shows how a magnetic-forming technique can likewise be used to pull metal. This is advantageous where it is impractical to exert a force from inside a hollow piece or from the reverse side of a large sheet. The figure shows the magnetic field lines in the two phases. First the primary circuit is energized, second, after interruption of that circuit, a secondary device becomes effective by closing a spark gap. This sophisticated technique is also developed for expanding cylinders. For making contact, bronze spring contacts instead of the spark gap are used. Idelberger [791] and Moll [791a] have written a survey of this modern technique in German.

I. Conversion of capacitively stored energy into acoustic impulses

1 Conversion by Electroacoustic Converters and Applications

a. *The Simple Sound Depth Finder*

Theoretically, it is possible to feed the energy of a storage capacitor into any electroacoustical converter for conversion into acoustic energy. A practical application that has gained general acceptance is in the field of depth sounding.

In this application the energy of a capacitor is made to actuate, by way of a mechanical switch or an ignitron, an electric or an electrostrictive oscillator. When an ignitron is used, the capacitor, in conjunction with the inductance of the oscillator winding, forms an oscillatory circuit. For that purpose it is necessary that either the electrical resonance frequency given by the capacitance and inductance coincide exactly with the acoustical frequency of the oscillator itself, or that it differ entirely therefrom. In the former case, the electrical circuit oscillates in unison with the mechanical system which is given by the magnetostrictive converter, in close coupling with the damping decrement of the converter. In the latter case, the energy output is ballistic, i.e. the oscillatory circuit transmits the energy as an impulse into the acoustic mechanical converter, which damps out as in a bell. The former case requires a switch that passes flux in both directions and has very accurate contact making qualities, avoiding bouncing. For the latter case the switch must pass current in only one direction for a short time, that is ballistically; a thyratron is the proper switch for that purpose.

It is possible to introduce from a capacitor discharge into an oscillatory wave train having the most common operating frequency between 15 and 30 kc, and a sequence rate of 10, a peak output of 1 kw. Higher energy values cannot be consumed by converters available today. Under this condition the approximate necessary capacitor energy for a 15 kc oscillator with about 1 millisec sounding duration at 1 kw, is 1 joule. If we include the efficiency factor, this becomes 1.5 joules. Because the winding consists of only a few turns, the required resonance frequency can be obtained with a few microfarads at 1–2 kv charging voltage. The permeability of the magnetostrictive converter changes to a large extent with magnetization so that an accurate computation of the necessary capacitance is hardly possible. For experimental determination the best procedure is to cover the oscillator with a thin layer of sand, and with light from the side, measure the height to which the sand grains bounce at the instant of capacitor discharge. The proper capacitance value, which is not very critical on account of damping, is effective when the bouncing height is at a maximum.

With electrostrictive converters, which in themselves represent capacitors, it is necessary to arrange an adaptation transformer between storage capacitor and converter, usually a barium titanate capacitor. The transformer should be loosely coupled, similarly to a band filter, and possess a similar resonance frequency at its primary and its secondary. If the coupling factor is, e.g., 0.1, the energy radiated into sea water is exactly replenished by coupling-in of the primary low damped oscillation, because the radiation into sea water also has a decrement between 0.05 and 0.1 within the frequency range referred to as commonly used. The sound concentrations, and therewith also the peak outputs, with barium titanate radiators are not higher than with magnetostrictive radiators; they lie at a maximum of 1–10 watts/sq cm. Thus it is of primary importance that the radiator operate in degassed water for prevention of cavitation, since when cavitation occurs the contact between converter and liquid is interrupted at a little more than 1–2 watts/sq cm, and after such occurrence no more energy radiation can be observed.

In recent years impulse radiators operating at high, mostly changeable, frequencies and for very short impulse durations have become known as defect finders for metals and poured concrete. For this purpose the capacitor discharge is used not directly but for triggering a high frequency generator having accurate frequency control by means of a quartz crystal. Smaller depth sounding devices on board ships today still operate mostly with direct capacitor discharges. More detailed information about the modes of operation of different depth sounding devices can be found in leaflets of the manufacturers.

b. *The Soniscope for Testing Material*

S. M. Breuning [233] applied an echoscope, called "Soniscope," for the maintenance of concrete structures. The Soniscope is an electric device for the nondestructive testing of materials. The instrument measures the time delay of a shock wave sent through the material, from which the velocity of propagation of sound waves can be computed. This velocity represents a measure of the quality of the material.

Testing techniques for various types of structures are discussed by Breuning. Tests on prestressed concrete beams loaded to failure show a close correlation of Soniscope readings with the behavior of the beam. The influence of reinforcing steel is shown by tests on a reinforced bridge pier. The detection of both cracks and deterioration in concrete is demonstrated by tests of several bridge members. A graphical presentation of Soniscope results on a bridge pier is then discussed.

The results of these tests indicate that the Soniscope can be a valuable supplement to visual inspection.

c. *The "Sonar Pinger" and "Sonar Thumper"*

Many instruments for studying the ocean use sound as a medium of transmission, since sound is the only form of radiation known to penetrate water for reasonable distances. Two underwater sound sources (pingers and thumpers) are described here in function and in application to oceanographic exploration.

One of the many problems in oceanographic exploration is that of getting information through the water over reasonable distances. In air, we have a choice of methods: light, infrared, radio waves, sound, etc. In water, however, we are left with only one choice, sound, since most other forms of radiation are absorbed within a few hundred feet. As an example of how well sound travels in water, consider that under the proper conditions (a small charge of dynamite exploded in the so-called "sound channel") underwater sounds can be heard over distances of thousands of miles.

Many instruments based upon underwater sound have been built for exploring the oceans. The most common is the echo sounder which measures water depth under a ship by bouncing a ping off the ocean floor. Some examples of other instruments using sound are free floating buoys that send out pings periodically so that they can be tracked by listening devices; submarine detection devices, both passive and active; submerged instruments that can be controlled from the surface by sending them commands acoustically, thus eliminating the necessary of electrical wires; submerged measuring instruments that telemeter information to the surface acoustically; etc. Underwater sound is a specialized field of study that is growing fast.

Sonar Pinger. J. Brackett Hersey, E. Edgerton, O. Raymond and G. Hayward developed an echoscope especially for oceanographic research, called "Sonar Pinger" [756].

Whenever a device is to be lowered in deep water to the ocean floor or to a given position above the floor, a device-to-bottom indicator is very useful. In many cases, such as when it is desired to lower a camera down 20,000 feet and suspend it 10 feet off the ocean floor, such an indicator is a necessity. Effects such as wire stretch, wire length meter reading inaccuracies, water depth measuring errors, and the fact that the wire never hangs straight down under a ship, combine to rule out the possibility of merely subtracting the wire length from water depth to get the device-to-bottom distance.

A simple solution to the problem is to mount a sonar pinger on the device to be lowered, such as on the camera shown in Fig. I1c-1. Periodically, the unit sends out a brief ping of sound which travels directly up to the ship as well as to the ocean floor where it is reflected. A sonar receiver on the ship is used to pick up the pings and they are displayed on a recorder or an oscilloscope so that the time difference between the direct and bottom-reflected pings can be monitored continuously. The time difference of two pings is a direct indication of the pinger-to-bottom distance since

$$D = \frac{VT}{2}$$

where D = pinger-to-bottom distance in feet
V = velocity of sound in water (approximately 5000 ft/sec)
T = time interval between direct and reflected signals in seconds.

For example, if the direct and reflected pings are one half second apart, the

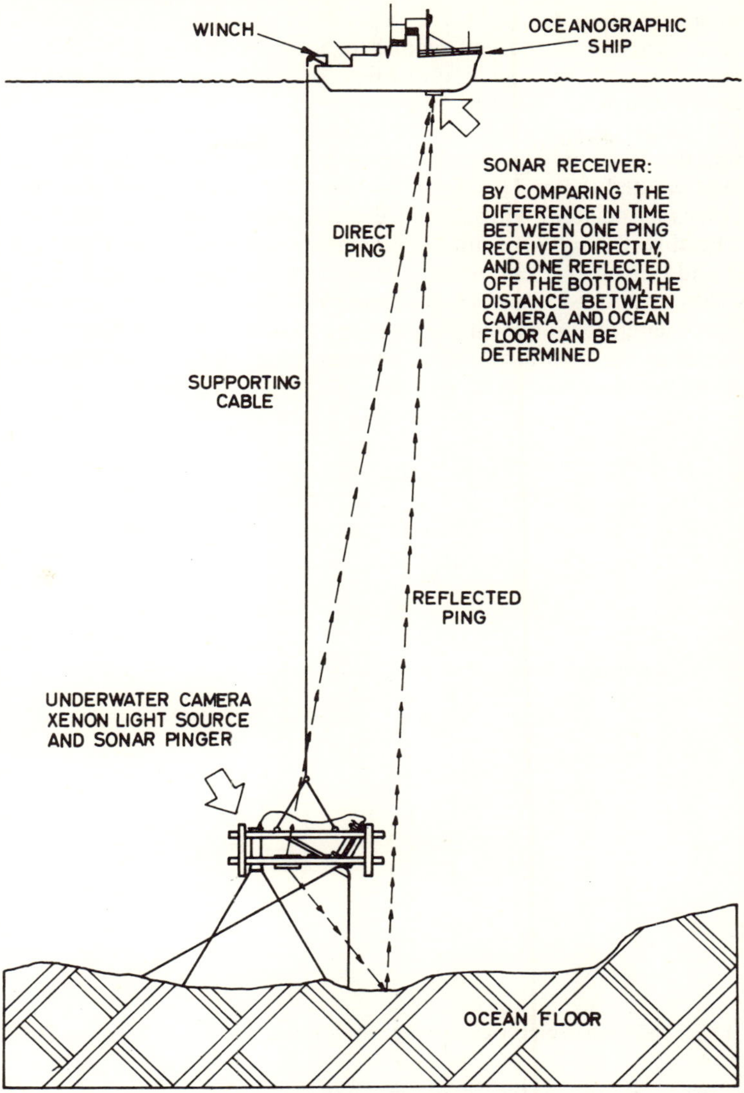

FIG. IIc-1. Deep sea camera positioning by use of a battery powered sonar pinger. No electrical connections to the ship are required. From Edgerton [756].

pinger-to-bottom distance is 1250 ft. If the pings are 2 millisec apart, the pinger is about 5 ft off bottom. The operator monitoring the pings is usually in continuous contact by telephone with the winch operator who is lowering the equipment. By telephone conversation they are able to keep the pinger, and any associated equipment, the desired distance above the ocean floor.

The operator often monitors the pings by three means simultaneously; by recorder, to give a permanent record for later study; by oscilloscope to give an instantaneous "enlarged" indication of distance; and by earphones to supplement the other two methods.

If the pinger has a ping rate of 1/sec and is more than 2500 feet off the bottom, the bottom echoes will arrive more than 1 sec behind the direct ping that created them. Therefore, one may not know with which direct ping the echo is associated. Thus, when the echoes are received one half second after the direct pings, seemingly indicating a distance of 1250 ft between pinger and bottom, the echoes may actually be arriving $1\frac{1}{2}$ sec after the direct pings with a pinger-to-bottom distance of 3750 ft. To overcome this "phase shift" problem, the pinger can be designed so that it omits a ping periodically. By observing the position of the "missing ping" in the direct and reflected ping display, the ambiguity is resolved.

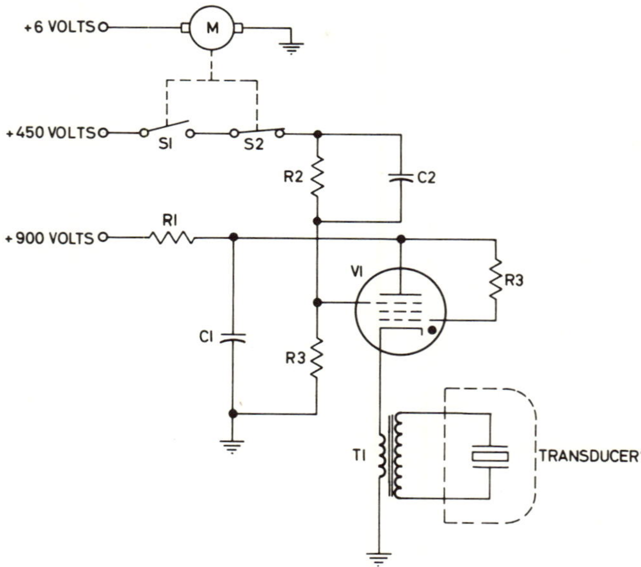

FIG. 11c-2. Sonar pinger circuit described in the text. From Edgerton [756].

The sonar pinger system consists of three basic units: a driver, a pulse transformer, and a transducer. The driver is housed in a pressure resistant casing. It is powered by a rechargeable silver-zinc battery capable of providing 16 hours of operation. A battery supplied transistor power supply generates a high dc voltage which charges a 4 μf capacitor C_1 as shown in Fig. 11c-2. The contacts S_1 of precision dc timer M close once each second to fire strobotron V_1, which discharges through the primary of pulse transformer T_1. Timer contacts S_2 open every 10 sec causing every tenth ping to be omitted to overcome the phase shift problem. The pulse transformer T_1 is not housed in a pressure resistant housing but in an oil filled resilient housing. The housing consists of a lucite tube closed at each end by a rubber stopper. The transformer has a turn ratio such that an 8 kv pulse is generated in the secondary. The secondary is connected to the transducer which contains a set of ADP crystals.

The crystals are housed in a castor oil filled aluminum casing with a flexible rubber diaphragm cover. The secondary of the pulse transformer and the transducer crystals form a tuned circuit which oscillates at a frequency of about 12 kc for a period of about 1/2 millisec. The 12 kc vibration of the crystals produces the sonar ping that is transmitted downward to the water through the oil and the rubber diaphragm. The 1/2 millisec duration provides adequate resolution for positioning the pinger within three feet. Since only a part of the incident energy is coherently reflected from the bottom, the direct signal returning to the ship would normally be about ten times stronger than the reflected signal if the transducer front-to-back ratio were unity. Hence the direct signal (the back lobe of the transducer pattern) is attenuated by a sheet of phenolic and a 1/2 inch aluminum plate, both built into the transducer housing. Thus both signals are received at about the same intensity.

The sonar pinger has been used extensively for deep sea photography to position underwater cameras the proper distance above the ocean floor. The pinger is mounted on the camera framework. Camera lowerings have been made more than 25,000 feet deep and the cameras have been kept suspended just a few feet above the ocean floor for over an hour in spite of ship drift, roll and changing bottom contour.

In camera work where an automatic electronic flash camera taking pictures every few seconds is used, the pinger can be wired so that it omits a ping only when the camera takes a picture. Thus the ping record will show the exact second at which a given picture was taken and the corresponding distance above bottom.

Whenever something must be lowered to the ocean floor in deep water, the sonar pinger can help bring precision to an operation which is otherwise a blind and groping one. It also gives the operators greatly increased confidence and helps insure safety of valuable equipment.

Sonar Thumper. Sound in the form of elastic waves is used in physical exploration on land for making surveys of the sediments and rock of the earth's crust. Until recently these waves have been produced mainly by explosives set off on or in the earth. As the waves travel downward, they are reflected at the interfaces between layers of material of different elastic impedance. Recording of these multiple echoes shows the geological structure of the earth in the vicinity of the charge. Many attempts have been made to use devices other than explosives as seismic pulse generators. Only during the past decade, however, have any such devices of wide utility been developed.

Although more seismic work has been done on dry land than at sea, seismic systems have also been developed for use in exploring the earth beneath the ocean. Fortunately, since water is such a good conductor of sound, it is not necessary to produce the sound right on the ocean floor, but it can be produced in the water near its surface. The pressure waves travel down through the water to the ocean floor and are reflected, as in usual echo sounding techniques. However, the waves produced also penetrate into the ocean floor and are reflected from the substrata. The type of echo sounders usually used for depth indication do not indicate sub-bottoms because their frequency is too high. Pulse generators for seismic work must have frequencies well below 2000 cps to obtain much penetration.

Explosives have been used extensively for marine seismic work; they can put more energy into the water and greater depth of penetration can be obtained than with any other means available today. They have drawbacks, however. They are dangerous to handle and use, and in some areas (such as congested harbors) they cannot be used. Although no capital equipment investment is needed, each shot costs money which can run into many thousands of dollars per survey. Explosives are essentially one shot devices in comparison to some of the newer seismic tools. Although the newer devices cannot obtain the depth of penetration possible with explosives, they can be actuated at sufficiently high repetition rates hour after hour to obtain continuous sub-bottom profiles as the ship moves along.

Some of the sound sources that have been developed for continuous sub-bottom profiling include "Sparkers" which use an electric spark in water, "Gas Poppers" which use an explosive gas mixture, and magnetostrictive types of devices. Each system has its own advantages and disadvantages. To supplement these systems, H. E. Edgerton, at the instigation of J. B. Hersey, started to work on the problem of developing a new device to put a large amount of electrical energy into the water in the form of a clean low frequency, repeatable pulse. The desired device would be simple, safe, reliable, rugged, and reasonably inexpensive. The device developed is called the Thumper.

The Thumper system utilizes a power supply cabinet which converts ship's power (110 or 220 volts ac) to high voltage dc. This is fed to a capacitor bank cabinet which contains ten large energy storage capacitors connected in parallel to give a total bank of 160 μf. A trigger circuit in the capacitor bank cabinet causes the capacitors to discharge into the transducer on command. The actual transducer consists of a flat aluminum plate 18 inches in diameter and 1/4 inch thick which is held against the face of a flat coil by means of a spring. Figure IIc-3 shows a cutaway drawing of the transducer. When a pulse of current flows through the coil, a magnetic field builds up around the coil and then collapses. The rise and fall of this field induce circular currents in the aluminum plate, and these in turn create a magnetic field opposing that of the coil. The results of this interaction is a force which drives the plate away from the coil.

The principle can be demonstrated by a simple experiment. Place a penny on a small flat coil. Charge a 1000 μf capacitor to 6 volts from a battery and then discharge the capacitor into the coil. The penny will jump into the air.

If the Thumper transducer is fired in air without its restraining bolt, the plate will rise some 30 feet in the air. In actual operation in water, most of the energy is radiated in the form of a pressure wave so that the plate travel is only about half an inch. The present efficiency of transformation from electrical energy to acoustic energy has been estimated to be about 10%. Peak acoustic pressures at a distance of 1 yd measure between one half and two million dynes per square centimeter.

The Thumper spectrum contains both "high" and "low" frequency components. High frequencies do not penetrate the ocean floor as well as do low frequencies; because of their shorter wavelength they instead tend to give a highly resolved picture of the bottom and any shallow structure which may be present. Conversely, the low frequency end of the spectrum does not have as much resolving power but penetrates farther into the bottom. By recording through adjustable filters on a two channel recorder, one channel can be set to show the high frequency bottom details and the other the deeper low frequency sub-bottom returns. Such a presentation is also useful in detecting the presence of spurious side echoes since these will show up in the high frequency channel and not in the low frequency one.

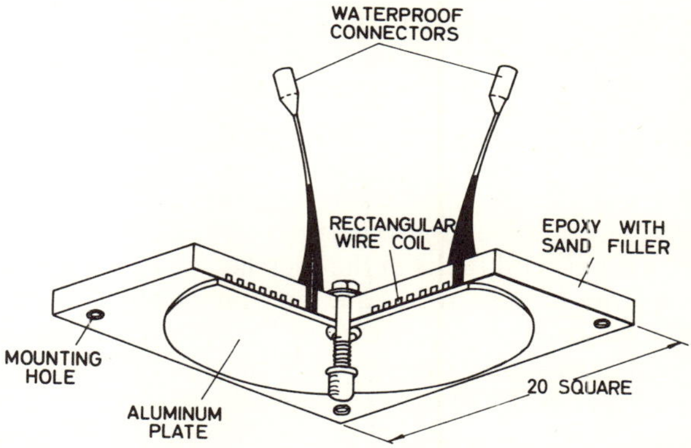

FIG. I1c-3. The sonar Thumper transducer. A pulse of current through the coil causes the aluminum plate to be repelled against the water, creating a pressure wave. From Edgerton [756].

Figure I1c-4 is a simplified wiring diagram of the system. The 110 or 220 volts ac from the ship's generator are stepped up to 4000 volts by transformer T_1. This is rectified by a full-wave arrangement using twenty-seven 400 volt diodes in each leg and then fed to the 10 capacitors which make up C_1. When it is desired to thump the device, an external keying contactor closes causing a capacitor in the strobotron trigger circuit to discharge through the primary of transformer T_2. A very high voltage is generated in the secondary of T_2 and is supplied to the trigger electrode of the spark gap shown on the right. This causes the air in the gap to ionize and thus become conductive. The charge in capacitors C_1 discharges through the spark gap which is in series with the transducer coil L_1. The current reaches a peak of about 1600 amp. The electrical energy stored in the capacitors is given by

$$W = \frac{CE^2}{2} = \frac{(160 \times 10^{-6})\,(4000)^2}{2} = 1280 \text{ watt-sec}$$

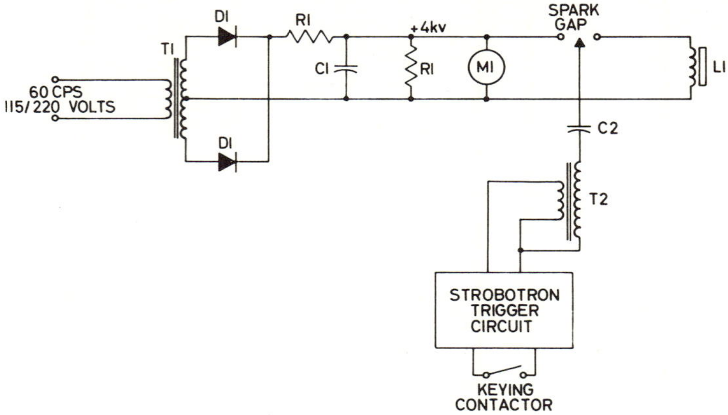

FIG. IIc-4. Sonar Thumper circuit. Keying contactor on recorder triggers spark gap, discharging C_1 into transducer L_1. From Edgerton [756].

In actual operation line losses and poor generator regulation usually reduce the energy to about 1000 watt-sec/pulse.

The echo train received from a single thump will contain reflections from any extensive rock strata and also energy scattered by smaller details such as sub-bottom boulders and faults in the rock layers. It is well known that single recordings are not subject to unique interpretation if the geometry of the situation is complex. Fortunately for the geophysicist, instrumental devices and techniques are available to help him interpret the received echoes. One of these is to move both the detector and the source while rapidly repeating the soundings. By this means extensive geological formations can be traced out and distinguished from local details. Successive soundings may be compared by means of some automatic correlating device or visually on a chart recorder. Such a presentation allows one to trace out the salient features of the underlying structure even in the presence of excessive noise and local side reflections. While a display of this type shows immediately the basic form of the underlying geology, the determination of exact values for the thickness of the various layers requires knowledge of the velocity of sound in these layers relative to that in water. This is a relatively simple operation in the case of the bottom sediment where samples can be obtained and a direct measurement made. The underlying rock layers present another problem, however, unless one is content to guess their composition by knowledge of the prevailing geology. One possibility is to follow the particular formation of interest for a distance in various directions. Often the recorder will indicate a spot where the overlying sediment layer becomes quite thin, and a sample of the material can often be obtained at this point. Another method of establishing the sound velocity in the various layers is to employ conventional seismic refraction techniques in connection with the Thumper sound source. Here the receiver is displaced from the source until refracted signals from the various layers are observed. For this method the receiving hydrophone can either be towed at the end of a long cable, or mounted on another ship, or a sonobuoy can be used to transmit return signals back to the ship by means of a radio link.

Echo returns from the Thumper can be recorded in various ways. They can be displayed on an oscilloscope and photographed there, but this has the disadvantage that no correlation techniques can readily be applied. They can be recorded on magnetic tape and analyzed later in the laboratory. This has the advantage that many different forms of analysis can be performed on the same data, especially if the tape is carefully made so as to include all frequencies present in the original returns.

d. *Method for Checking the Position of Fishing Nets*

Because of the spheric radiation diagram of spark sound sources, a method and system for checking the position of fishing nets became possible [234]. Also a combination with spark light sources is useful. The device relates to fishing gear and more particularly to a method and system for locating and checking the position of fishing nets by totally reflected sound having ultrasonic frequency or by totally reflected light signals.

In fishing with trawl nets, a type of fishing gear that has gained widespread acceptance, the net, having a center portion and two net wings, is towed at slow speed and at great distance behind the fishing vessel. In most cases the net is kept open by sheer boards ("otter boards") attached to the forward end of each wing. They are somewhat outwardly inclined to the direction of travel so that, when the net is towed, the two boards sheer in opposite directions and keep the mouth of the net open. In operation, when the quantity of catch increases and the net becomes heavier, the spacing of the sheer boards decreases so that their spacing is an indication of the state of catch. Thus it is very important for the fishermen to watch closely the position of the sheer boards which themselves travel under water and are not directly visible by sight. Optical observation—for instance, by means of buoys or floats attached to the wings of the net—is often very difficult since the distance between vessel and net is generally more than 1500 fathoms. Tracing by means of echo sounding is of interest, but it requires, powerful ultrasonic energy sources to assure that echo signals from the sheer boards are strong enough to be detected with certainty on board the fishing vessel.

In order to overcome these difficulties, a system has been devised which requires relatively little sounding energy and gives strong echo signals. The sheer boards and other strategic points of the fishing net are provided with triple reflectors or mirrors which at any inclination with respect to the transmitter give complete reflection of sound both in the direction of incidence and toward the transmitter. In addition it is possible to provide optical triple mirrors for total reflection of a beam of light for operation above the surface of the water.

In Fig. IId-1 the fishing vessel, indicated at 10, tows a net, 11, having a center bag portion, 12, and two net wings, 13 and 14. At the forward ends of the net wings there are secured inclined sheer boards, 15 and 16, sheering in opposite direction and keeping the net open. Substantially vertically above the

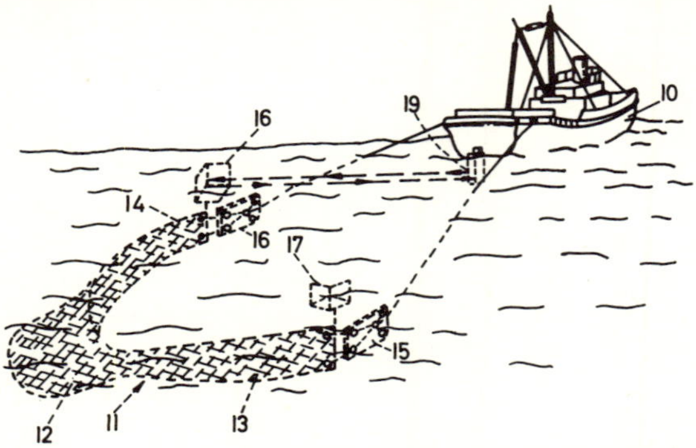

FIG. I1d-1. Diagrammatic representation of a system for underwater echo signaling for checking the position of fishing nets.

sheer boards, but still below the water surface, there can be seen fastened to the net the triple reflectors, 17 and 18, suitable for sound reflection.

Triple reflectors or mirrors, as is well known in the art, have three reflecting, substantially square planes, each perpendicular to the other two as the three sides adjacent to a corner of a cube. A ray of light or a beam of sound impinging upon the inner surface of one of these three planes is reflected in its direction of incidence so that one receives at the point of transmission always a total reflection from the triple mirror no matter what inclination it takes. Such total or complete reflection is very effective and relatively low transmitter energy will result in strong echo signals which surpass in intensity by far the interferences introduced by the water surface and make possible clear and definite observation.

The vessel, 10, carries an ultrasonic transmitter, 19, which beams an ultrasonic signal toward triple reflector, 17 or 18. When the ultrasonic beam strikes one of the reflectors, the beam is reflected in the direction of incidence, i.e. toward the transmitter, 19. Thus the location of the net, the position of the

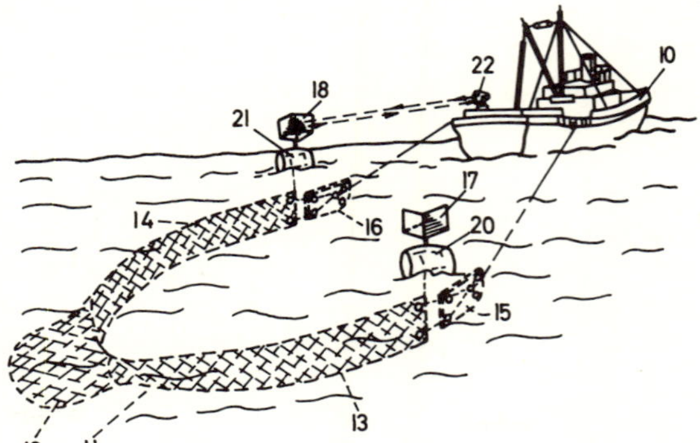

FIG. I1d-2. Diagrammatic representation of a system for light pulse reflector signaling for checking the position of fishing nets.

sheer boards with respect to each other, and therefrom the quantity of catch, can be determined.

Somewhat similar results to those with sound having ultrasonic frequency under water can be obtained by employing a light beam and optical triple mirrors above the surface of the water as depicted in Fig. IId-2. Here the triple mirrors, 17′ and 18′, are secured to floats or buoys, 20 and 21, extending to the water surface at points of the net located above the sheer boards. A searchlight, 22, mounted on the deck of the fishing vessel, sends a beam of pulse light toward the mirrors, and light incident on either one of the mirrors will be reflected in the direction of incidence in the same manner as sound under water.

In triple reflectors for sound having ultrasonic frequency the three planes forming the reflector are preferably constructed to render good total reflection against water. Such construction consists, for instance, of air filled, thin walled sheets of metal or other material having a considerably different wave resistance from that of water; for example, brass or steel.

2. Air Impulse Sound

a. *Theoretical Considerations on the Characteristic Sound Spectrum of Capacitor Discharges and Their Experimental Verification* [202]

Spontaneous electric discharges of capacitors are generally accompanied by an intense detonation-like noise, which can be imagined to originate as follows: The air which is permeated by the electric charge is subjected to a sudden temperature rise and thus develops a certain overpressure against the surrounding air. Consequently, there is a tendency for pressure equalization, and the compressed zone starts to expand in all directions. In the zones of slight overpressure the pressure wave proceeds in accordance with the differential equation of the sound field. Here a simple example may illustrate what slight temperature changes are sufficient to cause considerable sonar effects.

It is assumed that a spherical air volume of 1 cm diameter is subjected to a temperature rise of 10 °C in such a short time that no measurable volume expansion could occur during the heating-up interval. The pressure increase will then amount to 3.3×10^4 μbar. If we take the laws of sound propagation for granted, then an observer at 1 meter distance would measure a pressure change of 165 μbar. This impulse as such is comparatively short; however, it is prolonged by the many reverberations from the walls of the room and thus gains considerable intensity. If one recalls that normally the human ear is accustomed to sound pressure values in the region of 1 μbar, then he will see that the value of 165 μbar represents a considerable noise level.

The calculations made in [202] result in a time dependency of the sound pressure of

$$p = P_0\, e^{-\alpha t}$$

with

$$\alpha = \frac{1}{T} = \frac{3c}{r}\sqrt{1 + \left(\frac{c}{\omega r}\right)^2}$$

where

p = overpressure in μbar
P_0 = overpressure at $t = 0$
t = time
T = time constant
c = speed of sound
r = radius of a sphere in free atmosphere
$\omega = 2\pi f$
f = sound frequency

If one applies to this formula the rules of Fourier's analysis a relation results for the Fourier coefficients $A(\omega)$ which are a measure of the amplitudes of the partial tones.

$$A(\omega) = \frac{1}{\pi} \frac{1}{\sqrt{\omega^2 + \left(\frac{3c}{r}\right)^2 \cdot \left(\frac{c}{\omega r}\right)^2 + \left(\frac{3c}{r}\right)^2}}$$

The sound spectrum increases with the frequency and, as it progresses further, the gradient becomes less steep and finally comes to a halt at a maximum. Figure I2a-1 shows graphs of sound spectra of pressure equalization phenomena calculated for various spherical starting volumes. The pressure amplitude is indicated per cycle and related to the surface of the starting volume per sound pressure unit. The numbers beside the curves indicate the radius of the sphere.

The maximum will be found at that frequency where the wave length λ is 3.5-fold the radius of the sphere. There is one particularly noteworthy fact: At the higher frequencies the spectra of all spherical radii enter into the same boundary line.

The deductions applying to the pressure equalization process in free atmosphere can also be modified to suit two further characteristic cases. If the equalization process does not occur over the frequency dependent, sound resistance of a spherical radiator of 0 power but over the frequency independent sound resistance of an exponential funnel, then also the decaying constant becomes frequency independent. If it is assumed that the cross sectional area of the connected funnel is equal to the surface of the initial sphere then

$$\alpha = \frac{1}{T} = \frac{3c}{r}$$

If this value is inserted in the equation for $A(\omega)$, the result turns out to be characteristic spectrum of an e-function. The process is likewise illustrated in Fig. 12a-1 for various sphere diameters or initial volumes. Within the range $\alpha > \omega$ the sound pressure amplitude per cycle is constant. In the course of further progress, when $\alpha < \omega$ there occurs the same drop-off as formerly with $1/\omega$. The difference between the two curves is due to the spherical radiator's drop-off in sound resistance toward the low frequenceis. The value of α is apparently determined by the relation of the initial volume to the cross

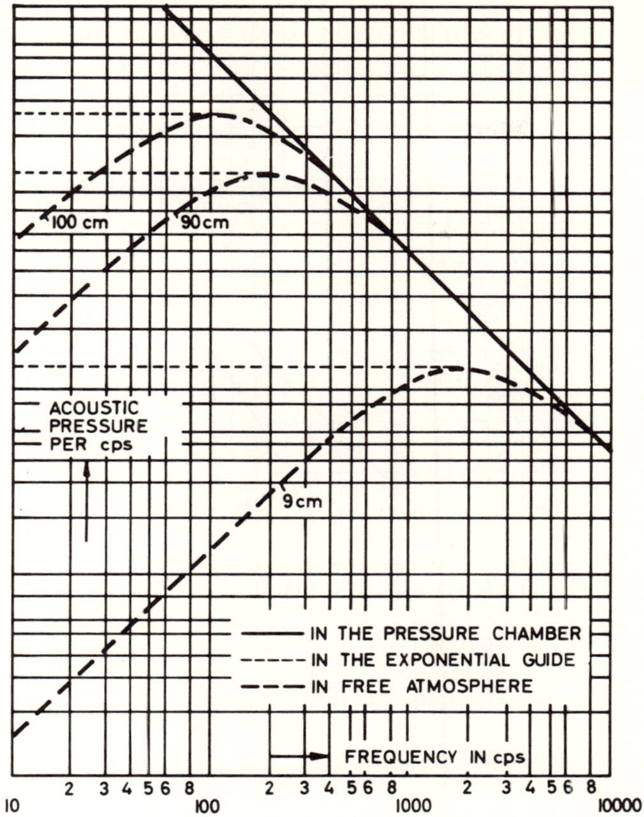

FIG. I2a-1. Graph of the calculated acoustic spectra in pressure equalization phenomena for various spherical starting volumes. The acoustic pressure amplitude is stated per cps per acoustic pressure unit in relation to the starting volume surface. The numbers at the curves state the radius of the sphere.

sectional opening area of the funnel. The radius of the assumed sphere is found by experimental results.

Another characteristic case is found when the capacitor discharge occurs within an enclosed space the relative size of which is small in comparison with the smallest wavelength. In such circumstances a pressure equalization will no longer be possible. In this case $\alpha = 0$. The amplitude of the partial tones, drops monotonously with increasing frequency (see Fig. 12a-1, drawn-out line). This fact is thoroughly comprehensible since at the moment of the capacitor discharge there occurs a pressure jump in the vessel which is of extremely long duration compared to the time scale applied here.

Figure I2a-2 shows the phase angles of partial tones during a pressure equalization in relation to the frequency for a spherical volume of $r = 5$ cm.

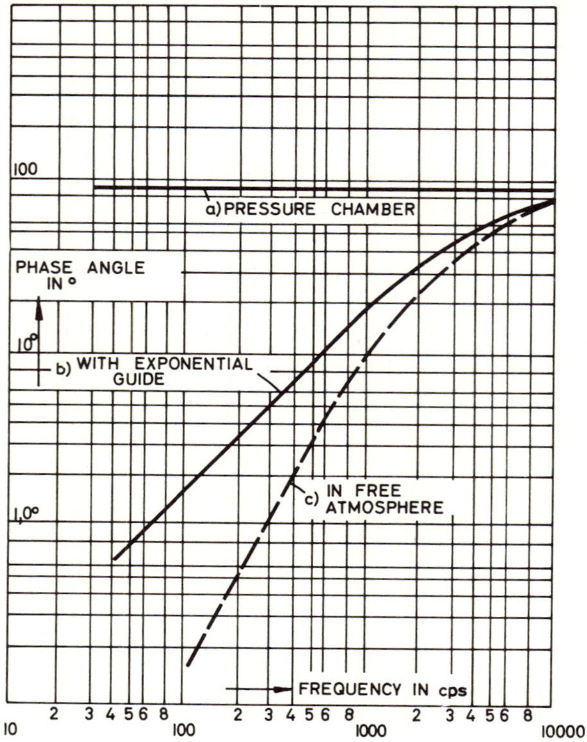

FIG. I2a-2. Phase angle of the partial tones during a pressure equalization phenomenon in relation to the frequency for a spherical volume of 5 cm radius. a–pressure chamber; b–with exponential horn; c–in free atmosphere.

Curve a relates to the pressure vessel, curve b to the exponential funnel and c to free atmosphere.

Of great importance also are the phase relations among individual partial tones since measuring results can be influenced by them, particularly during performance of the octave sifting analysis in the sound absorbing chamber. In the pressure chamber experiment all partial tones have the same initial phase of 90°. In the case of pressure equalization in free air the phase shifts from nearly 0° at the low frequencies up to 90° at high frequencies. A similar curve form results from a frequency independent time constant.

Figure I2a-3 shows measuring results obtained in the sound absorbing chamber at 30 cm distance between sound emitter and microphone for various sonar pulses. The peak sound pressure value per octave is indicated above the point of frequency concentration of the corresponding octave in the region from 37.5 cps to 12,000 cps. The sonar pulses originated from capacitor discharges at different electric energy levels and from the report of a Flobert pistol. The measuring results do not lend themselves to comparison with the theoretical values in Fig. I2a-1 since they relate to sound pressure per octave

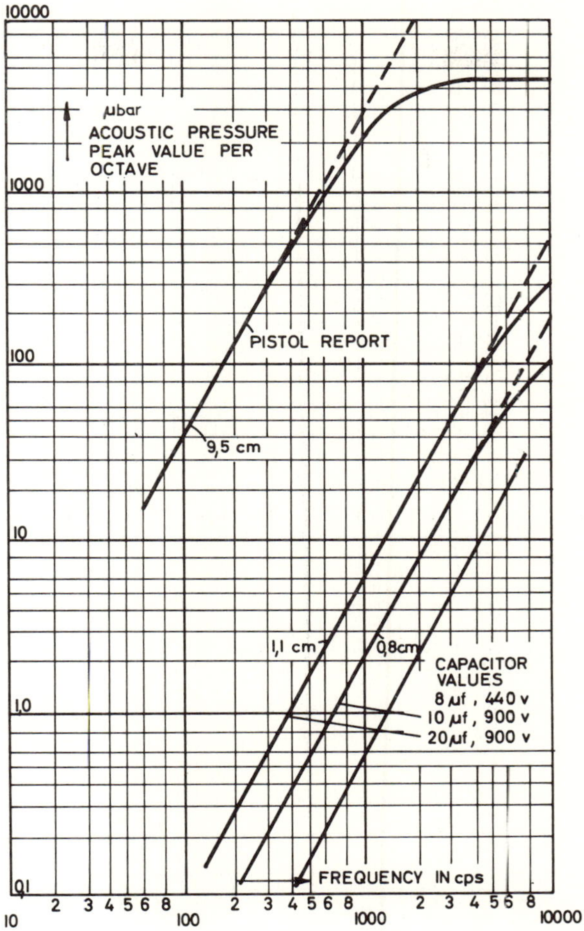

FIG. I2a-3. Measurements in a nonresonant room at 30 cm distance. Acoustic pressure peak value per octave in relation to the frequency for different reports. The centimeter indications at the curves indicate the size of the marginal radius determined from the spectral characteristic (at the curves 9.5 cm and 1.1 cm the references to the constant octaves have been omitted; hence, the much more erratic values.

at 30 cm distance, while the theoretically computed values relate to sound pressure per cps at the unit of sound pressure. In order to make possible a comparison of the measured curves with the computed curves of Fig. I2a-1, at least in their progress, they have been related to the sound pressure value per cps by dividing the measured sound pressure values by the bandwidth of the corresponding octave. This simple dividing procedure is applicable in this case since the phase shifting within one octave is only slight. In the octave

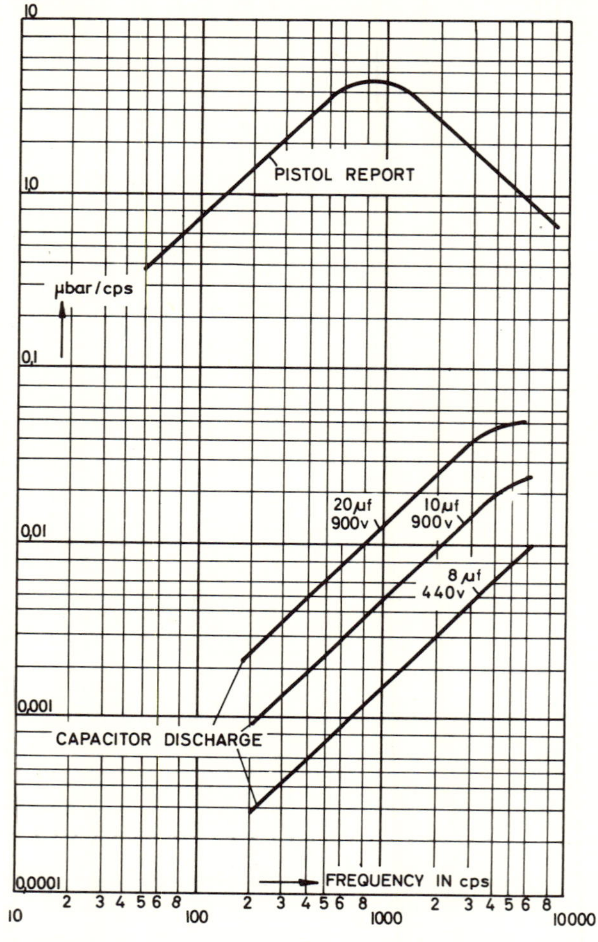

FIG. I2a-4. Graphs of the acoustic spectrum per cps for various reports related to 30 cm measuring distance calculated from the acoustic pressure values per octave measured in the nonresonant room.

embracing the maximum of the partial tone spectrum there is merely a phase shift of approximately 20°.

In all other cases only values of roughly 5° occur. The phase shiftings within the octave filter as such, particularly at the flanks, do not reflect on the curve shape since, as already mentioned, the error is identical in all octaves. The corrected curves are shown in Fig. I2a-4 and their shape is identical to that of the theoretically computed curves. Consequently, for the start of the sound wave we must assume a certain radius of the sphere. The measurements further indicate that the size of the sphere's radius depends on the energy released. Apparently the pressure wave first expands from its center of origin

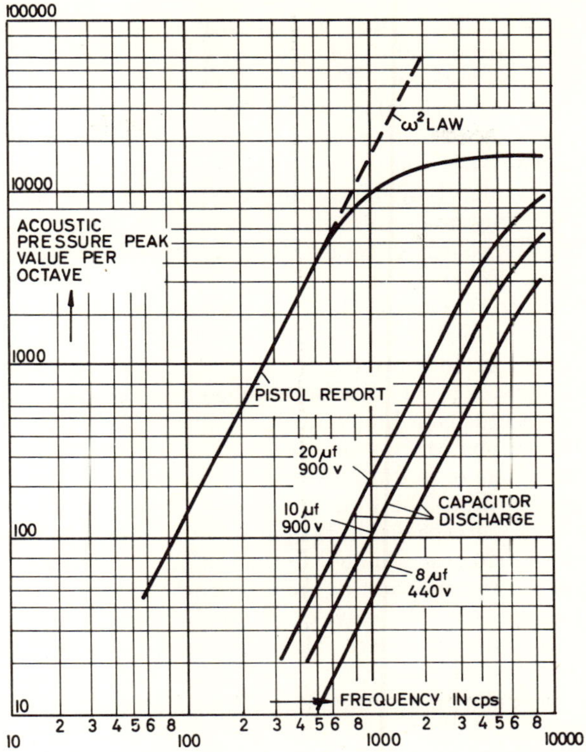

FIG. I2a-5. Measurements in a nonresonant room. Acoustic pressure peak value per octave related to the surface of the sphere, the marginal radius of which has been calculated from the spectral characteristic.

at supersonic speed until a radius is reached where the overpressure has dropped down to such a value that the linear acoustic laws begin to apply. The marked appearance of the separation point can be explained by assuming that at an adiabatic change of conditions between pressure and radius there exists the relation

$$\frac{P_1}{P_2} = c\left(\frac{r_1}{r_2}\right)^{4.2}$$

i.e. at a linear change of the radius the pressure change will proceed by the 4.2 power.

The increase of the radial boundary proportional to the energy released finally points to the conclusion that the same pressure must always prevail on the surface of the marginal sphere. In order to check this, the pressure values measured at a certain distance were reduced to the spherical radius determined from the spectrum. As can be seen from Fig. I2a-5, all curves tend to enter into an upper marginal value of approximately 15,500 μbar per octave, while in the ascending part the distance between the curves corresponds to the square of radial relations, as could be expected because of the changing radiation resistance. Thus the same starting pressure on the surface of the sphere with the critical radius has been proved for all cases in question.

The considerations relating to the pressure equalizing phenomena with regard to the pistol shot report can no doubt also be applied to other explosion phenomena, provided that the assumptions accepted in the derivation with regard to the steep gradient of the pressure jump will also be valid. But after all, it is in the very nature of explosion phenomena that a very steep pressure increase occurs. For instance, a sound analysis of the recorded acoustic spectrum of an internal combustion engine without exhaust showed the same mathematical interrelationship as in the above-discussed laws. The two-stroke engine seems to meet the theoretical conditions best because of its short exhaust stroke. It may also be assumed that the muzzle blast of heavier firearms will proceed according to identical physical laws.

To summarize, one can say that it is safe to assume that the overpressure of an acoustic sphere in free atmosphere is equalized over the sound resistance of the sphere's surface. The Fourier analysis of the equalizing process leads to a continuous sound spectrum; the partial tone intensity per cps at the lower frequency range increases similarly to the radiation resistance of a radiator of 0 power, that is, in proportion to the frequency up to a maximum occurring at that frequency where the wavelength of the partial tone has reached the 3.5-fold radius of the air sphere. Then follows a drop off of the partial tone intensity which is inversely proportional to the frequency. On double logarithmic coordinate paper the spectral curve becomes a bell-shaped curve, ascending symmetrically to the maximum. The size of the sphere's radius determining the shape of the spectral curve depends on the total energy which has been set free. However, independent of the total energy, the common value for the pressure jump on the sphere's surface is approximately 150,000 μbar. This seems to be the limit where the linear acoustic laws begin to apply, while up to this point the pressure equalization process proceeds at supersonic speed. From the pressure jump as such one can, if the equalization process is taken into account, calculate the sound energy, whereby the calculated results can be verified experimentally. The sound spectra of capacitor discharges can be used for determination of the frequency curves of microphones in free atmosphere. The measuring accuracy so far achieved is 1 db. If the capacitor discharge noise is released in the muzzle of an exponential funnel one can determine the radiation characteristic of the latter. Finally, the report of a pistol shot—which often is used for room acoustic measurements—has been for determining the frequency dependent characteristic of acoustic energy.

b. *Air Shock Waves in Traffic Safety*

This additional use was discovered by Bradfield. His ultrasonic blind guiding device utilizes a capacitor discharge of 1 to 1.5 μsec duration occurring 3 or 4 times per second at the focal point of a parabolic mirror. The transmitted "white" ultrasonic spectrum is reflected from an object and received by a crystal microphone located in a second hollow mirror. Devices of the same principle are useful in traffic safety control and are also used for automatic garage door openers.

The crack sound that emanates from an air spark is not low in its efficiency as might be assumed, since it is a strictly adiabatic sound radiation. The capacitor energy is converted in the spark plasma mostly into heat and only a small portion is lost by light radiation within the visible and invisible range. Thus the major portion of the energy remains as thermal energy and exerts itself, in accordance with the basic laws of adiabatic expansion, as mechanical and thus acoustic energy radiation. Only the facts that very high frequencies are immediately absorbed in air and other gases, due to nonideal behavior, and that a high percentage of the spark sound lies in the extremely high ultrasonic range, lead to the condition that a short distance away from the spark only low frequencies—and these with only low efficiencies in accordance with their modest Fourier participation—can be found in the "white" frequency band of the spark sound. The effect of this condition can be eliminated to a certain degree by utilizing an oscillatory circuit, the half-waves of which produce a sparkover each time so that, in accordance with the Fourier analysis, a main frequency is obtained that is twice the current cycle frequency and the sparkover frequency. This procedure, however, does not change the fact that every sparkover again produces high primary frequency portions which are subjected to the same lamentable high momentary absorption in the near vicinity of the spark.

Ultrasound generation by means of spark discharges belongs to the field of thermal ultrasonic generators. Exactly speaking, it is not proper to refer to ultrasonics when this method of sound generation is meant, because in connection with ultrasonics there is normally understood a definite frequency, whereas the property of the spark sound is to radiate a white spectrum and thus no frequency. Sound and "crack" have the same relationship as ultrasound and "shock wave sound" when the latter term is introduced with a corresponding meaning. With spark discharges it is possible to generate high shock sound peak outputs which could never be produced with conventional means, for instance, several hundred kilowatts peak outputs. In the following section some applications of shock sound and its generation are illustrated.

c. *Applications of Air Shock Waves*

In order to conclude the discussion about emission of shock sound in gases or atmospheric air, the construction of a radiator for crack sound waves in air will be illustrated. One of its uses is to scare birds.

A simple arrangement is shown in Fig. I2c-1. An electric capacitor, 1, is periodically charged and, after being charged it discharges itself through the electrodes, 2, with a cracking sound spark. The electrodes, 2, are so located that the issuing shock sound radiation can be projected in the desired direction as a narrow beam; for instance, at the focal point of a reflector, 3. For radiation in all directions, an electrode system as depicted in Fig. I2c-2 is suitable. The electric spark, fed from capacitor, 1, jumps the gap between the

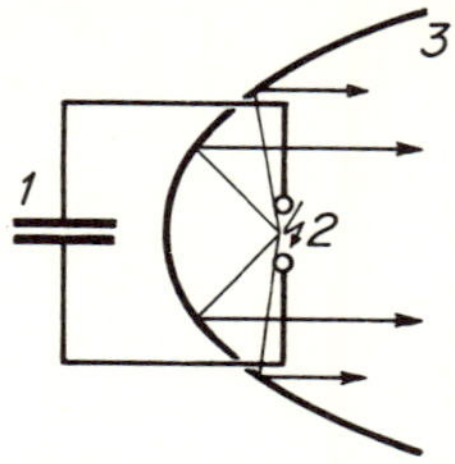

FIG. I2c-1. Electrical capacitor discharge by means of a sound spark gap for production of parallel directed, impulse sound radiation.

contact metal faces, 2, which are fastened to suitably curved metal sheets, 3, so that the shock sound originating from the spark travels in all radial directions as indicated by the arrows, 4. Since the contacts, 2, produce the electric spark in atmospheric air, they are subjected to strong oxidation. The most suitable electrodes for resisting oxidation are made of sintered metal, such as tungsten-silver.

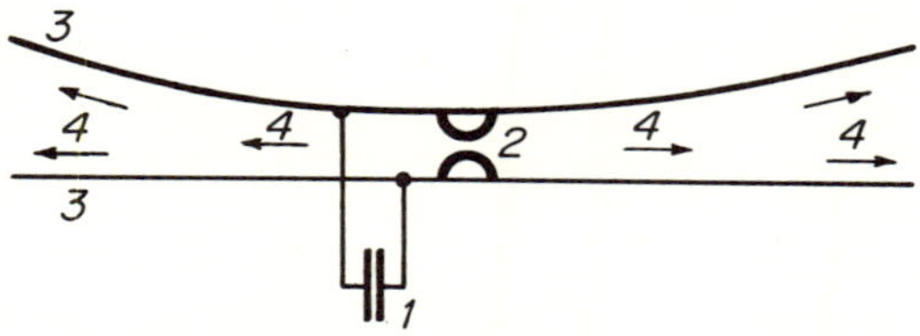

FIG. I2c-2. Electrodes ystem for all-around radially emerging disturbance sound creation.

For physiological crack sound production it is known by experience that 1 to 50 crack sparks per second will suffice. The charge of the capacitor must be at least 1/10 watt-sec. The electrode spacing is dependent on the maximum current value of the spark. The maximum current value in turn depends on the inductance lying in the discharge circuit or formed by it. The frequency components above 100 kc have such a limited radiation range in air that their production is uneconomical; the more so, since the share of higher frequencies becomes ineffective due to absorption. Thus, an inductance, 3, is inserted in the discharge circuit of capacitor 1 including electrodes 2 as indicated in Fig. I2c-3a. The value of this inductance must be such that, as illustrated in Fig. I2c-3b, the first half-wave with its duration t coincides with a half-wave of the desired emission frequency.

Example:

Desired electrical charge for one spark:	$\frac{1}{2}CU^2 = 1/4$ watt-sec
Capacitor value:	10,000 pf
Charging voltage:	7,000 volts
Inductance for critical damping:	100 kc 1/4000 h $= 250\ \mu$h
Charge of capacitor in amp-sec	$Q = CU = 7 \times 10^{-5}$ amp-sec
Spark resistance of the first half-wave:	$R_F = k \times l/Q = 16/7$ ohm for $k = 0.8 \times 10^{-3}$
Sparking distance for 7 kv:	assumed as $l = 0.2$ cm

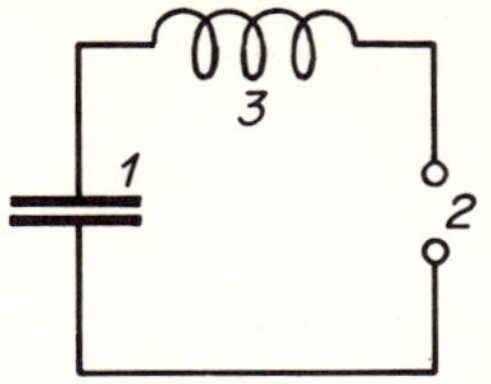

FIG. I2c-3a. Spark circuit with capacitor and line inductance.

The ohmic resistance of inductance 3, Fig. I2c-3a, must, for prevention of substantial losses, be lower than the spark resistance, which is 16/7 ohm in the present example. By applying a high frequency litz-wire coil it is possible to obtain with small outlay a coil resistance of 0.5 ohm, which is low enough not to weaken the crack sound noticeably, but to assist in releasing the full sound energy within the range of the effective frequency.

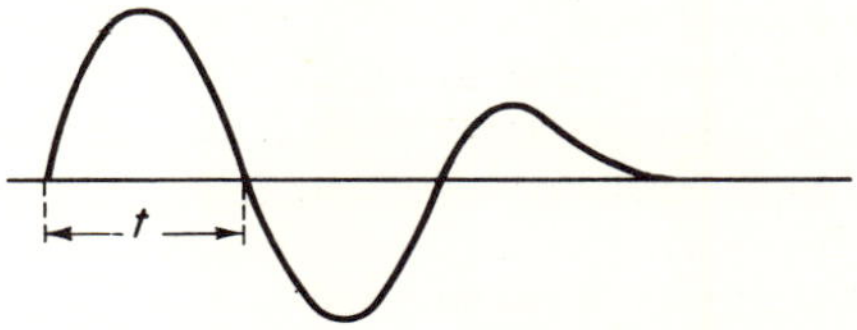

FIG. I2c-3b. Oscillogram of the discharge wave damped by the spark.

Charging of the capacitor, as shown by the example, requires only little electrical energy. For 50 sparks/sec the electrical energy would be 50 watts and including the nonavoidable charging losses this would require a power supply of 100 watts, whereas with a spark frequency of one/sec, which is usually still sufficient, the consumption is only 1 to 2 watts. If it is desirable to charge capacitor 1 from the ac power line, as illustrated in Fig. I2c-4 a high leakage transformer, 5, is interposed in the circuit. Its leakage inductance, 6, is by proper measures adjusted so that during each voltage half-wave there takes place only one capacitor charge up to the breakdown voltage of the spark gap. Transformers originally designed for neon tube illumination of the lowest commercial ratings are well suited for the purpose under consideration. For shielding high frequency oscillation from the secondary of such transformers, a small capacitor, 2, is connected in parallel to this winding whereby

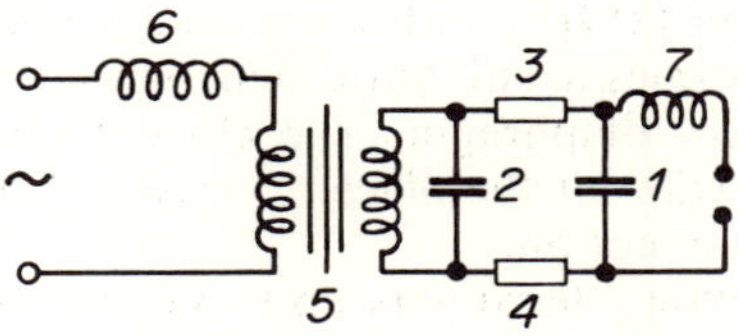

FIG. I2c-4. Equivalent circuit diagram for charging of a capacitor from the ac power line by means of a leakage transformer.

charging of capacitor 1 is effected through the charging resistance, 3 and 4. The voltage pulses at capacitor 1 thus become ineffective at the transformer so that completely radio-interference-free operation is assured.

3. Shock Sound by Underwater Capacitor discharges

a. *Spark Breakdown in Liquids*

Electrically conductive liquids such as salt solutions in water and alcohol can be used for enforcing a spark discharge by means of sufficiently great and quickly increased voltage differences between two metal electrodes. The electric potential gradients determine the breakdown voltage, which is a characteristic constant for the substance in question, just as, for instance, are conductivity and the dielectric constant.

For experimental purposes solutions of salts and acids were used mainly with low salt concentrations of 5%, where the conductivity as far as distilled water is concerned would change from 1 to 1000 and more. One or several capacitors in parallel were charged to high voltage potentials and the discharge was fed through a 2 mm copper conductor, which was interrupted at two places by spark gaps. Seen from the capacitor, the first spark gap was an air gap between brass spheres of 2 cm diameter with variable distance and the second spark gap was between brass spheres of 1.4 cm or 1 cm diameter at a certain distance. With this arrangement Heydweiller [217] found breakdown voltages of 200-800 kv/cm.

Similar to the direct sparkover in liquid mediums is the process with the electrolytic interrupter, which will be described because of the similarity of the phenomena involved. Two forms of the electrolytic interrupter must be distinguished, the Wehnelt interrupter and the Simon interrupter.

The Wehnelt interrupter [218] consists of a glass vessel filled with dilute sulphuric acid in which two electrodes are submerged. The positive electrode is a thin platinum wire, which protrudes only a few millimeters from an insulator. The negative electrode is formed by a lead plate. The interruption occurs when, at the small surface of the platinum wire, owing to the heavy current, a bubble of nonconductive vapor forms, caused by evaporation as well as electrolytic decomposition. The sudden interruption generates an overvoltage, which causes a sparkover in the gas bubble and therefore explodes the latter. The nonexplosive part of the gas bubble is blasted away by the explosion. The liquid then again contacts the wire and the voltage rise can start over again.

The Simon interrupter [219] operates in a similar manner. This interrupter has two large lead electrodes in sulphuric acid. They are separated from each other by a very narrow diaphragm. The current must pass through this diaphragm and thus creates at the place of greatest current density a vapor bubble which causes interruption.

By means of electrolytic interrupters up to several thousand interruptions per second can be attained. Every single interruption or disconnection is so sudden and so complete that a parallel capacitor becomes unnecessary with these interrupters. Also the electrolytic interrupters permit a certain control:

With the Wehnelt interrupter, for instance, the depth to which the platinum wire is submerged can be varied; with the Simon interrupter, control is effected by increasing or decreasing the diaphragm, e.g. by lowering a china cone more or less deeply into the diaphragm and thus partly closing the latter. Figure 13a-1 is a schematic diagram of the Wehnelt and Simon interrupters.

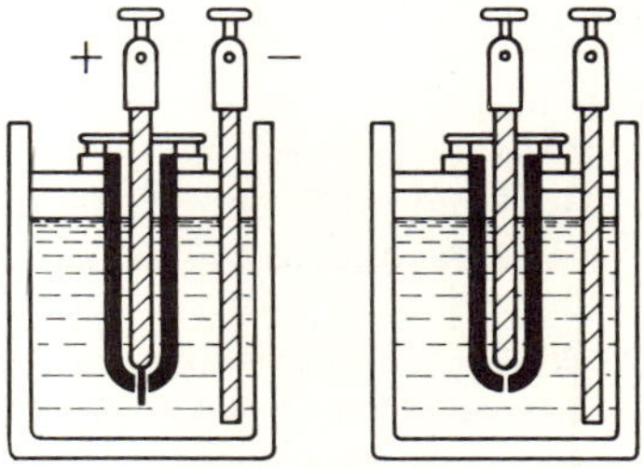

FIG. 13a-1. Schematic diagram of the Wehnelt and Simon interrupters.

L. R. Padberg [60] gives an incomplete compilation of a number of measurements made in cases where resonators were excited by fluid sparks. Further literature in this connection is given under [220, 229, and 101].

TABLE 13a-1

ACOUSTICAL AND MECHANICAL EFFICIENCY OF LIQUID SPARK GAPS (REPRODUCED FROM [53])

C(μf)	U(kv)	$\frac{1}{2}CU^2$ (mw-sec)	R	L(μh)	Weight (gm)	Throw height (cm)	Mech. work (cm-gm)	Mech. work (mw-sec)	Efficiency (%)
0.012	18.3	2000	—	0.2	2	130	260	26	1.3
0.012	18.3	2000	—	0.2	1.4	170	240	24	1.2
0.067	12.2	5000	—	0.2	5	100	500	50	1
0.067	12.2	5000	—	0.2	2	200	400	40	0.8
0.067	12.2	5000	—	0.2	2	250[a]	500	50	1
0.07	12.2	5200	—	0.2	20	25	500	50	0.96
0.07	12.2	5200	—	0.2	5	150	750	75	1.44
0.07	12.2	5200	—	0.2	2	200	400	40	0.8
0.07	12.2	5200	2.5	0.2	2	100	200	20	0.4
0.07	12.2	5200	—	3	5	100	500	50	0.95
0.07	12.2	5200	—	82	5	50	250	25	0.47

[a] Height to ceiling of room. Every one of above values is an average of 3–5 single measurements which scatter by $\pm$ 20%.

Investigations concerning the mechanical efficiency of liquid sparks were made by Früngel [53] on small weights thrown up by liquid sparks (see Table 13a-1). The results show efficiencies on the order of 1%, because it is impossible to establish mechanical-acoustic adaptation between weight and spark plasma. Since the conversion of energy is completed within one microsecond one obtains within this time shock sound intensities which create a weight acceleration as high as in the order of 10^5 watts, if the weight is concentrically located on its support. The drop in efficiency to one half by insertion of 2.5 ohms means that the resistance of the liquid spark, at the electrical data given, has about this value. According to Toepler's formula, the resistance of the series connected air spark gap of 3 mm is, with $k = 0.8 \times 10^{-3}$, about 0.24 ohm, thus negligibly small.

The resistance for the critically damped condition in the circuit (aperiodic limit resistance) lies, as calculated, at 3 ohm. The fact that the liquid spark resistance is very close to this favorable value, is important for rapid conversion of the energy.

The time from the sparkover until the first current maximum is, with a line inductance of 0.2 μh, about 0.2 μsec; and with 82 μh, about 4 μsec. The oscillations were oscillographically observed. The amplitude of the first half-wave is a triple of the second half-wave and then the oscillation decays rather slowly because with every passing of a current half-wave the damping spark resistance becomes lower. The process of mechanical energy transmission can take place only during the time when the spark has high resistance and takes considerable energy itself. In considering the influence of L the impression is gained that the discharge is slower than 0.2 μsec but faster than 4 μsec. At 1 μsec discharge the conversion of 75 milliwatt-sec would give the maximum mechanical output of 75 kw.

When the weight itself is used as counterelectrode and a thin aluminum foil is inserted for prevention of adhesive losses, the efficiency results remain the same as given above. By insertion of sheet aluminum there is no decrease up to a sheet thickness of 1–1.2 mm. When no weight is superposed, the mechanical energy exerts itself in the form of punch marks in the sheet. With higher sheet metal thicknesses the decrease is rather rapid, and with sheets of 2 mm the efficiency drops to less than 0.25% of the original values. A satisfactory explanation of this phenomenon could not be found. The propagation of the shock wave in aluminum appears to be considerable, the thin sheet is not in a position to develop the full sound hardness; it is caved in and transmits the impulse to the weight, whereas with a thicker medium, owing to the then fully pronounced acoustic-mechanical data, a reflection occurs at the aluminum-water surface which is not transferable to the weight. The time for sound to travel forward and return within the 1.2 mm material is about 0.5 μsec.

The punch-mark-like deformation of the thin sheet supports the assumption that the duration of the occurrence is about 1 μsec.

As measured by Schaaffs [28] the sound densities in the vicinity of a liquid spark are so high that the water may be compressed to a density of 2. Such steep pressure rises, up to amplitudes of 50,000 atm, usually cause plastic deformation in adjacent metal surfaces, which perhaps uses a considerable portion of the original acoustic peak output. Rust and Drubba [54] utilized the shock sound of underwater sparks with long rise time for development of an echo sounder. Bailitis [55] criticized their experimental arrangement and found that the sound progression given in milliseconds could have been caused by broad-faced electrodes and oscillation-like excitation of the same by the sparks, because in genuine liquid sparks one always finds only microsecond sound emissions.

b. *The Analogy Between Explosive Conversions and Shock Sound Discharges*

F. Früngel and H. Keller [56] calculated the analogy between explosives and sparks. The methods for producing shock sound in air, as discussed above, are different from those for shock sound in liquids, because liquids have densities that are higher by a factor of 10^3; they are, in contrast to air, practically noncompressible.

Two groups of sound producers can be distinguished:

The action of the first group is reversible. These devices radiate sound when electrical or mechanical energy is supplied, and they deliver similar mechanical or electrical energy when struck by sound. To this group belong magnetostrictive, electrostrictive, electrodynamic and electrostatic transmitters as well as those using quartz crystals.

The second group operates with irreversible processes. These devices cannot be used as microphones as is possible with devices of the first group. To this group belong explosives, which create sound because of chemical reaction. The irreversible emitters are distinguished in that the pressure amplitude can become so high that the medium in which they operate no longer has a constant sound velocity; the sound velocity is above normal in the vicinity close to the source. All the known methods of so-called sound optics which are of benefit in reversible emitters, such as application of lenses and mirrors, become useless, since the two prerequisites for their use, namely constant sound velocity and constant frequency, are not present. Interference considerations do not apply, since there are no minima subsequent to the peaks; instead, the pressure equalization proceeds very slowly. The irreversible sound producers have been useful only for producing single intensive sound shocks. Only very small sound emitters which produce sound not by chemical disintegration but by evaporation of the medium can be used for periodic operation, e.g. for low power signals. The Wehnelt interrupter belongs to the latter group. However, a useful application of this method has not yet been found despite many attempts to improve consistency and extend the life of equipment for that purpose.

During a high energy electrical sparkover in a liquid such as water, the sudden explosion of the evaporated and disintegrated liquid causes extraordinarily strong crack sound waves which often lead to destruction of the vessel in which the experiment is conducted. It is possible to convert up to 50% of the supplied electrical energy into mechanical energy which becomes effective in the sound field. By proper dimensioning of self-inductance and capacitance in the spark discharge circuit, the expansion progress of the spark, accompanied by an analogous pressure wave, can be influenced so that certain requirements with respect to impulse sequence and duration can be met.

The capacitor to be used must in each case have a rating such that its electrical energy $\frac{1}{2}CU^2$ is able to cover the demand of one impulse at the efficiency given by the selected parameters of the ultrasonic output circuit. It is possible to operate either with large capacitance and small voltage or *vice versa*, so that by predetermining the self-inductance and similar energy quantity in the capacitor, one can obtain a slow or a fast progression of the spark discharge respectively. Moreover, it is possible by variation of the self-inductance to render the occurrence of impulses periodic, aperiodic or variable. By selection of capacitance and self-inductance values one largely has control over the amplitude of the shock sound wave. Since—as is well known from oil circuit breakers for heavy loads in insulating liquids—the insertion of an auxiliary spark gap, also in water, causes the spark to extinguish very fast, it is possible to force a high number of sparks to occur within a unit of time without the formation of a continuous arc (cf. Wehnelt interrupter, described in Section 3a of this chapter). By operating in liquid under pressure it is possible to reach an even higher rate of sparks. The progression of the shock wave can also be influenced by the electrolytic conductivity of the medium; good conductivity (1–10% salt in water or a *p*H of 3–5) results in harder shocks than are obtained in, e.g., tap or drinking water. The following numerical example will illustrate the requirements for highest instantaneous impulse energy in supersonic sound production: Ultrasonic shock output in impulse operation, 10^4 watts max; dimensioning of the circuit 0.5 μf charging capacitance, 10 kv charging voltage, 0.5 μh selfinductance. With very large capacitor batteries and low selfinductance in the discharge circuit, the effect of the spark resembles very closely the detonation of a small quantity of explosives.

It is possible to point out an analogy between the phenomena of explosive detonations and liquid sparks in the calculation of super sound shocks of underwater sparks.

From practical experience it is known that the voltage and capacitance needed for production of underwater sparks of about 1 mm length amount to about 20 kv and 2×10^{-8} farads respectively. The energy necessary for such a spark becomes

$$\frac{CU^2}{2} = \frac{400 \times 10^6 \times 2 \times 10^{-8}}{2} = 4 \text{ watt-sec}$$

With 120 4-watt-sec sparks/sec at 60 cps obtainable directly through a transformer from an ac power line, we can assume an invested energy of about 500 watts. Using the physical laws of underwater detonations, we can make some basic disclosures about qualitative and quantitative values obtained in studying the physics of water sparks and their crack sound waves. Since 1 gram of explosive liberates about 1000 calories of energy (about 4000 watt-sec), the corresponding energy of a 4 watt-sec spark requires about 1 milligram of explosive; in volume that is a sphere of about 1 mm dia, assuming a specific gravity of 1.6 for the explosive. The temperature in the sphere volume after the detonation is about 2100°C. In answer to the question of how many calories must be introduced into a similar water volume to bring it to the same temperature, the following balance is given:

Assumed water volume: 0.6 cu mm (equivalent to a weight of 0.6×10^{-3} grams).
Heating of this water volume from 20°C to 341°C (critical temperature) requires 321 cal/grams $\times$ 0.6 $\times 10^{-3}$ grams = 190 $\times 10^{-3}$ calories

Conversion of this volume into steam: 539 cal/grams $\times$ 0.6 $\times 10^{-3}$ grams = 330 $\times 10^{-3}$ calories

Heating of the vapor from 341°C to 2100°C: 1759 cal/grams $\times$ 0.5 $\times$ 0.6 $\times 10^{-3}$ grams = 540 $\times 10^{-3}$ calories

(The factor 0.5 is the average specific heat of steam.)

Adding these calorie quantities, one obtains a total heat contribution of about 1 calorie which is equivalent to about 4 watt-sec. Thus the observed spark of a 4 watt-sec discharge obviously has an effect similar to that of the detonation of 1 milligram of explosive. Since the density of water is not very different from that of the explosive it is safe to assume that after the instant at which the gas bubble has reached the diameter of the explosive, the phenomena resulting from the underwater spark closely parallel those following the detonation. The question remaining open is whether the energy from the spark region is considerable during the time before that instant. It appears that, except for observable ultraviolet radiation, all energies remain within the spark bubble region and there is no reason to believe that more than one-tenth of the energy radiates farther. Even in the case of intensive spark discharges within gases, it is at present assumed that during the first phase of the occurrence the radiation losses are negligible [1]. Taking this for granted, it is permissible to deduce a rough quantitative conformity between the explosive phenomena and the spark in water, and since this is so, the experimental results gained with explosives will now be more fully illustrated:

As long as in the bubble region (explosive) pressures of 1–8 $\times 10^4$ atm still dominate, the water in the fringe region can be considered highly compressible (as in air; see Chapter I2a). Therefore, a steep compression wave develops in this region that travels away with ultrasonic velocity. From heavy blasting operations it is known that this compression wave as well as the following much weaker pressure waves contain about 15% of the total energy, whereas in smaller blastings a higher participation can be assumed, but perhaps never more than 30%. In the example given above with 100 sparks/sec there can be expected a sound participation of about 15–60 watts which is quite uniformly spherosymmetrically distributed. As mentioned above, the crack sound waves at first travel above sound velocity, but, after a few centimeters' travel distance, fall in step with and thereafter approximate closely the known laws of sound propagation. The blast pressure waves are roughly triangular waves having a very steep front so that the front range is smaller than one-tenth of the total wave length. Utilizing the measuring system chosen here, we can represent the amplitude of the respective pressure waves in the following form:

$$P_{max} \approx 540 \text{ [kp} \times \text{cm}^{-1} \times \text{cal}^{-1/3}] \, E^{1/3}/r$$

where E is the liberated energy and r is the distance from the center.

At a distance of 10 cm from the spark a pressure peak of about 50 atm can thus be expected. The duration of such a pressure wave becomes

$$t \approx 0.3 \times 10^{-6} \text{ [sec} \times \text{cal}^{-1/3}] E^{1/3}$$

In the example chosen $t = 0.3 \times 10^{-6}$ sec, and this time determines also the chief participating frequency as being 3×10^6 cps.

Thus, the application of the term "ultrasonic" seems to be proper here.

From the formulas it becomes apparent that the self-frequency which corresponds with the pressure wave duration varies with the third root of the spark energy which in an underwater spark can be increased only with difficulty and cannot be reduced at all. However, the amplitude can be changed very simply by distance variation. An important fundamental law of acoustics permits the calculation of the particle velocity amplitude as follows:

$$P = u \times c \times \rho$$

where P = sound pressure, u = particle velocity amplitude, c = sound velocity and ρ = density of the medium in cgs units. For a sound pressure of, e.g., 50 atm (about 50×10^6 dynes/sq cm) the formula gives a particle velocity amplitude of $u = 330$ cm/sec, a value that is very small against the sound velocity of $c = 15 \times 10^4$ cm/sec and thus justifies the acoustical mode of observation.

It has been mentioned above that 30 watts sound output would have to be reached but that this output is spherosymmetrically distributed. By utilization of a cone space having a 60° opening, only 1/30 of the total is available, that is, one watt. It is feasible to obtain concentration in a beam by employing a concave mirror of steel, but such concentration would hardly be possible above a factor of 5. Besides, reflection losses at intermediate media and transition surfaces must be considered.

The example gives a wave length of 0.5 mm in water; that means that a transition diaphragm must have at the maximum only 0.05 mm thickness and this fact demands that it can be installed only at a certain distance as otherwise it could easily be destroyed. A good energy utilization can be expected when the medium to be treated by ultrasound can be disposed all around the spark, that is, if it is a liquid or a quasi-liquid. Finally, consideration is given to the impulse values. With explosives the impulse per unit area has been observed to be.

$$I \approx 2.5 \times 10^2 \text{ (gram-sec}^{-1} \text{ cal}^{-2/3}) E^{2/3}/r \text{ (see [56])}$$

If the total impulse in the spherical shell is assumed to be constant (law of impulse preservation), the value in the denominator should be r^2. The region of the pressure wave proper and the following pressure wave tail region is negligible with respect to energy. But it is not negligible with respect to impulses in interaction with each other. The relation results as pointed out above. The total impulse of the pressure-wave spherical surface increases therefore in proportion to r.

In 1 mm distance the total impulse, in the example, assuming one milligram explosive, becomes

$$\frac{2.5 \times 10^2 \times 4 \times \pi}{0.1} \approx 3 \times 10^4 \frac{\text{cm} \times \text{grams}}{\text{sec}}$$

c. *Design of Long Lived Underwater Spark Sound Sources*

Because liquid sparks, the sound shocks of which depend on capacitance and self-inductance in a discharge circuit, are, with small self-inductance values, in their sound characteristic much like detonations of small explosive quantities, the material of the electrodes between which the sparks jump is subject to sputtering; and this the more so the harder the electrode material. H. H. Rust [54] tried to mount the electrodes on rubber in order to slow their destruction. The sound shock, however, acts first on the metal surface and then, a few milliseconds later, at the rear of the electrode, so that life expectancy was thereby only very little improved. A considerable increase in life expectancy resulted from an arrangement depicted in Fig. I3c-1 and devised by Bailitis and Früngel. The electrode, 1, consists of an elastic metal blade, e.g. a jig-saw blade, whereas the counterelectrode, 2, is in the form of a block of soft metal such as brass or copper. Adjustment of blade 1 relative to surface 2 can be effected by an adjustment screw, 3, which can be so constructed that it is accessible from the outside. The liquid in which spark formation takes place is indicated at 4. The light spring blade vibrates rhythmically with each sound shock in the sound field and, because its mass is very small, no destructive effects can attack it; whereas the counterelectrode, 2, with its large mass, is capable of taking the blow and consuming some plastic deformation work without ill effects.

A useful life of about 20,000 sparks can be obtained in such arrangement without electrode replacement, operating with 7–9 kv capacitor voltage, about 1 μh circuit inductance (1 meter coaxial cable) and capacitance values between 0.01 and 0.05 μf.

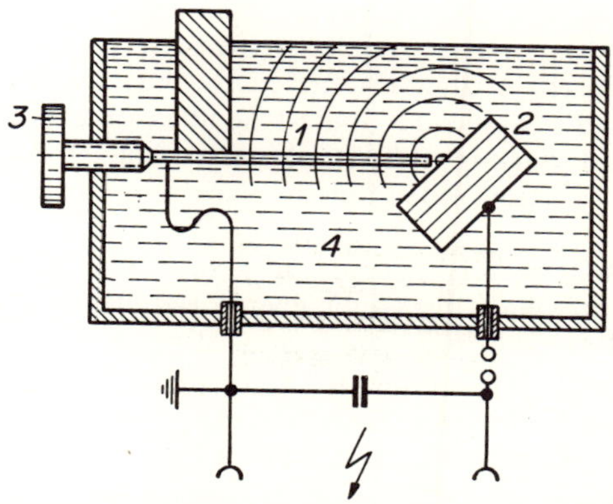

FIG. I3c-1. Functional diagram of a setup for producing underwater sparks. The capacitor discharges through the resilient electrode 1, which is adjustable by button 3 with respect to metal plate 2 within electrolyte 4. An auxiliary spark gap is to be provided in the discharge path.

To reduce the destructive effects on submerged electrodes of liquid sparks, the arrangement shown in Fig. I3c-2 is preferable. Here the pair of electrodes is designated by the reference numerals 1 and 2. Electrode 1 is the high potential electrode on which generally the high voltage pulses are imposed and it constitutes in this case a thin elastic metal blade. The counterelectrode, 2, generally connected to ground, constitutes a ductile metal block, e.g. brass or copper. Adjustment of electrode, 1, relative to the surface of electrode 2 is preferably made by an adjusting screw indicated at 3, which in a practical device will be arranged to be accessible from the outside. The liquid surrounding the electrodes and in which the sparkover is to occur is indicated at 4. In operation, the elastic metal electrode, 1, vibrates rhythmically during and after every sparkover so that destructive deformation and consumption of the high potential electrode is practically prevented, whereas the solid but ductile counterelectrode, 2, is capable of absorbing a certain amount of plastic deformation work.

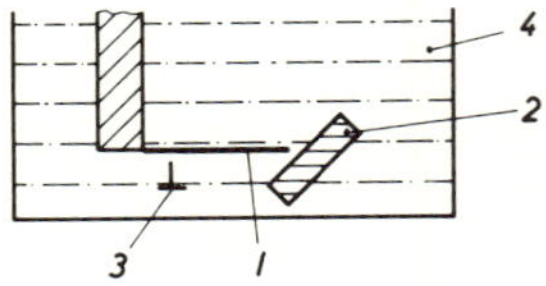

FIG. I3c-2. Diagrammatic representation of a spark gap within an electrolyte.

Fundamentally, liquid sparks as shock wave sources emit spherical radiation. For most useful purposes, however, it is desirable to operate with directional radiation giving a coherent ultrasonic wave front. To achieve this end, an attempt has been made to arrange the spark gap at the focal point of a parabolic mirror, but the results were not satisfactory, since the distances from the points of the reflecting surface to the receiving object were all of different length and resulted in loss of coherence of the wave front; the Fourier spectrum, and, consequently, the frequency content at the receiving point were different from that originated by the liquid spark. In order to avoid this effect and to achieve a phase-true ultrasonic front wave of parallel radiation, the liquid spark must be combined with a sonic lens in such a manner that not the refraction index but the various sonic velocities of the lens determine the focusing.

Figure I3c-3 shows the arrangement of a sonic lens, 5, relative to a liquid spark gap, 1–2 from which ultrasonic energy radiates in a spherical manner. It will be noted that the sonic lens, 5, consisting preferably of a plastic, for instance, the material known under the trade mark "Plexiglas," is set in front of the spark gap, 1–2. The sonic lens, 5, is so shaped that the radiation penetrating the lens has a coherent parallel wave front.

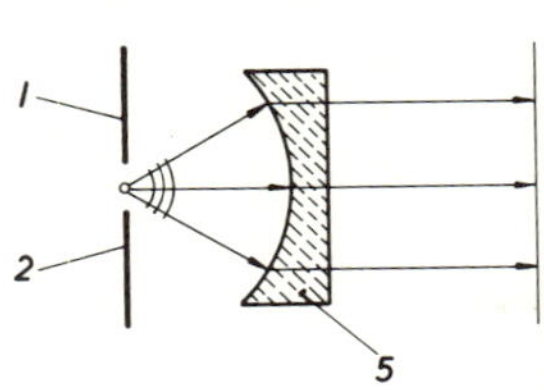

FIG. I3c-3. Arrangement of a sonic lens relative to a liquid spark gap.

The aforementioned elements can be combined into the single useful device shown in Fig. I3c-4. Here the liquid spark gap, 1–2, is adjustable by means of knob, 3, and is arranged within liquid 4 which, if desirable, may be subjected to a pressure above atmospheric pressure. A sonic lens, 5, forms the bottom of the sounding head or transducer from which parallel radiation of ultrasonic energy emanates. In order not to disturb the sonic field, the inner wall of the device is covered with sound absorbing material, 6, which detains the lateral portion of the spherical radiation. In a well known manner one of the electrodes, here electrode 2, receives the high voltage pulse by way of a lead-in insulator and an interposed auxiliary gaseous spark gap from a discharge capacitor (not shown).

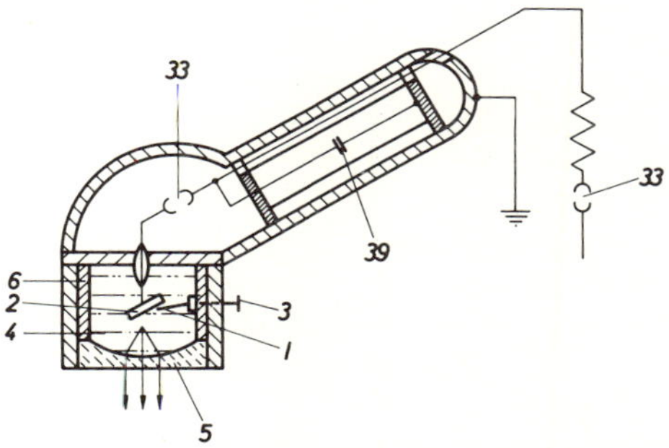

FIG. I3c-4. Diagrammatic representation of a spark gap device to be used as a source of sonic and ultrasonic energy waves within liquid media.

·The useful life of submerged spark gap devices can be considerably increased, up to one million discharges by an arrangement shown in Fig. I3c-5. As will be noted in this figure, the solid counterelectrode is completely omitted and the metallic container, 9, filled with an electrolytically conducting liquid, 4, serves as a substitute. The resilient high potential or main electrode used in previously described applications is here replaced by a thin metal electrode, e.g. a wire, which is enclosed in a yielding plastic or rubber sheath, 8. Such an insulated wire, 7 and 8, is preferably cut flat at its end.

In operation, when a normal pulse resulting from a capacitor discharge above a certain value, say about 1 watt-sec, is passed from the electrode, 7, through the electrolyte to the container, 9, a spontaneous heating with subsequent spark formation occurs at the point of contact between electrode, 7, and the electrolyte. The resulting shock wave of desirable magnitude progresses in all directions within a solid angle of 180° and acts only on the liquid so that destruction of electrodes cannot occur. The rearward radiating energy, covering the other 180° solid angle of the sphere, is absorbed to a considerable extent by the plastically deformable covering, 8, of the wire, 7. This covering, due to its reversible deformability with partly plastic energy absorption, suffers practically no deterioration. Only a small fraction of the total energy converted causes some sputtering of the metallic conductor, 7. It will be understood that a sonic lens as previously described can be incorporated in the arrangement of Fig. I3c-5 for obtaining directional radiation.

For best performance it is advisable to operate the arrangement as shown in Fig. I3c-5 with a capacitor potential of more than 5 kv and a capacitor capacitance higher than 0.02 μf. Foer xperimental work only, a suitable covering for the electrode wire is plasticized polyvinyl chloride.

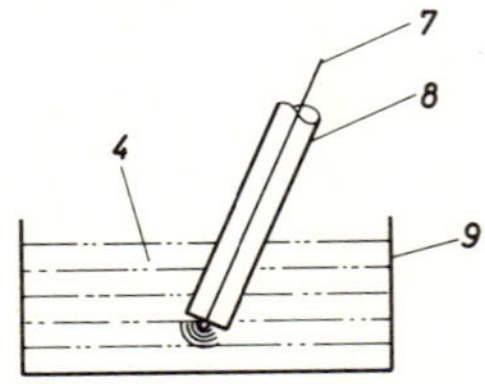

FIG. I3c-5. Modified spark gap device in which the high potential electrode is an insulated wire immersed in the electrolyte and the counter electrode is in the form of a metallic container holding the electrolyte.

In an arrangement according to Fig. I3c-5, having several unipolar spark electrodes in parallel connection the surprising observation has been made that the majoricy of the spark electrodes—and by proper dimensioning, even all of them—emit sparks simultaneously, whereby naturally the energy of each spark is correspondingly lower. Physically, this effect, which lends itself to many useful applications, can be explained thus: In *statu nascendi* all parallel connected spark gaps find similar initial conditions; that is, in all of them the electrolyte begins to decompose forming minute volumes of oxyhydrogen gas; then, as the rate of capacitor discharge energy increases, a multitude of small partial sparks develop. While these sparks, by nature, are of various sizes, it is possible to obtain sufficient uniformity for practical purposes. In perfecting this feature of multiple spark emission for practical application, the potential of the capacitor and other electrical constants of the discharge circuit must be selected in adaptation to the electrolyte in question, for instance, sea water. Dimensioning is thereby not particularly critical and a supply of, say, 1/100 joule discharge energy is sufficient to obtain spark

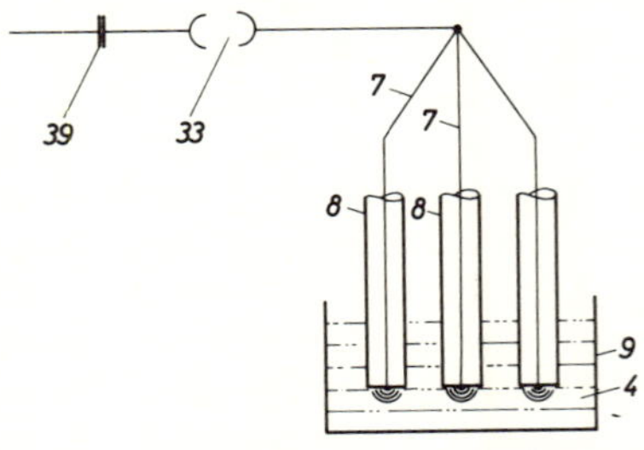

FIG. I3c-6. Arrangement wherein a number of electrodes of similar polarity and fed from the same current source are placed within a metallic container similar to the one shown in Fig. I3c-5.

discharges at a multitude of parallel-connected electrodes in fresh water as well as in salt (sea) water. In another arrangement, illustrated in Fig. I3c-6, several electrode systems, 7 and 8, are placed within the same body of electrolyte, 4, contained in vessel 9. This principle can also be practiced according to Fig. I3c-7 which indicates an arrangement for ultrasonic treatment of liquid media to achieve, for instance, special chemical or physical changes of the medium under treatment. For such treatment of solutions, the vessel, 9, is preferably in the form of a tube through which the solution flows in a desirable direction, for instance, as indicated by the arrow. The spark gap devices are inserted through the wall of tube 9 and may be in the form of conventional spark plugs or the like. Owing to this very simple arrangement, any suitable design of the vessel for ultrasonic treatment of solutions during through-flow can be devised. Depending on the character of the solution under treatment, it is advisable to select the electrode material correspondingly, and one will generally prefer iron for beverages and medicines, because iron emulsions are not poisonous and in every respect harmless.

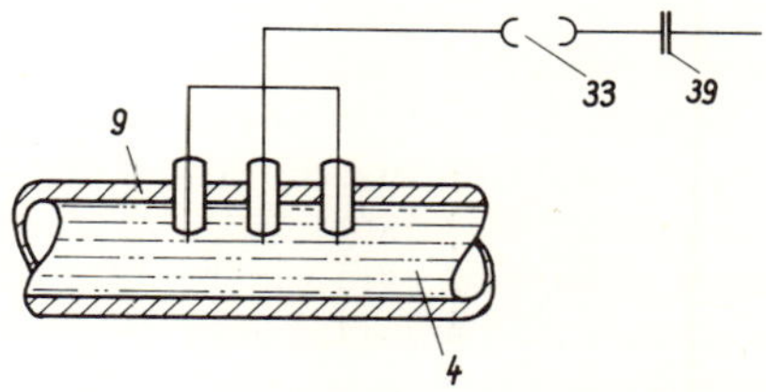

FIG. I3c-7. Group of electrodes inserted through the wall of a tube, whereby the tube through which the electrolyte is flowing serves as counterelectrode.

For creating strong ultrasonic fields by liquid sparks it has been discovered that sparkovers also occur at perforations in the insulation of conductor 7 in Fig. I3c-6. The spark, for its initiation, does not need to be in direct contact with the metallic conductor; it is sufficient that the electrolyte at certain points is particularly contracted and that through these minute cross sections of electrolyte, current lines are passed at greater density from a capacitor discharge. In this respect there exists a certain physical relationship with the well known Wehnelt interrupter, in which by gradual increase of current the electrolyte is heated up to evaporation and thus interrupts the circuit. In the present case, this is further developed in that chemically and thermally the electrolyte is momentarily transferred into the gaseous state, forming detonating or oxyhydrogen gas in certain cases, and that in this state of great dissociation it represents a remarkably good plasma for a sparkover.

The arrangement indicated in Fig. I3c-8 takes advantage of this phenomenon. Here the grounded counterelectrode is again connected with the conducting housing, 9, and it will be noted that the high potential electrode, 10, is surrounded by an insulating jacket, 11, which is perforated either only on one side or on both sides. When a discharge pulse passes, small spark discharges emanate out of the small perforations, which leads to a coherent shock wave emission having the sonic field characteristic of a magnetostrictive or piezoelectric transducer when excited by harmonic waves. It is hereby of importance to note that shock wave emission also occurs when no actual spark discharge takes place; strong contraction of current lines within the perforations suffices. The shock wave emission resulting from current line contraction without liquid spark formation is naturally much weaker, but for many purposes sufficient, and this type of shock wave generation has the advantage of nonaggressive action on electrodes, thus lending practically infinite life to the shock wave source.

FIG. I3c-8. Modification of a liquid spark gap, wherein the high potential electrode is a conductor surrounded by a perforated insulating jacket and the counterelectrode is the conducting container of the electrolyte.

Sparkless shock wave emission occurs also in the previously described arrangements shown in Fig. I3c-3 to 7. Therefore, it is possible to operate shock wave sources which are primarily designed to flash liquid sparks when subjected to strong capacitor discharge with weak capacitor discharges. When weak ultrasonic or sonic fields are sufficient for a certain purpose, this can be done in order to increase the useful life of the electrodes. Shock wave generation is caused by current line contraction in that the electrolytic medium is momentarily vigorously compressed and dilated by action of the current passing at high density through the contraction space.

Doubling of the shock wave effect can be achieved with an arrangement as shown in Fig. I3c-9, in which the insulation jacket, 11, on its rear side, the side from which no emission is desired, is made of foam rubber or other porous material so that at this portion it represents a reflector for shock wave effects. If so modified, liquid sparks occur only on the perforated side and the rearward emanating radiation, covering a solid angle of about 180°, is nearly completely reflected by the porous mass of foam rubber. In order to avoid interference phenomena in such an arrangement, the electrode, 12 (Fig. I3c-9), should be very thin, that is, extremely thin relative to λ, the wavelength of the ultrasonic radiation. The perforated side of jacket 11 consists preferably of a very thin perforated dielectricum so that the motion variance between direct and reflected waves is very small. Wavelength, as here referred to, is intended to mean the pertinent value of the main frequency content of the Fourier spectrum of a spark discharge. With an electrode of the construction diagrammatically indicated in Fig. I3c-9, it is possible to obtain a useful life of the electrode up to 10^7 sparks. After this upper limit has been reached, a noticeable burning takes place, giving rise to perforation or pore size increase. Instead of using rubber as a covering for the electrode, it is better to apply one of the recently developed plastics of leatherlike consistency, such as, for instance, the material known under the trade mark "Vulcaprene."

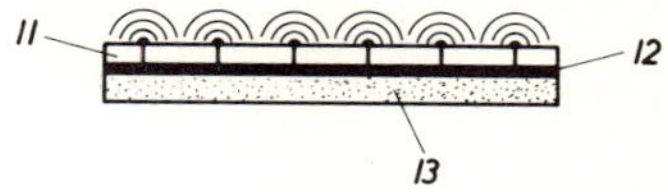

FIG. I3c-9. High potential electrode somewhat similar to the one indicated in Fig. I3c-8, but having an insulating jacket which on one side consists of foam rubber or the like which is not perforated and serves as a reflector for sonic radiation.

For precision problems in shock wave energy generation, where life of the electrodes is of secondary importance, it is possible to replace the flexible electrode covering by a suitable solid insulation, for example, solid synthetic resin or artificial horn, or Plexiglas. In such material the perforations can be shaped to the exacting requirements of the particular application in question. Figure I3c-10 shows a cross section through such a modified electrode, where numeral 12 designates again the metallic portion receiving the high voltage pulse, and 14 designates the plastic covering. The latter is provided with perforations having conical expansions in which the small spheric, ultrasonic waves develop at the points where the conical expansions join the cylindrical portions of the perforations.

FIG. I3c-10. Modification of a high potential electrode wherein the perforations in the electrode jacket have funnel-shaped enlargements toward the outside.

Another design is represented in Fig. I3c-11. Here a metal foil, 12, to be subjected to the electric pulse is provided with a perforated plastic foil, 11, of high durability. This combination is finally surrounded by a shock wave guide material, 14, preferably a hard plastic, but eventually even a metal may be suitable. In this manner the toughness and high burning resistance of foil 11 are combined with a solid material serving as precision shock wave guide.

FIG. I3c-11. Cross section through a flat, high potential electrode wherein the conducting element is tinfoil covered by a plastic layer over which in turn a sound guide material, for instance hard plastic or metal, is superposed.

It will be understood that with the aforedescribed design, the constructional possibilities for liquid spark devices are far from exhausted and that other designs will occur to other workers in this field. Each arrangement must be devised to meet a specific requirement. To obtain contraction or divergence of shock wave force fields, it is possible to curve the electrode combination as described in connection with Fig. I3-c11 in a manner indicated in Fig. I3c-12a and b, the former for converging and the latter for diverging force fields. The perforated plastic layer from which the shock waves emerge is shown at 11, the conducting metallic foil at 12, and the rearward foam rubber reflector at 13 (the latter may be omitted in some cases). Particularly in such an arrangement, shock wave force field convergence at the focal point can be achieved by making the surface from which the shock waves emerge spherical or ellipsoidal. Such a concentration of sound and/or supersound energy was hitherto unknown.

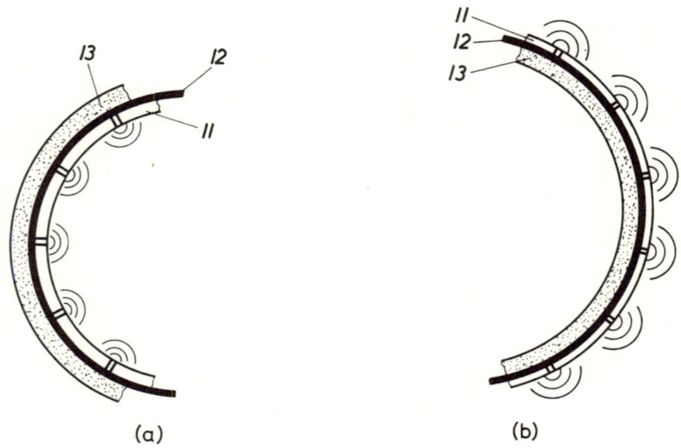

FIG. I3c-12. Modifications of high potential electrodes suitable for liquid spark gaps, whereby the arrangement indicated in Fig. I3c-12a serves for converging sonic radiation and the arrangement in Fig. I3c-12b for diverging sonic radiation.

A further modification with respect to the arrangement of a group of electrodes is shown in Fig. I3c-13a. Often the Fourier spectrum in its total width is not desirable, rather only a certain frequency band is to be transmitted. For this purpose the application of shock wave stairs is advantageous. These comprise metal foil members, 12, and reflectors, 13, the latter consisting of foam rubber or the like, whereby the spacing d of the reflector from the metal foil is the height of the stair step. The width of the step h is so chosen that it conforms to the desired shape of the shock wave radiation pattern. By superposition of the directly radiated wave over the echo wave of the adjacent stair step, one obtains, for instance, a three-stepped stair source having the amplitudes $a/2$, a, and $a/2$. The application of this principle of shock wave stairs makes possible the reception of such transmission by means of a conventional resonance receiver. Since a periodic sequence of pulses more nearly resembles a pure tone than a detonational crack as caused by single pulses, such construction is preferable for application in certain physiological fields, such as in a whale fishery for whale scaring and chasing. Figure I3c-13b shows a diagram of shock wave distribution amplitudes as they may be expected from an arrangement according to Fig. I3c-13a.

In multiple liquid spark devices having as an essential feature a perforated jacket surrounding the electrode, one finds that size and shape of the perfora-

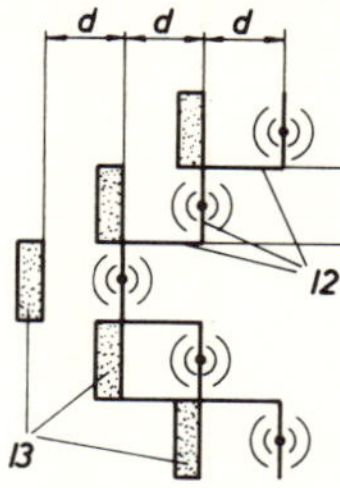

FIG. I3c-13a. Cross section through an arrangement of high potential electrodes in a staggered relation for preferred directional radiation of certain supersonic wave frequency bands.

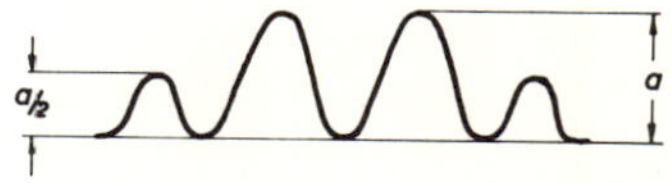

FIG. I3c-13b. Diagram of the shock waves realized by an arrangement according to Fig. I3c-13a.

tions is not critical and that considerable deformation of these openings will not diminish satisfactory operation. In fact, even bare metal portions of the electrode foil, 12, do not noticeably hamper sparkovers occurring at other points, since, presumably, the peak energy of the capacitor discharge is sufficiently high to maintain the current line concentration necessary for liquid spark initiation at a multitude of points. Giving due consideration to this finding, the author changed the small circular holes of the perforation, purposely provided electrode insulation having slots or slits and applied this in the form of overlapping collars or scales as indicated at 15 in Fig. I3c-14. The shock wave radiation emanates here at the abutting joints or at the slits in the overlap of the fish-scale-like jacket and is produced by a pure electrolytic spark without metallic counterelectrode, whereby the high potential electrode, 12, serves solely for electrical nourishment of the electrolyte. It will be understood that in all the designs described here it is possible to establish zero potential of the electrolyte by a grounded plate or another conducting member immersed therein, an important provision for using such devices on board ships, where the body of sea water itself serves as counterelectrode.

In the arrangement shown in Fig. I3c-14 it is generally of little consequence in how many spots the bare metal is exposed. The sparks develop even when a considerable number of bare spots are present, because there is always a surplus supply of current available from the capacitor. Only a small fraction of this current leaks through the bare surfaces, and there seems to exist a preference of the energy to be substantially consumed only where sparks actually detonate and give rise to shock wave radiation. A simplified variant, unlike the one with fish-scale-like insulation, as shown in Fig. I3c-14, is a design in which ribbons of rubber or other insulating material are wound around the electrode in abutting or overlapping fashion. Or, instead, the insulating jacket of the conductor may be of insulating tubing which is slipped over it with a certain clearance. Such tubing contacts the conducting part of the electrode only loosely, and may be slit longitudinally before or after assembly, whereafter the clearance is taken up so that the tubing overlaps at the slit.

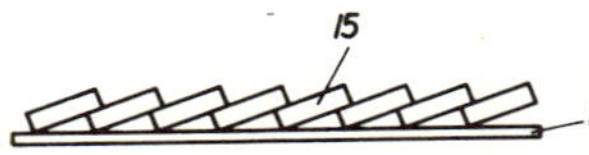

FIG. I3c-14. High potential electrode having an insulating jacket consisting of overlapping sections whereby the contracted current line flux emanates through the slits between the sections.

For a point source with long life expectancy, one design provides a construction in which the wire-shaped electrode, 7, is surrounded by a plastic insulation jacket as shown in Fig. I3c-15. The jacket is provided with an orifice, 17, facing the liquid, 4. The shock wave emanates from the orifice, 17, and by proper selection of the size of this orifice discharge can be made to occur either very rapidly with a dull full tone when the hole is large, or in the form of a hard dry detonation when the hole is small. In operation with a small hole, often only partial discharge of the capacitor takes place. In every form of construction it is possible to control the spark characteristic by varying the discharge circuit with respect to capacitance, self-inductance and charging voltage of the capacitor. In the system shown in Fig. I3c-15 the diameter of the orifice hole enters as an additional variable. Thus such devices can be further perfected as shown in Fig. I3c-16 which, with respect to life expectancy, is an improvement over the device shown in Fig. I3c-15. For control of the spark, the extremity of electrode 7 is pinnated and a tube, 18, is connected to the ball-shaped rubber jacket, 16, through which electrolyte under pressure may be pumped into the jacket. Continuous streaming of the electrolyte and replenishment of used-up material is thus assured. By virtue of the fact that it is possible to subject the liquid, 4, in the interior of the ball-shaped jacket, 16, to pressure, which causes an outlet stream through opening 17, the size of

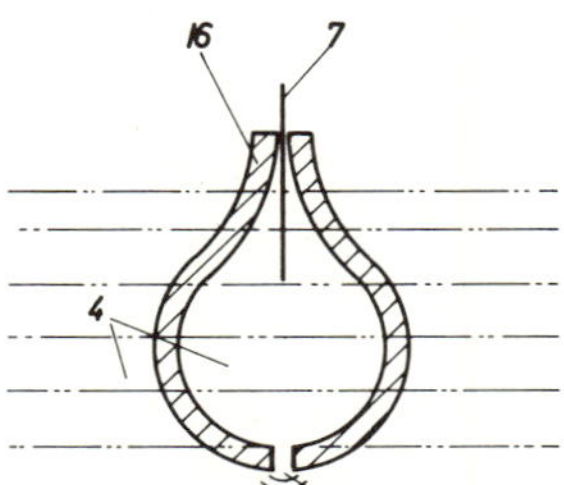

FIG. I3c-15. Diagrammatic representation of another modification of a high potential electrode surrounded by a flexible plastic bulb with an orifice through which communication with the main body of liquid is established and in which current line contraction occurs and ultrasonic energy radiation originates.

this opening, which is contracted by a spring ring, 19, can be varied by varying the pressure through control of flow in tube 18. At high pressure the outlet mouth, 17, opens somewhat wider, so that during spark discharge the spark comes to develop in a larger cross section and the sparks grow forceful and produce a hollow sound, thus radiating a substantial amount of sound energy at lower frequencies. On the other hand, if radiation at some higher frequency is desired, the pressure difference between the interior of the ball and the main body of liquid must be reduced by reducing the rate of inflow through tube 18 with the result that the outlet mouth, 17, contracts. Besides the fact that the characteristic of shock wave radiation can thus be changed, this arrangement makes it possible to compensate for burn-offs at the rim of the orifice, 17, by changing the internal pressure of the resilient jacket, 16. Such pressure adjustment can be derived from familiar external automatic means. Whenever desirable, a sonic reflector, 13, can be accommodated within the ball-shaped jacket, consisting, for instance, of a small foam rubber member; the reflector is beneficial in utilizing rearward radiating energy.

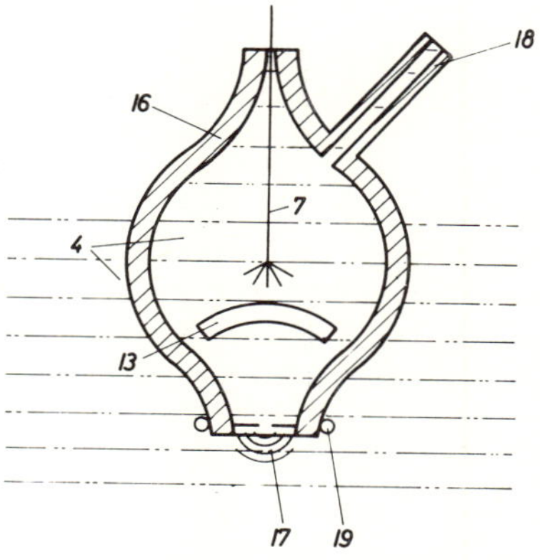

FIG. I3c-16. Modification of the arrangement shown in Fig. I3c-15, wherein the flexible plastic bulb is provided with a tubular inlet through which electrolyte under pressure is injected, thereby changing the size of the orifice for control of the magnitude of ultrasonic radiation, and a sonic reflector so arranged in this bulb as to reflect rearward emanating radiation.

To furnish the necessary supply of high voltage pulses for operating liquid spark devices, simple electric circuits can be employed. One such circuit is shown, by way of example, in Fig. I3c-17. The component parts of this circuit are an interrupter coil, 36, with self-interrupter, 35, primary and secondary winding, 36 and 37, and a current source, not shown. The secondary voltage pulses are passed through a dry-disc rectifier, 38, or a glow discharge valve for charging the pulse capacitor, 39, which in turn discharges through the auxiliary gaseous spark gap, 33, and the liquid spark gap within the liquid, 4.

A very simple control of spark sequence, which in the circuit of Fig. I3c-17 is arbitrary and not controllable, can be obtained by a modification as shown in Fig. I3c-18: The magnetic energy of the interrupter coil, 36, is critically measured so that its magnitude $\frac{1}{2}LI^2$ (L the inductance of the coil and I the current) is somewhat greater than the static energy $\frac{1}{2}CV^2$ (C the capacitance of the capacitor and V the voltage) required for the generation of the liquid spark. Thus, it is achieved that at each interruption of the interrupter, 35, the capacitor, 39, has a small surplus of charge leading to a breakdown of the auxiliary spark gap, 33, and to the firing of a liquid spark. The sparking frequency of the liquid spark in this circuit can be controlled in a simple manner by adjusting the interrupter frequency of the interrupter, 35.

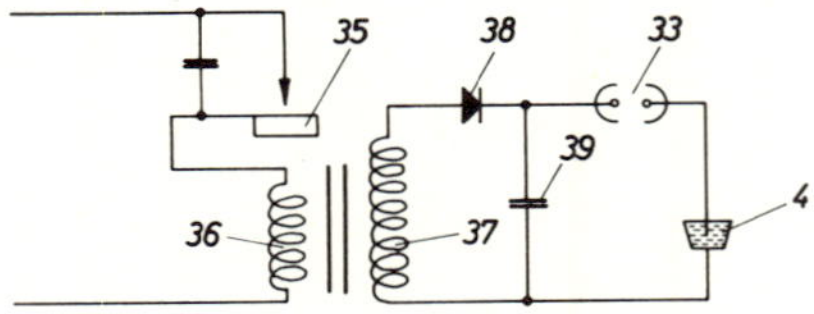

FIG. I3c-17. Schematic diagram of a preferred form of a power supply circuit for a liquid spark gap.

This circuit, in its principle, is a further development of the conventional Ruhmkorff's induction coil in that, instead of a very rapid interruption of the primary circuit with resulting high voltage pulses, here critical dimensioning and calculation of a predetermined magnetic minimum energy $\frac{1}{2}LI^2$ is effected. In this manner the contact opens the dependent-current circuit at the instant this energy surpasses the required static energy. The interrupter coils required for a circuit according to Fig. I3c-18 are of necessity physically larger than the coils employed in the well-known automobile ignition system.

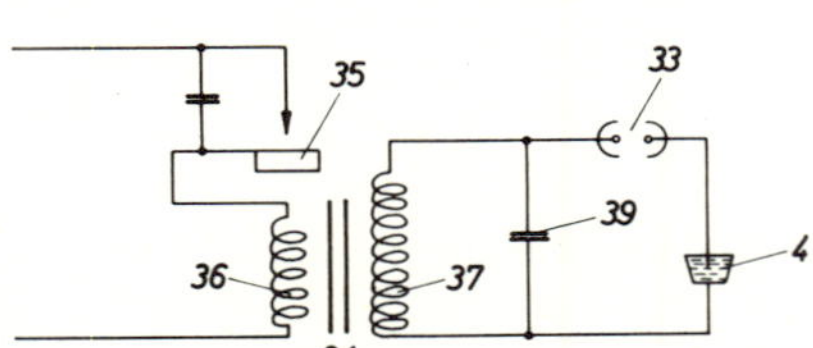

FIG. I3c-18. Schematic diagram of a modified form of supply circuit.

In some circuits, generally of the construction shown in Fig. I3c-17 and 18, it has been discovered that it is possible to eliminate the auxiliary gaseous spark gap entirely without materially reducing the usefulness of the system. When perfect interruption is assured, discharging of the capacitor takes place so rapidly—in less than about 1×10^{-5} sec—that the auxiliary gap becomes superfluous and thus the losses created by it due to spark energy consumption are eliminated. Without an auxiliary spark gap, the capacitor discharges spontaneously through the liquid spark gap, because the rearward discharge by way of the secondary winding of the coil would require a longer time. Therefore, it is advisable to connect the liquid spark gap directly, i.e. without auxiliary gap, to a discharge capacitor which in turn is connected with the secondary winding of an interrupter coil whose magnetic energy for breakdown of the liquid spark gap surpasses the required static capacitor energy.

Constructional details of a single point source shock wave radiator can be seen in cross section in Fig. I3c-19a. Here a chamber, 40, is formed by a metal housing, 41, and a metallic cover plate, 43. The housing, 41, is internally lined with an insulating material indicated at 42. Cover plate 43, which may extend over the housing to form a flange for mounting, is provided with an aperture, 44, behind which is located a diaphragm, 45, of a material of leather-like

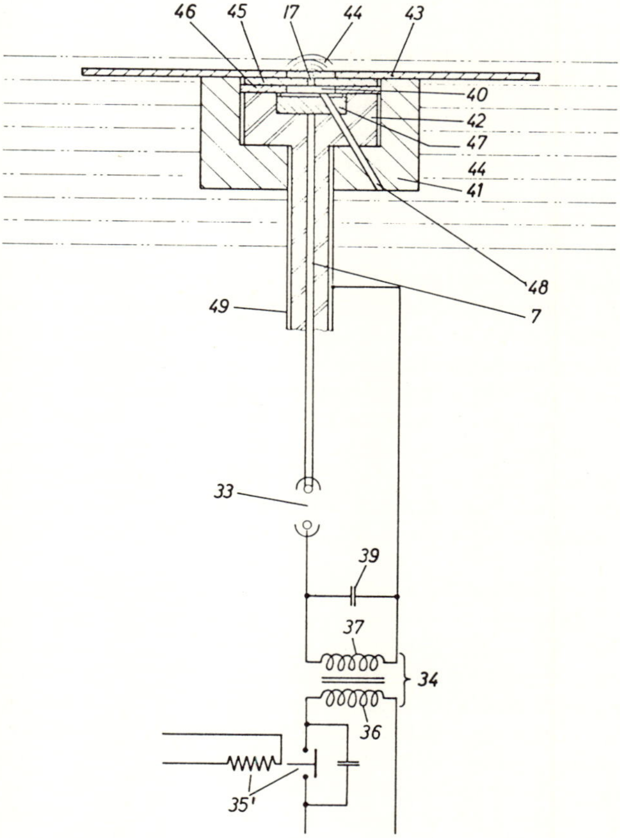

(a)

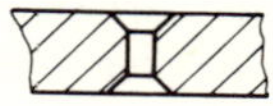

(b)

FIG. I3c-19a. Cross section through a shock wave radiator showing the construction in more detail.

FIG. I3c-19b. Enlarged sectional view of the orifice of Fig. I3c-19a.

consistency such as that known under the trade mark "Vulcaprene." Diaphragm 45 is provided with an orifice, 17, and is pressed against cover plate 43 by a resilient rubber gasket, 46, inserted between diaphragm 45 and insulation 42. It will be noted that insulation 42 extends through an opening in the rear wall of housing 41 and that the high potential electrode, 7, is embedded in the insulation at this point, thus projecting into chamber 40 where it terminates in a disc, 47, covering substantially the whole rear wall of this chamber. An enlarged cross section through orifice 17 is shown in Fig. I3c-19b where the funnel-like expansions on the inlet and outlet side will be noted. A hole, 48, is provided; it leads from the rear into chamber 40 and allows replenishment of the liquid therein, which may become exhausted by current line and liquid spark action. The circuit for feeding this shock wave radiator is indicated schematically in Fig. I3c-19a.

It will be noted that one side of the circuit is electrically connected to sheath 49 of the lead-in for the electrode 7 and thus grounded to the metallic housing of the device. The circuit is somewhat similar to the ones shown in Fig. I3c-17 and 18 with exceptions of the self-interrupter, 35, which is replaced here, as shown, by an externally controlled relay designated by reference numeral 35′. In operation, when the contacts of relay 35′ are momentarily closed by energizing from a control source (not shown), a current will flow through the primary, 36, of transformer 34 and induce a voltage pulse in the secondary, 37, of charging capacitor 39, which, upon passing the breakdown voltage of auxiliary spark gap 33 will send a burst of current through electrode 7 and into chamber 40. Since the inside of chamber 40 is insulated, the only conducting connection is through the electrolytic medium in orifice 17, where, on account of the small size of the orifice, considerable current line contraction disintegration of liquid into oxyhydrogen, and sparking occur. All of these effects, as previously explained, are sources of shock wave radiation. The counterelectrode in this device is the metallic cover, 43, which through housing 41 is in electric contact with sheath 49 to which the other terminal of the secondary circuit is connected. The flexible diaphragm, 45, in which orifice 17 is located lends considerable life to this shock wave radiator and can be readily replaced.

A particular field of usefulness is whale and seal hunting. Here shock waves are used for influencing the physiological behaviour of these sea mammalia. Most effective frequencies for this purpose lie in the range of 10^4 to 3×10^4 cps. Directional radiation in this frequency range can be realized by properly adapting shock wave stair arrangements as described in connection with Fig. I3c-13a. Assuming a sound velocity in water of 1435 meters/sec, the wavelength corresponding to 10 kilocycles (10^4 cycles) is 14 cm, and the one corresponding to 30 kilocycles (4×10^4 cycles) is 4.6 cm. Thus for affecting sea animals in their physiological behavior to achieve certain desirable reactions, e.g. to frighten them, the shock wave stairs for emitting such influencing waves must be set for wavelengths within the range of about 5 to 15 cm. Stairs with step heights and reflector spacing in the range between 1 and 10 cm have been found most suitable, and they may be placed in single or, for higher output, in multiple arrangement.

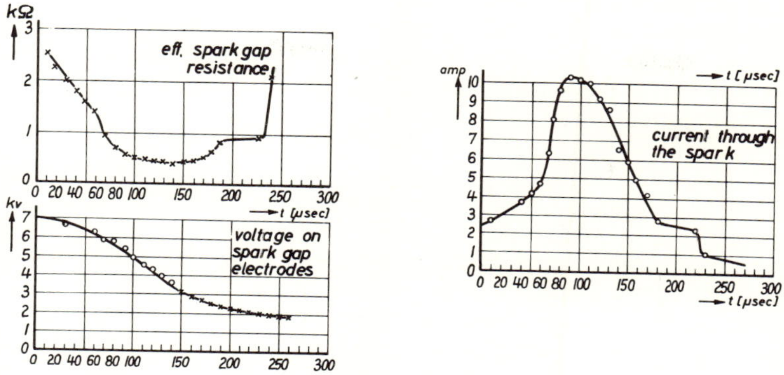

FIG. 13c-20. Resistance (top left), voltage (bottom left), and current (right) *versus* time in a shock sound head as shown in Fig. 13c-9 with about 10 joules discharge energy.

Plots taken in a test of a sound emitter depicted in Fig. 13c-9 and showing current, voltage, and resistance versus time are indicated in Fig. 13c-20. Such a multi-hole plate for technical ultrasonic treatment of chemical baths was designed for an average acoustic output of 200 watts and a shock sound peak of 300 kw (Fig. 13c-21). Its physical size is about 12 × 20 sq cm. In an endeavor to focus the spherical emission of the shock sound, schlieren optical investigations have been made into the influence on the sonic field of acoustic concave plexiglass and convex lenses of carbon tetrachloride. For that purpose the sound waves were photographed by a spark chamber lamp at a frame sequence frequency of 200,000/sec under control of an electronic retarding device [137]. Since the sound shocks have only positive pressure

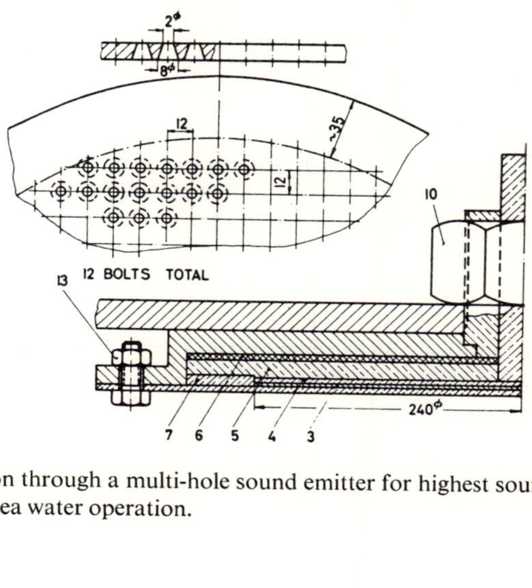

FIG. 13c-21. Section through a multi-hole sound emitter for highest sound shock energies, up to 1000 joules in sea water operation.

peaks and the negative equalization is very slow, interference considerations are absent. The sound waves can be radiated as desirable, either with linear fronts or concentrated in a focal point by sound-optical lenses. When two sound fronts penetrate each other, there occurs at the instant of interception a confusion region which must be contributed to the high pressure present. This interception region has no influence on the further propagation path of the waves, which is undisturbed.

With respect to flat high potential electrodes or shock wave radiators in the form of plates, as described in connection with Fig. I3c-9 to 11, it was found that, when the distance between the single holes of the perforation is very small relative to a preferred wavelength to be transmitted—according to Fourier all wavelengths are contained in a detonation—the total shock wave transmission can be considered as a uniform coherent directional radiation and the source itself as such a radiator. The radiation is spherical if only one radiator is applied. By simultaneous transmission from several radiators there exist interference phenomena among all directions which diverge sideways from the main radiation direction. For instance, it has been found that a square radiator plate having 100 holes equally spaced at distances of, say, 1 cm corresponds to an oscillator plate of 10λ diameter and shows a very pronounced directional radiation pattern. Such a characteristic is particularly useful for the construction of depth sounding systems utilizing step heights and reflector distances between 1 and 10 cm. These dimensions are most effective in directing wavelengths corresponding to frequencies in a range most suitable for depth sounding. For greater depths it is advisable to select the larger dimensions because there is less absorption of lower frequencies at greater depths. Because shock wave stairs and radiator plates can be fabricated at reasonable cost, it is now possible to give the radiators for depth sounding large dimensions with strong energy radiation, a construction which would have been too costly with former methods. The new means produce extremely strong directional radiation patterns making possible sounding of great depths. A further advantage lies in the fact that various frequencies are present in the shock waves produced so that, for instance, in the range of high frequencies, particularly good reflection may be obtained from fishes or fish shoals, whereas in the range of the lower radiated frequencies, particularly clear reflections from the sea bottom can be had.

It is therefore advisable to construct a depth sounding system so that any of the shock wave radiators described in connection with Fig I3c-2–10 can be used and to employ for the reception of this radiation several resonance receivers or an aperiodic receiver with electric band selection in such a manner that the receiver is tuned in each case for optimal reception. As is well known, the radiation pattern of an ultrasonic radiator or, generally speaking, an oscillator, is a function of the relation of the oscillator diameter to a particular wavelength. The greater the ratio of diameter to wavelength, the sharper is the focusing. Now in connection with the aforedescribed radiators it is impossible to speak of a specific wavelength. It is, rather, true that the highest frequencies with the shortest wavelengths that are still contained in the Fourier spectrum have the greatest ratio of diameter to wavelength and thus the most sharply focused radiation pattern. The lower the frequency, the wider becomes the solid angle of radiation, or the radiation cone, until at those frequencies whose wavelength is about equal to the radiator diameter, radiation takes place in conformity with a cosine distribution having no focusing direction at all. It is thus advisable to select the size of the radiator plate, or the size of the perforated portion of the insulating jacket, in according with the lowest frequency still suitable for a certain application, in order to obtain a desirable directional pattern; or, in case only a very high frequency portion such as a partial sector of the Fourier spectrum is desired, a size according to this frequency portion should be selected.

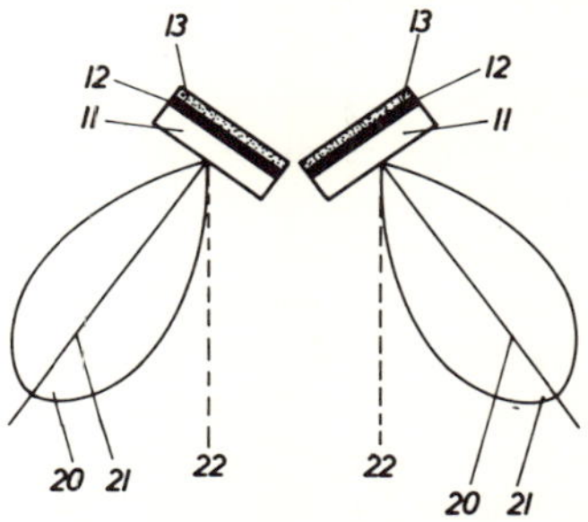

FIG. I3c-22. Diagrammatic representation of a system with two high potential electrodes of the construction shown in Fig. I3c-9 arranged at an angle to each other for preferred radiation in respective directions.

A special application is an improvement of depth sounding systems for making out fish shoals. The system is depicted in Fig. I3c-22. Two suitable shock wave radiation devices as described above are set at an angle with respect to each other. The metal plates, 12, of the high potential electrodes are subjected to the discharge pulses which cause shock waves to emanate from the perforated layer, 11. Reflectors, 13, of foam rubber of another reflecting material are preferably placed at the rear of the plates for higher efficiency. In operation, both radiators radiate shock waves in a predetermined frequency range according to the radiation pattern, 20, in directions, 21, normal to the planes of the plates. The shape of the radiation pattern can be altered by varying the spacing of the perforations in the plates and/or by varying the outlines of the plates. This arrangement is preferred for depth sounding when it is desirable to determine whether a fishing craft is at the right- or the left-hand side of a fish shoal because the radiator directed more nearly toward the fish shoal will give more pronounced signals. Generally, one radiator plate designed for a weakly focused radiation pattern, but cooperating with two sharply tuned receivers, will also give satisfactory results in directional orientation. For best results with two sharply focused radiators (Fig. I3c-19a) it is advisable to select the radiation patterns in conformity with the frequency band of the receivers so that the tangents, 22, of the pattern curves are parallel or diverging, in order to avoid disturbing interference areas between the two radiation fields. Still better, in order to obtain a clear distinction between the signals of the systems, they can be operated in alternating cycles, i.e., one is off while the other is on. In this way the respective image tubes, if cathode ray tubes are employed, need be rendered luminous only when the reflex of the pertinent depth sounder is expected. The usefulness of this arrangement can be further expanded through selective image reproduction of the frequency content of the transmitted energy by such well-known means as resonance receivers or filtering circuits. Such means are preferably employed when, for example, it is the objective to distinguish plankton from fishes and the latter from the bottom of the sea. It was found that the receiver frequencies from 30 to 50 kilocycles are most suitable for plankton and very small fishes; frequencies from 30 to 50 kilocycles, for medium size fishes, and very low frequencies for the sea bottom. With these means it is furthermore possible to ask several, for instance, 2 or 3 receivers about their sound reflexes, either successively or continually intermittently, by an automatically operating electronic switch or by means of relay circuits. Thus, it is possible to obtain in a system known in itself three different sounding diagrams upon a single fluorescent screen such as that indicated in Fig. I3c-23. On the fluorescent screen, 23, of a cathode-ray tube the sounding diagrams 24a, b, and c are indicated, of which 24a, for example, may be one for 10 kilocycles, 24b for 30 kilocycles, and 24c for about 80 kilocycles. The corresponding image sketches indicated in the figure are to represent at 24a the sea bottom, at 24b a strongly reflecting fish shoal and at 24c tiny reflexes of plankton and of smaller objects. This method is useful for echo sounding devices combined with several band selective receiving devices which, mechanically or electronically are successively or continually intermittently questioned and, by means of well-known circuits, give useful images upon one or several indicators.

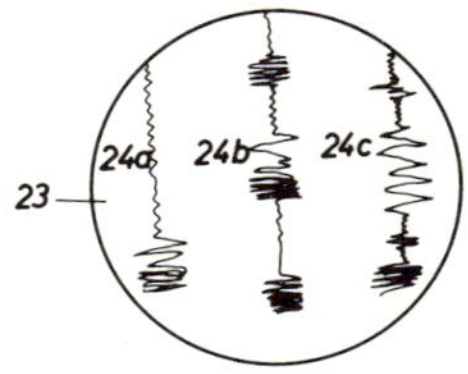

FIG. I3c-23. Luminescent screen with three different images reflected during a depth sounding operation with three different sounding frequencies.

On board ship, a radiator can be arranged to transmit according to an intensity pattern suitable for reaching for instance, 10 degrees to larboard as well as to starboard side, and to cooperate with two receivers with sharp receiving characteristics, of which one receives only the larboard and the other only the starboard reflexes. This arrangement thus provides a double depth sounding system comprising one transmitter and two oppositely arranged receivers with rapid characteristics which may actuate one common or two separate image devices.

When receivers of this kind are incorporated in a depth sounding system its usefulness can be further expanded by utilizing the acoustic as well as the optical phenomena, for instance, in a double-acting, acoustic-optical, underwater light barrier or signaling system of that kind. Also, in depth sounding by the reflex method an optical receiver is useful as an auxiliary; where very shallow depths are encountered, for instance. The light flash of the spark is better reflected, especially from a muddy bottom, than the sound. Moreover, this combination method lends itself as a reliable means for measuring sound velocities within liquids: The elapsed periods between transmission and arrival of the signal as light pulse and as sound pulse are measured and evaluated, and the results compared.

Besides being the source of ultrasonic and light radiation as explained in previous paragraphs a liquid spark, due to current line contraction, is also the origin of galvanic current fields passing through the electrolyte. This galvanic current passing in the form of pulses during each sparkover can also be made to serve as a signaling means if such action is desirable. However, if such galvanic current utilization is out of the question, or when it would disturb other galvanic communication within reach, it can substantially be confined by placing the grounded counterelectrode close to the sonic radiator, which, as will be remembered, is the high potential electrode of the discharge system. Thus galvanic field line disturbances in adjacent systems will be held to a minimum. On the other hand, if considerable galvanic disturbance is desired, either for the purpose of physiological influence or irritation of fish or in order to make galvanic current useful for signaling, the counterelectrode must be placed remote from the receiver, or the ship's hull itself may serve as counterelectrode.

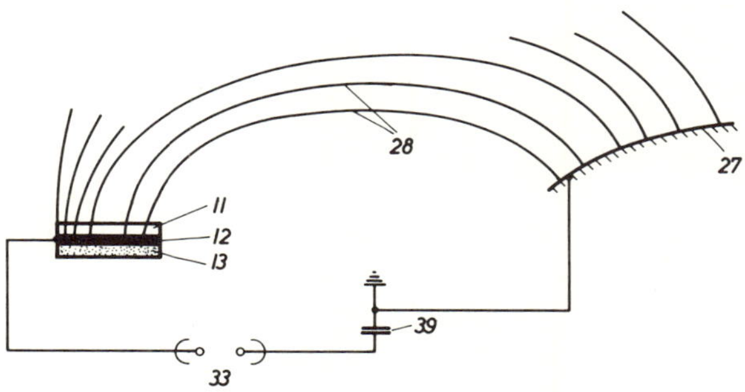

FIG. I3c-24. Arrangement of an ultrasonic pulse source with strong galvanic effects achieved by placing the counterelectrode, which may be a large metallic surface, remote from the high potential electrode.

Figure I3c-24 shows an arrangement with strong galvanic disturbing effect. A capacitor, 39, discharges by way of an auxiliary spark gap, 33, or directly through the high potential electrode, 12, which thus serves as an ultrasonic radiator. The grounded counterelectrode is remote therefrom, indicated as the mass of the ship's hull, 27, so that the elcetric field lines, 28, are long and thus detectable at greater distances. An enveloping arrangement as shown in the

Fig. I3c-25 confines the electric field lines. Here the ultrasonic pulse radiator is surrounded by a grounded envelope, 29, which is provided with perforations, 30, for emission of the ultrasonic radiation. The high voltage pulses are fed to electrode 12 through cable 31 and the insulated lead-in, 32. For weak sonic effects or galvanic fields the discharge can be made so small that no spark occurs. In that case, as previously mentioned, the dielectric stress on the surrounding liquid, for instance sea water, produces ultrasonic waves strong enough for moderate sounding depths when no sparkover occurs, and such sparkless operation prolongs the life of the radiator electrode by avoiding evaporation of electrode material.

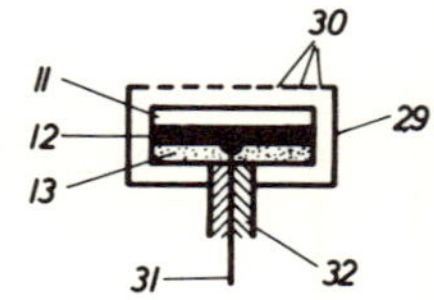

FIG. I3c-25. High potential electrode similar to the arrangement of Fig. I3c-24, but here surrounded by a grounded metallic envelope to suppress galvanic action.

An arrangement in which shock waves are utilized for beneficially influencing an electrochemical process is diagrammatically depicted in Fig. I3c-26. The direct current source, 50, is shown connected in a well known manner to electrodes, 51, and 52 immersed in electrolyte, 4, contained in cell 53. A shock wave source including a capacitor, 39, means for charging the same (not shown), and an auxiliary spark gap, 33, for causing discharges when the capacitor charging voltage reaches a predetermined value, is connected in parallel with the direct current source to the same electrodes, 51 and 52. In operation, when direct current flows at a certain density through the electrolyte, bursts of capacitor discharges in regular succession are superposed over this current flow through the electrolyte, producing high current line concentration and shock wave radiation at the electrodes. Such shock wave radiation, giving rise to high frequency vibration in the electrolyte, is beneficial for certain

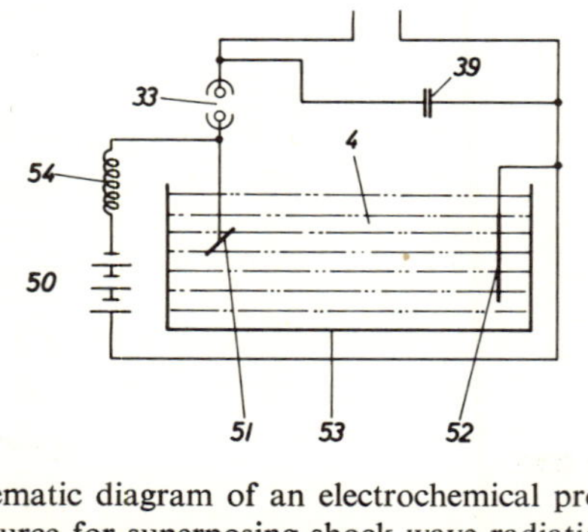

FIG. I3c-26. Schematic diagram of an electrochemical process circuit with connections to a shock wave source for superposing shock wave radiation over the electric field lines passing through the electrolyte.

electrochemical processes: for instance, it results in a more uniform deposit of the plating material on the electrode in electroplating. A choke coil, 54, is interposed in the direct current circuit to prevent the capacitor discharge pulses of the shock wave source from entering this circuit and to assure discharge through the electrolyte.

While by proper construction of shock producers such as so-called "hole generators" it is possible to command large energies with relatively slow discharge progression in submarine application, it is also feasible to build spark gaps of long life expectancy when employing so-called surface sparks and making one of the electrodes resilient, e.g., a helical spring, and the other very hard, e.g. of drill-rod steel. Such spark gaps have a steep pressure rise similar to that of direct underwater sparkover of higher voltage. The pressure needs, for its rise to the peak value, only 0.1 to 1 μsec, depending on the conductivity of the water [136]. Different forms of point and ring vibrators have been designed with which it is possible to produce at a shock sequence of 50/sec acoustically 4 joules/shock, i.e., 200 watts average output. Practically any desirable number of such shock sound sources can be operated in parallel for shaping of desirable sound patterns. The frequency spectrum can be influenced by configuration and can reach from 0 to one megacycle (see Section 4 of this chapter).

It must be understood that the system of Fig. I3c-26 is shown for example only, and that other arrangements for superposing shock wave radition upon an electrochemical process circuit can be made. It is, for instance, possible to connect the pulse discharge capacitor directly across the choke coil, 54, or to insert the secondary of a transformer into a direct current circuit whose primary is connected to receive the capacitor discharge pulse for shock wave generation.

4. Applications of High Intensity Shock Sound

a. *Systems for Probing Materials by Shock Waves*

A system for probing materials by passing shock wave fronts through them is described by Früngel [212].

The systems often used for probing material by ultrasonic signals operate with wave trains of high frequency vibration and use oscillatory circuits or high frequency generators to excite piezoelectric crystals, or magnetostrictive transducers from which mechanical high frequency waves are radiated into the material under test. The operator can observe the attenuation of the reflected signals and their characteristic velocity to draw conclusions about the internal structure of the material or, in other words, to detect hidden defects.

Such a source of energy for probing, inspecting and testing of materials, in contrast to conventional sources, produces no vibration, and for each signal, only a single impact becomes effective in the form of a compression wave. Impacts of this sort produce substantially only pressure and no tension in the material under test, and thus represent an asymmetrical phenomenon, whereas sonic and ultrasonic frequencies of the conventional character produce vibrations which are symmetrical in that compression and tension alternate.

The shock wave source is inexpensive and of rugged construction, and compression waves emanating therefrom act differently on the material under test and render in some cases more reliable information than ultrasonic vibrations. The source is an electric spark produced within a liquid which, at the instant of its occurrence, gives rise to a steep fronted elastic compression wave in the liquid. The compression wave is caused to beam into the material to be tested and can be received electrically at the opposite side of the sample under test by means of a quartz crystal or other piezoelectric receiver in a manner well known in the art.

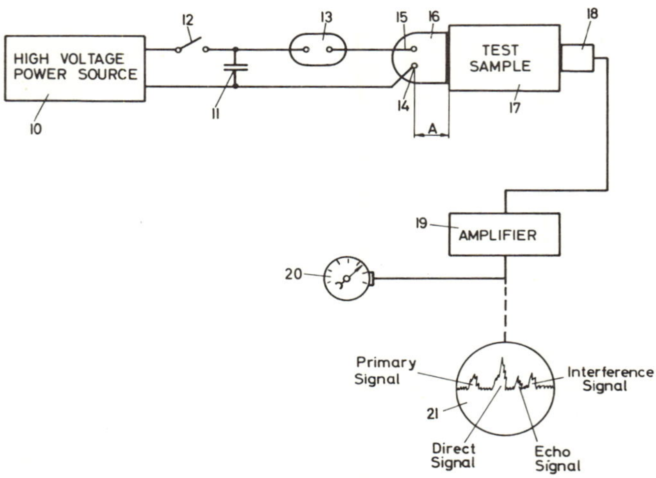

FIG. I4a-1. A system for probing material by shock waves.

In Fig. I4a-1 is indicated at 10 a high voltage power source charging a storage capacitor, 11, whenever switching means 12 is closed. A discharge circuit of capacitor 11 includes an auxiliary spark gap, 13, and a spark gap formed by electrodes, 14 and 15, immersed in a liquid such as water, enlcosed in a chamber, 16; the combination of these elements forms an emitter for elastic shock waves and a new source of energy for material probing. The auxiliary spark gap, 13, in air or a suitable gas serves for effecting at its breakdown, which is due to high capacitor potential, the extremely short timed discharges and, in turn, the steep fronted compression waves which would be considerably less pronounced if one depended on the liquid spark gap alone, particularly if the liquid were of the kind having a certain conductivity, for instance water. Or, in other words, without the auxiliary spark gap in series with the spark gap within liquid, current of the capacitor charge would start flowing off as soon as switch 12 was closed, because the liquid surrounding gap 14–15 has a certain conductivity and therefore this gap has a considerably lower resistance than the auxiliary gap in gas or air. It cannot have a pronounced accurate breakdown point at a certain potential. As pointed out above, flow-off of current, which starts as electrolytic flow through the liquid and gradually increases until finally a sparkover occurs, is undesirable because it represents a loss of energy and renders the instant of occurrence and the intensity of the spark discharge unreliable. Only with an auxiliary spark gap that can be set for an exactly predetermined breakdown potential, can closing of the discharge circuit be achieved at the instant when the capacitor charge reaches this breakdown potential. Such instantaneous closing of the discharge circuit when an exact quantity of discharge energy is available is needed to cause in liquid chamber 16 an immediate and rapid sparkover that is capable of giving rise to a steep compression wave.

Chamber 16 has a flat surface or a surface of another configuration suitable to be brought into close contact with the surface of a sample for testing, indicated at 17. Such test samples can be of metal or other construction materials, such as concrete blocks, whose internal structural conditions are of interest. At the opposite side of sample 17 there is arranged an acoustic-electric converter, 18, which preferably includes a piezoelectric crystal or the like. The converter changes the shock wave energy, after it has penetrated the sample, into weak electric signals which can be amplified by amplifier 19 and then either fed to an indicating instrument, 20, depicting the intensity of the attenuated impulse, or introduced into a cathode ray tube indicated as a screen of such a tube at 21. The single line connections shown between converter 18 and the instruments, etc., 19, 20 and 21, are only a diagrammatic illustration of conventional connections. The oscilloscope may be connected by a line with the discharge side of the capacitor, 11, for receiving a primary signal at the instant of its discharge. However, this connection is usually not needed since the discharge of capacitor 11 or the practically simultaneous sparkover at 13 transmits an impulse through the air into the amplifier, 19, which is usually not entirely shielded, or into oscilloscope 21, sufficient to produce a primary signal as indicated on the screen, 21.

From the amplitude of various signals and the shape of the pulse indicated on the screen, valuable information as to the structure of the sample can be had. The shock wave signal passing the test sample, 17, upon a sparkover between the electrodes, 14 and 15, corresponds to those obtained in conventional echo systems used for testing materials. The "primary signal" would correspond to the sparkover in the discharge gap, 13; the "direct signal" would correspond to the shock wave passing through the test sample, 17, and impinging upon the converter, 18. An "echo signal" would indicate the existence of a discontinuity or defect in the test sample, 17, caused by the interference of any of said signals with one another. An "echo signal" appears when energy transmitted through or past a defect of the test piece is reflected, at least partly, from its surface opposite the transmitter, 16, and is then reflected back by the flaw to the receiver, 18. In general, the means for receiving and evaluating the compression or shock wave signals emitted by the underwater spark source are substantially similar to those employed by supersonic vibration systems.

In operation, when switch 12 is closed, the capacitor is charged to a potential above the breakdown voltage of auxiliary spark gap 13, causing breakdown of the spark gap and, as a consequence thereof, also an abrupt and instant current impulse through spark gap 14–15 in liquid. The resulting spark discharge through liquid gives rise to a shock wave impulse which beams an elastic compression wave through the liquid, through the face of chamber 16 and, this being in close contact with the surface of test sample 17, into and through this sample. Depending on the internal structure of the sample, the impulse will be attenuated and recordings on the cathode ray screen will indicate corresponding comparable amplitudes, such as a direct signal, an echo signal, a primary signal, an interference signal, and so forth. The time of travel of the shock wave can be determined by the spacing between the signals on the screen as usual.

In experimental operation it has been discovered that, in a region close to the sparkover through liquid, the shock wave velocity is higher than sound velocity and may reach 10,000 meters/sec, (see Section 3b of this chapter) but that after a few centimeters of travel this speed decelerates to normal sound velocity. Therefore, such a system must have a space of the order of 5 cm between the spark gap in liquid and test sample so that the shock wave damps to normal sound velocity before entering the material under test. Such spacing is indicated at A in the drawing and can be obtained by a suitable length of chamber 16.

The system operating with compression shock waves produced by electric spark discharges through liquid resembles more closely natural phenomena, such as striking of a blow, and renders more accurate results than ultrasonic vibrations of high frequency which are generally employed in the pertinent art.

For measuring the expanding of sound traces in metal, W. Güth [230] combined a spark shock wave with spark illumination. By means of the integration method of "stationary phase" a theory of the propagation of impulses and similar sounds in a dispersive medium is developed and demonstrated on a circular metal cylinder. The theory is compared with experimental data obtained by schlieren methods. The possible use of the results to explain the formation of ripples on steel rails is discussed.

The chief requirement of the examination method for corroborating the above ideas will be—in view of the rather complex nature of the partly super-

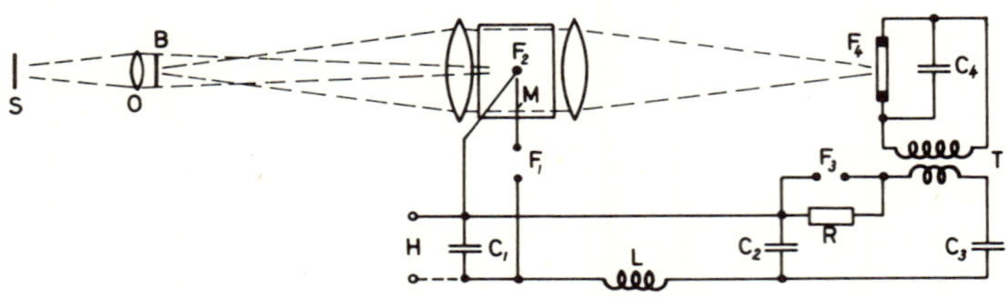

FIG. I4a-2. Experimental setup for the schlieren photographs. S–ground glass plate; O–lens; B–schlieren aperture; M–metal rod; F_2–shock wave spark gap; F_4–illumination spark gap; H–high voltage source; C_1 C_2 C_3–capacitors, F_1 F_3–auxiliary spark gap, L–inductance; R–resistor; T–pulse transformer; C_4–high voltage capacitor.

imposed multiple wave trains—the demand for simplification and clarity. That method was selected, therefore, which comes nearest to this requirement, namely the method of visual observation.

By means of an electric spark the steel rod submerged in a water filled vessel is activated. The waves traveling in the rod induce in the water shock waves according to Mach. These shock waves can be observed, that is, photographed by means of the Toepler Schlieren method. A second spark, which is triggered shortly after (approximately 10^{-4} sec) the shock induction as such, serves as illumination.

Figure I4a-2 shows a sketch of the schlieren arrangement with retarding circuit. By a high voltage source, H, the capacitors, C_1, C_2 and C_3, over the resistor, R, are charged; until the breakthrough voltage of the auxiliary spark gap, F_1, with the shock spark gap, F_2, which is in series, is reached. C_1 discharges over F_1 and F_2 and the metal rod, M, is thereby excited to produce a heavy shock wave at its lower end. By the discharge of C_1 the electric resonant circuit comprised by C_1, C_2 and the inductance L is agitated so

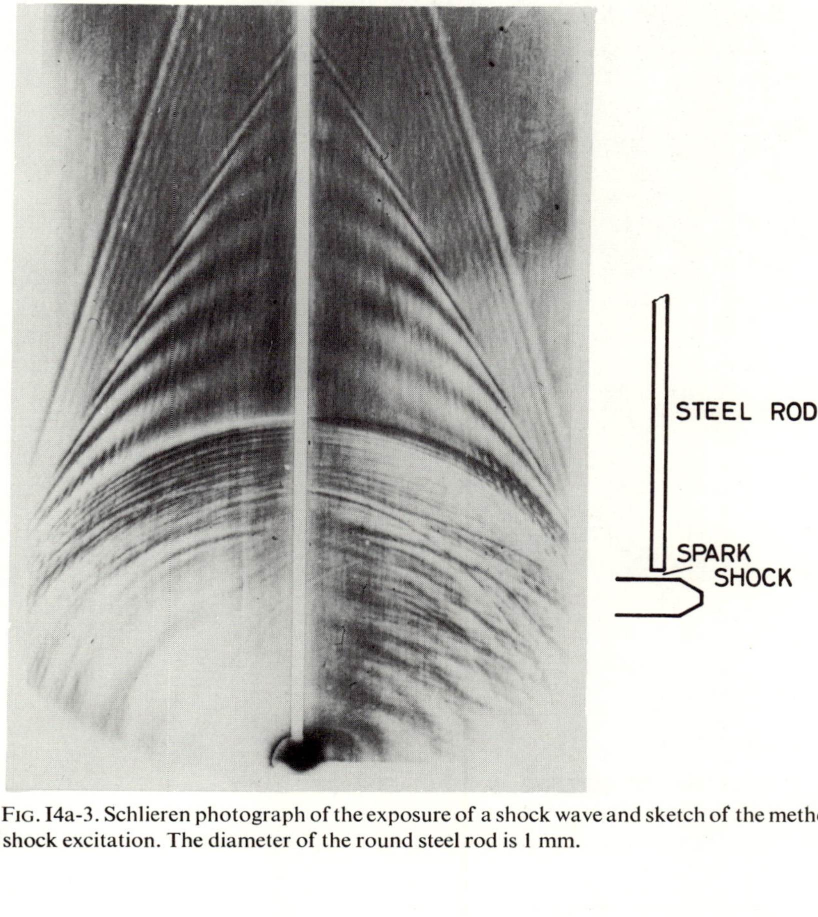

FIG. I4a-3. Schlieren photograph of the exposure of a shock wave and sketch of the method of shock excitation. The diameter of the round steel rod is 1 mm.

that the voltage on C_2 changes its polarity after a short interval (approximately 10^{-4} sec). R is so dimensioned that C_2 maintains its voltage throughout the whole oscillatory period. At the electrodes of the spark gap, F_3, there is built up a voltage resulting in a sparkover at F_3. This causes a current pulse to flow in the primary of the pulse transformer, T, which in turn induces in the secondary of T a high voltage. Thereby the capacitor, C_4, is charged and discharges itself over a spark gap, F_4, enclosed in a capillary tube, which serves for illumination purposes. By means of the lenses in front and behind the vessel, F_4 is depicted onto a rim of a schlieren head, B, and the metal rod in the vessel is depicted through a lens, O, onto the ground glass plate, S.

The electrical setup is similar to the delay circuit by Mach, offering, however, the advantage that one can operate at considerably lower voltages, which in turn permit the use of small capacitors and a smaller inductance.

The high voltage for the illumination spark gap, which is required for short

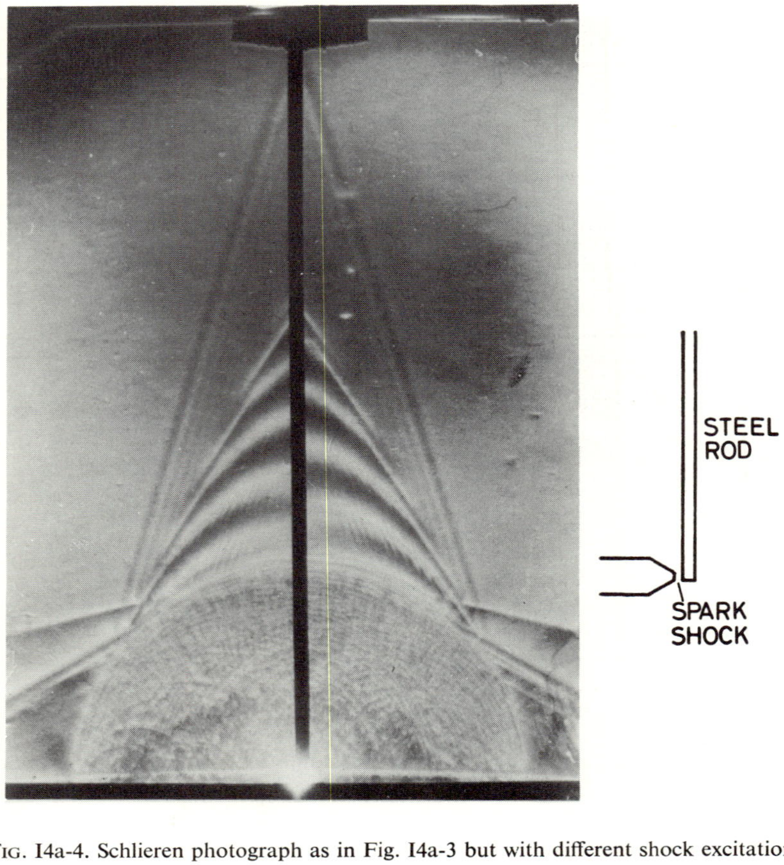

FIG. I4a-4. Schlieren photograph as in Fig. I4a-3 but with different shock excitation.

exposure times, can easily be supplied by the pulse transformer, T, the construction of which is very simple indeed. In this way the illumination circuit comprising the spark gap, F_4, the capacitor, C_4, and the wires between F_4 and C_4 can be designed with particularly low induction.

Photographs (see Fig. I4a-3, 4) were taken with the equipment described above. At the lower rim of the picture the shock pulse spark can be seen as a white dot. The end of the vertical steel rod is blurred out so that one cannot see the actual shock excitation. Separate switches beside the picture are provided to illustrate this point. The front of the diluted wave has already passed the upper part of the frame. One can clearly discern the first subsequent wave as almost linear parallel lines, the intensity of which first of all drops down to a minimum. The front of the wave train touches the spherical wave radiated from the shock spark in the water. Particularly in Fig. I4a-3 it appears black in the left-hand part of the frame and white in the right-hand part, which shows that the waveform as such is symmetrical. (Note that in all pictures the schlieren head is parallel to the rod so that the symmetrical waves traveling along the rod appear asymmetrical and the asymmetrical waves appear symmetrical.) The fishbone-like structure is caused by the bending stress wave. It can be recognized as being skew symmetrical. The rear-curved band stripes are interrupted by smaller lines, which originate from the wake of the dilatory below the just-mentioned amplitude minimum.

For better illustration Fig. I4a-3 and 4 were again reproduced as sketches in the left-hand part of Fig. I4a-5. The dilatory wave train, D, can only be followed from its back $x/c_0t = 1$ to the turning point 0.38. The subsequent part between 0.38 and 0.58 is not visible on Figs. I4a-3 and 4; therefore it

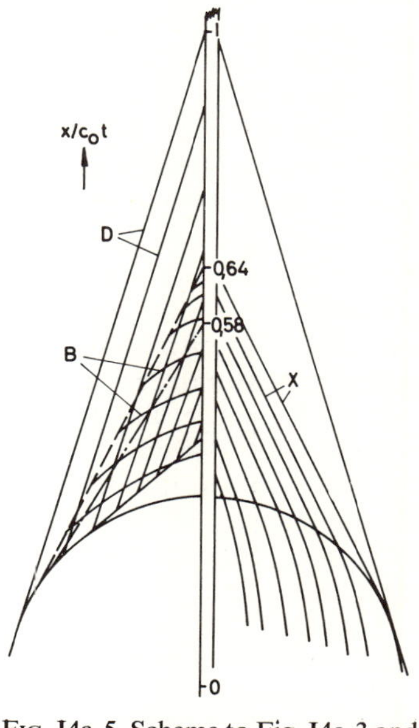

FIG. I4a-5. Scheme to Fig. I4a-3 and 4.

was also not drawn in Fig. 14a-5. Presumably, the high frequencies for formation of this part of the wave train are not contained in the pulse spectrum or they are utilized for forming other wave types.

The bent line, B, in Fig. I4a-5 represents the bend stress wave stripes. They travel from the rod to the dotted line where they turn and actually should approach asymmetrically toward the dash-dot line. For reasons mentioned above, these parts of the stripe can only be seen as a short portion in Fig. I4a-3; they are not visible at all in Fig. I4a-4 and therefore they were only indicated in Fig. I4a-5.

b. *Methodology of Sonic Measuring Procedures in Seismics* [207, 209]

Often short ultrasonic pulses are transmitted into the mineral sample by way of piezoelectric or magnetostrictive emitters. After passing through a predetermined path the pulses are detected by corresponding highly sensitive receivers, the so-called geophones. For continuous seismic velocity loggings in boreholes Vogel [210] has described a simple spark gap transmitter.

In V_t and V_l measurements the frequency of waves mostly ranges from several kilocycles up to several megacycles, in V_d measurements up to 25 kcps:

V_l = longitudinal velocity in an infinite media
V_t = transversal velocity in an infinite media
V_d = velocity of the dilatory wave within the rod

Coupling of transmitter and receiver to the mineral samples is effected mechanically, very often with the aid of a thin intermediate layer such as oil or putty. The emitter impulse and the signal arriving at the receiver are usually made visible, synchronized and amplified, on the screen of an oscilloscope together with a timing mark which in most cases will be quartz controlled. The high time resolution of an electronic beam oscilloscope permits transient time readings down to 0.02 μsec for a sample length of roughly 7 cm. Bradfield [211] gives a good general account of all points to be observed in such experiments, particularly with regard to connecting emitter and receiver to the sample, and also describes a pulse arrangement which permits measuring of transient times in the region of 0.01 μsec.

Figure I4b-1 is a block diagram of an ultrasonic seismic velocity logging arrangement for determination of V_d and E-modules. This arrangement has stood its test in several hundred measurements of borehole samples. By means of the magnetostrictive ultrasonic emitter an ultrasonic impulse of 22 kcps is transmitted 50 times per second into one end of the mineral sample and then tapped off at many measuring points at differing distances from the emitter. The distance of the measuring points is read on a scale with an accuracy of at least one millimeter. The trailing edge of the emitter pulse and the leading edge of the signal arriving at the geophone are made visible together with the timing marks of a quartz timer and the time lag is measured. The timing marks are placed at 5 μsec intervals, and every tenth mark is stressed, e.g. by a longer line. Figure I4b-2 shows the oscillogram for one of the mentioned measuring points.

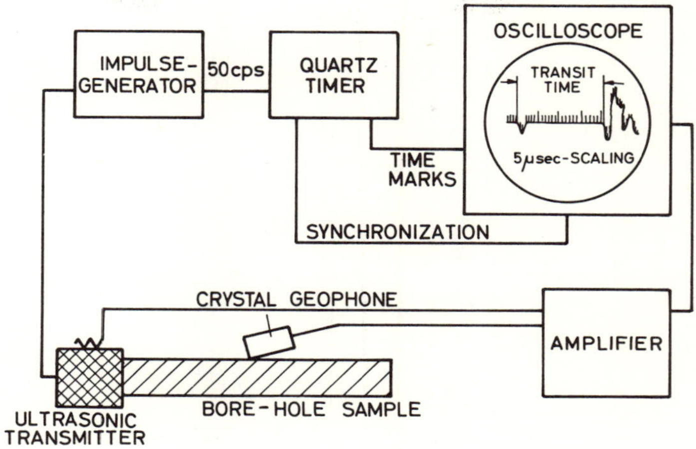

FIG. I4b-1. Block diagram for electronic measurements of traveling time.

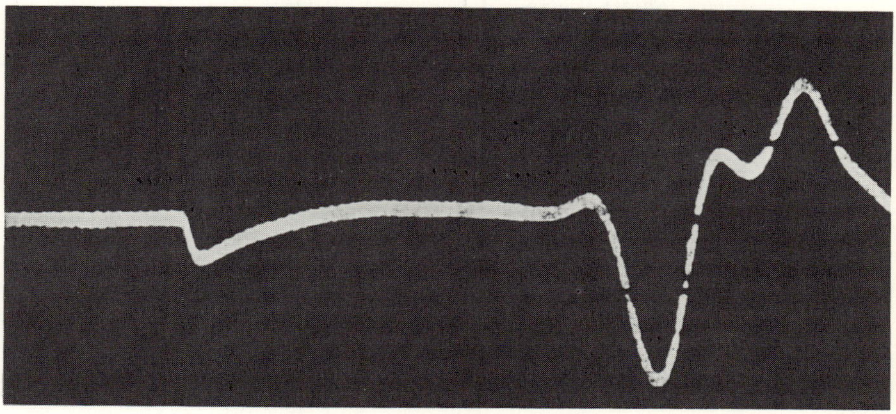

FIG. I4b-2. Ultrasound oscillogram of a sandstone core.

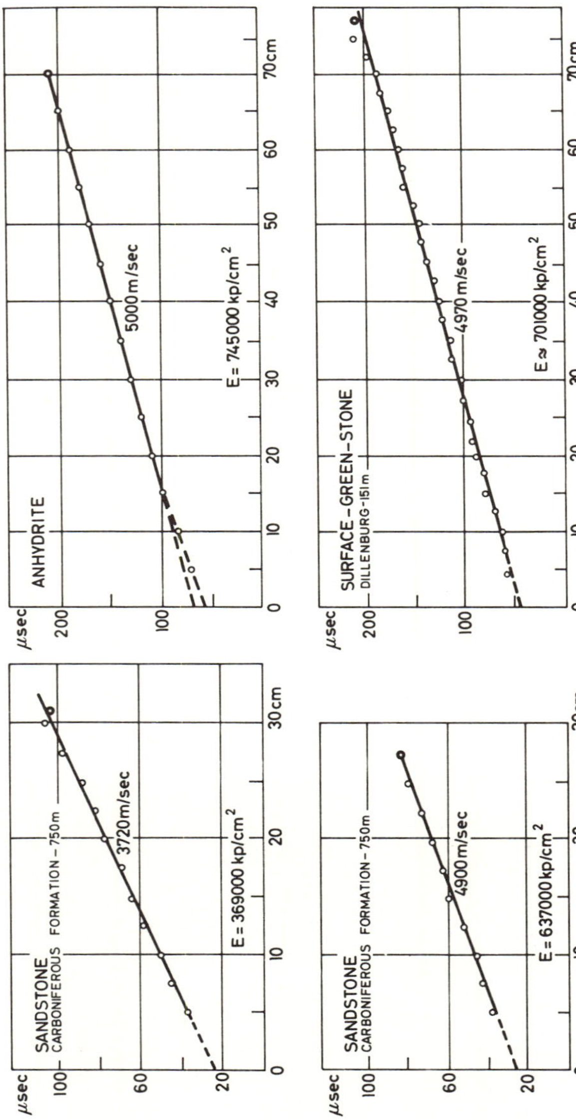

FIG. I4b-3. Traveling time curves of drilled cores of sandstone, anhydrite, and surface greenstone.

After determination of the transient times at many measuring points, a transient time trajectory is compiled, from which the velocity as such results. Some examples of such curves are shown in Fig. I4b-3.

Also by means of sparklight photography of the front waves one can determine the velocity of seismic waves. Here, a very instructive method for determining the wave velocities V_l and V_t in small samples at room temperature and atmospheric pressure was described by Erich Müller [216], who refers to work by Mach, Cranz, and Schardin [213] and publications of O. von Schmidt [214]. In this method the small mineral samples of 3 to 10 cm length are submerged in a liquid filled vessel and the headwaves appearing in the liquid (Mintrop waves) [215] are photographed in sparklight illumination according to the Toepler schlieren method. The angles α and β (see Fig. I4b-4) formed by the straight head waves and the plain surface of the samples are measured and calculated according to the relations

$$V_l = c/\sin \alpha \text{ and } V_t = c/\sin \beta$$

where c stands for the velocity of sound in the liquid. All other elastic constants can then be determined from V_l, V_t, and the mineral density ρ.

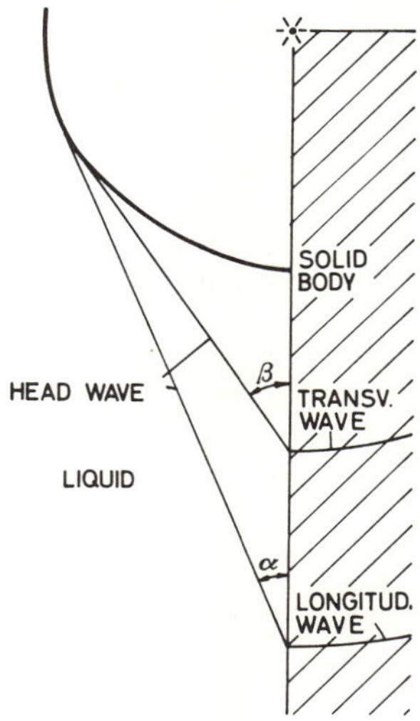

FIG. I4b-4. Spherical and head waves in the margin layer zone, mineral sample–liquid, after firing the acoustic spark (according to E. Müller [216]).

Figure I4b-5 shows a block diagram of the experimental setup. The shock wave spark, F_1, in the vessel, K, is to be followed by a subsequent illuminating flash, F_2, after an interval of approximately 10^{-5} sec (controllable by the inductance L) and with a duration of 10^{-7} sec. For this purpose the capacitors C_1 to C_3 are charged up to a voltage U of 20–30 kv until F_1 and the auxiliary spark gap, H, break through, while C_3 is temporarily insulated through the high resistance, W, whereas C_1 and C_2 go through an oscillating discharge.

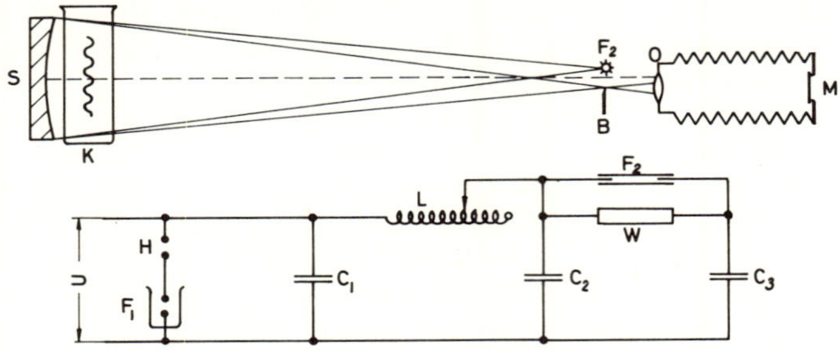

FIG. I4b-5. Diagram of the optical and electrical experimental arrangement (Toepler schlieren method and Mach circuitry) (according to E. Müller [216]).

After a half-cycle of the circuit LC_2 there occurs between the capacitor plates of C_2 and C_3—which formerly were on equal potential—a voltage of approximately $2U$, and now also C_3 discharges over the sharply defined spark gap F_2, the image of which is projected by the concave mirror, S, onto the schlieren edge, B. The camera lens, O, projects the image of K onto the ground glass, M. Some of the light is bent by the head waves, passes B and makes the schlieren visible on the ground glass M [216]. The mineral samples may have any shape; all that is required is one plane face, to which a thin wire is fastened for the sparkover.

Any changes of the elastic qualities cause a deformation of the normally straight line head wave and can thus be detected. Table I4b-1 shows some of

TABLE I4b-1

SOME MEASURED DATA BY E. MÜLLER [216]

Mineral, origin and depth	V_l (m/sec)	V_t (m/sec)	σ	E (10^{11}dyn/cm^2)	G (10^{11}dyn/cm^2)	$\varkappa$ (10^{-12} cm^2/dyn)
Granite	5200	3000	0.25	6.0	2.4	2.5
Bituminous rock	5400	3100	0.25	5.5	2.2	2.7
Toadstone	6400	3200	0.33	7.8	2.9	1.3
Eclogite	8000	4300	0.30	16.3	6.3	0.74
Anhydrite 2174 m ‖	5500	2400	0.38	4.7	1.7	1.5
Anhydrite, 2174 m ⊥	5150	2600	0.33	5.3	2.0	1.9
Anhydrite, 2180 m ‖	5700	2700	0.36	5.8	2.1	1.4
Anhydrite, 2180 m ⊥	5500	2600	0.36	5.4	2.0	1.6
Anhydrite	5610	2690	0.352	5.7	2.11	1.56
Cretaceous, 1460 m	2400	1350	0.27	1.1	0.45	12
Marly limestone, 1590 m	3100	1750	0.27	1.6	0.63	9
Chalky clay, 1721 m	3250	2000	0.20	2.3	0.97	8
Fine-grain sandstone, 2170 m	3500	2150	0.19	2.8	1.2	7

E. Müller's measuring results, and Table I4b-2 indicates the relationships between the mineral density ρ and the elastic constants for six samples of mottled sandstone.

TABLE I4b-2

DEPENDENCY OF MINERAL DENSITY ρ AND THE ELASTIC CONSTANTS OF SIX SAMPLES OF MOTTLED SANDSTONE BY E. MÜLLER [216]

Mottled Sandstone Borehole Rheden 5	ρ (g/cm^3)	V_l (m/sec)	V_t (m/sec)	σ	E (10^{11}dyn/cm^2)
1676–1677	2.58	4400	2200	0.33	3.3
2028–2030	2.66	4700	2500	0.30	4.3
1794–1797	2.66	4700	2600	0.28	4.6
1925–1927	2.71	4900	2600	0.30	4.8
1532–1534	2.79	5200	2900	0.27	6.0
1532–1542	2.82	5600	2900	0.32	6.3

c. *Experimental Work with Seismic Pulses Generated by Underwater Sparks*

In order to study the laws governing the expansion, reflection and refraction of seismic pulses, Howes, Tejada-Flores and Randolph [744] carried out model experiments in a water tank. Natural conditions were closely imitated. The pulse generation was effected by underwater sparks, recording on a cathode ray oscilloscope over barium-titanate hydrophones. The recordings which served to determine the expansion velocity of elastic waves in lime chalk show—besides the direct wave—a Rayleigh wave, another wave which proceded through the water and ran right around the tank, and a wave which was reflected from the water surface. By suitable change of experimental conditions the authors succeed in demonstrating the expansion, reflection and refraction of seismic pulses so that the solution of some problems related to these phenomena is facilitated in the interpretation of seismic recordings.

Cl. Behnke [207] has reported on a novel piece of equipment for generation of explosion sparks as a seismic impulse source. The apparatus works with impulse capacitor discharges by way of a fluid spark gap. The resultant ground motion is recorded on magnetic tape. A possibility for magnification of real existing and simultaneous diminution of arbitrary starts of oscillations is given by multiple recording of one upon another of the same seismic pulses on only one magnetic tape. There are various methods which are employed for seismic examination of upper ground layers particularly with regard to the generation of seismic pulses. In applied commercial geophysics one almost exclusively applies high energy explosive detonations in order to penetrate to depths of some ten kilometers. For measurements limited to lesser depth ranges, primers or blasting caps or even hammer blows are employed. However, for research and laboratory experimental work high explosives are disadvantageous, considering the complicated nature of high energy explosive regulations which have to be complied with, the cost of the explosives as such, etc. Experimenters therefore looked for new ways to generate the acoustic pulses required for seismic velocity logging purposes.

High voltage pulse discharges fed into a fluid spark gap offered themselves as a very convenient method for producing the acoustic pulses required. If the poles of a charged storage capacitor are connected to two electrodes facing each other in a liquid medium, a spontaneous sparkover will occur between the electrodes with simultaneous formation of a light arc. This sparkover is connected with an intense pressure wave which manifests itself by a loud detonation. If this sparkover is effected at a depth of several meters below ground a considerable shock wave is released. In this way impulses of steep gradients can be achieved.

Below an apparatus will be described by means of which shock waves with ranges of several hundred meters can be released on the surface of solid ground, while in water ranges of several kilometers can be achieved under favorable conditions.

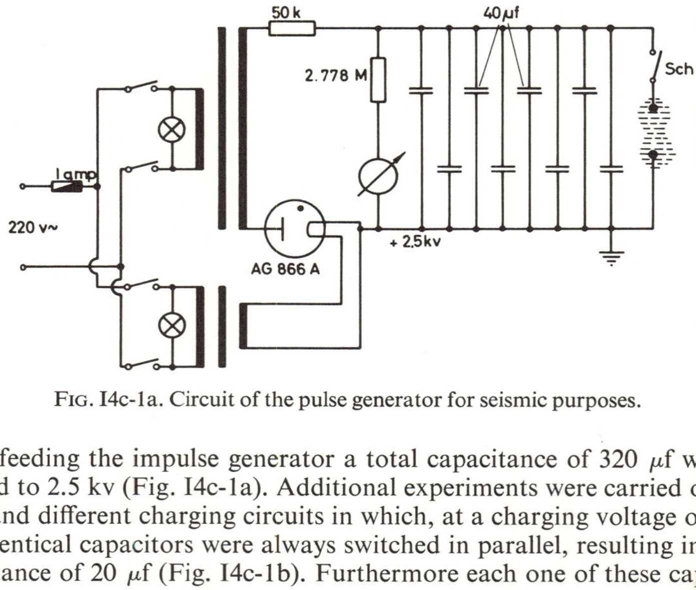

FIG. I4c-1a. Circuit of the pulse generator for seismic purposes.

For feeding the impulse generator a total capacitance of 320 μf was used charged to 2.5 kv (Fig. I4c-1a). Additional experiments were carried out with other and different charging circuits in which, at a charging voltage of 10 kv, four identical capacitors were always switched in parallel, resulting in a total capacitance of 20 μf (Fig. I4c-1b). Furthermore each one of these capacitors was bridged by a 10 megohm resistor for the following reasons.

1. The ohm resistances of such capacitors may well be equal but, due to production techniques, they are not identical. Through the parallel resistors an even voltage drop was achieved at every capacitor, whereby voltage overloads and sparkovers were avoided.

2. If for some reason or other no pulse discharge was released, then the capacitors would discharge themselves slowly over the parallel resistors, thus offering handling safety which is of utmost importance in high voltage work. The time constant of this automatic discharge was 16 minutes, which means that after approximately 1 hour the capacitor voltage dropped down to 50 volts, and the current flowing was less than 1 milliamp.

This circuitry arrangement was later discarded, mainly because it is simpler to build a setup with 2.5 kv operating voltage than to build and handle a 10 kv apparatus, and furthermore because the switch Sch (Fig. I4c-1a) (which is discussed in Chapter B, Section 6a, Impact Switch) could not be made for higher voltages than 2.5 kv.

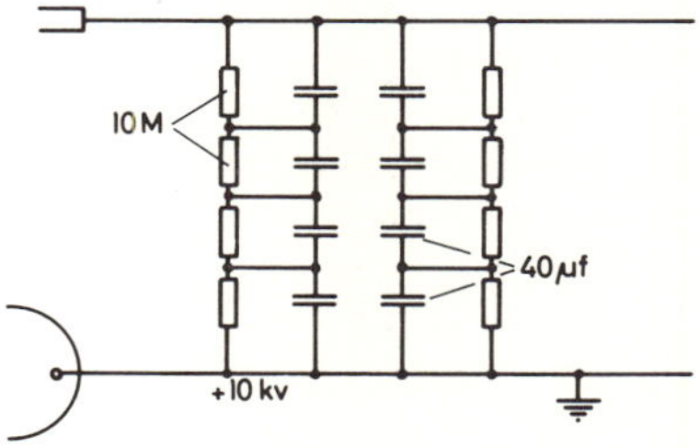

FIG. I4c-1b. Circuit of the pulse generator; series circuit of the capacitors.

The electric energy in the charged capacitor batteries, $(C/2) \times U^2$, where C = capacitance of the battery and U = charging voltage, is identical in both circuitries, viz. 1 kw-sec. The discharge time t and the electric energy $C/2 \times U^2$ result in an energy pulse of

$$N = \frac{C/2 \times U^2}{t} \text{ [watts]}$$

where t largely depends on C and the inductance L of the discharge circuit. By using paper capacitors the inductance of the capacitor battery could be kept low, at 0.0125 μh to which must be added the detrimental inductance of consumer leads amounting to approximately 0.15 μh. Assuming an aperiodic discharge the half-period will be

$$t = \pi\sqrt{L \times C}$$

and

$$N = \frac{C/2 \times U^2}{\pi \times \sqrt{LC}}$$

with L standing for the inductance of the capacitor battery with leads.

From the above there results a discharge time of

$$t \approx 2.27 \times 10^{-5} \text{ sec}$$

and an energy pulse of $N \approx 44$ megawatts. As soon as the switch, Sch, is closed, the total voltage is at the fluid spark gap, which will immediately sparkover with formation of an arc.

In order to effect the sparkover at several meters' depth in the ground, the spark electrodes were fitted into the point of a steel tube probe. In experiments carried out by Suhr [208] deformations at the probe and the electrodes occurred at energy levels as low as 25 watt-sec. For these reasons a probe was constructed having a 5 mm steel tube. Two stainless steel cylinders of 10 mm dia, with rounded tips face each other as electrodes (Fig. I4c-2). The electrode gap was steplessly variable. One of the electrodes is affixed to the probe's steel tip which is screwed into the lower end of the steel tube probe with the tube forming one current lead. The heavily insulated lead for the other electrode was in the middle of the hollow tube. The lower part of the tubular probe could be unscrewed and had 54 boreholes of 9 mm diameter to ensure good radiation of the energy to all sides. A tightly fitting rubber hose prevented escape of the water which could be filled in from above ground. At the probe's total length of 2.10 meters sparkovers could thus be effected up to a depth of 2 meters.

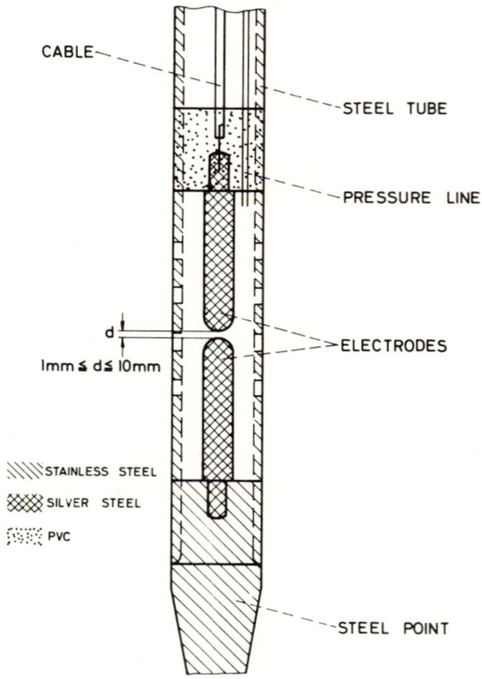

FIG. I4c-2. Cross section of spark sound probe.

According to work by Weizel [98] and Bailitis [55] a plasma, the so-called discharge space, originates over a fluid spark gap upon discharge of a capacitor. Depending on the properties of the discharge circuit the energy is transmitted into this discharge space either very rapidly or slowly. If the consumer's resistance is low and if also the leads have low inductance and the capacitor battery has a comparatively large capacitance as was the case in the apparatus described above, then the energy transport occurs much more quickly and more completely than would otherwise be the case.

Figure I4c-3 shows the profile of an underwater spark's acoustic impulse as determined by Bailitis, the pulse duration being 5 μsec. The acoustic impulse is particularly characterized by a very steep front gradient. The pressure increase up to 3/4 of the maximum value is effected within 0.2 to 0.3 μsec.

According to theoretical work by Früngel and Keller [56] (see Section 3b of this chapter) a quantity of 1 gram of TNT high explosive corresponds to an electric energy of approximately 4000 watt-sec. For the above-described pulse generator there thus results a comparative high explosive energy of 0.25 grams TNT per spark discharge.

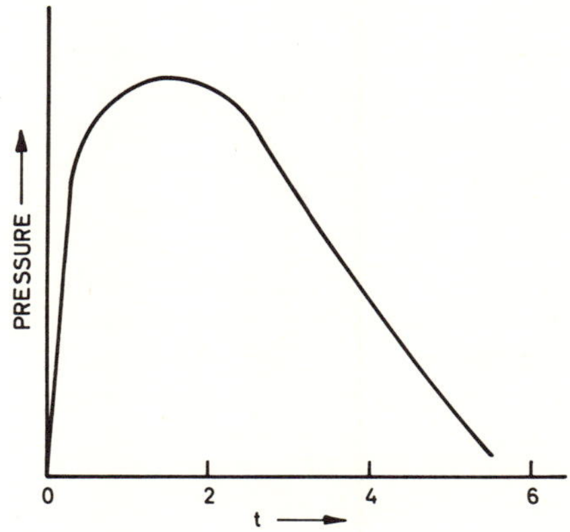

FIG. I4c-3. Pressure vs. time diagram of the emitted spark sound of the probe in Fig. I4c-2.

d. *Biological Application of Shock Sound—Sterilizing and Cleaning of Milking Machines.*

A useful application of shock waves produced by underwater sparks is a method and device for electrically sterilizing and cleaning. Früngel [231, 232] published a description of a device specifically for sterilizing and cleaning milking machines.

It is very difficult to remove milk residues which remain in the various elements of a milking machine after the milking procedure. Even water at higher temperatures to which dairy cleanser has been added will not clean completely, particularly the rubber or plastic-material hoses or tubes of the machine. As milk particles accumulate, various species of bacteria find nourishing living conditions therein so that they rapidly grow and multiply. When such bacteria are present in the tubes they can not be prevented from passing into the milk during the milking process. Thus it is impossible to produce milk of low bacteria count with machines whose tubes have been infested by such bacteria.

To obviate these difficulties it is possible to provide a method of electrically cleaning and sterilizing milking machines, vaccine apparatus, medical instruments and the like by means of electrical discharge sparks. The sparks are caused to jump across a liquid spark gap submerged in the cleaning water or solution, and passed through tubes and other elements to be sterilized, also immersed in the cleaning fluid, so that every portion of the surface of such elements is reached by the shock sound waves radiating from such liquid sparks.

A multiple-conductor cable, hereinafter called a discharge or shock sound cable, which carries at its free end a spark gap, can be inserted and pushed at slow speed through tubes, hoses and other elements to be cleaned and sterilized. Such a shock sound cable has a mouthpiece of plastic material which has a socket and a small hole in its face portion in which the electric spark discharges occur from which shock sound radiation emanates.

This cable mouthpiece is molded to exact dimension, without subsequent machining or finishing, from plastic material of high rupture resistance. It is possible also to provide for the cable mouthpiece of plastic material a protective sleeving of high mechanical deformation resistance, preferably consisting of high-grade steel, surrounding the mouthpiece except for its active face portion.

Additionally, a suitable power source and circuit for properly energizing the shock sound cable are needed. A storage capacitor is utilized; it is charged up to the breakdown potential of a spark gap in gas or open air. Upon breakdown of this spark gap the capacitor discharges instantly into the shock sound cable to produce the liquid spark. The circuitry between power source proper and discharge cable can take various forms.

In order to achieve the objects in view, it is of primary importance to produce effective shock wave radiation in an aqueous solution. For this purpose it was found necessary to force capacitor discharges through a small water capillary present in the small hole of the cable mouthpiece of detonation resisting plastic material. At the instant of discharge the water in the capillary evaporates and causes a pressure wave whose power is equal to the power of a detonation of about 10^{-9} grams of dynamite, so that at a distance of about 7 cm from the source a pressure of more than 1000 atm can be measured. On account of their small energy these shock waves are harmless to the human being, but they are able to destroy at short distance all microorganisms.

Intensive investigation and practical experience have shown that it is possible to sterilize milking machines and the like with such shock waves of sufficient intensity. Not only do they remove milk and fat remnants completely from the inner walls of the milk tubes, they also effect complete destruction of germs, viruses and bacteria at normal cleaning water temperatures and without chemical additives.

To convey a still better understanding of this method, a description of some designs may be given, with reference to Fig. I4d-1–14d-5.

For general orientation there is shown in Fig. I4d-1 a milking machine set having teat cups, 1, and milk tubes or hoses, 2, of rubber or the like which connect the teat cups to the milker head, 4. The inner walls of the milk tubes, indicated at 3, are particularly susceptible to infestation by bacteria, since the occurrence of surface ruptures and fine hairline cracks in these walls cannot be prevented. In some milking machine designs another rubber tube,

5, leads from the milker head to the milk can or milk container. Such a tube is another preferred hiding place for bacteria.

The shock wave source used is in the form of a cable as shown in a few preferred construction designs in Fig. I4d-2a, 2b, and 2c. The cable itself is of similar construction in each of these designs and comprises the outer insulating jacket, 6, which surrounds the outer, preferably tubular, conductor, 7; the insulation, 8, which is preferably polyethylene; and the center conductor, 9, which is preferably made of high grade steel. In the design shown in Fig. I4d-2a, (a cable as just described serves directly as shock wave source in that it is cut flat at its free end and the flat face of the center conductor, 9, serves as one electrode. The outer insulating jacket, 6, is cut back for a distance of about one to two centimeters so that a greater surface area of the outer conductor, 7, is exposed to the liquid when the cable is immersed therein. When capacitor discharges of sufficient potential are fed into such a cable immersed in liquid, a spark will jump across its face between center and outer conductor causing shock wave radiation therefrom.

In the design depicted in Fig. I4d-2b, the cable is provided with a head portion or mouthpiece which includes the center electrode, 10, in the form of a pin which is secured, preferably by welding, to the center conductor, 9, and a sleeve, 12, which forms the outer electrode and is secured to the outer conductor, 7, preferably also by welding. To assure long life in spite of continuous operation, both electrodes are preferably made of a high grade steel. The mouthpiece proper, 13, is formed by the insulation between electrodes 10 and 12, and is composed of detonation resisting plastic material, preferably of the kind known under the trade name "Vulkollan," a poly-addition product of a di-isocyanate with a polyester carrying free hydroxyl groups. In this construction, of course, the spark gap is formed between pin electrode 10 and sleeve electrode 12 across the detonation resisting plastic material, 13.

In the construction shown in Fig. I4d-2c the sleeve, 12, is replaced by a helically coiled wire, 12a, which in this case forms the outer electrode. This wire is preferably also of high grade steel and is conductively joined to outer conductor 7 by a tubular ring, 14, of a suitable metal. The center electrode, 10, is again a pin welded to conductor 9. The tubular outer conductor, 7, of the cable is preferably in the form of braided or interwoven wires as is customary in cable construction. The mouthpiece, 13, of Vulkollan is molded to exact dimensions in order to retain the smooth molded surfaces, and the small hole, 11, in its face is the point at which the water capillary previously mentioned evaporates when a discharge occurs. By experimenting with a mouthpiece construction as just described, and by connecting the positive terminal of the discharge source to the center conductor 9 with electrode 10, it has been found that, besides shock wave radiation, considerable traces of ozone are produced at each sparkover, which is of benefit in that ozone has a germ killing effect and renders after-sterilization.

The discharge energy, to which the cable of each of the types just described is subjected after immersion in an aqueous solution evaporates the small liquid volume in the spark gap and breaks immediately thereafter as an electric sparkover through this not yet expanded vapor space, thereby

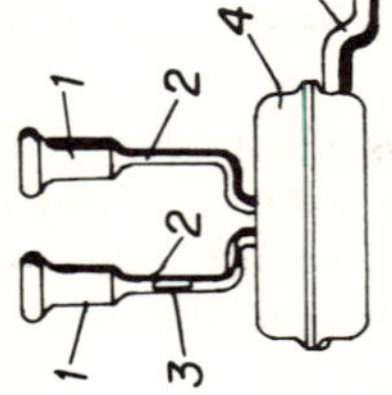

FIG. I4d-1

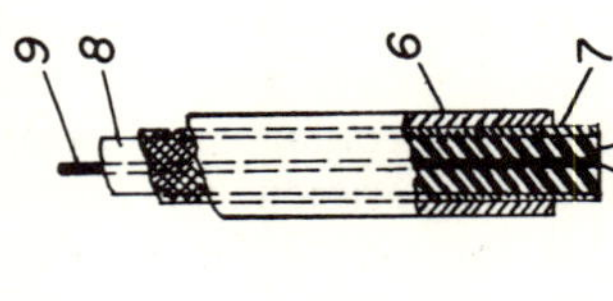

FIG. I4d-2a

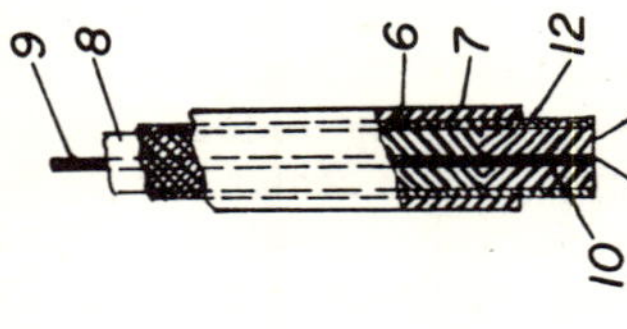

FIG. I4d-2b

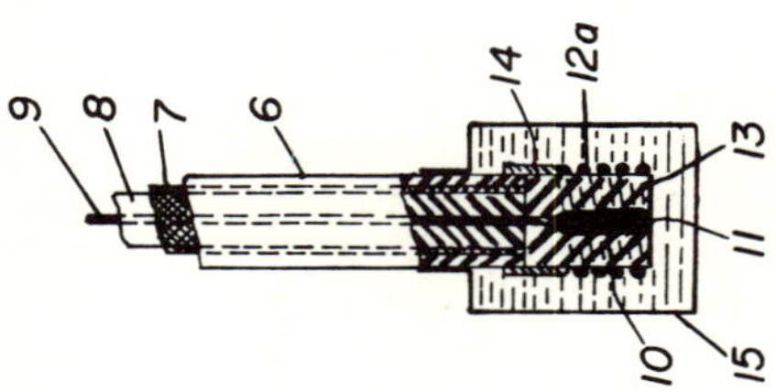

FIG. I4d-2c

FIG. I4d-2d

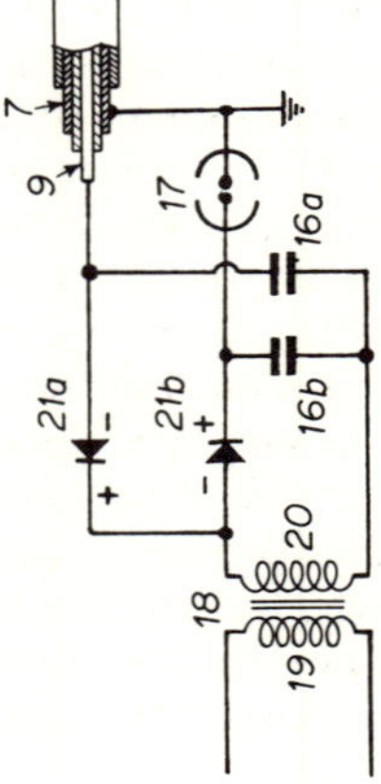

FIG. I4d-3a

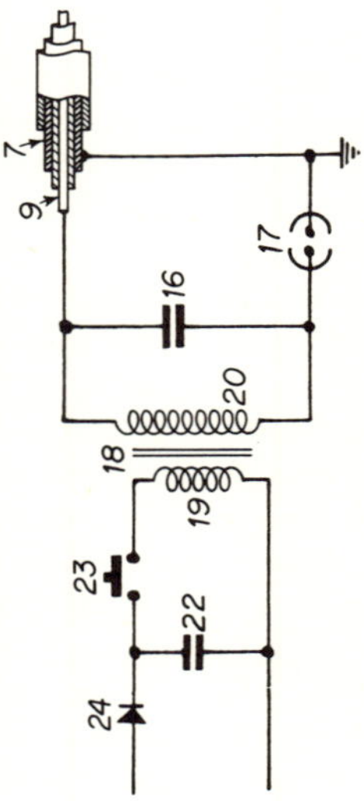

FIG. I4d-3b

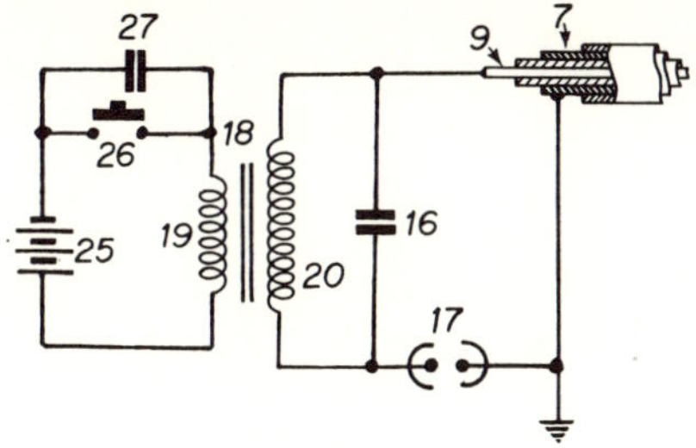

FIG. I4d-3c

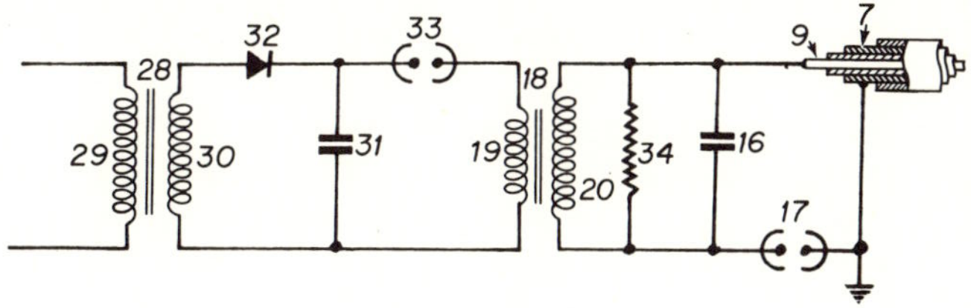

FIG. I4d-3d

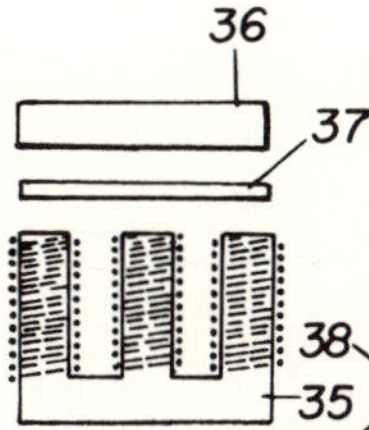

FIG. I4d-4

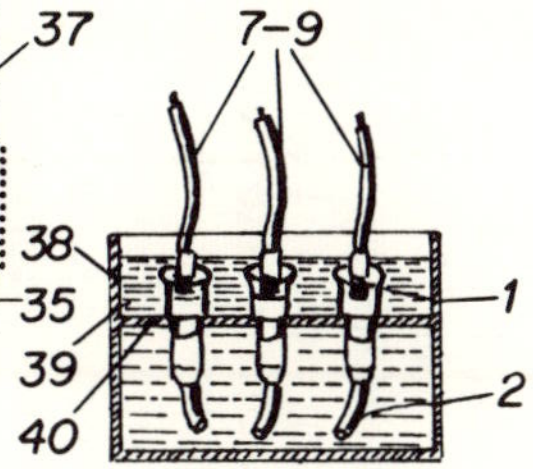

FIG. I4d-5

FIG. I4d-1. Diagrammatic view of a milking machine set to be cleaned and sterilized by the method and device described here.

FIG. I4d-2a–2c. Preferred designs of shock sound cables.

FIG. I4d-2d. A cross section through a shock sound cable taken above the mouthpiece. It may be a cross section through any one of the modifications shown in Fig. I4d-2a, b, or c.

FIG. I4d-3a–3d. Preferred designs of power supply circuits for energizing a shock sound cable.

FIG. I4d-4. Schematic view of a transformer construction as preferably employed in a power supply circuit.

FIG. I4d-5. Diagrammatic view of milking machine sets as suspended for cleaning in a special container.

develops a detonation like sound. In case the conductivity of the liquid is low, shock sound production is slow and the peak volume moderate. With increasing conductivity of the liquid, shock sound production becomes faster and peak volume greater.

To avoid confusion it should be remembered that shock sound waves differ from ultrasonic waves because they produce only a positive pressure peak without any oscillatory characteristic. The sterilizing effect, it can be assumed, is caused by the nearly infinitely steep wave front which, when it strikes a microorganism, destroys the latter completely.

For certain purposes it may be desirable to radiate shock sound waves into solutions that are not aqueous and do not have the proper conductivity for initiation of sparkovers. This, for instance, may be the case in serological studies and in vaccine production. To gain the benefit of shock wave action in such liquids of adverse conductivity, the mouthpiece portion of the shock sound cable can be enclosed by a water-containing envelope of thin plastic material as indicated at 15 in Fig. I4d-2c. If so enclosed, the sparkovers occur within the envelope, and shock sound radiates through its walls. It has been found that the film of plastic material does not hinder shock wave radiation, and, in such an arrangement, the outer surface of the envelope forms the source from which shock sound emanates. An envelope, 15, of modified form has the additional advantage that it lends itself to therapeutic treatment of the human body by shock sound waves instead of the conventional ultrasonic therapy. For that purpose, the plastic envelope which contains spark discharge water and the end portion of the discharge cable must be shaped to take the form of a traditional ultrasonic transducer head for medical uses, and should be applied to the skin with sliding motion and the assistance of contact liquid or cream. Shock sound therapy has the advantage of removing the danger of cavitation injuries that is present in ultrasonic treatments.

Figures I4d-3a, 3b, 3c and 3d, disclose some preferred examples of power supply circuits for shock sound cable operation. All arrangements shown are similar in this respect: a capacitor, 16, which is charged by some means to the breakdown potential of a spark gap, 17, discharges suddenly upon breakdown of this spark gap into the discharge or shock sound cable through conductors 7 and 9 and produces thereby a spark between the electrodes at the mouthpiece of the cable. The spark gap, 17, is of the gaseous type, but need not be sealed; it can also be in open air.

The arrangement depicted in Fig. I4d-3a is particularly suitable for discharge of considerable energy, and the capacitors, 16a and 16b, are connected for charging by the well known voltage doubling circuit from transformer 18, having primary winding, 19, and secondary winding, 20, by way of rectifiers, 21a and 21b. The discharge circuit is formed by the spark gap, 17, and shock sound cable, 7–9, which has the spark gap (not shown) at its free end. The primary, 19, of transformer 18 has leads to be connected to a commercial alternating current power line.

In the circuit shown in Fig. I4d-3b, the high voltage, again in the range between 4 and 8 kv, for charging the high potential capacitor 16, which is connected to the secondary, 20, of transformer 18, is produced in the following manner: A low-potential capacitor, 22, is connected to the primary, 19, of said transformer through a contact breaker, 23, whereby at each contact breaking a high potential voltage impulse is induced in the secondary, 20, of transformer 18; this instantly charges capacitor 16, which in turn, at breakdown of gaseous gap 17, abruptly discharges into the shock sound cable. Low potential capacitor 22 is kept charged from a commercial ac power line through rectifier 24. In this system an increase in liquid spark energy is gained by the fact that at the instant of breakdown of the spark gap not only the energy stored in capacitor 16, but in addition energy from the magnetic field of transformer 18 is released in the liquid spark. Such release of magnetic energy is impossible in the system of Fig. I4d-3a. The arrangement of Fig. I4d-3b is particularly suitable for yielding high ozone formation in the spark discharge space.

For setups of moderate size, a power supply circuit as shown in Fig. I4d-3c may be sufficient. The transformer, 18, is here actually a secondary type of induction coil which is fed from a battery, 25, by way of an interrupter, 26, with direct current in the same manner as in a motor car ignition system. The interrupter is bridged by a spark suppression capacitor, 27, and the secondary, 20, of the induction coil is connected across the high potential storage capacitor, 16. At each break of the interrupter, a high voltage impulse is induced in the secondary, 20, which charges capacitor 16 up to breakdown of gaseous gap 17 and subsequently discharges into the cable, 7–9, similarly to the system of Fig. I4d-3b.

Still another construction of a power supply system that has proved its value in reliable regularity of discharges at great energy conversion per spark is shown in Fig. I4d-3d. Here two transformers are employed and the transformer from whose secondary the high potential capacitor, 16, is charged is again designated by reference numeral 18. Another transformer, 28, whose primary, 29, is fed from a commercial power line, has its secondary connected to a capacitor, 31, through a rectifier, 32. Capacitor 31 discharges at breakdown of spark gap 33 into the primary, 19, of transformer 18. This discharge causes a high potential impulse in secondary, 20, which in turn charges high potential capacitor 16 to the breakdown point of gaseous spark gap 17, and finally initiates discharge of capacitor 16 into cable 7–9, whereby capacitor 16 and gap 17 assume the same functions as the corresponding circuit elements in the other power supply circuits described. A damping resistor, 34, having a resistance value suitable to substantially suppress circuit oscillation, is preferably connected across capacitor 16. Owing to spark gap 33 in the primary circuit of transformer 18, satisfactory functioning of this power supply system can also be achieved without spark gap 17 and this spark gap is thus considered optional. Capacitor 31 has from 3 to 5 times greater capacitance than capacitor 16, so that sufficient energy, stored as magnetic energy, is additionally available at the instant of discharge from transformer 18. Such energy reserve improves the shock sound curve of the discharge. The repetition rate of discharges in this system can be controlled by the resistance value of the secondary, 30, of transformer 28.

The transformers employed in the supply circuits just described can be constructed either with or without ferromagnetic cores. Figure I4d-4 is an exploded view of a transformer having a shell-type core. This transformer has the core, 35, and a separable yoke, 36. In such construction it is advantageous to insert a layer, 37, of rubber or the like between core and yoke. Such a layer suppresses the electrical and mechanical interactions of forces in the core and increases spark efficiency.

Figure I4d-5 finally shows the complete arrangement of a milking machine cleaning and sterilizing setup utilizing this method. It is advisable to use a suitable container, 38, which is filled with the cleaning solution, 39, to such a level that the teat cups and milk tubes of a milker set can be completely submerged therein. The cleaning fluid may be plain water, but cleaning additives suitable for dairy use may be dissolved in it. A supporting frame, 40, should be provided, in which the milking machine elements can be suspended to be readily accessible for the sterilizing and cleaning procedure. After this preparation, one or several shock sound cables as described above are inserted and passed through teat cups, milk tubes and other elements at slow speed, so that all portions of the inner surfaces and walls of such elements can be effectively reached by shock wave radiation emanating from the liquid sparks produced at the mouthpiece of the cable when the latter is energized by electric energy impulses. A repetition frequency of 5 to 15 sparks or shocks per second has been found most suitable. With such frequency the cleaning and sterilizing procedure of a whole milking machine set can be completed within a few minutes.

Practical tests and microscopic inspections have proved that not only lactic acid bacteria, but molds and viruses as well as other bacteria are destroyed up to complete sterilization, particularly in the milk tubes. Even colonies hiding in hair cracks are killed because shock sound radiates equally well through water and rubber. Thus when the steep shock front strikes the individual bacterium, it must kill it no matter whether it hides in a rubber crack, in a fat accumulation or in the cleaning solution.

Measurements with a probe microphone revealed at 7 cm distance from the shock sound source, pressures between 900 and 2,000 atm, depending on the capacitance of capacitor 16 and the charging voltage. From practical tests it appears that capacitors having capacitances between 0.05 and 0.1 μf at charging voltages between 4 and 8 kv are most suitable.

In order to force detonation-like capacitor discharges through the liquid spark gap in the mouthpiece of the discharge cable, the pH of the solution in which the present method is used must deviate slightly from 7.0. If it were exactly 7.0, the solution would be nonconducting and the shock sound cable would not be able to produce shock waves. But even a deviation of a few tenths from pH 7 is sufficient to create enough conductivity to assure shock wave generation. However, the more the pH deviates from 7 the stronger will be the detonation waves. Stronger detonation waves will have less duration on account of the higher current conductivity; however the total energy of each sound shock is practically the same, and no difference in the effect upon bacteria can be discerned. The sound of the detonations as heard by the ear varies, but the sterilizing effect is equally good whether the pH of the cleaning solution is 6, 6.5, 7.5, or 8. This explains the fact that suitable cleansers or detergents can be added to the water without diminishing the detonation-like sterilizing sound shock action in any way.

The shock sound cables can be modified and altered in their diameters to suit a particular application. For the milking machine application, the diameter of the steel sleeve, 12, forming the outer electrode is preferably 5 mm. For obtaining shock sound energy of higher intensity, the diameter can be increased. It is also possible to provide the cables with mouthpieces having noncircular cross sections, but the circular section has the advantage that it can be manufactured more economically.

Another application of shock sound is the treatment of seeds. It operates as follows: The seeds to be treated are placed in a container of plastic material, preferably in a water solution, and this container is then subjected to shock waves from several shock sound sources. The proper amplitude of the sound waves can be adjusted, e.g. by placing the sound source a certain distance from the plastic container, and the desired dose can be fixed by counting the shock impulses. Experiments have indicated that electrical values of 0.2 μf and 8 kv are particularly effective.

Seeds in a plastic container that had been subjected to from 10 to 40 shocks at a distance of 5 cm from such a sound source suffered no injury. The experimental seedlings (lettuce, radishes, sugar, beets, etc.) grown from these seeds exhibited, however, a higher germination rate and faster growth than those raised from untreated seeds. Moreover, shortening of the vegetation period could be observed.

When, however, the effective amplitude of the shock sound waves is increased by reducing the distance between seeds and sound source to about 1 cm, extraordinary cell-destroying effects become apparent in the seeds. The effects which are harmful to seeds can be utilized in other fields, e.g. for sterilization of the goods so treated. The form of the sound source should be adapted to its application.

An additional bactericidal effect of the shock sound source is obtained when the polarity of the sound source electrode is made positive. Directly adjacent to this electrode a large quantity of ozone is developed—about 1 gram of ozone for every 25 watt-hours. When, furthermore, an acid treatment solution is used, the time during which ozone, which actually has been produced as a byproduct, resides in the solution can be prolonged so that the presence of this ozone in itself enhances the sterilizing effect. However, in any case in which such additional ozone production is undesirable for some reason, it is possible to keep the ozone away from the treatment solution by a plastic or other thin membrane placed between the treated goods and the shock sound source. By means of such membranes it is furthermore possible to maintain desirable different pH values in a treated solution and in the solution where the shock sound originates.

e. *Treatment of Molten Metal by Shock Sound*

In the field of metallurgy an application of shock sound is also possible. A device for treatment of molten metal by shock waves has been made by public Früngel [223].

There have already been published various methods for degazing and forming a better texture by introducing ultrasonic energy into a bath of molten metal. None of these methods could gain widespread recognition because there was not available an ultrasonic transducer that could safely be inserted directly into molten metal. It was necessary to insert long intermediate members, such as rods of tungsten or porcelain, between the ultrasonic transducer and molten metal, and these caused considerable loss of ultrasonic energy, particularly on account of high absorption at vibration frequencies not natural to the inserted members. The alternative was to excite the crucible containing the molten metal by a suitable transducer, and this requires great amounts of ultrasonic energy due to the considerable size of the crucible and the thickness of its lining. Furthermore, an ultrasonic transducer is hard to apply to a crucible of conventional form.

It is the purpose of this section to describe a simple device and a method for employing ultrasonic vibration in connection with conventional foundry and casting equipment at little additional expense.

This ultrasonic treatment is effected as the molten metal is on its way between crucible and mold, i.e. directly in the pouring ladle; and it therefore consists in providing a special ladle having an ultrasonic generator or transducer in its long handle. A preferred example of this method is described here with reference to Fig. I4e-1, which is a diagrammatic view of such a pouring ladle.

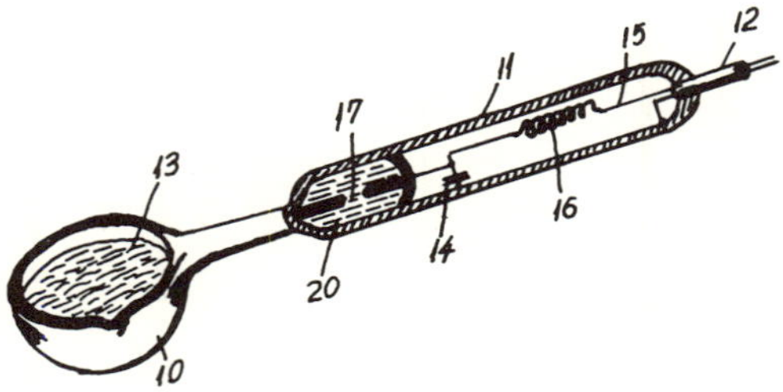

FIG. I4e-1. Treatment of molten metal by shock sound of sparks.

The ladles, 10, shown in their conventional form as half spherical bowls, are provided with long handles, 11, to which electrical connections are made in the form of cables, 12. Pouring ladles are generally not provided with a refractory inner lining, but are coated with a graphite preparation which is a good conductor for ultra sound. Therefore, it is readily possible to subject the molten metal, 13, in the ladle to ultrasonic vibration which can be generated in the long handle of the ladle itself. Such ultrasonic generating means may include, as shown in Fig. I4e-1, a storage capacitor, 14, to which one conductor, 15, of cable 12 is connected through an impedance, 16. The other terminal of capacitor 14 and the other conductor of the cable are grounded to the handle and complete a circuit to which an external source of voltage impulses can be connected for charging capacitor 14. Impedance 16 serves to limit the charging rate. A discharge circuit of capacitor 14 includes a spark gap, 17, preferably arranged in a sealed chamber, 20, filled with water or other liquid. When the peaks of the capacitor charging pulses are higher than the breakdown voltage of the spark gap, sparks will jump through the gap and cause ultrasonic shock waves in the liquid, imparting in turn ultrasonic vibration to the contents of the ladle.

f. *A High Current Switch Using an Underwater Spark*

The underwater spark can be used to actuate a high current switch for capacitor discharges. Bodenseher [867] has developed a device which moves a contact foil as a part of the high current switch with a speed which raises suddenly to 500 meters/sec or 0.5 mm/μsec. The jitter is less than 100 nanosec. Figure I4f-1 shows the operating principle.

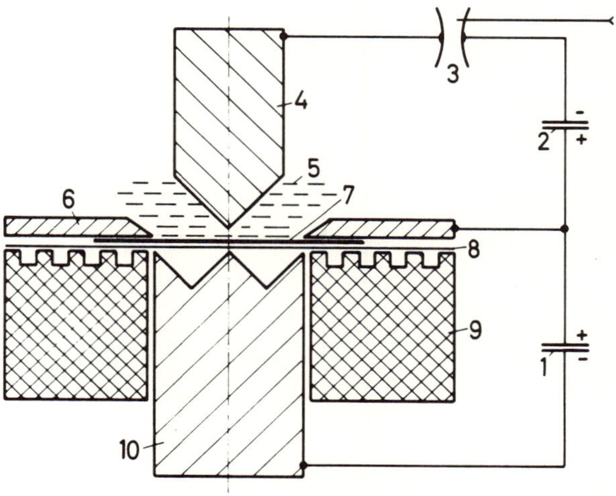

FIG. I4f-1. Metal foil switch for high current discharges. 1–Main capacitor battery 3μf, 25 kv, 200 kcps ringing frequency; 2–auxiliary capacitor; 3–ignition spark gap; 4–electrode; 5–water; 6–flange for mounting the sheet metal; 7–sheet metal of 0.25 mm aluminium foil; 8–Mylar foil; 9–insulating electrode holder; 10–copper electrode of the main discharge circuit.

The high current switch consists of a solid electrode, 10, and a second electrode, 7, which is made of thin sheet metal. The electrode, 10, is held by the insulator, 9. The metal sheet, 7, is pressed against the insulator, 9, by means of the flange, 6, The insulating material between 7 and 9 is a very thin Mylar foil, 8. For closing the contacts, 7 and 10, of the switch the metal sheet, 7, is moved by means of the pressure wave of an underwater spark between the metal sheet, 7, and conic electrode, 4. The water itself is indicated at 5.

For starting the main discharge of the capacitor, 1, through the electrode, 7–10, another capacitor, 2, is discharged by means of the triggered spark gap, 3.

Then the underwater spark occurs between 4 and 7 through the water, 5. The pressure wave of the underwater spark immediately destroys the thin Mylar foil and presses the metal sheet, 7, with high power (this results in extremely small electrical resistance) against the electrode, 10. If the auxiliary triggering circuit, consisting of capacitor, 2, spark gap, 3, and electrodes, 4 and 7, is designed for low inductance, a resonance frequency of 2.5 megacycles is available. The contacting procedure has a total duration of less than 100 nanosec and the jitter in the case of switching 240 kamp is below 100 nanosec too.

All parts of the described heavy current switch are easily replaced. The main advantage of this indirect fired gap may be that owing to the applied Mylar foil undesired breakdown rarely occurs.

5. The Photography of Discharge Sound Shocks and Its Technical Application

a. *High Speed Photography of Shock Waves in Water*

Fundamentally, shock wave sources of the type referred to emit spherical radiation. Since, however, for most useful purposes it is desirable to operate with directional radiation giving a coherent wave front, attempts have been made to arrange the spark at the focal point of a concave mirror. The results of such focusing were not successful because the distances from the points

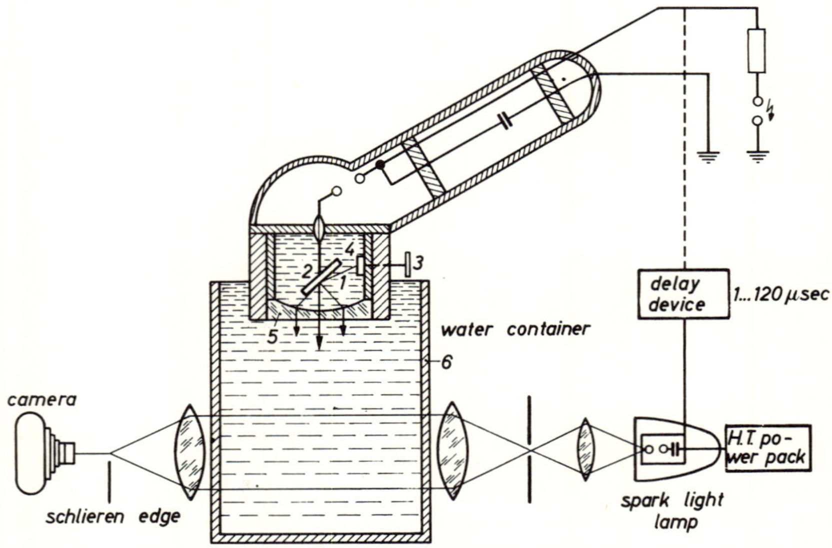

FIG. I5a-1. Underwater spark gap in the form of a sounding head shown in a test arrangement for schlieren-optical photography of spark sound waves by spark light. 1–resilient metal blade as electrode; 2–counter electrode, 3–adjusting screw, 4–liquid, 5–sonic lens, 6–wall covered with absorbing material.

FIG. I5a-2a. Sound shock in water (schlieren photo) during start at the surface.

of the reflecting surface to the receiving object were all different and this led to loss of coherence of the wave front. Thus the Fourier spectrum and consequently the frequency content were different at the receiving point from at the source.

Improved focusing was obtained with an acoustic or sonic lens. With such a lens, the difference between the sound velocity in the lens material and, for example, water effects the desired emergence of a coherent-front wave at the plane side of the lens. Interference considerations are out of place since the shock has practically an infinitely flat minimum that does not interfere. The principle just discussed was materialized in an arrangement, illustrated in Fig. I5a-1, that very closely resembles in its form a medical sounding head. Herein the liquid spark gap is again 1–2 and is adjustable by button 3. The spark gap is immersed in liquid, 4, which, if desirable, can be under pressure.

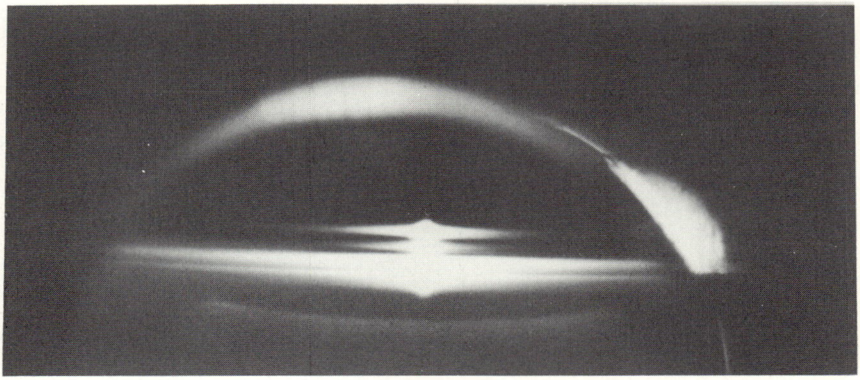

FIG. I5a-2b. The same sound shock about 5 μsec later. The slightly arced form is clearly visible.

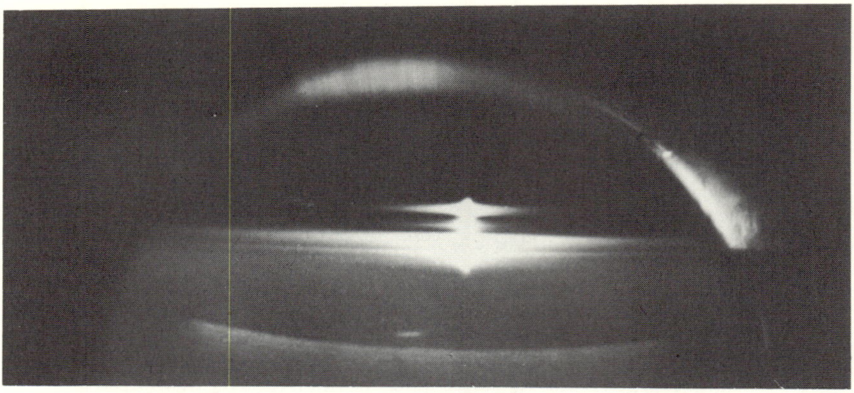

FIG. I5a-2c. The same sound shock about 15 μsec later. It can be clearly seen that it has hardly changed its profile against Fig. I5a-2b.

The sonic lens, 5, is arranged in the bottom of the sounding head and serves for focusing the radiation in a parallel sided beam. In order to prevent disturbances of the sound field, the inner wall of the shell is covered with an absorbing material, 6, that consumes the lateral radiation. The high voltage impulses are impressed on electrode 2 by way of a high tension insulator and a series spark gap from the charging capacitor.

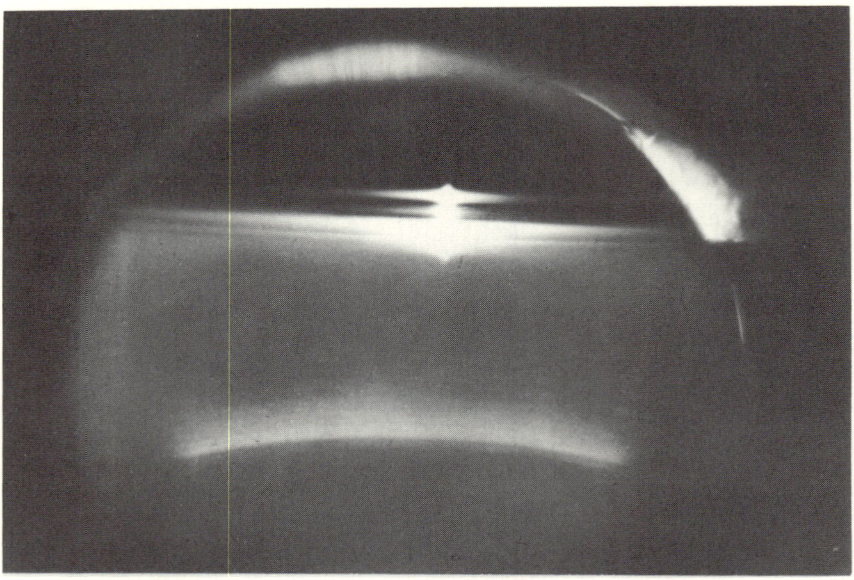

FIG. I5a-2d. The sound shock returns after reflection from the bottom of the vessel. Observe the mirror image of the wave profile as against Figs. I5a-2a–2c.

Instead of using plexiglass as construction material for the lens it is also possible to build a lens by confining a liquid that has a sound velocity differing from that of water under pressure between 2 thin plastic diaphragms. It is possible to construct a biconvex liquid lens with selectable curvature radius and desirable refraction power. Carbon tetrachloride (specific gravity 1.58/cu cm and sound velocity 920 meters/sec) is well suited as a lens liquid. The effect of shock wave straightening and also of concentrating with different lenses becomes apparent from the schlieren photographs made with such apparatus and published in a paper of Früngel and Thorwart [59].

FIG. I5a-2e. The same reflected sound shock. The profile of the wave remained unchanged during the entire time. Thus the shock traveled as a coherent "package" through the liquid owing to the fact that interference was impossible.

For taking the pictures the shock sound source with lens was operated in water in a 20 liter vessel with plane glass walls. The electric shock released, at the instant of the discharge, a tube controlled delay device having a variable delay range between 1 and 120 μsec. The initial impulse, given after the delay time, triggered a flash lamp of 1 μsec flash duration. The whole arrangement is indicated in the lower portion of Fig. I5a-1. At triggering, an exposure spark occurs and between each picture and the subsequent one the delay is advanced by a repeated constant value, e.g. by 5 μsec. Thus the exposure series of Figs. I5a-2a–2e corresponds to a frame sequence frequency of 200,000/sec.

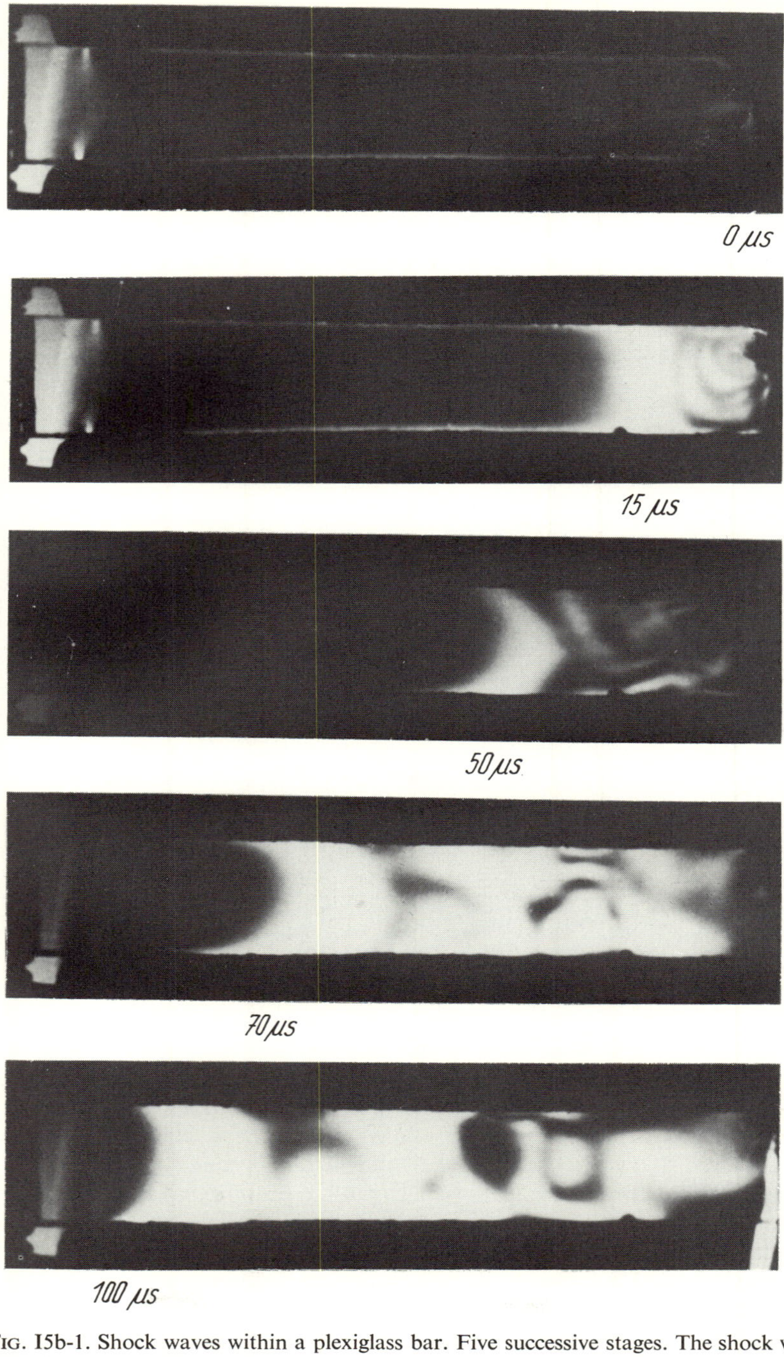

FIG. I5b-1. Shock waves within a plexiglass bar. Five successive stages. The shock wave was produced by a liquid spark.

b. *Shock Waves in Stress Analysis and Schlieren Photography*

The previous sections have shown that a shock wave of a liquid spark resembles a microdetonation. An advantage over actual explosions, however, lies in the fact that such shock waves can be repeated practically as often as desirable.

In strain optics it is often necessary to study a certain shock wave image after various delay durations. Figure I5b-1 shows such a series of recordings. A plexiglass bar is pushed from the right side by means of a steeply rising liquid spark. Simultaneously therewith a light flash of 1 μsec duration is released by way of a delay device. Now, the capacitor discharge action, incidental with a shock wave release, is arbitrarily repeated and between subsequent releases the delay time is changed so that the travel time of the shock wave between the instant of release and the instant of recording increases with each subsequent release. Thus the subsequent recordings in the illustration show the shock wave after 0, 15, 50, 70, and 100 μsec travel times. It is each time another shock wave, but owing to similar energy output in each shock and similar point of action without destruction to the object, it is practically possible to take a whole film of recordings at various instants of delay by this single frame method [59].

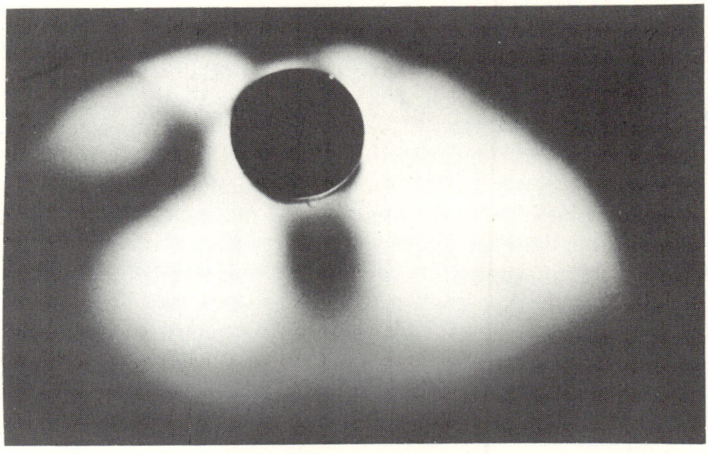

FIG. I5b-2. Shock wave pattern in plexiglass. One sound shock struck a hole (large black spot). The image of the hole is still visible in images of subsequent shock wave advances, but becomes slowly blurred as it travels.

Analogously to this method, Fig. I5b-2 shows the occurrence in a circular perforated plexiglass disc. The shock, introduced from above at the center of the picture, expands circularly. The hole in the plate travels along with the shock wave as a stress-free shadow figure and is only gradually filled with the mechanical shock force strains [139].

The "Strobokin" method about to be described affords the possibility of photographic analysis of dynamic phenomena in liquids and of optical strain models. With the help of short time exposures and a spark shock wave source, Früngel, Bailitis and Thorwart delivered the paper [59].

Optical strain models of transparent material have been used in strain optics for a long time. Shock waves and rapid changes in strain conditions spread rapidly into the medium to be observed. In the method to be discussed here, such spreading strains are photographed by monochromatic light flashes of 0.5 μsec duration. Using the schlieren method, the progress of shock waves within liquids can also be investigated by this method. A spark shock wave source has been designed for producing shock waves having an exactly measured energy content within liquid and solid plastic phantoms in periodic succession. The "high-frequency flashing source," described elsewhere in this volume (Früngel), is utilized as a light source for this kind of investigation of dynamic shock waves.

The examination of dynamic shock phenomena in gaseous, liquid and solid media is limited by two fundamental requirements:

(1) The need to use a shock wave source capable of delivering the shortest possible intensive sound pulse with steep front.

(2) The need to use a light source of great intensity and the shortest possible flash duration.

A detonation is much used as a shock wave source meeting the above conditions, but the liquid spark has now proved very suitable as a new form of shock wave source. The discharge of a capacitor across a liquid spark gap behaves as a sound which is similar to the firing of an explosive charge. The pressure amplitude of such a spark wave is very high; the discharge of a capacitor of 0.25 μf charged to 7 kv gave a pressure amplitude of 900 kp/sq cm at a distance of 1 cm. The pressure rise to one third of its maximum value occurred in about 0.1–0.2 μsec. If the liquid volume is bounded by a sound transmitting solid boundary, this steep sound pulse can be transmitted into solid or gaseous media. The capacitor discharge occurs via a triggering spark gap within the liquid spark; this renders possible the control of the liquid spark by the triggering spark which is itself controllable. Thus, periodic and very exactly spaced sound pulses are obtained. The control of the illuminating flash occurs simultaneously with the control of the sound pulses, and with the aid of a delay line a controllable delay of the instant of illumination can be obtained. In this way an advancing shock wave can be observed at any desired instant.

By using a periodic shock source and continuous variation of the delay time (i.e. instant of observation) every phase of the dynamic process can be rendered visible and may be photographed. If the individual pictures so obtained are arranged in a suitable series, the actual process can be shown cinematographically.

As an intense light source, a spark discharge in an argon atmosphere is used—the same as in the commercially available Früngel stroboscope and high frequency flash apparatus. These light flashes have a duration in the visible spectral region of 1 μsec. When monochromatic light in the blue spectral region is used for optical strain photographs an even shorter illumination time is achieved, 0.5 μsec.

For the examination of shock waves of high propagation velocities (in solid media up to 2000 meters/sec) such short illumination times are a necessary condition for obtaining sharp pictures, as well as for observing shock waves in liquid or gases by schlieren or shadow methods. The high ignition speed of the Früngel light source facilitates observation of any desired phase of the motion.

Similar pictures as derived from this method are shown in Figs. I5a-2a–2e.

c. *The Underwater Spark as Flash Source with Quenching by Shock Wave*

The underwater spark has been known for many years in the field of spectroscopy as a source of light in the continuous ultraviolet spectrum [57]. Relatively weak discharges in fast succession were used for that purpose. The usefulness of single heavy current discharges from a 6 μf capacitor, charged to 25 kv, for a flash light source of high light intensity was investigated by H. C. Early and E. A. Martin (University of Michigan, Ann Arbor) [58]. In their experiments a wire of about 2/100 mm dia was suspended between two electrodes at a spacing of 1.5 cm so that the discharge of the capacitor resulted first in an explosion of the wire. Through the channel thus formed and in the space thus closely defined, the main quantity of the capacitor discharge now passed.

From measurements of the refractive index which changes with the pressure wave of the underwater discharge, and the density of the water, it could be concluded that a pressure of about 5000 atm was created in the cylindrical pressure wave issuing from the discharge channel. In the light emanating from the underwater spark, no line structure could be observed between 3400 and 6900 Å. Under the assumption that the radiation is that of a blackbody the resulting light temperature figures to above 30,000°K. In the photographically effective range, the light output radiated into the whole space lies at about 430,000 watts.

The duration of the light impulse was about 24 μsec, for many purposes, too long. By a clever move it was possible to reduce the duration to about 2 μsec: To that end, the pressure wave of the underwater spark was forced to shatter a glass plate so that the sand-fine splinters blocked the light path again to a large extent. The whole arrangement is called a "shatter shutter." That the quantity of light radiated from a heavy current underwater spark is sufficient for many purposes is evidenced by the experiment in which a perforated diaphragm of 1.5 mm dia was placed in front of the spark gap and gave practically a point light source. The amount of light emanating from this opening sufficed to record a shadow image of an object on a film at 8.7 meters distance from the hole diaphragm. There was no lens in the path of the light.

J. Material working by high-frequency capacitor discharges (spark erosion)

1. Facts and Purposes of the Mechanism of Spark Erosion

The fact that a liquid spark can be given desired mechanical and dynamical properties through the design of the discharge circuit was made public in Germany in 1943 by Früngel [76] in a patent description. In spectrochemistry for metal analysis particularly, the storage capacitor spark discharge with slow oscillation due to a standardized inductivity is used to transport particles of matter from the metallic surface into the spark with subsequent ionization in the spark itself. This method is applied in all spark analytical methods.

In modern fully automatic equipment of this kind (such as that manufactured by the Beckmann Corp.) the spark circuit is electronically stabilized; that is to say, spark energy and frequency are constant (the latter is determined by the uniformity of the line frequency). Behind the spectral analyzer a number of multipliers are adjusted to the exact place of importance in the spectral length indicating the metal characteristic. These multipliers operate directly onto the electrometer. Such a unit prints fully automatically the total quantitative result for, e.g., given steel sample within a few minutes. Aging of the photo-electric devices is kept constant by subjecting them to intermediate radiation from a xenon ultraviolet lamp, so that no regeneration can occur in the working intervals and cause false readings.

These automatic units prove that the transport of particles of matter by spark discharges is an extremely constant phenomenon with energy being the determining factor. The above-mentioned analysis results show a consistency of 1% and better. Quantitative research into the quantities of metal transported in the electric spark was recently carried out by Bunge [77], who used zinc cathodes in his experiments. The standard work in this field of research was performed by Kaiser and Wallraff [78]. The specific evaporation loss on two cathodes is 5×10^{-5} gram per coulomb. The evaporation loss is proportional to the total electricity which has flowed through the spark gap. When the discharge oscillates, then only every half wave to which the electrode in question is a cathode counts. From such information, reasonable estimates can be derived as to the life of spark gaps used as switches and for purposes of emitting light pulses. Unfortunately, the consistency for tungsten is not yet sufficiently known.

In the case of underwater shock wave sources with liquid sparks, where material destruction is caused by acoustic dynamic effects, as described in Chapter I, spark erosion is far greater than the cathode disintegration. It can only be eliminated by making that part of the spark discharge which generates the most substantial part of the acoustic energy into a sliding spark on nonmetallic materials, and by starting it not directly at the metal surface, but in underwater spark gaps (see Chapter I, Section 3).

2. Electro-Erosion Machining of Metals

A. L. Livshits published an extensive paper wherein he described the electro-erosion machining of metals [270].*

a. *Dependent Pulse Generators*

Dependent pulse generators, mainly of sign-reversal current and voltage pulses, form the basis of the electro-spark machining process as described in [61-78].

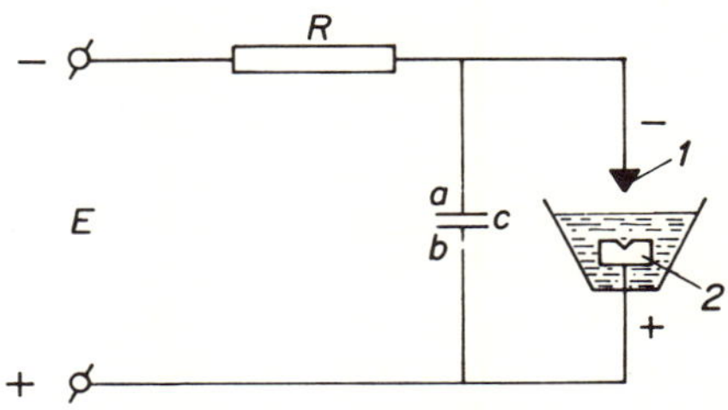

Fig. J2a-1. Diagrammatic circuit of *RC* type relaxation generator. By permission from "Electro-Erosion Machining of Metals," A. L. Livshits, Butterworths, London, 1960.

In 1943, B. R. and N. I. Lazarenko proposed the use of the ordinary relaxation (capacitance discharge) *RC* type of generator circuit for machining metals (Fig. J2a-1). The capacitor C is charged through the resistor R, from the dc source E. When the capacitor voltage reaches the value U_0 required to form conducting bridges, a discharge occurs through the circuit b-2-1-a, in which are included the tool electrode, 1, and the work piece, 2. The tool electrode forms the cathode and the work piece the anode, i.e. direct polarity is used. Since the resistance *R* of the charging circuit is considerably more than that of the discharge circuit, discharge takes place very much more rapidly than charging so that a large amount of power is released in the erosion gap in a very short interval of time. After discharge the capacitor C is recharged and the cycle is repeated.

Relaxation generators, in view of the presence of unavoidable inductance in the leads, produce sign-reversal pulses, i.e. in addition to the forward half-wave, the pulse contains a smaller reverse half-wave. This leads to reduced stability of the process and to additional tool wear. The frequency

* Chapter J, Sections 2a and 2b is reprinted from A. L. Livshits, "Electro-Erosion Machining of Metals," Butterworths, London, 1960, pp. 14–20, 51, 74–81.

and parameters of the current pulses depend on the open circuit voltage U_0 of the discharge, the resistance R of the charging circuit, and the capacitance of capacitor C. The starting voltage is fixed and maintained by the automatic feed control, but the latter can respond only to the mean value of U_0 or the voltage drop corresponding to it in R. Therefore U_0 can vary over a very wide range from the set value. When U_0 is low (or when the distance between the electrodes is small), the pulse frequency is correspondingly increased, and the amplitude and duration decreased; when, on the other hand, U_0 is high (or the distance between the electrodes large), the frequency falls, the amplitude increases, and the duration of the pulses falls slightly. Since the primary variations in U_0 are the direct cause of changes in the physical conditions in the erosion gap, there is an obvious relationship between the parameters of the pulses generated and the conditions in the erosion gap, which justifies classifying relaxation generators, in the dependent group. The disadvantage of dependent generators is the passage across the erosion gap of pulses of variable duration, amplitude, and frequency, which do not correspond to the optimum conditions for metal removal for the given values of R, C and E.

The resistance of the charging circuit controls the pulse frequency and the over-all amount of energy supplied from the source to the working gap. As the resistance R is decreased, the pulse recurrence frequency increases and the pulse interval decreases, since the charging cycle for the capacitor is shortened. At some critical value R_{cr}, near to the resistance of the discharge circuit, the interval between pulses disappears and, in place of the interrupted pulse current across the erosion gap, interrupted dc flows. Since this does not satisfy the basic condition for electro-erosion machining, i.e. pulse supply energy, the process of metal removal ceases. Since, for a given supply voltage, the value of R determines the power which can be drawn from the supply, the value R_{cr} will correspond to the limiting value of power P_{cr} to which the relaxation generator is limited in the production of pulse currents. To this limit of power there corresponds a limit in the rate of metal removal—which cannot be exceeded in the most powerful electro-spark equipment*—of about 600–900 cu mm/min at frequencies of 100–300 pulses per second, in copy drilling work.

The transformation to pulses of the most commonly available ac at mains frequency, by relaxation generators, is carried out in a three-stage circuit: mains ac–dc–pulse current. The presence of three stages of transformation, each of which involves a loss of energy (particularly in the stage of transforming dc to pulse current), the use of sign-reversal pulses instead of unidirectional, and the marked variations in the pulse parameters even under steady conditions, lead to a drastic reduction in the efficiency of relaxation generators, increased power consumption in the process and a reduction in the accuracy of machining.

Thus dependent relaxation pulse generators possess a number of substantial

* The Russian Lenkarz machine is powered by a type RC generator of about 16 kw at 700 μf, and has a metal removal rate under particularly favorable conditions of 900 cu mm/min.

disadvantages (production of sign-reversal pulses in place of the required unidirectional pulses, nearness to the limits of dc which prevents increase of output), which give rise to high energy consumption in electro-spark machining and to rapid tool wear. The advantages of relaxation generators are their simplicity, relatively small size for small power outputs. Moreover, it must be borne in mind that they can also produce, by simple means, high frequency pulses at the lowest power outputs.

These positive features of dependent relaxation generators at present determine the rational field of application, in equipment designed solely for low power conditions (drilling small diameter holes in fuel equipment, fine screens, etc.), and in cases where efficiency is not important and the considerable tool wear can be directly compensated by renewing the working part of the electrode; for example, by feeding electrode wire through a guide.

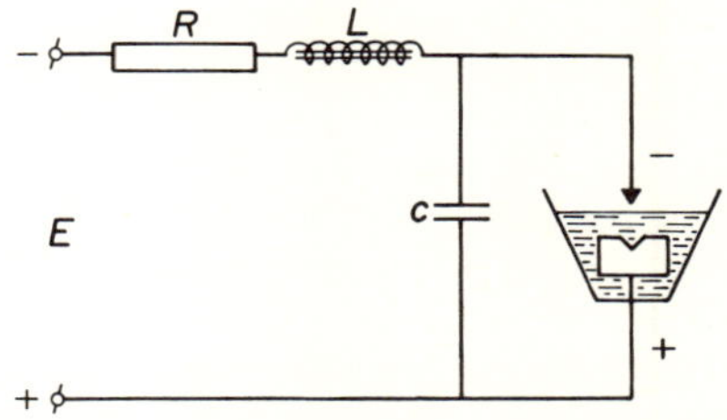

FIG. J2a-2. Diagrammatic circuit of *RCL* type relaxation generator. By permission from "Electro-Erosion Machining of Metals," A. L. Livshits, Butterworths, London, 1960.

Efforts to remove some of the faults of type *RC* relaxation generators led to the introduction of a number of new elements in the circuit. One of the useful proposals was the inclusion in the charging circuit of an inductor (Fig. J2a-2). The effect of the inductor is to cause oscillatory charging of the capacitor. The conditions for this are represented by the inequality

$$\epsilon^2 = \frac{4L}{R^2C} > 1$$

where ϵ is a dimensionless parameter of the circuit,
L is the inductance in henries,
C is the capacitance in farads of the charging circuit, and
R is the resistance in ohms of the charging circuit.

When the above condition is satisfied, the capacitor C is charged to a higher voltage than the source. Under steady conditions the maximum voltage across the capacitor cannot exceed twice the supply voltage.

The optimum conditions correspond to charging the capacitor to a voltage equal to that of the supply. This produces an increase in the amplitude and energy of the current pulses, and consequently also in output. The efficiency of a type *RCL* generator, in comparison with that of a type *RC*, is 20–25% greater.

The value of the inductance added to the circuit is found from the relationship

$$L = 6.25R^2C$$

Further increase in the value of the inductance, beyond this value, is of no use, since the efficiency is practically unaffected, whilst the size of the inductor increases. The inductor is made in the form of a cored coil, with a winding designed to carry the effective charging current, and the required number of turns for the given value of L. The laminated magnetic circuit is open, leaving a gap.

When there is a sudden short circuit, a reverse voltage is set up across the capacitor which can be about 15 times the supply voltage. Thus for a supply $E = 200$ volts, a short circuit current of 30 amps, $L = 0.12$ h and $C = 100\ \mu f$, the reverse voltage is 3,750 volts, which endangers the equipment insulation and could cause a capacitor breakdown. To avoid this it is essential to select a ratio between R and C, for a given value of L, such that the reverse voltage due to a short circuit does not exceed the permitted maximum for the type of capacitor in use. Thus, for example, if the supply voltage is $E = 220$ volts, it is necessary to use capacitors for working voltage of $2E = 400$ volts. For [the Russian] Type KBGM capacitors, which are the nearest standard type, the working voltage is 600 volts, and the maximum test voltage, 1,800 volts. Thus, for this type of capacitor, the resistance R must be such that the reverse voltage factor does not exceed $1800/220 = 8$. The relationship between R and C is usually selected so that the reverse voltage does not exceed 4–5 times E.

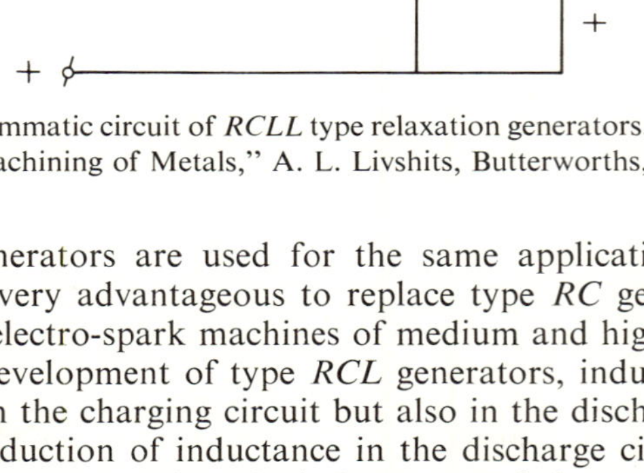

FIG. J2a-3. Diagrammatic circuit of *RCLL* type relaxation generators. By permission from "Electro-Erosion Machining of Metals," A. L. Livshits, Butterworths, London, 1960.

Type *RCL* generators are used for the same applications as type *RC* generators. It is very advantageous to replace type *RC* generators by type *RCL* in existing electro-spark machines of medium and high power.

As a further development of type *RCL* generators, inductors were introduced not only in the charging circuit but also in the discharge circuit (Fig. J2a-3). The introduction of inductance in the discharge circuit alone is ineffective. On the contrary, when the inductance L_2 is increased considerably, the rate of metal removal falls and tool wear increases. However, when inductances L_1 and L_2 are included simultaneously, the performance of the generator is improved. This phenomenon can be explained as follows:

The inductance L_1 (steel cored coil) in the charging circuit increases the initial discharge voltage and hence the unit pulse energy and the charging circuit efficiency. The inductance (coreless) in the discharge circuit increases the duration of the discharge, but at the same time somewhat reduces the pulse amplitude, since the impedance of this circuit is increased. Increase in the duration of the discharge increases the initial store of energy in the magnetic field of the coil L_1 as the current in the charging circuit increases during discharge. This increase in the stored energy in the coil in turn leads to an increase in the initial voltage and amplitude of the discharge current. Thus, the decrease in the amplitude, due to the inclusion of inductance in the discharge circuit, is counterbalanced by the higher initial discharge voltage, due to the increased stored energy in the coil. In all, the pulses across the working gap will have a longer duration at about the same amplitude, i.e. the pulse energy is increased. The resultant increase in rate of metal removal, compared with type *RCL* generators, is about 10–15%. This is accompanied by a reduction in relative tool wear of about 10–15%. The electrode polarity, as for type *RC* and *RCL* generators, is negative.

Type *RCLL* generators are used for the same applications as type *RCL*. Since the addition of inductance to the discharge circuit is not difficult it is advantageous in general to use the *RCLL* circuit, particularly for machining certain alloys and for multihead circuits operating a large number of tools simultaneously. The inductor L_2 in this case constitutes the wear and cutting rate control and can maintain the required stability over the entire set of electrodes.

Another type of generator is the *RCR* relaxation generator, which is of fundamental interest since with its use in 1948, M. M. Pisarevskii [270], for the first time machined metal using unidirectional pulses of increased duration and achieved the effect of considerably reducing tool wear. This type of generator is still dependent, but in contrast to the type *RC* generator, the discharge circuit includes a resistance r, additional to the existing circuit resistance r_c (Fig. J2a-4). The value of r is selected so that the capacitor discharge becomes aperiodic. * The condition for aperiodic discharges is given by the following relationship between the parameters of the discharge

* Aperiodic pulses can be obtained in a relaxation generator by another method. If the initial discharge voltage is reduced to a value near to the voltage drop across the erosion gap in an arc spark discharge, the reverse half-wave is reduced almost to the point of disappearance. This is explained by the fact that, when the initial discharge voltage is reduced, the discharge current falls.

Reduction of the current causes the erosion gap resistance to increase, which is equivalent to adding a further resistance to the circuit, and causes the oscillatory current to become aperiodic.

However, at the same time, the source voltage remains near to the voltage drop in the erosion gap, and the capacitor almost stops charging and becomes idle. This apparently explains the uselessness of including a capacitor in the so-called low voltage electro-spark (more accurately electro-contact) method of machining in a liquid medium on dc.

circuit:

$$r_c + r \approx 2K\sqrt{\frac{L_r}{C}}$$

where L_r is the self-inductance of the discharge circuit in henries,
C is the capacitance in farads, and
K is a coefficient accounting for the effect of the erosion gap.

In a linear circuit without an erosion gap, $K = 1$. The erosion gap has a resistance which varies with current, and one cannot use normal methods to determine the critical resistance at which the discharge becomes aperiodic. A coefficient K is therefore introduced in the formula, which, as more detailed analysis shows, is equal to about 0.80–0.85. In other words, the critical resistance of the circuit is smaller with the erosion gap than without. As the inductance of the circuit increases, or the capacitance decreases, there is an increase in the amount of resistance which must be added to obtain unidirectional aperiodic pulses. Since, as the capacitance decreases and the resistance increases, the pulse energy rapidly falls, and consequently also the rate of metal removal, it is necessary in type *RCR* generators to increase the capacitance considerably and keep the critical resistance correspondingly low. Type *RCR* generators in use have 2–3 times the capacitance of type *RC* generators of medium power; the total capacitance reaches 700–1000 μf.

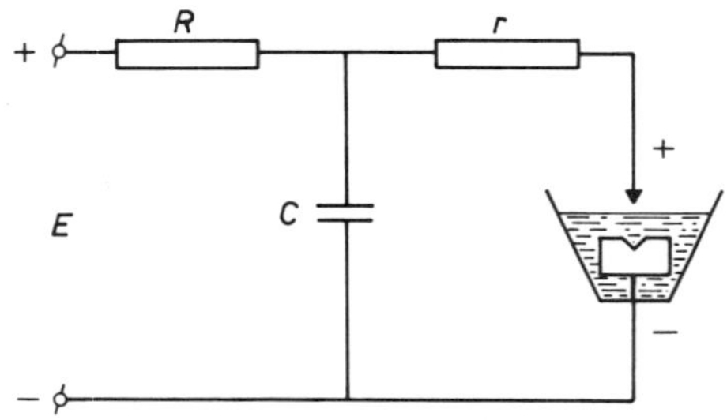

FIG. J2a-4. Diagrammatic circuit of *RCR* type relaxation generator. By permission from "Electro-Erosion Machining of Metals," A. L. Livshits, Butterworths, London, 1960.

The increase in capacitance and the introduction of ohmic resistance produces aperiodic pulses which are longer in duration and smaller in amplitude than in electro-spark machines with type *RC* generators. The discharge becomes predominantly arc type. As a result the thermal balance of the electrodes is altered and it is found, for example, that copper electrode tools wear to a much smaller extent when connected as the anode; the workpiece then forms the cathode, i.e. it has the opposite polarity to that used in electro-spark machines.

The main advantage of *RCR* generators is the reduction of tool wear, which is several times smaller than in electro-spark machines using dependent type *RC*, *RCL* or *RCLL* generators.

The important disadvantage of type *RCR* generators, in comparison with type *RC* and *RCL* generators are as follows:

(1) Considerable narrowing of the limit of continuous dc. As the discharge circuit resistance increases, the transition from pulse to continuous discharge occurs at higher resistance and lower currents in the charging circuit. Since the discharge circuit resistance is many times greater with *RCR* generators than with *RC* or *RCL* generators, this transition occurs considerably earlier. This defect prevents increased power and output rates being obtained in machines with type *RCR* generators.

(2) A marked drop in efficiency of the discharge circuit, due to losses in the added resistance. The maximum rate of metal removal is reduced by 5–10 times with *RCR* generators, due to the losses just mentioned.

(3) Low pulse recurrence frequency, due to the need to introduce high capacitances. The surface quality is consequently poorer than is obtained on electro-spark machines without added resistance.

(4) Large size of the capacitor bank.

The rational field of application for type *RCR* generators is indicated by their ability to maintain low tool wear, which is particularly important in machining parts of great accuracy and also in drilling deep holes with stepped electrodes. The dependent *RCR* generator of unidirectional pulses forms a transitional type between dependent generators of sign-reversal pulses (electro-spark method) and independent generators of unidirectional pulses (electro-pulse method).

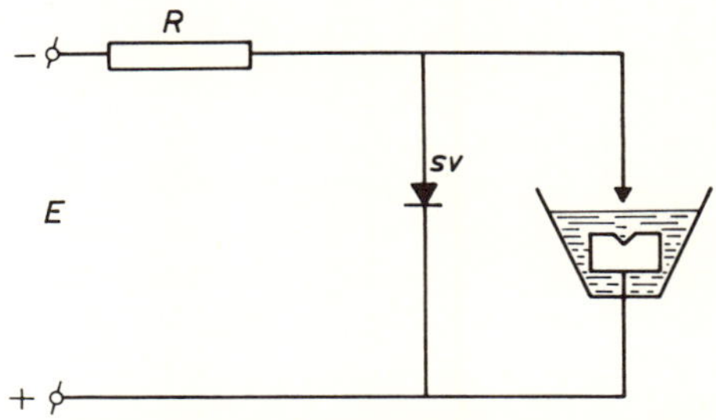

FIG. J2a-5. Diagrammatic circuit of *RV* type relaxation generator. By permission from "Electro-Erosion Machining of Metals," A. L. Livshits, Butterworths, London, 1960.

Efforts to prevent the passage between tool and workpiece of currents in the wrong direction have led to numerous attempts to solve the problem by means of various types of basic circuit.

One of these circuits has already been described (Fig. J2a-4). In this circuit the reverse half-wave is suppressed and at the same time the pulse duration is increased.

Circuits are known, in which a rectifier is connected in series in the discharge circuit. However, they have not been adopted, because there are no rectifiers available with sufficiently high efficiency, bearing in mind the low voltage, high current and relatively high frequency.

Figure J2a-5 shows the circuit of a type *RV* dependent generator which reduces or eliminates the reverse half-wave of current and voltage without affecting the duration of the forward half-wave. Use is made in this circuit of the self-capacitance of semiconductors (e.g. selenium) in the rectifier. An element of the working surface of a selenium rectifier has a self-capacitance of the order of 0.013 μf/sq cm [273]. If the selenium rectifier unit, designed for a reverse voltage equal to the supply voltage, is connected to oppose the latter, then during the charging period the self-capacitance of the rectifier will be charged, the rectifier being shunted by its very high resistance in the nonconducting direction.

During the last 2 years additional sophisticated developments have been made. The patent [951] describes, for instance, a device which automatically controls the distance between the electrode and the work piece which is to be machined by fully transistorized motion control.

b. *Electro-Spark Machines*

The controlled factor in electro-spark machines with dependent relaxation pulse generators is normally the current in the charging circuit which is functionally related to the discharge circuit current. The maximum value of the mean charging current of electro-spark machines does not exceed 40–50 amp in the largest machines. Consequently, the entire capacity range of electro-spark machines can be handled by two sizes of controller, differing in control motor capacity and magnetic amplifier size. Figure J2b-1 shows a circuit for a controller with parameters suitable for an electro-spark machine with a current range of 1 to 60 amp. The current windings, W_T, of the differential amplifier are connected into the charging circuit and the voltage windings, W_s, across the erosion gap (Fig. J2b-2 and 3).

The resistance and number of turns in coils W_s must be selected so as to have no influence on the discharge circuit (high resistance winding designed for low currents). The feed speed is set by adjusting the voltage on coil W_r. It is possible to place a series coil in the discharge circuit, but this increases the inductance of the circuit, which is only permissible when *RCLL* type pulse generators are used.

Electro-spark machines are suitable for spark machining-through of blind holes of arbitrary shape in any conducting materials, preferably those difficult to machine. The sinking of forging and stamping dies and also press tools, and the boring of holes in hardened steels and hard metals can be carried out on the machine. Costing should first be carried out carefully on any given operation, since it is not always most economical to use this machine. The relative efficiency increases when machining parts in difficult-to-machine alloys, when the part has a complicated shape, or when the operation does not lend itself to conventional machining.

Figure J2b-4 gives a general view of the machine. The cast, box section, pedestal 1 forms the base of the machine and carries the frame, 2. The transformer-rectifier set is placed inside the pedestal. The frame is welded from angle sections and faced with steel sheets.

On the front face of the frame are fitted the guide, 3, which is of swallow-tail design and on which the carriage, 4, travels, the supporting table, 5, the transverse and longitudinal transverse slides, 6 and 7, and the plates, 8, on which is fixed the tank of working fluid.

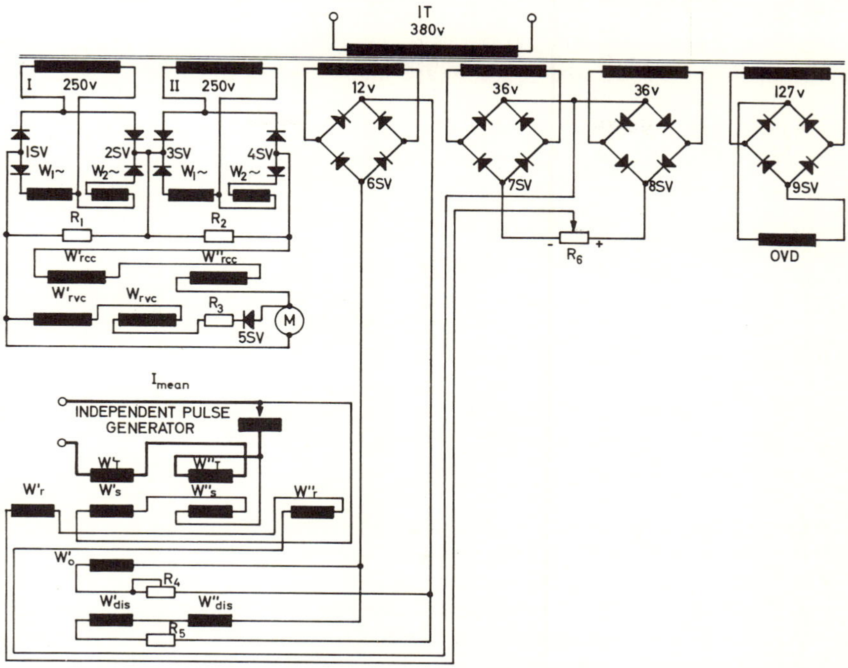

FIG. J2b-1. Circuit for an automatic feed controller using a differential magnetic amplifier. The circuit parameters are designed for a feed motor capacity of 25 watts at 110 volts. R_1–560 ohms, 40 milliamp; R_2–350 ohms, 400 milliamp; R_3–300 ohms, 100 milliamp; R_4–500 ohms, 200 milliamp; R_5–700 ohms, 100 milliamp; R_6–500 ohms, 200 milliamp.

Number of plates in the arms of rectifiers: 1SV–16 (type VS 45); 2 SV–16 (type VS 45); 3 SV–16 (type VS 45); 4 SV–16 (type VS 45); 5 SV–6 (type VS 45); 6 SV–1 (type VS 45); 7 SV–2 (type VS 45); 8 SV–2 (type VS 45); 9 SV–6 (type SV 18).

OVD–motor excitation winding. By permission from "Electro-Erosion Machining of Metals," A. L. Livshits, Butterworths, London, 1960.

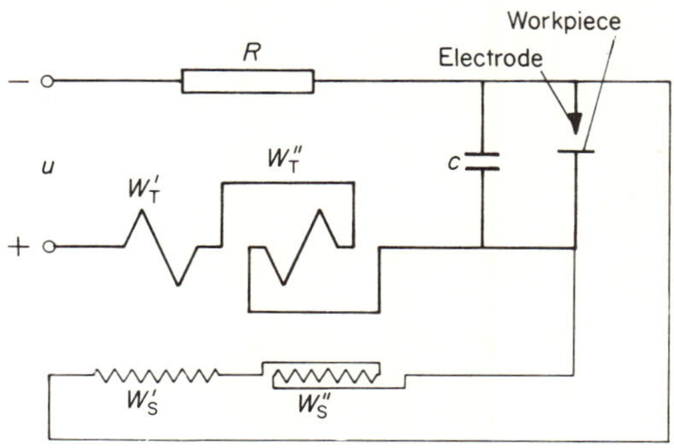

FIG. J2b-2. Controller circuit in electro-spark machines. By permission from "Electro-Erosion Machining of Metals," A. L. Livshits, Butterworths, London, 1960.

Inside the front face of the frame are fitted the capacitors and on the side faces the instrument panel, 9, the throttle, and pump, 10, for passing fluid through the electrode; the resistors are placed on the lid of the frame. The rear of the frame is filled by door 11, on which is mounted the electrical connection panel.

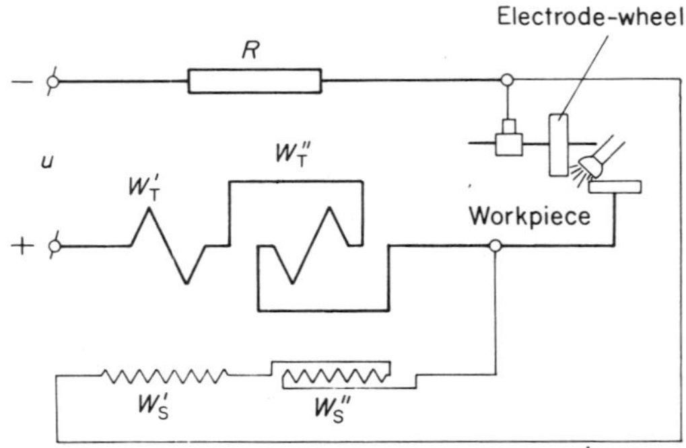

FIG. J2b-3. Controller circuit in anode mechanical machines. By permission from "Electro-Erosion Machining of Metals," A. L. Livshits, Butterworths, London, 1960.

On the upper part of transverse slide 6 is fitted the feed control, consisting of electric motor 12 and reducing gear 13, transmitting the feed motion through a leadscrew and nut to the tool slide, 14. Slide 14 ends in shank 15 with a No. 2 Morse cone to fix the arbor with a clamp for the tool electrode. The table is lowered into the fluid tank by means of electric motor 16, through a single stage reducing gear. The carriage and slides then move along the guide into a position at which the liquid takes up the required level over the work piece.

Gases and vapors are removed by means of the pipework, 17, running around the rim of the tank and having a telescopic connection, 18, to exhaust the gases directly from the working zone. The rim exhaust system is connected through special pipework to the shop ventilation system. Valves, 19, and 20 are provided to drain out used fluid, and the tank can be cleaned of slime through the side door, 21.

The electrical equipment of the machine (see Fig. J2b-5) consists of a source of dc (a dependent, relaxation type, pulse generator), the feed control, and the control instrumentation and safety devices.

The source of dc is a 3-phase selenium rectifier of the Larionov type. It is supplied via transformer Tr from mains at 380/220 volts (star connected transformer) or at 220/127 volts (delta connected transformer). The rectifier output is 4.4 kw, i.e. 220 volts at 20 amp and 5.5 kw is taken from the mains at a rectifier efficiency of 75%.

The machine uses a type *RCLL* dependent, relaxation-type pulse generator, the principles of which were described in Chapter J, Section 2a.

The generator discharge circuit consists of banked resistors R_1–R_6 and capacitors C_1–C_6 and an inductance, comprising two steel cored coils Dr_1

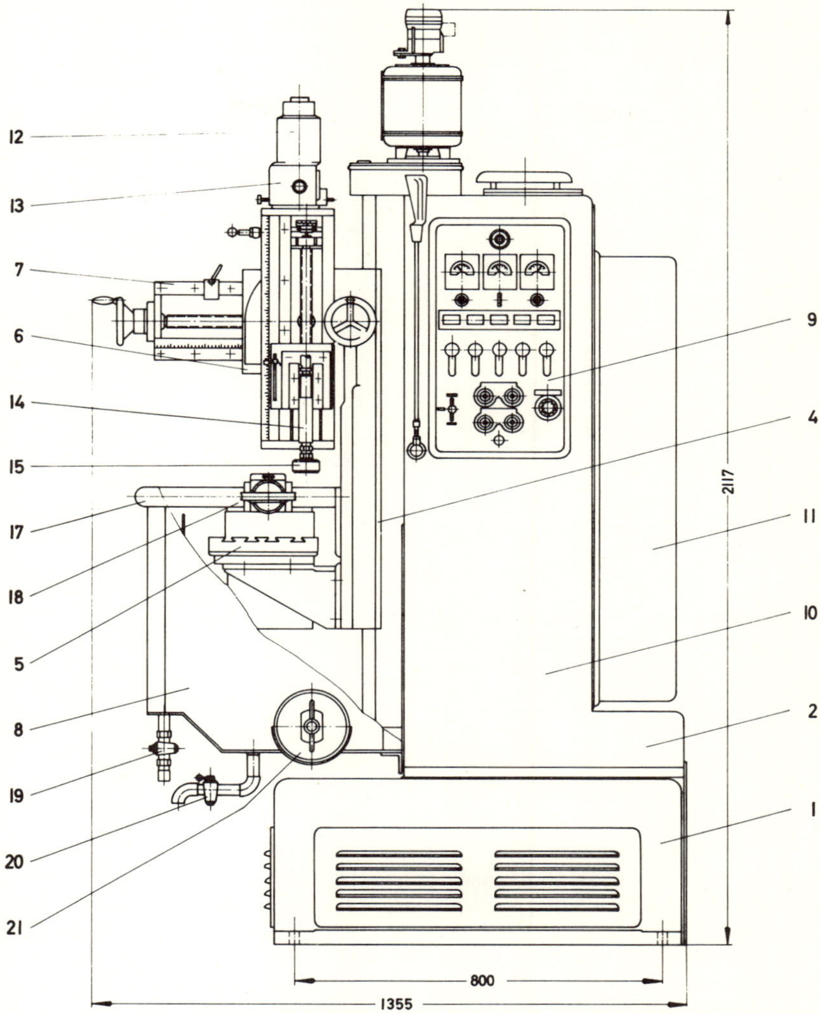

FIG. J2b-4. General view of type 4A722 electro-spark copy drilling machine. By permission from "Electro-Erosion Machining of Metals," A. L. Livshits, Butterworths, London, 1960.

and Dr_2 common to all settings. The total capacitance available is 200. 25 μf, the short circuit current 33 amp, and the nominal mean current 20 amp.

The feed control is connected through a very simple bridge circuit without intermediate amplification. The armature of the shunt wound electric motor, 4M, is supplied with the difference between two voltages: voltage E taken from potentiometer PP-1 and voltage U_r, across the current limiting resistance and coils Dr_1 and Dr_2. Voltage U_r is taken from resistance R_6 together with

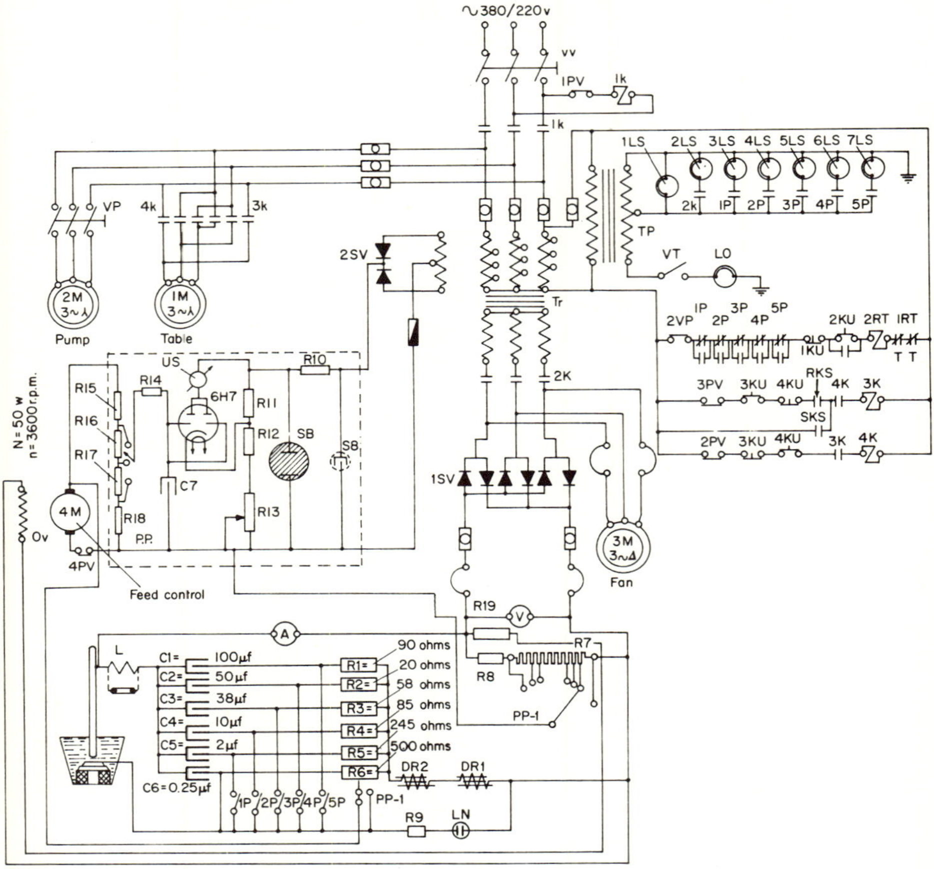

FIG. J2b-5. Outline of electrical circuit of type 4A722 machine, T_r, power transformer; TP, stepdown transformer; 1SV power rectifier; 2SV, supply rectifier for speed indicating circuit; 1P–5P, setting selector; C_1–C_6, working circuit capacitors; R_1–R_6 current limiting resistances; 1PV–4PV, starting switches; OV, feed motor excitation winding; 1K–4K, magnetic starters; LO, local lighting lamp; LN, neon lamp. By permission from "Electro-Erosion Machining of Metals," A. L. Livshits, Butterworths, London, 1960.

any of the parallel resistances R_1–R_5 in the circuit, through the double potentiometer *PP*-1.

When erosion is in progress $E > U_r$ and electric motor 4M rotates to bring the electrodes closer. At short circuit $E < U_r$ and the sign of the armature voltage in motor 4M is reversed, so that it withdraws the electrodes. Part of the supply voltage is dropped across resistor R_8, so that the nominal operating voltage for the motor (110 volts) is not exceeded.

The speed of the feed motion is determined by the slider potentiometer, PP-1. To determine the optimum feed speed, the machine is equipped with a speed indicator which operates as follows: The armature voltage on motor 4M, which is approximately proportional to its speed of rotation, is supplied, via the filter circuit R_{14}–C_7 to the grid of valve 6H7. A current flows in the anode circuit of this valve which is proportional to the feed speed, and this is shown on ammeter US, which is graduated in velocity units. To improve the accuracy of this device the anode circuit is supplied from a supplementary rectifier, 2SV, via voltage stabilizer SB and potentiometer R_{11}–R_{13}. The adjustment of the slider on potentiometer PP-1 gives the required range of feed speeds. When altering the position of the slider on PP-1, i.e. the values of voltage E and feed speed, the operator follows the movement of the needle on the meter, US. At the optimum feed speed for the machining conditions, the needle shows a maximum deflection. Further movements of the slider on PP-1 in the same direction lead to the needle indicating a lower feed speed. When the needle movement changes direction the optimum feed speed has been reached. Machining conditions are selected by pressing buttons 1P to 5P, switching in parallel the resistor-capacitor banks. At the same time the signal lamps, 2LS to 7LS, light up. In all, the machine has 32 electrical settings, producing sign-reversal pulses at recurrence frequencies within the approximate range of 250 to 1300 pulses/sec.

The table drive is powered by a reversible electric motor, 1M. The limiting table positions are fixed by means of end limit switches. To eliminate the effects of inertia when the table approaches its lower limiting position, the motor, 1M, is fitted with a speed control relay, RKS, so that the motor is breaked by reversal of connections.

Fluid is pumped through the hollow electrode by a pump driven by electric motor 2M.

Brief technical characteristics of a model 4A722 machine (Fig. J2b-4 and 5):

Machining area, sq mm	
minimum	2
optimum	1000, circular
Depth of penetration (maximum), mm	50
Rate of removal of metal, on optimum area, cu mm/min	525
Maximum work piece dimensions (length × width × height), mm	300 × 200 × 150
Maximum work piece weight, kg	50
Table dimensions (length × width), mm	400 × 250
Distance from table surface to highest liquid level mm	210
Maximum relative movement of tool and table, mm	
longitudinal	260
transverse	165
vertical	285
Over-all machine dimensions (length × width × height), mm	1355 × 1023 × 2177
Weight of machine, kg	900
Working fluid	transformer oil, or No. 2 spindle oil

Type of pulse generator	*RCLL* dependent relaxation
Nominal power required from the mains, kw	5.5
Rectifier voltage, volts	220
Nominal mean current, amp	20
Total available capacitance, μf	200.25
Control system	bridge type without amplifier

From its technical characteristics this copy-piercing machine is one of the best examples in the class of medium power electro-spark machines. However, the faults inherent in the dependent, relaxation type of generator, namely, rapid electrode wear, low productivity, etc., make it worthwhile to modernize the electrical side by modifying the machine for working on the electro-pulse principle. If a pulse generator requiring an input of 4–5 kw, i.e. 1 kw less than the machine normally uses, is fitted to the 4A722, removal rates of 1,200 cu mm/min, i.e. 2.3 times faster, can be obtained. At the same time the electrode wear rate is reduced from 5 to 20 times and the unit energy consumption from 2 to 3 times. This example shows the advantage to be gained by using an independent generator of unidirectional pulses.

A semiautomatic machine for boring atomizer jet bodies is manufactured in Russia by Mintransmash. The boring operation is one of the most highly developed and widely used applications of spark machining.

The advantage of replacing mechanical drilling by electro-spark piercing is that the jets can be pierced after the parts have been heat treated, eliminating the formation of oxide in the holes and the necessity of cleaning them out. Scrap jet bodies due to drill breakages are reduced and the process is easily automated.

Holes can be bored on electro-spark machines with diameters down to 0.1 mm, and even less in some cases, and this is difficult to do mechanically.

Figure J2b-6 shows schematically an electro-spark machine for fine-hole drilling. The voltage from an *RC*-type pulse generator is supplied to the jet body, 9, and the precision drawn, brass, electrode wire. The wire is vibrated axially in the direction of drilling. Spindle 3 is suspended in steel membranes, 2 and 6, fixed to the frame along the outer diameter by means of rings. Electromagnet 1 causes the spindle to vibrate at 100 cps with an amplitude of a few hundredths of a millimeter. Electrode wire is fed (as the hole deepens and the wire is eroded) by means of rolls, 5, connected to the automatic feed control. Collars, 4, give the wire a preliminary straightening, and final straightening occurs in guide 7. Jet body, 9, in the paraffin-filled tank, 10, can be automatically rotated to the given angle under the piercing head. Completion of each hole is signaled by electrical contact between the electrode wire and auxiliary electrode 12, which is insulated from the jet body by bush 11. Contact leads to the operation of a relay, actuating the withdrawal and workpiece indexing mechanism.

For drilling fine holes in jet bodies various designs of machines have been developed at the Leningrad carburetor factory. However, the degree of automation in these machines is lower than in the machine just described.

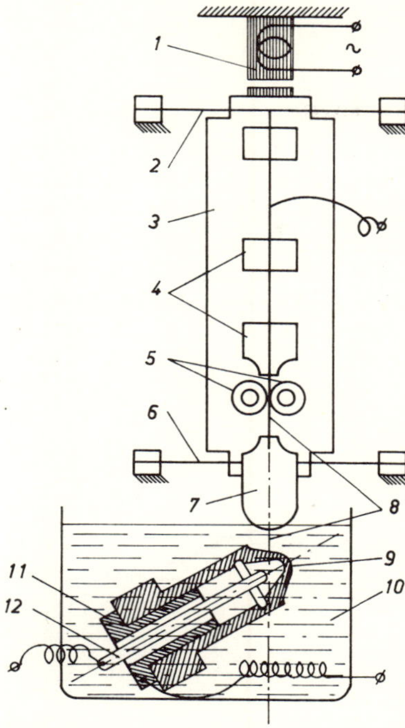

FIG. J2b-6. Outline of machine for drilling holes in jet bodies. By permission from "Electro-Erosion Machining of Metals," A. L. Livshits, Butterworths, London, 1960.

There is at present an experimental design for a semi-automatic machine for drilling several jet bodies simultaneously.

Another semi-automatic machine for grinding the working cone of jet bodies is also manufactured by Mintransmash. This machine is intended for electro-spark grinding to remove the eccentricity of the working cone of jet bodies in relation to the cylindrical fitting hole. The principle of this operation is shown in Fig. J2b-7.

Negative voltage pulses from a dependent relaxation-type (*RC*) generator are supplied to slide 9, which vibrates in the direction of arrow B, i.e. parallel to the walls of the jet body cone. Rolls, 10 and 12, on the slide feed the brass or copper electrode wire 11 through guide, 13. The positive side of the pulse generator is connected to contact 4 and the jet body 3. Transformer oil is fed through the tube, 8 and tank, 7, communicating with the inside of the jet body; excess oil drains out through the tube, 6.

The jet body, 3, held against the hollow bush, 5, by thrust bearing, 2, and compression spring, 1, is rotated as shown by arrow. The end of the electrode wire, approaching the surface of the jet body cone, forms a series of discharges as the jet body rotates, leading to metal being removed.

As the wire is eroded the rolls, 10 and 12, operated by an electromagnetic nonreversible, lead screw control, feed the wire by the amount required to balance its erosion and the removal of metal from the cone surface.

The use of a nonreversible, lead screw control [272] is necessary because the normal reversible control would copy the irregularities of the surface being ground.

The technology, design and construction of this semiautomatic machine provided an example of the rational application of the electro-spark process, successfully replacing the existing abrasive grinding process. The latter entails the use of small, rapid grinding, abrasive wheels, driven by air turbines at speeds up to 60,000 rpm. The mechanical and electrical parts of this machine are safer and simpler to use than existing machines for similar operations.

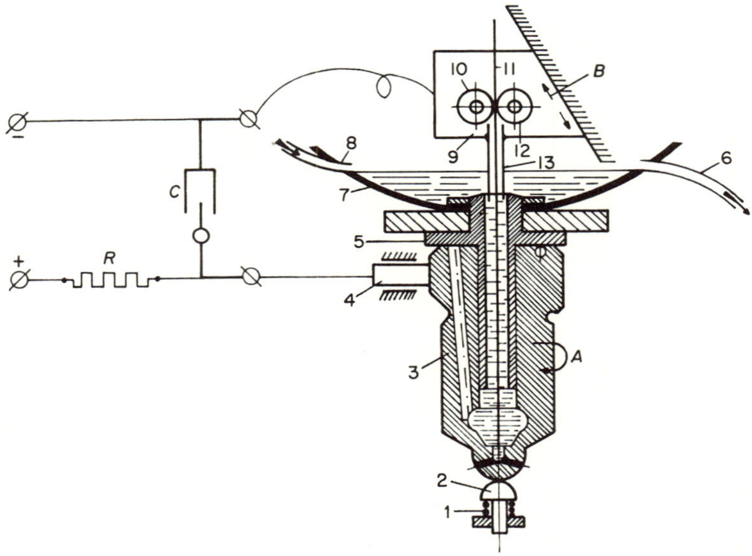

FIG. J2b-7. Outline of method of grinding jet body cones. By permission from "Electro-Erosion Machining of Metals," A. L. Livshits, Butterworths, London, 1960.

The electro-pulse method appeared later than the other variants of the electro-erosion process and at present no production models of electro-pulse machines have yet been developed.

A group of Russian manufacturers had completed as early as 1953 the design of a special, two spindle, electro-pulse machine for machining large numbers of slots in coal centrifuge screens. In 1953–54 a study was made in ENIMS of electro-pulse equipment for recovering broken cutting tools using a low frequency pulse generator. On the basis of these studies a method was developed at the Odessa Polytechnic Institute in 1952–54 for the production of needle files, and an experimental electro-pulse machine was constructed with a mercury jet generator circuit breaker. There are at ENIMS and at the Lenin Polytechnic Institute at Kharkov a number of experimental electro-pulse machines with rotary pulse generators, on the basis of which the batch production of new types of electro-pulse machine has been worked out and planned.

On the basis of the above-mentioned results Russian factories have developed special electro-pulse machines for the production of plane Nichrome resistor elements and the boring of communicating channels in deep recesses. At ENIMS in 1956, an experimental basic type of universal copy piercing electro-pulse machine was designed (Model 473) with a rotary pulse generator.

A series of universal copy piercing machines will be built up, basically in the form of electro-pulse machines supplied by rotary generators of unidirectional voltage pulses. The electro-pulse machines generally are suitable for copy piercing work. Electro-pulse machines for other operations (slitting, grinding, etc.) are still in the development stage.

3. Shaping Metals by Electrical Explosion Shock Wave

Electrical explosions are being employed in a development aimed at low cost shaping of hard-to-form metals like titanium, stainless steel and tungsten. The work may make possible savings of millions of dollars in the costly working of these difficult metals.

The controlled explosions create shock waves of such force that intricate contours can be blown into the metals in millionths of a second. Under development at the General Electric Company's General Engineering Laboratory [452], the new process is described as "capacitor discharge electro-spark forming." No TNT, dynamite, or other chemical explosives are used. The explosions are induced entirely through a buildup of electrical energy—available at about 35,000 volts at present, but eventually to go to 100,000 volts or more.

Made to occur under water, the electrical blasts create high intensity shock waves which are directed against the metal pieces to be formed. Upon impact of the shock wave, the metal instantly takes the shape of the die into which it is blown. The die is evacuated to remove air pockets that could cause the piece to have surface irregularities.

Electro-spark forming promises substantial savings through elimination of storage and handling problems associated with chemical explosives. Facilities where such work is performed need not be located remote from other manufacturing operations.

The new process is expected to have particular significance in the manufacture of aircraft and missile parts fabricated of exotic metals which generally are difficult to form by conventional processes. Such metals include titanium, niobium, tungsten, molybdenum, and certain steels and beryllium alloys. Using the electro-spark technique, the engineering laboratory has thus far successfully shaped titanium, niobium, beryllium-copper and the more difficult grades of stainless steel. It is believed the new process can also be used in forming tungsten and molybdenum, although the laboratory has not tested these two exceptionally hard-to-form metals.

The forming is done at room temperatures and no preheating of the metals is needed to make them ductile. The pieces take on the shape and surface texture of the die into which they are exploded and require minimum finish machining after they are removed. The laboratory has formed pieces up to 10 inches in diameter and 3/32 inch thick by the new process, but the forming of much larger pieces is considered entirely feasible. There is present planning for electro-spark forming of missile sections 10 feet or more in diameter and at least an inch in thickness, although this would not be done by the laboratory.

By the underwater high-voltage spark discharge a shock wave is created which forms parts on the inner contours of a closed epoxy die.

Each die has its own electrodes and initiating wire. A water-filled metal blank, sealed at the bottom with a rubber wafer, is placed in the die. The coverplate with its electrodes and initiating wire is then attached.

When the unit is fired, the wire vaporizes and expands against the surrounding water. The water's inertia causes a pressure buildup of about 15 to 20 atm. [955].

An advantage of the new technique* is that it can be operated indoors, and by nontechnical personnel. Once electrodes are connected and the power setting made, the operator merely has to press a release button to form a part.

The over-all unit consists of a portable metal box containing capacitors, switching circuits, a power-regulating Variac, and charging and discharging switches.

The new forming technique grew out of research which studied the behavior of shock waves in explosives forming; it is not yet available to industry in general. Eventual applications are expected to include the expanding and sizing of various hollow shapes.

Experts believe that high-voltage forming—as well as explosive and pneumatic forming—will find a place in many sheet metal fabricating shops. Similar techniques may also be developed using magnetic and ultrasonic sources of energy. For further literature see [956–958.]

---

* Developed by Convair of the United States, and known as "Hi-Vo-Pac."

Bibliography[1]

1.* Ambrosen, J. Apparat zur integrierenden Messung von Röntgenstrahlung oder radioaktiver Strahlung. German Patent C6019 VIII c/21 (June, 1952).

2.* Kurchatov, I. V. *Nucleonics* **14,** 37–42 (1956)

3. Früngel, F. Luftgekühlte, aus mehreren Teilkondensatoren zusammengesetzte Speicherkapazität. German Patent 901,087 (Jan. 7, 1954).

4.* VDI (Ver. Deut. Ing.). *Wärmetech. Arbeitsmappe, Arbeitsbl.* **B5/B6** (1953/1954).

5.* Liebendörfer, K., *Jahr. Chem. Ind.* No. 7 (1958).

6. Früngel, F. Leuchtdichten intensiver Funkenentladungen. *Optik* **3,** 128-136 (1948).

7a.* Kroebel, W. and Stutzer, G. Die Anwendung des Multivibratorprinzips für Zählrohrverstärker. *Z. angew. Physik* **6** (1), 14-19 (1954).

7b. Früngel, F., Thorwart, W., and Mahnau, H. Funkenblitzgerät mit 140,000 Blitzen/sec., Massnahmen zur Verkürzung der Blitzdauer. *Physik. Verhandl.* **8** (7), 191-192 (1957).

8.* Hübner, R. Das charakteristische Verhalten von Glückathodenröhren. *Elektron. Rundschau* (1955).

9.† Kretzmann, R. Beispiele für die industrielle Anwendung gasgefüllter Elektronenröhren. *VDI* (*Ver. Deut. Ing.*) *Z.* **91** (18), 457-62 (1949).

10.* Johnson, J. E. Description of mobile 250 kilovolts D.C. and Impulse set for cable testing and fault locating. Published report from Philadelphia Electric Company, Philadelphia (1957).

11. Früngel, F. and Vogler, H. Röntgenanlagen mit Impulstransformator. *Fortsch. Gebiete Röntgenstrahlen. Röntgenpraxis* **71,** 4 (1949).

12.* Strigel, R. "Elektrische Stossfestigkeit" p.104. Springer, Berlin (1955).

13.* Raether, H. Über eine gasionisierende Strahlung einer Funkenentladung. *Z. Physik* **110**, 611-624 (1938).

14. Früngel, F. Die Kontrolltechnik schneller Bewegungsvorgänge. *Z. angew. Physik* **6** (4), 183-192 (1954).

15.* Marx, E. Hochfrequenzerzeugung durch Blasfunkenstrecken. *VDI* (*Ver. Deut. Ing.*) *Z.* **18** (Sep. 1949).

16.* Brinkmann, K., and Marderwald, E. Dielektrische Trocknung der Papierisolation von Hochspannungskabeln. *Elektrotech. Z.* **73** (14), 449-450 (1952).

17.† Fürth H. P., and Waniek R. W. Production and use of high transient magnetic fields. *Rev. Sci. Instr.* **27,** 4 (1956).

[1] *Key to symbols:* *These references have been abstracted in the text. †These references have been paraphrased in the text. ‡These references have been quoted directly or very slightly paraphrased.

18.* VDE 0450/XI. 39, Leitsätze für Erzeugung und Verwendung von Stossspannung für Prüfzwecke. (VDE = Verband Deutscher Elektrotechniker e.V. VDE-Verlag, Berlin).

19.* Binder, L. "Die Wanderwellenvorgänge auf experimenteller Grundlage," pp. 24 ff. Springer, Berlin (1928).

20. Früngel, F. An ignition system for internal combustion engines. British Patent 699, 144 (Jan. 19, 1951).

21. Früngel, F. Das impulsoptische Überholmeldegerät. *Z. angew. Physik* **5** (3)' 102-104 (1953).

22.* Baatz, H. Blitzeinschlag-Messungen in Freileitungen. *Elektrotech. Z.* **72,** 7 (Apr. 1951).

23.* Fletcher, R. C. Impulse breakdown in the 10^{-9} sec range of air at atmospheric pressure. *Phys. Rev.* **76** (10), 1501-1511 (1949).

24.* Steenbeck, M. *Wiss. Veröffentl. Siemenswerke* **17,** 1 (1938).

25.* Slack, Ch. M., and Ehrke, L. F. *J. Appl. Phys.* **12,** 165 (1941); Slack, Ch. M., and Dickson, D. C., Jr. *Proc. I.R.E.* **35,** 600 (1947.)

26.* Trump, J. G. and van de Graaf, R. J. *J. Appl. Phys.* **18,** 327 (1947).

27.* Schaaffs, W., and Trendelenburg, F. Röntgeneographische Untersuchung der beim dielektrischen Funkenüberschlag auftretenden Schallwellen. *Z. Naturforsch.* **3a** (12), 656-668 (1948).

28.* Schaaffs, W. Röntgenblitzröhren und ihre Anwendung. *Z. angew. Physik* **1** (10), 462-73 (1949).

29.* Elle, D. Ein Röntgenblitzrohr mit Kunststoffisolierung und einer Doppelanode. *Z. angew. Physik* **8** (2), 81-85 (1956).

30.* Schaaffs, W., and Herrmann, K. H. Aufbau und technische Eigenschaften von Hochvakuum-Röntgenblitzröhren. *Z. angew. Physik* **6** (1), 23-35 (1954).

31.* Marx, E. *Elektrotech. Z.* **46,** 1298 (1925).

32.* Elle, D. Eine neue Anwendung des Faraday-Effekts zur Herstellung eines Kurzzeitverschlusses für Reihenbildvorderlichtaufnahmen für die Röhrenblitzphotographie. *Z. angew. Physik* **6** (2), 49-54 (1954).

33.* Landolt-Börnstein "Physikalisch-chemische Tabellen" 5th ed., II pp. 1332 and I Erg. pp. 721 (1923).

34.* Thomer, G. Wirkungsweise und Anwendung eines Doppel-Röntgenblitzes. *Z. angew. Physik.* **5** (6), 217-221 (1953).

35.* Glasoe, G. N., and Lebacqz, J. V. "Pulse Generators," p. 338. McGraw-Hill, New York (1948).

36.* Rompe, R., and Weizel, W. *Z. Physik* **122,** 630-39 (1944).

37.† Thorwart, W. Photoanalyse mit Kurzzeitphotographie und Zeitdehnerfilm. Kurzzeitphotostudien von Strömungsvorgängen. Dynamische Spannungsoptik und aerodynamische Strömungsbilder mit Funkenblitzen. *Photo-Technik und Wirtschaft* **9,** (1958); *Kinotechnik* **9** (1958).

38. Früngel, F. Hochfrequenzzeitlupen mittels Funkenblitzen. *Explosivstoffe* **6,** 4 (1958).

39. Früngel, F. Verfahren und Blitzlichtgerät zur Erzeugung extrem kurzer Lichtblitze. German Patent 1,027,309 (March 28, 1955).

40. Früngel, F. Ein neues Hochfrequenz-Blitzgerät für lange Blitzserien (25,000/sec) sowie Anwendung gesteuerter Einzelblitze in der Spannungsoptik. *Z. angew. Physik* **8,** 86-90 (1956).

41.* Brasch *Anschlusstherapie* **4,** 505-520 (1932).

42.* Elger, H.Eigenschaften von Hochspannungsleitungen für Hochfrequenzübertragungen und für Fehlerortung mit drahtverbundenen Funktmessgeräten. *Arch. elek. Übertragung* **1,** 195-204 (1947).

43.* Jäger, H. and Krone, H. H. Zum Entwicklungsstand der Elektrozaungeräte. *Landtech. Forsch.* **4** (2), 1-7 (1954).

44.* Hoffmann W., and Schumacher G. Schweissversuche mittels Kondensator Stossentladung. *Wiss Abhandl. Tech. Hochsch. Braunschweig* **5**, 122-131 (1953).

45.* Zdralek, A., and Wrana, J. *Elektrotech. Z.* pp. 579-580 (1935).

46.* Hudec, E. Verfahren zur Einstellung der Phasenlage einer erzwungenen Kippschwingung zu ihrer Steuerspannung. German Patent 705,537 (May 2, 1941).

47.* Presser, E. Verfahren zur elektrischen Widerstandschweissung mit Hilfe einer Gleichstromquelle. German Patent 224,879 (Sept. 21, 1909).

48.* Anon. DC-Welding with a transformer. "The Electrician," pp. 371-372 (1941).

49.* Brunst, W. "Das elektrische Widerstandsschwiessen," p. 152. Springer, Berlin (1952).

50. Früngel, F., and Thorwart, W. Kondensator-Impulsschweissung. *VDI (Ver. Deut. Ing.) Z.* **96,** (1), 18-24 (1954).

51.† Thorwart, W. Mikro-Induktionshärtung mittels hochfrequenten Impulsen. *VDI (Ver. Deut. Ing.) Z.* **95** (11 12) 341-44 (1953).

52.* Bradfield, G. Obstacle detection using ultrasonic waves in air. *Electronic-Eng.* **21,** pp. 464-68 (1949).

53. Früngel, F. Zum mechanischen Wirkungsgrad von Flüssigkeitsfunkten. *Optik* **3,** 124-127 (1948).

54.* Rust, H. H., and Drubba, H. Praktische Anwendung des Unterwasserfunkens als Impuls-Schallgeber für die Echolotung. *Z. angew. Physik* **5,** 251-252 (1953).

55.† Bailitis, E. Der Schallimpuls eines Flüssigkeitsfunkens. *Z. angew. Physik* **9,** 429 (1957).

56. Früngel, F., and Keller, H. Stoss-Schallquellen, Grundlagen und Analogie zu Sprengstoffumsetzungen. *Z. angew. Physik.* **9,** 145 (1957).

57.* von Keussler, V. *Spectrochim. Acta* **4,** 366 (1951); *Umschau* **51,** 88 (1951).

58.* Early, H. D., and Martin, A. *Trans. Am. Inst. Elec. Engrs.* **74,** 788 (1955).

59. Früngel, F., Bailitis, E., and Thorwart, W. Photographic analysis of dynamic phenomena in liquids and of optical strain models with the help of short-time exposures and a new shock-wave source. *Proc. 3rd. Intern. Congr. on High Speed Photography, London,* 1956, pp. AB 160-162. Butterworths, London.

60.* Padberg, L. R. Underwater-spark sound source. *Confidential* **VI,** 105-112. U.S. Patent Office, Washington, D.C. (1956).

61.* Meshujewa, A. W., Stekolnikow, I. S., and Effendijew, A. S. Untersuchungen des Anfangsstadiums eines Funkens. *Zhur. Tekh. Fiz.* **20** (3), 308 (1950).

62.* Stekolnikow, L. S. Untersuchung des Anfangsstadiums der Entladung bei sehr kleinen Elektrodenzwischenräumen. *Nachr. Akad. Wiss. UdSSR, Abt. tech. Wiss.* **7,** 985 (1950).

63.* Abramson, I. S., Gogetschkori, N. M., Drabkina, S. I., and Mandelstamm, S. L. Über die Kanal Funkenentladung. *Zhur. Elektrotek. Fiz.* **17** (10), 862 (1947).

64.* Drabkina, S. I. Zur Theorie der Ausbildung einer Funkenentladung. *Zhur. Elektrotek. Fiz.* **21** (4), 473 (1951).

65.* Gogetschkori, N. M. Experimentelle Untersuchungen des Kanals einer Funkenentladung. *Zhur. Elektroket. Fiz.* **21** (4), 493 (1951).

66.* Spiwak, G. W., and Lukazkaja, R. A. Elektronenmiskroskopie geringer Vergrösserung beim Vorhandensein von Gasen bei Atmosphärendruck und bei Unterdruck. *Nach. Akad. Wiss. UdSSR, Reihe Physik* **15** (4), 434 (1951).

67.* Spiwak, G. W., and Kardash, I. *J. Physics* **10,** 252 (1946).

68.* Spiwak, G. W., and Lukazkaja, R. A. Elektronenoptische Effekte bei atmosphärischem Druck. *Nachrbl. Moskauer Univ.* **4,** 37 (1948).

69.* Spiwak, G. W., Kardash, I. *J. Physics* **9,** 447 (1945).

70.* Kapzow, N. A. "Elektrische Erscheinungen in Gasen und im Vakuum." Gostekhizdat, Moscow (1947).

71.* Mandelstamm, S. L. "Einführung in die Spektralanalyse." Gostechisdat, Moscow (1946).

72.* Mandelstamm, S. L., and Raiski, S. M. Über den Mechanismus der elektrischen Erosion von Metallen. *Nachr. Akad. Wiss. UdSSR, Reihe Physik.* **13,** (5), 549 (1949).

73.* Pfeilstiker, K. Der Niederspannungsfunke und seine Anwendung zur Analyse von Halogenen und anderen schwer erregbaren Elementen. *Spektrochim. Acta* **1** (5), 424 (1940).

74a.*Solotych, B. N. Über einige Gesetzmässigkeiten der elektrischen Funkenerosion in einem flüssigen dielektrischen Medium. Dissertation NII MEP (1947).

74b. Früngel, F. Verfahren zur Erzeugung von Ultraschallstössen höchster Leistung. German Patent 914,590 (July 5, 1954).

75.* Solotych, B. N. Physikalische Grundlagen der Elektrofunkenbearbeitung von Metallen. *Schriftenreihe Verlag. Technik* **175** (1955).

76. Früngel, F. Ultraschallvergaser für Kraftstoffgemische. German Patent 826,088 (Jan. 7, 1952).

77.* Bunge, K. Materialbau und Funkenentladungen. *Forschungsber. Wirtschafts und Verkehrsministeriums No.* **209** (1955).

78.* Kaiser H., and Walraff A. *Ann. Physik* **34,** 297-340 (1939).

79.* Bergmann, L. "Der Ultraschall," p. 46. Hirzel, Stuttgart (1954).

80.* Toepler M. *Ann. Physik* [4] **27,** 1043-1050 (1908).

81.* Fünfer, E. Einige experimentelle Untersuchungen der elektrischen und optischen Vorgänge beim Funkendurchschlag in Gasen. *Z. angew. Physik* **1** (7), pp. 295-304 (1949).

82.† Luy, H., and Schade, R. A method for the production of front-light pictures at rates up to 10,000 pictures per second. *Proc. 2nd Intern. Congr. on High Speed Photography* pp. 59-63. Dunod, Paris (1956).

83.* Aumont, R., and Vodar, B. Some observations on guided sparks in compressed gases. *Proc. 2nd. Intern. Congr. on High Speed Photography* pp. 39-42. Dunod, Paris (1956).

84.* Laporte, M. The electronic flash tubes and their applications. *Proc. 2nd. Intern. Congr. on High Speed Photography* pp. 1-7. Dunod, Paris (1956).

85.* Chestermann, W. D., and Glegg, D. R. Short duration spark discharges in xenon. *Proc. 2nd Intern. Congr. on High Speed Photography* pp. 8-18. Dunod, Paris (1956).

86.* Carrol, J. S. *Electr. Manufacturing* **39,** 47 (1947).

87.* Edgerton, H. E. *Phot. Soc. Am. J.* **13,** 439-441 (1947).

88.* Many, W. G. Electronic Photoflash Construction. *Phot. Age.* **2,** 8-12 (1947).

89.* Schafer, H. J. *J. Soc. Mot. Pic. Eng.* **53,** 524-544 (1949).

90.* Taylor, H. H., and Waldram, J. M. *J. Sci. Instr.* **10** 278-389 (1953).

91.* Raether, H. *Z. Physik* **110,** 611-624. (1938).

92.* Raether, H. *Z. Physik* **112,** 464-489 (1939).

93.* Schwieker, Z. *Z. Physik* **116,** 562-575 (1940).

94. Früngel, F. High frequency flash lamp. *Proc. 3rd Intern. Congr. on High Speed Photography, London,* 1956, pp. AW 13-17. Butterworths, London, (1956).

95.* Glaser, G. Zur Lichtemission stromstarker Funkenentladungen. *Optik* **7**, 33-53 (1950).

96.* Fletcher, R. C. Impulse breakdown in the 10^{-9} sec range of air at atmospheric pressure. *Phys. Rev.* **76** (10), 1501-1511 (1949).

97.* Raether, H., and Kerner, H. Uber die Elektronenemission von kalten Metalloberflächen bei mittleren Feldstärken. *Z. angew. Physik.* **6** (5), 212-13 (1954).

98.* Weizel, W. Berechnung des Ablaufs von Funken mit Widerstand und Selbstinduktion im Stromkreis. *Z. Physik* **135**, 639-57 (1953).

99. Früngel, F. Zeitlicher und spektraler Ablauf intensiver Funkenentladungen. Forschungsbericht No. 1856 (1944) im ZWB (Zentrale für das wissenschaftlich Berichtswesen, Berlin).

100.† Thorwart, W. Spitzenspektrum von Lichtimpulsquellen. ZWB (Zentrale für das wissenschaftliche Berichtswesen, Berlin, 1944).

101.* Glaser, G. Zur Lichtemission stromstarker Funktenenladungen. *Optik* **7**, 33-53 (1950).

102.* Rompe, R., and Weizel, W. *Z. Physik* 122, 639 ff. (1944).

103.* Forsythe, W. E., ed. "Measurement of Radiant Energy," McGraw-Hill, New York (1937).

104. Früngel, F. Stand der Signalübertragung mit kurzzeitigen Lichtimpulsen. *Physik. Verhandl.* **3** (2), 42 (1962).

105. Früngel, F. Das impulsoptische Überholmeldegerät. *Z. angew. Physik* **5** (3), 102-104 (1953).

106. Früngel, F. Ein neues Auto-Überholgerät nach dem impulsoptischen Prinzip. *Optik* **10**, 404-408 (1953).

107.* Klumb, H., and Koch, B. Verwendung von UV-Strahlen. Forschungsbericht No. 5198 in der ZWB (Zentrale für das wissenschaftliche Berichtswesen, Berlin).

108.* Klumb, H. Die Verwendbarkeit von UV-Strahlen zur Übertragung von Nachrichtensignalen, Messsignalen und Fernbetätigungssignalen. Vorbericht des Inst. für Fernmeldetechnik der Technischen Akademie dier Luftwaffe (1940).

109.* Gaertner, H. Stand und Aussichten der thermischen Ortung im OKH (Oberkommando des Heeres) Bericht No. 148. "Forschungsarbeiten über ultrarotempfindliche Strahlungsempfänger" (1940).

110.* Meyer, A. E. H., and Seitz, E. O. "Ultraviolette Strahlen," p. 132. Verlag W. de Gruyter, Berlin (1949).

111.* Olbers, W. Grundlagen der objektiven Wolkenhöhenmessung. *Tech. Mitt. Instrumentenwesens deut. Wetterdienstes* [*N. F.*] 1 (1955).

112. Früngel, F. Funkenstrecke für Stroboskopie. German Patent No. 826,369 (May 31, 1950).

113. Früngel, F. Optisches Durchstrahluntsverfahren für stark absorbierende Medien. German Patent No. 836,410 (Aug. 23, 1951).

114.* Altrogge, W. *Elektrotech. Z.* **5**, 112 (1952).

115.* Weizel, W. Forschungsbericht No. 103, Wirtschafts-Ministerium, Nordrhein-Westfalen, Germany (1954).

116.* Franck, S. "Messentladungstrecken (Ionenstrecken)." Springer, Berlin (1931).

117.* Weicker, W. *Elektrotech. Z.* **32**, 436 (1911).

118.* Peek, F. W. "Dielectric Phenomena," 3rd ed., New York, 1929.

119.* Kuske, A. "Verfahren der Spannungsoptik." Deutscher Ingenieur-Verlag, Düsseldorf (1950).

120.* Hennig-G. "Dauermagnet-Technik" p. 75. Franzis, Munich (1952).

121.* Marx, E. "Hochspannungs-Praktikum," p. 94. Springer, Berlin (1952).

122.* Pelier, P. M. "Elektrische Messung mechanischer Grössen." Springer, Berlin (1956).

123.* Naslin, P., and Vivie, J. "Photographie et cinematographie ultrarapides." Dunod, Paris (1956).

124.* Gänger, B. "Der elektrische Durchschlag von Gasen," p. 452. Springer, Berlin (1953).

125.* Schumann, W. O. "Elektrische Durchbruchfeldstärke von Gasen," p. 35 Springer, Berlin (1923).

126.* Kluge, W. Thyratrone. *Elektrotech. Z.* **63,** 204-207 (1942).

127.* Költer, J. Breitbandverstärkung mit Gegenkopplung. *Hausmitteilungen der Fernseh AG.* **2,** 6.

128.* Wanjukow, M. P., and Mak, A. A. *Opt. i. Spektroskopiya* **1,** 642-649 (1956).; Ures, M. A., and Parasinskaja, H. B. *ibid.* **4,** 90-92 (1957); Marshak, I. S. *Zhur. Nauchn. i Prikl. Fotogr. i Kinematogr.* **1** (5), 366-368 (1956). Laborbericht des Labors für Blitzlichtröhren des Moskauer Glühlampenwerkes. *Zhur. Eksptl. i Teoret. Fiz.* **16,** 703-717, 718-727 (1946).

129. Früngel, F. Verfahren zur Erzeugung von aperiodischen Höchstspannungsimpulsen. German Patent 892,629 (Oct. 8, 1953).

130. Früngel, F. Vorrichtung zur Erzeugung von steilen Stossspannungen. German Patent 881,211 (July 23, 1953).

131. Früngel, F. Anordnung zur Erzeugung von hohen Stossspannunsgen. German Patent 881,835 (July 2, 1953).

132. Früngel, F. Gebereinrichtung zur Erzeugung von Signalen. German Patent 1,019,940 (Feb. 10, 1950).

133. Früngel, F. Höchstdrucklampe. German Patent 890,990 (Sept. 24, 1953).

134. Früngel, F. Entladungsgefäss, insbesondere für Lichtblitz-Stroboskope. German Patent 856,995 (Nov. 27, 1952).

135. Früngel, F. Mit Kondensatorentladungen betriebene Stoss-Schallquelle für elektrolytisch leitende Medien. Austrian Patent 195,135 (Jan. 25, 1958).

136. Früngel, F. Metode til elektrisk sterilisering, til emulgering og til rensing av melkemaskiner. Norwegian Patent 93,484 (March 24, 1958).

137. Früngel, F. Ein neues Hochfrequenz-Blitzgerät für lange Blitzserien (25000/sec.) sowie Anwendungen gesteuerter Einzelblitze in der Spannungsoptik. *Z. angew. Physik.* **8** (2), 86-90 (1956).

138. Früngel, F., and Bailitis, E. Aufbau und Arbeitsweise leistungssstarker Stoss-Schallerzeuger mit weissem Ultraschallspektrum. *Physik. Verhandl.* **7,** (6), 186-187 (1956).

139a. Früngel, F., Bailitis, E., and Thorwart, W. Photographic analysis of dynamic phenomena in liquids and of optical strain models with the help of short-time exposures and a new shock-wave source. *Z. angew. Physik* **4** (1957).

139b. Früngel, F., Bailitis, E., and Thorwart, W. *Proc. 3rd Intern. Congr. on High Speed Photography,* 1956, *London* pp. AB 160-162. Butterworths, London (1956).

140.* Knight, B., Herbert, L., and Maddison, R. C. *Proc. Inst. Elec. Engrs.* (*London*) **106,** Pt. A., No 2947 (1959).

141.* Bird, M. A. *Proc. Inst. Elec. Engrs.* (*London*) **106,** Pt. A, No. 2879 (1959).

142.* Carruthers, R. *Proc. Inst. Elec. Engrs.* (*London*) **106,** Pt. A, No. 2915 (1959).

143.* Robson, A. E., and Hancox, R. *Proc. Inst. Elec. Engrs.* (*London*) **106,** Pt. A, No. 2948 (1959).

144.* Payne, M., and Kaufman, S *Proc. Inst. Elec. Engrs.* (*London*) **106,** Pt. A, No. 2900 (1959).

145.* Butt, E. P., Carruthers, R., Mitchell, J. T. D., Pease, R. S., Thonemann, P. C., Bird,

M. A., Blears, J., and Harthill, E. R. *Proc. Inst. Elec. Engrs.* **106,** Pt. A, No. 2901 (1959).

146.* Fitch, R. A., and McCormick, N. R. *Inst. Elec. Engrs. Brit. Nuclear Energy Conf.* (April, 1959).

147.* Hartill, E. R. *Proc. Inst. Elec. Engrs.* **106,** Pt. A, No. 2876 (1959).

148.* Fry, D. W. *Proc. Inst. Elec. Engrs.* No. 2903 (1959): Lecture on British Nuclear Energy Conference Dec. 4, 1958.

149.* Pahl, M. *Z. Naturforsch.* **14a,** 239-246 (1959).

150.* Pahl, M., Weimer, U. *Z. Naturforsch.* **13a,** No. 1 (1958).

151.* Pahl, M., Weimer, U. *Z. Naturforsch.* **13a,** No. 9 (1958).

152. Früngel, F., Thorwart, W., and Mahnau, H. Funkenblitzgerät mit 140,000 Blitzen/sec. Massnahmen zur Verkürzung der Blitzdauer. *Physik. Verhandl.* **8** (7), 191-192 (1957).

153. Früngel, F. Steigerung der Funkenfolgefrequenz für Photographie und Hochfrequenzfilm. *Physik. Verhandl.* **6** (4), 62-63 (1955).

154. Früngel, F. Optisches Durchstrahlungsverfahren für stark absorbierende Medien. German Patent 836,410 (April 10, 1952).

155.* Weske, J. R. Spec. publ. of University of Maryland: Tech. Note BN-128 AFORSTN 57304 AD 154 214 (1958).

156.* Bomelburg, H. J. Spec. publ. of University of Maryland: Tech. Note BN-68 OSR-TN 56-38, ASTIA No. AD80549 (1957).

157.* Herzog, F. A new experimental approach to the analysis of compressor performance. Spec. publ. of University of Maryland: No. USAF BN-90 AFOSR-TH-57-3 AD 115035 (Jan., 1957).

158a.*Thorwart, W. Schneller als Elektronenblitz. *Photomagazin* **11** (1958).

158b.†Thorwart, W. Photoanalyse mit Kurzzeitphotographie und Zeitdehnerfilm. Kurzzeitphotostudien von Strömungsvorgängen. Dynamische Spannungsoptik und aerodynamische Strömungsbilder mit Funkenblitzen. *Kino-Technik* **9,** 254-258 (1958).

158c.*Thorwart, W. Photographische Zeitmikroskopie mit Funkenblitzen. *Naturwiss. Rundschau* **12** (6), 233-35 (1959).

159. Früngel, F. Verfahren zur Erzeugung extrem kurzer Lichtblitze. German Patent 1,044,969 (Nov. 10, 1956).

160. Früngel, F., and Hands, G. Die Signalübertragung mit Lichtimpulsen. *Arch. elek. Übertragung* **13,** pp. 121-131, 176-183 (1959).

161. Früngel, F. Stand der Signalübertragung mit kurzzeitigen Lichtimpulsen. *Physik Verhandl.* **3,** 2, (1952).

162.* Koch, B. *Optik* **5,** 258-273 (1949).

163.* Middleton, K. W. E. "Vision through Atmosphere." University of Toronto Press, Toronto, Canada (1952).

164. Früngel, F. Höchstdrucklampe. German Patent 890,990 (Sept. 24, 1953).

165. Früngel, F. Verfahren zur Verlagerung der Spektralausbeute einer Quecksilberdampf- oder Edelgas-Höchstdrucklampe in gewünschte Gebiete. German Patent 967,196 (Oct. 24, 1957).

166. Früngel, F. Bandverstärker für Empfänger von Funkenlicht-Signalen. German Patent 1,008,620 (April 20, 1954).

167. Früngel, F. Fotoelektrischer Empfänger für kurze Lichtimpulse, insbesondere für impulsoptische Überholmelder. German Patent 1,048,199 (April 20, 1954).

168.* Smith, S. *Astrophys. J.* **61,** 186 ff (1925).

169. Früngel, F. Spannungsvervielfacherschaltung für Hochspannungs-Prüfgeräte. German Patent 1,011,989 (April 20, 1954).

170. Früngel, F. Hochspannungs-Prüfgeräte mit hyperbolischer U-I-Kennlinie. *Elektro-Welt* **4,** (C3), 67-68 (1959).

171. Früngel, F. Vorrichtung zum Messen und Dosieren von Elektrizitätsmengen. German Patent 971,719 (Aug. 30, 1951).

172.* Fischer, H. Simple submicrosecond light source with extreme brightness. *J. Opt. Soc. Am.* **47** (11), 981-984 (1957).

173.* Fischer, H. Funkenkanäle extremer Leistung. *Physik. Verhandl.* **6,** 177 (1955).

174. Freise, W., and Früngel, F. Zur Messung des Verlustwinkels von grossen einseitig geerdeten Kapazitäten. *Elektrotech. Z.* **80** (10), 296-300 (1959).

175.† Zuckermann, W. A., and Awdejenko, A. L. *Zhur. Techk. Fiz.* **7,** 185 (1942.)

176.† Zuckermann, W. A. *Vorträge Akad. Wiss. USSR* **40,** 7 (1943).

177.† Zuckermann, W. A. *Vorträge Akad. Wiss. USSR* **53,** 4 (1946).

178.† Strachowski, G. M., and Zuckermann, W. A. *Izvest. Akad. Nauk. SSSR, Ordel. Tekh. Nauk.* **3,** (1946).

179.† Zuckermann, W. A., and Manakowa, M. A. *Zhur. Tekh. Fiz.* **27,** 391 (1957)

180.† Lukaschw, A. A. *Pribory i tekhn. eksperim.* **4,** 56 (1957).

181.† Sjusin, W. P., Manakowa, M. A., and Zuckermann, W. A. *Pribory i tekhn. eksperim.* **1,** 85 (1958).

182.‡ Dummer, G. W. A. "Fixed Capacitors." Pitman, London (1956).

183.‡ Dummer, G. W. A. "Fixed Capacitors" pp. 149-152. Pitman, London (1956).

184.‡ von Tersch, L. W. and Swago, A. W. "Recurrent Electrical Transients," pp. 18-27, 320-332. Prentice-Hall, Englewood Cliffs, New Jersey (1953).

185.‡ Lebacqz, J. V. and White, H. J. The rotary spark gap. *In* "Pulse Generators," pp. 275-292. McGraw-Hill, New York (1948).

186.‡ Germeshausen, K. J. The hydrogen thyratron. In "Pulse Generators," pp. 336-344. McGraw-Hill, New York (1948).

187.‡ Dillinger, J. R. General considerations for gap design. In "Pulse Generators," pp. 316-332. McGraw-Hill, New York (1948).

188.* Chace, W. G., and Moore, H. K. "Exploding Wires," pp. 197. Plenum Press, New York (1959).

189.‡ Dillinger, J. R. The three-electrode fixed spark gap. In "Pulse Generators," pp. 332-335. McGraw-Hill, New York (1948).

190.† Goldberg, S., and Germeshausen, K. J. On the design and measurement of hydrogen thyratron modulator characteristics. A summary interim report of findings under Signal Corps Development contract DA 36-029 sc 15372 (1953).

191. Früngel, F. Quick-Responsive spark gap device. U.S. Patent 2,886,737 (May, 1959).

192.‡ White, H. J., Gillette, P. R. and Lebacqz, J. V. The pulse-forming network. *In* "Pulse Generators." pp. 201-213. McGraw-Hill, New York (1948).

193.* Hawley, W. G. "Impulse Voltage Testing," pp. 32-34. Chapman & Hall, London (1959).

194.† Hauser, S. M., and Quan, H. Applying the Kerr cell to nanosecond. *Electronics* pp. 57-59 (Aug., 1961).

195.† McFarlane, H. B. Spark gaps for fast high voltage switching. *Electronics* pp 72-73 (July, 1959)

196. Früngel, F. Quenching spark gaps as trigger elements in high speed cinematography. *Proc. 5th Intern. Congr. on High Speed Photography, Washington*, 1960. pp. 469-472. Society of Mot. Pic. and Tel. Eng., New York (1962).

197.* Dessau-Righ. Die elektrischen Wellen. In "Die Telegraphie ohne Draht," p. 177. Vieweg, Braunschweig (1907).

198.* Zenneck, J. Die technischen Löschfunkenstrecken. "Lehrbuch der drahtlosen Telegraphie," pp. 222-226. Springer, Berlin (1925).

199.* Wien, M. Über eine Methode zur Erzeugung schwach gedämpfter elektrischer Schwingungen. *Physik. Z.* **9** (2), 49 (1908).

200.* Boas, H. Löschfunkenstrecken für enge Koppelung. "Jahrbuch der drahtlosen Telegraphie," p. 563 (1912).

201.* Weichart, F. Funkenstrecken: (a) Knall-oder Knarrfunkenstrecken; (b) Löschfunkenstrecken; (c) Funkenstrecken mit rotierenden Elektroden. "Jahrbuch der drahtlosen Telegraphie," pp. 420-422 (1928).

202.† Weber, W. Das Schallspektrum von Knallfunken und Knallpistolen mit einem Beitrag über Anwendungsmöglichkeiten in der elektroakustischen Messtechnik. *Z. Akustik* **4,** 373-391 (Nov., 1939).

203.* McFarlane, W. The sound radiation from a condenser discharge. *Phil. Mag.* [7] **7,** 117 (July, 1934).

204.* Geiger, H. and Scheel, U. K. in "Handbuch der Physik" (H. Geiger and K. Scheel, eds.), Vol. VIII, Chap. 14, p. 635. Springer, Berlin (1926).

205.* Schindelin, W. Beiträge zur Raumakustik. *Ann. Physik* [4] **52,** 129 (1920).

206.* von Békésy, G. Über die mechanische Frequenzanalyse einmaliger Schwingungsvorgänge und die Bestimmung der Frequenzabhängigkeit von Übertragungssystemen und Impedanzen mittels Ausgleichsvorgängen. *Z. Akustik* **3** (5), 217 (1937).

207.† Behnke, A. Über die Speicherung mehrerer durch Flüssigkeitsfunken erzeugter seismischer Impulse auf einem einzigen magnetischen Tonträger. *Z. Geophysik* **26** (1), 24-31 (1960).

208.* Suhr, G. Eine neue Apparatur zur Messung des vertikalen Gradienten der seismischen Wellengeschwendigkeit. Dissertation, University of Hamburg (1956).

209.† Baule, H., and Müller, E. Messung elastischer Eigenschaften von Gesteinen. *In* "Handbuch der Physik" (S. Flügge, ed.), Vol. 47, pp. 180-189 Springer. Berlin (1956/57).

210.* Vogel, C. B. A seismic velocity logging method. *Geophysics* **17,** 3 (1952).

211.* Bradfield, G. Precise measurement of velocity and attenuation using ultrasonic waves. *Nuovo Cimento 7,* Suppl. 2, pp. 162-181 (1950).

212. Früngel, F. System for probing materials by shock wave signals. U.S. Patent 2,946,217 (May, 1955).

213.* Schardin, H. Das Töplersche Schlierenverfahren. *VDI-Forsch.* Ausgabe B, p. 367 (1934).

214.* von Schmidt. O. Über Knallwellenausbreitung in Flüssigkeiten und festen Körpern. *Z. Physik* **39,** 868 (1938); Über Kopfwellen in der Seismik. *Z. Geophysik* **15,** 141 (1939).

215.* Mintrop, L. 100 Jahre physikalische Erdbebenforschung und Sprengseismik. *Naturwissenschaften* **34,** 257, 289 (1947); Über Anwendung des seismischen Verfahrens im Erdölbergbau und ihre wirtschaftlichen und wissenschaftlichen Auswirkungen. *Öl und Kohle* p. 269 (1943).

216.† Müller, E. Eine neue Messmethode zur Bestimmung der Wellengeschwindigkeiten in Gesteinsproben. *Naturwiss. Rundschau* **41** (4), 85 (1954).

217.† Heydweiler, A. Über die dielektrische Festigkeit leitender Flüssigkeiten. *Z. Instrumentenk.* **12,** 377 (1892); *Ann. Physik* [3] **48,** 110 (1893).

218.* Wehnelt, A. *Elektrotech. Z.* **20,** 76 (1899); *Ann. Physik* [3] **68,** 233 (1899).

219.* Simon, H.Th *Ann. Physik* [3] **68,** 860 (1899).

220.* Anderson, V. C. Wide band sound scattering in deep scattering layers. *Scripps Inst. Oceanog. Contribs.* Reference 53-36 (1953).

221.* Burhorn, F., *et al.* Temperature measurements of water stabilized high power arcs. *Z. Physik* **131** (1), 28-40 (Dec. 1951).

222.* Duddell, W. On rapid variation in the current through the direct-current arc. *J. Inst. Elec. Engrs. (London)* **30,** 232-283 (1900/1901).

223. Früngel, F. Device for ultrasonic treatment of molten metal. U.S. Patent 2,820,263 (Jan., 1958).

224.* Glaser, G. About the generation of very high temperatures in high power sparks. *Z. Naturforsch.* **6a** (11), 706-713 (1953).

225.* Loeb, L. B. "Mechanism of the Electric Spark." Stanford Univ. Press, Stanford, California (1941).

226.* Loeb, L. B. Spark breakdown in uniform fields. Office of Naval Research, Dept. of Navy, Washington, D.C. (1954).

227.* Peters, T. Temperature and radiation measurements of water stabilized high pressure arcs. *Z. Physik* **135** (5), 573-592 (1953).

228.* Wilcox, R. H. Factors affecting the operation of spark noise generators. Naval Research Laboratory, Report 4307 (Feb., 1954).

229.* Wood, A. B. "A Textbook of Sound." 2nd ed. Macmillan, New York, (1941).

230.† Güth, W. Die Leitung von Schallimpulsen in Metallstäben. *Acustica* **5,** 35-43 (1955).

231. Früngel, F. Method and apparatus for sterilizing and cleaning articles, particularly milking machines, by shock sound radiation. Australian Patent 228,608 (1960).

232. Früngel, F. Method and device for electrically sterilizing and cleaning milking machines or the like. U.S. Patent 2,931,947 (1960).

233.† Breuning, S. M., and Bone, A. J. Soniscope applied to maintenance of concrete structures. *Joint Highway Research Project, Research Rep.* **5,** 210-216 (1954).

234. Früngel, F. Method and system for checking the position of fishing nets. U.S. Patent 2,789,382 (1957).

235.* Cheesman, W. J. Dynamic testing of concrete with the soniscope apparatus. *Proc. Highway Research Board* **29,** 176-183 (1949).

236.* Jones, R. Testing of concrete by ultrasonic-pulse technique. *Proc. Highway Research Board* **32,** 258-275 (1953).

237.* Leslie, J. R. Pulse technique applied to dynamic testing. *Proc. Am. Soc. Testing Materials* **50,** 104-116 (1950).

238.* Leslie, J. R., and Cheesman, W. J. An ultrasonic method of studying deterioration and cracking in concrete structures. *J. Am. Concrete Inst.* **21** (1), 17-36 (1949).

239.* Meyers, R. C. Dynamic testing of concrete pavements with the soniscope. *Proc. Highway Research Board* **31,** 234-245 (1952).

240.* Whitehurst, E. A. Soniscope tests concrete structures. *J. Am. Concrete Inst.* **22,** No. 26 (Feb. 1951); and *Proc. Am. Concrete Inst.* **47,** 433-444 (1951).

241.* Thomson, E. "Electric Welding", A paper read at the States Meeting of Franklin Institute, held March 16, 1887. *Franklin Inst.* **123,** No. 737.

242.* van Nostrand, D. (Publisher). *Eng. Mag.* **35,** 5 (Nov., 1886).

243.* de Fodor, E. "Die elektrische Schweissung und Lothung." Hartlebens-Verlag, Athens, (1892).

244. Früngel, F. Method for high-frequency soldering or induction hardening. British Patent 716,298 (1952).

245. Früngel, F. Method and device for high-frequency soldering and induction hardening. U.S. Patent 2,799,760 (1952).

246. Früngel, F. Method and device for cooling electric resistance welding machines. U.S. Patent 2,827,546 (1955).

247. Früngel, F. Method of an apparatus for electric resistance welding. British Patent 814,608 (1955).

248. Früngel, F. An ignitron system for internal combustion engines. British Patent 699,144 (1950).

249. Früngel, F. Long-lived impulse transformer. U.S. Patent 2,942,214 (1957).

250. Früngel, F. Method and system for operating x-ray tubes. U.S. Patent 2,899,562 (1959) (German patent priority 1948).

251.* Zuckermann, W. A., and Manakowa, M. A. Röntgenblitzquellen zur Untersuchung schnellverlaufender Vorgänge. *Proc. 4th Intern. Congr. Kurzzeitphotographie und Hochfrequenzkinematographie, Cologne,* 1958 pp. 118-122 Helwich, Darmstadt, Germany (1958).

252.* Thomer, G., and Stenzel, A. Versuche zur Röntgenblitzkinematography mit weicher Strahlung. *Proc. 4th Intern. Congr. Kurzzeitphotographie und Hochfrequenzkinematographie, Cologne,* 1958 pp. 112-117 Helwich, Darmstadt, Germany (1959).

253.* Vollrath, K. Measurement of spark gap deionisation. *Proc. 2nd Intern. Congr. on High-Speed Photography* pp. 31-38. Dunod, Paris (1956).

254.* Stenzel, A., and Vollrath, K. Electronic cine-camera with a large number of frames. *Proc. 2nd Intern. Congr. on High-Speed Photography* pp. 55-58. Dunod, Paris (1956).

255. Früngel, F. "Impulstechnik," pp. 374 ff. Akademische Verlagsges., Leipzig, Germany (1960).

256.* Thomer, G. Recent developments in the field of x-ray flash and motion picture equipments. *Proc. 2nd Intern. Congr. on High-Speed Photography,* pp. 72-77. Dunod, Paris (1956).

257.* Conn, W. M. "Die technische Physik der Lichtbogenschweissung". Springer, Berlin (1959).

258.† Thomer, G., and Schall, R. Application of image intensifier in flash radiography. *Proc. 5th Intern. Congr. on High-Speed Photography, Washington,* 1960. pp. 176-178. Soc. of Motion Pictures and Television Engineers, New York (1962).

259.† Stenzel, A., and Thomer, G. X-ray flash cinematography up to 12,000 images/sec. *Proc. 5th Intern. Congr. on High-Speed Photography, Washington,* 1960, pp. 173-175. Soc. of Motion Picture and Television Engineers, New York (1962).

260.* Thomer, G., and Stenzel, A. Versuche zur Röntgenblitz-Kinematographie mit weicher Strahlung. *Proc. 4th Intern. Congr. Kurzzeitphotographie und Hochfrequenzkinematographie, Cologne* 1958, pp. 112-117. Helwich, Darmstadt, Germany (1959).

261.* Edgerton, H. E. The past, present and future of high speed photography. *PSA Journal* **13,** 437-440 (July 1947).

262. Patzke, H. G., Früngel, F., and Thorwart, W. High-speed photography of rapid air currents and shock waves by means of high-frequency high-voltage sparks. *Proc. 5th Intern. Congr. on High-Speed Photography, Washington,* 1960, pp. 498-502. Soc. of Motion Picture and Television Engineers, New York (1962).

263. Früngel, F. High speed slow motion pictures by means of spark flashes. *Explosivstoffe* **10,** 236-246. (1958)

264. Früngel, F., and Alberti, H. X-ray flash high speed cinematography of small objects. *Proc. 5th Intern. Congr. on High-Speed Photography, Washington,* 1960, pp. 170-172. Soc. of Motion Picture and Television Engineers, New York (1962).

265.† Rieck, J. Ein Tageslicht-Zeitraffer mit Kurzzeitverschluss. *Proc. 4th Intern. Congr. Kurzzeitphotographie und Hochfrequenzkinematographie, Cologne,* 1958 pp. 300-303. Helwich, Darmstadt, Germany (1959).

266.* Schall, R. Feinstrukturaufnahmen in ultrakurzen Zeiten mit dem Röntgenblitzrohr. *Z. angew. Physik* **3,** 83-88 (1950).

267.* Schaaffs, W. Die Erzeugung von Röntgenblitzinterferenzen von polykristallinem Material. *Z. Naturforsch.* **5a,** 631-632 (1950).

268.* Teves, M. C., and Tol, T. Elektronische Verstärkung von Röntgenbildern. *Phillips Tech. Rundschau* **14,** 37-47 (1952).

269.† Gow, J. D., and Ruby, L. Simple pulsed neutron source based on crossed filde trapping. *Rev. Sci. Instr.* **30** (5), 315-317 (1959).

270.† Livshits, A. L. "Electro-Erosion Machining of Metals," pp. 14-20, 51, 74-81. Butterworths, London (1960).

271.* Höhne, E. Elektroerosion und elektrolytische Maschinen. Induktions—Wärmeanlagen—Ultraschallgeräte. *VDI* (*Ver. Deut. Ing.*) *Z.* **102** (33), 1641-1646 (1960).

272.* Gutkin, B. G. "Avtomatizatsiya elektoriskrovyka i anodno-mekhanicheskikh stankov." (Automation of electro-spark and anodo-mechanical machine tools.) Mashgiz (1952).

273.† Levinson, A. Z. "Poluprovodnikovye vypryamiteli." (Semi-conductor Rectifiers). Gosenergoizdat (1948.)

274.‡ Conn, W. M. Combination of electrically exploded wires and electric arc. *In* "Conference on Extremely High Temperatures," pp. 29-41. Wiley, New York (1958).

275.† Conn, W. M. Studium zum Mechanismus von elektrischen Drahtexplosionen. *Z. angew. Physik* **7** (11), 539-554 (1955).

276.* Chace, W. C. Technical memo to the Director, Geophysics Research Directorate, Air Force Cambridge Research Center, AFCRC, GRC, TM 57-5 (1956).

277.† Thackeray, D. P. C. Current developments in the production and assessment of high intensity discharges. *Proc. 4th Intern. Congr. Kurzzeitphotographie und Hochfrequenzkinematographie, Cologne*, 1958, pp. 123-129. Helwich, Darmstadt, Germany (1959).

278.* Schardin, H., and Fünfer, E. Grundlagen der Funkenkinematographie. *Z. Physik* **4,** 224 (1952).

279.* von Ardenne, M. "Tabellen der Elektronenphysik, Ionenphysik und Übermikroskopie." *Vol.* 2, p. 916. Deutscher Verlag der Wissenschaftten, Berlin (1956).

280.* Larche, K. *Z. Physik* **136,** 74 (1953).

281.* Hills, H. F. *Proc. 3rd Intern. Congr. on High-Speed Photography, London,* 1956 p. 279. Butterworths, London (1956).

282.* Tawil, E. P. *Proc. 3rd Intern. Congr. on High-Speed Photography, London,* 1956 p. 9. Butterworths, London (1956).

283.* Edgerton, H. E. *Proc, 3rd Intern. Congr. on High-Speed Photography, London,* 1956 p. 51. Butterworths, London (1956).

284.* Friedland, S.S., Katzenstein, J., and Zatzick, M. R. *J. Geophys. Res.* **61,** 415 (1956).

285.* Hey, J. S. unpublished work.

286.* Thackeray, D. P. C. *J. Sci. Instr.* **35,** 206 (1958).

287.* Thackeray, D. P. C. *J. Sci. Instr.* **35,** 248 (1958).

288.* Thackeray, D. P. C. *Nature* **180,** 913 (1957).

289.* Fischer, H. *J. Opt. Soc. Am.* **51,** 543 (1961).

290.* Moden, J. C. F.V.R.D.E. exhibit, Physical Society (London), Exhibition Catalogue p. 7 (1958)

291.* Berglung, S., Nilsson, R., Ohlin, P., Siegbahn, K., Sundström, T., and Svennerstedt, S. *Nucl. Instr.* **1,** 233 (1957).

292.* Thumwood, R. F. Flash x-rays; a lecture given to the Institute of Physics, London (April, 1958).

293.* Claesson, S., and Lindqvist, L. *Arkiv Kemi* **12,** 1 (1957).

294.* Saxe, R. F. *Brit. Appl. Phys.* **7,** 336 (1956).

295.* Saxe, R. F. *Electronics* **31** 58 (1958).

296.* Müller, W. Der Ablauf einer elektrischen Drahtexplosion mit Hilfe einer Kerrzellen-Kamera. *Z. Physik* **149** (4), 397-411 (1957).

297.* Lawrence, E. O., and Dunnington, F. G. *Phys. Rev.* **35,** 396 (1930)

298.* Clark, G. L., Holhauser, D. F., and von Foerster, H. M. A.F.O.S.R. Report TN 57-114 (March, 1957).

299.* Maddock, I. Ultra-high speed oscillography. Lectures given to the British Institution of Radio Engineers, London (Jan., 1958).

300.* Ferranti Ltd., Chadderton, England Commercial literature. Industrial paper.

301.* 20th Century Electronics Ltd. Exhibit Physical Society (London), exhibition catalogue p. 18 (1958).

302.* Gerharz, R. *Z. angew. Math. und Physik* (*Basle*) 7, 529 (1956).

303.* Kazan, B. *R.C.A. Rev.* **19,** 19 (1958).

304.* Malan, D. J. *Geophys. pura e appl.* (*Milan*) **38,** 250 (1957).

305.* Courtney-Pratt, J. S., and Thackeray, D. P. C. *Actes 2^e Congr. Intern. Photographie et Cinematographie Ultrarapides*, 1954 p. 163 Dunod, Paris (1954/1956).

306.* Beeson, E. J. G. *B.T.H. Activities* **27,** 131 (1956).

307.* Thackeray, D. P. C. *Proc. 3rd Intern. Congr. on High-Speed Photography, London*, 1956 p. 21. Butterworths, London (1956).

308.* Griffin, W. C. *J. Soc. Motion Picture and Television Engrs.* **66,** 127 (1957).

309.* Courtney-Pratt, J. S., and Edgerton, H. E. *Repts. Progr. Phys.* **20,** 379 (1957).

310a.* Edgerton, H. E. Shock wave photography of large subjects in daylight. *Rev. Sci. Instr.* **29** (2), 171-172 (1958).

310b.† Edgerton, H. E., and Germeshausen, K. J., A microsecond still camera *J. Soc. Motion Picture and Television Engrs.*, **5,** 42-50 (1954).

311.* Conn, W. M. *Bull. Am. Phys. Soc.* **2,** 376 (1957).

312.* Conn, W. M. *Naturwissenschaften* **45,** 6 (1958).

313.* Conn, W. M. "Die Technische Physik der Lichtbogenschweissung." Springer, Berlin (1958).

314.† Zernow, L., Woffinden, G., and Kreyenhagen, K. N. High-speed cinemicrography of electrically exploded tungsten and molybdenum wires. *Phot. Sci. and Engin.* **4,** (1), 31-36 (1960).

315. Liebing, L. Multiple Kerr-cell system with square shuttering characteristic. *Proc. 5th Intern. Congr. on High-Speed Photography, Washington*, 1960 pp. 138-140. Soc. Motion Picture and Television Engineers, New York (1962).

316.* Schall, R., and Thomer, G. Röntgenblitzaufnahmen von Stosswellen in festen, flüssigen und gasförmigen Medien. *Z. angew. Physik* **3** (2), 41-44 (1951).

317.* Thomer, G. Wirkungsweise und Anwendung eines Doppel-Röntgenblitzrohres. *Z. angew. Physik.* **5** (6), 217-221 (1953).

318.* Cullington, E. H., Chace, W. G., and Morgan, R. L. Geophysics Research Directorate, Air Force Cambridge Research Center, Rept. AFCRC-TR 55-227 (1955).

319.* Müller, W. Zur Anwendung des Kerr-Zellenverschlusses bei Schlierenaufnahmen. *Proc. 4th Intern. Congr. Kurzzeitphotographie und Hochfrequenzkinematographie, Cologne*, 1958, pp. 225-229. Helwich, Darmstadt, Germany (1959).

320.* Müller, W. *Z. Physik* **149,** 397-411 (1957).

321.* Rieck, J., and Elle, D. *Askania Warte* **12** (1954).

322.* Rieck, J. Kinematographische Anlage zur Unterdrückung der durch Tageslichtschwankungen möglichen Deckungsunterschiede der photographischen Schicht bei Zeitraffer-Filmaufnahmen. German Patent 946,867 Aug. 9, (1956).

323.‡ Meyer-Waarden, P. F. "Basic Principles of Electrical Fishing," pp. 3-51. Food and Agriculture Organization of the United Nations, Rome (1957).

324.* Andrew, F. J., Kersey, L. R., and Johnson, P. C. An investigation of the problem of guiding downstream migrant salmon at dams. Bulletin 8, p. 65. International Pacific Salmon Fisheries Commission, New Westminster, British Columbia, Canada (1955).

325.* Applegate, V. C., Smith, B. R., and Nielsen, W. L. Use of electricity in the control of sea lampreys. Spec. Sci. Rept. on Fisheries No. 92, p. 52. U.S. Department of the Interior Fish and Wildlife Service, Washington, D.C. (1952).

326.* Bary, B. McK. The effect of electric fields on marine fishes. Marine Research No 1, p. 32. Scottish Home Department, Edinburgh (1956).

327.* Burnet, A. M. R. Studies on the ecology of the New Zealand freshwater eels. I. The design and uses of an electric fishing machine. *Australian J. Marine and Freshwater Research* (*Melbourne*) **3,** (2), 111-125 (1952).

328.* Cattley, J. G. Your guide to electrical fishing. *World Fishing* (*London*) Pt. I: **4,** 125-127; Pt. II: **4,** 166-169; Pt. III: **5,** 202-205 (1955).

329.* Denzer, H. W. The significance of the water conductivity for electrical fishing in inland waters. *Fischwirt* (*Hamburg*) **2** (11), 411-415 (1952).

330.* Denzer, H. W. Present advances in electrical fishing in inland waters. *Fischereiwelt* (*Hamburg*) **4** (10), 301-305 (1954).

331.* Denzer, H. W. The influence of the bottom conductivity on the fishing effect of electrical fishing. *Allgem. Fischerei Z.* (*Munich*) **80** (8), 144-145 (1955).

332.* Denzer, H. W. Electrical fishing. *In* "Handbuch der Binnenfischerei Mitteleuropas," Vol. 5, Suppl., pp. 142-233. Schweizerbart' sche Verlagsbuchhandlung, Stuttgart (1956).

333.* Dickson, W. Marine electrical fishing. *World Fishing* (*London*) **3** (4), 148-151 (1954).

334.* Haier, U. The electrotechnical principles of electrical fishing in the sea. *Arch. Fischereiwissenschaft* (*Braunschweig*) **5** (3/4), 132-144 (1954).

335.* Halsband, E. Experiments on the significance of the law polar stimulation for the reactions of fish in the electrical field. *Arch. Fischereiwissenschaft* (*Braunschweig*) **5** (3/4), 144-152 (1954).

336.* Halsband, E. Experiments on the narcotizing impulse threshold values of various fresh water fishes. *Arch. Fischereiwissenschaft* (*Braunschweig*) **5** (1/2), 45-53 (1955).

337.* Halsband, E. The narcotizing impulse threshold values of various marine fishes. *Arch. Fischereiwissenschaft* (*Braunschweig*) **6** (3/4), 223-228 (1955).

338.* Halsband, E. Experiments on the influence of different types of currents on the metabolism of animals. *Arch. Fischereiwissenschaft* (*Braunschweig*) **6** (5/6), 350-361 (1955).

339.* Halsband, E. The relation between the stimulus intensity and the duration of the single impulses during the electrical stimulation of fishes. *Arch. Fischereiwissenschaft* (*Braunschweig*) **7,** (1/2), 74-81 (1956).

340.* Haskell, D. C., McDougal, J. and Freeman, R. Two back-pack fish shockers *N. Y. Fish and Game J.* **1,** 65-74 (1954).

341.* Hösl, A. The electrotechnical principles of electrical fishing. *Fischwirt* (*Hamburg*) **5** (8), 235-244 (1955).

342.* Holzer, W. An absolute stimulus tension in fishes. *Arch. ges. Physiol.* **229,** 153-172 (1932).

343.* Kreutzer, C. The physiological principles of electrical fishing in the sea. *Arch. Fischereiwissenschaft* (*Braunschweig*) **2** (1/2), 10-14 (1950).

344.* Kreutzer, C. Tuna caught with electric hook and line. *Fischereiwelt* (*Hamburg*) **3** (10), 160-161 (1951).

345.* Kreutzer, C. Electrical fishing. *Elektrotech. Z.* **5,** 1-4 (1954).

346.* Kuhn, O., Schulze, J., and Spieker, D. Experiments on the cause of the galvanotaxic reactions of fishes. *Experientia* **11** (8), 1-7 (1955).

347.* Larsen, K. Electrifishing of sea trout for stripping. *Medd. Danmarks Fiskeri Havundersøgelser* (*Copenhagen*) **1** (6), 1-2 (1954).

348.* Larsen, K. Fish population analyses in some small Danish trout streams by means of D.C. electrofishing, with special reference to the populations of trout (*Salmo trutta* L.) *Medd. Danmarks Fiskeri Havundersøgelser* (*Copenhagen*) **1** (10), 1-69 (1955).

349.* Leduc, St. *Ann. electro-biol. radiol.* (*Paris*) **3/4** (March-April, 1900).

350.* McLean, A. L., and Nielsen, W. L. Directing the movement of fish with electricity. Spec. Sci. Rept. Fisheries No. 93, 24 pp. U.S. Department of the Interior, Fish and Wildlife Service, Washington, D.C. (1953).

351.* Meyer-Waarden, P. F. Electricity and fishing. *Orion* (*Munich*) **8** (3/4), 91-99 (1953).

352.* Meyer-Waarden, P. F. Does electronarcosis influence the vitality and growth of fish. *Fischwirt* (*Hamburg*) **3** (7), 225-228 (1953).

353.* Meyer-Waarden, P. F. New ways of electrical fishing. *Elektronik* (*Würzburg*) **4** (7), 159-165 (1955).

354.* Moore, W. H. A new type of electrical fish catcher. *J. Animal Ecology* **23** (2), 373-375 (1954).

355.* Reidel, F. The effect of electrical current, as applied in electrical fishing on the gonads of fishes. *Z. Fischerei* (*Berlin*) [N. S.] **3** (1/2/3), 1-53 (1954).

356.* Sheminsky, Fe., Sheminsky, Fr., and Bukatsch, F. Electrobiology. Special reprint from "Tabulae Biologicae," Vol. 19, 198 pp. The Hague, Netherlands (1941).

357.* Schiemenz, F., and Schönfelder, A. Fishing with electricity. *Z. Fischerei* (*Berlin*) **25**, 161-187 (1927).

358.* Skoskiewicz, Janina, and Janusz. "The Newest Fishing Methods in the U.S.S.R.", pp. 57. Fachbuch, Leipzig (1954).

359.* Trefethen, P. S. Exploratory experiments in guiding salmon fingerling by narrow D.C. electric field. Spec. Sci. Rept. Fisheries No. 158, 42 pp. U.S. Department of the Interior, Fish and Wildlife Service, Washington, D.C. (1955).

360.* Tschernigin, N. F. A new fishing method (in Russian). *Fishing Industry* (*Moscow*) **2,** 22 (1952).

361.* Applegate, V. C., Macy, P. T., and Harris, V. E. Selected bibliography on applications of electricity in fishery science. Spec. Sci. Rept. Fisheries No. 127, 55 pp. U.S. Department of the Interior, Fish and Wildlife Service, Washington D.C. (1954).

362.* Denzer, H. W. Electrical fishing. *In* "Handbuch der Binnenfischerei Mitteleuropas," Vol. 5, Suppl. pp. 142-233. Schweizerbart' sche Verglasbuchhandlung, Stuttgart (1956).

363.* Scheminsky, Fe., Scheminsky, Fr., and Bukatsch, F. Electrobiology. Special reprint from "Tabulae Biologicae," Vol. 19, pp. 198. The Hague, Netherlands (1941).

364.* Dethloff, J., and Matusche, H. Elektro-Fischerei auf hoher See. *VDI* (*Ver. Deut. Ing.*) *Z.* **14,** 3 (1960).

365. Früngel, F. Electric impulse lamp as transmitter for light-flash signalling system. British Patent 742,396 (1955).

366. Früngel, F. Electric impulse lamp as transmitter for light-flash signalling system. U.S. Patent 2,749,482 (June, 1956) (German priority, 1951).

367. Früngel, F. Receiver for light flash signalling system. U.S. Patent 2,941,084 (June, 1960) (German priority, 1951).

368. Früngel, F. System for signalling by light impulses. U.S. Patent 2,921,184 (Jan., 1960) (German priority, 1951).

369. Früngel, F. Receiver for light-flash signalling system. British Patent 746,030 (March, 1956) (German priority, 1951).

370.† Borkowski, H. Wolkenhöhenmesser. Bestimmung der Höhe der Wolkenuntergrenze zu jeder Tages- und Jahreszeit. *Deut. Flugtechnik* **4** (7), 210-213 (1960).

371.* Baude, A., La télémetrie des nuages au moyen d'impulsions lumineuses. *Radio Française* **10,** 3-17 (1949).

372.* Baude, A. Conditions électronique de réalisation d'un radar à ondes lumineuses. *L'onde électrique*. pp. 44-48, 90-101 (1951).

373.* Moles, F. J. The cloud range meter. *Gen. Elec. Rev.* No. 4 (1946).

374.* Bureau, R. Altimétrie des nuages par impulsions lumineuses. *Météorol.* (July/ Sept., 1946).

375.* Hansen, W. W. Surveying by pulsed light radar. *Electronics* (July, 1948).

376. Früngel, F. Ein impulsoptisches Transmissometer für Registrierung der Normsichtweite zwischen etwa 40 m und 5 km (Skopograph). *Arch. Meteorol., Geophys. u. Bioklim.* **B10,** 253-262 (1960).

377. Früngel, F. Der SKOPOGRAPH ein Flugplatz-Transmissometer mit Impulslicht. *In* "Physik der Atmosphäre", Vol. 33, 1/2, pp. 36-56 Akademische Verlagsges, Frankfurt (1960).

378. Früngel, F. Ein impulsoptischer Wolkenhöhenmesser für Registrierung von Wolkenhöhen bis 1000 m bzw. 3000 m CEILOGRAPH. *Archi. Meteorol. Geophys. u. Bioklim.* **10** (2), 244-251 (1960).

379.* Beutell, R. G., and Brewer, R. G. Instruments for the measurement of the visual range. *J. Sci. Instr.* **26,** 357-359 (1949).

380.* Ruppersberg, G. H., Eigene Neuentwicklung von Sichtregistriergeräten, Sicht-Kolloquium, Darmstadt (1959).

381.* Foitzik, L. Sichtmessgeräte. *Ann. Met.* **5,** 215-222 (1952).

382.* Douglas, C. A. Development of a transmissometer for determining visual-range. Devel. Rept. No. 47, U.S. Department of Commerce, Washington, D.C. (1945).

383.* Douglas, C. A. Final-Reports. Air Force-Navy-Civil Landing Aids Experimental Station. Arcata, California (1949).

384.* Douglas, C. A. Final-Approach visibility studies. U.S. Weather Bureau. Air Navigation Development Board. Project 4, 14, Washington, D.C. (1953/1954.)

385.* Foitzik, L. Die Sichweite und ihre Registrierung. Jahrbuch 1955 der Wissenschaftlichen Gesellschaft für Luftfahrt (WGL) pp. 112-116. Vieweg, Braunschweig (1955).

386.* Foitzik, L. Bestimmung der Schrägsicht nach einer Kopmonenten-Methode. Sicht-Kolloquium, Darmstadt, 1959.

387.* Stevens, N. B., Horman, M. H., and Dodd, E. E. The Determination of atmospheric transmissivity by backscatter of a pulsed light system. Riverside Research Lab., Motorola Inc., Riverside, California (1957).

388.* Middleton, W. E. K. "Vision through the Atmosphere," pp. 199, "Univ. of Toronto Press, Toronto, Canada (1952).

389.* Marshak, I. S., and Feigenbaum, M. G. Vergleich der Wirksamkeit von Impulslichtquellen und Gleichlichtlampen bei Sichtweiten im Nebelbereich (Opt. Institut der Universität Moskau). *Svetotekhnika* **3,** 17-22 (1959).

390. Früngel, F. Ein impulsoptisches Transmissometer mit Funkenlicht. Sicht-Kolloquium, Darmstadt (1959).

391.† Meek, J. M. The mechanism of growth of spark discharges. *Proc. Intern. Symposium on Electrical Discharges in Gases, Delft, Netherlands* (April, 1955).

392.† Fischer, L. H. Formative time lags of spark breakdown. *Proc. Intern. Sypmosium on Electrical Discharges in Gases, Delft, Netherlands* (April, 1955).

393.† Craggs, J. D. Some properties of spark channels. *Proc. Intern. Symposium on Electrical Discharges in Gases, Delft, Netherlands* (April, 1955).

394.† von Engel, A. Internal radiation and nature of discharges. *Proc. Intern. Symposium on Electrical Discharges in Gases, Delft, Netherlands* (April, 1955).

395.† Warren, R. W., and Loeb, L. B. Field measurements in glow discharges with a refined electron beam probe and automatic recording. *Proc. Intern. Symposium on Electrical Discharges in Gases, Delft, Netherlands* (April, 1955).

396.† Morgan, C. G., and Sommerville, J. M. Time studies of ionization processes. *Proc. Intern. Symposium on Electrical Discharges in Gases, Delft, Netherlands* (April, 1955).

397.† Raether, H. The electron avalanche and its development. *Proc. Intern. Symposium on Electrical Discharges in Gases, Delft, Netherlands* (April, 1955).

398.† Massey, H. S. W. Fundamental primary processes in gas discharges. *Proc. Intern. Symposium on Electrical Discharges in Gases, Delft, Netherlands* (April, 1955).

399.† Liebing, L. Über das Triggern von Kabelfunkstrecken. *Z. angew. Physik* **13** (4), 201-202 (1961).

400.* Middleton, W. E. K. The effect of the angular aperture of telephotometer. *J. Opt. Soc. Am.* **39**, 576-581.

401.† Kash, S. W. Experiments at Lockheed Missile Systems Division. *In* "Magnetohydrodynamics," pp. 92-93. Stanford Univ. Press, Stanford, California (1957).

402.† Kolb, A. C. Production of high energy plasma by magnetically driven shock waves. *Phys. Rev.* **107** (2), 345-350 (1957).

403.* Kolb, A. C. *Phys. Rev.* **107,** 1197 (1957).

404.‡ Janes, G. S., and Patrick, R. M. The production of high temperature gas by magnetic acceleration. *In* "Conference on Extremely High Temperatures." pp. 3-10. Wiley, New York (1958).

405.‡ Korneff, T., Nadig, F. H., and Bohn, Ll. Plasma acceleration experiments. *In* "Conference on Extremely High Temperatures," pp. 197-207. Wiley, New York (1958).

406.‡ Bostick, W. H. Plasma motors. *In* "Conference on Extremely High Temperatures," pp. 169-179. Wiley, New York (1958).

407.‡ Post, R. F. Experimental research on high temperature plasmas. *In* "Symposium of Plasma Dynamics," pp. 1-7. Pergamon, London (1960).

408.† Blackman, V. H., and Niblett, B. Experiments using a hydromagnetic shock tube. *In* "The Plasma in a Magnetic Field," pp. 87-98. Stanford Univ. Press, Stamford, California (1958).

409.‡ Kash, S. W., Gauger, J., Starr, W., and Vali, U. Velocity measurements in magnetically driven shock tubes. *In* "The Plasma in a Magnetic Field," pp. 99-109. Stanford Univ. Press, Stanford, California (1958).

410.‡ Scott, F. R., Basman, W. P., Little, E. M., and Thomson, D. B. Magnetic channeling of a strong shock. *In* "The Plasma in a Magnetic Field," pp. 110-116. Stanford Univ. Press, Stanford, California (1958).

411.* Fowler, R. G., Atkinson, W. R., Compton, W. D., and Lee, E. J. *Phys. Rev.* **88,** 137 (1952).

412.* Landshoff, R. K. M. (Editor). "The Plasma in a Magnetic Field." Stanford Univ. Press, Stanford, California (1958).

413.* Landshoff, R. K. M. (Editor) "Magnetohydrodynamics." Stanford Univ. Press, Stanford, California (1957).

414.* Gilmore, F. R. Equilibrium composition and thermodynamic properties of air to 24,000° K. RAND Report 1543 (Aug., 1955).

415.* Broadwell, J. E. A simple model of the nonequilibrium dissociation of a gas in couette and boundary layer flow. Douglas Aircraft Co. Rept. SM 22888 (Aug., 1957.)

416.* Rossow, V. J. On flow of electrically conducting fluids over a flat plate in the presence of a transverse magnetic field. NACA TN 3971 (1957).

417.* Kaplun, S., Lagerstrom, P. A., and Kaplun, S. A symptotic expansion of Navier stokes solutions for small Reynolds numbers. Low Reynolds number flow past a circular cylinder and note on the preceding two papers. *J. Math. Mech.* **6** (5), 585-606 (Sept., 1957).

418.* Chester, E. Stokes flow in a conducting fluid. *J. Fluid Mech.* **3** (3), 304-308 (Dec., 1957).

419.† Nolan, Ph. High intensity fractional microsecond light sources. *J. Soc. Motion Picture and Television Engrs.* **70,** 632-633 (1961).

420.* Petschek, H. E. Magnetically driven shock waves. Experiments at AVCO, "Magnetohydrodynamics." Stanford Univ. Press Stanford, California (1957).

421.* Petschek, H. E., and Byron, S. *Ann. Phys.* (*N.Y.*) **1,** 270-315 (1957).

422.* Keck, J. C., Kivel, E., and Wentink, T. Emissivity of high temperature air. AVCO Research Report 8 (1957).

423.* Bostick, W. H. *Phys. Rev.* **104,** 292 (1956); **106,** 104 (1957).

424.* Bostick, W. H., and Twite, O. A. *Nature* **179,** 214 (1957).

425.* Harris, E., Theus, R., and Bostick, E. H. *Phys. Rev.* **105,** 46 (1957).

426.* Millsaps, K. The linear acceleration of large masses by electrical means. Hollmann Air Development Center, Operations Research Office, Tenchical Memorandum No. 3.

427.* Morozov, A. I. *Soviet Phys. JETP* (*Engl. Transl.*) **5,** 215 (1957).

428.* Artsimovich, L., Chuvatin, S., Lukjanov, S., and Podgorny, I. Electrodynamical acceleration of plasma bunches. *Soviet Phys. JETP* (*Engl. Transl.*) **33,** 3 (1957).

429.* Elassser, W. M. Dimensional relations in magnetohydrodynamics *Phys. Rev.* **95,** 1 (1954).

430.* Grad, H., and Rose, M. H. Notes on magnetohydrodynamics, Number II. AEC Research and Development Report, New York Univ. (1956).

431.* Spitzer, L. "Physics of fully Ionized Gases." Wiley (Interscience), New York (1956).

432.* Chapman, S., and Cowling, T. G. The mathematical theory of non-uniform gases. Cambridge Univ. Press, London and New York (1952).

433.* Stratton, T. A. "Electromagnetic Theory." McGraw-Hill, New York (1941).

434.* Bond, J. W. Structure of a shockfront in argon. *Phys. Rev.* **105,** 1683 (1957).

435.* Petschek, H. E., and Byron, S. Approach to equilibrium ionization behind strong shockwaves in argon. Cornell Univ. Press, Ithaca, New York (1955).

436.* Lin, S. C., Resler, E. L., and Kantrowitz, A. Electrical conductivity of highly ionized argon produced by shock waves. *J. Appl. Phys.* **26,** 95 (1955).

437.* Cousins, S. W., and Ware, A. A. Pinch effect oscillations in a high current toroidal ring discharge. *Proc. Phys. Soc.* (*London*) **64,** 159 (1951).

438.* Kurtchatov, L. V. On the possibility of producing thermonuclear reactions in a gas discharge. *Nucleonics* **14** (6), 36 (1956).

439.* Burkhardt, L. C., Dunaway, R. E., Mather, J. W., Philips, J. A., Sawyer, G. A., Stratton, T. F., Stovall, E. J., and Tuck, J. L. Pinch effect. *J. Appl. Phys.* **28,** 519 (1957).

440.* Kolb, A. C. "Lockheed Symposium on Magnetohydrodynamics." Stanford Univ. Press, Stanford, California (1956).

441.* Kolb, A. C. *Bull. Am. Phys. Soc.* [2] **2,** p. 47 (1957).

442.* Fowler, R. G., Goldstein, and Clotfelder. *Phys. Rev.* **82,** p. 879 (1951).

443.* Fowler, R. G., Atkinson, W. R., and Marks. *Phys. Rev.* **87,** 966 (1952).

444.* Fowler, R. G., Atkinson, W. R., Clotfelder, and Lee, R. J. University of Oklahoma, Research Institute, Project Report 1952 (unpublished).

445.* Spitzer, L. "Physics of Fully Ionized Gases." Wiley (Interscience), New York (1956).

446.* Tucker, D. G. The synchrodyne receiver. *Electronic Eng.* **19,** 241, 276 (1947).
447.* Bohm, D., and Aller, L. *Astrophys. J.* **105,** 131 (1947).
448.* Cohen, Spitzer, L., and Routly. *Phys. Rev.* **80,** 230 (1950).
449.* Gasirowicz, Neumann, and Riddel. *Phys. Rev.* **101,** 922 (1956).
450.* Industrial paper, Field Emmission Corp., McMinnville, Oregon; Fexitron flash x-ray system, Bulletin S 210 5-15-61, T-10-15-60, S 10-15-60, S 4-18-60.
451.* Rosenbluth, M., MacDonald, and Judd. *Phys. Rev.* (to be published).
452.† Demaris, A. J. Shaping metals by shock wave. *Exporter* **3** (1) 9 (Spring 1961). Publ. by Intern. General Electric Co.
453.* Watermann, M. N., Reid, K. M., and Taylor, G. J. (Editors). "IES Lighting Handbook," 3rd ed. pp. 8-41, 46. Publ. by the Illuminating Engineering Society. New York (1959).
454.† Tawil, E. New development in "guided" air sparks. *Proc. 3rd Intern. Congr. on High Speed Photography, London,* 1956 pp. 148-152. Butterworths, London (1956).
455.* Fayolee, P., and Naslin, P. J. *J. Soc. Motion Picture Television Engrs.* **1** (60), 603 (1953).
456.† Laporte, M. High power electronic flash lamps for cinematography. *Proc. 3rd Intern. Congr. on High-Speed Photography, London,* 1956 pp. G31-32 Butterworths, London, (1956).
457.* Anon. Description of instruments. Lewis Flight Propulsion Laboratory, National Advisory Committee for Aeronautics, Cleveland, Ohio, NACA RM E50K27.
458.† Beams, J. W., Kuhlhau, A. R., Lapsley, A. C., McQueen, J. H., Snoddy, L. B., and Whitehead, W. D. Spark light source of short duration. *J. Opt. Soc. Am.* **37** (10), 868-869 (1947).
459.* Beams, J. W. *Phys. Rev.* **35,** 24 (1930).
460.* Snoddy, L. B., Dietrich, J. R., and Beams, J. W. *Phys. Rev.* **52,** 739 (1937).
461.* Street, J. C., and Beams, J. W. *Phys. Rev.* **38,** 416 (1931).
462.† Edgerton, H. E. Submicrosecond flashes of light. *Proc. 4th Intern. Kongr. für Kurzzeitphotographie und Hochfrequenzkinematographie, Cologne,* 1958 pp. 91-97. Helwich, Darmstadt, (1959).
463.* Talbot, W. J. F. On the production of instantaneous images. *Phil. Mag.* **34,** 73-77 (Jan., 1852).
464.* Mach, E. A review of the results reported in the 1887, 1888 and 1889 issues of the Akademie der Wissenschaften in Wien. *Nature* **11** (2), 250-251 (July, 1890).
465.* Boys, C. V. Electric spark photography of flying bullets. *Nature* **11** (7), 415-421 (1893); pp. 165-182, 440-446 (1893). *Smithsonian Rep.*.
466.* Cranz, C. "Experimental Ballistics," Vol. 3, Chapter 9. Springer, Berlin (1927).
467.* Bull, L. *Compt. rend. acad. sci.* 755 (1904).
468.* Worthington, A. M. "Splashes" Longmans Green, London (1901).
469.* Looms, J. S. T., and North, R. J. Short duration spark light sources for photography of high speed airflow. *Proc. 3rd Intern. Congr. on High Speed Photography, London,* 1956 pp. AX 18-AX 22. Butterworths, London (1956).
470.* Holder, D. W., North, R. L., and Wood, G. P. Schlieren methods and interferometer methods. *AGARDograph No.* 23.
471.* Fischer, H. Afterglow-tails and stability of high density nanosecond arc channels. *Proc. 6th Intern. Symposium on Ionization Phenomena in Gases, Paris* (July, 1963).
472.* Schardin, H., and Struth, W. Neuere Ergebnisse der Funkenkinematographie. *Z. tech. Physik* **18** (11), 474-477 (1937).
473.* Edgerton, H. E. Double flash microsecond silhouette photography. *Rev. Sci. Instr.* **23** (10), 532-533 (1952).

474.* Edgerton, H. E., and Cathou, J. Y. Xenon flashtube of small size. *Rev. Sci. Instr.* **27** (10), 821-825 (1956).

475.* Tawil, E. P. Guided air sparks. *Proc. 3rd Intern. Congr. on High Speed Photography, London,* 1956 pp. Y 148-152. Butterworths, London, (1956).

476.* Schardin, H., and Fünfer, E. Grundlagen der Funkenkinematographie. *Z. angew. Physik* **4** (5), 185-199 (1952); **4** (6), 224-238 (1952).

477.* Stenzel, A., Minimum inductance circuit with circular Ba Ti O_3-plate, C = 40000$\mu\mu$F. electrical energy = 2 Joule, light duration 100 mμs. Note tech. 17a/57. L.R.S.L.

478.* Struth, W. Neue Verfahren der Kurzzeitphotographie und Hochfrequenzkinematographie. *Kinotechnik* **12** (9), 230-248 (1958).

479.† North, R. J. A Cranz-Schardin high speed camera for use with a hypersonic shock tube. *Proc. 4th Intern. Kongr. für Kurzzeitphotographie und Hochfreauenzkinematographic, Cologne,* 1958 pp. 151-155. Helwig, Dormstadt (1959).

480.* Fischer, H., and Mansur, L. C. *In* "Conference on Extremely High Temperatures." Wiley, New York (1958).

481.* Fischer, H. *Proc. High Temperatures Conf. Berkeley* (June, 1956).

482.* Sauer, J. *Z. Physik* **143,** 44 (1955).

483.† Hills, H. F. Layout and details of the multiple spark range at Cambridge. *Proc. 3rd Intern. Congr. on High-Speed Photography, London,* 1956 pp. T 112-117. Butterworths, London (1956).

484.* Cranz, C., and Schardin, H. *Z. Physik* **56,** 147 (1929).

485.* Adams, C. A. M. O. S. Report (1940). Not published.

486. Früngel, F. Controlled, electric sparking system for stroboscope and the like. British Patent 719,356 (Dec., 1954).

487. Früngel, F. High frequency spark device. U.S. Patent 2,906,924 (Sept., 1959).

488. Früngel, F. High intensity general purpose stroboscope. Industrial paper, Impulsphysik GmbH, Hamburg, Germany.

489. Früngel, F. Stroboscopic light source. British Patent 719,296 (Dec., 1954).

490. Früngel, F. Stroboscopic light source. U.S. Patent 2,703,374 (March, 1955).

491.† Osterwalder, J. Ein neuer Prüfstand für Kennlinien-Messungen und Kavitations-Beobachtungen an Wasserturbinen und Pumpen. *Escher-Wyss-Mitt.* No. 2 (1956).

492.* Rint, C. *In* "Handbuch für Hochfrequenz- und Elektrotechniker," p. 236 Vol. 2, Verlag für Radio-Foto-Kino-Technik, Berlin (1953).

493. Früngel, F. Strobokin-Ultra-Rapid Flasher. Industrial paper. Impulsphysik GmbH, Hamburg, Germany.

494.† Bomelburg, H. J., Herzog, J., and Weske, J. R. The electric spark method for quantitative measurements in flowing gases. *Z. Flugswissenschaft* **7** (11), 322-329 (1959).

495.* Angerer, E. V. "Technische Kunstgriffe by physikalischen Untersuchungen" Vieweg, Braunschweig, 1924.

496.* Wortmann, F. X. Eine Methode zur Beobachtung und Messung von Wasserströmungen. *Z. angew. Math. u. Physik* **5,** 210-206 (1953).

497.* Geller, E. W. An electrochemical method of visualizing the boundary layer. *J. Aeronaut. Sci.* **22,** 869-870 (1955).

498.* Townend, H. C. H. Hot wire and spark shadowgraphs of the air flow through an air screw. *Phil. Mag.* [7] **14,** 700-712 (1932).

499.* Townend, H. C. H. Visual and photographic methods of studying boundary layer flow. ARC 1803 (1937).

500.* Goldstein, S. "Modern Developments in Fluid Dynamics," pp. 290-292. Oxford Univ. Press (Clarendon), London and New York (1950).

501.* Yoichi Saheki. On the measurement of wind velocity distribution by the electric spark method. *Hokaido Fac. Eng. Mem.* **8,** 185-190 (1947). Translation by E. Hope, Scientific Intelligence Division, Defense Research Board, Canada (1950).

502.* Weske, J. R. Ein Beitrag zur Untersuchung stetiger und unstetiger dreidimensionaler Strömungsfelder in Turbomaschinen. *Z. angew. Math. u. Physik* **9b,** 721-724 (1958)

503.* Rudinger, G., and Somers, L. M. Behaviour of small regions of different gases carried in accelerated gas flows. SQUID-Rep. Cal. 76 P (1958).

504. Früngel, F., Thorwart, W., and Patzke, H. G. Tracing of rapid air currents and shock waves by means of high frequency high voltage sparks. *Proc. 5th Intern. Congr. on High Speed Photography Washington, D.C.* 1960, pp. 498-502. Society of Motion Picture and Television Engineers (1962).

505.† Vogel, J. Präzisionsaufnahmen schneller Luftströmungen durch quartzgesteuerte Funkenblitzfolgen. *Proc. 4th Intern. Kongr. für Kurzzeitphotographie und Hochfrequenzkinematographie, Cologne,* 1958 pp. 267-273. Helwich, Darmstadt (1959).

506. Früngel, F. "Impulstechnik," pp. 376-385. Akademischer Verlagsges., Leipzig (1960).

507. Früngel, F. Stosstransformatoren zur Erzeugung von Hochspannungsimpulsen. *Elektrotech. Z.* **81,** 355-360 (1960).

508. Früngel, F. Anordnung zur Erzeugung von starken Funkenentladungen sehr kurzer Dauer. German Patent 887,545 (Aug., 1953).

509. Früngel, F. Anordnung zur Erzeugung von starken Funkenentladungen sehr kurzer Dauer. Japanese Patent 233,700 (Dec., 1954).

510.* Rumpf, H. Versuche zur Bestimmung der Teilchenbewegung in Gasstrahlen des Beanspruchungsmechanismus in Strahlmühlen. *Chem. Ing-Tech.*-**36** (5), 335-342 (1960). Verlag Chemie, Weinheim.

511. Früngel, F. Bewegungsaufnahmen rascher Luftströmungen und Stosswellen durch hochfrequente Hochspannungsfunken. Wissenschaftliche Gesellschaft für Luftfahrt Jahrbuch 1960. Vieweg, Braunschweig (1960).

512. Früngel, F. Methoden der photographischen Erfassung schneller Bewegungsvorgänge in Strahl-, Strömungs- und Werkstoff-Forschung. *Motortech. Z.* **22,** 5 (May, 1961).

513. Früngel, F., and Thorwart, W. Bewegungsanalyse mit Funktenlich. *Physik. Blätter* **17** (3), 72-78 (1961).

514.* Thorwart, W. Photographische Analyse einer unsichtbaren Welt. *Photo-Technik. u. Wirtschaft* **4,** 102-107 (1961).

515.* Bardocz, A., Vörös, T., and Vanjek, U. M. Zeitlicher Verlauf der Verbreiterung von Spektrallinien und der lone konzentration in Funkenentladungen. *Z. angew. Physik* **14,** 581-584 (1962).

516.‡ Chesterman, W. D. "The Photographic Study of Rapid Events," pp. 105-113, 116-117. Oxford Univ. Press, (Clarendon) London and New York (1951).

517.† Rudinger, G., and Somers, L. M. A simple schlieren system for two simultaneous views of a gas flow. *J. Motion Picture and Television Engnrs.* p. 107 (Oct., 1957).

518.* Picard, C. *La recherche aeronautique* **32,** 15 (1953).

519.* Veret, C. *La recherche aeronautique* **29,** 3 (1952).

520.* Holder, D. W., and North, R. J. Optical methods for examining the flow in high speed wind tunnels. AGARDOgraph, Part I, No. 23, Schlieren-methods (1956).

521.* Hett, J. H. A high speed stereoscopic schlieren system. *J. Motion Picture and Television Engrs.* **56,** 214-218 (1951).

522.* Altrogge, W. Ignition spark system. U.S. Patent 2,904,723 (1958).

523.* Chace, W. G., and Moore, H. K. "Exploding Wires," Plenum Press, New York (1959).

524.* Zernow, D., and Woffinden, G. The cinemicroscopic observation of exploding wires. "Exploding Wires," pp. 170-185. Plenum Press, New York (1959).

525.* Miller, W. Studies of exploding wire phenomenon by use of exploding schlieren photography "Exploding Wires," pp. 186-208. Plenum Press, New York (1959).

526.* Nadig, F. H., Bohn, J. L. and Korneff, T. High speed framing camera for photographing exploding wire phenomena. "Exploding Wires." pp. 345-364. Plenum Press, New York (1959).

527.* Holtzworth, R. E., and Hinz, D. J. Exploding wire backlighting for the study of detonation, shock and shaped charges. Office of Technical Service, Washington, D.C.

528.* Müller, W. Der Ablauf einer elektrischen Drahtexplosion mit Hilfe der Kerrzellen-Kamera. *Z. Physik* [4], **149,** 397-411 (1957).

529.* Anderson, J. A. The spectrum of electrically exploded wires. *Astrophys. J.* **51,** 37-48 (Jan., 1920).

530.* Anderson, J. A., and Smith, S. General characteristics of electrically exploded wires. *Astrophys. J.* **64,** 295-314 (1925).

531.* Schaaffs, W. Beobachtungen an elektrischen Drahtexplosionen. *Z. angew. Physik* **11,** 63-65 (1959).

532.* Marshall, F. R., and Zarem, A. M. A multiple Kerr-cell camera. *Rev. Sci. Instr.* **21,** 514-519 (1950).

533.* Holtmann, A. E. J., and Prime, H. A. The operation and photographic characteristics of Kerr-cell type of electro-optical shutter. *Proc. Phys. Soc.* (*London*) **63B,** 561-572 (Aug,. 1950).

534.* Quinn, H. F., McKay, W. B., and Bourque O. J. A Kerr-cell camera and flash illumination unit for ballistic photography. *J. Appl. Phys.* **21,** 995-1001 (Oct., 1950).

535.* Pugh, E. M., von Heine-Geldern, R., Foner, S., and Mutschler, E. C. Kerr-cell photography of high speed phenomena. *J. Appl. Phys.* **22,** 487-493 (April, 1951).

536.* Froome, K. D. An electronically operated Kerr-cell shutter. *Phot. J.* **92B,** 158-161 (Sept./Oct.)., 1952).

537.* Sultanoff, M. A 0.1 microsecond Kerr-cell shutter. Phot. Eng. **5,** 80-90 (1954).

538.* Nichloson, W. Q., and Ross, I. Kerr-cell shutter has submicrosecond speed. *Electronics* **28,** 171-173 (June, 1955).

539.* Senior, D. A. The Kerr-cell, a high speed electro-optical shutter. Pts. I and II. *Instruments Practice* **11,** 338-341 (April, 1957); **11,** 471-476 (May, 1957).

540.* Zarem, A. M., Marshall, F. R. and Hauser, S. M. Millimicrosecond Kerr cell camera shutter. *Rev. Sci. Instr.* **29,** 1041-1044 (Nov., 1958).

541.* Müller, W. Der Ablauf einer elektrischen Drahtexplosion mit Hilfe der Kerr-Zellen-Kamera. *Z. Physik*, [4] **149,** 397-411 (1957).

542.* Jenkins, J. A., and Chippendale, R. A. High speed photography by means of the image converter. *Phillips Tech. Rev.* **14,** 213-225 (Feb., 1953).

543.* Hogan, A. W. Use of image converter tube for high speed shutter action. *Proc. I.R.E.* **29,** 268-270 (March, 1951).

544.* Turnock, R. C. The performance of image converters as high speed shutters. *Proc. I.E.E.* **98,** 635-641 (Oct., 1951).

545.* Jenkins, J. A., and Chippendale, R. A. Some new image converter tubes and their applications. *Electronic Eng.* **24,** 302-307 (July, 1952).

546.* Courtney-Pratt, J. S. Image converter tubes and their application to high speed photography. *Phot. J.* **92B,** 137-148 (Sept./Oct. 1952).

547.* Chippendale, R. A. Image converter techniques applied to high speed photography. *Phot. J.* **92,B** 149-157 (Sept./Oct., 1952).

548.* Gibson, F. C., Bowser, M. L., Ramaley, C. W., and Scott, F. H. Image converter for studies of explosive phenomena. *Rev. Sci. Instr.* **25,** 173-176 (Feb., 1954).

549.* Braunstein, S. Image converter at Aberdeen. *Indust. Phot.* **5,** 30-31, 67 (July, 1956).

550.* Mullard Ltd. Image converter and dissector tubes for use in high-speed photography. *J. Sci. Instr.* **34,** 78, (Feb., 1957). (Manufactured by Mullard Ltd.).

551.* Linden, B. R., and Snell, P. A. Shutter Image converter tubes. *Proc. I.R.E.* **45,** 513-523 (April, 1957).

552.* Stoudenheimer, R. G., and Moor, J. C. An image converter tube for high speed photographic shutter service. *RCA Rev.* **18,** 322-331 (Sept., 1957).

553.* Duckworth, G. W. Experimental RCA tube for high speed photography. *Phot. Sci. Eng.* **1** (3), 138 (Jan., 1958).

554.* Reshovsky, E. E. Photo instrumentation. The PTI image converter camera. *Photo Methods for Industry* **1,** 20 (July, 1958).

555.* Drosd, R. D., Liddiard, T. P., and Singleton, B. N. High speed photography with the I.P. 25 A image converter tube. PB 106 734, Office of Tech. Service, U.S. Dept. of Commerce, Washington, D.C. (1951).

556.* Jenkins, J. A., and Chippendale, R. A. Application of image converter to high speed photography. *J. Brit. Inst. Radio Eng.* **11,** 505-517 (Nov., 1951).

557.* Laurence, R. F. and, Shaftan K. Some current image converter practice. *Phot. Eng.* **3** (4), 189-206 (1952).

558.* Maninger, R. C., and Butenbach, R. W. Millimicrosecond photography with image converter camera. *I. R. E. WESCON Convention Record* **1** (5), 36-42 (1957).

559.* Beachamp, K. G., and Tyrrell, A. J. An image converter equipment for the observation of self-luminous gaseous discharges. *Electronic Eng.* **31,** 384-389 (July, 1959).

560.* Strong, C. L. How to make extremely energetic sparks for high speed photography and other purposes. *Sci. American* **197,** 148-150, 152, 154, 156-58 (Nov., 1957).

561.* Wilson, M. R., and Hiemenz, R. J. High speed multiple spark light source. *Rev. Sci. Inst.* **29** 949-51 (Nov., 1958).

562.* Simonds, B. T., Shafer, P. E., and Moulton, J. F. Method for producing multiple sparks at high frequencies across a single gap. P.B. 120 922, Office of Tech. Service, Washington, D.C.

563.* Osterstrom, G. E. Knocking combustion observed in a spark ignitron engine with direct and schlieren high speed motion pictures and pressure records. U.S. National Advisory Committee for Aeronautics, Tech. Note 1614 (June, 1948).

564.* Adams, G. K. A repetitive spark source for shadow and schlieren photography. *J. Sci. Instr.* **28,** 379-384 (Dec., 1951).

565.* Quale, P. P. Single spark photography and its application to some problems in ballistics. *Nature* **115,** 767-770 (May, 1925).

566.* Smith, J. E. Spark photography as a means of measuring rate of explosion. *Phys. Rev.* **25,** 870-876 (June, 1925).

567.* Woodbridge, D. B., and Palmer, A. E. Some application of spark photography. *J. Opt. Soc. Am.* **16,** (2), 125-133 (1928).

568.* Schardin, H. Die Grenzen der Hochfrequenzkinematographie. *Kinotechnik* **14,** 41-45 (Feb., 1932).

569.* Cranz, C., and Schardin, H. Fortschritte auf dem Gebiet der Hochfrequenzkinematographie. *VDI* (*Ver. Deut. Ing.*) *Zeitschrift* **79,** 1075-1079 (1935).

570.* Payman, W., and Woodhead, D. W. The pressure wave sent out by an explosive. III. Spark photographs with permitted explosives. Safety in Mines Res. Board Paper Paper No. 88 (1934).

571.* Chiles, J. A., Jr. A photographic study of the Vacuum spark discharge. *J. Appl. Phys.* **8,** 622-626 (Sept., 1937).

572.* Rothrock, A. M., and Spencer, R. G. A photographic study of combustion and knock in a spark ignition engine. U.S. National Advisory Committee for Aeronautics Report No. 622, 21 pp. Lab. Langley Field, Virginia (1938).

573.* Miller, C. D. Study of high speed photography of combustion and knock in a spark ignition. U.S. National Advisory Committee for Aeronautics Report No. 727, 23 pp. (1942).

574.* Miller, C. D., and Olsen, H. L. Identification of knock in NACA High speed photographs of combustion in a spark ignition engine. U.S. National Advisory Committee for Aeronautics Report No. 761 (1943).

575.* Miller, C. D. Relation between spark ignition engine knock detonation waves and autoignition as shown by high speed photography. U.S. National Advisory Committee for Aeronautics. WartimeReport No. E 238 (1947). (P.B. 96 955 issued in 1949).

576.* Miller, C. D. Roles of detonation waves and autoignition in spark ignition engine knock as shown by photographs taken at 40,000 and 200,000 frames per second. *Trans. Soc. Automotive Eng.* **1**, 98-143 (1947). Abridged in: *J. Soc. Automotive Eng.* **54**, 34-35 (Oct., 1946).

577.* Allen, J. E., and Craggs, J. D. High current spark channels. *Brit. J. Appl. Phys.* **5**, 446-453 (Dec., 1954).

578.* Lintin, D. R., and Wooding, E. R. Investigation of the ignition of a gas by an electric spark. *Brit. J. Appl. Phys.* **10**, 159-166 (Apr., 1959).

579.* Oosterkamp, W. J. X-ray photographs with extremely short exposure times. *Philips Tech. Rev.* **5**, 22-25 (Jan., 1940).

580.* Ehrke, L. F., and Slack, C. M. Ultra-speed radiography. *Photo-Tech.* **3**, 53-55 (Jan., 1941).

581.* Slack, C. M. One-millionth-second radiography. *J. Phot. Soc. Am.* **11**, 302-306 (Sept., 1945).

582.* Strakhovskii, G. M., and Tsukerman, V. A. Methods of synchronization for very brief x-ray exposure. *Izvest. Akad. Nauk. S.S.S.R., Otdel Tekh. Nauk.* **3**, 371-384 (1946) (In Russian).

583.* Dickson, D. C., and Slack, C. M. One-millionth-second radiography and its applications. *Proc. I.R.E.* **36**, 600-606 (June, 1947).

584.* Clark, J. C. Flash radiography applied to ordnance problems. *J. Appl. Phys.* **20**, 363-70 (April, 1949).

585.* Pollitt, C. G. Flash radiography. *Phot. J.* **90B**, 104-107. (July/Aug., 1950).

586.* Criscuolo, E. L., and O'Connor, D. T. The development of a fine focus flash x-ray tube *Rev. Sci. Instr.* **24**, 944-948 (Oct., 1953).

587.* Dapoigny, J., Kieffer, J., and Vodbar, B. Les ondes de choc en milieu dense. II-Méthodes expérimentales et quelques résults de mésures obtenus par la méthode de radiographie ultrarapide. *J. recherches C.N.R.S.* **31**, 260-270 (June, 1955).

588.* Dapoigny, J., Kieffer, J., and Vodar, B. Ondes de choc engendrées dans l'argon liquide et leur etude par radiographie ultrarapide. *J. phys. radium* **8**, 733-734 (Aug./Sep., 1955).

589.* Thomer, G. Photographie und Kinematographie mit Röntgenblitzen. *Phot. u. Wissenschaft* **7** (1), p. 5-8 (1958).

590.* Knight, H. T., and Vanable, D. Apparatus for precision flash radiography of shock and denotation waves in gases. *Rev. Sci. Instr.* **29**, 92-98 (Feb., 1958).

591.* Herrmann, K. H. Röntgenblitzuntersuchungen an Funkenstosswellen in Gasen. *Z. angew. Physik* **10**, 349-356 (Aug., 1958).

592.* Elle, D. Eine neue Anwendung des Faraday-Effekts zur Herstellung eines Kurzzeitverschlusses für Reihenbildvorderlichtaufnahmen und für die Röhrenblitzphotographie. *Z. angew. Physik* **6** (2), 49-54 (1954).

593.* Edgerton, H. Photography of very early states of nuclear explosions. *J. Soc. Motion Picture and Television Eng.* **68,** 77-79 (Febr., 1959).

594.† Fraser, R. P., and Dombrowski, N. Technique for the study of movement. *J. Imp. Coll. Chem. Eng. Soc.* **11,** 131-151 (1957).

595.† Vajnstejn, E. E. and Korolev, V. V. Générateurs à impulsions d'excitation de spectres et leur utilisation pour l'analyse spectrale quantitative de silicates. *Chem. anal.* **42** (10), 479-489 (Oct., 1960).

596.* Marshak, I. S. Elektrische Charakteristika röhrenförmiger Impulslampen bei Speisung aus organischen Energiequellen. *In* "Symposium on Vacuum-Technique Materials," No. 22, pp. 27-31. Gossenergoizdat, Moscow (1960).

597.† Wdowtschenko, R. G., Marshak, I. S., and Nikolskaja, S. A. Abhängigkeit lichtgebender röhrenförmiger Impulslampen von Kapillarlampen, Kathodenmaterial und Kapillargrösse. *In* "Symposium on Vacuum-Technique Materials," No. 21. Gossenergoizdat, Moscow (1959).

598.* Gelstein, M. I. Prinzipien der Berechnung von Kreisdiagrammen über Kaltkathoden-Thyratrons der Type TH 3B. *In* "Symposium on Vacuum-Technique Materials," No. 21 Gossenergoizdat, Moscow (1959).

599.* Marshak, I. S. Röhrenartige Gasentladungslampen kondensierter Funkengasentladungen. *In* "Uspekhi Nautschoi Photographii," Vol. VI, pp. 16-26. Akad. Nauk. SSSR, Moscow (1959). (In Russian). German translation available.

600.* Marshak, I. S. Koeffizient vom Wirkungsgrad von Impulslampen. *Svetotekhnika* **1,** 17 (1957).

601.* Marshak, I. S. Elektrische Charakteristika röhrenförmiger Impulslampen. *Svetotekhnika* **6,** 22 (1957).

602.* Marshak, I. S. Fotometrische Methoden von Impulslichtquellen. *In* "Symposium on Vacuum-Technique Materials "No. 10. Gossenergoizdat, Moscow (1957).

603.* Korablew, L. N. "Neue anwendbare Lampen mit Kaltkathoden in Impuls-Apparaturen." Gossenergoizdat, Moscow (1956).

604.* Flood, W. The design of cold cathode valve circuits. *Electronic Eng.* No. 10 (1956).

605.† Drechsel, D. Investigations of the variations with time of the intensity of spectrallines produced by light-flashes of very short duration. *Z. Vakuum-Technik* **9,** 7 (Oct., 1960).

606.* Conn, W. M. *J. Opt. Soc. Am.* **41,** 445 (1951).

607.* Plateau, J. *Mem. Acad. Roy. Bruxelles* 16 (1843); 23 (1850); 30 (1857); 31 (1859); *Smithsonian Inst. Repts.* 207 (1864); 285 (1865).

608.* Kleen, W. *Ann. Physik* [5] **11,** 579 (1931).

609.* Gow, Ruby, Smith, and Wilcox The trapping of charged particles in axially symmetrical systems of electric and magnetic fields. UCRL-8156 (Jan., 1958).

610.* Rechen, J. B., and Jordan, J. C. *J. Sci. Instr.* **28,** 584 (1957).

611.‡ Bostick, W. H. Elementary theory of pulse transformers. *In* "Pulse Generators," pp. 499-526. McGraw Hill, New York (1948).

612.* Slater, J. C. "Microwave Transmission," p. 98 McGraw-Hill, New York, (1942).

613.† Marshak, I. S., and Shchoukin, L. I. Physical and technical parameters of flashtubes. *J. Soc. Motion Picture and Television Engrs.* **70** (3), 169-176 (1961).

614.* Marshak, I. S. Tubular gas-discharge lamps with a condensed spark gas discharge. *In* "Symposium on Vakuum-Technique Materials," No. p. 3. Gossenergoizdat (1955). (In Russian).

615.* Marshak, I. S. Some physical and technical aspects of gas discharges in flashtubes. *Proc. 3rd Intern. Congr. on high Speed Photography, London*, 1956. Butterworth, London (1956).

616.* Marshak, I. S. Efficiency of tubular flash lamps. *Svetotekhnica* **1,** 17 (1957). (In Russian.)

617.* Marshak, I. S. Electric characteristics of tubular flash lamps. *Svetotekhnica* **6,** 22 (1957). (In Russian.)

618.* Marshak, I. S., and Vassilyev, V. I. Load limits for tubular flash lamps. *In* "Symposium on Vakuum-Technique Materials," No. 14, p.19 Gossenergoizdat, Moscow. (1958) (In Russian.)

619.* Marshak, I. S. Flash tubes for high speed photography. *Proc. 4th Intern. Congr. für Kurzzeitphotographie und Hochfrequenzkinematographie, Cologne,* 1958 pp. 130-135. Helwich, Darmstadt (1959).

620.* Marshak, I. S., Vassilyev, V. I., Mironova, A. L., Ivanov, V. P., and Vdovtshenko, R. G. New flash tubes. *Uspekhi Naootshnoy Fotografi* **6,** 43 (1959). (In Russian.)

621.* Vdovtshenko, R. G., Marshak, I. S., and Nikolskaya, S. A. Light output dependence of tubular flash lamps on the tube length, cathode material and the pre-electrode volumes. *In* "Symposium on Vacuum-Technique Materials," No. 21, p. 17. Gossenergoizdat, Moscow (1959). (In Russian.)

622.* Vodovatov, B. M., and Epstein, M. I. Spectral characteristics of tubular flash lamps. *Uspekhi Naootshnoy Fotografi* **6,** 35 (1959). (In Russian.)

623.* Carlson, F. E., and Pritchard, D. A. Characteristics and application of flash-tubes. *Illum. Eng.* **42,** 235 (1948).

624.* Grebennikov, O. F. A faster method of high speed cinematography. *Uspekhi Naootshnoy Fotografi* **6,** 145 (1959). (In Russian.)

625.* Vanewkov, M. P., Dobretzov, A. F., Issayenko, V. I., and Mak, A. A. A powerful pulsed light source. *Uspekhi Naootshnoy Fotografi* **6,** 53 (1959). (In Russian.)

626.† Edgerton, H. E., Tredwell, J., and Cooper, K. W., Jr. Submicrosecond flash sources. *J. Motion Picture and Television Engrs.* **70** (3), 177-188 (1961).

627.* Fayolle, P., and Naslin, P. Simple electronic devices for high speed photography and cinematography. *J. Soc. Motion Pictures and Television Engrs.* **60,** 603-626 (1953).

628.* Looms, J. S. T., and North, R. J. Short duration spark light sources. *Proc. 3rd Intern. Congr. on High Speed Photography, London,* 1956 p. 62. Butterworths, London (1956).

629.* Prescott, R. An introduction to spark shadowgraph techniques. *Phot. Eng.* **3,** 121 (1952).

630.* Schardin, H., and Fünfer, J. Grundlagen der Funkenkinematographie. *Z. angew. Physik* **4,** 185-199, 224-238 (1952).

631.* Tawil, E. New developments in "guided air sparks." *Proc. 3rd. Intern. Congr. on High Speed Photography, London,* 1956 pp. Y 148-152. Butterworths, London (1956).

632.* Vanyukov, M. P., and Mack, A. A. High intensity pulsed light sources. *Uspekhi Fiz. Nauk* 137-155 (1958).

633.* Edgerton, H. E. Submicrosecond flashes of light. *Proc. 4th Intern. Congr. für Kurzzeitphotographie und Hochfrequenzkinematographie, Cologne,* 1958. pp. 91-97. Helwich, Darmstadt (1959).

634.* Cooper, K. W., Jr. An investigation of spark discharge phenomena in the millimicrosecond region. M. S. Thesis, Elec. Eng. Dept., Massachusetts Inst. Technol. (1959.)

635.* Fischer, H. Simple submicrosecond light souruce with extreme brightness. *J. Opt. Soc. Am. pp.* 981-984 (Nov., 1957);

636.* *AGARDOgraph* No. 22 (Nov., 1956).

637.* Tredwell, J. The guided air spark as a light source. M. S. Thesis, Elec. Eng. Dept., Massachusetts Inst. Technal. (1960).

638.† Abella, I. D., and Cummins, H. Z. Thermal tuning of ruby optical maser. *J. Appl. Phys.* **32** (6), 1177-1178 (1961).

639.* Schawlow, A. L. and Townes, C. H. *Phys. Rev.* **112,** 1940 (1958).
640.* Maiman, T. H. *Nature* **187,** 493 (1960).
641.* Collins, R. J., Nelson, D. F., Schawlow, A. L., Bond, W., Garrett, C. G. B., and Kaiser, W. *Phys. Rev. Letters* **5,** 303 (1960).
642.* Schawlow, A. L., and Devlin, G. E. to be published.
643.* Gibson, K. S. *Phys. Rev.* **8,** 38 (1916).
644.* Sugano, S., and Tanabe, Y. *J. Phys. Soc. Japan* **13,** 880 (1958).
645.† Schawlow, A. L. Optical masers. *Sci. American* **204,** (6), 52-61 (June, 1961).
646.* Trümper, J. Parallelplattenzähler mit elektronisch gesteuerter Funkenzündung für Teilchenbahnbestimmungen. *Atomkernenergie* **5,** 4 (1960).
647.* Ruthenglen, J. G., and Paterson, J. M. Accuracy of track location in spark chambers. *Rev. Sci. Instr.* **32** (5), 519-523 (1961).
648.* Meyer, D. I., and Terwilliger, K. M. A 14 × 14 × 7-in. thin plate spark chamber. *Rev. Sci. Instr.* **32** (5), 512-518 (1961).
649.* Fischer, J., and Zorn, G. T. Observations on pulsed spark chambers. *Rev. Sci. Instr.* **32** (5), 499-511 (1961).
650.* Roberts, A. Developments of the spark chamber: a review. *Rev. Sci. Instr.* **32** (5), 482-484 (1961).
651.* Talbot, W. H. F. Spark photography. Brit. Patent 13,664 (June, 1851).
652.* Lord Rayleigh. Electricity on colliding water drops. *Proc. Roy. Inst.* (Feb., 1891).
653.* Worthington, A. M. *Proc. Roy. Inst.* (Oct., 1894); *Proc. Roy. Soc.* p. 250 (1895).
654.* Wood, R. W. *Phil. Mag.* [5] **48,** 218 (1899).
655.* Wood, R. W. *Phil. Mag.* [5] **50,** 148 (1900).
656.* Boys, C. V. *Nature* **47,** 415, 440 (1893).
657.* Beams, J. W. *J. Appl. Phys.* **8,** 795 (1937).
658.* Bull, L. *Rev. opt.* **13,** 9 (1934).
659.* Dvorak, *Ann. Physik* [3] **9,** 502 (1880).
660.* Mach, E. *Wiener Reports* **78,** 467 (1878). **95,** 764 (1887); **98,** 41, 1303, 1310, 1327, 1333 (1889); **105,** 605 (1896).
661.* Quayle, P. P. *Sci. Papers* of *Natl. Bur. Standards* (*U.S.*) No. 508 (1925).
662.* Quale, P. P. Photography of bullets in flight. *J. Franklin Inst.* **193,** 627-640 (1922).
663.* Foucault, L. *Ann. observatoire imp. Paris* **5,** 197 (1859).
664.* Toepler, A. *Ann. Physik und Chem.* **128,** 126 (1866); **131,** 33 (1867).
665.* Cranz, C. "Lehrbuch der Ballistik" Vol. 3, p. 259, 264. Springer, Berlin (1927).
666.* Schardin, H. *Ergeb. exact. Naturw.* **20** (1941).
667.* Barnes, N. F. *Ergeb. exact. Naturw.* **34,** 775 (1944).
668.* Drewell, P. *Z. tech. Physik.* **16,** 614 (1936).
669.* Slack, C. M., and Arnott, E. G. F. Report on enclosed pressure gaps. NDRC 14-150, Westinghouse Electric Corp. (1942).
670.* Goucher, F. S., Haynes, J. R., Depp, W. A., and Ryder, E. J. Spark gap switches for radar. *Bell System Tech. J.* (Oct., 1946).
671.† Marshak, I. S. *Optika i Spektroskpoiya* **10,** 803 (1961).
672.* (1) Anderson, J. A., and Smith, S. *Astrophys. J.* **64,** 295 (1926).
673.* (2) Amramson, I. S., and Marshak, I. S. *Zhur. Elektrotekh. Fiz.* **12,** 632 (1942).
(3) Sobolew, H. H. *Zhur. Elektrotekh. Fiz.* **17,** 986 (1947).
(4) Heine, R., Geldern, E. M., and Foner, P. S. *Phys. Rev.* **79,** 230 (1950).
(5) Conn, W. M. *J. Opt. Soc. Am.* **41,** 445 (1951).
(6) Kwarzhawa, I. F., Plutto, A. A., Schernow, A. A., and Bondarenko, W. W. *Zhur. Elektrotekh. Fiz.* **30,** p. 42 (1956).

(7) O'Day, M., Chace, W. G., and Cullington, E. H. *Proc. 3rd congr. intern. fenom ioniz. gas Milano*, 1957 p.784.

(8) Müller, W. *Z. Physik* **149,** 397 (1957).

(9) Wanjukowi, M. P., and Mak, A. A. *Uspekhi fys. nauk* **66,** 301 (1958).

(10) David, E. *Z. Physik* **150,** 162 (1958).

(11) Chace, W., and Moore, H. K. "Exploding Wires." Plenum Press, New York (1959).

(12) Marshak, I. S. *Uspekhi Naootshnoy Fotografi* **6,** 19 (1959).

(13) Marshak, I. S. *Zhur. Eksptl. i Teoret. Fiz.* **16,** 703, 718 (1946).

(14) Andreew, S. I., and Wangukow M. O. Vortrag auf dem II. Reichsbund über Hochgeschwindigkeits-Photographie, Moscow (1960).

(15) Marshak, I. S., and Shukin, L. I. Vortrag auf dem II. Reichsbund über Hochgeschwindigkeits-Photographie, Moscow (1960).

(16) *Svetotekhnika* **9,** 32 (1960).

674.* Goucher, F. S., Haynes, J. R., and Ryder, E. J. High power series gaps having sintered iron sponge mercury cathodes. NDRC 14-488. Bell Tel. Laboratories (Oct., 1945).

675.* Dillinger, J. R. General characteristics of aluminium cathode. Type series spark gaps. RL Report No. 682-3 (Nov., 1945).

676.† Stitch, M. L., Woodbury, E. J., and Morse, J. H. Optical ranging system uses laser transmitter. *Electronics* p. 51-53 (April, 1961).

677a.† Marx, E. Arc converter blast spark gap. U.S. Patent 2,128,884 (1938).

677b.† Arnott, E. G. F. Development of series spark gaps. NDRC 14-327 Westinghouse Electric Corp. (Aug., 1944).

678.* Fluharty, R. G., and Dillinger, J. R. Dissipation in series spark gaps and voltage current relationships during the discharge. RL Report No. 682-1 (June, 1945).

679.* Dillinger, J. R. Operation of sintered iron sponge mercury cathode type series gaps at 5 microsecond conditions. RL Report No. 682-5 (Nov., 1945); Line-type modulator and HP10V magnetron operation at 6 megawatts. RL Report No. 682-6 (Nov., 1945).

680a.* Lewis, I. A. D., and Wells, F. H. "Millimicrosecond Pulse Techniques," 2nd ed. pp. 115-132. Pergamon, London (1959).

680b.* Wells, F. H. Millimicrosecond pulse circuits. *J. Brit. Instr. Radio Engrs.* **11,** 491-503 (1951).

681.* Espley, D. C. Generation of very short pulses. *J. Inst. Elec. Engrs.* **93** (Pt. IIIA), 314-315 (1946).

682.* Espley, D. C., Cherry, E. C., and Levy, M. M. The pulse testing of wide-band networks. *J. Instr. Elec. Engrs.* **93** (Pt. IIIA), 1176-1187 (1946).

683.* Glasoe, G. N., and Lebacqz, J. V. "Pulse Generators," M. I. T. Radiation Laboratory Series No. 5. McGraw-Hill, New York (1948).

684.* Craggs, J. D., and Meek, J. M. "High Voltage Laboratory Technique." Butterworths, London (1954).

685.* White, E. L. C. The use of delay networks in pulse formation. *J. Inst. Elec. Engrs.* **93** (Pt. IIIA), 312-314 (1946).

686.* Brown, J. T. L., and Pollard, C. F Mercury contact relays. *Electronic Eng.* **66,** 1106-1109 (1947).

687.* Brown, J. T. L., and Pollard, C. E. Balanced polar mercury contact relay. *Bell System Tech. J.* **32,** 1392-1411 (1953).

688.* Moody, N. F., McLusky, G. J. R., and Deigth, M. O. Millimicrosecond pulse technique. *Electronic Eng.* **24,** 214-219, Part I (1952).

689.* Garwin, R. L. A pulse generator for the millimicrosecond range. *Rev. Sci. Instr.* **21,** 903-904 (1950).

690.* Schultz, C. W., and Smith, G. B. Pulse calibrator for wide-band oscilloscopes. *Tele-Tech.* **14** (12), 80-81 (1955).

691.* Fish, V. A coaxial mercury relay for fast pulse generation. Univ. of Calif. Radiation Lab. Report 3062 (1955).

692.* Whitby, H. C. Improvement in or relating to switch mechanism. U.S. Patent 731,668.

693.* Ives, R. L. The relay oscillator and related devices. *J. Franklin Inst.* pp. 242, 243-277 (1946).

694.* Brown, D. R. An interim report on tests to study the long term stability of braided r.f. cables. R.A.E. Tech. Note No. RAD 476 (1950).

695.* Finch, T. R. An impulse generator electronic switch for visual testing of wide-band networks. *Proc. Inst. Radio Engrs.* **38,** 657-661 (1950).

696.* Dorn, C. G. Fast rise pulse generator with high pulse repetition frequency. *Re. Sci. Instr.* **27,** 263-264 (1956).

697.* Rockett, F. Special purposer elays gain new uses. *Electronics* p. 150-56 (Feb., 1955).

698.* Boothby, L. W. Coaxial switch. U.S. Patent 2,690,540 (1954).

699.* Volz, A. H. High speed magnetically operated coaxial switch. *Bell Lab. Record* **34,** 305-307 (1956).

700.* Birnbaum, M. A method for the measurement of ionization and deionization times of thyratron tubes. *Trans. Am. Inst. Elec. Engrs.* **67,** 209-214 (1948).

701.* Chance, J. C. R. Note on the ionization and de-ionization times of gas-filled thyratrons. *J. Sci. Instr.* **23,** 50-52 (1946).

702.* de Knight, H. The development of mercury vapour thyratrons for radar modulator service. *J. Inst. Elec. Engrs.* **93** (Pt. IIIA), 949-962 (1946).

703.* de Knight, H. Hot cathode thyratrons, practical studies of characteristics. *Proc. Inst. Elec. Engrs.* **96,** 361-378, 379-381 (1949).

704.* Mullin, C. J. Initiation of discharge in arcs of thyratron type. *Phys. Rev.* **70,** 401-405 (1946).

705.* Romanowitz, H. A., and Dow, W. G. Statistical nature and physical concepts of thyratron de-ionization time. *Trans. Am. Inst. Elec. Engrs.* **69,** 368-379 (1950).

706.* Webster, E. W. Note on the ionization time of an argon filled relay. *J. Sci. Instr.* **24,** 299-301 (1947).

707.* Wittenberg, H. H. Thyratron in radar modulator service. *RCA Rev.* **10,** 322-327 (March, 1949).

708.* Appel, H., and Fünfer, E. Firing delay and build-time in the thyratron. *Z. angew. Physik* **8,** 322-327 (1956).

709.* Chantereau, J., and Leduc, P. Hydrogen thyratrons. *Rev. Tech. C.F.T.H.* **22,** 45-64 (April, 1956).

710.* Anon Thyratron recovery time. *Mullard Tech. Commun.* **2,** 290-298 (1956).

711.* Charles, D., and Warnecke, R. J. The hydrogen thyratron. *Ann. radio elect.* **10,** 256-302 (1955).

712.* Yu, Y. P., Christaldi, P. S., and Kallmann, H. E. Progress report on millimicrosecond oscillography. *Proc. Natl. Electronics Conf.* **6,** 360-372 (1950).

713.* Fowler, C. S. A narrow pulse generator. *Wireless Eng.* **27,** 265-269 (1950).

714.* de Knight, H., and Hooker, O. N. The hot cathode hydrogen filled thyratron. B.T.H. Research Lab. Rep. No. RLP 203.

715.* Heins, H. The hydrogen thyratron. *Instruments* **19,** 211, 250 (April, 1946).

716.* Woodford, J. B., and Williams, E. M. The initial conduction interval in high speed thyratrons. *J. Appl. Phys.* **23,** 722-724 (1952).

717.* Heard, H. G. 20 kilovolt delta-function generator. *Rev. Sci. Instr.* **25,** 454-458 (1954).

718.* Pearce, S. F., and Smith, D. C. G. A pulse generator producing 0,01μ sec pulses. Rept. Brit. Elec. Res. Assoc. M/T 113 (1951).

719.* Naray, Z. Pulse generator for producing positive and negative mμ sec-pulses. *Acta. Phys Hung.* **5,** 175-187 (1955).

720.* Edwards, W. D. High voltage saw-tooth and rectangular-wave pulse generator. *Electronic Eng.* **26,** 36-39 (1954).

721.* Querzoly, R. A generator of rectangular 0,2 to 10μ sec pulses. *Rend. ist. super sanita* **15** (Pt. 3), 162-167 (1953).

722.* Goodman, D. H., Sloan, D. H., and Trau, E. A high voltage high speed square wave surge generator. *Rev. Sci. Instr.* **23,** 766-767 (1952).

723.* Fletcher, R. C. Impulse breakdown in the 10^{-9} sec range of air at atmospheric pressure. *Phys. Rev.* **76** (10), 1501-1511 (Nov., 1949).

724.* Fletcher, R. C. Production and measurement of ultra high speed impulses. *Rev. Sci. Instr.* **20,** 861-869 (1949).

725.* Craggs, J. D., Haine, M. E., and Meek, J. M. The development of triggered spark gaps for high power modulators. *J. Inst. Elec. Engrs.* **93,** (Pt. 111A), 963 (1946).

726.* Wilson, R. R. Very short time lag of sparking. *Phys. Rev.* **50,** 1082-1088 (1936).

727.* Bell Laboratory Staff "Radar System and Components." Van Nostrand, Princeton, New Jersey (1949).

728.* Reintjes, J. F. "Principles of Radar." McGraw-Hill, New York (1946).

729.* Ridenour, L. N. "Radar System Engineering." McGraw-Hill, New York (1947).

730.* Drechsel, D. Investigation of the variations with time of the intensity of spectral lines produced by light-flashes of very short duration. *Vakuum-Technik* **9** (7), 185-190 (Oct., 1960).

731.* Pahl, M., and Weimer, U. Zur Massenspektrometrie an Glimmentladungen. *Z. Naturforsch.* **9,** 745-53 (1958).

732.* Pahl, M. and Weimer, U. Zur Bildung von He_2^+ und Ne_2^+ in der stationären positiven Niederdrucksäule. *Z. Naturforsch.* **9,** 754-757 (1958).

733.* Pahl, M., and Weimer, U. Massenspektrometrische Untersuchungen über die Bildung von $(HeNe)_2^+$ in der positiven Säule. *Z. Naturforsch* **12a** (11), 926-931 (1957).

734.* Pahl, M., and Weimer, U. Die Reaktion $He_2^+ + Ne \rightarrow Ne^+ + 2\,He$ in der positiven Säule. *Z. Naturforsch.* **13a** (1), 50-51 (1958).

735.* Bemerl, W., and Fetz, H. Das extrem kurzwellige UV-Licht einer Edelgasentladung. *Naturwissenschaften* **43** (12), 272-73 (1956).

736.* Bemerl, W., and Fetz, H. Über die zeitlich verzögerte Lichtemission einer Edelgaskorona. *Naturwissenschaften* **45** (16), 381-82 (1958).

737.* Bemerl, W., and Fetz, H. Untersuchungen des photoelektrisch wirksamen UV-Liches einer in H_2 und O_2 brennenden Korona-Entladung. *Z. angew. Physik* **8** (9) 424-429 (1956).

738.* Bemerl, W., and Fetz, H. Über das extrem kurzwellige UV-Licht einer zylinder symmetrischen Edelgaskorona. *Z. Physik* **153,** 458-467 (1959).

739.* Schmidt-Tiedemann, K. J. Elektronennachlieferung in den Gasen N_2, H_2 und O_2 bei niedrigen Drucken. (pd $\approx$ 50 Torr. cm) *Z. Physik* **150,** 299-310 (1958).

740.* Bemerl, W., and Fetz, H. Über das extrem kurzwellige UV-Licht einer zylindersymmetrischen Edelgaskorona. *Z. Physik* **153,** 468-83 (1959).

741.* Pahl, M. Zur Bildung von Molekülionen in stationären Edelgasentladungen. *Z. Naturforsch.* **14a** (3), 240-246 (1959).

742. Früngel, F. Bewegungsaufnahmen rascher Luftströmungen und Stosswellen

durch hochfrequente Hochspannungsfunken. Jahrbuch 1960. Wissenschaftliche Gesellschaft Luftfahrt. Vieweg, Braunschweig (1960).

743.† Edgerton, H. Xenon flash lamp design. *Proc. 2nd Intern. Conf. on Quantum Electronics*. Office of Naval Research, University of California (1962).

743a.* Edgerton, H. E. Circus action in color. *Nat. Geograph. mag.* **93,** No. 3, pp. 305 ff. (1948).

743b.* Edgerton, H. E. Characteristics of a flashtube for aerial night reconnaissance. *Photo. Eng.* **6,** No. 2, 110-115 (1955).

744.† Howes, E. T., and Tejada-Flores, L. H., and Randolph, L. Seismische Modelluntersuchungen. *J. Acoust. Soc. Am.* **25** (5), 915-921 (1953).

745.† Edgerton, H. Shock wave photography of large subjects in daylight. *Rev. Sci. Instr.* **29** (2), 171-172 (Feb., 1958).

746.* Edgerton, H. Travelling wave oscilloscope. Industrial Paper, Edgerton-Germeshausen, & Grier, Inc., Boston. Data Sheet 7071.

747.* Sharpe, J. A review of the EMI development of photomultiplier tubes. Industrial paper, EMI Electronics Ltd., Ruislip, England (Feb., 1958).

748.† STL-image converter camera. Industrial paper STL-Products, a Div. of Space Technol. Lab., Los Angeles.

749.† Slow neutrons detectors NE 400 and NE 401. Industrial paper, Bull. 13, 1959, Nukleonik, Solingen, Germany.

750.* Sun, K. H., Malmberg, P. R. and Pecjak, F. A. *Nucleonics* **14** (7), 46 (1956). Nukleonic, Solingen, Germany.

751.† Patzke, H. G. A simple cinematographic high speed attachment. Pending German Patent Application (1961).

752.† von Ardenne, M. "Duoplasmatron" Ion source with auxiliary magnetic field. Industrial paper, Vakutronik, Dresden, Germany.

753.† von Ardenne, M. "Tables of Electron Physics, Ion Physics and Electron Microscopy," Vol. I, pp. 544-557.

754. Früngel, F. Ein neues Verfahren zum Weichmachen von Kupferdrähten. *Fachzeitschrift "Draht"* **12** (6), 259-260 (1961). Prost & Meiner, Coburg, Germany.

755.† Spitzer, L. "Physics of Fully Ionized Gases," p. 42. Wiley (Interscience), New York (1956).

756.† Hersey, J. B., Edgerton, H., Raymond, O., and Hayward, G. Surface sonar thumper. Inst. Soc. Am. Conf. Reprint 21-60, Fall Instrument-Automation Conf. and Exhibit, New York (Sept., 1960).

757.* Ron, A. An investigation of the sonar transducer. M. S. Thesis, Massachusetts Inst. Technol. (May, 1959).

758.* Edgerton, H. E. Uses of sonar in oceanography. *Electronics* (June, 1960).

759.* Edgerton, H., and Cousteau, J. Y. Underwater camera positioning by sonar. *Rev. Sci. Instr.* (Dec., 1959).

760.* Hersey, J. B. Tests of the Edgerton thumper in vineyard sound from Atlantis. W.H.O.I. technical memorandum No. 1-60 (March, 1960).

761.† Fischer, H. Millimicrosecond light source with increased brightness. *J. Opt. Soc. Am.* **51** (5), 543-547 (May, 1961).

762.* Castle, J., and Webb, J. R. *Phot. Eng.* **4,** 51 (1954).

763.† Edgerton, H. Multiflash microsecond flash photography. *Research Film* **2,** 3 (July, 1956).

764.* Kerns, Q. H., Kirsten, F. A., and Cox, G. C. *Rev. Sci. Instr.* **30,** 31 (1959).

765.* Porter, G. A., and Woodling, E. R. *J. Sci. Instr.* **36,** 147 (1959).

766.* Elmore, W. C., and Sands, M. "Electronics," p. 139. McGraw-Hill, New York (1949).

767.* Fischer, H. AFCR-TN 59-185 (May, 1959).

768.* Edgerton, H. E. Xenon flash tube. Industrial paper, Edgerton. Germeshausen, Grier. Data Sheet FX-6A.

769.* Eichinger, R., and Krempl. H. Über den Einfluss des Entladungsgases auf den zeitlichen Verlauf des Stromes, der Spannung und der Spektrallinienintensität bei Funkenentladungen. *Z. angew. Physik* **12** (10), 461-469 (1960).

770.* Krempl, H., and Späth, H. Untersuchungen über die Statistik der Schwankung von Linienintensitäten bei periodischer Funkenentladung. Verlag Stahleisen. Düsseldorf, No. E2115 Vol. 8 (1960).

771.* Goes, O. W. Registrierung der Sichtweite auf einem Flughafen. *Physik der Atmosphäre* **33,** 1/2 (1960).

772.* Grosskurth, K. Etwas über die Möglichkeit, Nebel mit Lichtstrahlen zu durchdringen. *Signal und Draht* **5,** 73-76 (May, 1954).

773.* Koch, B. Die Durchlässigkeit der bodennahnen Atmosphäre im ultravioletten und sichtbaren Spektralbereich *Optik* **5,** 258-273 (1949).

774.* Nichols, J. W., *et al.* An automatic fog warning instrument. *Electronic Eng.* pp. 519-523 (1957).

775.* Curcio, J. A., Knestrick, G. L., and Drummeter, L. F. A preliminary study on the correlation of atmospheric transmission with backscattering. Naval Res. Lab. Washington, Rept. 4818 (Aug., 1956).

776.* Curcio, J. A., Knestrick, G. L., and Cosden, T. H. Further studies on the correlation of backscattering with atmospheric transmission. Naval Res. Lab. Washington, Rept. 5143 (June, 1958).

777.* Horman, M. H. Measurement of atmospheric transmissivity using backscattered light from a pulsed light beam. *J. Opt. Soc. Am.* **51** (6), 681-691 (1961).

778.* Ruppersberg, G. H. "Neuere Versuche zur Registrierung der Sichtweite auf Flughäfen ,"pp. 230-236. Wissenschaftliche Gesellschaft Luftfahrt, Vieweg, Braunschweig, (1959).

779. Früngel, F. Long lived impulse transformer. U.S. Patent 2,942,214 (1960).

780.† Hadfield, D., *et al.* Electronic magnetizing and demagnetizing *Electronic and Radio Eng.* pp. 242-245 (July, 1959).

781.† Chase, N., Hankin, N., and Webb, F. Instrumentation for exploding wire research. *Electronics* pp. 43 (July, 1960).

782.* Hauser, S. M. Final Report on WADD contract AF 33 (616)-669. Electro-Optical Systems Inc., Pasadena, California (June, 1960).

783.* Goss, W. C. *J. Soc. Motion Picture and Television Engrs.* **69,** 889 (1960).

784.* Baugh, C. W., and Ogland, J. W. Cathodoluminescent optical maser pumping. Symposium of Optical Society of America, Spring Meeting (1962).

785. Früngel, F. "Impulstechnik," pp. 329-341. Akademische Verlagsges., Leipzig (1960).

786. Früngel, F. The relationship between laser-radar equipment and sparklight signal systems in a critical view. Laser Symposium, The Hague (April, 1962).

787. Früngel, F. Discussions at the Meeting "Sichtmessverfahren" at Düsseldorf, Germany, held by Gesellschaft für Ortnung und Navigation (Dec., 1961).

787a. Alberti H., and Früngel, F. Das Widerstandsschweissen mittels transformierter Kondensatorentladung. *Blech* **9,** 3-10 (September, 1961).

787b.† Chace, W. G. Exploding wires and their uses. *New Scientist* **16,** 386-388 (May, 1963).

787c.* Fünfer, E., Kegel, W. H., Kronast, B., and Kunze, H. J. Local determination of plasma parameters in a θ-pinch plasma by light scattering experiments. *Arbeitsber. Inst. Plasmaphysik Garching bei München* Sect. VIII, 5.

787d.* Wellinger, K. Die technologischen Verfahren der Explosiv-Umformung von Metallen mittels detonierender Sprengstoffe. *Industrieanzeiger* (*Essen*) No. 58, 23-31 (July, 1963).

788.* Mahnau, H. Ein Indikator zur Zielort- und Kaliberbestimmung von Kernwaffen-Explosionen. *Zivilschutz* **27,** No. 6 (1963).

789. Früngel, F. Magnet-Charger. Industrial paper, Impulsphysik GmbH, Hamburg-Rissen.

790.* Thompsen Diskussionsbeiträge aus dem Bericht über das 11. Aachener Werkzeugmaschinen-Kolloquium 1962. *Industrieanzeiger* No. 72 (Sept., 1962).

791.* Idelberger, K. Hochleistungsumformung von Blechen und Profilen mit Magnethämmern *Blech* **4,** 191-195 (1963).

791a.* Idelberger, K. Elektromagnetisches Umformen von Blechen und Hohlkörpern aus Stahl. *Tech. Rundschau* No. 4, Pt. 2 (Jan. 1964).

792.* Beeson, E. J. G. The control of high power xenon flashes by magnetic switching. *AEI Eng.* pp. 134-138 (May-June, 1963).

793.† McDermott, J. E. Fiber optics: designing in a new light. Maser Optics Inc., Boston, Massachusetts.

794.* McKellar, A. C. Operation of xenon arc discharge lamps. *Elec. Rev.* **30,** 313-316 (Aug., 1963).

795.* Brewer, L. Gases and gas-condensed phase reactions. University of California, Berkeley, No. 9 (Sept., 1963).

796.* Diamond, J. J. Materials in the condensed state. *Nat. Bur. Standards* (*U.S.*) No. 3 (July-Sept., 1963).

797.* Nanosecond Light Source. Industrial paper, Impulsphysik GmbH, Hamburg-Rissen.

798.* Unknown Phenomena? STROBOKIN Precision Analysis. Industrial paper, Impulsphysik GmbH, Hamburg-Rissen.

799.† Fischer, H., Gallagher, C. C., and Tandy, P. Improved nanosecond light sources. *6th Intern. Congr. on High-Speed Photography, The Hague*, pp. 152-157 (Sept., 1962).

800.* Girard, M. J. M. Quelques applications particulières de l'éclair électronique. *Bull. Soc. franç. Electriciens* (May, 1962).

801.* Remuuth, H. Normale und hochfrequenz-kinematographische Analyse des Spinnvorganges. *Sonderdruck Z. ges. Textilind.* **65,** No. 6, 435-444 (1963).

802.* Frechinger, H. Kurzzeitphotographie und Hochfrequenzkinematographie als Prüf- und Forschungsmethoden der Textiltechnologie. *Sonderdruck Spinner Weber Textilveredlung* 81, 542-547 (1963).

803.* Bogen, B., and Conrads, H. A Vacuum sliding spark for the excitation of an intense Li III spectrum. *Arbeitsber. Inst. Plasmaphysik Jülich* (1963).

804.* Marshak, I. S. Very high-power tubular xenon lamps without ballast. A new kind of effective light source for high-speed cinematography. *Proc. 6th Intern. Congr. on High-Speed Photography, The Hague* (Sept., 1962).

805. Früngel, F. Hochgeschwindigkeits-Kinematographie mit kaltem Funkenlicht. *Physik. Verhandl.* No. 1 (1963).

806.* Patzke, H. -G. Funktions- und Fehleranalyse hochbeanspruchter Kinematiken mittels Hochgeschwindigkeitskinematographie. *Sonderdruck Feinwerktech.* **67,** No. 3 (1963).

807.* Röder, G., and Früngel, F. Ein neues Verfahren zur Zündung von Funkenstrecken und Lichtbögen. *Physik. Verhandl.* No. 2/3 (1963).

808.* Thorwart, W. Mikroskopische Zeitdehneraufnahmen. *Research Film* **4,** No. 2 (1961).

809.* Thorwart, W. Photographische Verfahren zur Analyse kürzester Zeiten. *Z. Instrumentenk.* **71,** No. 4/5 (1963).

810.* Foitzik, R., and Reizuch, H. Funkenstrecken für höchste Impulsströme. *Siemens-Z.* No. 4 (April, 1962).

811.* Fischer, H. Nanosecond Light Source. Industrial paper, Impulsphysik GmbH, Hamburg-Rissen (1964).

812.* Ascoli-Bartoli, U. and Martellucci, S. Schlieren Photography of an Orthogonal Pinch. *Nuovo Cimento* [10] **27,** No. 2 (1963).

813.* Frommhold, L. Die Erzeugung sehr kurzer und intensiver UV-Licht impulse. *Z. Naturforsch.* **18a,** No. 5 (1963).

814.† Isomet Bulletin Isomet electro-optic light modulators. Isomet Bull. 1201 New Jersey Palisades Park, (1963).

815.† Burkhalter, J. H., Martin Company Techniques and applications of Q-switching of several laser materials. Martin Co., Orlando, Florida (1964).

816. Früngel, F. Das Laser-Prinzip, Grundlagen und thermische Anwendung. *Schweisstech. Tagung Deut. Verbandes Schweisstechnik e. V.* (June, 1963).

817.* Armed Services, Technical Information Agency Soviet gas laser research. Review of open literature. AID Work Assignment No. 14, Report 62-100 (July, 1962).

818.† Fairbanks, R., Sr., and Martin, R. L. Aspects of laser beam welding. Presented before National Electronics Conference, Chicago (Oct., 1962).

819.* Hughes Aircraft Co., Culver City, California, Ein Entfernungsmesser mit Laser. *Elektrotech. Z.* **22,** 7 (1963).

820.† Vogel, Sy., and Dulberger, L. H. Lasers: devices and systems—Part I. Maser Optics Inc. Boston, Mass. (Oct., 1961).

821.† Maguire, Th. Microwelding laser or electron beam? *Electronics* (July 5, 1963).

822.* Rhodes, M. B., Keedy, D. A., and Stein, R. S. The use of a laser as a light source for photographic light scattering from polymer films. Polymer Research Inst., Univ. of Massachusetts, Amherst.

823.* Fine, S., Klein, E., Scott, R., Seed, R., and Roy, A. Some effects of laser radiation on the skin of the Syrian hamster. *Life Sciences* **1,** 30-53 (Dec., 1962).

824.* Sage, St. J. High efficiency optical maser parameters. *Appl. Opt.* **1,** No. 2, 173-179 (March, 1962).

825.* Schröter, F. Fragen der Anwendbarkeit des Lasers für Navigationszwecke. *Elektronische Rundschau* **5,** 242-247 (1963).

826.* Sirons, J. Lasers for aerospace weaponry. Tech. Documentary Rep. No. ASD-TDR-62-440 (May, 1962).

827.† Fünfer, E., Kronast, B., and Kunze, H.-J. Experimental results on light scattering by a θ-pinch plasma using a ruby laser. *Physics Letters* **5,** No. 2 (June, 1963).

828.* Compton, M. E., Jr. Surveying with the velocity of light. *Am. Congr. on Surveying and Mapping, Washington D.C.* pp. 22-24 (March, 1954).

829.* Ellis, A. T., and Fourney, M. E. Application of a ruby laser to high speed photography. *Proc. IEEE* **51,** No. 6 (June, 1963).

830.* CBS Industrial paper. CBS Laboratories, Stamford, Conneticut (1962).

831a.* Hintz, E., and Beerwald, H. The three electrode spark gap and its application for the simultaneous switching of many parallel spark gaps. *Proc. 4th Intern. Conf. on Ionization Phenomena in Gases, Uppsala, Sweden* (1959).

831b.* Mason, J. F. Military electronics. Anti-guerrilla gear. *Electronics* pp. 83-85 (June 1, 1964).

832.* Industrial paper TRG 604. Microwelder Electro-Optical Prod. Sec., Melville, New York (1964).

833. Früngel, F. Intensitätsverhältnisse des Rückstrahllichtes in der Atmosphäre und des Objekt-Reflexlichtes bei Verwendung eines Lasers als Entfernungsmesser. Unpublished report.

834. Früngel, F. Laser Kerr Q-switch. Industrial paper, Impulsphysik GmbH, Hamburg-Rissen.

835.* Narris, T. J. High-Speed photographs of laser-induced heating automatic step-size control. *IBM Journal* pp. 341-344 (Oct., 1963).

836.* Braun, G. Bestrahlung von Zellen mit Hilfe eines Rubin-Lasers. *Atompraxis* No. 4, 158-160 (1963).

837.* Hearne, M. R., and Le Fèvre, R. J. W. Molecular polarisability: the molar Kerr constants of five sydnones. *J. Chem. Soc.* p. 4753 (1962).

838.* Knestrick, G., Cosden, T. H., and Curcio, J. A. Atmospheric scattering co-efficients in the visible and infrared regions. *J. Opt. Soc. Am.* **52,** No. 9 (Sept., 1962).

839.* Klein, W. Pulse voltage measuring. *Mitteilungsblätter Strahlungsmessgeräte* No. 7, 311 (April, 1962).

840.* Beer, F. Impulsoptische Messgeräte im Dienste der Luftfahrt *VDI* (*Ver. Dent. Ing.*) *Z.* **104,** No. 24, 1245-1246 (Aug., 1962).

841.† Bock, O. H., and Clemens, P. L. Aerodynamic measurements in a hypervelocity gun range. *Electronics* No. 3, 33-37 (1961).

842.* Kasten, Fr. I. Theorie der horizontalen Sichtweite nicht selbstleuchtender Objekte unter bedecktem Himmel. *Beitr. Physik Atmosphäre* **34,** No. 3/4, 234-258 (1962).

843.† Dreeskamp, H. and Burton, M. Measurement of fast luminescence decay times. *Phys. Rev. Letters* **2.** No. 2 (1959).

844.* Anderson, L. K. Photodiode detection. Symposium on Optical Masers, Poly-tech. Inst. of Brooklyn (April, 1963). New York

845. Früngel, F. High-speed photography using a high-frequency spark source and a Kerr cell in combination. *J. Soc. Motion Picture Television Engrs.* **71,** 93 (Feb., 1962).

846. Früngel, F. Automatic weather-station. Industrial paper, Impulsphysik GmbH, Hamburg-Rissen (March, 1963).

847.* Helmchen, G. Die Entwicklung der Stosspannungs technik. *Elektrotech. Z.* **84A** (4), 107-113 (1963).

848.* Hansen, W. W. Surveying with pulsed-light radar. *Electronics* 76-79 (July, 1948).

849.* Balchan, A, S. A radiographic study of shock-loaded iron. *J. Appl. Phys.* **34,** No. 2, 242-245 (1963).

850.† Field Emission Corp., Flash X-ray systems. Field Emmission Corp., Mc-Minnville, Oregon.

851.† Fischer, H. and Rüppel, W. B. Afterglow-tails of high density nanosecond arc channels. *Proc. 6th Conf. Intern. Phénomènes d'Ionisation dans les Gaz* (May, 1963).

852.* Schall, R. *Proc. 6th Intern. Kongr. für Kurzzeitphotographie. Kino-Technik* **16,** No. 12 (1962).

853.* Stuetzer, O. M. Gas bubbles in a charged liquid. *J. Appl. Phys.* **34,** No. 4, 958-963 (1963).

854.* Stuetzer, O. M. Pressure analysis of conduction in liquid. *Physics of Fluids* **6,** No. 2, 190-194 (1963).

855.* Stuetzer, O. M. Electrohydrodynamic precipitator. *Rev. Sci. Instr.* **33,** No. 11, 1171-1177 (Nov., 1962).

856. Früngel, F., Köhler, H., and Reinhard, H. P. The illumination of the CERN 2-meter hydrogen bubble chamber. *Appl. Opt.* **2,** 1017-1024 (October, 1963).

857.‡ Courtney-Pratt, J. S. Lenticular plate multiple picture shadowgraph recording. *J. Soc. Motion Picture Television Engrs.* 70, 710-715 (Sept., 1961).

858.* McLaughlin, J. W. A method of improving contrast when copying dissected images. *Proc. 5th Intern. Congr. High-Speed Photography* pp. 227-230 (1962).

859.* Fischer, H. Ultra-Kurzzeit-Aufnahmen der Grenzschichtstrukturen schneller Strömungsvorgänge. *Chem.-Ing. Tech.* **34,** 118-119 (1962).

860.* Gallant, H. Untersuchung über Kavitationsblasen. *Oesterreich. Ing.-Z.* **5,** No. 3, 74-83 (1962).

861.† Wichmann, H. G. Elektrische Antriebe für Raumfahrzeuge. *Elektrotech. Z.* **84** No. 8 (April, 1963).

862.* Gnoth, B. Doppelimpulsgenerator für das Nanosekundengebiet. *Elektronik* No. 3, 85-88 (1963).

863.* Ebeling, D. Modulation von Impulslasern. *Physik. Verhandl.* **3,** No. 7 (1963).

864.* Thorwart, W. Kombinierte Framing-Smear-Aufnahmen mit Trommelkamera bei Zündprozessen in Gasgemischen. *Physik. Verhandl.* **3,** No. 7 (1963).

865.* Suarez, J. F. Ein Verzögerungsgenerator hoher Genauigkeit mit konventioneller Schaltung. *Physik. Verhandl.* **3,** No. 7 (1963).

866. Früngel, F. Jittermessungen im ns-Bereich. *Physik. Verhandl.* **3,** No. 7 (1963).

867.† Bodenseher, H. Ein schneller Folienschalter für elektrische Stossentladungen. *Vortrag Deut. Physikertagung in Hamburg* (1963).

868.* Verbraak, C. A., and De Graaf, J. G. A. On the Mechanism of brittle fracture in steel. *Proc. 6th Intern. Congr. on High-Speed Photography, The Hague* pp. 529-533 (Sept., 1962).

869.* Becker, H. Das Widerstands-Kondensatorimpuls- und Hochfrequenzschweissen an Platal, einem mit Kunststoff beschichteten, kaltgewalzten Stahlblech. *Schweisstechnik* **35**· (Deutscher Verlag für Schweisstechnik, Düsseldorf).

870. Früngel, F., and Patzke, H. -G. Das Laserprinzip, Grundlagen und thermische Anwendung. *Die elektrische Ausrüstung* No. 2 (April, 1964). Vogel-Verlag, Würzburg.

871.* Chanas, M. J. Procédé d'aimantation par décharge de condensateurs. *Rev. Gen. Electricité* **51,** No. 9, 419-423 (1942).

872.* Sosoka, J. Control pulse polarity for weld strength. *Elektronics Mag.* (April 13, 1962).

873.* Sackmann, W. Untersuchungen über die Anwendbarkeit der Wirbeltheorie auf die Ermittlung der Gestalt instationärer Flammenfronten in Spaltrohren. Dissertation, Technische Hochschule Aachen (Nov., 1962).

874.* Below, F. Vergleich der Detailerkennbarkeit des menschlichen Auges bei Kurzzeit- und Gleichlichtbeleuchtung. Diplomarbeit, Universität Hamburg (1963).

875.* Conrads, H. A Sliding spark with low inductance as a light source for the vacuum ultraviolet. *Ber. Kernforschungsanlage, Jülich* (Dec., 1962).

876.* Conrads, H. and Hartwig, H. Untersuchungen über die Reproduzierbarkeit der Lichtemission einer Gleitfunkenstrecke im Hochvakuum. *Ber. Kernforschungsanlage, Jülich* (April, 1963).

877.* Lunartron Electronics Ltd. Argon jet spark light source. Luton, Beds (1963).

878.* MacKellar, Operation of xenon arc discharge lamps. *Elec. Rev.* 313-316 (Aug., 1963).

879.* Sherin, R. J. Ein leistungsfähiger Transistor-Spannungswandler für Elektronenblitzgeräte. *Electronics* **33,** No. 4, 57 (1960).

880.* Langeberg, K. Elektronenblitzgeräte für ultrakurze Belichtungszeiten. *Elektronik, Funkschau* No. **21,** 527-528 (1960).

881.* Schardin, H., and Schall, R. Der Leistungsstand der Kurzzeitphotographie *Umschau.* No. 13, 396-398 (1959); No. 14, 419-421 (1959); No. 15, 456-457 (1959).

882.* Fritz, J. C. Die Automatik in der Schweisstechnik und verwandten Gebieten. *Fachz. "Automatik"* **5,** No. 4 119-125 (1961).

883. Früngel, F. Device for heating a ground covering. German Patent 3,047,701 (July, 1962).

884.* Non-destructive testing in electrical engineering. *Conf. Inst. Elec. Engrs.* (*London*) pp. 815, 874-875, No. II (Nov., 1961).

885.* Hughes, T. P. *Nature* **194,** 268 (1962).

886.* Schrade, H. *Proc. 5th Intern. Conf. on Ionization Phenomena in Gases., München* Vol. 2 (1961).

887.* Marshak, I. S., Kirsanov, V. P., and Gavanin, V. A. Die Leuchtdichte von Röhren- und Kugelimpulslampen. Izvest. Akad. Nauk SSSR **13,** 276-279 (1962) (In Russian).

888.* Marshak, I. S. Starkstrom-Impuls-(Funken) Entladungen in Gasen, angewendet in Impulslichtquellen. Akad. Nauk. SSSR, Moscow, No. 77, pp. 229-286 (1962).

889.* Spreadbury, F. G. Impulse Methods of Magnetization. *Permanent Magnets* PA 228507, pp. 230-231 (1949).

890.* Marshak, I. S. Präzise Impuls-(Funken) Entladungen in Gasen, angewendet in Impulslichtquellen. Akad. Nauk SSSR, Moscow, No. 77, pp. 229-286 (1962).

891.* Marshak, I. S. Impulsnye Istochniki Svyeta. Akad. Nauk SSSR, Moscow, No. 3, pp. 5-19 (1962).

892.* Kirsanov, V. P., and Marshak, I. S. Rabota katoda v. impulsnikh lampakh. *Elektronika* No. 12, pp. 66-72 (1960).

893.* Marshak, I. S., Zhilteov, V. P., and Shukin, L. I. O predelnoi chastotye uspishek impulsnikh lampakh. *Svetotekhnika* **11,** 13-17 (1961).

894. Früngel, F. Neue Methoden und Anwendungen der Kurzzeit-Photographie. *Proc. 6th Intern. Congr. on High-Speed Photography, The Hague* (Sept., 1962).

895. Früngel, F., Thorwart, W., and Suarez, J. F. A system for motion analysis cameras with triggering by high frequency pulses. *Proc. 6th Intern. Congr. on High-Speed Photography, The Hague*, pp. 367-370 (Sept., 1962).

896.* Raether, L. F. H. Eine Untersuchung von Funkenlichtquellen im nahen Ultraviolett, mit 10^{-8} s Leuchtdauer. *Proc. 6th Intern. Congr. on High-Speed Photography, The Hague* pp. 124-128 (Sept. 1962).

897.* Flanders, A. E. The universal transmission line (UXL). *Elec. Design News* (Nov., 1963).

898.* Cooley, R. A., (International Electronics Lab.) The theory and practice of explosive metal forming. *Am. Machinist*, pp. 127-130 (June 15, 1959).

899.* Clark, G. L., Hickey, J. J., Kingsley, R. J., and Wuerker, R. F. Exploding wire driven shock waves. STL Products, El Segundo, California (Dec. 1961).

900. Früngel, F., and Alberti, H. Neue Ergebnisse der Hochfrequenz-Röntgenblitz Kinematographie kleiner Objekte. *Proc. 6th Intern. Congr. on High-Speed Photography, The Hague*, pp. 195-198 (Sept., 1962).

901.* *Proc. Intern. Conf. Progr. in Very High Pressure Research, New York* (1961).

902.* Harvey, G. W. Verfahren zur Verformung von in fester Form vorliegenden leitenden insbesondere metallischen Werkstücken. German Patent 1,122,188 (Aug., 1959).

903.* Clark, G. L. and Wuerker, R. F. Plasmoid rotation in a theta pinch. STL Products, El Segundo, California (Oct., 1961).

904.* Zernow, L., Wright, F., Jr., and Woffinden, G. High speed-cinematographic studies of electrically exploded metal films. *Exploding Wires* **2,** 245-262 (1962).

905. Früngel, F. On the utilization of light pulses for ceilometers. Industrial paper, Impulsphysik GmbH, Hamburg-Rissen (June, 1959).

906.† Hansen, W. W. Surveying with pulsed-light radar. *Electronics* pp. 76-79 (July, 1948).

907.* Space Technology Lab. Inc., An introduction to Space Technology Laboratories Inc. STL Products, El Segundo, California (1961).

908. Patzke, H. G., and Früngel, F. 3-D high-speed photography. *Proc. 6th Intern. Congr. on High-Speed Photography, The Hague* pp. 250-257 (September, 1962).

909.† Tektronix Inc. Abridged Catalog. Beaverton, Oregon (1963).

910.* Schwab, A. Ueber die Anstiegszeiten von Koronaentladungsimpulsen in einer Spitze-Platte-Funkenstrecke. *Elektrotech. Z.* **84A,** 484 (1963).

911.* Warren-Becker, P., and Becker, St. E. "Bibliography on Masers, 1954-1961"; "Bibliography on Lasers, 1962." Oak Ridge, Tennessee (1963).

912. Früngel, F. Super flash equipment "Giant Lamp". Industrial paper, Impulsphysik GmbH, Hamburg-Rissen (Nov., 1963).

913. Früngel, F. Eine ausdruckende 15 MHz-Zählschaltung. *Physik. Verhandl. VDPG* **3,** No. 7 (1963).

914.* Kollmann, E. Eine Impulsverzögerungsschaltung mit extrem kleinem Jitter und weitem Arbeitsbereich. *Physik. Verhandl. VDPG* **3,** No. 7 (1963).

915.* Röder, G. Lebensdauer und Helligkeit von Funkenentladungslampen hoher Energie. *Physik. Verhandl. VDPG* **3,** No. 7 (1963).

916.* Wenzel, J. (Microwave Research Inst.) Prediction of Q-switched laser energy output. *Proc. Symposium on Optical Masers* Vol. 13 (April, 1963).

917.* Mallory, W. R., and Tittel, K. F. High repetition rate pulsed lasers. *Proc. Symposium on Optical Masers* Vol. 13 (April, 1963).

918. Früngel, F. Laser-Blitze in der High-Speed-Photography. *Proc. 6th Intern. Congr. on High-Speed Photography, The Hague* pp. 52-56 (Sept., 1962).

919. Früngel, F. Stereoskopische Funkenblitzserien. Dreidimensionale Bewegungsaufnahmen instationärer Strömungen durch hochfrequente Hochspannungsfunken. *Proc. 6th Intern. Congr. on High-Speed Photography, The Hague* pp. 622-628 (Sept., 1962).

920.* DVL-Institut für Turbulenzforschung, Berlin Untersuchung der Zerfallprozesse bei der Zerstäubung von Flüssigkeiten unter besonderer Berücksichitgung des Einflusses der Turbulenz *Sonderdruck Tatigkeitsber. Anlass des* 50 *jährigen Bestehens der DVL* (April, 1962).

921.* Fister, W. Versuche zur Erfassung der Strömungsverhältnisse an Radiallaufrädern. Industrial Research Report, Gute Hoffnungshütte, Duisburg.

922.* Merrill, H. J. "Light and Heat Sensing. Sixth Agard Avionics Panel Meeting, Paris, July, 1962." Pergamon Press, London (1963).

923.* Beard, M. Instrumentation and high-speed photography. *J. Soc. Motion Picture Television Engrs.* **2,** (1960-1963).

924.* Schoeller & Hoesch GmbH, Kondensatorpapiere. Papierfabrik Schoeller & Hoesch GmbH, Gernsbach.

925. Früngel, F. Zum Mechanismus der elektrostatischen Ladungsbildung beim Zerblasen fallender Tropfen. Sonderdruck aus: *Ber. Deut. Wetterdienstes* No. 91. Die Meteorologische Tagung in Hamburg (1962).

926. Früngel, F. Methoden der Transmissions-und Streulichtmessung an grossen Luftvolumina mittels Impulslicht, Nebelwarn- und Sichtweitenmessgeräte mit sehr grosser Messbasis. *Sonderdruck Ber. Deut. Wettendierstes* No. 91: Die Meteorologische Tagung in Hamburg (1962).

927.* Schoeller & Hoesch GmbH, Kondensatorpapier. Papierfabrik Schoeller & Hoesch, GmbH Gernsbach.

928.* Hoh, S. R. Ferroelectric energy converters. *Attis Congr. Sei. Rassegna Intern. Elettronica Nucl. Rome* (1960).

929.* Cleveland, F. J. & Co. Light control arrangement. British Patent 947,568 (Jan., 1964).

930. Früngel, F. Arrangement and method for electrical impulse welding. U.S. Patent 3,089,947 (May, 1963).

931.* Marshak, I. S. "Impulsnyl Istochniki Svyeta." Moscow, 1963.

932.* De Carli, Formation of diamond by explosive shock. Reports of the Poulter Laboratories, Stanford Research Inst., Stanford, California (1961).

933.* Vondenbusch, A. Schaltungskenngrössen von Kondensator-Stossstromerzeugern. *Elektrotech. Z.* **84A,** No. 14 (1963).

934.* Early, H. C., and Walker, R. C. Economics of multimillion-joule inductive energy storage. *Trans. Am. Inst. Elec. Eng.* **76,** 320-325 (1957).

935.* Beans, J. W. In this application, the Kerr-cell was utilized as a light modulator valve. *J. Opt. Soc. Am.* **5,** 597 (1926).

936.* Zarem, A. M., Marshall, F. R., and Hauser, S. M. Millimicrosecond Kerr-cell shutter. *Rev. Sci. Instr.* **29,** No. 11, 1041-1044 (1958).

937.* Hauser, S. M., and Quan, H. Applying the Kerr-cell to nanosecond photography. *Electronics* (Aug., 1961).

938.* Hauser, S. M. Multiple frame Kerr-cell camera. *Res./Develop. Mag.* (April, 1962).

939.* Brower, D. F. Magnetic-pulse forming. General Atomic Div., General Dynamics Corp., San Diego, California (1962).

940.* Brower, D. F. Metalforming with pulsed magnetic fields. General Atomic Div., General Dynamics Corp., San Diego, California (Oct., 1962).

941.* Brower, D. F. General Dynamics Magneform machine shapes metal parts with 300 kilogauss pulse. *Am. Machinist/Metalworking Manufacturing* (Dec., 11., 1961).

942.* Löb, H. Neue Raketenantriebe. *Kerntechnik* **5,** p. 475-498 (1963).

943.† Collapsing foil capping shutter. Electro-Optical Intsruments Inc., Monrovia, California (1963).

944.* Furth, H. P., and Waniek, R. W. New ideas on magnetic forming *Am. Machinist/Metalworking Manufacturing* (Jan. 22, 1962).

945.* Algar, L. E., Marshall, J. E., and Russell, C. R. Ein deuteriumgefülltes 200-MW-Thyratron neuester Entwicklung. *Intern. Elektronische Rundschau* No. 11 (Nov., 1963).

946.* Cook, K. G., and Isaacs, G. C. Deuterium filled thyratrons. *Brit. J. Appl. Phys.* **9,** 497-498 (1958).

947.* Armstrong, R. J., and Nicholls, N. S. Deuterium filled thyratrons. *Brit. J. Appl. Phys.* **9,** 498-499 (1958).

948.* Cranet, J., and Guman, W. The application of the pinch process for space propulsion. *Z. Flugwiss.* **10,** 83-93 (March, 1962).

949.† High fleld pulsed electromagnets. Advanced Kinetics, Inc., Costa Mesa, California.

950.† Design principles. Advanced Kinetics, Inc., Costa Mesa, California.

951.* Page, White & Farrer, Improvements in or relating to Spark Erosion Machines. British Patent 941,272 (Nov., 1963).

952.* Höcker, K. H., Kluge, W., and Schrade, H. Impulsentladung in dichten Plasmen. *Sonderdruck Ann. Physik* [7] **11,** No. 1-6 (1963).

953.* Thielo, W. Die Untersuchung eines Stickstoffplasmas hoher Teilchendichte mit magnetischen Sonden. *Sonderdruck Z. Naturforsch.* **18a,** No. 7 (1963).

954.* Geiger, W. Stationäre Strömungen chemisch reagierender Gase und Plasmen. *Z. Naturforsch.* **18a,** No. 12 (1963).

955.* *Am. Machinist/Metalworking Manufacturing* p. 104. (April, 18 1960).

956.* Pipher, F. C., Rardin, G. N., and Richter, W. L. High energy rate metal forming. Final Tech. Eng. Rep. PB 171896, Lockheed Aircraft Corp.; obtainable from Office of Technical Services, Washington, D.C. (1960).

957.* Peckner, D. High energy rate forming process—their present and future. *Materials Design Eng.* **52,** No. 1, pp. 87-96 (1960).

958.* Du Mond, T. C. Forming metals at high velocities. *Metal Progr.* **74,** No. 5, pp. 68-76 (1958).

959.† New Danish visibility measuring systems introduced. Impulsphysics Association e. V. *The Aeroplane and Commercial Aviation News* (Jan., 1964).

960.* Gaulder, C. F. K. Portable laser has high pumping efficiency. *Electronics Products* (Dec., 1961). pp. 26-28

961. Früngel, F. High frequency spark device. U.S. Patent 2,906,924 (September, 1959)

962.* Sirons, J. A. Laser pumping technique using an internal pumping source. U.S. Patent 3,102,920 (Sept., 1963).

963.* Miles, P. A., and Edgerton, H. E. Optically efficient ruby laser pump. *J. Appl. Phys.* **32,** 740-741 (1961).

964.* Collins, R. J., and Kisliuk, P. Control of population inversion in pulsed optical masers by feedback modulation. *J. Appl. Phys.* **33,** No. 6, 2009-2011 (June, 1962).

965.* Stitch, M. L., Woodbury, E. J., and Morse, J. H. Optical ranging system uses laser transmitter. *Electronics* pp. 51-53 (April, 21, 1961).

966.* "Uspekhi Nautschoi Photographii," Vol. VI. Akad. Nauk SSSR, Moscow (1959).

967.* Comes, F. I. Lichtquellen für das Vakuum-Ultraviolett *Z. Instrumentenk.* **68,** 69-76 (1960).

968.* Rheindorf, H. H. Versuche zur schnellen magnetischen Kompression. Diplomarbeit, Technische Hochschule Aachen (1960).

969.* Romand, J., and Balloffet, G. *J. Phys. Rad.* **16,** pp. 489-490. (1955)

970.* Bockasten, K. A study of CIII by means of a sliding vacuum spark. *Arkiv. Fys.* **9,** Paper 30, 457-481 (1955).

971.* Mandel'stam, S. L. *Zhur. Eksptl. Teoret. Fiz.* **37,** No. 2, 248-252 (1960).

972.* Fischer, H., Von Thuena, P., and Rose, H. *J. Appl. Opt.* **1,** 373 (1962); see *Appl. Spectroscopy* **2,** to be published.

973.* Elwert, G. *Z. Naturforschung* **7a,** 432 (1952).

974.* Pahl, M. Massenspektrometrische Untersuchungen an der positiven Säule in Ar, Ar-He- und Ar-H_2-Gemischen. *Z. Naturforsch.* **18a,** No. 12, 1276-1288 (1963).

975.* Kleber, W. Der Mechanismus der elektrostatischen Lackzerstäubung. *Plaste und Kautschuk* **10,** No. 7 (1963).

976.* Kyser, J. B. Tracer-spark technique for velocity mapping of hypersonic flow fields. Technical Notes. *AIAA. J.* **2,** 393-394 (Feb., 1964).

977.* Rukop, H. Messungen im elektromagnetischen Sepktrum des Wassers mit wenig gedämpften, durch Stosserregung hervorgebrachten Schwingungen von 65 bis 20 cm Wellenlänge. *Ann. Physik* [4] **42,** 491 (1913).

978.* Solimene, N. TRG-134-TR-7, Appendix C (July, 1962); TRG-134-QTR-1, Appendix I (Sept., 1962); TRG-134-QTR-3, Appendices A and B (March, 1963); TRG-134-QTR-4, Appendix D (June, 1963); under Contract AF49 (638)-673.

979.* Wagner, W. G. and Lengyel, B. A. *J. Appl. Phys.* **34,** 2040 (1963).

980.* Vuylsteke, A. A. *J. Appl. Phys.* **34,** 1615 (1963).

981.† Daly, R., and Sims, S. D. An improved method of mechanical Q-switching using TIR. TRG. Inc., Melville, New York (1963).

982.* Stitch, M. L. Optical maser. British Patent 946,199 (July, 1962).

983.* Caskey, D. L. Hair-trigger operation of a neodymium laser. M.I.T. Department of Electrical Engineering, Cambridge, Massachusetts (Jan., 1964).

984.† McClung, F. J., and Hellwarth, R. W. Giant optical pulsation from ruby. *J. Appl. Phys.* **33,** 828-29 (March, 1962).

985.* McClung, F. J. and Hellwarth, R. W. Characteristics of giant optical pulsations from ruby. *Proc. IRE* **51,** 46 (1963).

986.* Solemine, N. Rate equation approach to the performance of Q-switches laser. *Proc. 3rd Intern. Conf. on Quantum Electronics, Paris* (1963).

987.* Wenzel, J. Pulsed laser performance prediction. *Ohio State Univ. Symposium on Lasers, Columbus, Ohio* (Nov., 1963).

988.* Soay, W. R., Congleton, R. S., Dobratz, B. E., and Ng, W. K. Dynamic limitations on the attainable inversion in ruby lasers. *Proc. 3rd Intern. Conf. on Quantum Electronics, Paris* (1963).

989.* Maiman, L. H. Stimulated optical emmission in fluorescent solids. *Phys. Rev.* **123,** 1145 (1961).

990.* Stitch, M. D. Power output characteristics of a ruby laser. *J. Appl. Phys.* **32,** 1994 (1961).

991.* Wenzel, J. Pump power absorption variation in cylindrical and composite rod lasers. General Electric Corp., Advanced Electronics Center, Ithaca, New York (to be published).

992.* Wagner, W. G., and Lengyel, B. A. Evaluation of the giant pulse in a laser. Hughes Research Laboratories, Malibu, California [to appear in *J.Appl. Phys.* (1963)].

993.* Wittke, J. P. Effects of elevated temperature on the fluorescence and optical maser action in ruby. *J. Appl. Phys.* **37,** No. 7, 2333 (1962).

994.* World Literature Service; *Laser/Maser International* **1,** No. 1 (April, 1964).

995.* Butayeva, F. A., and Fabrikant, V. A. A medium with negative absorption coefficient. *In* "Issledovaniya po Eksperimentalnoy i Teoreticheskoy Fizike, Sbornik" (Studies of Experimental and Theoretical Physics, Collection of Articles.) pp. 62-70. Akad. Nauk SSSR, Fiz. Inst. P. N. Lebedeva. Moskow, 1959.

996.* Townes, C. H. Optical masers and their possible applications to biology. *Biophys. J.* **2,** No. 2, Part 2 (March, 1962).

997.* Courtney-Pratt, J. S. *J. Soc. Motion Picture Television Engrs.* **70,** 509 (1961).

998.* Dacey, G. C. Ruby laser as a microsurgical instrument. *Science* **141,** No. 3575 (July, 1963).

999.* Post, R. F., and Shiren, H. S. *Phys. Rev.* **78,** 80 (1950).

1000.* Swank, R. K., and Buck, W. L. *Rev. Sci. Instr.* **26,** 15 (1955).

1001.* Gläser, F. *Z. Naturforsch.* **11a,** 1030 (1956).

1002.* Sangster, R. C., and Irvine, J. W. *J. Chem. Phys.* **24,** 670 (1956).

1003.* Williams, E. M., Woodford, J. B., Jr., and Smith, R. E. Recent developments in the theory and design of electric spark machine tools. Inst. of Technol. Pittsburg, Pennsylvania. Presented at Am. Inst. Elec. Engrs. General Meeting, New York (May, 1954).

1004.* Kohlrausch, F. "Praktische Physik," Vol. 1, p. 19; Vol. 2, p. 710. Teubner, Stuttgart, (1955, 1956).

1005.† Wachsmann, F., and Dimotsis, A. "Graphs and Tables for Radiotherapy," p. 30. Hirzel, Stuttgart (1957).

1006.‡ Technical Notes in *J. Opt. Soc. Am.* **54,** 281 (1964).

Author Index

Manufacturers Index

Subject Index